VOLUME FOUR HUNDRED AND SIXTY-THREE

METHODS IN ENZYMOLOGY

Guide to Protein Purification, 2nd Edition

METHODS IN ENZYMOLOGY

VOLUME FOUR HUNDRED AND SIXTY-THREE

METHODS IN ENZYMOLOGY

Guide to Protein Purification, 2nd Edition

EDITED BY

RICHARD R. BURGESS

McArdle Laboratory for Cancer Research
University of Wisconsin-Madison
Madison, Wisconsin, USA

MURRAY P. DEUTSCHER

Department of Biochemistry and Molecular Biology
University of Miami School of Medicine
Miami, FL, USA

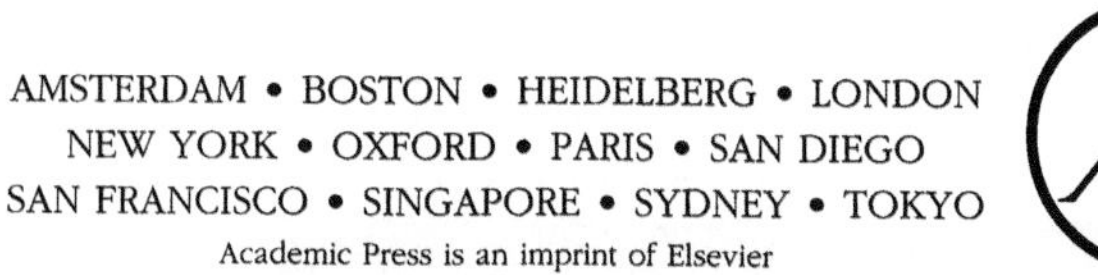
AMSTERDAM • BOSTON • HEIDELBERG • LONDON
NEW YORK • OXFORD • PARIS • SAN DIEGO
SAN FRANCISCO • SINGAPORE • SYDNEY • TOKYO
Academic Press is an imprint of Elsevier

Academic Press is an imprint of Elsevier
525 B Street, Suite 1900, San Diego, CA 92101-4495, USA
30 Corporate Drive, Suite 400, Burlington, MA 01803, USA
32 Jamestown Road, London NW1 7BY, UK

First edition 1990
Second edition 2009

ISBN: 978-0-12-374536-1 (hardback)
ISBN: 978-0-12-374978-9 (paperback)
ISSN: 0076-6879

Contents

Contributors xix
Preface xxv
Volumes in Series xxvii

1. Why Purify Enzymes? **1**
Arthur Kornberg (with preface by Murray Deutscher)

Section I. Developing Purification Procedures **7**

2. Strategies and Considerations for Protein Purifications **9**
Stuart Linn

1. General Considerations 10
2. Source of the Protein 15
3. Preparing Extracts 16
4. Bulk or Batch Procedures for Purification 17
5. Refined Procedures for Purification 18
6. Conclusions 19

References 19

3. Use of Bioinformatics in Planning a Protein Purification **21**
Richard R. Burgess

1. What You Can Learn from an Amino Acid Sequence 22
2. What You Cannot yet Predict 25
3. Conclusion 26

References 27

4. Preparing a Purification Summary Table **29**
Richard R. Burgess

1. Introduction 29
2. The Importance of Footnotes 32
3. The Value of an SDS–Polyacrylamide Gel Analysis on Main Protein Fractions 32
4. Some Common Mistakes and Problems 32

Section II. General Methods for Handling Proteins and Enzymes 35

5. Setting Up a Laboratory 37

Murray P. Deutscher

1. Supporting Materials 38
2. Detection and Assay Requirements 39
3. Fractionation Requirements 40

6. Buffers: Principles and Practice 43

Vincent S. Stoll and John S. Blanchard

1. Introduction 43
2. Theory 44
3. Buffer Selection 45
4. Buffer Preparation 48
5. Volatile Buffers 49
6. Broad-Range Buffers 50
7. Recipes for Buffer Stock Solutions 50

References 56

7. Measurement of Enzyme Activity 57

T. K. Harris and M. M. Keshwani

1. Introduction 58
2. Principles of Catalytic Activity 58
3. Measurement of Enzyme Activity 64
4. Formulation of Reaction Assay Mixtures 69
5. Discussion 71

Acknowledgments 71
References 71

8. Quantitation of Protein 73

James E. Noble and Marc J. A. Bailey

1. Introduction 74
2. General Instructions for Reagent Preparation 75
3. Ultraviolet Absorption Spectroscopy 80
4. Dye-Based Protein Assays 83
5. Coomassie Blue (Bradford) Protein Assay (Range: 1–50 μg) 85
6. Lowry (Alkaline Copper Reduction Assays) (Range: 5–100 μg) 86
7. Bicinchoninic Acid (BCA) (Range: 0.2–50 μg) 88
8. Amine Derivatization (Range: 0.05–25 μg) 89

9. Detergent-Based Fluorescent Detection (Range: 0.02–2 μg) 91
10. General Instructions 91
Acknowledgment 94
References 94

9. Concentration of Proteins and Removal of Solutes 97
David R. H. Evans, Jonathan K. Romero, and Matthew Westoby

1. Chromatography 98
2. Electrophoresis 103
3. Dialysis 104
4. Ultrafiltration 107
5. Lyophilization 113
6. Precipitation 116
7. Crystallization 118
References 118

10. Maintaining Protein Stability 121
Murray P. Deutscher

1. Causes of Protein Inactivation 121
2. General Handling Procedures 122
3. Concentration and Solvent Conditions 122
4. Stability Trials and Storage Conditions 123
5. Proteolysis and Protease Inhibitors 124
6. Loss of Activity 125

Section III. Recombinant Protein Expression and Purification 129

11. Selecting an Appropriate Method for Expressing a Recombinant Protein 131
William H. Brondyk

1. Introduction 132
2. *Escherichia coli* 133
3. *Pichia pastoris* 135
4. Baculovirus/Insect Cells 136
5. Mammalian Cells 138
6. Protein Characteristics 139
7. Recombinant Protein Applications 143
8. Conclusion 144
References 144

12. Bacterial Systems for Production of Heterologous Proteins **149**

Sarah Zerbs, Ashley M. Frank, and Frank R. Collart

1. Introduction 150
2. Heterologous Protein Production Using *Escherichia coli* 150
3. Planning a Bacterial Expression Project 151
4. Evaluation of Project Requirements 152
5. Target Analysis 152
6. Cloning 153
7. Preparation of T4 DNA Polymerase-Treated DNA Fragments 155
8. Expression in the *E. coli* Cytoplasm 156
9. Expression of Cytoplasmic Targets in *E. coli* 157
10. Analysis of Heterologous Protein Expression in *E. coli* 157
11. Small-Scale Expression Cultures in Autoinduction Media Protocol 160
12. Periplasmic Expression of Proteins 160
13. Expression of Periplasmic Targets in *E. coli* 161
14. Small-Scale Osmotic Shock Protocol 162
15. Alternative Bacterial Systems for Heterologous Protein Production 164
16. Alternative Vector and Induction Conditions 165
17. Production Scale 166

Acknowledgment 166

References 166

13. Expression in the Yeast *Pichia pastoris* **169**

James M. Cregg, Ilya Tolstorukov, Anasua Kusari, Jay Sunga, Knut Madden, and Thomas Chappell

1. Introduction 170
2. Other Fungal Expression Systems 170
3. Culture Media and Microbial Manipulation Techniques 171
4. Genetic Strain Construction 172
5. Gene Preparation and Vector Selection 174
6. Transformation by Electroporation 176
7. DNA Preparation 176
8. Examination of Strains for Recombinant Protein Production 178
9. Assay Development—The Yeastern Blot 182
10. Posttranslational Modification of the Recombinant Protein (Proteinases and Glycosylation) 184
11. Selection for Multiple Copies of an Expression Cassette 185

References 187

14. Baculovirus–Insect Cell Expression Systems **191**
Donald L. Jarvis
1. Introduction 192
2. A Brief Overview of Baculovirus Biology and Molecular Biology 193
3. Baculovirus Expression Vectors 195
4. Baculovirus Expression Vector Technology—The Early Years 196
5. Baculovirus Expression Vector Technology—Improved 198
6. Baculovirus Transfer Plasmid Modifications 198
7. Parental Baculovirus Genome Modifications 200
8. The Other Half of the Baculovirus–Insect Cell System 210
9. A New Generation of Insect Cell Hosts for Baculovirus Expression Vectors 212
10. Basic Baculovirus Protocols 214
References 218

15. Recombinant Protein Production by Transient Gene Transfer into Mammalian Cells **223**
Sabine Geisse and Cornelia Fux
1. Introduction 224
2. HEK293 and CHO Cell Lines Commonly Used in TGE Approaches 224
3. Expression Vectors for HEK293 and CHO Cells 226
4. Cultivation of HEK293 Cells and CHO Cell Lines in Suspension 228
5. Transfection Methods 228
6. Conclusions 234
Acknowledgments 234
References 235

16. Tagging for Protein Expression **239**
Arun Malhotra
1. Introduction 240
2. Some Considerations When Designing a Tagged Protein 241
3. Protein Affinity Tags 245
4. Solubility Tags 249
5. Removal of Tags 251
6. Conclusions 253
Acknowledgment 254
References 254

17. Refolding Solubilized Inclusion Body Proteins 259

Richard R. Burgess

1. Introduction 260
2. General Refolding Consideration 262
3. General Procedures 262
4. General Protocol 263
5. Comments on this General Procedure 264
6. Performing a Protein Refolding Test Screen 271
7. Other Refolding Procedures 275
8. Refolding Database: Refold 277
9. Strategies to Increase Proportion of Soluble Protein 277
10. Conclusion 279

References 279

Section IV. Preparation of Extracts and Subcellular Fractionation 283

18. Advances in Preparation of Biological Extracts for Protein Purification 285

Anthony C. Grabski

1. Introduction 286
2. Chemical and Enzymatic Cell Disruption 287
3. Mechanical Cell Disruption 290
4. Concluding Remarks 293
5. Procedures, Reagents, and Tips for Cell Disruption 293

References 301

19. Isolation of Subcellular Organelles and Structures 305

Uwe Michelsen and Jörg von Hagen

1. Introduction 306
2. Extraction and Prefractionation of Subproteomes 308

References 327

Section V. Purification Procedures: Bulk Methods 329

20. Protein Precipitation Techniques 331

Richard R. Burgess

1. Introduction 332
2. Ammonium Sulfate Precipitation 332
3. Polyethyleneimine Precipitation 337

4. Other Methods 341
5. General Procedures When Fractionating Proteins by Precipitation 341
References 342

21. Affi-Gel Blue for Nucleic Acid Removal and Early Enrichment of Nucleotide Binding Proteins 343

Murray P. Deutscher

1. A Representative Protocol 344
Reference 345

Section VI. Purification Procedures: Chromatographic Methods 347

22. Ion-Exchange Chromatography 349

Alois Jungbauer and Rainer Hahn

1. Introduction 349
2. Principle 351
3. Stationary Phases 353
4. Binding Conditions 355
5. Elution Conditions 361
6. Operation of Ion-Exchange Columns 363
7. Example: Separation of Complex Protein Mixture 366
8. Example: High-Resolution Separation with a Monolithic Column 367
References 370

23. Gel Filtration 373

Earle Stellwagen

1. Principle 373
2. Practice 374

24. Protein Chromatography on Hydroxyapatite Columns 387

Larry J. Cummings, Mark A. Snyder, and Kimberly Brisack

1. Introduction 388
2. Mechanisms 389
3. Chemical Characteristics 392
4. Purification Protocol Development 396
5. Packing Laboratory-Scale Columns 397
6. Process-Scale Column Packing 399
7. Applications 401
References 402

25. Theory and Use of Hydrophobic Interaction Chromatography in Protein Purification Applications **405**

Justin T. McCue

1. Theory 406
2. Latest Technology in HIC Adsorbents 408
3. Procedures for Use of HIC Adsorbents 409

References 413

Section VII. Purification Procedures: Affinity Methods **415**

26. Affinity Chromatography: General Methods **417**

Marjeta Urh, Dan Simpson, and Kate Zhao

1. Introduction 418
2. Selection of Affinity Matrix 419
3. Selection of Ligands 423
4. Attachment Chemistry 429
5. Purification Method 433

References 435

27. Immobilized-Metal Affinity Chromatography (IMAC): A Review **439**

Helena Block, Barbara Maertens, Anne Spriestersbach, Nicole Brinker, Jan Kubicek, Roland Fabis, Jörg Labahn, and Frank Schäfer

1. Overview on IMAC Ligands and Immobilized Ions 440
2. IMAC Applications 444
3. Conclusions 467

Acknowledgments 468

References 468

28. Identification, Production, and Use of Polyol-Responsive Monoclonal Antibodies for Immunoaffinity Chromatography **475**

Nancy E. Thompson, Katherine M. Foley, Elizabeth S. Stalder, and Richard R. Burgess

1. Introduction 476
2. Polyol-Responsive Monoclonal Antibodies 477
3. Conclusions 492

Disclosure 493

References 493

Section VIII. Purification Procedures: Electrophoretic Methods **495**

29. One-Dimensional Gel Electrophoresis **497**

David E. Garfin

1. Background 498
2. Polyacrylamide Gels 500
3. Principle of Method 501
4. Procedure 502
5. Detection of Proteins in Gels 508
6. Marker Proteins 510
7. Molecular Weight Determination 511
8. Preparative Electrophoresis 511

References 513

30. Isoelectric Focusing and Two-Dimensional Gel Electrophoresis **515**

David B. Friedman, Sjouke Hoving, and Reiner Westermeier

1. Introduction 516
2. Materials 527
3. Methods 528

References 538

31. Protein Gel Staining Methods: An Introduction and Overview **541**

Thomas H. Steinberg

1. Introduction 542
2. General Considerations 543
3. Instrumentation: Detection and Documentation 545
4. Total Protein Detection 545
5. Phosphoprotein Detection 556
6. Glycoprotein Detection 557

References 559

32. Elution of Proteins from Gels **565**

Richard R. Burgess

1. Introduction 565
2. Elution of Proteins from Gels by Diffusion 566
3. Replacing the SDS Gel with a Reverse Phase HPLC 570
4. Electrophoretic Elution 570
5. Conclusion 571

References 571

33. Performing and Optimizing Western Blots with an Emphasis on Chemiluminescent Detection **573**

Alice Alegria-Schaffer, Andrew Lodge, and Krishna Vattem

1. Western Blotting 574
2. Types of Western Blots 575
3. Detection Methods 579
4. The Chemiluminescence Signal 583
5. Common Problems and their Explanations 588
6. Blotting and Optimization Protocols using Chemiluminescent Substrates 593

References 598

Section IX. Purification Procedures: Membrane Proteins and Glycoproteins 601

34. Detergents: An Overview **603**

Dirk Linke

1. Introduction 604
2. Detergent Structure 604
3. Properties of Detergents in Solution 605
4. Exploiting the Physicochemical Parameters of Detergents for Membrane Protein Purification 612
5. Detergent Removal and Detergent Exchange 613
6. Choosing the Right Detergent 613
7. Conclusions 615

Acknowledgments 616
References 616

35. Purification of Membrane Proteins **619**

Sue-Hwa Lin and Guido Guidotti

1. Introduction 619
2. Preparation of Membranes 620
3. Solubilization of Native Membrane Proteins 622
4. Purification of Membrane Proteins 625
5. Detergent Removal and Detergent Exchange 628
6. Expression and Purification of Recombinant Integral Membrane Proteins 628

References 629

36. Purification of Recombinant G-Protein-Coupled Receptors **631**

Reinhard Grisshammer

1. Introduction 632
2. Solubilization: General Considerations 633
3. Purification: General Considerations 634
4. Solubilization and Purification of a Recombinant Neurotensin Receptor NTS1 637
5. Analysis of Purified NTS1 641
6. Conclusions 642

Acknowledgments 642
References 642

37. Cell-Free Translation of Integral Membrane Proteins into Unilamelar Liposomes **647**

Michael A. Goren, Akira Nozawa, Shin-ichi Makino, Russell L. Wrobel, and Brian G. Fox

1. Introduction 648
2. Overview of Cell-Free Translation 650
3. Expression Vectors 651
4. Gene Cloning 652
5. PCR Product Cleanup 655
6. Flexi Vector and PCR Product Digestion Reaction 656
7. Ligation Reaction 657
8. Transformation Reaction 658
9. Purification of Plasmid DNA 659
10. Preparation of mRNA 660
11. Preparation of Liposomes 661
12. Wheat Germ Translation Reaction 662
13. Purification by Density Gradient Ultracentrifugation 665
14. Characterization of Proteoliposomes 667
15. Considerations for Scale-Up 669
16. Isotopic Labeling for Structural Studies 669
17. Conclusions 670

Acknowledgments 670
References 670

Section X. Characterization of Purified Proteins 675

38. Determination of Protein Purity 677

David G. Rhodes and Thomas M. Laue

1. Composition-Based and Activity-Based Analyses 680
2. Electrophoretic Methods 681
3. Chromatographic Methods 685
4. Sedimentation Velocity Methods 687
5. Mass Spectrometry Methods 687
6. Light Scattering Methods 688

References 689

39. Determination of Size, Molecular Weight, and Presence of Subunits 691

David G. Rhodes, Robert E. Bossio, and Thomas M. Laue

1. Introduction 692
2. Chemical Methods 695
3. Transport Methods 698
4. Scattering Methods 716
5. Presence of Subunits 719

References 721

40. Identification and Quantification of Protein Posttranslational Modifications 725

Adam R. Farley and Andrew J. Link

1. Introduction 726
2. Enrichment Techniques for Identifying PTMs 731
3. Nitrosative Protein Modifications 740
4. Methylation and Acetylation 741
5. Mass Spectrometry Analysis 744
6. CID versus ECD versus ETD 747
7. Quantifying PTMs 750
8. Future Directions 756

Acknowledgments 758

References 758

Section XI. Additional Techniques 765

41. Parallel Methods for Expression and Purification 767

Scott A. Lesley

1. Introduction 767
2. Strategies Based on End-Use 768
3. Parallel Cloning Strategies for Creating Expression Constructs 771
4. Small-Scale Expression Screening to Identify Suitable Constructs 774
5. Analytical Testing of Proteins for Selection 778
6. Large-Scale Parallel Expression 780
7. Conclusion 783

Acknowledgments 783
References 784

42. Techniques to Isolate O_2-Sensitive Proteins: [4Fe–4S]-FNR as an Example 787

Aixin Yan and Patricia J. Kiley

1. Introduction 788
2. Anaerobic Isolation of 4Fe-FNR 790
3. Characterization of $[4Fe–4S]^{2+}$ Cluster Containing FNR 799
4. Summary 803

References 803

Section XII. Concluding Remarks 807

43. Rethinking Your Purification Procedure 809

Murray P. Deutscher

1. Introduction 809

44. Important but Little Known (or Forgotten) Artifacts in Protein Biochemistry 813

Richard R. Burgess

1. Introduction 814
2. SDS Gel Electrophoresis Sample Preparation 814
3. Buffers 816
4. Chromatography 817
5. Protein Absorption During Filtration 818
6. Chemical Leaching from Plasticware 819
7. Cyanate in Urea 819

References 820

Author Index 821
Subject Index 835

Contributors

Alice Alegria-Schaffer
Thermo Fisher Scientific, Pierce Protein Research, Rockford, Illinois, USA

Marc J. A. Bailey
Nokia Research Centre - Eurolab, University of Cambridge, Cambridge, United Kingdom

John S. Blanchard
Department of Biochemistry, Albert Einstein College of Medicine, Bronx, New York

Helena Block
QIAGEN GmbH, Qiagen Strasse 1, Hilden, Germany

Robert E. Bossio
Department of Chemistry, University of Michigan–Dearborn, Dearborn, Michigan, USA

Nicole Brinker
QIAGEN GmbH, Qiagen Strasse 1, Hilden, Germany

Kimberly Brisack
Bio-Rad Laboratories, Inc., Hercules, California, USA

William H. Brondyk
Genzyme Corporation, Framingham, Massachusetts, USA

Richard R. Burgess
McArdle Laboratory for Cancer Research, University of Wisconsin-Madison, Madison, Wisconsin, USA

Thomas Chappell
Biogrammatics, Inc., Carlsbad, California, USA

Frank R. Collart
Biosciences Division, Argonne National Laboratory, Lemont, Illinois, USA

James M. Cregg
Keck Graduate Institute of Applied Life Sciences, Claremont, California, USA, and Biogrammatics, Inc., Carlsbad, California, USA

Larry J. Cummings
Bio-Rad Laboratories, Inc., Hercules, California, USA

Murray P. Deutscher
Department of Biochemistry and Molecular Biology, University of Miami School of Medicine, Miami, Florida, USA

David R. H. Evans
Process Biochemistry, Biopharmaceutical Development, Biogen Idec. Inc., Cambridge, Massachusetts, USA

Roland Fabis
QIAGEN GmbH, Qiagen Strasse 1, Hilden, Germany

Adam R. Farley
Department of Biochemistry, Vanderbilt University School of Medicine, Nashville, Tennessee, USA

Katherine M. Foley
McArdle Laboratory for Cancer Research, University of Wisconsin-Madison, Madison, Wisconsin, USA

Brian G. Fox
Department of Biochemistry, Dr. Brian Fox Lab, University of Wisconsin-Madison, Madison, Wisconsin, USA

Ashley M. Frank
Biosciences Division, Argonne National Laboratory, Lemont, Illinois, USA

David B. Friedman
Proteomics Laboratory, Mass Spectrometry Research Center, Vanderbilt University, Nashville, Tennessee, USA

Cornelia Fux
Novartis Pharma AG, Department of NBx/PSD: Scale up, Basel, Switzerland

David E. Garfin
Chemical Division, Research Products Group, Bio-Rad Laboratories, Incorporated, Richmond, California

Sabine Geisse
Novartis Institutes for BioMedical Research, Department of NBC/PPA, Basel, Switzerland

Michael A. Goren
Department of Biochemistry, Dr. Brian Fox Lab, University of Wisconsin-Madison, Madison, Wisconsin, USA

Anthony C. Grabski
Department of Research and Development, Semba Biosciences, Inc., Madison, Wisconsin, USA

Reinhard Grisshammer
Department of Health and Human Services, National Institutes of Health, National Institute of Neurological Disorders and Stroke, Rockville, Maryland, USA

Guido Guidotti
Department of Molecular and Cellular Biology, Harvard University, Cambridge, Massachusetts, USA

Rainer Hahn
Department of Biotechnology, University of Natural Resources and Applied Life Sciences, Vienna, Austria

T. K. Harris
Department of Biochemistry and Molecular Biology, Miller School of Medicine, University of Miami, Miami, Florida, USA

Sjouke Hoving
Novartis Institutes of BioMedical Research, Basel, Switzerland

Donald L. Jarvis
Department of Molecular Biology, University of Wyoming, Laramie, Wyoming, USA

Alois Jungbauer
Department of Biotechnology, University of Natural Resources and Applied Life Sciences, Vienna, Austria

M. M. Keshwani
Department of Biochemistry and Molecular Biology, Miller School of Medicine, University of Miami, Miami, Florida, USA

Patricia J. Kiley
Department of Biomolecular Chemistry, University of Wisconsin, Madison, Wisconsin, USA

Jan Kubicek
QIAGEN GmbH, Qiagen Strasse 1, Hilden, Germany

Anasua Kusari
Keck Graduate Institute of Applied Life Sciences, Claremont, California, USA

Jörg Labahn
Institute of Structural Biology (IBI-2), Research Center Jülich, Jülich, Germany

Thomas M. Laue
Department of Biochemistry and Molecular Biology, University of New Hampshire, Durham, New Hampshire, USA

Scott A. Lesley
Genomics Institute of the Novartis Research Foundation, San Diego, California, USA

Sue-Hwa Lin
Department of Molecular Pathology, University of Texas, Houston, Texas, USA

Andrew J. Link
Department of Microbiology and Immunology, Vanderbilt University School of Medicine, Nashville, Tennessee, USA

Dirk Linke
Department I, Protein Evolution, Max Planck Institute for Developmental Biology, Tübingen, Germany

Stuart Linn
Department of Molecular and Cellular Biology, University of California, Berkeley, California, USA

Andrew Lodge
Thermo Fisher Scientific, Pierce Protein Research, Rockford, Illinois, USA

Knut Madden
Biogrammatics, Inc., Carlsbad, California, USA

Barbara Maertens
QIAGEN GmbH, Qiagen Strasse 1, Hilden, Germany

Shin-ichi Makino
Department of Biochemistry, Dr. Brian Fox Lab, University of Wisconsin-Madison, Madison, Wisconsin, USA

Arun Malhotra
Department of Biochemistry and Molecular Biology, University of Miami Miller School of Medicine, Miami, Florida, USA

Justin T. McCue
Biogen Idec Corporation, Cambridge, Massachusetts, USA

Uwe Michelsen
Merck KGaA, Darmstadt, Germany

James E. Noble
Analytical Science, National Physical Laboratory, Teddington, Middlesex, United Kingdom

Akira Nozawa
Department of Biochemistry, Dr. Brian Fox Lab, University of Wisconsin-Madison, Madison, Wisconsin, USA

David G. Rhodes
Lipophilia Consulting, Storrs, Connecticut, USA

Jonathan K. Romero
Process Biochemistry, Biopharmaceutical Development, Biogen Idec. Inc., Cambridge, Massachusetts, USA

Frank Schäfer
QIAGEN GmbH, Qiagen Strasse 1, Hilden, Germany

Dan Simpson
Promega Corporation, Madison, Wisconsin, USA

Mark A. Snyder
Bio-Rad Laboratories, Inc., Hercules, California, USA

Anne Spriestersbach
QIAGEN GmbH, Qiagen Strasse 1, Hilden, Germany

Elizabeth S. Stalder
McArdle Laboratory for Cancer Research, University of Wisconsin-Madison, Madison, Wisconsin, USA

Thomas H. Steinberg
McArdle Laboratory for Cancer Research, University of Wisconsin–Madison, Madison, Wisconsin, USA

Earle Stellwagen
Department of Biochemistry, University of Iowa, Iowa City, Iowa

Vincent S. Stoll
Department of Biochemistry, Albert Einstein College of Medicine, Bronx, New York

Jay Sunga
Keck Graduate Institute of Applied Life Sciences, Claremont, California, USA

Nancy E. Thompson
McArdle Laboratory for Cancer Research, University of Wisconsin-Madison, Madison, Wisconsin, USA

Ilya Tolstorukov
Keck Graduate Institute of Applied Life Sciences, Claremont, California, USA, and Biogrammatics, Inc., Carlsbad, California, USA

Marjeta Urh
Promega Corporation, Madison, Wisconsin, USA

Krishna Vattem
Thermo Fisher Scientific, Pierce Protein Research, Rockford, Illinois, USA

Jörg von Hagen
Merck KGaA, Darmstadt, Germany

Reiner Westermeier
Gelcompany GmbH, Paul-Ehrlich-Strasse, Tübingen, Germany

Matthew Westoby
Process Biochemistry, Biopharmaceutical Development, Biogen Idec. Inc., San Diego, California, USA

Russell L. Wrobel
Department of Biochemistry, Dr. Brian Fox Lab, University of Wisconsin-Madison, Madison, Wisconsin, USA

Aixin Yan
School of Biological Sciences, The University of Hong Kong, Hong Kong, Hong Kong SAR

Sarah Zerbs
Biosciences Division, Argonne National Laboratory, Lemont, Illinois, USA

Kate Zhao
Promega Corporation, Madison, Wisconsin, USA

PREFACE

Protein biochemistry continues to be an essential part of modern biological research. With the huge increase in genome sequencing, the sequences of most studied organisms are available, facilitating gene cloning, protein production, and the study of protein properties, structure, and function. Tremendous advances in heterologous expression of recombinant proteins have greatly increased our ability to produce proteins of interest. However, a protein must still be purified and characterized. This is particularly important if the gene is unknown.

Much has happened since the First Edition of the Guide to Protein Purification (Methods in Enzymology, Volume 182) was published in 1990. Major changes have occurred, including the explosion of genomics and proteomics, high-throughput expression/purification for the Protein Structure Initiative, protein target-based high throughput screening of chemicals to find new pharmaceuticals, a growing need to express and purify membrane proteins, the use of purification and solubilization tags, the systematic testing of refolding conditions to facilitate and optimize protein refolding, mass spectrometry analysis of post-translational modifications, and expression of recombinant proteins in non-bacterial hosts. These changes have necessitated significant advances in techniques, materials, reagents and equipment.

In the Second Edition of the Guide to Protein Purification, we have tried to identify the areas of greatest change in the last 20 years and to present authoritative reviews and methods that reflect the current best practice in each area of protein purification. First and foremost, we wanted these chapters to be educational and highly useful to the reader, and included typical best use protocols, cautions, technical tips, and limitations. While many of the chapters are by experts in academia, a significant number are by experts in certain areas in the biotechnology and pharmaceutical industry. We feel that high-level scientists in companies that use, develop, sell, or provide technical support for products in technical areas are particularly valuable in writing such chapters. We have encouraged such chapters and tried to ensure that they are even-handed and not focused entirely on the products of a particular company.

This Guide is a self-contained volume covering all the important procedures for purifying proteins as well as the more specialized techniques.

We hope that the Guide will be an invaluable resource and reference text for researchers new to protein purification as well as for the more experienced researchers, and that it will find an important place in every protein biochemistry laboratory.

RICHARD R. BURGESS
MURRAY P. DEUTSCHER

Methods in Enzymology

Volume I. Preparation and Assay of Enzymes
Edited by Sidney P. Colowick and Nathan O. Kaplan

Volume II. Preparation and Assay of Enzymes
Edited by Sidney P. Colowick and Nathan O. Kaplan

Volume III. Preparation and Assay of Substrates
Edited by Sidney P. Colowick and Nathan O. Kaplan

Volume IV. Special Techniques for the Enzymologist
Edited by Sidney P. Colowick and Nathan O. Kaplan

Volume V. Preparation and Assay of Enzymes
Edited by Sidney P. Colowick and Nathan O. Kaplan

Volume VI. Preparation and Assay of Enzymes *(Continued)*
Preparation and Assay of Substrates
Special Techniques
Edited by Sidney P. Colowick and Nathan O. Kaplan

Volume VII. Cumulative Subject Index
Edited by Sidney P. Colowick and Nathan O. Kaplan

Volume VIII. Complex Carbohydrates
Edited by Elizabeth F. Neufeld and Victor Ginsburg

Volume IX. Carbohydrate Metabolism
Edited by Willis A. Wood

Volume X. Oxidation and Phosphorylation
Edited by Ronald W. Estabrook and Maynard E. Pullman

Volume XI. Enzyme Structure
Edited by C. H. W. Hirs

Volume XII. Nucleic Acids (Parts A and B)
Edited by Lawrence Grossman and Kivie Moldave

Volume XIII. Citric Acid Cycle
Edited by J. M. Lowenstein

Volume XIV. Lipids
Edited by J. M. Lowenstein

Volume XV. Steroids and Terpenoids
Edited by Raymond B. Clayton

Volume XVI. Fast Reactions
Edited by Kenneth Kustin

Volume XVII. Metabolism of Amino Acids and Amines (Parts A and B)
Edited by Herbert Tabor and Celia White Tabor

Volume XVIII. Vitamins and Coenzymes (Parts A, B, and C)
Edited by Donald B. McCormick and Lemuel D. Wright

Volume XIX. Proteolytic Enzymes
Edited by Gertrude E. Perlmann and Laszlo Lorand

Volume XX. Nucleic Acids and Protein Synthesis (Part C)
Edited by Kivie Moldave and Lawrence Grossman

Volume XXI. Nucleic Acids (Part D)
Edited by Lawrence Grossman and Kivie Moldave

Volume XXII. Enzyme Purification and Related Techniques
Edited by William B. Jakoby

Volume XXIII. Photosynthesis (Part A)
Edited by Anthony San Pietro

Volume XXIV. Photosynthesis and Nitrogen Fixation (Part B)
Edited by Anthony San Pietro

Volume XXV. Enzyme Structure (Part B)
Edited by C. H. W. Hirs and Serge N. Timasheff

Volume XXVI. Enzyme Structure (Part C)
Edited by C. H. W. Hirs and Serge N. Timasheff

Volume XXVII. Enzyme Structure (Part D)
Edited by C. H. W. Hirs and Serge N. Timasheff

Volume XXVIII. Complex Carbohydrates (Part B)
Edited by Victor Ginsburg

Volume XXIX. Nucleic Acids and Protein Synthesis (Part E)
Edited by Lawrence Grossman and Kivie Moldave

Volume XXX. Nucleic Acids and Protein Synthesis (Part F)
Edited by Kivie Moldave and Lawrence Grossman

Volume XXXI. Biomembranes (Part A)
Edited by Sidney Fleischer and Lester Packer

Volume XXXII. Biomembranes (Part B)
Edited by Sidney Fleischer and Lester Packer

Volume XXXIII. Cumulative Subject Index Volumes I-XXX
Edited by Martha G. Dennis and Edward A. Dennis

Volume XXXIV. Affinity Techniques (Enzyme Purification: Part B)
Edited by William B. Jakoby and Meir Wilchek

VOLUME XXXV. Lipids (Part B)
Edited by JOHN M. LOWENSTEIN

VOLUME XXXVI. Hormone Action (Part A: Steroid Hormones)
Edited by BERT W. O'MALLEY AND JOEL G. HARDMAN

VOLUME XXXVII. Hormone Action (Part B: Peptide Hormones)
Edited by BERT W. O'MALLEY AND JOEL G. HARDMAN

VOLUME XXXVIII. Hormone Action (Part C: Cyclic Nucleotides)
Edited by JOEL G. HARDMAN AND BERT W. O'MALLEY

VOLUME XXXIX. Hormone Action (Part D: Isolated Cells, Tissues, and Organ Systems)
Edited by JOEL G. HARDMAN AND BERT W. O'MALLEY

VOLUME XL. Hormone Action (Part E: Nuclear Structure and Function)
Edited by BERT W. O'MALLEY AND JOEL G. HARDMAN

VOLUME XLI. Carbohydrate Metabolism (Part B)
Edited by W. A. WOOD

VOLUME XLII. Carbohydrate Metabolism (Part C)
Edited by W. A. WOOD

VOLUME XLIII. Antibiotics
Edited by JOHN H. HASH

VOLUME XLIV. Immobilized Enzymes
Edited by KLAUS MOSBACH

VOLUME XLV. Proteolytic Enzymes (Part B)
Edited by LASZLO LORAND

VOLUME XLVI. Affinity Labeling
Edited by WILLIAM B. JAKOBY AND MEIR WILCHEK

VOLUME XLVII. Enzyme Structure (Part E)
Edited by C. H. W. HIRS AND SERGE N. TIMASHEFF

VOLUME XLVIII. Enzyme Structure (Part F)
Edited by C. H. W. HIRS AND SERGE N. TIMASHEFF

VOLUME XLIX. Enzyme Structure (Part G)
Edited by C. H. W. HIRS AND SERGE N. TIMASHEFF

VOLUME L. Complex Carbohydrates (Part C)
Edited by VICTOR GINSBURG

VOLUME LI. Purine and Pyrimidine Nucleotide Metabolism
Edited by PATRICIA A. HOFFEE AND MARY ELLEN JONES

VOLUME LII. Biomembranes (Part C: Biological Oxidations)
Edited by SIDNEY FLEISCHER AND LESTER PACKER

VOLUME LIII. Biomembranes (Part D: Biological Oxidations)
Edited by SIDNEY FLEISCHER AND LESTER PACKER

VOLUME LIV. Biomembranes (Part E: Biological Oxidations)
Edited by SIDNEY FLEISCHER AND LESTER PACKER

VOLUME LV. Biomembranes (Part F: Bioenergetics)
Edited by SIDNEY FLEISCHER AND LESTER PACKER

VOLUME LVI. Biomembranes (Part G: Bioenergetics)
Edited by SIDNEY FLEISCHER AND LESTER PACKER

VOLUME LVII. Bioluminescence and Chemiluminescence
Edited by MARLENE A. DELUCA

VOLUME LVIII. Cell Culture
Edited by WILLIAM B. JAKOBY AND IRA PASTAN

VOLUME LIX. Nucleic Acids and Protein Synthesis (Part G)
Edited by KIVIE MOLDAVE AND LAWRENCE GROSSMAN

VOLUME LX. Nucleic Acids and Protein Synthesis (Part H)
Edited by KIVIE MOLDAVE AND LAWRENCE GROSSMAN

VOLUME 61. Enzyme Structure (Part H)
Edited by C. H. W. HIRS AND SERGE N. TIMASHEFF

VOLUME 62. Vitamins and Coenzymes (Part D)
Edited by DONALD B. MCCORMICK AND LEMUEL D. WRIGHT

VOLUME 63. Enzyme Kinetics and Mechanism (Part A: Initial Rate and Inhibitor Methods)
Edited by DANIEL L. PURICH

VOLUME 64. Enzyme Kinetics and Mechanism
(Part B: Isotopic Probes and Complex Enzyme Systems)
Edited by DANIEL L. PURICH

VOLUME 65. Nucleic Acids (Part I)
Edited by LAWRENCE GROSSMAN AND KIVIE MOLDAVE

VOLUME 66. Vitamins and Coenzymes (Part E)
Edited by DONALD B. MCCORMICK AND LEMUEL D. WRIGHT

VOLUME 67. Vitamins and Coenzymes (Part F)
Edited by DONALD B. MCCORMICK AND LEMUEL D. WRIGHT

VOLUME 68. Recombinant DNA
Edited by RAY WU

VOLUME 69. Photosynthesis and Nitrogen Fixation (Part C)
Edited by ANTHONY SAN PIETRO

VOLUME 70. Immunochemical Techniques (Part A)
Edited by HELEN VAN VUNAKIS AND JOHN J. LANGONE

VOLUME 71. Lipids (Part C)
Edited by JOHN M. LOWENSTEIN

VOLUME 72. Lipids (Part D)
Edited by JOHN M. LOWENSTEIN

VOLUME 73. Immunochemical Techniques (Part B)
Edited by JOHN J. LANGONE AND HELEN VAN VUNAKIS

VOLUME 74. Immunochemical Techniques (Part C)
Edited by JOHN J. LANGONE AND HELEN VAN VUNAKIS

VOLUME 75. Cumulative Subject Index Volumes XXXI, XXXII, XXXIV–LX
Edited by EDWARD A. DENNIS AND MARTHA G. DENNIS

VOLUME 76. Hemoglobins
Edited by ERALDO ANTONINI, LUIGI ROSSI-BERNARDI, AND EMILIA CHIANCONE

VOLUME 77. Detoxication and Drug Metabolism
Edited by WILLIAM B. JAKOBY

VOLUME 78. Interferons (Part A)
Edited by SIDNEY PESTKA

VOLUME 79. Interferons (Part B)
Edited by SIDNEY PESTKA

VOLUME 80. Proteolytic Enzymes (Part C)
Edited by LASZLO LORAND

VOLUME 81. Biomembranes (Part H: Visual Pigments and Purple Membranes, I)
Edited by LESTER PACKER

VOLUME 82. Structural and Contractile Proteins (Part A: Extracellular Matrix)
Edited by LEON W. CUNNINGHAM AND DIXIE W. FREDERIKSEN

VOLUME 83. Complex Carbohydrates (Part D)
Edited by VICTOR GINSBURG

VOLUME 84. Immunochemical Techniques (Part D: Selected Immunoassays)
Edited by JOHN J. LANGONE AND HELEN VAN VUNAKIS

VOLUME 85. Structural and Contractile Proteins (Part B: The Contractile Apparatus and the Cytoskeleton)
Edited by DIXIE W. FREDERIKSEN AND LEON W. CUNNINGHAM

VOLUME 86. Prostaglandins and Arachidonate Metabolites
Edited by WILLIAM E. M. LANDS AND WILLIAM L. SMITH

VOLUME 87. Enzyme Kinetics and Mechanism (Part C: Intermediates, Stereo-chemistry, and Rate Studies)
Edited by DANIEL L. PURICH

VOLUME 88. Biomembranes (Part I: Visual Pigments and Purple Membranes, II)
Edited by LESTER PACKER

Volume 89. Carbohydrate Metabolism (Part D)
Edited by Willis A. Wood

Volume 90. Carbohydrate Metabolism (Part E)
Edited by Willis A. Wood

Volume 91. Enzyme Structure (Part I)
Edited by C. H. W. Hirs and Serge N. Timasheff

Volume 92. Immunochemical Techniques (Part E: Monoclonal Antibodies and General Immunoassay Methods)
Edited by John J. Langone and Helen Van Vunakis

Volume 93. Immunochemical Techniques (Part F: Conventional Antibodies, Fc Receptors, and Cytotoxicity)
Edited by John J. Langone and Helen Van Vunakis

Volume 94. Polyamines
Edited by Herbert Tabor and Celia White Tabor

Volume 95. Cumulative Subject Index Volumes 61–74, 76–80
Edited by Edward A. Dennis and Martha G. Dennis

Volume 96. Biomembranes [Part J: Membrane Biogenesis: Assembly and Targeting (General Methods; Eukaryotes)]
Edited by Sidney Fleischer and Becca Fleischer

Volume 97. Biomembranes [Part K: Membrane Biogenesis: Assembly and Targeting (Prokaryotes, Mitochondria, and Chloroplasts)]
Edited by Sidney Fleischer and Becca Fleischer

Volume 98. Biomembranes (Part L: Membrane Biogenesis: Processing and Recycling)
Edited by Sidney Fleischer and Becca Fleischer

Volume 99. Hormone Action (Part F: Protein Kinases)
Edited by Jackie D. Corbin and Joel G. Hardman

Volume 100. Recombinant DNA (Part B)
Edited by Ray Wu, Lawrence Grossman, and Kivie Moldave

Volume 101. Recombinant DNA (Part C)
Edited by Ray Wu, Lawrence Grossman, and Kivie Moldave

Volume 102. Hormone Action (Part G: Calmodulin and Calcium-Binding Proteins)
Edited by Anthony R. Means and Bert W. O'Malley

Volume 103. Hormone Action (Part H: Neuroendocrine Peptides)
Edited by P. Michael Conn

Volume 104. Enzyme Purification and Related Techniques (Part C)
Edited by William B. Jakoby

VOLUME 105. Oxygen Radicals in Biological Systems
Edited by LESTER PACKER

VOLUME 106. Posttranslational Modifications (Part A)
Edited by FINN WOLD AND KIVIE MOLDAVE

VOLUME 107. Posttranslational Modifications (Part B)
Edited by FINN WOLD AND KIVIE MOLDAVE

VOLUME 108. Immunochemical Techniques (Part G: Separation and Characterization of Lymphoid Cells)
Edited by GIOVANNI DI SABATO, JOHN J. LANGONE, AND HELEN VAN VUNAKIS

VOLUME 109. Hormone Action (Part I: Peptide Hormones)
Edited by LUTZ BIRNBAUMER AND BERT W. O'MALLEY

VOLUME 110. Steroids and Isoprenoids (Part A)
Edited by JOHN H. LAW AND HANS C. RILLING

VOLUME 111. Steroids and Isoprenoids (Part B)
Edited by JOHN H. LAW AND HANS C. RILLING

VOLUME 112. Drug and Enzyme Targeting (Part A)
Edited by KENNETH J. WIDDER AND RALPH GREEN

VOLUME 113. Glutamate, Glutamine, Glutathione, and Related Compounds
Edited by ALTON MEISTER

VOLUME 114. Diffraction Methods for Biological Macromolecules (Part A)
Edited by HAROLD W. WYCKOFF, C. H. W. HIRS, AND SERGE N. TIMASHEFF

VOLUME 115. Diffraction Methods for Biological Macromolecules (Part B)
Edited by HAROLD W. WYCKOFF, C. H. W. HIRS, AND SERGE N. TIMASHEFF

VOLUME 116. Immunochemical Techniques
(Part H: Effectors and Mediators of Lymphoid Cell Functions)
Edited by GIOVANNI DI SABATO, JOHN J. LANGONE, AND HELEN VAN VUNAKIS

VOLUME 117. Enzyme Structure (Part J)
Edited by C. H. W. HIRS AND SERGE N. TIMASHEFF

VOLUME 118. Plant Molecular Biology
Edited by ARTHUR WEISSBACH AND HERBERT WEISSBACH

VOLUME 119. Interferons (Part C)
Edited by SIDNEY PESTKA

VOLUME 120. Cumulative Subject Index Volumes 81–94, 96–101

VOLUME 121. Immunochemical Techniques (Part I: Hybridoma Technology and Monoclonal Antibodies)
Edited by JOHN J. LANGONE AND HELEN VAN VUNAKIS

VOLUME 122. Vitamins and Coenzymes (Part G)
Edited by FRANK CHYTIL AND DONALD B. MCCORMICK

Volume 123. Vitamins and Coenzymes (Part H)
Edited by Frank Chytil and Donald B. McCormick

Volume 124. Hormone Action (Part J: Neuroendocrine Peptides)
Edited by P. Michael Conn

Volume 125. Biomembranes (Part M: Transport in Bacteria, Mitochondria, and Chloroplasts: General Approaches and Transport Systems)
Edited by Sidney Fleischer and Becca Fleischer

Volume 126. Biomembranes (Part N: Transport in Bacteria, Mitochondria, and Chloroplasts: Protonmotive Force)
Edited by Sidney Fleischer and Becca Fleischer

Volume 127. Biomembranes (Part O: Protons and Water: Structure and Translocation)
Edited by Lester Packer

Volume 128. Plasma Lipoproteins (Part A: Preparation, Structure, and Molecular Biology)
Edited by Jere P. Segrest and John J. Albers

Volume 129. Plasma Lipoproteins (Part B: Characterization, Cell Biology, and Metabolism)
Edited by John J. Albers and Jere P. Segrest

Volume 130. Enzyme Structure (Part K)
Edited by C. H. W. Hirs and Serge N. Timasheff

Volume 131. Enzyme Structure (Part L)
Edited by C. H. W. Hirs and Serge N. Timasheff

Volume 132. Immunochemical Techniques (Part J: Phagocytosis and Cell-Mediated Cytotoxicity)
Edited by Giovanni Di Sabato and Johannes Everse

Volume 133. Bioluminescence and Chemiluminescence (Part B)
Edited by Marlene DeLuca and William D. McElroy

Volume 134. Structural and Contractile Proteins (Part C: The Contractile Apparatus and the Cytoskeleton)
Edited by Richard B. Vallee

Volume 135. Immobilized Enzymes and Cells (Part B)
Edited by Klaus Mosbach

Volume 136. Immobilized Enzymes and Cells (Part C)
Edited by Klaus Mosbach

Volume 137. Immobilized Enzymes and Cells (Part D)
Edited by Klaus Mosbach

Volume 138. Complex Carbohydrates (Part E)
Edited by Victor Ginsburg

Volume 139. Cellular Regulators (Part A: Calcium- and Calmodulin-Binding Proteins)
Edited by Anthony R. Means and P. Michael Conn

Volume 140. Cumulative Subject Index Volumes 102–119, 121–134

Volume 141. Cellular Regulators (Part B: Calcium and Lipids)
Edited by P. Michael Conn and Anthony R. Means

Volume 142. Metabolism of Aromatic Amino Acids and Amines
Edited by Seymour Kaufman

Volume 143. Sulfur and Sulfur Amino Acids
Edited by William B. Jakoby and Owen Griffith

Volume 144. Structural and Contractile Proteins (Part D: Extracellular Matrix)
Edited by Leon W. Cunningham

Volume 145. Structural and Contractile Proteins (Part E: Extracellular Matrix)
Edited by Leon W. Cunningham

Volume 146. Peptide Growth Factors (Part A)
Edited by David Barnes and David A. Sirbasku

Volume 147. Peptide Growth Factors (Part B)
Edited by David Barnes and David A. Sirbasku

Volume 148. Plant Cell Membranes
Edited by Lester Packer and Roland Douce

Volume 149. Drug and Enzyme Targeting (Part B)
Edited by Ralph Green and Kenneth J. Widder

Volume 150. Immunochemical Techniques (Part K: *In Vitro* Models of B and T Cell Functions and Lymphoid Cell Receptors)
Edited by Giovanni Di Sabato

Volume 151. Molecular Genetics of Mammalian Cells
Edited by Michael M. Gottesman

Volume 152. Guide to Molecular Cloning Techniques
Edited by Shelby L. Berger and Alan R. Kimmel

Volume 153. Recombinant DNA (Part D)
Edited by Ray Wu and Lawrence Grossman

Volume 154. Recombinant DNA (Part E)
Edited by Ray Wu and Lawrence Grossman

Volume 155. Recombinant DNA (Part F)
Edited by Ray Wu

Volume 156. Biomembranes (Part P: ATP-Driven Pumps and Related Transport: The Na, K-Pump)
Edited by Sidney Fleischer and Becca Fleischer

Volume 157. Biomembranes (Part Q: ATP-Driven Pumps and Related Transport: Calcium, Proton, and Potassium Pumps)
Edited by Sidney Fleischer and Becca Fleischer

Volume 158. Metalloproteins (Part A)
Edited by James F. Riordan and Bert L. Vallee

Volume 159. Initiation and Termination of Cyclic Nucleotide Action
Edited by Jackie D. Corbin and Roger A. Johnson

Volume 160. Biomass (Part A: Cellulose and Hemicellulose)
Edited by Willis A. Wood and Scott T. Kellogg

Volume 161. Biomass (Part B: Lignin, Pectin, and Chitin)
Edited by Willis A. Wood and Scott T. Kellogg

Volume 162. Immunochemical Techniques (Part L: Chemotaxis and Inflammation)
Edited by Giovanni Di Sabato

Volume 163. Immunochemical Techniques (Part M: Chemotaxis and Inflammation)
Edited by Giovanni Di Sabato

Volume 164. Ribosomes
Edited by Harry F. Noller, Jr., and Kivie Moldave

Volume 165. Microbial Toxins: Tools for Enzymology
Edited by Sidney Harshman

Volume 166. Branched-Chain Amino Acids
Edited by Robert Harris and John R. Sokatch

Volume 167. Cyanobacteria
Edited by Lester Packer and Alexander N. Glazer

Volume 168. Hormone Action (Part K: Neuroendocrine Peptides)
Edited by P. Michael Conn

Volume 169. Platelets: Receptors, Adhesion, Secretion (Part A)
Edited by Jacek Hawiger

Volume 170. Nucleosomes
Edited by Paul M. Wassarman and Roger D. Kornberg

Volume 171. Biomembranes (Part R: Transport Theory: Cells and Model Membranes)
Edited by Sidney Fleischer and Becca Fleischer

Volume 172. Biomembranes (Part S: Transport: Membrane Isolation and Characterization)
Edited by Sidney Fleischer and Becca Fleischer

VOLUME 173. Biomembranes [Part T: Cellular and Subcellular Transport: Eukaryotic (Nonepithelial) Cells]
Edited by SIDNEY FLEISCHER AND BECCA FLEISCHER

VOLUME 174. Biomembranes [Part U: Cellular and Subcellular Transport: Eukaryotic (Nonepithelial) Cells]
Edited by SIDNEY FLEISCHER AND BECCA FLEISCHER

VOLUME 175. Cumulative Subject Index Volumes 135–139, 141–167

VOLUME 176. Nuclear Magnetic Resonance (Part A: Spectral Techniques and Dynamics)
Edited by NORMAN J. OPPENHEIMER AND THOMAS L. JAMES

VOLUME 177. Nuclear Magnetic Resonance (Part B: Structure and Mechanism)
Edited by NORMAN J. OPPENHEIMER AND THOMAS L. JAMES

VOLUME 178. Antibodies, Antigens, and Molecular Mimicry
Edited by JOHN J. LANGONE

VOLUME 179. Complex Carbohydrates (Part F)
Edited by VICTOR GINSBURG

VOLUME 180. RNA Processing (Part A: General Methods)
Edited by JAMES E. DAHLBERG AND JOHN N. ABELSON

VOLUME 181. RNA Processing (Part B: Specific Methods)
Edited by JAMES E. DAHLBERG AND JOHN N. ABELSON

VOLUME 182. Guide to Protein Purification
Edited by MURRAY P. DEUTSCHER

VOLUME 183. Molecular Evolution: Computer Analysis of Protein and Nucleic Acid Sequences
Edited by RUSSELL F. DOOLITTLE

VOLUME 184. Avidin-Biotin Technology
Edited by MEIR WILCHEK AND EDWARD A. BAYER

VOLUME 185. Gene Expression Technology
Edited by DAVID V. GOEDDEL

VOLUME 186. Oxygen Radicals in Biological Systems (Part B: Oxygen Radicals and Antioxidants)
Edited by LESTER PACKER AND ALEXANDER N. GLAZER

VOLUME 187. Arachidonate Related Lipid Mediators
Edited by ROBERT C. MURPHY AND FRANK A. FITZPATRICK

VOLUME 188. Hydrocarbons and Methylotrophy
Edited by MARY E. LIDSTROM

VOLUME 189. Retinoids (Part A: Molecular and Metabolic Aspects)
Edited by LESTER PACKER

Volume 190. Retinoids (Part B: Cell Differentiation and Clinical Applications)
Edited by Lester Packer

Volume 191. Biomembranes (Part V: Cellular and Subcellular Transport: Epithelial Cells)
Edited by Sidney Fleischer and Becca Fleischer

Volume 192. Biomembranes (Part W: Cellular and Subcellular Transport: Epithelial Cells)
Edited by Sidney Fleischer and Becca Fleischer

Volume 193. Mass Spectrometry
Edited by James A. McCloskey

Volume 194. Guide to Yeast Genetics and Molecular Biology
Edited by Christine Guthrie and Gerald R. Fink

Volume 195. Adenylyl Cyclase, G Proteins, and Guanylyl Cyclase
Edited by Roger A. Johnson and Jackie D. Corbin

Volume 196. Molecular Motors and the Cytoskeleton
Edited by Richard B. Vallee

Volume 197. Phospholipases
Edited by Edward A. Dennis

Volume 198. Peptide Growth Factors (Part C)
Edited by David Barnes, J. P. Mather, and Gordon H. Sato

Volume 199. Cumulative Subject Index Volumes 168–174, 176–194

Volume 200. Protein Phosphorylation (Part A: Protein Kinases: Assays, Purification, Antibodies, Functional Analysis, Cloning, and Expression)
Edited by Tony Hunter and Bartholomew M. Sefton

Volume 201. Protein Phosphorylation (Part B: Analysis of Protein Phosphorylation, Protein Kinase Inhibitors, and Protein Phosphatases)
Edited by Tony Hunter and Bartholomew M. Sefton

Volume 202. Molecular Design and Modeling: Concepts and Applications (Part A: Proteins, Peptides, and Enzymes)
Edited by John J. Langone

Volume 203. Molecular Design and Modeling: Concepts and Applications (Part B: Antibodies and Antigens, Nucleic Acids, Polysaccharides, and Drugs)
Edited by John J. Langone

Volume 204. Bacterial Genetic Systems
Edited by Jeffrey H. Miller

Volume 205. Metallobiochemistry (Part B: Metallothionein and Related Molecules)
Edited by James F. Riordan and Bert L. Vallee

VOLUME 206. Cytochrome P450
Edited by MICHAEL R. WATERMAN AND ERIC F. JOHNSON

VOLUME 207. Ion Channels
Edited by BERNARDO RUDY AND LINDA E. IVERSON

VOLUME 208. Protein–DNA Interactions
Edited by ROBERT T. SAUER

VOLUME 209. Phospholipid Biosynthesis
Edited by EDWARD A. DENNIS AND DENNIS E. VANCE

VOLUME 210. Numerical Computer Methods
Edited by LUDWIG BRAND AND MICHAEL L. JOHNSON

VOLUME 211. DNA Structures (Part A: Synthesis and Physical Analysis of DNA)
Edited by DAVID M. J. LILLEY AND JAMES E. DAHLBERG

VOLUME 212. DNA Structures (Part B: Chemical and Electrophoretic Analysis of DNA)
Edited by DAVID M. J. LILLEY AND JAMES E. DAHLBERG

VOLUME 213. Carotenoids (Part A: Chemistry, Separation, Quantitation, and Antioxidation)
Edited by LESTER PACKER

VOLUME 214. Carotenoids (Part B: Metabolism, Genetics, and Biosynthesis)
Edited by LESTER PACKER

VOLUME 215. Platelets: Receptors, Adhesion, Secretion (Part B)
Edited by JACEK J. HAWIGER

VOLUME 216. Recombinant DNA (Part G)
Edited by RAY WU

VOLUME 217. Recombinant DNA (Part H)
Edited by RAY WU

VOLUME 218. Recombinant DNA (Part I)
Edited by RAY WU

VOLUME 219. Reconstitution of Intracellular Transport
Edited by JAMES E. ROTHMAN

VOLUME 220. Membrane Fusion Techniques (Part A)
Edited by NEJAT DÜZGÜNEŞ

VOLUME 221. Membrane Fusion Techniques (Part B)
Edited by NEJAT DÜZGÜNEŞ

VOLUME 222. Proteolytic Enzymes in Coagulation, Fibrinolysis, and Complement Activation (Part A: Mammalian Blood Coagulation Factors and Inhibitors)
Edited by LASZLO LORAND AND KENNETH G. MANN

Volume 223. Proteolytic Enzymes in Coagulation, Fibrinolysis, and Complement Activation (Part B: Complement Activation, Fibrinolysis, and Nonmammalian Blood Coagulation Factors)
Edited by Laszlo Lorand and Kenneth G. Mann

Volume 224. Molecular Evolution: Producing the Biochemical Data
Edited by Elizabeth Anne Zimmer, Thomas J. White, Rebecca L. Cann, and Allan C. Wilson

Volume 225. Guide to Techniques in Mouse Development
Edited by Paul M. Wassarman and Melvin L. DePamphilis

Volume 226. Metallobiochemistry (Part C: Spectroscopic and Physical Methods for Probing Metal Ion Environments in Metalloenzymes and Metalloproteins)
Edited by James F. Riordan and Bert L. Vallee

Volume 227. Metallobiochemistry (Part D: Physical and Spectroscopic Methods for Probing Metal Ion Environments in Metalloproteins)
Edited by James F. Riordan and Bert L. Vallee

Volume 228. Aqueous Two-Phase Systems
Edited by Harry Walter and Göte Johansson

Volume 229. Cumulative Subject Index Volumes 195–198, 200–227

Volume 230. Guide to Techniques in Glycobiology
Edited by William J. Lennarz and Gerald W. Hart

Volume 231. Hemoglobins (Part B: Biochemical and Analytical Methods)
Edited by Johannes Everse, Kim D. Vandegriff, and Robert M. Winslow

Volume 232. Hemoglobins (Part C: Biophysical Methods)
Edited by Johannes Everse, Kim D. Vandegriff, and Robert M. Winslow

Volume 233. Oxygen Radicals in Biological Systems (Part C)
Edited by Lester Packer

Volume 234. Oxygen Radicals in Biological Systems (Part D)
Edited by Lester Packer

Volume 235. Bacterial Pathogenesis (Part A: Identification and Regulation of Virulence Factors)
Edited by Virginia L. Clark and Patrik M. Bavoil

Volume 236. Bacterial Pathogenesis (Part B: Integration of Pathogenic Bacteria with Host Cells)
Edited by Virginia L. Clark and Patrik M. Bavoil

Volume 237. Heterotrimeric G Proteins
Edited by Ravi Iyengar

Volume 238. Heterotrimeric G-Protein Effectors
Edited by Ravi Iyengar

VOLUME 239. Nuclear Magnetic Resonance (Part C)
Edited by THOMAS L. JAMES AND NORMAN J. OPPENHEIMER

VOLUME 240. Numerical Computer Methods (Part B)
Edited by MICHAEL L. JOHNSON AND LUDWIG BRAND

VOLUME 241. Retroviral Proteases
Edited by LAWRENCE C. KUO AND JULES A. SHAFER

VOLUME 242. Neoglycoconjugates (Part A)
Edited by Y. C. LEE AND REIKO T. LEE

VOLUME 243. Inorganic Microbial Sulfur Metabolism
Edited by HARRY D. PECK, JR., AND JEAN LEGALL

VOLUME 244. Proteolytic Enzymes: Serine and Cysteine Peptidases
Edited by ALAN J. BARRETT

VOLUME 245. Extracellular Matrix Components
Edited by E. RUOSLAHTI AND E. ENGVALL

VOLUME 246. Biochemical Spectroscopy
Edited by KENNETH SAUER

VOLUME 247. Neoglycoconjugates (Part B: Biomedical Applications)
Edited by Y. C. LEE AND REIKO T. LEE

VOLUME 248. Proteolytic Enzymes: Aspartic and Metallo Peptidases
Edited by ALAN J. BARRETT

VOLUME 249. Enzyme Kinetics and Mechanism (Part D: Developments in Enzyme Dynamics)
Edited by DANIEL L. PURICH

VOLUME 250. Lipid Modifications of Proteins
Edited by PATRICK J. CASEY AND JANICE E. BUSS

VOLUME 251. Biothiols (Part A: Monothiols and Dithiols, Protein Thiols, and Thiyl Radicals)
Edited by LESTER PACKER

VOLUME 252. Biothiols (Part B: Glutathione and Thioredoxin; Thiols in Signal Transduction and Gene Regulation)
Edited by LESTER PACKER

VOLUME 253. Adhesion of Microbial Pathogens
Edited by RON J. DOYLE AND ITZHAK OFEK

VOLUME 254. Oncogene Techniques
Edited by PETER K. VOGT AND INDER M. VERMA

VOLUME 255. Small GTPases and Their Regulators (Part A: Ras Family)
Edited by W. E. BALCH, CHANNING J. DER, AND ALAN HALL

VOLUME 256. Small GTPases and Their Regulators (Part B: Rho Family)
Edited by W. E. BALCH, CHANNING J. DER, AND ALAN HALL

Volume 257. Small GTPases and Their Regulators (Part C: Proteins Involved in Transport)
Edited by W. E. Balch, Channing J. Der, and Alan Hall

Volume 258. Redox-Active Amino Acids in Biology
Edited by Judith P. Klinman

Volume 259. Energetics of Biological Macromolecules
Edited by Michael L. Johnson and Gary K. Ackers

Volume 260. Mitochondrial Biogenesis and Genetics (Part A)
Edited by Giuseppe M. Attardi and Anne Chomyn

Volume 261. Nuclear Magnetic Resonance and Nucleic Acids
Edited by Thomas L. James

Volume 262. DNA Replication
Edited by Judith L. Campbell

Volume 263. Plasma Lipoproteins (Part C: Quantitation)
Edited by William A. Bradley, Sandra H. Gianturco, and Jere P. Segrest

Volume 264. Mitochondrial Biogenesis and Genetics (Part B)
Edited by Giuseppe M. Attardi and Anne Chomyn

Volume 265. Cumulative Subject Index Volumes 228, 230–262

Volume 266. Computer Methods for Macromolecular Sequence Analysis
Edited by Russell F. Doolittle

Volume 267. Combinatorial Chemistry
Edited by John N. Abelson

Volume 268. Nitric Oxide (Part A: Sources and Detection of NO; NO Synthase)
Edited by Lester Packer

Volume 269. Nitric Oxide (Part B: Physiological and Pathological Processes)
Edited by Lester Packer

Volume 270. High Resolution Separation and Analysis of Biological Macromolecules (Part A: Fundamentals)
Edited by Barry L. Karger and William S. Hancock

Volume 271. High Resolution Separation and Analysis of Biological Macromolecules (Part B: Applications)
Edited by Barry L. Karger and William S. Hancock

Volume 272. Cytochrome P450 (Part B)
Edited by Eric F. Johnson and Michael R. Waterman

Volume 273. RNA Polymerase and Associated Factors (Part A)
Edited by Sankar Adhya

Volume 274. RNA Polymerase and Associated Factors (Part B)
Edited by Sankar Adhya

Volume 275. Viral Polymerases and Related Proteins
Edited by Lawrence C. Kuo, David B. Olsen, and Steven S. Carroll

Volume 276. Macromolecular Crystallography (Part A)
Edited by Charles W. Carter, Jr., and Robert M. Sweet

Volume 277. Macromolecular Crystallography (Part B)
Edited by Charles W. Carter, Jr., and Robert M. Sweet

Volume 278. Fluorescence Spectroscopy
Edited by Ludwig Brand and Michael L. Johnson

Volume 279. Vitamins and Coenzymes (Part I)
Edited by Donald B. McCormick, John W. Suttie, and Conrad Wagner

Volume 280. Vitamins and Coenzymes (Part J)
Edited by Donald B. McCormick, John W. Suttie, and Conrad Wagner

Volume 281. Vitamins and Coenzymes (Part K)
Edited by Donald B. McCormick, John W. Suttie, and Conrad Wagner

Volume 282. Vitamins and Coenzymes (Part L)
Edited by Donald B. McCormick, John W. Suttie, and Conrad Wagner

Volume 283. Cell Cycle Control
Edited by William G. Dunphy

Volume 284. Lipases (Part A: Biotechnology)
Edited by Byron Rubin and Edward A. Dennis

Volume 285. Cumulative Subject Index Volumes 263, 264, 266–284, 286–289

Volume 286. Lipases (Part B: Enzyme Characterization and Utilization)
Edited by Byron Rubin and Edward A. Dennis

Volume 287. Chemokines
Edited by Richard Horuk

Volume 288. Chemokine Receptors
Edited by Richard Horuk

Volume 289. Solid Phase Peptide Synthesis
Edited by Gregg B. Fields

Volume 290. Molecular Chaperones
Edited by George H. Lorimer and Thomas Baldwin

Volume 291. Caged Compounds
Edited by Gerard Marriott

Volume 292. ABC Transporters: Biochemical, Cellular, and Molecular Aspects
Edited by Suresh V. Ambudkar and Michael M. Gottesman

Volume 293. Ion Channels (Part B)
Edited by P. Michael Conn

VOLUME 294. Ion Channels (Part C)
Edited by P. MICHAEL CONN

VOLUME 295. Energetics of Biological Macromolecules (Part B)
Edited by GARY K. ACKERS AND MICHAEL L. JOHNSON

VOLUME 296. Neurotransmitter Transporters
Edited by SUSAN G. AMARA

VOLUME 297. Photosynthesis: Molecular Biology of Energy Capture
Edited by LEE MCINTOSH

VOLUME 298. Molecular Motors and the Cytoskeleton (Part B)
Edited by RICHARD B. VALLEE

VOLUME 299. Oxidants and Antioxidants (Part A)
Edited by LESTER PACKER

VOLUME 300. Oxidants and Antioxidants (Part B)
Edited by LESTER PACKER

VOLUME 301. Nitric Oxide: Biological and Antioxidant Activities (Part C)
Edited by LESTER PACKER

VOLUME 302. Green Fluorescent Protein
Edited by P. MICHAEL CONN

VOLUME 303. cDNA Preparation and Display
Edited by SHERMAN M. WEISSMAN

VOLUME 304. Chromatin
Edited by PAUL M. WASSARMAN AND ALAN P. WOLFFE

VOLUME 305. Bioluminescence and Chemiluminescence (Part C)
Edited by THOMAS O. BALDWIN AND MIRIAM M. ZIEGLER

VOLUME 306. Expression of Recombinant Genes in Eukaryotic Systems
Edited by JOSEPH C. GLORIOSO AND MARTIN C. SCHMIDT

VOLUME 307. Confocal Microscopy
Edited by P. MICHAEL CONN

VOLUME 308. Enzyme Kinetics and Mechanism (Part E: Energetics of Enzyme Catalysis)
Edited by DANIEL L. PURICH AND VERN L. SCHRAMM

VOLUME 309. Amyloid, Prions, and Other Protein Aggregates
Edited by RONALD WETZEL

VOLUME 310. Biofilms
Edited by RON J. DOYLE

VOLUME 311. Sphingolipid Metabolism and Cell Signaling (Part A)
Edited by ALFRED H. MERRILL, JR., AND YUSUF A. HANNUN

VOLUME 312. Sphingolipid Metabolism and Cell Signaling (Part B)
Edited by ALFRED H. MERRILL, JR., AND YUSUF A. HANNUN

VOLUME 313. Antisense Technology (Part A: General Methods, Methods of Delivery, and RNA Studies)
Edited by M. IAN PHILLIPS

VOLUME 314. Antisense Technology (Part B: Applications)
Edited by M. IAN PHILLIPS

VOLUME 315. Vertebrate Phototransduction and the Visual Cycle (Part A)
Edited by KRZYSZTOF PALCZEWSKI

VOLUME 316. Vertebrate Phototransduction and the Visual Cycle (Part B)
Edited by KRZYSZTOF PALCZEWSKI

VOLUME 317. RNA–Ligand Interactions (Part A: Structural Biology Methods)
Edited by DANIEL W. CELANDER AND JOHN N. ABELSON

VOLUME 318. RNA–Ligand Interactions (Part B: Molecular Biology Methods)
Edited by DANIEL W. CELANDER AND JOHN N. ABELSON

VOLUME 319. Singlet Oxygen, UV-A, and Ozone
Edited by LESTER PACKER AND HELMUT SIES

VOLUME 320. Cumulative Subject Index Volumes 290–319

VOLUME 321. Numerical Computer Methods (Part C)
Edited by MICHAEL L. JOHNSON AND LUDWIG BRAND

VOLUME 322. Apoptosis
Edited by JOHN C. REED

VOLUME 323. Energetics of Biological Macromolecules (Part C)
Edited by MICHAEL L. JOHNSON AND GARY K. ACKERS

VOLUME 324. Branched-Chain Amino Acids (Part B)
Edited by ROBERT A. HARRIS AND JOHN R. SOKATCH

VOLUME 325. Regulators and Effectors of Small GTPases (Part D: Rho Family)
Edited by W. E. BALCH, CHANNING J. DER, AND ALAN HALL

VOLUME 326. Applications of Chimeric Genes and Hybrid Proteins (Part A: Gene Expression and Protein Purification)
Edited by JEREMY THORNER, SCOTT D. EMR, AND JOHN N. ABELSON

VOLUME 327. Applications of Chimeric Genes and Hybrid Proteins (Part B: Cell Biology and Physiology)
Edited by JEREMY THORNER, SCOTT D. EMR, AND JOHN N. ABELSON

VOLUME 328. Applications of Chimeric Genes and Hybrid Proteins (Part C: Protein–Protein Interactions and Genomics)
Edited by JEREMY THORNER, SCOTT D. EMR, AND JOHN N. ABELSON

Volume 329. Regulators and Effectors of Small GTPases (Part E: GTPases Involved in Vesicular Traffic)
Edited by W. E. Balch, Channing J. Der, and Alan Hall

Volume 330. Hyperthermophilic Enzymes (Part A)
Edited by Michael W. W. Adams and Robert M. Kelly

Volume 331. Hyperthermophilic Enzymes (Part B)
Edited by Michael W. W. Adams and Robert M. Kelly

Volume 332. Regulators and Effectors of Small GTPases (Part F: Ras Family I)
Edited by W. E. Balch, Channing J. Der, and Alan Hall

Volume 333. Regulators and Effectors of Small GTPases (Part G: Ras Family II)
Edited by W. E. Balch, Channing J. Der, and Alan Hall

Volume 334. Hyperthermophilic Enzymes (Part C)
Edited by Michael W. W. Adams and Robert M. Kelly

Volume 335. Flavonoids and Other Polyphenols
Edited by Lester Packer

Volume 336. Microbial Growth in Biofilms (Part A: Developmental and Molecular Biological Aspects)
Edited by Ron J. Doyle

Volume 337. Microbial Growth in Biofilms (Part B: Special Environments and Physicochemical Aspects)
Edited by Ron J. Doyle

Volume 338. Nuclear Magnetic Resonance of Biological Macromolecules (Part A)
Edited by Thomas L. James, Volker Dötsch, and Uli Schmitz

Volume 339. Nuclear Magnetic Resonance of Biological Macromolecules (Part B)
Edited by Thomas L. James, Volker Dötsch, and Uli Schmitz

Volume 340. Drug–Nucleic Acid Interactions
Edited by Jonathan B. Chaires and Michael J. Waring

Volume 341. Ribonucleases (Part A)
Edited by Allen W. Nicholson

Volume 342. Ribonucleases (Part B)
Edited by Allen W. Nicholson

Volume 343. G Protein Pathways (Part A: Receptors)
Edited by Ravi Iyengar and John D. Hildebrandt

Volume 344. G Protein Pathways (Part B: G Proteins and Their Regulators)
Edited by Ravi Iyengar and John D. Hildebrandt

Volume 345. G Protein Pathways (Part C: Effector Mechanisms)
Edited by Ravi Iyengar and John D. Hildebrandt

VOLUME 346. Gene Therapy Methods
Edited by M. IAN PHILLIPS

VOLUME 347. Protein Sensors and Reactive Oxygen Species (Part A: Selenoproteins and Thioredoxin)
Edited by HELMUT SIES AND LESTER PACKER

VOLUME 348. Protein Sensors and Reactive Oxygen Species (Part B: Thiol Enzymes and Proteins)
Edited by HELMUT SIES AND LESTER PACKER

VOLUME 349. Superoxide Dismutase
Edited by LESTER PACKER

VOLUME 350. Guide to Yeast Genetics and Molecular and Cell Biology (Part B)
Edited by CHRISTINE GUTHRIE AND GERALD R. FINK

VOLUME 351. Guide to Yeast Genetics and Molecular and Cell Biology (Part C)
Edited by CHRISTINE GUTHRIE AND GERALD R. FINK

VOLUME 352. Redox Cell Biology and Genetics (Part A)
Edited by CHANDAN K. SEN AND LESTER PACKER

VOLUME 353. Redox Cell Biology and Genetics (Part B)
Edited by CHANDAN K. SEN AND LESTER PACKER

VOLUME 354. Enzyme Kinetics and Mechanisms (Part F: Detection and Characterization of Enzyme Reaction Intermediates)
Edited by DANIEL L. PURICH

VOLUME 355. Cumulative Subject Index Volumes 321–354

VOLUME 356. Laser Capture Microscopy and Microdissection
Edited by P. MICHAEL CONN

VOLUME 357. Cytochrome P450, Part C
Edited by ERIC F. JOHNSON AND MICHAEL R. WATERMAN

VOLUME 358. Bacterial Pathogenesis (Part C: Identification, Regulation, and Function of Virulence Factors)
Edited by VIRGINIA L. CLARK AND PATRIK M. BAVOIL

VOLUME 359. Nitric Oxide (Part D)
Edited by ENRIQUE CADENAS AND LESTER PACKER

VOLUME 360. Biophotonics (Part A)
Edited by GERARD MARRIOTT AND IAN PARKER

VOLUME 361. Biophotonics (Part B)
Edited by GERARD MARRIOTT AND IAN PARKER

VOLUME 362. Recognition of Carbohydrates in Biological Systems (Part A)
Edited by YUAN C. LEE AND REIKO T. LEE

VOLUME 363. Recognition of Carbohydrates in Biological Systems (Part B)
Edited by YUAN C. LEE AND REIKO T. LEE

VOLUME 364. Nuclear Receptors
Edited by DAVID W. RUSSELL AND DAVID J. MANGELSDORF

VOLUME 365. Differentiation of Embryonic Stem Cells
Edited by PAUL M. WASSAUMAN AND GORDON M. KELLER

VOLUME 366. Protein Phosphatases
Edited by SUSANNE KLUMPP AND JOSEF KRIEGLSTEIN

VOLUME 367. Liposomes (Part A)
Edited by NEJAT DÜZGÜNEŞ

VOLUME 368. Macromolecular Crystallography (Part C)
Edited by CHARLES W. CARTER, JR., AND ROBERT M. SWEET

VOLUME 369. Combinational Chemistry (Part B)
Edited by GUILLERMO A. MORALES AND BARRY A. BUNIN

VOLUME 370. RNA Polymerases and Associated Factors (Part C)
Edited by SANKAR L. ADHYA AND SUSAN GARGES

VOLUME 371. RNA Polymerases and Associated Factors (Part D)
Edited by SANKAR L. ADHYA AND SUSAN GARGES

VOLUME 372. Liposomes (Part B)
Edited by NEJAT DÜZGÜNEŞ

VOLUME 373. Liposomes (Part C)
Edited by NEJAT DÜZGÜNEŞ

VOLUME 374. Macromolecular Crystallography (Part D)
Edited by CHARLES W. CARTER, JR., AND ROBERT W. SWEET

VOLUME 375. Chromatin and Chromatin Remodeling Enzymes (Part A)
Edited by C. DAVID ALLIS AND CARL WU

VOLUME 376. Chromatin and Chromatin Remodeling Enzymes (Part B)
Edited by C. DAVID ALLIS AND CARL WU

VOLUME 377. Chromatin and Chromatin Remodeling Enzymes (Part C)
Edited by C. DAVID ALLIS AND CARL WU

VOLUME 378. Quinones and Quinone Enzymes (Part A)
Edited by HELMUT SIES AND LESTER PACKER

VOLUME 379. Energetics of Biological Macromolecules (Part D)
Edited by JO M. HOLT, MICHAEL L. JOHNSON, AND GARY K. ACKERS

VOLUME 380. Energetics of Biological Macromolecules (Part E)
Edited by JO M. HOLT, MICHAEL L. JOHNSON, AND GARY K. ACKERS

VOLUME 381. Oxygen Sensing
Edited by CHANDAN K. SEN AND GREGG L. SEMENZA

Volume 382. Quinones and Quinone Enzymes (Part B)
Edited by Helmut Sies and Lester Packer

Volume 383. Numerical Computer Methods (Part D)
Edited by Ludwig Brand and Michael L. Johnson

Volume 384. Numerical Computer Methods (Part E)
Edited by Ludwig Brand and Michael L. Johnson

Volume 385. Imaging in Biological Research (Part A)
Edited by P. Michael Conn

Volume 386. Imaging in Biological Research (Part B)
Edited by P. Michael Conn

Volume 387. Liposomes (Part D)
Edited by Nejat Düzgüneş

Volume 388. Protein Engineering
Edited by Dan E. Robertson and Joseph P. Noel

Volume 389. Regulators of G-Protein Signaling (Part A)
Edited by David P. Siderovski

Volume 390. Regulators of G-Protein Signaling (Part B)
Edited by David P. Siderovski

Volume 391. Liposomes (Part E)
Edited by Nejat Düzgüneş

Volume 392. RNA Interference
Edited by Engelke Rossi

Volume 393. Circadian Rhythms
Edited by Michael W. Young

Volume 394. Nuclear Magnetic Resonance of Biological Macromolecules (Part C)
Edited by Thomas L. James

Volume 395. Producing the Biochemical Data (Part B)
Edited by Elizabeth A. Zimmer and Eric H. Roalson

Volume 396. Nitric Oxide (Part E)
Edited by Lester Packer and Enrique Cadenas

Volume 397. Environmental Microbiology
Edited by Jared R. Leadbetter

Volume 398. Ubiquitin and Protein Degradation (Part A)
Edited by Raymond J. Deshaies

Volume 399. Ubiquitin and Protein Degradation (Part B)
Edited by Raymond J. Deshaies

Volume 400. Phase II Conjugation Enzymes and Transport Systems
Edited by Helmut Sies and Lester Packer

Volume 401. Glutathione Transferases and Gamma Glutamyl Transpeptidases
Edited by Helmut Sies and Lester Packer

Volume 402. Biological Mass Spectrometry
Edited by A. L. Burlingame

Volume 403. GTPases Regulating Membrane Targeting and Fusion
Edited by William E. Balch, Channing J. Der, and Alan Hall

Volume 404. GTPases Regulating Membrane Dynamics
Edited by William E. Balch, Channing J. Der, and Alan Hall

Volume 405. Mass Spectrometry: Modified Proteins and Glycoconjugates
Edited by A. L. Burlingame

Volume 406. Regulators and Effectors of Small GTPases: Rho Family
Edited by William E. Balch, Channing J. Der, and Alan Hall

Volume 407. Regulators and Effectors of Small GTPases: Ras Family
Edited by William E. Balch, Channing J. Der, and Alan Hall

Volume 408. DNA Repair (Part A)
Edited by Judith L. Campbell and Paul Modrich

Volume 409. DNA Repair (Part B)
Edited by Judith L. Campbell and Paul Modrich

Volume 410. DNA Microarrays (Part A: Array Platforms and Web-Bench Protocols)
Edited by Alan Kimmel and Brian Oliver

Volume 411. DNA Microarrays (Part B: Databases and Statistics)
Edited by Alan Kimmel and Brian Oliver

Volume 412. Amyloid, Prions, and Other Protein Aggregates (Part B)
Edited by Indu Kheterpal and Ronald Wetzel

Volume 413. Amyloid, Prions, and Other Protein Aggregates (Part C)
Edited by Indu Kheterpal and Ronald Wetzel

Volume 414. Measuring Biological Responses with Automated Microscopy
Edited by James Inglese

Volume 415. Glycobiology
Edited by Minoru Fukuda

Volume 416. Glycomics
Edited by Minoru Fukuda

Volume 417. Functional Glycomics
Edited by Minoru Fukuda

Volume 418. Embryonic Stem Cells
Edited by Irina Klimanskaya and Robert Lanza

VOLUME 419. Adult Stem Cells
Edited by IRINA KLIMANSKAYA AND ROBERT LANZA

VOLUME 420. Stem Cell Tools and Other Experimental Protocols
Edited by IRINA KLIMANSKAYA AND ROBERT LANZA

VOLUME 421. Advanced Bacterial Genetics: Use of Transposons and Phage for Genomic Engineering
Edited by KELLY T. HUGHES

VOLUME 422. Two-Component Signaling Systems, Part A
Edited by MELVIN I. SIMON, BRIAN R. CRANE, AND ALEXANDRINE CRANE

VOLUME 423. Two-Component Signaling Systems, Part B
Edited by MELVIN I. SIMON, BRIAN R. CRANE, AND ALEXANDRINE CRANE

VOLUME 424. RNA Editing
Edited by JONATHA M. GOTT

VOLUME 425. RNA Modification
Edited by JONATHA M. GOTT

VOLUME 426. Integrins
Edited by DAVID CHERESH

VOLUME 427. MicroRNA Methods
Edited by JOHN J. ROSSI

VOLUME 428. Osmosensing and Osmosignaling
Edited by HELMUT SIES AND DIETER HAUSSINGER

VOLUME 429. Translation Initiation: Extract Systems and Molecular Genetics
Edited by JON LORSCH

VOLUME 430. Translation Initiation: Reconstituted Systems and Biophysical Methods
Edited by JON LORSCH

VOLUME 431. Translation Initiation: Cell Biology, High-Throughput and Chemical-Based Approaches
Edited by JON LORSCH

VOLUME 432. Lipidomics and Bioactive Lipids: Mass-Spectrometry–Based Lipid Analysis
Edited by H. ALEX BROWN

VOLUME 433. Lipidomics and Bioactive Lipids: Specialized Analytical Methods and Lipids in Disease
Edited by H. ALEX BROWN

VOLUME 434. Lipidomics and Bioactive Lipids: Lipids and Cell Signaling
Edited by H. ALEX BROWN

VOLUME 435. Oxygen Biology and Hypoxia
Edited by HELMUT SIES AND BERNHARD BRÜNE

VOLUME 436. Globins and Other Nitric Oxide-Reactive Protiens (Part A)
Edited by ROBERT K. POOLE

VOLUME 437. Globins and Other Nitric Oxide-Reactive Protiens (Part B)
Edited by ROBERT K. POOLE

VOLUME 438. Small GTPases in Disease (Part A)
Edited by WILLIAM E. BALCH, CHANNING J. DER, AND ALAN HALL

VOLUME 439. Small GTPases in Disease (Part B)
Edited by WILLIAM E. BALCH, CHANNING J. DER, AND ALAN HALL

VOLUME 440. Nitric Oxide, Part F Oxidative and Nitrosative Stress in Redox Regulation of Cell Signaling
Edited by ENRIQUE CADENAS AND LESTER PACKER

VOLUME 441. Nitric Oxide, Part G Oxidative and Nitrosative Stress in Redox Regulation of Cell Signaling
Edited by ENRIQUE CADENAS AND LESTER PACKER

VOLUME 442. Programmed Cell Death, General Principles for Studying Cell Death (Part A)
Edited by ROYA KHOSRAVI-FAR, ZAHRA ZAKERI, RICHARD A. LOCKSHIN, AND MAURO PIACENTINI

VOLUME 443. Angiogenesis: *In Vitro* Systems
Edited by DAVID A. CHERESH

VOLUME 444. Angiogenesis: *In Vivo* Systems (Part A)
Edited by DAVID A. CHERESH

VOLUME 445. Angiogenesis: *In Vivo* Systems (Part B)
Edited by DAVID A. CHERESH

VOLUME 446. Programmed Cell Death, The Biology and Therapeutic Implications of Cell Death (Part B)
Edited by ROYA KHOSRAVI-FAR, ZAHRA ZAKERI, RICHARD A. LOCKSHIN, AND MAURO PIACENTINI

VOLUME 447. RNA Turnover in Bacteria, Archaea and Organelles
Edited by LYNNE E. MAQUAT AND CECILIA M. ARRAIANO

VOLUME 448. RNA Turnover in Eukaryotes: Nucleases, Pathways and Analysis of mRNA Decay
Edited by LYNNE E. MAQUAT AND MEGERDITCH KILEDJIAN

VOLUME 449. RNA Turnover in Eukaryotes: Analysis of Specialized and Quality Control RNA Decay Pathways
Edited by LYNNE E. MAQUAT AND MEGERDITCH KILEDJIAN

VOLUME 450. Fluorescence Spectroscopy
Edited by LUDWIG BRAND AND MICHAEL L. JOHNSON

VOLUME 451. Autophagy: Lower Eukaryotes and Non-Mammalian Systems (Part A)
Edited by DANIEL J. KLIONSKY

VOLUME 452. Autophagy in Mammalian Systems (Part B)
Edited by DANIEL J. KLIONSKY

VOLUME 453. Autophagy in Disease and Clinical Applications (Part C)
Edited by DANIEL J. KLIONSKY

VOLUME 454. Computer Methods (Part A)
Edited by MICHAEL L. JOHNSON AND LUDWIG BRAND

VOLUME 455. Biothermodynamics (Part A)
Edited by MICHAEL L. JOHNSON, JO M. HOLT, AND GARY K. ACKERS (RETIRED)

VOLUME 456. Mitochondrial Function, Part A: Mitochondrial Electron Transport Complexes and Reactive Oxygen Species
Edited by WILLIAM S. ALLISON AND IMMO E. SCHEFFLER

VOLUME 457. Mitochondrial Function, Part B: Mitochondrial Protein Kinases, Protein Phosphatases and Mitochondrial Diseases
Edited by WILLIAM S. ALLISON AND ANNE N. MURPHY

VOLUME 458. Complex Enzymes in Microbial Natural Product Biosynthesis, Part A: Overview Articles and Peptides
Edited by DAVID A. HOPWOOD

VOLUME 459. Complex Enzymes in Microbial Natural Product Biosynthesis, Part B: Polyketides, Aminocoumarins and Carbohydrates
Edited by DAVID A. HOPWOOD

VOLUME 460. Chemokines, Part A
Edited by TRACY M. HANDEL AND DAMON J. HAMEL

VOLUME 461. Chemokines, Part B
Edited by TRACY M. HANDEL AND DAMON J. HAMEL

VOLUME 462. Non-Natural Amino Acids
Edited by TOM W. MUIR AND JOHN N. ABELSON

VOLUME 463. Guide to Protein Purification, 2nd Edition
Edited by RICHARD R. BURGESS AND MURRAY P. DEUTSCHER

PREFACE TO CHAPTER 1

It is with great admiration that we reprint in its entirety Chapter 1 of the first edition of this volume, "Why Purify Enzymes" by the late Arthur Kornberg. Arthur was one of the giants of 20th century biochemistry and his forte was enzymology. Upon reading the chapter, one can appreciate why this was so. His passion for identifying an enzyme activity and purifying it to homogeneity so that many of its properties could be directly studied become clearly evident from Arthur's words. Yet, he was clearly mindful of what he called the enzyme's "social face," its interactions with other cellular components, an attribute that has gained more prominence as we have come to understand the importance of cell organization.

There have been many advances in the 20 years since the chapter was written, particularly in the ease with which a protein can be purified if its gene is known. Nevertheless, the admonition to "purify, purify, purify" remains relevant when one is studying the catalytic properties of an enzyme, elucidating the structure of a protein, or identifying possible regulatory factors. Protein purification remains of prime importance if one endeavors to understand the enzyme responsible for a newly discovered cellular reaction or the proteins involved in a cellular process. With this in mind, the lessons to be learned from reading, or re-reading, Arthur's wonderfully lucid and pertinent chapter will be extremely rewarding.

MURRAY P. DEUTSCHER

CHAPTER ONE

Why Purify Enzymes?

Arthur Kornberg✝

"Don't waste clean thinking on dirty enzymes" is an admonition of Efraim Racker's which is at the core of enzymology and good chemical practice. It simply says that detailed studies of how an enzyme catalyzes the conversion of one substance to another is generally a waste of time until the enzyme has been purified away from the other enzymes and substances that make up a crude cell extract. The mixture of thousands of different enzymes released from a disrupted liver, yeast, or bacterial cell likely contains several that direct other rearrangement of the starting material and the product of the particular enzyme's action. Only when we have purified the enzyme to the point that no other enzymes can be detected can we fell assured that a single type of enzyme molecule directs the conversion of substance A to substance B, and does nothing more. Only then can we learn how the enzyme does its work.

The rewards for the labor of purifying an enzyme were laid out in a series of inspirational papers by Otto Warburg in the 1930s. From his laboratory in Berlin-Dahlem came the discipline and many of the methods of purifying enzymes and with those the clarification of key reactions and vitamin functions in respiration and the fermentation of glucose. Warburg's contributions strengthened the *classic approach* to enzymology inaugurated with Eduard Büchner's accidental discovery, at the turn of this century, of cell-free conversion of sucrose to ethanol. One tracks the molecular basis of cellular function—alcoholic fermentation in yeast, glycolysis in muscle, luminescence in a fly, or the replication of DNA—by first observing the phenomenon in a cell-free system. Then one isolates the responsible enzyme (or enzymes) by fractionation of the cell extract and purifies it to homogeneity. Then one hopes to learn enough about the structure of the enzyme to explain how it performs its catalytic functions, responds to regulatory signals, and is associated with other enzymes and structures in the cell.

By a reverse approach, call it *neoclassical*, especially popular in recent decades, one first obtains a structure and then looks for its function. The protein is preferably small and stable, and has been amplified by cloning or is commercially available. By intensive study of the protein and homologous proteins, one hopes to get some clues to how it functions. As the popularity

✝ Deceased

Methods in Enzymology, Volume 463
ISSN 0076-6879, DOI: 10.1016/S0076-6879(09)63001-9

of the neoclassical approach has increased, so has there been a corresponding decrease in interest in the classical route: pursuit of a function to isolate the responsible structure.

Implicit in the devotion to purifying enzymes is the faith of a dedicated biochemist of being able to reconstitute in a test tube anything a cell can do. In fact, the biochemist with the advantage of manipulating the medium: pH, ionic strength, etc., by creating high concentrations of reactants, by trapping products and so on, should have an easier time of it. Another article of faith is that everything that goes on in a cell is catalyzed by an enzyme. Chemists sometimes find this conviction difficult to swallow.

On a recent occasion I was told by a mature and well-known physical chemist that what fascinated him most in my work was that DNA replication was catalyzed by enzymes! This reminded me of a seminar I gave to the Washington University chemistry department when I arrived in St. Louis in 1953. I was describing the enzymes that make and degrade orotic acid, and began to realize that my audience was rapidly slipping away. Perhaps, they had been expecting to hear about an organic synthesis of erotic acid. In a last-ditch attempt to retrieve their attention, I said loudly that every chemical event in the cell depends on the action of an enzyme. At that point, the late Joseph Kennedy, the brilliant young chairman, awoke: "Do you mean to tell us that something as simple as the hydration of carbon dioxide (to form bicarbonate) needs an enzyme?" The Lord had delivered him into my hands. "Yes, Joe, cells have an enzyme, called carbonic anhydrase. It enhances the rate of that reaction more than a million fold."

Enzymes are awesome machines with a suitable level of complexity. One may feel ill at ease grappling with the operations of a cell, let alone those of a multicellular creature, or feel inadequate in probing the fine chemistry of small molecules. Becoming familiar with the personality of an enzyme performing in a major synthetic pathway can be just right. To gain his intimacy, the enzyme must first be purified to near homogeneity. For the separation of a protein species present as one-tenth or one-hundredth of 1% of the many thousands of other kinds in the cellular community, we need to devise and be guided by a quick and reliable assay of its catalytic activity.

No enzyme is purified to the point of absolute homogeneity. Even when other proteins constitute less than 1% of the purified protein and escape detection by our best methods, there are likely to be many millions of foreign molecules in a reaction mixture. Generally, such contaminants do not matter unless they are preponderantly of one kind and are highly active on one of the components being studied.

Only after the properties of the pure enzyme are known is it profitable to examine its behavior in a crude state. "Don't waste clean thinking on dirty enzymes" is sound dogma. I cannot recall a single instance in which I begrudged the time spent on the purification of an enzyme, whether it led to the clarification of a reaction pathway, to discovering new enzymes, to acquiring a unique analytical reagent, or led merely to greater expertise with purification procedures. So, purify, purify, purify.

Purifying an enzyme is rewarding all the way, from first starting to free it from the mob of proteins in a broken cell to having it finally in splendid isolation. It matters that, upon removing the enzyme from its snug cellular niche, one cares about many inclemencies: high dilution in unfriendly solvents, contact with glass surfaces and harsh temperatures, and exposure to metals, oxygen, and untold other perils. Failures are often attributed to the fragility of the enzyme and its ready denaturability, whereas the blame should rest on the scientist for being more easily denatured. Like a parent concerned for a child's whereabouts and safety, one cannot leave the laboratory at night without knowing how much of the enzyme has been recovered in that day's procedure and how much of the contaminating proteins still remain.

To attain the goal of a pure protein, the cardinal rule is that the ratio of enzyme activity to the total protein is increased to the limit. Units of activity and amounts of protein must be strictly accounted for in each manipulation and at every stage. In this vein, the notebook record of an enzyme purification should withstand the scrutiny of an auditor or bank examiner. Not that one should ever regard the enterprise as a business or banking operation. Rather, it often many seem like the ascent of an uncharted mountain: the logistics like those of supplying successively higher base camps. Protein fatalities and confusing contaminants may resemble the adventure of unexpected storms and hardships. Gratifying views along the way feed the anticipation of what will be seen from the top. The ultimate reward of a pure enzyme is tantamount to the unobstructed and commanding view from the summit. Beyond the grand vista and thrill of being there first, there is no need for descent, but rather the prospect of even more inviting mountains, each with the promise of even grander views.

With the purified enzyme, we learn about its catalytic activities and its responsiveness to regulatory molecules that raise or lower activity. Beyond the catalytic and regulatory aspects, enzymes have a social face that dictates crucial interactions with other enzymes, nucleic acids, and membrane surfaces. To gain a perspective on the enzyme's contributions to the cellular economy, we must also identify the factors that induce or repress the genes responsible for producing the enzyme. Tracking a metabolic or biosynthetic enzyme uncovers marvelous intricacies by which a bacterial cell gears enzyme production precisely to its fluctuating needs.

Popular interest now centers on understanding the growth and development of flies and worms, their cells and tissues. Many laboratories focus on the aberrations of cancer and hope that their studies will furnish insights into the normal patterns. Enormous efforts are also devoted to AIDS, both to the virus and its destructive action on the immune system. In these various

Figure 1.1 Personalized license plate expressing a commitment to enzymology.

studies, the effects of manipulating the cell's genome and the actions of viruses and agents are almost always monitored with intact cells and organisms. Rarely are attempts made to examine a stage in an overall process in a cell-free system. This reliance in current biological research on intact cells and organisms to fathom their chemistry is a modern version of the vitalism that befell Pasteur and that has permeated the attitudes of generations of biologists before and since.

It baffles me that the utterly simple and proven enzymologic approach to solving basic problems in metabolism is so commonly ignored. The precept that discrete substances and their interactions must be understood before more complex phenomena can be explained is rooted in the history of biochemistry and should by now be utterly commonsensical. Robert Koch, in identifying the causative agent of an infectious disease, taught us a century ago that we must first isolate the responsible microbe from all others. Organic chemists have known even longer that we must purify and crystallize a substance to prove its identity. More recently in history, the vitamin hunters found it futile to try to discover the metabolic and nutritional roles of vitamins without having isolated each in pure form. And so with enzymes it is only by purifying enzymes that we can clearly identify each of the molecular machines responsible for a discrete metabolic operation. Convinced of this, one of my graduate students expressed it in a personalized license plate (Fig. 1.1).

ACKNOWLEDGMENT

This article borrows extensively from "For the Love of Enzymes: The Odyssey of a Biochemist," Harvard University Press, 1989.

SECTION ONE

DEVELOPING PURIFICATION PROCEDURES

CHAPTER TWO

Strategies and Considerations for Protein Purifications

Stuart Linn

Contents

1. General Considerations 10
   1.1. Properties and sensitivities of the protein 10
   1.2. For what is the protein to be used 10
   1.3. Assays 11
   1.4. What should be added to the suspension, storage, and assay buffers 13
   1.5. Storage of protein solutions 14
   1.6. Contaminating activities 15
2. Source of the Protein 15
   2.1. Preliminary studies to obtain sequence information 15
   2.2. Overexpressed protein as the source material 16
3. Preparing Extracts 16
4. Bulk or Batch Procedures for Purification 17
5. Refined Procedures for Purification 18
   5.1. High-capacity steps 18
   5.2. Intermediate-capacity steps 19
   5.3. Low-capacity steps 19
6. Conclusions 19
References 19

Abstract

Prior to embarking upon the purification of a protein, one should begin by considering what the protein is to be used for. In particular, how much of the protein is needed, what should be its state of purity, and must it be folded correctly and associated with various other peptides or cofactors. Using such criteria, an appropriate assay should be chosen and a procedure be planned taking into account the source of the protein, how it is to be extracted from the source, and what agents the protein ought to be exposed to or ultimately be stored in.

Department of Molecular and Cellular Biology, University of California, Berkeley, California, USA

Methods in Enzymology, Volume 463
ISSN 0076-6879, DOI: 10.1016/S0076-6879(09)63002-0

One is often surprised at the time necessary to develop an appropriate protein purification procedure relative to the time required to clone a gene or to accumulate information with the purified protein. There are an overwhelming number of options for protein purification steps, so forethought is necessary to expedite the tedious job of developing the purification scheme, or to avoid having to redesign it upon attempting to use the protein. This chapter points out general considerations to be undertaken in designing, organizing, and executing the purification, while subsequent chapters of this volume supply more specific options and technical details.

1. General Considerations

1.1. Properties and sensitivities of the protein

If known, one must consider whether the protein is soluble, and, if not, what agents might help to solubilize it. Is the protein labile at high or low concentrations? Is the protein sensitive to high or low salt concentrations, high or low temperatures, high or low pH, or oxidation, particularly by oxygen. A few preliminary experiments in this regard might be well worth the effort so as to learn what reagents ought to be present during the purification, what agents must be avoided (e.g., oxygen), and under what conditions the protein ought to be stored.

1.2. For what is the protein to be used

The required amount of purified protein may vary from a few micrograms needed for a cloning or identification endeavor to several kilograms required for industrial or pharmaceutical applications. Therefore, a very major consideration is the amount of material required. One should be aware of the scale-up that will be necessary and the final scheme should be appropriate for expansion to those levels. There are very real limitations to scaling up a procedure which are brought about not only by considerations of cost and availability of facilities but also by physical constraints by such factors as chromatographic column size limits or electrophoresis (ohmic) heating. As outlined below, individual steps of the procedure should flow from high-capacity/low-cost techniques toward low-capacity/high-cost ones. In some cases, alternative procedures may be required: for example, one to obtain microgram quantities for sequencing and cloning and a second to produce kilogram amounts of the cloned and overproduced material in its native state. The protein chemist should anticipate such variable needs and remain flexible for adopting new procedures as the need arises.

A corollary to the amount is the concentration of protein necessary. Concentration comes into play not only for applications of the protein but also for detection or assay of the amount and activity of the protein. Concentration techniques such as those covered in Chapter 9 often must be included after an intermediate or the final step of purification.

Other considerations include whether the protein must be active (as an enzyme, a regulatory protein, or an antibody, for example). If so, must it be folded correctly into its normal native configuration? (Refolding of denatured- or overexpressed proteins may alter precise kinetic properties, for example, if subtle, but stable alternative configurations are assumed.) Or, folding may not be an issue, for example, if the protein is to be utilized for sequence information or identification. The techniques employed during a procedure should be as gentle as are necessary. But, whenever possible, some of the harsher, but often spectacularly successful procedures such as those that involve extremes of pH, organic solvents, detergents, or hydrophobic or strong affinity chromatographic media should also be tried.

How free from contaminants should a protein be and, in particular, which contaminants? For example, proteins used as pharmaceuticals must be ultra-pure in all respects, whereas polymerases must be free of enzymes that degrade the polymer product of the polymerase or proteins that add or remove protein modifications must be free of their counterparts.

A consideration that is currently receiving a great deal of attention is that of the appropriate subunit content of a multimeric protein. What are the subunits of the desired protein? Is an isolated complex identical in content to that within the starting material? And, how might this content change with the growth state, growth conditions, or storage of the source material? These considerations are further exacerbated by the possibility that a protein in question might exist simultaneously in several or many different complexes or states of aggregation. Likewise, it might exist simultaneously with alternative posttranslational modifications. In either or both of these instances, the distribution of the protein among these various alternatives might be extremely sensitive to stresses imposed upon the cells used as the source material prior to purification. One should take these possibilities into account, particularly when the protein seems to partition into several peaks during purification. These variables may preclude an easy or successful purification scheme. A case in point is p53 (the product of the human TP53 gene) for which a satisfactory purification of active protein has yet to be achieved, in spite of its importance as the "guardian of the human genome," because of its numerous patterns of associated peptides, alternative modifications, and states of aggregation.

1.3. Assays

Classically, a most important consideration is the development of appropriate assays for the protein. The success of the purification is often dependent on this. Six considerations are relevant: sensitivity, accuracy, precision, linearity, substrate availability, and cost as measured in time and money.

Sensitivity can be the limiting factor as the protein becomes diluted into column effluents, etc. Before formulating a step, dilution and losses ought to be estimated so that the ability to detect and utilize the protein will be possible.

Accuracy and precision are often compromised, but clearly these factors must be controlled to the extent that the assay is reliable for assuring reproducibility and the recovery of sufficient material.

The linear range of the assay must be measured both with respect to time and to protein amount. As a procedure progresses, these ranges ought to be reevaluated, particularly for activity assays. For example, removal of protein contaminants might affect the limits if an inhibitor or stimulator were removed. Or, the limits could be altered if the highly purified protein becomes less stable during assay incubations.

Substrate availability and cost refer to the practicality of the assay: Is enough substrate available to perform the entire purification without interruption? Can the assay be performed simultaneously on a reasonable number of fractions in a reasonable amount of time? Stopping to prepare more substrate or skimping on material and/or the number of fractions assayed usually compromises the purification's success. On the other hand, assays at some steps might be compromised, for example, by omitting specificity controls at later stages of purification or assaying alternate or combined chromatography fractions.

It is often tempting not to use assays for activity, but to purify a gel band or an antigen instead. Although this approach is appropriate in some instances, particularly when an activity is not being sought, it is to be strongly discouraged when activity is in fact what is desired. It cannot be emphasized strongly enough that an activity assay is necessary to obtain optimal yields of unaltered activity, be it one associated with an enzyme, an antibody, a hormone, a regulatory protein, or a protein that modifies the secondary or tertiary structure of a polymer.

An inherent part of an assay is the assay conditions. Optimal conditions must be determined empirically, but, beware: optimal conditions often change with purification due to removal of interfering inhibitory or stimulatory factors, lower overall protein concentrations, etc. Moreover, always take into account materials added with the substrate or with the protein storage, suspension or dilution buffers as these may change the pH and ionic strength or otherwise affect the reliability of the assay.

A final comment pertains to the protein assay. While the goals also include simplicity, linearity, reproducibility, specificity, and reliability, accuracy is generally compromised, as no commonly used assay is absolute with regard to all proteins. With crude fractions, color reactions are probably best. While the Bradford Method (Bradford, 1976) is the simplest of these, in our laboratory we find it to be unreliable with crude fractions from animal cells or when detergents are present. Protein assays can and often must by changed as the purification progresses. For example, for column effluents, ultraviolet absorption may be utilized: it is simple, sensitive, does not consume the material, and can even be continually monitored with a flow detector. For extremes of sensitivity, wavelengths between 210 and 230 nm can be utilized (Tombs *et al.*, 1959; Waddell, 1956). Beware, however, that blanks must be adjusted to take into account the UV-absorption of materials in the solutes and that nonprotein contaminants, particularly nucleic acids, may have significant UV-absorption. Remember, protein assays are done to follow the extent of purification and to monitor reproducibility; they are not absolute. In the rare cases where absolute values must be obtained, such assays as nitrogen determinations are necessary (see Chapter 8).

1.4. What should be added to the suspension, storage, and assay buffers

A common query is, "why is the protein suspended in *x*?" And its answer is often, "if I change the components, I don't know what will happen." The obvious lesson is to add something only with good reason in the first place. Even more of a problem in this regard pertains to proteins obtained commercially in complex buffers, the origin of which may be obscure and, even worse, the content of which may be proprietary and not disclosed. In this instance, replacement with a known buffer cocktail may be advisable prior to embarking upon a study of the protein's properties or its application in complex experiments (see Chapter 9).

Solutes are added usually to improve stability, prevent the growth of microorganisms, reduce the freezing point, or keep the protein in solution. Table 2.1 lists several classes and examples of such materials, but the reader is referred to chapters later in this volume for comprehensive discussions of them.

It is well worth the effort to carry out stability studies (e.g., heat inactivation or storage trials) in order to learn how to maintain a stable, soluble protein. Three notes of caution: (1) optimal storage conditions may change with purification; (2) storage conditions may alter peptide–peptide interactions; and (3) optimal storage conditions need not relate to optimal conditions for activity. Indeed, materials that stabilize a protein might inhibit it when added to activity mixes. Of course, the latter situation must be considered when utilizing the protein—interfering substances will have to be removed or "diluted out" for utilization of the protein.

In our experience, reducing agents are particularly effective for storage of bacterial enzymes that often derive from a reducing environment, whereas enzymes from animals often take kindly to surfactants or protease inhibitors. Fungal proteins also respond to protease inhibitors. Optimal pH and salt concentrations vary widely.

Table 2.1 Additions to protein solutions

Class	Examples	Purpose
Buffer	Potassium phosphate, Tris–HCl, sodium acetate, Hepes–NaOH	Stability
Salt	KCl, NaCl, $(NH_4)_2SO_4$	Stability
Ionic detergents	Sodium deoxycholate, sodium cholate	Stability, solubility
Nonionic detergents	Triton X-100, Nonidet P-40, Tween 20, Brij 35	Stability
Glycerol, sucrose		Stability; unfrozen storage below 0 °C
Sodium azide		Bacteriostatic
Metal chelators	EDTA (ethylenediaminetetraacetic acid), EGTA (ethylene glycol bis (*β*-aminoethylether) *N*,*N*′-tetraacetic acid)	Stability
Reducing agents	*β*-Mercaptoethanol (BME), dithiothreitol (DTT), TCEP	Stability; reduce disulfide bonds
Ligands	Mg^{2+}, ATP, phosphate	Stability
Protease inhibitors	PMSF (phenylmethylsolfonyl fluoride), TPCK (*N*-tosyl-L-phenylalanine chloromethyl ketone), TLCK (N^{α}-*p*-tosyl-L-lysine chloromethyl ketone)	Stability

In a converse context, what might be absolutely necessary to eliminate from buffers and resins? Special precautions could be necessary to deal with heavy metals (e.g., the use of ultra-pure reagents and double-distilled water) or with oxygen, for which the purging of all buffers with N_2 and/or working in an anoxic glove box might be required. Keep in mind that a seemingly minor amount of contaminating heavy metal or O_2 in a liter of dialysis buffer can be a relatively large amount in molar terms versus a few micrograms of purified protein being dialyzed.

Special precautions, which must be taken for large protein complexes, are noted in Chapter 10.

1.5. Storage of protein solutions

Most proteins prefer to be stored at the lowest temperature possible. If not frozen, 0 °C (on ice, rather than in a refrigerator in order to minimize degradation or denaturation and bacterial or fungal growth) or − 20 °C with sufficient glycerol or sucrose present to avoid freezing. If frozen, storage above liquid nitrogen or at −70 °C is usually best and "frost-free" freezers ought to be avoided if possible due to the possibility of repeated thawing of small volumes during the thaw cycles. Allocation into small aliquots avoids losses brought about by multiple freezing/thawing and thawing should be as rapid as possible to minimize exposure of the protein to concentrated solute in partially thawed solutions.

Concerning the containers utilized for purified proteins, particularly at low concentrations, plastic rather than glass should be used. In our experience, polypropylene is superior to polyethylene or polystyrene and other clear plastics. In rare instances, we have had to precoat storage tubes with an inert protein prior to their use. Be sure to have tight-fitting caps if storage is in "frost-free" freezers. These and other precautions are discussed in detail in Chapter 10.

1.6. Contaminating activities

Often proteins need not or cannot be obtained in a pure state, but particular interfering activities (e.g., nucleases, proteases, phosphatases) must not be present. In our experience, attempting to purify one activity against one or more others by utilizing multiple types of assays, as a criterion of purity is extremely frustrating. Instead, purifying so as to optimize yield of the specific protein and minimize total protein content (optimize specific activity) with selective choice of fractions only at the last or penultimate step is more likely to be successful.

Finally, to avoid an extremely embarrassing—though surprisingly common—error, be certain that the activity that you have purified is associated physically with the protein species being characterized. Genetic experiments or physical criteria that confirm that these two entities comigrate under independent procedures (e.g., sedimentation, nondenaturating gel electrophoresis) are most strongly encouraged.

2. Source of the Protein

2.1. Preliminary studies to obtain sequence information

If gene or peptide sequence is not available and a low-abundance protein is to be purified in order to obtain such information, careful consideration of the source material (e.g., organism, tissue, culture cell type) is worth the time and effort, and trial extracts from a number of sources should be explored. These should be tested for content of the protein (per gram of starting material or per unit cost), the starting specific activity if relevant, and the stability of the protein. If relevant, the classical microbiological approach of isolating microorganisms with unique growth requirements can lead to unexpected success.

The cost and availability of the source material, particularly if a largely scaled-up preparation might be desirable in the future, should be considered. Genetic knowledge and technology for the organism or cell type ought to be available should regulatory characterization and/or gene sequence manipulations be envisioned for the future.

2.2. Overexpressed protein as the source material

With the recent dramatic technological improvements in gene manipulations, one usually expects ultimately to overexpress a protein, whether for in-depth characterization or for reagent use. For overexpression, is a bacterial, fungal, insect cell, or mammalian system best? (see Chapters 11–15). What special precautions are necessary for each source? Will the protein be appropriately posttranslationally modified and will such modifications be easily identifiable? Will the protein be obtained in a biologically relevant state of aggregation? If a heterologous source is to be utilized (e.g., expression of a mammalian protein in insect cells), might an endogenous homologue cause any problems? Should a mutant host cell be utilized (e.g., one lacking a similar protein or an interfering activity such as a nuclease or protease)?

All of the subunits of a heteromultimeric protein often must be overexpressed together in order to obtain the protein in its "native" state. To accomplish this, it is necessary to learn the efficiency of expression of each of the peptide constructs in order that the peptides are produced in roughly the desired stoichiometric amounts. This is often ultimately not difficult to accomplish, but a good deal of effort must be expended in learning the conditions to differentially control such joint expression systems by variation of transfection levels or promoter efficiencies as the case may be.

If tags are to be attached to a protein to aid in purification and/or detection, will they interfere with subsequent studies? Where in the protein should the tags be placed? If on the N-terminus, prematurely terminated peptides will contaminate affinity-purified material. If on the C-terminus, functional or binding properties of that part of the protein might be compromised (see Chapter 16 for discussion of this important topic).

3. Preparing Extracts

Preparation of extracts is discussed in details in Section IV, so only general considerations are noted here. In our experience, the manner in which cells are disrupted has a profound, but unpredictable effect on the yield and quality of the protein preparation. Trials of alternative procedures are clearly necessary.

Thought should always be given to scaling up of the preparation. Will the volumes or time required become unreasonable? Can a subsequent clarification of the extract be conveniently scaled up? In general, volumes should be kept to a minimum—for example, extracts should be as concentrated as possible. If relevant and convenient, tissue, cell type, or organelle fractionation is almost always worthwhile prior to general disruption.

The buffer that favors the most efficient disruption may not prove to be one that favors stability or a subsequent step, and, in extreme cases, the buffer might need immediate alteration or a subsequent step might immediately be necessary upon completion of the disruption protocol. For example, hypotonicity, chelators, or pH extremes might aid in total release of the desired protein, but labialize the protein in the resulting extract. Or, an overproduced protein might not remain soluble in the disruption medium. Parenthetically, such buffer modifications are also often necessary during subsequent purification protocols in which, for example, it is not uncommon to collect chromatography fractions into tubes containing additional buffer constituents, or to dialyze individual fractions immediately as they are eluted.

4. Bulk or Batch Procedures for Purification

These procedures are almost always utilized early in the purification as they are usually most effective in removing nonprotein material and are amenable to the relatively large volume and amounts of material that exist in earlier stages of the preparation. A great deal of effort went into designing these steps in the early days of protein chemistry, and much frustration, time, and money can often be avoided by including some of these "old-fashioned" procedures.

Section V details many of these procedures. Gentle procedures including "salting out" or phase partition with organic polymers are most common. More drastic methods such as heat, extremes of pH, or phase partition with organic solvents might be particularly effective with stable proteins, though subtle forms of damage may be difficult to foresee or to detect. In addition, one might consider the use of high-capacity ion-exchange resins either added as a slurry to crude material or used in columns for batch elution. The time expended in developing and optimizing these early steps is always worthwhile—even removal of half of the contaminating protein may dictate the feasibility of a subsequent step from both cost and technical considerations.

5. Refined Procedures for Purification

Once the bulk methods have yielded a protein preparation that is reasonably free of nucleic acids, polysaccharides, and lipids, the preparation becomes amenable to the more interesting and often spectacular procedures, which have been developed in recent years. The general strategy is to proceed from high- to low-capacity procedures and to attempt to exploit specific affinity materials whenever possible.

Applications, specific examples, and technical details for these procedures are discussed in Sections VI and VII and will not be described here. As a general consideration, in proceeding from one procedure to the next, one ought to reduce as much as possible the necessity for dialysis and concentration. Hence, procedures that separate by size and shape can simultaneously remove salt or otherwise modify the buffer. Procedures utilizing high-capacity resins can concentrate proteins as well as purify them, or resins from which the desired protein elutes at relatively low salt concentrations can be directly followed with a resin to which the protein binds at this concentration. Also, some steps (e.g., sedimentation through gradients of sucrose or glycerol) may leave the protein in a medium that might be ideal for long-term storage, and appropriate (or inappropriate) for use directly for studies or a subsequent purification step. Finally, interchanging the order of the steps of a procedure can, and often does, have a profound effect on the success of a purification scheme.

Two notes of caution: it has been our experience that rapid dilution of a fraction that has a high concentration of salt, glycerol, or sucrose often results in a loss of activity. If this phenomenon is experienced, the use of dialysis or gel exclusion chromatography may be a useful alternative. Also, during a purification scheme, the protein solution sometimes becomes cloudy. This solution must be clarified prior to chromatography to avoid unacceptably low flow rates. In our experience, the insoluble material rarely contains native protein.

Some procedures which cannot be effectively scaled up (e.g., sedimentation or HPLC/FPLC) can be carried out with small aliquots of the preparation if left to the final stages. In fact, in some cases the utilization of such aliquots is desirable as the less purified fractions might be more stable to long-term storage.

5.1. High-capacity steps

Generally, these include ion-exchange resins or very general affinity agents such as dyes or glass. When used in the initial stages of a purification scheme with large amounts of material, a particular type of ion-exchange resin can often be successfully reutilized for purification at a later stage (especially if the pH is changed) or for concentration.

5.2. Intermediate-capacity steps

These might include the hydrophobic resins for which long chromatographic times reduce activity yields. Many affinity agents bound to inert chromatography scaffolds (e.g., bulk DNA, simple oligonucleotides, antibodies, or ligands of a protein) fall into this class. In these instances, thought and effort must be given to finding conditions and/or materials that can successfully elute the protein without affecting its subunit structure or destabilizing or inactivating it. Free, unbound ligand often is successful, although it may be expensive and/or difficult to unbind from the protein. Finally, gel filtration is considered as a step with intermediate capacity.

5.3. Low-capacity steps

Affinity steps utilizing valuable ligands such as substrate analogs, complex polynucleotide sequences, monoclonal antibodies, or lectins might be included here. Also included are electrophoresis methods, including isoelectric focusing, as precipitation of the protein may be a problem with moderate amounts of protein. Ultracentrifugation steps are usually limited to small volumes of material and HPLC/FPLC steps are also often limited by cost and/or capacity of the column. Small hydrophobic columns might be successfully utilized when larger ones of the same resin result in activity loss.

6. Conclusions

Although the development and execution of protein purification procedures are often difficult and frustrating processes, the rewards are great. Moreover, given the continual addition of new technology, high-quality commercial materials utilized for purification procedures, and genetically altered sources of proteins, the future bodes well for ever-simpler procedures accompanied by ever-greater rewards.

REFERENCES

Bradford, M. M. (1976). A rapid and sensitive method for quantitation of microgram quantities of protein utilizing the principle of protein-dye binding. *Anal. Biochem.* **72,** 248–254.

Tombs, M. P., Souter, F., and MacLagan, N. F. (1959). The spectrophotometric determination of protein at 210 μm. *Biochem. J.* **73,** 167–171.

Waddell, W. J. (1956). A simple ultraviolet spectrophotometric method for the determination of protein. *J. Lab. Clin. Med.* **48,** 311–314.

CHAPTER THREE

Use of Bioinformatics in Planning a Protein Purification

Richard R. Burgess

Contents

1. What You Can Learn from an Amino Acid Sequence 22
    1.1. Molecular weight of the polypeptide chain 22
    1.2. Charge versus pH/titration curve, isoelectric point 22
    1.3. Molar extinction coefficient 23
    1.4. Cysteine content 23
    1.5. Secondary structure 24
    1.6. Stability 24
    1.7. Hydrophobicity and membrane-spanning regions 24
    1.8. Sequence similarity suggests homology and possible cofactor affinity 24
    1.9. Potential modification sites 25
    1.10. Solubility on overexpression in *E. coli* 25
2. What You Cannot yet Predict 25
    2.1. Three-dimension structure; shape, surface features 26
    2.2. Multisubunit features; homomultimers, heteromultimers? 26
    2.3. Precipitation properties 26
3. Conclusion 26
    3.1. Protein bioinformatic resources 27
    3.2. Purification in the denatured state 27
References 27

Abstract

Now that many hundreds and even thousands of whole genomes have been sequenced, it is rare to be studying a target protein whose amino acid sequence is not known. However, it is still often necessary to obtain large amounts of the target protein for a variety of purposes including structural studies, drug discovery, enzymology, protein biochemistry, and industrial application. It would seem that knowing the amino acid sequences would make it much easier

McArdle Laboratory for Cancer Research, University of Wisconsin-Madison, Madison, Wisconsin, USA

Methods in Enzymology, Volume 463
ISSN 0076-6879, DOI: 10.1016/S0076-6879(09)63003-2

to design an effective purification of that protein. We examine in this chapter what you can and cannot predict from an amino acid sequence and conclude that protein purification is still largely an empirical science.

1. What You Can Learn from an Amino Acid Sequence

Often there are two common scenarios that a protein biochemist faces these days: (1) you know the protein and at least some of its functions; and (2) you have identified a gene and its protein through transcriptional profiling or proteomic studies by two-dimensional gel analysis or mass spectroscopy, but it is of unknown function. In either case, you know the sequence and want to purify it. Here are some of the types of information you can learn from its sequence that might help you in designing a protein purification procedure. Many of these properties can readily be obtained using sequence analysis software as described below.

1.1. Molecular weight of the polypeptide chain

It is very straightforward to calculate the molecular weight (MW) of a polypeptide chain protein by multiplying the MW of an amino acid in a polypeptide (the MW of an amino acid −18 Da) by the number of times that amino acid occurs in the polypeptide. Remember to add an extra 18 Da to the total to account for the free amino and carboxy termini. Of course, one cannot know, except by N-terminal sequencing or mass spectroscopy, whether the N-terminal initiating methionine has been removed or not.

1.2. Charge versus pH/titration curve, isoelectric point

Assuming a p*K*a for each ionizable group in a protein (the p*K*a is the pH at which that residue is half ionized), one can calculate the total charge as a function of pH of a given protein from its amino acid sequence. The pH at which the net charge is zero is defined as its isoelectric point or p*I*. This information is helpful in deciding which ion exchange chromatography resin to use. For example, if you assume that the protein is a monomer and not modified to change the charge, then an acidic protein (one that is negatively charged at pH 7) is likely to bind to a positively charged anion exchange resin such as MonoQ while a basic protein (one that is positively charged at pH 7) is likely to bind to a negatively charged cation exchange resin such as MonoS. There are exceptions to this generalization if the charge on the surface is not uniformly distributed. A protein could have a positive patch and a separate negative patch and be able to bind to both anion and cation exchange columns under the same conditions. To the first approximation, the greater the charge on a protein at the pH of the column buffer, the tighter it will bind and the higher the salt needed to elute it from the column resin. Of course, if the protein is part of a multiprotein complex, then you have no idea what its ion exchange column binding properties will be.

Finally, since a protein is generally least soluble at its p*I*, one can consider an isoelectric precipitation step (assuming that the protein is not in a stable complex with another protein or proteins) (see Chapter 20).

1.3. Molar extinction coefficient

At a wavelength of 280 nm, all of the absorbance of an unmodified protein is due to the absorption of its amino acids tryptophan, tyrosine, and cysteine. Gill and von Hippel (1989) and Pace *et al.* (1995) have described similar methods to estimate the molar absorption/extinction coefficient of a protein given its amino acid composition. This involves some assumptions, based on experimental data, on the average proportion of the tryptophan and tyrosine that are exposed on the surface versus buried in the interior. This method is perhaps the most useful practical method of determining the concentration of a purified protein. For example, if the protein has six tryptophans, seven tyrosines, and no cysteines, the molar extinction coefficient ($\varepsilon_{280\ \mathrm{nm}}$) would be $(6 \times 5500) + (7 \times 1490) + (0 \times 125) = 43{,}430$. The absorption ($A_{280\ \mathrm{nm}}$) of a 10^{-5} *M* solution would be 0.43. A somewhat more useful number is the $A_{280\ \mathrm{nm}}$ of a 1 mg/ml solution of that protein. This is sometimes called $E^{1\ \mathrm{mg/ml}}_{280\ \mathrm{nm}}$. You divide the molar extinction coefficient by the MW (e.g., if the protein above is 30,000 Da, then 43,430/30,000 = 1.45). Then a careful measurement of the $A_{280\ \mathrm{nm}}$ with appropriate buffer absorption controls will give you the protein concentration (e.g., if the measured $A_{280\ \mathrm{nm}}$ is 0.75, then the protein concentration of the measured protein solution is 0.75/1.45 = 0.52 mg/ml).

It should be stressed that this method is not valid if the protein contains any contaminating nucleic acid, additional 280 nm-absorbing moieties (such as bound heme, iron–sulfur [Fe–S] centers, or nucleotide substrates or cofactors), or fluorescent modifications (such as with green fluorescent protein).

1.4. Cysteine content

I always look to see if a protein I am trying to purify contains cysteine. If not, then I can omit reducing agents such as dithiothreitol (DTT) from my buffers. If the protein is from *E. coli* and has cysteines, then I assume that there are no disulfides in the native protein since the environment in the cytoplasm is reducing making it unlikely that the protein is naturally in an oxidized condition unless it is located in the periplasm. In general when my protein has cysteines, I include DTT in my buffers to prevent unwanted intra- or intermolecular disulfide formation. It is harder to predict whether proteins from other sources and that contain multiple cysteines are likely to form disulfides.

1.5. Secondary structure

There are several quite reasonable methods to predict regions of the protein that are likely to be in the form of alpha helical or beta strand secondary structures. However, this knowledge is rarely of use in designing a protein purification scheme.

1.6. Stability

It is possible, using ProtParam (see below) to estimate the *in vivo* half-life and the instability index of a protein from its sequence. The *in vivo* half-life is calculated based on the N-end rule (Varshavsky, 1997) and the N-terminal amino acid of the protein in question. Approximate half-life estimates are given for the protein expressed in mammalian cells, yeast and *E. coli*. The instability index provides an estimate of the stability of your protein *in vitro* and is based on the analysis of dipeptide occurrences in your protein compared to those of a set of test proteins that are known to be unstable or stable (Guruprasad *et al.*, 1990). This information might be useful since a protein predicted to be unstable *in vitro* might warrant special care in maintaining low temperatures during purification and perhaps in the use of protease inhibitors.

1.7. Hydrophobicity and membrane-spanning regions

It is possible to analyze sequence information to determine regions that are particularly hydrophobic or hydrophilic. One such method is that of Kyte and Doolittle (1982) that allows one to plot the hydropathy value along a sequence. This can allow one to recognize potential membrane-spanning regions and, for a protein of unknown function, predict that it is a membrane protein. A membrane-spanning region is usually a stretch of 23 hydrophobic amino acids that form an alpha helix. If you know that the protein of interest is a membrane protein, it will certainly affect how you design a purification scheme.

1.8. Sequence similarity suggests homology and possible cofactor affinity

If your protein is of unknown function, it is possible to do a search to determine if it is highly similar to other known proteins in the protein database. If the sequence similarity is great enough to infer homology and if your protein is related to other homologous family members that have been studied, then perhaps you know, for example, that it usually occurs as a homodimer. This will help in predicting its behavior on gel filtration column chromatography. This information might also be used to devise a suitable assay for your protein.

Again, as above, sequence can identify if it belongs to a protein family all of which are known to bind to a particular cofactor or substrate. For example, if it is a member of the AAA ATPase family, it is likely that it will bind to an affinity column that has an immobilized ATP analog. Such an affinity purification step can aid greatly in a purification scheme (see Chapter 26).

1.9. Potential modification sites

It is now possible to identify in amino acid sequences, short amino acid stretches or "motifs" that are commonly sites of posttranslational modification. A few of these motifs are: glycosylation site (N**X**S or N**X**T); biotinylation site (AM**K**M); zinc finger (metal binding site) (F/YX**C**X_{2-4} **C**X_3FX_5LX_2**H**X_{3-4}**H**X_5), heart muscle protein kinase recognition site (RRA**S**V). This information can be highly useful. For example, if the protein is glycosylated, it might bind to a lectin affinity column (see Chapter 35). Many posttranslational modifications become handles by which a protein can be fractionated away from many other proteins. The problems are that one cannot predict if the site is on the surface of the protein of interest or if it is in fact modified and to what extent.

1.10. Solubility on overexpression in *E. coli*

Several studies have proposed that one can predict from the protein sequence the likelihood that a protein, if overproduced in *E. coli* will be soluble or form insoluble inclusion bodies (Idicula-Thomas and Balaji, 2005; Wilkinson and Harrison, 1991). Obviously, this is useful information, but is most easily obtained by simply overproducing a protein, breaking the cells, centrifuging the lysate to separate soluble from insoluble, and then analyzing the two resulting fractions by SDS–PAGE. The fact that the proportion of an overproduced protein that is soluble can be significantly altered by manipulating cell growth conditions suggests that solubility predictions based on sequence may be useful but have severe limitations.

2. What You Cannot yet Predict

Unfortunately, most of the information needed to design a purification scheme is not predictable from a protein's amino acid sequence. Some of the most critical problems are listed below.

2.1. Three-dimension structure; shape, surface features

While a huge amount of time and effort has gone into finding a practical method of predicting three-dimensional structure from amino acid sequence, it is presently not yet possible to do so with any confidence. In the case where a target protein shows high similarity to a protein whose structure has been determined, it is possible to "thread" the sequence into the known structure and come up with a reasonable approximation of the structure of the target protein.

Without accurate structural information, one cannot predict shape, or detailed surface features such as hydrophobic patches, charge distribution, and antigenic sites. Therefore, one cannot easily predict behavior on hydrophobic interaction chromatography (HIC) or ion exchange chromatography. Since the shape affects the Stokes radius, one cannot predict behavior during gel filtration chromatography (a spherical protein will appear smaller than an asymmetric or cigar-shaped protein of the same MW during gel filtration).

2.2. Multisubunit features; homomultimers, heteromultimers?

Even if one could accurately predict the three-dimensional structure of a protein, it would still be impossible to predict if it exists in solution as a multimer (e.g., a hexamer) or as a monomer. That lack of knowledge precludes any reasonable prediction of its behavior on sizing columns or on ion exchange columns. Even more problematic is the fact that many proteins exist as parts of multisubunit complexes and one cannot predict whether a given protein will exist as a complex whose purification properties may be determined largely by its binding partners.

2.3. Precipitation properties

Different proteins vary in their solubility properties in ways that are not yet understood. Therefore, it is not presently possible to predict what ammonium sulfate concentration to use in precipitating a given protein from solution. This will be discussed further in Chapter 20.

3. Conclusion

I think it is very clear that one cannot yet use amino acid sequence to predict the behavior of a given protein in most of the primary fractionation methods used by protein purifiers. By far the best approach still is to fractionate an extract by ammonium sulfate precipitation, by ion exchange and gel filtration chromatography and determine how the protein of interest behaves. Therefore, although we have tremendous knowledge of protein sequence, protein purification is still very empirical! Don't be afraid to simply do the experiment!

3.1. Protein bioinformatic resources

One of the most used sites for obtaining sequence data and using it to compute various physicochemical properties of proteins is the ProtParam feature of ExPASy (Expert Protein Analysis Software) Web site (http://www.expasy.org/tools/protparam). ProtParam calculates many of the protein parameters including MW, theoretical p*I*, amino acid composition, atomic composition, extinction coefficient, estimated half-life, instability index, aliphatic index, and grand average of hydropathicity (Gasteiger *et al.*, 2005).

This is very simple to use. Go to ProtParam tools, enter a Swiss-Prot/TrEMBL accession number (AC) or a sequence identifier (ID), or you can paste your own sequence in the box, hit button that says "compute parameters," and print out the results.

One commercial package available through DNASTAR (http://www.dnastar.com), called Lasergene, version 8.0, contains a section on protein sequence analysis called "Protean." This allows sequence analysis similar to ProtParam and in addition, predicts antigenicity, surface probability, maps proteolytic digestion sites, and has very nice graphical displays.

3.2. Purification in the denatured state

After all that was discussed above, it is only fair to point out that one can do much better at predicting purification properties of a given protein is you are willing to carry out the purification in the denatured state (Knuth and Burgess, 1987). If you do not have to worry about possible multimeric states, then one can simply denature in urea, fractionate somewhat predictably by ion exchange and gel filtration, and then refold the protein to its native structure (see Chapter 17).

REFERENCES

Gasteiger, E., Hoogland, C., Gattiker, A., Duvaud, S., Wilkins, M. R., Appel, R. D., and Bairoch, A. (2005). Protein identification and analysis tools on the ExPASy server. *In* "The Proteomics Protocols Handbook", (J. M. Walker, ed.), pp. 571–607. Humana Press, Totowa.

Gill, S. C., and von Hippel, P. H. (1989). Calculation of protein extinction coefficients from amino acid sequence data. *Anal. Biochem.* **182,** 319–326.

Guruprasad, K., Reddy, B., and Pandit, M. W. (1990). Correlation between stability of a protein and its dipeptide composition: A novel approach for predicting *in vivo* stability of a protein from it primary sequence. *Protein Eng.* **4,** 155–161.

Idicula-Thomas, S., and Balaji, P. V. (2005). Understanding the relationship between the primary structure of proteins and its propensity to be soluble on overexpression in *E. coli. Protein Sci.* **14,** 582–592.

Knuth, M. W., and Burgess, R. R. (1987). Purification of proteins in the denatured state. *In* "Protein Purification: Micro to Macro", (R. R. Burgess, ed.), pp. 279–305. Alan R. Liss, New York.

Kyte, J., and Doolittle, R. F. (1982). A simple method for displaying the hydrophobic character of a protein. *J. Mol. Biol.* **157,** 105–132.

Pace, C. N., Vajdos, F., Fee, L., Grimsely, G., and Gray, T. (1995). How to measure and predict the molar absorption coefficient of a protein. *Protein Sci.* **11,** 2411–2423.

Varshavsky, A. (1997). The N-end rule pathway of protein degradation. *Genes Cells* **2,** 13–28.

Wilkinson, D. L., and Harrison, R. G. (1991). Predicting the solubility of recombinant proteins in *E. coli. Biotechnology* **9,** 443–448.

CHAPTER FOUR

Preparing a Purification Summary Table

Richard R. Burgess

Contents

1. Introduction 29
2. The Importance of Footnotes 32
3. The Value of an SDS–Polyacrylamide Gel Analysis on Main Protein Fractions 32
4. Some Common Mistakes and Problems 32

Abstract

Once a protein purification scheme has been developed, the purification, characterization, and use/structure of a target protein are usually published. It is highly desirable to present the major steps in the purification and the corresponding features of the protein at each step summarized in the form of a purification summary table. In considering whether to repeat a published protein purification, a reader needs this information to evaluate the purification, and to decide if it is worth following or if it needs major modifications. In this chapter, I discuss the main characteristics of a useful purification summary table and point out common mistakes and problems I see in many such tables.

1. Introduction

As an executive editor and editor-in-chief of the journal, *Protein Expression and Purification*, I have reviewed on the order of 100 protein purification papers a year for over 18 years. It is remarkable how many manuscripts I receive where there is either no purification summary or one that is severely lacking in necessary information and accuracy. The essentials of a reasonable purification table are illustrated by the example below.

Suppose one set out to purify an enzyme from the bacterium *E. coli* starting with 10 g of wet weight cell pellet from a 4-l culture (10 g of wet weight cells typically would contain about 2 g of dry weight and about

McArdle Laboratory for Cancer Research, University of Wisconsin-Madison, Madison, Wisconsin, USA

Methods in Enzymology, Volume 463
ISSN 0076-6879, DOI: 10.1016/S0076-6879(09)63004-4

1200 mg of total protein). The cells are lysed by sonication to give a crude lysate and the debris is removed by centrifugation to give a crude extract. A 45–50% saturated ammonium sulfate cut was prepared. The 50% saturated ammonium sulfate pellet was dissolved in buffer and diluted to low salt and applied to a DEAE anion exchange column. The column was washed at low salt and then eluted with a linear salt gradient from 0.1 to 0.6 *M* NaCl, the peak of activity eluting at about 0.25 *M* NaCl. The peak was pooled and applied to a Sephacryl S-300 gel filtration column and eluted with a buffer at constant salt (isocratically). The fractions of peak activity were pooled and shown by SDS–PAGE with Coomassie blue staining to be a single band. The specific activity of the final material is the same as that of a known pure reference sample. The main fractions were all assayed for enzyme activity and protein determinations were carried out. The resulting data are given in Table 4.1.

A purification summary table should allow a reader to evaluate easily the procedure and readily detect particularly effective and ineffective purification steps. It should be easy to see if large losses occurred at a particular step.

A suitable table will contain the following columns:

1. *Major steps in the purification.* These typically include steps like:
   Crude lysate (the result of cell or tissue disruption). This step is often omitted since assays may be difficult but it is useful and even essential when much of the expressed recombinant target protein is in an insoluble inclusion body.
   Crude extract (the lysate after any insoluble material has been removed by centrifugation)

Table 4.1 A typical purification summary table

Step	Total protein (mg)[b]	Total activity (units)[c]	Specific activity (units/mg)	Yield (%)	Purity (%)
Crude lysate[a]	1200	120	0.10	100	0.8
Crude extract	1000	110	0.11	92	0.9
Ammonium sulfate 45–55% cut[d]	180	75	0.42	62	3.4
DEAE column (pooled peak)	24	60	2.5	50	20
Sephacryl column (pooled peak)	3.6	46	12.5	38	100

[a] From 10 g of wet weight *E. coli* cell pellet (from 4 l of bacterial culture).
[b] Protein concentration determined by Bradford assay using BSA as a standard protein.
[c] Enzyme activity measured as described in the methods section.
[d] Crude extract material that is soluble at 45% but precipitates at 55% saturated ammonium sulfate.

Ammonium sulfate cut
Pooled peak from an ion exchange column
Pooled peak from a gel filtration column
Pooled peak from an affinity column
Concentrated and dialyzed final product
Solubilized inclusion bodies. This step and the following two are often used in the case where an expressed recombinant protein is produced as insoluble inclusion bodies; see Chapter 17.
Washed inclusion bodies
Refolded, centrifuged, and concentrated material

2. *Amount of total protein (mg).* This is usually determined by a standard protein assay. Most commonly these days a Bradford dye-binding assay or a bicinchoninic acid (BCA) assay is used (see Chapter 8). It is important to indicate in the methods section what protein is used as the protein standard (typically BSA). Once the protein is purified it can also sometimes be quantified by measuring its absorbance at 280 nm and the use of an appropriate molar extinction coefficient.
3. *Amount of target protein or total activity (mg or units).* If there is a suitable enzyme assay for the target protein, it should be carried out on material from each major step. If the protein is not an enzyme or there is no quantitative assay, and if the protein is visible on a Coomassie blue stained SDS–PAGE, then often the stained gel is scanned and the amount of protein in the target protein band is determined. In other words, purity is determined and multiplied by the total protein to give an estimate of the total target protein.
4. *Specific activity (units/mg).* If enzyme activity assays are possible, then the total activity (units) is divided by the total protein (mg) to give specific activity as units/mg.
5. *Overall yield (%).* The yield at a step in the procedure is the amount of target (either total target protein or total activity) at that step divided by the amount of target in the first step (defined as 100%).
6. *Purity of target protein (%).* Purity is often determined by scanning a stained SDS–PAGE and measuring the amount of the stain associated with the target band as a fraction of the stain associated with all the bands on the gel. If one has a reliable assay, then if the final material is pure, its specific activity can be used to define purity. For example, if an earlier step has a specific activity 10% of the final pure material, then the purity at that step would be 10%.
7. *Relative or fold purification.* This is not essential since it can be calculated from the other values above, but it is often useful. This is merely setting the initial purity at a value of one and then giving the purity at each step relative to that of the first step. For example, in Table 4.1, the final step represents an overall fold purification of 125.

2. The Importance of Footnotes

Every protein and purification is different and footnotes are needed to help the reader understand what has been done. There should be a footnote that indicates the amount of raw material used in the preparation being summarized. For recombinant protein expressed in bacteria, for example, one should always give the number of grams of wet weight cell pellet used in the preparation. It is also useful to know the volume of the bacterial culture used, but that in itself is not enough since, depending on the growth media and conditions, the yield of wet weight cell pellet can range from 1 to 80 g/L. Another useful footnote indicates how the protein amount was determined (e.g., sometimes a Bradford assay is used on the early steps, but absorbance and extinction coefficient is used on the final product).

3. The Value of an SDS–Polyacrylamide Gel Analysis on Main Protein Fractions

I find that an SDS–polyacrylamide gel image is a very valuable complement to the purification summary table. If the same samples that represent the various steps in the purification table are also shown in a gel photo, then it is particularly easy to see the progress of the various fractionation steps toward production of a purified final target protein. The most useful gels are ones on which an equal proportion of the material at each step is loaded on the gel lanes.

4. Some Common Mistakes and Problems

1. *Use too many significant figures.* This is one of my pet peeves. I suspect that the concept of significant figures is no longer taught, because I find that a good 75% of the purification papers I review give protein amounts like 235.052 mg and yields like 46.72%. Just because a calculator or computer can divide two numbers and give one the result to eight figures does not mean that value is what one should enter into the table. Just remember that most protein quantification methods or enzyme assays are not accurate to better than 5–10%. When one writes 23.47 mg, one implies that it is not 23.46 or 23.48, but 23.47. In other words, one is implying that it is accurate to one part in over 2000 when it is not even

clear that it is accurate to one part in 20 (is it 22, 23, or 24?). No numbers should be given to more than three significant figures and in general most percentages can be given to two significant figures. Also, remember that any number resulting from the division of two numbers each accurate to $\pm$ 10% will only be accurate to $\pm$ 20%.

2. *Calculate values erroneously.* Remarkably many tables contain simple arithmetic errors. All numbers should be checked and rechecked before submission of a manuscript.
3. *Use step yields instead of overall yields.* A step yield is the yield from a single step in the purification procedure, that is, the amount of target protein or activity after that step compared to that in the previous step of the procedure. A series of four fractionation steps might all give 60% step yields, but the overall yield is $(0.6)^4 = 0.6 \times 0.6 \times 0.6 \times 0.6 = 0.13$ or 13%. Overall yield is more useful. A procedure that gives an overall yield of a few percent may be due to a very lengthy and difficult purification of a rare or unstable protein, but more likely it indicates that the procedure has not been optimized very well.
4. *Calculate yield as yield of total protein.* Often I see a table in which the yield given is yield of total protein, rather than target protein. This is relatively useless information. Yield is always recovery of total target protein or activity.
5. *Write up and try to publish the first purification that gives any product.* There is a tendency for readers, especially inexperienced readers, to assume that a published purification is the result of many cycles of improvement and optimization, and represents the best way to purify the protein. This is very often not true. Many times, it is just a series of steps, often chosen arbitrarily that happens to result in some product. One should look very carefully at a purification to assess if it is a procedure that is worth trying to follow if one wants to purify some of the target protein. This is why a proper purification summary table is so valuable. If huge losses occur at a particular fractionation step, if the overall yield is very low, if the final purity is not high, or if similar fractionation steps are used several times, then perhaps the procedure should merely be used as a beginning point in designing a better, more effective purification.
6. *What to do when purifying a protein where a fusion partner is cleaved off during the procedure?* Very often a recombinant protein is expressed as a fusion with another protein or tag that aids in its folding or purification (see Chapter 16). Since most often the desired final product is the target protein without the fusion partner or tag, especially for structural studies, the fusion partner must be cleaved off by one of several specific proteases. Let us say that the target is 20 kDa and the fusion partner is 40 kDa, so the fusion protein expressed is 60 kDa. At the step where the fusion protein is cleaved, the yield of target protein seems to decrease by 67% even if all of it is recovered. How is this indicated in the summary table?

I suggest that the column on target protein amount contain two numbers; the mg of fusion protein and the calculated mg of the final target protein in parenthesis. That way at the step where the cleavage has occurred, one can continue giving just the mg of the cleaved product and the theoretical amount of cleaved product in the first step can be used to calculate the overall yield.

SECTION TWO

GENERAL METHODS FOR HANDLING PROTEINS AND ENZYMES

CHAPTER FIVE

Setting Up a Laboratory

Murray P. Deutscher

Contents

1. Supporting Materials 38
   1.1. Glassware and plasticware 38
   1.2. Chemicals 38
   1.3. Disposables 38
   1.4. Small equipment and accessories 39
   1.5. Equipment and apparatus 39
2. Detection and Assay Requirements 39
3. Fractionation Requirements 40

The aim of this chapter is to provide some general information on the basic equipment, chemicals, and supplies that should be present in any laboratory undertaking protein purification. Details relevant to individual pieces of equipment, information on apparatus and chemicals for specialized applications, useful vendors, etc., can be found in chapters throughout this volume.

Although any laboratory engaged in protein purification may have many types of equipment, chemicals, and supplies, all these materials basically fall into three categories, those used for fractionation, those needed for detection and assay, and those that I call supporting materials. The supporting materials (e.g., tubes, pipets, baths, stirrers, timers, salts, buffers, and much more) are common to every biochemical laboratory. They are generally the least costly, used most frequently, required in largest numbers, and are the most essential. It is natural in setting up a laboratory to focus on the large, expensive apparatus, but in practice, available funds should first go to ensuring an adequate supply of supporting materials. (It obviously makes no sense to buy a sophisticated fraction collector, and not to have enough tubes.) Obtain the necessary amount of glassware, chemicals, disposables, etc., for the number of people who will be working in the laboratory. A representative (but not complete) list follows.

Department of Biochemistry and Molecular Biology, University of Miami School of Medicine, Miami, Florida, USA

Methods in Enzymology, Volume 463
ISSN 0076-6879, DOI: 10.1016/S0076-6879(09)63005-6

1. Supporting Materials

1.1. Glassware and plasticware

Tubes, beakers, flasks, bottles, cylinders, funnels, and pipets in a wide range of sizes (disposable materials are often most useful)

Transfer (Pasteur) pipets
Pipettor tips
Baking dishes
Plastic containers
Large carboys and jars
Ice buckets

1.2. Chemicals

High-grade distilled H_2O
Salts (generally chlorides or acetates)
Sodium and potassium phosphates
Enzyme-grade ammonium sulfate
Tris and other organic buffers
EDTA
Acids and bases
Reducing agents (2-mercaptoethanol, dithiothreitol, glutathione)
Protease inhibitors
Detergents
Glycerol

1.3. Disposables

Dialysis tubing
Concentrators/fractionators
Weighing paper and boats
Filter paper
pH paper
Aluminum foil
Glass wool
Syringes and needles
Sterile gloves
Marking tape and pens
Paper wipes and towels
Razor blades

1.4. Small equipment and accessories

Burners and flints
Timers (including a stopwatch)
Vortex mixers
Magnetic stirrers and stirring bars of various sizes
Variable pipettors of various sizes
Forceps, scissors, spatulas
Small tools
Hot plate
Homogenizers
Thermometers

1.5. Equipment and apparatus

Refrigerator
Freezer (− 20 °C) and, if funds available, a low-temperature freezer
Water baths (shaking and standing)
Balances (top loading and analytical)
pH meter, electrode, and standard solutions
Oven
Pump
Microfuge

Accessibility to a cold room, autoclave, ice machine, lyophilizer, dry ice, or liquid N_2.

If funds still remain after filling the above list, obtain other items directly relevant to protein purification (although some of these could also be considered supporting materials), that is, those necessary for detection and assay and for fractionation. In these areas, some of the equipment could be quite costly and sophisticated. A great deal of thought should be given to the planned usage of such equipment to determine your actual needs. In some cases it might be essential to have the item in your immediate laboratory. However, in others, if only occasional use is contemplated, you might get by with nearby access to the piece of equipment. With limited funds, and the cost of some equipment, a priority list is very helpful. In some instances, duplicating a frequently used item may be more advantageous than purchasing a new piece of equipment that will only be used infrequently. Thus, in my experience, a lab actively engaged in protein purification never has enough fraction collectors, columns, and gel electrophoresis apparatus.

2. Detection and Assay Requirements

Probably, the most important detection device in the laboratory is the spectrophotometer. It can be used for determining protein concentrations, measuring the growth of bacterial cultures, as well as for a variety of enzymatic and colorimetric assays. The spectrophotometer should be equipped with both UV and visible optics and cover the range from about 200 to 800 nm to be of most use. Both glass and quartz cuvettes are necessary to cover the visible and UV range, respectively. It is often useful to have one set of microcuvettes for analysis of small volumes (~0.2 ml). Disposable cuvettes are available, and are best for measuring cell growth.

Most enzymatic assays rely on either spectrophotometry or the use of radioisotopes. In the latter case, a scintillation counter is a necessity. The use of a scintillation counter means that the supplies, chemicals, and other accessories needed for preparing radioactive samples will also be required. Scintillation counters are quite costly, and often are shared among several laboratories. Likewise, a phosphorimager and a fluoroimager are essential for gel assays, but are often shared. If the use of radioactive material is contemplated, radiation monitors, shielding, and other precautions will be needed as well.

Two other detection devices that often come in handy are a conductivity meter and a refractometer. These are used to measure salt gradients on chromatographic columns, and sucrose, glycerol, or CsCl centrifugal gradients.

3. Fractionation Requirements

Protein purification means protein fractionation. What distinguishes a protein purification laboratory from the usual biochemistry or molecular biology laboratory is largely the number and types of fractionation apparatus and materials available. Subsequent chapters will discuss these items in detail; they will be mentioned here only briefly.

Probably, the most frequently used piece of equipment in the laboratory is the centrifuge. The workhorse of the protein purification laboratory is the refrigerated high-speed centrifuge which attains speeds up to 20,000 rpm. The usefulness of such a centrifuge is directly related to the presence in the laboratory of a wide variety of rotors and centrifuge tubes and bottles. Rotors are available that hold as small a volume as a few milliliters per tube to ones that hold 500-ml bottles. The large rotors are invaluable for handling the large volumes of extracts often encountered in early steps of a protein purification.

In instances in which one wants to remove or prepare subcellular organelles, access to an ultracentrifuge is desirable. Instruments are available that can process reasonably large volumes at speeds as high as 80,000 rpm, and smaller machines on the market can go even faster. The availability of this instrumentation has greatly reduced the time required to prepare microsomal or high-speed supernatant fractions. In view of the cost of these machines, and their relatively infrequent use in most cases, they are often shared among laboratories.

The advent of many microanalytical techniques has also made the minifuge or microcentrifuge an essential item. Though not really a fractionation apparatus in a protein purification laboratory, it is a necessary addition. In this regard, the larger centrifuges are also frequently used for assays of various types, rather than only for fractionation purposes.

In order to isolate proteins, a means of rupturing cells is required. Various apparatuses are available for this purpose, including hand-held and motor-driven homogenizers, blenders, sonicators, pressure cells, etc. These will be discussed in detail in other chapters. In general, it is desirable to have a variety of the less costly items in individual laboratories, with the remainder available as shared equipment.

Column chromatography is a primary protein purification method in use in most laboratories. Every laboratory involved in protein fractionation should have available a large supply of columns of various lengths and diameters in anticipation of every conceivable need, since they will arise during the course of developing purification schemes. Columns are available in various degrees of sophistication (and cost). We have found that simple, gravity-flow, open-top columns fitted with stoppers and syringe needles, or tubing, for fluid inlet and control, are satisfactory for most chromatographic procedures. However, for many purposes, prepacked columns and pressurized FPLC systems may be the preferred route.

A dependable fraction collector is one of the most important pieces of equipment in the laboratory. Failure of a fraction collector may result in the loss of several month's work. In this instance, extra money spent on a good, versatile machine is a wise investment. Instruments able to handle a large number of tubes, of various sizes, in different collection modes, are the most useful. Many different types of fraction collectors are available. Careful analysis of the various models, and matching to anticipated requirements, is good practice prior to purchase. A suitable strategy for many laboratories would be the purchase of one of the more sophisticated instruments for special needs, and one or more of the less costly, simple machines for routine use.

A number of other accessories to column chromatography are useful, if not essential. These include a peristaltic pump, various sizes of gradient makers, and a UV monitor. Gradient makers can be homemade from flasks or bottles, if necessary. Following the protein elution profile during a chromatographic run provides important information. This can be done by determining the absorbance of individual fractions with a spectrophotometer, or automatically with an in-stream UV monitor. Dual-wavelength models with different size flow cells are the most versatile (and also most costly).

Finally, every laboratory should also have on hand a basic supply of chromatographic gels and resins. These should include an anion and cation exchanger, various porosity gel filtration media, hydroxyapatite, a hydrophobic gel, and probably an immobilized dye resin.

No protein purification laboratory is complete without the presence of gel electrophoresis equipment. These items are used to monitor a purification procedure or for fractionation itself. Generally, a vertical slab-gel apparatus with various-sized spacers and combs is satisfactory for most applications. An electrophoresis power supply unit is also required. If only one is to be purchased, a regulated constant-current (0–50 mA) constant-voltage (0–200 V) model is useful. In addition to the chemicals necessary for preparing gels, several protein standards and staining dyes should also be obtained. The instrumentation used for gel electrophoresis can also be used for isoelectric focusing.

The equipment, chemicals, and supplies mentioned in this chapter should allow you to enter the field of protein purification. As you read through this volume, and actually begin to purify proteins, many other useful items will become apparent.

CHAPTER SIX

BUFFERS: PRINCIPLES AND PRACTICE[1]

Vincent S. Stoll *and* John S. Blanchard

Contents

1. Introduction 43
2. Theory 44
3. Buffer Selection 45
4. Buffer Preparation 48
5. Volatile Buffers 49
6. Broad-Range Buffers 50
7. Recipes for Buffer Stock Solutions 50
References 56

1. INTRODUCTION

The necessity for maintaining a stable pH when studying enzymes is well established (Good and Izawa, this series; Johnson and Metzler, this series). Biochemical processes can be severely affected by minute changes in hydrogen ion concentrations. At the same time, many protons may be consumed or released during an enzymatic reaction. It has become increasingly important to find buffers to stabilize hydrogen ion concentrations while not interfering with the function of the enzyme being studied. The development of a series of N-substituted taurine and glycine buffers by Good *et al.* (1966) has provided buffers in the physiologically relevant range (6.1–10.4) of most enzymes, which have limited side effects with most enzymes (Good *et al.*, 1966). It has been found that these buffers are nontoxic to cells at 50 m*M* concentrations and in some cases much higher (Ferguson *et al.*, 1980).

Department of Biochemistry, Albert Einstein College of Medicine, Bronx, New York
[1] Reprinted from *Methods in Enzymology*, Volume 182 (Academic Press, 1990)

Methods in Enzymology, Volume 463
ISSN 0076-6879, DOI: 10.1016/S0076-6879(09)63006-8

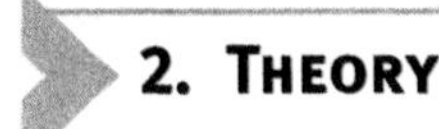

2. Theory

The observation that partially neutralized solutions of weak acids or weak bases are resistant to pH changes on the addition of small amounts of strong acid or strong base leads to the concept of "buffering" (Perrin and Dempsey, 1974). Buffers consist of an acid and its conjugate base, such as carbonate and bicarbonate, or acetate and acetic acid. The quality of a buffer is dependent on its buffering capacity (resistance to change in pH by addition of strong acid or base), and its ability to maintain a stable pH upon dilution or addition of neutral salts. Because of the following equilibria, additions of small amounts of strong acid or strong base result in the removal of only small amounts of the weakly acidic or basic species; therefore, there is little change in the pH:

$$\mathrm{HA(acid)} \rightleftharpoons \mathrm{H^+} + \mathrm{A^-(conjugate\ base)} \tag{6.1}$$

$$\mathrm{B(base)} + \mathrm{H^+} \rightleftharpoons \mathrm{BH^+(conjugate\ acid)} \tag{6.2}$$

The pH of a solution of a weak acid or base may be calculated from the Henderson–Hasselbalch equation:

$$\mathrm{pH} = \mathrm{p}K_a' + \log\frac{[\text{basic species}]}{[\text{acidic species}]} \tag{6.3}$$

The pK_a of a buffer is that pH where the concentrations of basic and acidic species are equal, and in this basic form the equation is accurate between the pH range of 3–11. Below pH 3 and above pH 11 the concentrations of the ionic species of water must be included in the equation (Perrin and Dempsey, 1974). Since the pH range of interest here is generally in the pH 3–11 range, this will be ignored.

From the Henderson–Hasselbalch equation an expression for buffer capacity may be deduced. If at some concentration of buffer, c, the sum $[A^-] + [HA]$ is constant, then the amount of strong acid or base needed to cause a small change in pH is given by the relationship:

$$\frac{d[B]}{d\mathrm{pH}} = 2.303\left\{\frac{[K_a'c[H^+]]}{(K_a' + [H^+])^2} + [H^+] + \frac{K_w}{H^+}\right\} \tag{6.4}$$

In this equation, K_w refers to the ionic product of water, and the second and third terms are only significant below pH 3 or above pH 11. In the pH range of interest (pH 3–11), this equation yields the following expression:

$$\beta_{max} = \frac{2.303c}{4} = 0.576c, \tag{6.5}$$

which represents a maximum value for d[B]/dpH when pH = pK_a. The buffer capacity of any buffer is dependent on the concentration, c, and may

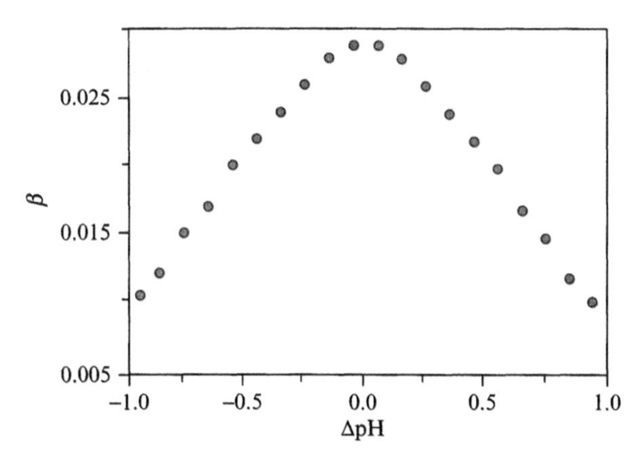

Figure 6.1 Buffer capacity (β) versus ΔpH over the range $\pm$ 1 pH unit of the pK_a for HEPES (0.05 *M*). Points calculated using Eq. (6.5), and data from Perrin and Dempsey (1974).

be calculated over a buffer range of $\pm$ 1 pH unit around the p*K* to determine the buffer capacity, as shown in Fig. 6.1 for one of the Good buffers, HEPES. It can be seen that the buffer capacity is greatest at its p*K*, and drops off quickly 1 pH unit on either side of the p*K*. In practice, buffers should not be used beyond these values.

3. Buffer Selection

There are many factors that must be considered when choosing a buffer. When studying an enzyme one must consider the pH optimum of the enzyme, nonspecific buffer effects on the enzyme, and interactions with substrates or metals. When purifying a protein, cost becomes an important consideration, as does the compatibility of the buffer with different purification techniques. Table 6.1 lists a wide variety of buffers covering a broad pH range.

Determining the pH optimum of a protein is a first step in determining the best buffer to employ (Blanchard, this series). Since the buffering capacity is maximal at the p*K*, buffers should be used close to this value. When determining the pH optimum for an enzyme, it is useful to use a series of related buffers that span a wide pH range. Once an optimal pH has been approximated, different buffers within this pH range can be examined for specific buffer effects.

The Good buffers have been shown to be relatively free of side effects. However, inorganic buffers do have a high potential for specific buffer effects. Many enzymes are inhibited by phosphate buffer, including carboxypeptidase, urease, as well as many kinases and dehydrogenases (Blanchard, this series). Borate buffers can form covalent complexes with mono- and oligosaccharides, the ribose moieties of nucleic acids, pyridine nucleotides, and other *gem*-diols. Tris and other primary amine buffers may form Schiff base adducts with aldehydes and ketones.

Table 6.1 Selected buffers and their p*K* values at 25 °C

Trivial name	Buffer name	pK_a	dpK_a/dt
Phosphate (pK_1)	–	2.15	0.0044
Malate (pK_1)	–	3.40	–
Formate	–	3.75	0.0
Succinate (pK_1)	–	4.21	– 0.0018
Citrate (pK_2)	–	4.76	– 0.0016
Acetate	–	4.76	0.0002
Malate	–	5.13	–
Pyridine	–	5.23	– 0.014
Succinate (pK_2)	–	5.64	0.0
MES	2-(*N*-Morpholino) ethanesulfonic acid	6.10	– 0.011
Cacodylate	Dimethylarsinic acid	6.27	
Dimethylglutarate	3,3-Dimethylglutarate (pK_2)	6.34	0.0060
Carbonate (pK_1)	–	6.35	– 0.0055
Citrate (pK_3)	–	6.40	0.0
Bis–Tris	[Bis(2-hydroxyethyl)imino]tris (hydroxymethyl)methane	6.46	0.0
ADA	*N*-2-Acetamidoiminodiacetic acid	6.59	– 0.011
Pyrophosphate	–	6.60	–
EDPS (pK_1)	*N*,*N*′-Bis(3-sulfopropyl) ethylenediamine	6.65	–
Bis–Tris propane	1,3-Bis[tris(hydroxymethyl) methylamino]propane	6.80	–
PIPES	Piperazine-*N*,*N*′-bis(2-ethanesulfonic acid)	6.76	– 0.0085
ACES	*N*-2-Acetamido-2-hydroxyethanesulfonic acid	6.78	– 0.020
MOPSO	3-(*N*-Morpholino)-2-hydroxypropanesulfonic acid	6.95	– 0.015
Imidazole	–	6.95	– 0.020
BES	*N*,*N*-Bis-(2-hydroxyethyl)-2-aminoethanesulfonic acid	7.09	– 0.016
MOPS	3-(*N*-Morpholino) propanesulfonic acid	7.20	0.015
Phosphate (pK_2)	–	7.20	– 0.0028
EMTA	3,6-Endomethylene-1,2,3,6-tetrahydrophthalic acid	7.23	–
TES	2-[Tris(hydroxymethyl) methylamino]ethanesulfonic acid	7.40	– 0.020

Table 6.1 *(continued)*

Trivial name	Buffer name	pK_a	dpK_a/dt
HEPES	*N*-2-Hydroxyethylpiperazine-*N*′-2-ethanesulfonic acid	7.48	−0.014
DIPSO	3-[*N*-Bis(hydroxyethyl)amino]-2-hydroxypropanesulfonic acid	7.60	−0.015
TEA	Triethanolamine	7.76	−0.020
POPSO	Piperazine-*N*,*N*′-bis(2-hydroxypropanesulfonic acid)	7.85	−0.013
EPPS, HEPPS	*N*-2-Hydroxyethylpiperazine-*N*′-3-propanesulfonic acid	8.00	–
Tris	Tris(hydroxymethyl) aminomethane	8.06	−0.028
Tricine	*N*-[Tris(hydroxymethyl)methyl] glycine	8.05	−0.021
Glycinamide	–	8.06	−0.029
PIPPS	1,4-Bis(3-sulfopropyl)piperazine	8.10	–
Glycylglycine	–	8.25	−0.025
Bicine	*N*,*N*-Bis(2-hydroxyethyl)glycine	8.26	−0.018
TAPS	3-{[Tris(hydroxymethyl)methyl] amino}propanesulfonic acid	8.40	0.018
Morpholine	–	8.49	–
PIBS	1,4-Bis(4-sulfobutyl)piperazine	8.60	–
AES	2-Aminoethylsulfonic acid, taurine	9.06	−0.022
Borate	–	9.23	−0.008
Ammonia	–	9.25	−0.031
Ethanolamine	–	9.50	−0.029
CHES	Cyclohexylaminoethanesulfonic acid	9.55	0.029
Glycine (pK_2)	–	9.78	−0.025
EDPS	*N*,*N*′-Bis(3-sulfopropyl) ethylenediamine	9.80	–
APS	3-Aminopropanesulfonic acid	9.89	–
Carbonate (pK_2)	–	10.33	−0.009
CAPS	3-(Cyclohexylamino) propanesulfonic acid	10.40	0.032
Piperidine	–	11.12	–
Phosphate (pK_3)	–	12.33	−0.026

Buffer complexation with metals may present additional problems. In this respect, inorganic buffers can prove problematic in that they may remove, by chelation, metals essential to enzymatic activity (e.g., Mg^{2+} for kinases,

Cu^{2+} or Fe^{2+} for hydroxylases). Release of protons upon chelation or precipitation of metal–buffer complexes may also be a potential problem. Where metal chelation presents a problem, the Good buffers are useful since they have been shown to have low metal-binding capabilities (Good *et al.*, 1966).

Once a suitable buffer has been found (noninteracting, with an appropriate p*K*), a concentration should be chosen. Since high ionic strength may decrease enzyme activity, the buffer concentration should be as low as possible (Blanchard, this series). A reasonable way to determine how low a concentration may be used is to examine the properties (reaction rate or protein stability) at a low (10–20 m*M*) concentration of buffer. The pH prior to, and an adequate time after, addition of protein should not vary more than $\pm$0.05 pH. If the pH changes too drastically (greater than $\pm$0.1 pH unit), then the buffer concentration should be raised to 50 m*M*. In cases where protons are consumed or released stoichiometrically with substrate utilization, pH stability becomes increasingly important.

Buffers may be made up in stock solutions, then diluted for use. When stock solutions are made, it should be done close to the working temperature, and in glass bottles (plastic bottles can leach UV-absorbing material) (Perrin and Dempsey, 1974). Buffers have temperature-sensitive p*K* values, particularly amine buffers. The carboxylic acid buffers are generally the least sensitive to temperature, and the Good buffers have only a small inverse temperature dependence on p*K*. The effects of dilution of stock solutions, or addition of salts, on pH should be checked by measurement of the pH after addition of all components.

Choosing a buffer for protein purification requires some special considerations. Large amounts of buffer will be needed for centrifugation, chromatographic separations, and dialysis, which makes cost a concern. Tris and many inorganic buffers are widely used since they are relatively inexpensive. Although buffers like Tris are inexpensive, and have been widely used in protein purification, they do have disadvantages. Tris is a poor buffer below pH 7.5 and its p*K* is temperature dependent (a solution made up to pH 8.06 at 25 °C will have a pH of 8.85 at 0 °C). Many primary amine buffers such as Tris and glycine (Bradford, 1976) will interfere with the Bradford dye-binding protein assay. Some of the Good buffers, HEPES, EPPS, and Bicine, give false-positive colors with Lowry assay.

Spectroscopic measurement of enzyme rates is a commonly applied method. It may be important to use a buffer that does not absorb appreciably in the spectral region of interest. The Good buffers, and most buffers listed in Table 6.1, can be used above 240 nm.

4. Buffer Preparation

Once a suitable buffer has been chosen it must be dissolved and titrated to the desired pH. Before titrating a buffer solution, the pH meter must be calibrated. Calibration should be done using commercially available pH standards, bracketing the desired pH. If monovalent cations interfere, or are being investigated, then titration with tetramethylammonium hydroxide can be done to avoid mineral cations. Similarly, the substitution of the most commonly used counteranion, chloride, with other anions such as acetate, sulfate, or glutamate, may have significant effects on enzyme activity or protein–DNA interactions (Leirmo *et al.*, 1987). Stock solutions should be made with quality water (deionized and double-distilled, preferably) and filtered through a sterile ultrafiltration system (0.22 μm) to prevent bacterial or fungal growth, especially with solutions in the pH 6–8 range. To prevent heavy metals from interfering, EDTA (10–100 μM) may be added to chelate any contaminating metals.

5. Volatile Buffers

In certain cases, it is necessary to remove a buffer quickly and completely. Volatile buffers make it possible to remove components that may interfere in subsequent procedures. Volatile buffers are useful in electrophoresis, ion-exchange chromatography, and digestion of proteins followed by separation of peptides or amino acids. Most of the volatile buffers (Table 6.2) are transparent in the lower UV range except for the buffers

Table 6.2 Types of systems for use as volatile buffers[a]

System	pH range
87 ml glacial acetic acid + 25 ml 88% HCOOH in 11 l	1.9
25 ml 88% HCOOH in 1 l	2.1
Pyridine–formic acid	2.3–3.5
Trimethylamine–formic acid	3.0–5.0
Triethylamine–formic (or acetic) acid	3–6
5 ml pyridine + 100 ml glacial acetic acid in 1 l	3.1
5 ml pyridine + 50 ml glacial acetic acid in 1 l	3.5
Trimethylamine–acetic acid	4.0–6.0
25 ml pyridine + 25 ml glacial acetic acid in 1 l	4.7
Collidine–acetic acid	5.5–7.0
100 ml pyridine + 4 ml glacial acetic acid in 1 l	6.5
Triethanolamine–HCl	6.8–8.8
Ammonia–formic (or acetic) acid	7.0–10.0
Trimethylamine–CO_2	7–12
Triethylamine–CO_2	7–12
24 g NH_4HCO_3 in 1 l	7.9
Ammonium carbonate–ammonia	8.0–10.5
Ethanolamine–HCl	8.5–10.5
20 g $(NH_4)_2CO_3$ in 1 l	8.9

[a] From Perrin and Dempsey (1974).

containing pyridine (Perrin and Dempsey, 1974). An important consideration is interference in amino acid analysis (i.e., reactions with ninhydrin). Most volatile buffers will not interfere with ninhydrin if the concentrations are not too high (e.g., triethanolamine less than 0.1 *M* does not interfere).

6. Broad-Range Buffers

There may be occasions where a single buffer system is desired that can span a wide pH range of perhaps 5 or more pH units. One method would be a mixture of buffers that sufficiently covers the pH range of interest. This may lead to nonspecific buffer interactions for which corrections must be made. Another common approach is to use a series of structurally related buffers that have evenly spaced p*K* values such that each p*K* is separated by approximately ± 1 pH unit (the limit of buffering capacity). The Good buffers are ideal for this approach since they are structurally related and have relatively evenly spaced p*K* values. As the pH passes the p*K* of one buffer it becomes nonparticipatory and therefore has no further function. These nonparticipating buffer components may show nonspecific buffer effects as well as raising the ionic strength with potential deleterious effects. A detailed description of buffer mixtures which provide a wide range of buffering capacity with constant ionic strength is available (Ellis and Morrison, this series).

7. Recipes for Buffer Stock Solutions

1. Glycine–HCl buffer (Sorensen, 1909a,b) stock solutions
   - A: 0.2 *M* solution of glycine (15.01 g in 1000 ml)
   - B: 0.2 *M* HCl

   50 ml of A + *x* ml of B, diluted to a total of 200 ml:

x	*pH*	*x*	*pH*
5.0	3.6	16.8	2.8
6.4	3.4	24.2	2.6
8.2	3.2	32.4	–

2. Citrate buffer (Lillie, 1948) stock solutions
   - A: 0.1 *M* solution of citric acid (21.01 g in 1000 ml)
   - B: 0.1 *M* solution of sodium citrate (29.41 g $C_6H_5O_7Na_3 \cdot 2H_2O$ in 1000 ml)

   x ml of A + *y* ml of B, diluted to a total of 100 ml:

x	*y*	*pH*
46.5	3.5	3.0
43.7	6.3	3.2
40.0	10.0	3.4
37.0	13.0	3.6
35.0	15.0	3.8
33.0	17.0	4.0
31.5	18.5	4.2
28.0	22.0	4.4
25.5	24.5	4.6
23.0	27.0	4.8
20.5	29.5	5.0
18.0	32.0	5.2
16.0	34.0	5.4
13.7	36.3	5.6
11.8	38.2	5.8
9.5	41.5	6.0
7.2	42.8	6.2

3. Acetate buffer (Walpole, 1914) stock solutions
   - A: 0.2 *M* solution of acetic acid (11.55 ml in 1000 ml)
   - B: 0.2 *M* solution of sodium acetate (16.4 g of $C_2H_3O_2Na$ or 27.2 g of $C_2H_3O_2Na \cdot 3H_2O$ in 1000 ml)

   x ml of A + *y* ml of B, diluted to a total of 100 ml:

x	*y*	*pH*
46.3	3.7	3.6
44.0	6.0	3.8
41.0	9.0	4.0
36.8	13.2	4.2
30.5	19.5	4.4
25.5	24.5	4.6
14.8	35.2	5.0
10.5	39.5	5.2
8.8	41.2	5.4
4.8	45.2	5.6

4. Citrate–phosphate buffer (McIlvaine, 1921) stock solutions
   - A: 0.1 *M* solution of citric acid (19.21 g in 1000 ml)
   - B: 0.2 *M* solution of dibasic sodium phosphate (53.65 g of $Na_2HPO_4 \cdot 7H_2O$ or 71.7 g of $Na_2HPO_4 \cdot 12H_2O$ in 1000 ml)

   x ml of A + *y* ml of B, diluted to a total of 100 ml:

x	*y*	*pH*
44.6	5.4	2.6
42.2	7.8	2.8
39.8	10.2	3.0
37.7	12.3	3.2
35.9	14.1	3.4
33.9	16.1	3.6
32.3	17.7	3.8
30.7	19.3	4.0
29.4	20.6	4.2
27.8	22.2	4.4
26.7	23.3	4.6
25.2	24.8	4.8
24.3	25.7	5.0
23.3	26.7	5.2
22.2	27.8	5.4
21.0	29.0	5.6
19.7	30.3	5.8
17.9	32.1	6.0
16.9	33.1	6.2
15.4	34.6	6.4
13.6	36.4	6.6
9.1	40.9	6.8
6.5	43.6	7.0

5. Succinate buffer (G. Gomori, unpublished observation) stock solutions
   - A: 0.2 *M* solution of succinic acid (23.6 g in 1000 ml)
   - B: 0.2 *M* NaOH

   25 ml of A + *x* ml of B, diluted to a total of 100 ml:

x	*pH*	*x*	*pH*
7.5	3.8	26.7	5.0
10.0	4.0	30.3	5.2
13.3	4.2	34.2	5.4
16.7	4.4	37.5	5.6
20.0	4.6	40.7	5.8
23.5	4.8	43.5	6.0

6. Cacodylate buffer (Plumel, 1949) stock solutions
   - A: 0.2 *M* solution of sodium cacodylate (42.8 g of $Na(CH_3)_2AsO_2 \cdot 3H_2O$ in 1000 ml)
   - B: 0.2 *M* NaOH

   50 ml of A + *x* ml of B, diluted to a total of 200 ml:

x	pH	x	pH
2.7	7.4	29.6	6.0
4.2	7.2	34.8	5.8
6.3	7.0	39.2	5.6
9.3	6.8	43.0	5.4
13.3	6.6	45.0	5.2
18.3	6.4	47.0	5.0
13.8	6.2		

7. Phosphate buffer (Sorensen, 1909a,b) stock solutions
   - A: 0.2 *M* solution of monobasic sodium phosphate (27.8 g in 1000 ml)
   - B: 0.2 *M* solution of dibasic sodium phosphate (53.65 g of $Na_2HPO_4 \cdot 7H_2O$ or 71.7 g of $Na_2HPO_4 \cdot 12H_2O$ in 1000 ml)

   x ml of A + y ml of B, diluted to a total of 200 ml:

x	y	pH	x	y	pH
93.5	6.5	5.7	45.0	55.0	6.9
92.0	8.0	5.8	39.0	61.0	7.0
90.0	10.0	5.9	33.0	67.0	7.1
87.7	12.3	6.0	28.0	72.0	7.2
85.0	15.0	6.1	23.0	77.0	7.3
81.5	18.5	6.2	19.0	81.0	7.4
77.5	22.5	6.3	16.0	84.0	7.5
73.5	26.5	6.4	13.0	87.0	7.6
68.5	31.5	6.5	10.5	90.5	7.7
62.5	37.5	6.6	8.5	91.5	7.8
56.5	43.5	6.7	7.0	93.0	7.9
51.0	49.0	6.8	5.3	94.7	8.0

8. Barbital buffer (Michaelis, 1930) stock solutions
   - A: 0.2 *M* solution of sodium barbital (veronal) (41.2 g in 1000 ml)
   - B: 0.2 *M* HCl

   50 ml of A + x ml of B, diluted to a total of 200 ml:

x	pH	x	pH
1.5	9.2	22.5	7.8
2.5	9.0	27.5	7.6
4.0	8.8	32.5	7.4
6.0	8.6	39.0	7.2
9.0	8.4	43.0	7.0
2.7	8.2	45.0	6.8
17.5	8.0		

Solutions more concentrated than 0.05 *M* may crystallize on standing, especially in the cold.

9. Tris(hydroxymethyl)aminomethane (Tris) buffer (Hayaishi, this series) stock solutions
   - A: 0.2 *M* solution of tris(hydroxymethyl)aminomethane (24.2 g in 1000 ml)
   - B: 0.2 *M* HCl

   50 ml of A + *x* ml of B, diluted to a total of 200 ml:

x	*pH*
5.0	9.0
8.1	8.8
12.2	8.6
16.5	8.4
21.9	8.4
26.8	8.0
32.5	7.8
38.4	7.6
41.4	7.4
44.2	7.2

10. Boric acid–borax buffer (Holmes, 1943) stock solutions
    - A: 0.2 *M* solution of boric acid (12.4 g in 1000 ml)
    - B: 0.05 *M* solution of borax (19.05 g in 1000 ml; 0.2 *M* in terms of sodium borate)

    50 ml of A + *x* ml of B, diluted to a total of 200 ml:

x	*pH*	*x*	*pH*
2.0	7.6	22.5	8.7
3.1	7.8	30.0	8.8
4.9	8.0	42.5	8.9
7.3	8.2	59.0	9.0
11.5	8.4	83.0	9.1
17.5	8.6	115.0	9.2

11. 2-Amino-2-methyl-1,3-propanediol (Ammediol) buffer (Gomori, 1946) stock solutions
    - A: 0.2 *M* solution of 2-amino-2-methyl-1,3-propanediol (21.03 g in 1000 ml)
    - B: 0.2 *M* HCl

    50 ml of A + *x* ml of B, diluted to a total of 200 ml:

x	pH	x	pH
2.0	10.0	22.0	8.8
3.7	9.8	29.5	8.6
5.7	9.6	34.0	8.4
8.5	9.4	37.7	8.2
12.5	9.2	41.0	8.0
16.7	9.0	43.5	7.8

12. Glycine–NaOH buffer (Sorensen, 1909a,b) stock solutions
    - A: 0.2 *M* solution of glycine (15.01 g in 1000 ml)
    - B: 0.2 *M* NaOH

    50 ml of A + x ml of B, diluted to a total of 200 ml:

x	pH	x	pH
4.0	8.6	22.4	9.6
6.0	8.8	27.2	9.8
8.8	9.0	32.0	10.0
12.0	9.2	38.6	10.4
16.8	9.4	45.5	10.6

13. Borax–NaOH buffer (Clark and Lubs, 1917) stock solutions
    - A: 0.05 *M* solution of borax (19.05 g in 1000 ml; 0.02 *M* in terms of sodium borate)
    - B: 0.2 *M* NaOH

    50 ml of A + x ml of B, diluted to a total of 200 ml:

x	pH
0.0	9.28
7.0	9.35
11.0	9.4
17.6	9.5
23.0	9.6
29.0	9.7
34.0	9.8
38.6	9.9
43.0	10.0
46.0	10.1

14. Carbonate–bicarbonate buffer (Delory and King, 1945) stock solutions
    - A: 0.2 *M* solution of anhydrous sodium carbonate (21.2 g in 1000 ml)
    - B: 0.2 *M* solution of sodium bicarbonate (16.8 g in 1000 ml)

    x ml of A + y ml of B, diluted to a total of 200 ml:

x	*y*	*pH*
4.0	46.0	9.2
7.5	42.5	9.3
9.5	40.5	9.4
13.0	37.0	9.5
16.0	34.0	9.6
19.5	30.5	9.7
22.0	28.0	9.8
25.0	25.0	9.9
27.5	22.5	10.0
30.0	20.0	10.1
33.0	17.0	10.2
35.5	14.5	10.3
38.5	11.5	10.4
40.5	9.5	10.5
42.5	7.5	10.6
45.0	5.0	10.7

REFERENCES

Blanchard, J. S. (this series). Vol. 104, p. 404.
Bradford, M. M. (1976). *Anal. Biochem.* **22,** 248.
Clark, W. M., and Lubs, H. A. (1917). *J. Bacteriol.* **2,** 1.
Delory, G. E., and King, E. J. (1945). *Biochem. J.* **39,** 245.
Ellis, K. J., and Morrison, J. F. (this series). Vol. 87, p. 405.
Ferguson, W. J., *et al.* (1980). *Anal. Biochem.* **104,** 300.
Gomori, G. (1946). *Proc. Soc. Exp. Biol. Med.* **62,** 33.
Gomori, G. Unpublished observations.
Good, N. E., and Izawa, S. (this series). Vol. 24, p. 53.
Good, N. E., Winget, G. D., Winter, W., Connolly, T. N., Izawa, S., and Singh, R. M. M. (1966). *Biochemistry* **5,** 467.
Hayaishi, O. (this series). Vol. 1, p. 144.
Holmes, W. (1943). *Anat. Rec.* **86,** 163.
Johnson, R. J., and Metzler, D. E. (this series). Vol. 22, p. 3.
Leirmo, S., Harrison, C., Cayley, D. S., Burgess, R. R., and Record, M. T. (1987). *Biochemistry* **26,** 2095.
Lillie, R. D. (1948). Histopathologic Technique. Blakiston, Philadelphia, PA.
McIlvaine, T. C. (1921). *J. Biol. Chem.* **49,** 183.
Michaelis, L. (1930). *J. Biol. Chem.* **87,** 33.
Perrin, D. D., and Dempsey, B. (1974). Buffers for pH and Metal Ion Control. Chapman & Hall, London.
Plumel, M. (1949). *Bull. Soc. Chim. Biol.* **30,** 129.
Sorensen, S. P. L. (1909a). *Biochem. Z.* **21,** 131.
Sorensen, S. P. L. (1909b). *Biochem. Z.* **22,** 352.
Walpole, G. S. (1914). *J. Chem. Soc.* **105,** 2501.

CHAPTER SEVEN

Measurement of Enzyme Activity

T. K. Harris *and* M. M. Keshwani

Contents

1. Introduction 58
2. Principles of Catalytic Activity 58
   2.1. Chemical kinetics 58
   2.2. Basic enzyme kinetics 62
3. Measurement of Enzyme Activity 64
   3.1. Continuous assays 65
   3.2. Discontinuous assays 67
4. Formulation of Reaction Assay Mixtures 69
5. Discussion 71
Acknowledgments 71
References 71

Abstract

To study and understand the nature of living cells, scientists have continually employed traditional biochemical techniques aimed to fractionate and characterize a designated network of macromolecular components required to carry out a particular cellular function. At the most rudimentary level, cellular functions ultimately entail rapid chemical transformations that otherwise would not occur in the physiological environment of the cell. The term *enzyme* is used to singularly designate a macromolecular gene product that specifically and greatly enhances the rate of a chemical transformation. Purification and characterization of individual and collective groups of enzymes has been and will remain essential toward advancement of the molecular biological sciences; and developing and utilizing enzyme reaction assays is central to this mission. First, basic kinetic principles are described for understanding chemical reaction rates and the catalytic effects of enzymes on such rates. Then, a number of methods are described for measuring enzyme-catalyzed reaction rates, which mainly differ with regard to techniques used to detect and quantify concentration changes of given reactants or products. Finally, short commentary is given toward formulation of reaction mixtures used to measure enzyme activity. Whereas a comprehensive treatment of enzymatic reaction assays is not within

Department of Biochemistry and Molecular Biology, Miller School of Medicine, University of Miami, Miami, Florida, USA

Methods in Enzymology, Volume 463
ISSN 0076-6879, DOI: 10.1016/S0076-6879(09)63007-X

the scope of this chapter, the very core principles that are presented should enable new researchers to better understand the logic and utility of any given enzymatic assay that becomes of interest.

1. Introduction

One primary goal of biological sciences is to deduce the molecular bases of all chemical processes that take place in living organisms. It is well established that the vast majority of biochemical reactions are governed by protein gene products called enzymes. In this sense, enzymes behave as finely tuned chemical catalysts, which greatly enhance reaction rates with both temporal and spatial resolution. Thus, elucidation of biochemical pathways involving biosynthesis, modification, and degradation of macromolecular, metabolic, and signaling molecules remains ultimately tied to experimental methods for purification, reconstitution, and direct demonstration of enzymes to catalyze specific chemical reactions. The focus of this chapter is to explain the most fundamental principles that must be considered in developing effective methods for measuring the progress of enzyme-catalyzed reactions. First, kinetic formulation is described for chemical and enzymatic reaction processes. Next, instructive commentary is given on various experimental methods that are commonly used for measuring enzyme activity; and consideration is given toward numerous procedures and conditions that must be optimized.

2. Principles of Catalytic Activity

2.1. Chemical kinetics

Chemical kinetics involves the experimental study of reaction rates in order to infer about the kinetic mechanisms for chemical conversion of reactants (R) into products (P) (Fig. 7.1) (House, 2007; Laidler, 1987). For any given chemical reaction, (i) the *mechanism* refers to the sequence of elementary steps by which overall chemical change occurs and (ii) an *elementary step* refers to the passing of a reactant or reaction complex through a single-transition state to a chemical form with a defined and detectable lifetime. If chemical conversion of reactants (R) to products (P) involves more than one elementary step, then the chemical structures that exist after each elementary step preceding final product formation are defined as reaction *intermediates* (I). Figure 7.1 also shows that each elementary step leading to product is characterized by a forward microscopic rate constant (e.g., k_{+1} and k_{+2}). If a particular elementary step is reversible, then a reverse rate

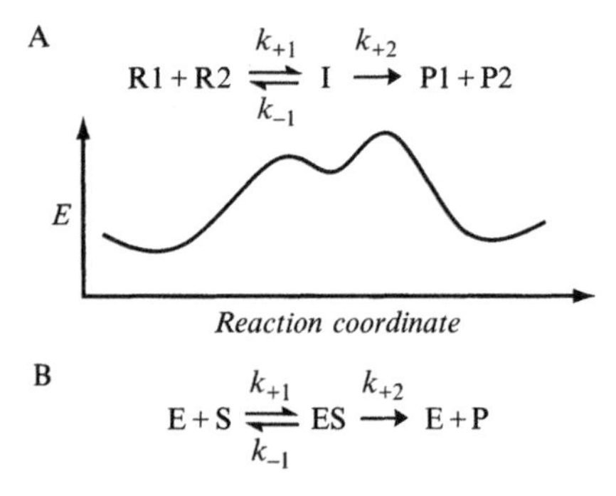

Figure 7.1 (A) Basic kinetic mechanism for chemical conversion of two molecular reactants (R1 and R2) into two molecular products (P1 and P2). In this mechanism, an intermediate complex (I) is reversibly formed from reactants (k_{+1} and k_{-1}) but is irreversibly converted to products (k_{+2}). (B) Basic kinetic mechanism for enzyme (E)-catalyzed conversion of a molecular substrate (S) into a molecular product (P). In this mechanism, an ES intermediate complex is reversibly formed (k_{+1} and k_{-1}) but is irreversibly converted to product (k_{+2}). In this case, the enzyme (E) is regenerated and can undergo subsequent catalytic cycles. For both mechanisms, a free-energy diagram depicts typical energy changes that occur along the reaction coordinate leading to intermediates and products. Ground-state energies for reactants, intermediates, and products are depicted as valleys, whereas transition-state energies for chemical changes are depicted by peaks.

constant is designated (e.g., k_{-1}). The principle of microscopic reversibility states that only one transition state exists for reversible chemical conversion between the species associated with one elementary step (i.e., the transition state for the forward reaction is the same as for the reverse reaction).

For any given unidirectional conversion within an elementary step, a *rate law* is used to describe the mathematical relationship between the reaction rate or velocity and the concentration of each given reactant. For a given chemical reaction step occurring in solution, the instantaneous reaction *velocity* (v) is defined as the derivative of chemical *concentration* with respect to *time* (t); and it is expressed either as (i) a decreasing concentration of a given reactant ($v = -d[\mathrm{R}]/dt$) or (ii) an increasing concentration of a given product ($v = +d[\mathrm{P}]/dt$). The standard international units of reaction velocity are $\mathrm{M\ s^{-1}}$. The rate law for a single-unidirectional step shows the instantaneous velocity to depend on variable reactant concentration values according to parameter values of (i) a proportionality constant (k) and (ii) the exponent of each concentration term (Table 7.1). The proportionality constant is defined as the forward *microscopic rate constant* for a given chemical conversion, and its dimensions depend on the overall *molecularity* or *order* of the reaction. For the rate laws defined in Table 7.1, the *molecularity* with respect to a given reactant is given by the exponent in the given concentration term and represents the stoichiometric quantity of reactant involved in the reaction step. The kinetic *order* of a simple chemical reaction

Table 7.1 Relationship between rate law, order, and the rate constant, k

Rate law[a]	Order	Units of k
$v = k$	Zero	$M\ s^{-1}$
$v = k[R1]$	First order with respect to R1 First order overall	s^{-1}
$v = k[R1]^2$	Second order with respect to R1 Second order overall	$M^{-1}\ s^{-1}$
$v = k[R1][R2]$	First order with respect to R1 First order with respect to R2 Second order overall	s^{-1}

[a] For each rate law, the units of the reaction velocity (v) will always be $M\ s^{-1}$. M represents moles per liter ($mol\ l^{-1}$).

is usually the same as molecularity, but it is important to emphasize that order refers to the exponent value that relates the experimentally measured reaction rate's dependence on reactant concentration. The overall molecularity or order of a reaction step is the sum of all the concentration term exponents.

In order to more fully understand the dependence of instantaneous velocity on reactant concentration, we will first consider a simple *first-order* reaction in which one molecule of reactant R forms one molecule of product P ($R \rightarrow P$) according to the rate law given by Eq. (7.1).

$$v = \frac{-d[R]}{dt} = \frac{d[P]}{dt} = k[R] \tag{7.1}$$

In Eq. (7.1), the instantaneous reaction velocity ($v = -d[R]/dt = +d[P]/dt$) is directly proportional to the concentration of R raised to the first power. If the velocity for either loss of R or gain of P was measured immediately after initiation of the reaction (Fig. 7.2A), then a secondary plot of this *initial velocity* versus different initial or starting concentrations of [R] would be linear with a slope equal to the first-order rate constant, k (s^{-1}) (Fig. 7.2B). For such analyses, great care must be taken to ensure measurement of initial velocities. With increasing times past initiation of the reaction, the measured velocity decreases proportionately with depletion of [R]. Thus, it is common and good practice to measure initial velocities for either loss of reactant or gain of product at times corresponding to $\leq 5\%$ of product conversion.

To better illustrate the decreasing reaction velocity with time, the differential rate Eq. (7.1) can be integrated with respect to [R] to yield the first-order or single-exponential decay function given by Eq. (7.2). The limits of integration correspond to the initial concentration of reactant

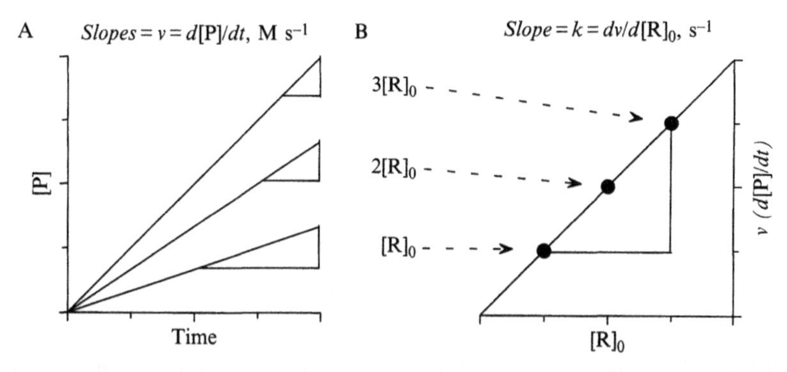

Figure 7.2 Determination of a first-order rate constant from measuring the initial velocity of product formation at differing concentrations of a reactant. (A) Initial velocities of product formation (slopes $= v = d[\mathrm{P}]/dt$, M s^{-1}) are measured for different starting concentrations of reactant (i.e., $[\mathrm{R}]_0$, $2[\mathrm{R}]_0$, or $3[\mathrm{R}]_0$). (B) The measured initial velocities are then plotted versus the different starting concentrations of reactant. The slope of this dependence yields the first-order rate constant for conversion of reactant into product (slope $= k = dv/d[\mathrm{R}]_0$, s^{-1}).

$$\int_{[\mathrm{R}]_0}^{[\mathrm{R}]} \frac{d[\mathrm{R}]}{[\mathrm{R}]} = -k\int_0^t dt$$

$$\ln\left(\frac{[\mathrm{R}]}{[\mathrm{R}]_0}\right) = -kt$$

$$[\mathrm{R}] = [\mathrm{R}]_0\,\mathrm{e}^{-kt} \qquad (7.2)$$

when the reaction is initiated (i.e., $[\mathrm{R}] = [\mathrm{R}]_0$ at $t = 0$) and the concentration of reactant at a given time after the reaction has started. Figure 7.3A shows the relationship between measuring *instantaneous velocity* (i.e., $v = -d[\mathrm{R}]/dt$, where Δt is very small) according to Eq. (7.1) and measuring overall *time progress* (i.e., either [R] or [P] versus t, where Δt is very large) according to Eq. (7.2) for the simple first-order conversion of reactant to product. The derivative in Eq. (7.1) is simply the slope of the tangent for the concentration curve at a specific time.

If no product is present at $t = 0$, the sum of reactant and product concentrations at any time must be equal to $[\mathrm{R}]_0$ (i.e., $[\mathrm{R}] + [\mathrm{P}] = [\mathrm{R}]_0$). Using this relationship, ($[\mathrm{R}]_0 - [\mathrm{P}]$) may be substituted for [R] in Eq. (7.2) and rearranged to yield the first-order rate Eq. (7.3) that describes the corresponding increasing product concentration with time (Fig. 7.3B).

$$[\mathrm{P}] = [\mathrm{R}]_0(1 - \mathrm{e}^{-kt}) \qquad (7.3)$$

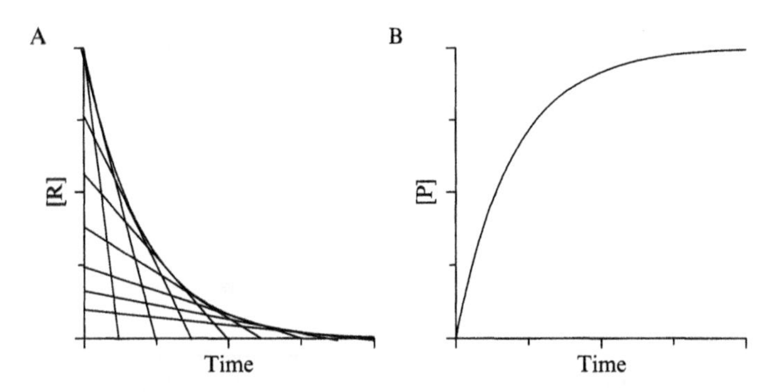

Figure 7.3 (A) Time progress for decreasing reactant concentration [R] in a first-order reaction. It can be seen that the instantaneous velocity at any given reactant concentration (lines) decreases with decreasing reactant concentration [R]. (B) Corresponding time progress curve for increasing product concentration [P] in the same first-order reaction.

2.2. Basic enzyme kinetics

Measurement of *enzyme* activity is actually a measurement of *catalytic* activity (Cook and Cleland, 2007; Cornish-Bowden, 1995; Segel, 1975). An enzyme or catalyst participates in chemical reactions by increasing the reaction velocity without itself appearing in the end products. In this case, an enzyme (E) first combines with or *binds* one or more chemical reactants or *substrates* (S) (Fig. 7.1B, k_{+1} and k_{-1}). The resulting enzyme–substrate (ES) complex is an intermediate species from which catalytic conversion of substrate to product (P) takes place, ultimately releasing the intact enzyme so that it may catalyze a subsequent reaction (Fig. 7.1B, k_{+2}). Since the k_{+2} proportionality constant typically comprises all processes or chemical steps involving conversion of the ES complex to release of products, it is a "pseudo" first-order rate constant (s^{-1}). It is more commonly referred to as the *turnover number*, k_{cat}, which is the maximum number of substrate molecules converted to product per active site per unit time. Thus, the instantaneous enzyme-catalyzed reaction velocity ($v = +d[P]/dt$) is directly proportional to the concentration of the ES complex according to k_{cat} in Eq. (7.4).

$$v = \frac{d[P]}{dt} = k_{cat}[ES] \tag{7.4}$$

Equation (7.4) is rendered impractical to the experimentalist, because the differential rate expression that describes the time-dependent concentration of ES complex includes terms for both its formation ($k_{+1}[E][S]$) and decay ($k_{-1}[ES] + k_{cat}[ES]$) according to Eq. (7.5). Furthermore,

$$\frac{d[ES]}{dt} = k_{+1}[E][S] - k_{-1}[ES] - k_{cat}[ES] \tag{7.5}$$

Equation (7.5) requires expressions for the time-dependent concentrations of E and S. To overcome the difficulty of integrating all differential rate expressions, the very useful *steady-state approximation* is applied to the concentration of ES. In cases where the reaction velocity is measured to be approximately constant over a time interval (e.g., measuring the initial velocity), then it follows that the concentration of ES does not vary appreciably. Thus, Eq. (7.5) is equated to zero and rearranged to obtain the Eq. (7.6) expression for [ES] in terms of [E] and [S].

$$[\mathrm{ES}] = \frac{k_{+1}[\mathrm{E}][\mathrm{S}]}{k_{-1} + k_{\mathrm{cat}}} = \frac{[\mathrm{E}][\mathrm{S}]}{K_{\mathrm{m}}} \tag{7.6}$$

In Eq. (7.6), the rate constants are grouped to form a composite constant $[K_{\mathrm{m}} = (k_{-1} + k_{\mathrm{cat}})/k_{+1}]$, which is referred to as the *Michaelis constant.* When the *initial velocity* of an enzyme-catalyzed reaction is measured under steady-state conditions, the time-dependent [S] in Eq. (7.6) is approximated by its initial concentration, $[\mathrm{S}]_0$, since the amount of product formed is substantially less than the amount of remaining substrate (i.e., $[\mathrm{S}] \approx [\mathrm{S}]_0$ since $[\mathrm{P}] \ll [\mathrm{S}]$). Since the time-dependent [E] cannot be approximated by its initial concentration, $[\mathrm{E}]_0$, the law of conservation of mass is used, whereby $[\mathrm{E}] = [\mathrm{E}]_0 - [\mathrm{ES}]$ is further substituted into Eq. (7.6) to yield Eq. (7.7), which is rearranged to give the Eq. (7.8) expression for [ES]. Now, this expression is substituted back into Eq. (7.4) to yield the familiar *Michaelis–Menten* Eq. (7.9),

$$[\mathrm{ES}] = \frac{([\mathrm{E}]_0 - [\mathrm{ES}])[\mathrm{S}]_0}{K_{\mathrm{m}}} \tag{7.7}$$

$$[\mathrm{ES}] = \frac{[\mathrm{E}]_0[\mathrm{S}]_0}{K_{\mathrm{m}} + [\mathrm{S}]_0} \tag{7.8}$$

$$v = \frac{d[\mathrm{P}]}{dt} = k_{\mathrm{cat}}[\mathrm{ES}] = \frac{k_{\mathrm{cat}}[\mathrm{E}]_0[\mathrm{S}]_0}{K_{\mathrm{m}} + [\mathrm{S}]_0} = \frac{V_{\mathrm{max}}[\mathrm{S}]_0}{K_{\mathrm{m}} + [\mathrm{S}]_0} \tag{7.9}$$

which shows the initial velocity of product formation to be directly proportional to the total concentration of enzyme, $[\mathrm{E}]_0$. Equation (9) further shows the initial velocity of product formation to vary in a hyperbolic-dependent manner with regard to the initial concentration of substrate according to the composite *Michaelis constant,* K_{m}. Here, the $K_{\mathrm{m}} = (k_{-1} + k_{\mathrm{cat}})/k_{+1}$ values represents the substrate concentration at which half-maximum initial velocity is attained, and V_{max} ($\mathrm{M\ s^{-1}}$) $= k_{\mathrm{cat}}[\mathrm{E}]_0$ represents the maximum initial velocity of product formation. In the limiting case where catalytic conversion to products is much slower than dissociation of the substrate (i.e., $k_{cat} \ll k_{-1}$), the K_m value approximates the true dissociation constant for the ES complex, $K_d = k_{-1}/k_{+1}$; and the system is described as being in rapid equilibrium.

If the measured initial velocity is then normalized to amount of enzyme in the reaction mixture, then Eq. (7.9) takes the form of Eq. (7.10), whereby the *activity of the enzyme* is defined

$$\text{Activity}\,(\text{s}^{-1}) = \frac{v}{[\text{E}]_0} = \frac{k_{cat}[\text{S}]_0}{K_m + [\text{S}]_0} \tag{7.10}$$

as the instantaneous enzyme-catalyzed reaction velocity normalized to the amount of enzyme ($v/[\text{E}]_0$). In this case, enzyme activity is measured as a function of substrate concentration to obtain values of the *turnover number*, k_{cat}, and the *Michaelis constant*, K_m. The ratio of k_{cat}/K_m ($\text{M}^{-1}\,\text{s}^{-1}$) is termed the *specificity constant*, which is an apparent second-order rate constant that refers to the properties and reactions of the free enzyme leading to the first irreversible step.

It is important to point out that the above activity definition pertains to cases in which the enzyme active site *concentration* (moles per volume) in the reaction mixture is known (e.g., assay of a purified homogeneous form of the enzyme). When enzyme activity measurements are carried out in the presence of other proteins such as crude cell lysates or partially purified enzyme preparations, the measured initial velocity is normalized to the *total protein concentration* in the assay. In this case, total protein concentration is measured and expressed as weight per volume. As such, when the molar concentration units of initial velocity ($\text{mol l}^{-1}\,\text{s}^{-1}$) are divided by weight concentration units of total protein (g l^{-1}), the volume terms cancel so that *specific enzyme activity* is expressed as the number of moles of product converted per unit time per weight of protein. Specific activity expressed in this manner is most often reported as $\mu\text{mol min}^{-1}\,\text{mg}^{-1}$. When one *unit* (U) is defined as conversion of 1 μmol of substrate per minute, then enzyme activity is also commonly expressed as units per milligram of protein (U mg^{-1}). The following sections describe a number of initial considerations and experimental methods for proper development of enzymatic assays.

3. Measurement of Enzyme Activity

In order to measure an enzyme's catalytic activity, it is essential to first identify the chemical changes involving conversion of substrate(s) to product(s). To date, the textual accounting of the incredible variety of enzymatic activities is most typically managed by grouping enzymes according to the types of chemical reactions they catalyze (e.g., oxidation–reductions, group transfers, eliminations, isomerizations, rearrangements, condensations, carboxylations, etc.) (Frey and Hegeman, 2007). Further, considerations should take into account whether mechanistically similar chemistry (i) involves either small molecule substrates or larger macromolecular proteins or nucleic acids, (ii) occurs readily in solution or requires membrane-bound components, and (iii) proceeds to any extent in the reverse direction. In any case, the enzymatic assay is designed around the ability to distinguish between the physicochemical properties of a given substrate with respect to those of a given product in a quantifiable manner (Eisenthal and Danson, 2002; Rossomando, 1990). To a large degree, similar methods of measuring activity have been applied to various enzymes categorized within such subgroups; and *it would be prudent to first consider devising assays on the basis of those already well established for either (i) homologous enzymes from different organisms or (ii) different enzymes that catalyze a related chemical reaction.* The central theme among all enzyme activity measurements is to *ensure measurement of initial velocities.* To reiterate this all important point, it is common and good practice to measure initial velocities for either loss of reactant or gain of product at times corresponding to $\leq$5% of product conversion. This gives the best approximation for the initial velocity at the starting or initial substrate concentration.

3.1. Continuous assays

After surveying the literature for applicable enzymatic assays, one may find a number of suitable methods, which may be initially distinguished according to whether a "continuous" or a "discontinuous" assay is employed. Continuous assays are defined by the ability to continuously monitor either disappearance or appearance of a given substrate or product, respectively; and they most often rely on spectroscopic techniques such as electronic ultraviolet–visible absorption and fluorescence emission. For absorption spectroscopy, molar absorbance extinction coefficients (ε) can be gravimetrically determined for any number of compounds with conjugated bond systems. According to the Lambert–Beer relationship ($A = \varepsilon cl$), absorbance (A, unitless) is directly proportional to (i) the molar extinction coefficient (ε, $M^{-1}\,cm^{-1}$), (ii) the concentration of the compound (c, M), and (iii) the path length of the cuvette (l, cm) used for the measurement (Segal, 1976). Thus, the molar change in concentration with time or initial velocity ($v = -d[S]/dt$ or $d[P]/dt$) is directly calculated from the slope relating the change in spectroscopic signal at a designated wavelength with respect to time (dA/dt) on dividing by εl; and the *enzyme activity* is ultimately obtained by dividing the initial velocity by either the enzyme molar concentration (M) or the weight concentration of total protein in the assay (mg ml^{-1}) according to either Eq. (7.11) or (7.12), respectively.

$$\text{Activity}\,(\text{s}^{-1}) = \frac{v}{[\text{E}]_0} = \frac{(dA/dt, \text{s}^{-1})}{(\varepsilon, \text{M}^{-1}\,\text{cm}^{-1})(l, \text{cm})([\text{E}]_0, \text{M})} \tag{7.11}$$

$$\begin{aligned}\text{Activity}\,(\mu\text{mol}\,\text{min}^{-1}\,\text{mg}^{-1}) &= \frac{v}{[\text{E}]_0} \\ &= \frac{(dA/dt, \text{min}^{-1})}{(\varepsilon, \text{ml}\,\mu\text{mol}^{-1}\,\text{cm}^{-1})(l, \text{cm})([\text{E}]_0, \text{mg}\,\text{ml}^{-1})}\end{aligned} \tag{7.12}$$

Although ultraviolet–visible absorption spectroscopy is (i) convenient for continuous monitoring and (ii) accurate with respect to determining concentration changes, it is limited with respect to both (i) the number of ESs and products that can be detected and (ii) the range of concentrations that can be accurately detected. Alternatively, fluorescence and phosphorescence spectroscopies can be used for continuously monitoring reaction progress; and in most cases provide significantly enhanced sensitivity. However, it must be pointed out that the concentration of the fluorophore cannot be calculated from the measured fluorescence emission by application of a universal constant equivalent to the molar extinction coefficient. Rather, *relative* changes in fluorescence emission with time are compared. Regardless, fluorescence and phosphorescence spectroscopies are highly suited for use in high-throughput platforms utilized for screening very large numbers of compounds for effects on enzyme activity. For a large number of enzyme drug targets (e.g., particularly protein kinases and proteases), artificial peptide substrates have been chemically modified to contain one or more small molecule fluorophores; and such substrates undergo fluorescence emission changes occur on either phosphorylation or cleavage. Before employing fluorescence- or phosphorescence-based assays, researchers should thoroughly review literature pertaining to such principles, as numerous properties must be considered (e.g., fluorescence lifetime, quantum yields, and inner-filter effects) (Lakowicz, 2004).

Finally, it should be pointed out that numerous "continuous" enzymatic assays can and have been developed for reactions involving substrate–product pairs with no spectroscopic properties. In these cases, either one or more additional enzymes are included in the reaction mixture, which serve to "couple" a given product to a reaction that does yield a desirable spectroscopic property (Eisenthal and Danson, 2002). For the most part, such *coupled continuous* assays have been applied to metabolic enzymes for which a given product may either be oxidized or reduced by chromogenic coenzymes such as NADH. If the product is not a direct substrate for such a reaction, then an additional enzyme may be included to convert the first product to a second product that undergoes the chromogenic reaction. In any case, it must be ensured that the reaction mixture contains adequate amounts of all additional components to ensure that the reaction velocity is solely determined by the enzyme of interest. In other words, the measured reaction velocity should be (i) dependent only on the target substrate(s) and enzyme concentration and (ii) independent or zero order with respect to the concentration of each and every component added to facilitate the coupled reaction.

3.2. Discontinuous assays

Discontinuous assays are defined by the *inability* to continuously and *selectively* monitor concentration changes of either substrate or product, such as when a given substrate–product pair exhibits either similar or no spectroscopic properties. Rather, the enzymatic reaction must be manually stopped or "quenched" at different times. The quenched samples are then subjected to some method whereby the product can be efficiently separated from the substrate so that the change in concentration of either can be determined for each individual time point. In setting up a discontinuous enzymatic assay, it is crucial to devise methods for both (i) efficiently stopping or "quenching" an enzyme-catalyzed reaction at designated time points and (ii) efficiently separating the large amount of unreacted substrate ($\geq$95%) from the very small amount of formed product ($\leq$5%) required for measuring initial velocities. These two methodological points must be considered as a pair, since the "quenched" reaction solution conditions must be amenable to subsequent procedures required for fractionation of substrate and product molecules. Since a large number of enzymes can be rapidly inactivated under a wide variety of conditions (e.g., addition of acid, base, or metal chelators), substrate–product fractionation is foremost considered.

High-performance liquid chromatography (HPLC) offers extensive ranges in both resolution capabilities and detection sensitivity (McMaster, 2007). For example, a great variety of mixtures containing either small molecules or macromolecules can be resolved on the basis of differential retention when passed over a column containing a given stationary phase (e.g., ion-exchange, size-exclusion, or reversed-phase chromatography). In addition, HPLC systems can be fixed with any number of detectors so that a compound with a given spectroscopic property can be detected as it elutes from the column, which results in a spectroscopic peak. To gauge the amount of product formed, the area of the resolved product peak is integrated and compared with a standard curve prepared from integrated peak areas obtained for range of product concentrations. In cases where the designated product does not exhibit any spectroscopic properties, one may consider possibilities of either (i) chemically modifying the substrate to contain a spectroscopic active compound or (ii) modifying the product in the quenched reaction mixture. In the latter case, it must be established that essentially all product undergoes modification and that the modified product can be readily resolved from any remaining unreacted spectroscopic active compound. The determination of enzyme activity is carried out exactly as in Eqs. (7.11) and (7.12) except that dA/dt is obtained as $\Delta A/\Delta t$ (i.e., a single-point determination rather than a slope determined from many points).

In cases when a given substrate–product pair exhibits no useful spectroscopic properties, the amount of product at a given time point in a discontinuous assay can be determined by radiometric analysis (Eisenthal and Danson, 2002). In fact, radiometric assays provide the highest degree of sensitivity, as well as exhibit the highest degree of broad scale utility. Almost any conceivable enzyme substrate can be synthesized to contain a radioactive isotopic atom (or radioisotope) in a position that undergoes transfer from the parent substrate, which can be detected in very small amounts. The most commonly used radioisotopes for enzymatic assays are ^{3}H, ^{14}C, ^{32}P, and ^{35}S. In all four cases, the unstable nucleus undergoes slow first-order decay to form a stable isotope with an atomic number one higher and an atomic weight identical to the original radioisotope through a process called beta-particle emission. Scintillation counting of beta-particle emission gives very sensitive determination of the amount of radioactivity present in a sample, which is directly quantified in counts per minute (cpm). The half-life ($t_{1/2}$) of a radioisotope is the time required for half of the original number of atoms to decay. Half-life values of ^{3}H ($t_{1/2} = 12.3$ years), ^{14}C ($t_{1/2} = 5700$ years), ^{32}P ($t_{1/2} = 14.3$ days), and ^{35}S ($t_{1/2} = 87.1$ days) have been well determined. From these values, it can be seen that the amount of radioactivity emitted by ^{3}H and ^{14}C does not appreciably change over the time course of routine laboratory work; whereas the amount of radioactivity emitted by ^{32}P and ^{35}S significantly decreases over such time.

Commercial manufacturers incorporate such radioisotopes into designated positions of target molecules to yield radiolabeled substrate molecules, which have a defined specific (radio)activity (SA) (Segal, 1976). The specific activity refers to the amount of radioactivity (cpm) per unit amount of molecule (mol). Typically, one adds a small amount of the highly radioactive "hot" substrate molecule into a much larger amount of the nonradioactive or "cold" substrate. A small volume of a known concentration is then subjected to scintillation counting to determine the specific (radio)activity of the substrate (SA^{S}, cpm/mol^{-1}). By controlling the mixture of hot and cold substrate molecules, a very wide range of specific (radio)activities can be generated so that enzyme activity measurements are possible over an equal range of substrate concentrations. For example, a radiolabeled substrate with $SA = 1000$ cpm mmol^{-1} would enable efficient detection for transfer of label to form 1 mmol of product (1000 cpm), whereas generation of a radiolabeled substrate with SA = 1000 cpm pmol^{-1} would enable efficient detection of 1 pmol of product!

Most commonly, ion-exchange methods are used in which the product form of the molecule is retained on solid support (e.g., ion-exchange resin, paper, or disk), while the substrate form is not retained. Other methods of separation may include either selective precipitation, solvent extraction, or gel electrophoresis. Before carrying out the enzyme-catalyzed reaction, it is essential to establish that (i) the radiolabeled product is highly resolved from radioactive substrate and (ii) the amount of resolved product closely approximates the amount subjected for resolution. Typically, one finds a point of compromise, whereby a certain minimum amount of "background" substrate radiation is tolerated in the medium or location where product is resolved. For most accurate analysis, the medium containing the purified product is directly analyzed by scintillation counting (e.g., ion-exchange resin, paper, or disk). Although selected gel regions after electrophoresis may be removed for direct counting, it has been more common practice to quantify spatial radioactivity across an entire gel using scintillation plates and computer densitometry analysis.

The amount of enzyme-catalyzed product (mol) detected in the aliquot removed for analysis is calculated by dividing the amount of radioactivity detected in the product fraction (cpm) by the specific (radio)activity of the substrate (cpm/mol^{-1}); the concentration of product, [P], is obtained by dividing by the volume of the aliquot (l); and the initial velocity, $d[\mathrm{P}]/dt$, is obtained by further dividing [P] by the time of the reaction (s). Thus, *enzyme activity* is ultimately obtained by dividing the initial velocity by either (i) the enzyme molar concentration according to Eq. (7.13) or (ii) the total protein concentration according to Eq. (7.14).

$$\text{Activity}\,(\mathrm{s}^{-1}) = \frac{d[\mathrm{P}]}{dt[\mathrm{E}]_0} = \frac{\text{cpm}}{(\mathrm{SA^S}, \mathrm{cpm}/\mu\mathrm{mol}^{-1})(\mathrm{vol}, \mu\mathrm{l})(t, \mathrm{s})([\mathrm{E}], \mu\mathrm{mol}/\mu\mathrm{l}^{-1})} \quad (7.13)$$

$$\text{Activity}\,(\mu\mathrm{mol\,min^{-1}\,mg^{-1}}) = \frac{d[\mathrm{P}]}{dt[\mathrm{E}]_0} = \frac{\text{cpm}}{(\mathrm{SA^S}, \mathrm{cpm}/\mu\mathrm{mol}^{-1})(\mathrm{vol}, \mu\mathrm{l})(t, \mathrm{min})([\mathrm{E}], \mu\mathrm{g}/\mu\mathrm{l}^{-1})} \quad (7.14)$$

4. Formulation of Reaction Assay Mixtures

Having established effective methods for resolution, detection, and quantification of reaction species, it is next time to consider preparation of the reaction mixture itself. For continuous spectroscopic assays, total volumes should be selected to accommodate cuvettes made for the given spectrometer. For visible absorption ($\geq$325 nm), 1 ml plastic disposable cuvettes are most widely used due to convenience. Since plastic cuvettes increasingly absorb light of lower wavelengths, quartz cuvettes should be used in this region. Of increasing availability and utility are fluorometers fitted with microplate readers, which simultaneously obtain accurate emission readings from a large number of individual small reactions ($\leq$50 μl) on a single microtiter plate. For discontinuous enzyme assays, a total volume of the reaction mixture should be selected in which several ($\geq$3) designated fixed amount aliquots can be removed at varying times for reaction quenching and product analysis.

For all enzyme assays, stock concentrations of suitable buffer, substrate(s), and enzyme are prepared so that dilution of each component to its desired concentration can be achieved by adding volumetric amounts that sum to less than the total volume. The remaining volume is occupied by water. The buffer component should have a pK_a value $\pm$ 1 of the desired reaction pH, and its concentration should be chosen to exceed any other component that has ionizable groups. The stock buffer solution may also contain additional compounds that may be necessary to maintain enzyme stability and activity such as salts (e.g., NaCl or KCl), osmolytes (e.g., polyethylene glycol, glycerol, or sucrose), and reducing agents (e.g., 2-mercaptoethanol, dithiothreitol, or TCEP). It is very important to include such components in the preparation of the stock buffer, and the final stock buffer mixture should be titrated to its desired pH at the temperature in which the reaction will be carried out. Pay special attention to use of (i) nitrogen containing buffers (such as Tris) whose pK_a values are very sensitive to temperature and (ii) phosphate buffers whose pK_a values are very sensitive to ionic strength. In fact, dilution of phosphate buffer will change its pH value, so the pH of its stock concentration must be adjusted accordingly.

It is best to generate the reaction mixture by first adding nonenzyme components in the following order: (i) designated volume of water, (ii) designated volume of stock buffer mixture, and (iii) designated volume of stock substrate. It is best to prepare a stock concentration of enzyme so that only a relatively small amount ($\leq$10% volume) is added to bring the mixture to the designated total volume and initiate the reaction. Immediately, after the enzyme has been added to initiate a continuous assay, the cuvette(s) or microtiter plate must be inserted into a spectrometer that is programmed to begin collecting absorption or emission data at the designated wavelength(s). For discontinuous assays, one must be prepared to remove a precise volume of the reaction mixture and mix with a precise volume of the quench reagent; and this must be repeated at precise times. In all enzyme assays, one should always formulate an additional control reaction mixture that does not contain enzyme to determine the "background" amounts of either (i) noncatalyzed product formation or (ii) signal retained from incomplete resolution. Any such background amounts must be subtracted from the signal detected in the presence of enzyme.

5. Discussion

In this chapter, the most fundamental principles of measuring enzyme reaction rates were presented, and it is in no way a complete treatment on the subject. Nevertheless, clear understanding of these principles is absolutely prerequisite toward either (i) effective utilization of well-developed assays or (ii) efficient development of assays for newly discovered enzymes. Over the past *century*, countless research articles, review articles, book chapters, and even entire books have reported on the development, utilization, and interpretation of enzyme activity measurements. The selection of books and articles cited in this article provide more comprehensive treatments of the various topics related to measuring enzyme activity.

ACKNOWLEDGMENTS

This work was supported by NIGMS, National Institutes of Health Grant GM69868 to T. K. H. and a Maytag Fellowship to M. M. K.

REFERENCES

Cook, P. F., and Cleland, W. W. (2007). Enzyme Kinetics and Mechanism. Garland Science, Hamden, CT.

Cornish-Bowden, A. (1995). Fundamentals of Enzyme Kinetics. Portland Press, Ltd., London, UK.

Eisenthal, R., and Danson, M. (2002). Enzyme Assays: A Practical Approach. Oxford University Press, Inc., New York, NY.

Frey, P. A., and Hegeman, A. D. (2007). Enzymatic Reaction Mechanisms. Oxford University Press, Inc., New York, NY.

House, J. E. (2007). Principles of Chemical Kinetics. 2nd edn. Academic Press, Burlington, MA.

Laidler, K. J. (1987). Chemical Kinetics. 3rd edn. Prentice Hall, Upper Saddle River, NJ.

Lakowicz, J. R. (2004). Principles of Fluorescence Spectroscopy. 2nd edn. Springer Science+Business Media, Inc., New York, NY.

McMaster, M. C. (2007). HPLC, a Practical Users Guide. 2nd edn. John Wiley & Sons, Inc., Hoboken, NJ.

Rossomando, E. F. (1990). Measurement of enzyme activity. *Methods Enzymol.* **182,** 38–49.

Segel, I. H. (1975). Enzyme Kinetics. John Wiley & Sons, Inc., New York, NY.

Segal, I. H. (1976). Biochemical Calculations. 2nd edn. John Wiley & Sons, Inc., New York, NY.

CHAPTER EIGHT

Quantitation of Protein

James E. Noble* *and* Marc J. A. Bailey†

Contents

1. Introduction 74
2. General Instructions for Reagent Preparation 75
3. Ultraviolet Absorption Spectroscopy 80
   3.1. Ultraviolet absorbance at 280 nm (Range: 20–3000 μg) 80
   3.2. Method 81
   3.3. Comments 81
   3.4. Ultraviolet absorbance at 205 nm (Range: 1–100 μg) 82
   3.5. Calculation of the extinction coefficient 82
4. Dye-Based Protein Assays 83
   4.1. Protein concentration standards 83
5. Coomassie Blue (Bradford) Protein Assay (Range: 1–50 μg) 85
   5.1. Reagents 85
   5.2. Procedure 85
   5.3. Comments 86
6. Lowry (Alkaline Copper Reduction Assays) (Range: 5–100 μg) 86
   6.1. Reagents 87
   6.2. Procedure 87
   6.3. Comments 88
7. Bicinchoninic Acid (BCA) (Range: 0.2–50 μg) 88
   7.1. Reagents 88
   7.2. Procedure 89
   7.3. Comments 89
8. Amine Derivatization (Range: 0.05–25 μg) 89
   8.1. Reagents 90
   8.2. Procedure 90
   8.3. Comments 90
9. Detergent-Based Fluorescent Detection (Range: 0.02–2 μg) 91

* Analytical Science, National Physical Laboratory, Teddington, Middlesex, United Kingdom
† Nokia Research Centre - Eurolab, University of Cambridge, Cambridge, United Kingdom

Methods in Enzymology, Volume 463

ISSN 0076-6879, DOI: 10.1016/S0076-6879(09)63008-1

10. General Instructions 91
10.1. Cuvettes 91
10.2. Microwell plates 92
10.3. Interfering substrates 93
Acknowledgment 94
References 94

Abstract

The measurement of protein concentration in an aqueous sample is an important assay in biochemistry research and development labs for applications ranging from enzymatic studies to providing data for biopharmaceutical lot release. Spectrophotometric protein quantitation assays are methods that use UV and visible spectroscopy to rapidly determine the concentration of protein, relative to a standard, or using an assigned extinction coefficient. Methods are described to provide information on how to analyze protein concentration using UV protein spectroscopy measurements, traditional dye-based absorbance measurements; BCA, Lowry, and Bradford assays and the fluorescent dye-based assays; amine derivatization and detergent partition assays. The observation that no single assay dominates the market is due to specific limitations of certain methods that investigators need to consider before selecting the most appropriate assay for their sample. Many of the dye-based assays have unique chemical mechanisms that are prone to interference from chemicals prevalent in many biological buffer preparations. A discussion of which assays are prone to interference and the selection of alternative methods is included.

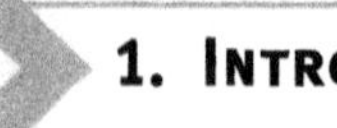

1. Introduction

The quantity of protein is an important metric to measure during protein purification, for calculating yields or the mass balance, or determining the specific activity/potency of the target protein. Various platforms and methods are available to quantitate proteins and will be described elsewhere in this volume; however, for this chapter, we will concentrate on spectrophotometric assays of protein in solution that do not require either enzymatic/chemical digestion or separation of the mixture prior to analysis.

The spectrophotometric assays described are UV absorbance methods and dye-binding assays using colorimetric and fluorescent-based detection. In comparison to other methods, these assays can be run at a high throughput, using inexpensive reagents with equipment found in the majority of biochemical laboratories. These spectrophotometric assays require an appropriate protein standard or constituent amino acid sequence information to make a good estimate of concentration. The choice of method used to determine the concentration of a protein or peptide in solution is dependent on many factors that will be discussed. The majority of methods require a soluble analyte such as peptides, proteins, posttranslationally modified protein (e.g., glycosylated), or chemically modified proteins (e.g., PEGylated). For common protein purification procedures, the flow chart in Fig. 8.1 describes the process of selecting the most appropriate assay, based on key criteria.

Where other protein purification techniques are available or complex buffer systems are present in the sample, refer to Table 8.1, or other reviews (Olson and Markwell, 2007) for assay selection. Other criteria that need to be considered when selecting an assay include:

- *Sample volume*: The amount of material available to analyze, typically fluorescent-based assays display the best sensitivity and dynamic range (see Fig. 8.2). Microplate assays (by using lower assay volumes, thereby less protein sample) show improved sensitivity, typically up to 10-fold when compared with cuvette-based assays.
- *Sample recovery*: If the sample is limited, a nondestructive method, for example, UV spectroscopy may be more appropriate.
- *Throughput*: If multiple samples are to be analyzed, a microplate-compatible rapid one step assay should be considered.
- *Robustness*: The absorbance-based dye-binding assays appear to display enhanced repeatability and robustness when compared to fluorescent assays.
- *Chemical modification*: Covalent modification, for example glycosylation (de Moreno *et al.*, 1986; Fountoulakis *et al.*, 1992) or PEGylation (Noble *et al.*, 2007), can interfere with specific assays.
- *Protein aggregation*: The solubility of a protein in solution, often a problem for membrane proteins, or proteins prone to aggregation can alter the expected response for many assays.

Other protein quantitation methods are becoming more commonly employed in biochemistry laboratories due to automation, regulatory, and sensitivity requirements. Alternative methods not detailed in this chapter include isotope dilution mass spectrometry (ID IC MS/MS) (Burkitt *et al.*, 2008), Kjeldahl nitrogen method, amino acid analysis (Ozols, 1990), gravimetric determination (Blakeley and Zerner, 1975), immunological, and quantitative gel electrophoresis with fluorescent staining.

2. General Instructions for Reagent Preparation

For the methods detailed, reagents should be used at the highest purity available and dyes should be obtained at spectroscopy grade where available. Ideally deionized, filtered water should be used at a minimum quality of

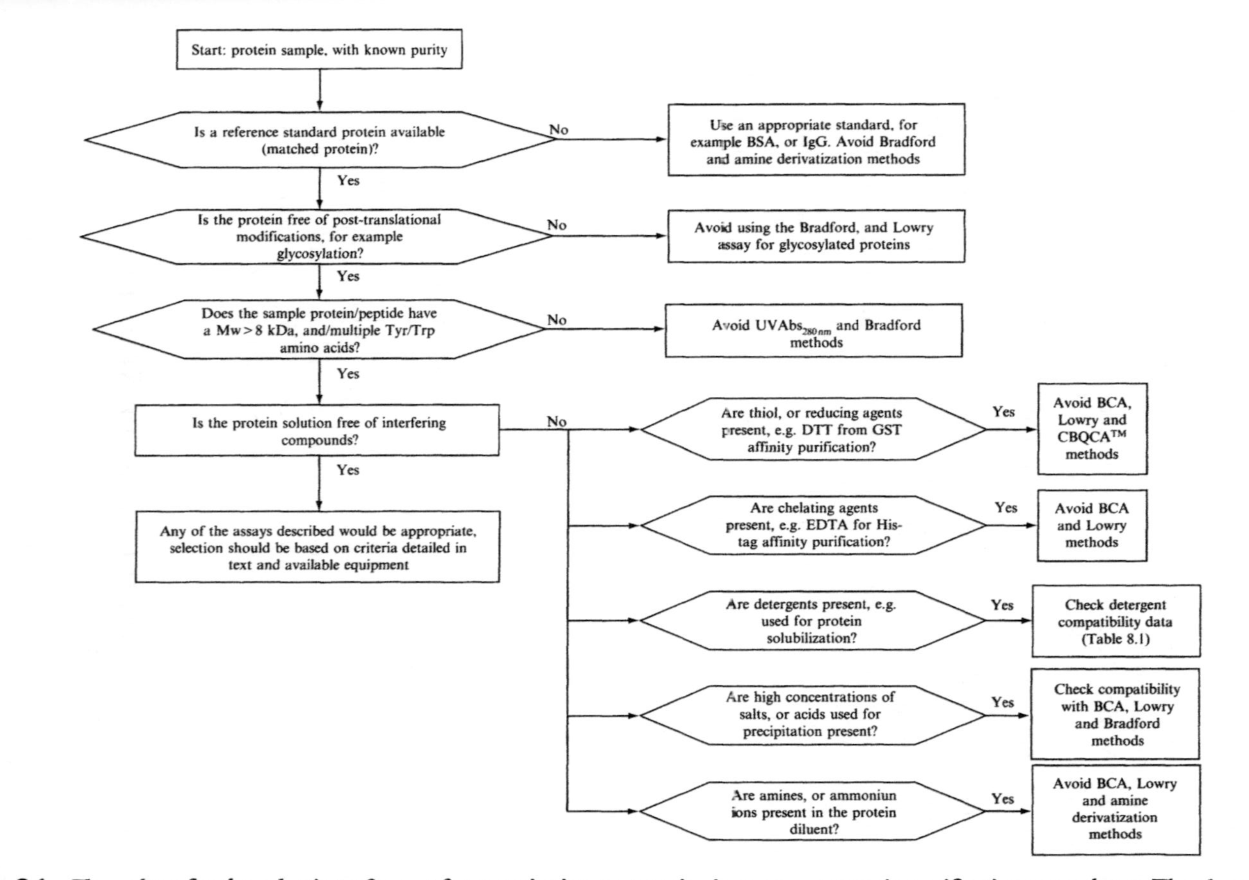

Figure 8.1 Flow chart for the selection of assays for quantitation or proteins in common protein purification procedures. The chart assumes that the sample for analysis is relatively pure, that is the analyte for quantitation is the major component, for example fractions from affinity chromatography, or extraction from inclusion bodies. The "reference standard protein" refers to a standard that is the same protein that is being quantitated in the same, or similar matrix that is "matched."

Table 8.1 Substance compatibility table

Substance	Compatible concentration[a]					
	BCA[b]	Lowry[c]	Bradford[d]	Amine derivatization[e]	Fluorescent detergent[f]	UV $Abs_{280\ nm}$[g]
Acids/bases						
HCl	0.1 *M*	na	0.1 *M*	na	10 m*M*	>1 *M*
NaOH	0.1 *M*	na	0.1 *M*	na	10 m*M*	>1 *M*
Perchloric acid	<1%	>1.25%	na	na	na	10%
Trichloroacetic acid	<1%	>1.25%	na	na	na	10%
Buffers/salts						
Ammonium sulfate	1.5 *M*	>28 m*M*	1.0 *M*	10 m*M*	10–50 m*M*	>50%
Borate	10 m*M*	Undiluted	Undiluted	Undiluted	Undiluted	Undiluted
Glycine	1 m*M*	1 m*M*	100 m*M*	–	na	1 *M*
HEPES	100 m*M*	1 m*M*	100 m*M*	na	10–50 m*M*	na
Imidazole	50 m*M*	25 m*M*	200 m*M*	na	na	na
Potassium chloride	<10 m*M*	30 m*M*	1.0 *M*	na	20–200 m*M*	100 m*M*
PBS	Undiluted	Undiluted	Undiluted	Undiluted	Undiluted	Undiluted
Sodium acetate	200 m*M*	200 m*M*	180 m*M*	na	na	na
Sodium azide	0.2%	0.5%	0.5%	0.1%	10 m*M*	na
Sodium chloride	1.0 *M*	1.0 *M*	5.0 *M*	na	20–200 m*M*	>1 *M*
Triethanolamine	25 m*M*	100 m*M*	na	na	na	na
Tris	250 m*M*	10 m*M*	2.0 *M*	10 m*M*	na	0.5 *M*
Detergents						
Brij 35	5%	0.031%	0.125%	na	na	1%
CHAPS	5%	0.0625%	5%	na	na	10%

(*continued*)

Table 8.1 *(continued)*

	Compatible concentration[a]					
Substance	BCA[b]	Lowry[c]	Bradford[d]	Amine derivatization[e]	Fluorescent detergent[f]	UV $Abs_{280\ nm}$[g]
Deoxycholic acid	5%	625 μg/ml	0.05%	na	na	0.3%
Nonidet P-40	5%	0.016%	0.5%	na	na	na
SDS	5%	1%	0.125%	–	0.01–0.1%	0.1%
Triton X-100	5%	0.031%	0.125%	na	0.001%	0.02%
Tween-20	5%	0.062%	0.062%	0.1%	0.001%	0.3%
Reducing agents						
Cysteine	na	1 m*M*	10 m*M*	na	na	na
DTT	1 m*M*	0.05 m*M*	5–1000 m*M*	0.1 m*M*	10–100 m*M*	3 m*M*
2-Mercaptoethanol	0.01%	1 m*M*	1.0 *M*	0.1 m*M*	10–100 m*M*	10 m*M*
Thimerosal	0.01%	0.01%	0.01%	na	na	na
Chelators						
EDTA	10 m*M*	1 m*M*	100 m*M*	na	5–10 m*M*	30 m*M*
EGTA	na	1 m*M*	2 m*M*	na	na	na
Solvents						
DMSO	10%	10%	10%	na	na	20%
Ethanol	10%	10%	10%	na	na	na
Glycerol	10%	10%	10%	10%	10%	40%

Guanidine–HCl	4.0 *M*	0.1 *M*	3.5 *M*	na	na	na
Methanol	10%	10%	10%	na	na	na
PMSF	1 m*M*	1 m*M*	1 m*M*	na	na	na
Sucrose	40%	7.5%	10%	10%	10–500 m*M*	2 *M*
Urea	3.0 *M*	3.0 *M*	3.0 *M*	na	na	>1.0 *M*
Miscellaneous						
DNA	0.1 mg	0.2 mg	0.25 mg	na	50–100 μg/ml	1 μg

Values relate to the maximum concentration of interfering compound within the protein sample that does not result in significant loss in assay performance. The guide is an updated version of that prepared by Stoscheck to inform of any issues related to assay interference. Concentrations were obtained from product inserts and references (Bradford, 1976; Peterson, 1979; Smith *et al.*, 1985; Stoscheck, 1990), where there is not a consensus of values a range is given. Changing the protein-to-dye ratios, or formulation of many of the dye-based assays can alter the maximum concentration of compound permissible. Interfering compounds have been selected to represent those commonly encountered in protein purification and enzymology.

[a] na indicates the reagent was not tested. A blank indicates that the reagent is not compatible with the assay at the reagent concentrations analyzed. A figure preceded by (<) or (>) symbols indicates the tolerable limit is unknown but is respectively, less than or greater than the amount shown.

[b] Figures indicate the concentration in a 0.1-ml sample using a final reaction volume of 2.1 ml.

[c] Figures indicate the concentration in a 0.2-ml sample using a final reaction volume of 1.3 ml.

[d] Figures indicate the concentration in a 0.05-ml sample using a final reaction volume of 1.55 ml.

[e] Figures indicate interference concentrations with the CBQCA™ assay (You *et al.*, 1997) in a 90-μl sample using a final reaction volume of 100 μl.

[f] Figures indicate interference concentrations with the NanoOrange™ and Quant-iT™ assays ((Hammer and Nagel, 1986) and Quant-iT™ product insert) in a 40- and 20-μl sample (respectively) using a final reaction volume of 200 μl.

[g] Figures indicate the concentration of the chemical that does not produce an absorbance of 0.5 over water (Stoscheck, 1990).

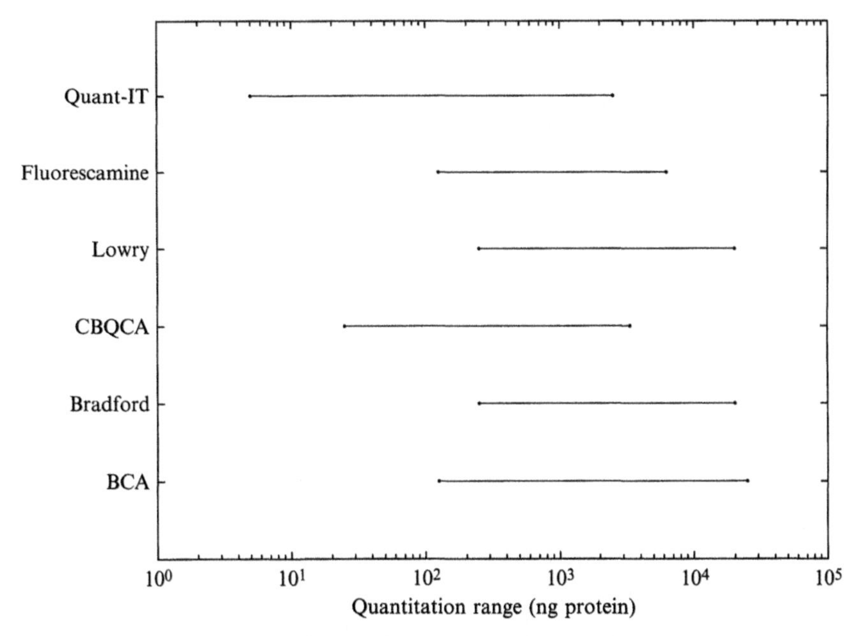

Figure 8.2 The markers designate the upper and lower values for the quantitation range of dye-based protein assay performed in a microplate format. The quantitation range was defined as the range of protein amounts (ng) that displayed good precision and did not show any deviation from the fitted response curve. Figure used with permission from Noble *et al.* (2007).

18 $M\Omega$ cm and a total organic carbon of below 6 ppb. All buffer preparations should be filtered using 0.2 μm filtration (Millipore, Sartorius) devices upon preparation to remove bacteria and fines. If precipitation occurs during storage, the reagent should be discarded, unless stated in the method.

3. Ultraviolet Absorption Spectroscopy

3.1. Ultraviolet absorbance at 280 nm (Range: 20–3000 μg)

Proteins display a characteristic ultraviolet (UV) absorption spectrum around 280 nm predominately from the aromatic amino acids tyrosine and tryptophan. If the primary sequence contains no or few of these amino acids then this method will give erroneous results. Quartz crystal cuvettes are routinely used for measurement as plastic materials can leach plasticizers, and are not UV transparent. Similarly, buffer components with strong UV absorbance such as some detergents especially Triton X-100 should be avoided (Table 8.1) and "blank" samples should be measured using the sample buffer solution but with no protein present. UV absorbance is routinely used to give an estimate of protein concentration but if the molar extinction coefficient of the protein is known then the Beer–Lambert law can be used to accurately quantitate amount of protein by UV absorbance, assuming the protein is pure and contains no UV absorbing nonprotein components such as bound nucleotide cofactors, heme, or iron–sulfur centers.

Beer–Lambert (molar absorption coefficient):

$$A = a_{\mathrm{m}} cl \tag{8.1}$$

where a_{m} is the molar extinction coefficient, c the concentration of analyte, and l the path length in cm.

3.2. Method

For the measurement of a protein with unknown extinction coefficient, using a protein standard:

1. Add blank buffer to a clean quartz cuvette and use to zero the spectrophotometer.
2. Either using a fresh identical cuvette or replace the buffer with the sample, then measure the absorbance at 280 nm. If the signal is outside the linear range of the instrument (typically an absorbance greater than 2.0), then dilute the protein in buffer and remeasure.
3. After measurement of the sample remeasure the blank buffer to correct for any instrument drift.
4. Determine the unknown concentrations from the linear standard response.

3.3. Comments

The determination of the absorbance coefficient for a protein is discussed below but if a stock of the protein at known concentration is available then this can be used as a standard. Very rough estimates can be made from the relationship that if the cuvette has a path length of 1 cm, and the sample volume is 1 ml then concentration (mg/ml) = absorbance of protein at 280 nm.

Light scattering from either turbid protein samples or particles suspended in the sample with a comparable size to the incident wavelength (250–300 nm) can reduce the amount of light reaching the detector leading to an increase in apparent absorbance. Filtration using 0.2 μm filter units (that do not adsorb proteins), or centrifugation can be performed prior to analysis to reduce light scattering. Corrections for light scattering can be performed by measuring absorbance at lower energies (320, 325, 330, 335, 340, 345, and 350 nm), assuming the protein does not display significant absorbance at these wavelengths. A log–log plot of absorbance versus wavelength should generate a linear response that can be extrapolated back to 280 nm, the resulting antilog of which will give the scattering contribution at this wavelength (Leach and Scheraga, 1960).

Nucleic acids absorb strongly at 280 nm and are a common contaminant of protein preparations. A pure protein preparation is estimated to give a ratio of A_{280} to A_{260} of 2.0 while, if nucleic acid is present, the protein concentration can be derived by the following formula (Groves *et al.*, 1968).

$$\text{Protein concentration (mg/ml)} = 1.55A_{280} - 0.76A_{260} \qquad (8.2)$$

3.4. Ultraviolet absorbance at 205 nm (Range: 1–100 μg)

The peptide bond absorbs photons at a maximum wavelength below 210 nm. However, the broad absorption peak of the peptide bond allows measurements at longer wavelengths, which can have many practical advantages in terms of instrumentation and measurement accuracy. Due to interference from solvents and components of biological buffers, absorbance at 214 and 220 nm is often used as an alternative to measure proteins and peptides.

The large number of peptide bonds within proteins can make $\text{Abs}_{205\ \text{nm}}$ measurements more sensitive and display less protein-to-protein variability than $\text{Abs}_{280\ \text{nm}}$ measurements. Most proteins have extinction coefficients at $\text{Abs}_{205\ \text{nm}}$ for a 1-mg/ml solution of between 30 and 35; however, an improved estimate can be obtained using Eq. (8.3) that takes into account variations in tryptophan and tyrosine content of the protein to be quantitated (Scopes, 1974). Absorbance at 205 nm is used to quantitate dilute solutions, or for short path length applications, for example, continuous measurement in column chromatography, or for analysis of peptides where there are few, if any aromatic amino acids.

$$\varepsilon_{205\ \text{nm}}^{1\ \text{mg/ml}} = 27.0 + 120 \times \left(\frac{A_{280\ \text{nm}}}{A_{205\ \text{nm}}}\right) \qquad (8.3)$$

3.5. Calculation of the extinction coefficient

The extinction coefficient (ε) at a set wavelength describes the summation of all the photon absorbing species present within the molecule at a defined wavelength; the molar extinction coefficient is defined in Eq. (8.1). The extinction (absorption) coefficient is commonly expressed either in terms of molarity ($M^{-1}\ cm^{-1}$) or as a percentage of the mass $\varepsilon^{1\%}$ ($\%^{-1}\ cm^{-1}$), where $\varepsilon^{1\%}$ is defined as the absorbance value of a 1% protein solution.

Deviation from experimentally derived values for ε, and those derived by sequence data can be due to the influence of salts and buffers within the protein sample. The absorbance spectra from various amino acids are environmentally sensitive; therefore, ε derived for a protein in a set buffer may not be the same for another buffer system if gross changes in pH (tyrosine ionization at pH 10.9), or solvent polarity (denaturing agents) occur.

To determine ε_{280}, the amino acid composition or sequence of the protein is required. From the protein sequence, ε_{280} can be calculated from first principles using a standard formula (Gill and von Hippel, 1989), which has been refined (Pace *et al.*, 1995). Such models use the absorption coefficients for specific amino acids (Trp, Tyr, and disulfide bond) to generate a good estimate of ε_{280} where these amino acids are in abundance. However, where there is a low abundance of these amino acids (e.g., insulin), the model can display deviations of up to 15% from that determined by physical methods (Pace *et al.*, 1995). Physical (empirical) methods to determine extinction coefficient include amino acid analysis (AAA) via acid hydrolysis and chromatographic separation of resulting amino acids (Sittampalam *et al.*, 1988) and Kjeldahl and gravimetric analysis (Kupke and Dorrier, 1978).

4. Dye-Based Protein Assays

Methods to prepare the established (nonproprietary) protein quantitation assays are described. These reagents can be economically prepared in bulk and stored for prolonged periods. The majority of such assays are available from commercial suppliers such as Sigma-Aldrich, Bio-Rad, Novagen, and Pierce. It should be noted that suppliers can have different preparations of such reagents and these can perform differently with specific proteins. The use of commercial reagents can improve the long-term repeatability and performance of the assay and for microplate-based assays is reported to reduce issues with dye precipitation after long-term storage of reagents (Stoscheck, 1990). The majority of the spectrophotometric protein quantitation methods described can be adapted to a microplate format (typically 96-well plate), we have highlighted where changes in the assay formulations are required.

4.1. Protein concentration standards

The ideal protein standard to use in a quantitative assay is the exact same protein in a matched matrix/solution that has been assigned using a higher order method, for example AAA (Sittampalam *et al.*, 1988) or gravimetric analysis (Blakeley and Zerner, 1975). Gravimetric analysis is prone to errors due to the extensive dialysis and drying to remove water and salts from commercial preparations. Prepared standards should be redissolved at a high concentration in water and stored at -20 °C for long-term storage.

In practice, there is not always a matched protein standard available; however, some commercially available standards may be suitable for use, the most common being BSA, bovine gamma globulins, or immunoglobulins (used for antibody quantitation). The use of a BSA standard is known to give misleading results in many assays, especially those methods that are sensitive to the protein sequence, that is where the signal is generated by specific amino acids (Fig. 8.3). Assays with a low protein sequence dependence will give better estimates when BSA calibration is compared to AAA assignment (Alterman *et al.*, 2003). AAA assignment quantitates the amount

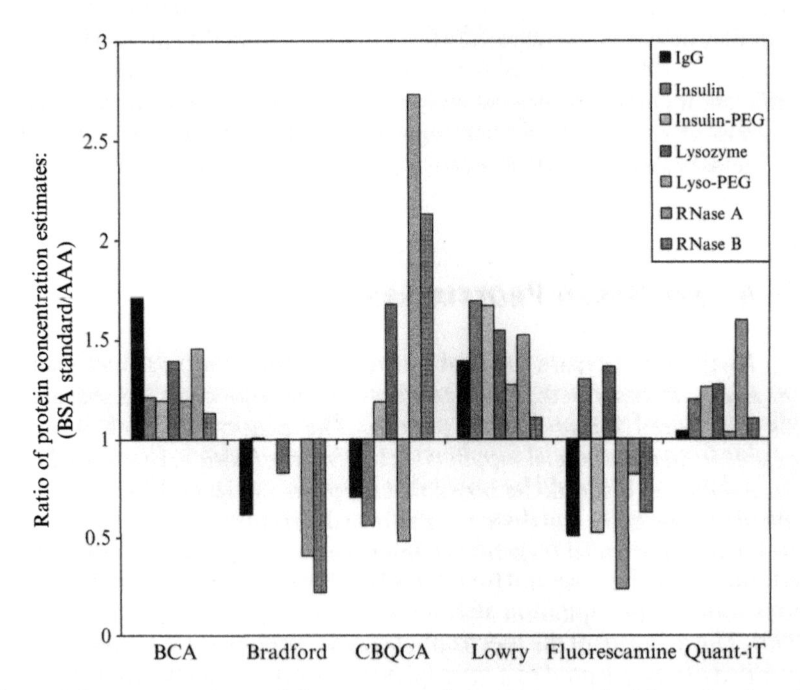

Figure 8.3 A comparison of the accuracy of the BSA standard in estimating the concentration of a protein using dye-based protein quantitation assays. AAA was used to determine the concentration of the model proteins; from these estimates a calibration curve for each protein was prepared using the dye-based assays. In the same plate, a calibration curve using the BSA standard (Pierce; concentration defined by manufacturer) was also prepared and the response of this was compared to that of the model proteins to see how well the BSA standard estimated the true concentration of the model proteins. The "ratio of concentration estimations" refers to the concentration of protein derived using the BSA standard when compared to the "true" value using AAA, where a ratio of 1 indicates the two methods gave the same value. The variation "% CV" associated with the dye-based protein concentration assays ranged from 2% to 8%, dependent on the assay. AAA concentration assignment typically displayed 5% CV values, dependent on the protein analyzed. Figure adapted with permission from Noble *et al.* (2007).

of specific amino acids present following protein hydrolysis and separation, using peptide sequence information the amount of target protein can then be calculated.

5. Coomassie Blue (Bradford) Protein Assay (Range: 1–50 μg)

The Bradford assay encompasses various preparations of the dye Coomassie Brilliant Blue G-250 used for protein quantitation purposes, and was first described by Bradford (1976). The basic mechanism of the assay is the binding of the dye at acidic pH to arginine, histidine, phenylalanine, tryptophan and tyrosine residues (de Moreno *et al.*, 1986), and hydrophobic interactions (Fountoulakis *et al.*, 1992). The exact mechanism is however still not fully understood (Sapan *et al.*, 1999). Upon binding protein, a metachromatic shift from 465 to 595 nm is observed due to stabilization of the anionic form of the dye. The majority of the observed signal is due to the interaction with arginine residues, resulting in the wide protein-to-protein variation characteristic of Bradford assays (Fig. 8.3).

5.1. Reagents

Dissolve 100 mg Coomassie Brilliant Blue G-250 in 50 ml of 95% ethanol and add 100 ml of 85% phosphoric acid while stirring continuously. When the dye has dissolved dilute to 1 l in water. The reagent is stable for up to a month at room temperature; however, for long-term storage keep at 4 °C, if precipitation occurs filter before use.

5.2. Procedure

1. Prepare standards in the range 100–1500 μg/ml in a Bradford-compatible buffer. For more dilute samples the sensitivity can be extended by increasing the ratio of sample to reagent volumes (Micro Bradford assay: 1–25 μg/ml). If the ratio of the sample to dye is too high, the pH of the reaction mixture could increase leading to higher background responses.
2. Add the standard and unknown samples to disposable cuvettes (plastic disposable cuvettes and microplates should be used as the dye sticks to various surfaces).
3. Allow the Bradford reagent to warm to room temperature. Add 1 ml of the dye solution to 25 μl of the protein sample, mix and incubate for 10 min at room temperature.

4. Measure the absorbance at 450 and 595 nm (for filter-based instruments a range from 570 to 610 nm can be used without significant loss of assay performance).
5. Plot either the 595 nm data or for improved precision at lower response values the ratio 595 nm/450 nm. The standard response curve can be fit to a polynomial response, from which unknown protein estimates can be calculated.

5.3. Comments

The advantages of the Bradford assay include the ease of use, sensitivity and low cost of the reagents. For microplate-based assays the reagent volumes can be decreased giving a total volume of 300 μl. Due to the path of the light source on the majority of microplate spectrophotometers, it is recommended to use commercial sources of Bradford reagent that are less predisposed to precipitation during prolonged storage.

We have observed significant variation in response between various commercial suppliers of Bradford preparations (Noble *et al.*, 2007). This appears to be most pronounced when analyzing low-molecular-weight proteins or peptides. Indeed the assay is reported to display a molecular weight cutoff "threshold"; requiring a certain number of residues for full signal development (de Moreno *et al.*, 1986). Changes in the formulation of the Bradford reagent are reported to change the response generated from specific proteins; therefore, care should be taken when comparing Bradford data from different suppliers or preparations (Chan *et al.*, 1995; Friedenauer and Berlet, 1989; Lopez *et al.*, 1993; Read and Northcote, 1981).

The Bradford assay is sensitive to interferences from various reagents detailed in Table 8.1 that include most ionic and nonionic detergents and glycosylated proteins. If precipitation of the reaction mixture occurs, for example hydrophobic or membrane proteins, the reaction can be supplemented with 1 *M* NaOH at 5–10% (v/v) to aid solubilization.

6. Lowry (Alkaline Copper Reduction Assays) (Range: 5–100 μg)

The Lowry assay (Lowry *et al.*, 1951) and other preparations with enhanced assay performance are based on a two-step procedure. Initially, the Biuret reaction involves the reduction of copper (Cu^{2+} to Cu^{+}) by proteins in alkaline solutions, followed by the enhancement stage, the reduction of the Folin–Ciocalteu reagent (phosphomolybdate and phosphotungstate) (Peterson, 1979) producing a characteristic blue color with absorbance maxima at 750 nm. The assay displays protein sequence variation, as color development is due not only to the reduced copper–amide bond complex but also to tyrosine, tryptophan, and to a lesser extent cystine, cysteine, and histidine residues (Peterson, 1977; Wu *et al.*, 1978).

The Lowry assay has been modified to reduce its sensitivity to interfering agents, increase the dynamic range and increase the speed and resulting stability of the color formation (Peterson, 1979). There are many commercial sources of the modified Lowry assay (Roche, Pierce, Bio-Rad, and Sigma), but different preparations may not give equal responses when using the same standard, dilution buffer, or interfering compounds.

6.1. Reagents

6.1.1. Folin and Ciocalteu's reagent

The preparation of this reagent has been described (Lowry *et al.*, 1951); however, the solution can be obtained from commercial sources (Sigma). Mix 10 ml of Folin–Ciocalteu's Phenol reagent to 50 ml of water.

6.1.2. Copper sulfate reagent

100 mg $CuSO_4 \cdot 5H_2O$ and 200 mg of sodium tartrate dissolved in 50 ml of water. Dissolve 10 g of sodium carbonate into 50 ml of water, then pour slowly while mixing to the copper sulfate solution, prepare fresh daily.

6.1.3. Alkaline copper reagent

Mix one-part copper sulfate solution, one-part 5% SDS (w/v) and two-parts 3.2% sodium hydroxide (w/v). This solution can be stored at room temperature for up to 2 weeks, discard if a precipitate forms.

6.2. Procedure

1. To 1 ml of sample and protein standards ~ 5–100 μg/ml, add 1 ml of the alkaline copper reagent, mix and allow to stand for 10 min.
2. Add 0.5 ml of Folin–Ciocalteu's reagent mix, vortex thoroughly and incubate for 30 min.
3. After incubation vortex again and measure the absorbance at 750 nm. Absorbance can be read from 650 to 750 nm depending on the availability of appropriate filters (microplate readers), or if the signal is too high, without significant loss in assay performance. Lowry is not an endpoint assay, so samples should be staggered to obtain more accurate estimates.
4. The response observed will be linear over a limited range of standards. Polynomial, exponential, and logarithmic models can be used to fit the data to extend the dynamic range of the response curve.

6.3. Comments

The Lowry method above can be adapted to a microplate format by reducing the volume of reactants added, resulting in a dynamic range ~50–500 μg/ml. The Lowry assay has been largely superseded by the BCA assay due to sensitivity, linearity, and improved methodology.

The Lowry protein assay is sensitive to many interfering compounds (Table 8.1), which may not generate a linear response (making extrapolations of interfering data complex). Formation of precipitates can occur with detergents, lipids, potassium ions, and sodium phosphate.

7. Bicinchoninic Acid (BCA) (Range: 0.2–50 μg)

The BCA assay replaces the Folin–Cioalteu's reagent as described for the Lowry method with Bicinchoninic acid that results in a protein assay with improved sensitivity and tolerance to interfering compounds (Smith *et al.*, 1985). The BCA reaction forms an intense purple complex with cuprous ions (Cu^{+}) resulting from the reaction of protein and alkaline Cu^{2+}. The residues that contribute to the reduction of Cu^{2+} include the cysteine, cystine, tryptophan, tyrosine, and the peptide bonds (Smith *et al.*, 1985). The chemical reaction is temperature dependent with different functional groups displaying a different reactivity at elevated temperatures, which result in less protein variability (Wiechelman *et al.*, 1988). At elevated temperatures (60 °C compared to 37 °C), more color formation is observed due to the higher reactivity of tryptophan, tyrosine, and peptide bonds.

Most of the commercial preparations are formulated close to the original preparation described by Smith *et al.* (1985), which is described in the following subsections. Variations have been employed to improve the sensitivity of the assay and can be obtained from commercial sources (Pierce, Novagen). The sample-to-working reagent ratio can be varied to maximize signal, or reduce assay interference, typically ratios of 8–20-fold excess of BCA working reagent are added to the protein sample.

7.1. Reagents

Reagent A: 1 g sodium bicinchoninate, 2 g Na_2CO_3, 0.16 g sodium tartrate, 0.4 g NaOH, and 0.95 g $NaHCO_3$, made up to 100 ml and the pH adjusted to 11.25 with either solid or concentrated NaOH.

Reagent B: 0.4 g $CuSO_4 \cdot 5H_2O$ dissolved in 10 ml water. Both reagent A and B are stable indefinitely at room temperature.

The working solution is prepared by mixing 100 parts of reagent A with two parts reagent B to form a green solution that is stable for up to a week.

7.2. Procedure

1. Cuvette analysis can be performed with 50–150 μl of protein and 3 ml of BCA working reagent, whereas microplate assay can use 25 μl of protein and 200 μl of BCA working reagent, that is a lower reagent to protein ratio.
2. Incubate the sample and standards ~ 5–250 μg/ml at either 37 or 60 °C for 30 min (longer incubations at 37 °C will improve protein-to-protein variability) and allow the sample to equilibrate to room temperature before reading. Microplates should be covered during incubation to avoid evaporation of the sample. For cuvette analysis at 37 °C, samples should be staggered to ensure equal incubation times.
3. Measure absorbance at 562 nm, for filter-based plate readers wavelengths in the range of 540–590 nm can be used instead without a significant loss in assay performance.
4. The BCA assay will produce a linear response over a wide concentration range; however, to extend the dynamic range of the data analysis a quadratic response can be used to model the data.

7.3. Comments

A microbased BCA assay can be used to improve the sensitivity of the procedure (1–25 μg/ml). The microbased assay uses a more concentrated working solution and can be prone to precipitation; again commercial sources of this modified BCA assay are available (Pierce). The BCA assay is sensitive to either copper chelators (e.g., EDTA) or reagents that can also reduce Cu^{2+} (e.g., DTT), a summary of the maximum tolerances can be found in Table 8.1.

8. Amine Derivatization (Range: 0.05–25 μg)

Amine-labeling "derivatization" using various fluorescent probes is a common technique to quantitate amino acid mixtures in AAA. The same technique can be used to quantitate proteins and peptides containing either lysine or a free N-terminus, both of which need to be accessible to the dye. Upon reaction with amines, the dyes display a large increase in fluorescence that for part of the dynamic range will generate a linear response with increasing protein concentration. Three dyes that have been used to quantitate proteins, or amino acids in a microplate format include *o*-phthalaldehyde (OPA) (Hammer and Nagel, 1986), Fluorescamine (Lorenzen and Kennedy, 1993), and 3-(4-carboxybenzoyl)quinoline-2-carboxyaldehyde (CBQCA™) (Asermely *et al.*, 1997; Bantan-Polak *et al.*, 2001; You *et al.*, 1997).

Fluorescamine reacts directly with the amine functional group, whereas OPA and CBQCA™ require the addition of a thiol (2-mercaptoethanol) or cyanide (CBQCA™). A cuvette-based format is described for the OPA assay, which can be converted to a microplate format by adjusting the volume of reactants and NaOH.

8.1. Reagents

OPA stock: Dissolve 120 mg of *o*-phthalaldehyde (high purity grade from Sigma or Invitrogen) in methanol, then dilute to 100 ml in 1 *M* boric acid, pH 10.4 (pH adjusted with potassium hydroxide). Add 0.6 ml of polyoxyethylene (23) lauryl ether and mix. The stock is stable for 3 weeks at room temperature.

8.2. Procedure

1. At least 30 min before analysis, add 15 μl of 2-mercaptoethanol to 5 ml of OPA stock, this reagent is stable for a day. Protect all fluorescent samples and reactions from light at all times.
2. Protein standards (0.2–10 μg/ml) and unknown samples need to be adjusted to a pH between 8.0 and 10.5 before analysis. Mix 10 μl of test sample with 100 μl of OPA stock (supplemented with 2-mercaptoethanol) and incubate at room temperature for 15 min.
3. Add 3 ml of 0.5 N NaOH and mix.
4. Read fluorescence at excitation 340 nm and emission from 440 to 455 nm in a fluorescent cuvette.
5. The relationship between protein concentration and fluorescence should be linear over the dynamic range of the assay and can be used to estimate unknown samples.

8.3. Comments

All three dyes offer improved sensitivity and dynamic range when compared with absorbance-based protein quantitation assays. OPA is generally preferred over fluorescamine due to its enhanced solubility and stability in aqueous buffers.

The use of amine-derivatization agents for protein quantitation is limited as the assay displays a large protein-to-protein variability due to variation in the number of lysine residues in proteins, requiring the need for a "matched" standard. Assay interference from glycine and amine containing buffers, ammonium ions, and thiols common in many biological-buffering systems limit the application of such assays (Table 8.1). The reproducibility of the assay is dependent on the pH of the reaction, protein samples that contain residual acids, for example from precipitation steps could reduce the rate of amine derivatization (You *et al.*, 1997).

A noncovalent amine reactive dye epicocconone can also be used for total protein assays in solution (Sigma), for which the mechanism has been reported (Bell and Karuso, 2003; Coghlan *et al.*, 2005).

9. Detergent-Based Fluorescent Detection (Range: 0.02–2 μg)

The development of fluorescent probes whose quantum yields are enhanced significantly when binding at the detergent–protein interface have been used to quantitate proteins within gels and in solution-based assays (Daban *et al.*, 1991; Jones *et al.*, 2003). Two commercial preparations of these assays are available NanoOrange™ and Quant-iT™ (Invitrogen); however, limited independent testing of the respective reagents prevents a full critical analysis. The NanoOrange™ assay is limited by the need to heat samples to 90 °C to denature the proteins thereby reducing the protein-to-protein variability (Jones *et al.*, 2003). Both assays are sensitive to detergents and high salt concentrations (Table 8.1), which presumably disrupt the protein-dye-detergent interface. The Quant-iT™ assay displays good sensitivity and dynamic range compared to other dye-based assays (Fig. 8.2), and a relatively low protein-to-protein variability (Fig. 8.3).

10. General Instructions

The choice of measurement format used will depend on the throughput, sensitivity, and precision required of the assay, and concerns about assay interferences that can be reduced by dilution in cuvette-based assays. From our experience, both in industry and academia, plate-based assays are replacing cuvette assays due to increases in throughput. For all spectrophotometric techniques, the instrument should be warmed up for 15 min prior to measurement and any calibration programs run before sample analysis. Samples and reagents should be equilibrated to room temperature before analysis to avoid condensation on optical surfaces.

10.1. Cuvettes

Traditionally, cuvettes have been used for the majority of spectrophotometric protein assays. Quartz cuvettes can be costly, therefore glass cuvettes are preferred; however, both of these may have to be washed between measurements to remove dye and adsorbed protein. Disposable plastic cuvettes are available and can be used to increase the throughput where many samples have to be measured, or the reagent is prone to sticking to the cuvette surface, for example Bradford reagent. Staggering of sample analysis is especially important if the signal is not stable, or does not run to completion within the time frame of the assay, for example BCA or Lowry assays. The best precision is obtained from a two-beam instrument incorporating a reference cell to account for instrument drift. Replacing the cuvette in the holder between each measurement due to cleaning, or the use of disposable cuvettes can result in changes in alignment, resulting in significant changes in amount of light reaching the detector. This is especially important if low-volume cuvettes are being used where the transmission window is reduced in size. Care should also be taken with low-volume cuvettes to ensure the sample covers the entire transmission window.

Care should be taken when handling and cleaning cuvettes. Prevent fingerprints from contaminating the transmitting surfaces. Cuvettes should be washed with either water or an appropriate solvent between runs and dried using a stream of nitrogen gas. If smearing of the transmitting surface is observed, the cuvette can be rewashed in water, ethanol, and finally acetone, or removed using ethanol and lintless lens tissue. If protein deposition is a recurring problem, cuvettes can be washed overnight in nitric acid and thoroughly washed before use.

10.2. Microwell plates

The majority of protein assays have been adapted for use in microwell plates, typically 96-well plates to enhance speed, throughput and lower sample and reagent usage. Many of the commercial fluorescent assays are specifically designed for plate formats. The plate reader format also offers the advantage of being able to read multiple samples within a short period (typically 25 s) reducing potential timing differences in reactions that do not go to completion, or are unstable.

Protein UV measurements can be made in a plate format; however, the effective path length can be difficult to calculate due to meniscus formation for concentrated protein solutions (many commercial plate readers can estimate effective path-length and thereby improving protein quantitation calculations). Quartz 96-well plates tend to be expensive, difficult to clean and prone to scratches that can affect light transmission.

Care should be taken in the preparation of protein assays in plate formats. The use of lower volume samples (down to 5 μl for some assays) can increase the relative pipetting errors of high viscosity solutions. Well-to-well contamination should be avoided by using fresh pipette tips for each sample and reagent. Regular calibration of the instrument should be performed using either optical standards or solid phase fluorescent standard plates (Matech) to ensure equal transmission/light detection from all wells.

Many of the 96-well plates conform to a standard geometry; however, in our experience, it is worth analyzing plate geometry in the plate-reader, especially if a different plate supplier is used to ensure equal illumination, and detection for fluorescent-based measurements. Plate-based assays can also be more sensitive to sample precipitation (common in the Lowry and Bradford assays) when compared to cuvette-based assay due to the detection geometry.

Recently, spectrophotometers that can measure low microliter samples (typically 1–2 μl), without the need for a cuvette or microplates have become commercially available (Tecan and Thermo Scientific), further minimizing sample usage.

10.3. Interfering substrates

Interfering substances for many protein preparations will be variable from batch-to-batch and can be difficult to adequately control for when standards are formulated differently. The choice of assay used should take into account interfering contaminants in the protein preparation, either used as stabilizers or as a result of purification that cannot be replaced, or substituted with a suitable alternative, for example reducing agents or chelators. Inclusion of an interfering substance can be accommodated using a matched standard; however, this can result in loss of dynamic range and poor assay performance and is therefore not recommended. The concentration or amount of interfering substance that can be tolerated is often quoted with the assay instructions; however, this can be dependent on the formulation of the assay, the maximum tolerated concentrations are summarized in Table 8.1.

Interfering substances can be removed prior to concentration determination, however, this adds additional steps to the procedure and can often result in dilution, or incomplete recovery of the original sample leading to errors in the concentration estimate. Changes in sample recovery can be compensated for by comparing the recovery of the standard that has been subjected to interference removal steps.

Precipitation of protein followed by separation and resuspension probably offers the most accurate method to remove interfering substances where they cannot be avoided. Buffer components, detergents, and lipids can be removed by precipitating the protein with trichloroacetic acid (TCA), perchloric acid (PCA), or acetone (Olson and Markwell, 2007); however Triton X-100 can coprecipitate with TCA and PCA.

In addition to precipitation techniques, specific interferences can be removed through chemical treatment, for example reducing agents (iodoacetic acid treatment), lipids through chloroform extraction, volatility, or neutralization of strong acids/bases.

ACKNOWLEDGMENT

We thank A. Hills for comments and help in preparing this review.

REFERENCES

Alterman, M., Chin, D., Harris, R., Hunziker, P., Le, A., Linskens, S., Packman, L., and Schaller, J. AAARG2003 study: Quantitation of proteins by amino acid analysis and colorimetric assays. Poster available for download from www.abrf.org (http://www.abrf.org/ResearchGroups/AminoAcidAnalysis/EPosters/aaa2003_poster_print.pdf).

Asermely, K. E., Broomfield, C. A., Nowakowski, J., Courtney, B. C., and Adler, M. (1997). Identification of a recombinant synaptobrevin-thioredoxin fusion protein by capillary zone electrophoresis using laser-induced fluorescence detection. *J. Chromatogr. B Biomed. Sci. Appl.* **695,** 67–75.

Bantan-Polak, T., Kassai, M., and Grant, K. B. (2001). A comparison of fluorescamine and naphthalene-2, 3-dicarboxaldehyde fluorogenic reagents for microplate-based detection of amino acids. *Anal. Biochem.* **297,** 128–136.

Bell, P. J., and Karuso, P. (2003). Epicocconone, a novel fluorescent compound from the fungus epicoccumnigrum. *J. Am. Chem. Soc.* **125,** 9304–9305.

Blakeley, R. L., and Zerner, B. (1975). An accurate gravimetric determination of the concentration of pure proteins and the specific activity of enzymes. *Methods Enzymol.* **35,** 221–226.

Bradford, M. M. (1976). A rapid and sensitive method for the quantitation of microgram quantities of protein utilizing the principle of protein-dye binding. *Anal. Biochem.* **72,** 248–254.

Burkitt, W. I., Pritchard, C., Arsene, C., Henrion, A., Bunk, D., and O'Connor, G. (2008). Toward Systeme International d'Unite-traceable protein quantification: From amino acids to proteins. *Anal. Biochem.* **376,** 242–251.

Chan, J. K., Thompson, J. W., and Gill, T. A. (1995). Quantitative determination of protamines by coomassie blue G assay. *Anal. Biochem.* **226,** 191–193.

Coghlan, D. R., Mackintosh, J. A., and Karuso, P. (2005). Mechanism of reversible fluorescent staining of protein with epicocconone. *Org. Lett.* **7,** 2401–2404.

Daban, J. R., Bartolome, S., and Samso, M. (1991). Use of the hydrophobic probe Nile red for the fluorescent staining of protein bands in sodium dodecyl sulfate-polyacrylamide gels. *Anal. Biochem.* **199,** 169–174.

de Moreno, M. R., Smith, J. F., and Smith, R. V. (1986). Mechanism studies of coomassie blue and silver staining of proteins. *J. Pharm. Sci.* **75,** 907–911.

Fountoulakis, M., Juranville, J. F., and Manneberg, M. (1992). Comparison of the Coomassie brilliant blue, bicinchoninic acid and Lowry quantitation assays, using non-glycosylated and glycosylated proteins. *J. Biochem. Biophys. Methods* **24,** 265–274.

Friedenauer, S., and Berlet, H. H. (1989). Sensitivity and variability of the Bradford protein assay in the presence of detergents. *Anal. Biochem.* **178,** 263–268.

Gill, S. C., and von Hippel, P. H. (1989). Calculation of protein extinction coefficients from amino acid sequence data. *Anal. Biochem.* **182,** 319–326.

Groves, W. E., Davis, F. C. Jr., and Sells, B. H. (1968). Spectrophotometric determination of microgram quantities of protein without nucleic acid interference. *Anal. Biochem.* **22,** 195–210.

Hammer, K. D., and Nagel, K. (1986). An automated fluorescence assay for subnanogram quantities of protein in the presence of interfering material. *Anal. Biochem.* **155,** 308–314.

Jones, L. J., Haugland, R. P., and Singer, V. L. (2003). Development and characterization of the NanoOrange protein quantitation assay: A fluorescence-based assay of proteins in solution. *Biotechniques* **34,** 850–854, 856, 858 *passim.*
Kupke, D. W., and Dorrier, T. E. (1978). Protein concentration measurements: The dry weight. *Methods Enzymol.* **48,** 155–162.
Leach, S. J., and Scheraga, H. A. (1960). Effect of light scattering on ultraviolet difference spectra. *J. Am. Chem. Soc.* **82,** 4790–4792.
Lopez, J. M., Imperial, S., Valderrama, R., and Navarro, S. (1993). An improved Bradford protein assay for collagen proteins. *Clin. Chim. Acta* **220,** 91–100.
Lorenzen, A., and Kennedy, S. W. (1993). A fluorescence-based protein assay for use with a microplate reader. *Anal. Biochem.* **214,** 346–348.
Lowry, O. H., Rosebrough, N. J., Farr, A. L., and Randall, R. J. (1951). Protein measurement with the Folin phenol reagent. *J. Biol. Chem.* **193,** 265–275.
Noble, J. E., Knight, A. E., Reason, A. J., Di Matola, A., and Bailey, M. J. (2007). A comparison of protein quantitation assays for biopharmaceutical applications. *Mol. Biotechnol.* **37,** 99–111.
Olson, B. J., and Markwell, J. (2007). Assays for determination of protein concentration. *Curr. Protoc. Protein Sci.* Chapter 3, Unit 3.4.
Ozols, J. (1990). Amino acid analysis. *Methods Enzymol.* **182,** 587–601.
Pace, C. N., Vajdos, F., Fee, L., Grimsley, G., and Gray, T. (1995). How to measure and predict the molar absorption coefficient of a protein. *Protein Sci.* **4,** 2411–2423.
Peterson, G. L. (1977). A simplification of the protein assay method of Lowry et al. which is more generally applicable. *Anal. Biochem.* **83,** 346–356.
Peterson, G. L. (1979). Review of the Folin phenol protein quantitation method of Lowry, Rosebrough, Farr and Randall. *Anal. Biochem.* **100,** 201–220.
Read, S. M., and Northcote, D. H. (1981). Minimization of variation in the response to different proteins of the Coomassie blue G dye-binding assay for protein. *Anal. Biochem.* **116,** 53–64.
Sapan, C. V., Lundblad, R. L., and Price, N. C. (1999). Colorimetric protein assay techniques. *Biotechnol. Appl. Biochem.* **29**(Pt 2), 99–108.
Scopes, R. K. (1974). Measurement of protein by spectrophotometry at 205 nm. *Anal. Biochem.* **59,** 277–282.
Sittampalam, G. S., Ellis, R. M., Miner, D. J., Rickard, E. C., and Clodfelter, D. K. (1988). Evaluation of amino acid analysis as reference method to quantitate highly purified proteins. *J. Assoc. Off. Anal. Chem.* **71,** 833–838.
Smith, P. K., Krohn, R. I., Hermanson, G. T., Mallia, A. K., Gartner, F. H., Provenzano, M. D., Fujimoto, E. K., Goeke, N. M., Olson, B. J., and Klenk, D. C. (1985). Measurement of protein using bicinchoninic acid. *Anal. Biochem.* **150,** 76–85.
Stoscheck, C. M. (1990). Quantitation of protein. *Methods Enzymol.* **182,** 50–68.
Wiechelman, K. J., Braun, R. D., and Fitzpatrick, J. D. (1988). Investigation of the bicinchoninic acid protein assay: Identification of the groups responsible for color formation. *Anal. Biochem.* **175,** 231–237.
Wu, A. M., Wu, J. C., and Herp, A. (1978). Polypeptide linkages and resulting structural features as powerful chromogenic factors in the Lowry phenol reaction. Studies on a glycoprotein containing no Lowry phenol-reactive amino acids and on its desialylated and deglycosylated products. *Biochem. J.* **175,** 47–51.
You, W. W., Haugland, R. P., Ryan, D. K., and Haugland, R. P. (1997). 3-(4-Carboxybenzoyl)quinoline-2-carboxaldehyde, a reagent with broad dynamic range for the assay of proteins and lipoproteins in solution. *Anal. Biochem.* **244,** 277–282.

CHAPTER NINE

Concentration of Proteins and Removal of Solutes

David R. H. Evans,* Jonathan K. Romero,* *and* Matthew Westoby[†]

Contents

1. Chromatography 98
   1.1. Gel filtration 99
   1.2. Ion-exchange chromatography 99
   1.3. Reversed phase chromatography 100
   1.4. Hydrophobic and affinity chromatography 100
   1.5. Manufacturing-scale chromatographic applications 102
2. Electrophoresis 103
3. Dialysis 104
   3.1. Concentration and affinity binding applications 107
4. Ultrafiltration 107
   4.1. Ultrafiltration membranes 109
   4.2. Ultrafiltration devices 110
   4.3. Purification applications 112
5. Lyophilization 113
6. Precipitation 116
7. Crystallization 118
References 118

Abstract

The dramatic advances in recombinant DNA and proteomics technology over the past decade have supported a tremendous growth in biologics applied to diagnostics, biomarkers, and commercial therapeutic markets. In particular, antibodies and fusion proteins have now become a main focus for a broad number of clinical indications, including neurology, oncology, and infectious diseases with projected increase in novel first-class molecules and biosimilar entities over the next several years. In line with these advances are the improved analytical, development, and small-scale preparative methods

* Process Biochemistry, Biopharmaceutical Development, Biogen Idec. Inc., Cambridge, Massachusetts, USA
† Process Biochemistry, Biopharmaceutical Development, Biogen Idec. Inc., San Diego, California, USA

Methods in Enzymology, Volume 463
ISSN 0076-6879, DOI: 10.1016/S0076-6879(09)63009-3

employed to elucidate biologic structure, function, and interaction. A number of established methods are used for solvent removal, including lyophilization, reverse extraction, solute precipitation, and dialysis (solvent exchange), ultrafiltration, and chromatographic techniques. Notably, advances in the miniaturization and throughput of protein analysis have been supported by the development of a plethora of microscale extraction procedures and devices that exploit a wide array of modes for small-scale sample preparation, including the concentration and desalting of protein samples prior to further analysis. Furthermore, advances in process handling and data monitoring at microscale have dramatically improved complex control and product recovery of small quantities of biologics using techniques such as lyophilization and precipitation. In contrast, the efficient concentration of feed streams during preparative chromatography has been enhanced by improvements to protein binding capacity achieved through advanced bead and ligand design.

The objective of solvent removal may be to prepare or concentrate solutes for analysis, or to facilitate their production or modification. Here, we describe the most recent advances in these techniques, particularly focusing on improved capabilities for bench-scale preparative methods.

1. Chromatography

Over the past two decades, the innovation of chromatographic techniques for protein concentration and desalting has been driven by the increasing application of proteomics technologies. Proteomics generally requires an enrichment of proteins from small amounts of complex mixtures including the selective concentration of specific species. The high sensitivity of detection and detailed biochemical characterization provided by mass spectrometry (MS) and antibody-based assays has found utility in diverse applications including clinical diagnostics, biologic drug characterization, and basic research into protein structure–function. These applications require the analysis of drug targets and biomarkers present at low amounts in biological fluids, or the acquisition of detailed information on the structure and posttranslational modification of proteins expressed from heterologous systems. Samples for proteomics analysis are routinely generated in aqueous solutions that may contain a range of nonvolatile salts, solvents, dyes, detergents, chaotropes, DNA/RNA, and lipids. These contaminants can impair the performance of sensitive analytical techniques, notably electrospray ionization (ESI) matrix-assisted laser desorption/ionization (MALDI) MS, where they cause a detrimental effect on resolution and sensitivity by interfering with sample ionization, adduct formation, and ion-source fouling. As a result, an excess of microscale solid phase extraction procedures and devices have been developed by multiple vendors to concentrate target proteins and purify them from contaminating molecules with high yield.

1.1. Gel filtration

Gel filtration matrices have been widely used for desalting proteins under nondenaturing conditions. Unlike small molecules, macromolecular proteins are excluded from the pores of these resins and are desalted during flow. Typically, a bed volume of 4–20 times the volume of the sample and a column length-to-diameter ratio of between 5 and 10 provide optimal resolution. A minimum sample dilution of 1.25-fold is generally obtained during desalting but, with an appropriate selection of eluant, protein losses can be minimal even with sample concentrations in the μg/ml range. Polymeric carbohydrate-based sephadex media offered by various vendors have been extensively used for desalting and a range Bio-Gel P acrylamide-based matrices (Bio-Rad, Hercules, CA) permit the user to tailor the gel exclusion limit of the matrix to the molecular weight of the target protein. Added convenience is provided by Zeba Spin columns (Thermo/Pierce, Rockford, IL) which permit rapid desalting of protein samples of molecular weight >7000 Da by centrifugation without the requirement for column equilibration or gravity flow fractionation. Through appropriate choice of column size, sample volumes in the range 2–4000 μl can be accommodated with high recovery at protein concentrations as low as 25 μg/ml and with 95% retention of molecules of molecular weight <1000 Da. The Zeba Spin matrix is also available in 96-well plate format facilitating high-throughput nondenaturing desalting of multiple protein samples in parallel.

1.2. Ion-exchange chromatography

An alternative to gel filtration for protein desalting is the use of mixed bed ion-exchange (IEX) resins. Large bead size and surface modification of mixed bed IEX sorbents can permit effective capture of small anions and cations (<1000 Da) with minimal retention of high-molecular-weight protein molecules. Desalting may be achieved in batch mode or via a packed bed of the matrix. However, an undesirable consequence of this technique is that the exchange of ions that occurs on a mixed bed of conventional IEX sorbents can cause acidification of the protein sample and precipitation. In contrast, ion retardation matrices, which have been used to desalt milk and whey, undergo salt uptake by absorption rather than by ion exchange and may therefore achieve the required protein desalting without significant pH change. A commercially available example is AG11 A8 resin (Bio-Rad, Hercules, CA). Similar to gel filtration, a drawback to these techniques is that the protein sample is diluted during the desalting process and may require subsequent concentration.

1.3. Reversed phase chromatography

Reversed phase (RP) chromatography is a popular technique for the separation, desalting, and concentration of proteins, in part because the sample is concentrated in a small volume of volatile solvent that can be removed by evaporation. With the advent of routine analysis by MS, a broad array of RP extraction formats have been commercially developed improving the speed and convenience of the technique. These formats include microspin columns, capillaries, micropipette tips, and plates for rapid desalting of as little as femtomole quantities of proteins with recovery in microlitre volumes of eluate. Typically, optimal protein binding is achieved at pH <4 in the presence of an ion-pairing agent such as 0.1–1.0% TFA. Reversible binding may be promoted by the presence of an organic component, typically 15% acetonitrile or methanol, or a chaotropic agent such as guanidine hydrochloride. Dilution of contaminating detergents may be required to prevent interference with protein binding to the sorbent. Desalting may require the resin to be washed, for example with 5% methanol, 0.1% TFA, prior to elution while selective elution may be achieved by experimenting with alternative ion-pairing agents. The choice of RP sorbent is based on protein size, with aliphatic C4, C8, C12, and C18 chemistries being routinely employed. Examples of micropipette-based products are ZipTips (Millipore, Billerica, MA) which are packed with RP beads or NuTips (Glygen, Columbia, MD) in which the interior walls of the tip are coated with the binding material thereby reducing backpressure and minimizing nonspecific interactions with the tip surfaces. For high-throughput desalting of multiple samples in parallel, RP resin can also be obtained in 96-well plates, an example being the ZipPlate format (Millipore, Billerica, MA).

1.4. Hydrophobic and affinity chromatography

In addition to the commonly used gel filtration, IEX, and RP sorbents, additional chemistries are available as batch resin or in microscale device format that support various modes of chromatography for protein concentration and desalting. These include hydrophobic, hydrophilic, metal chelating, and protein affinity resins. Notably, hydrophilic media provide a useful alternative to RP chemistry when the protein sample is suspended in an organic solvent and removal of contaminating salts or detergents is required. An example is PAGE-Prep media (Thermo/Pierce, Rockford, IL) that can be used to desalt proteins prior to SDS–PAGE analysis when bound to the resin in the presence of 50% dimethyl sulfoxide. Alternatively, Glygen (Columbia, MD) offers hydrophilic media in a micropipette tip format (NuTip). Similar to RP media, a potential disadvantage of hydrophilic resins is that the protein is generally recovered in a denatured form. In contrast, an important use of immobilized metal affinity chromatography

(IMAC) sorbents and, more recently, media-containing TiO_2 has been to selectively concentrate and desalt phosphopeptides prior to MS analysis. This has found particular application in the area of signal transduction research where the detection of low-abundance phosphopeptides is critical to the understanding of cellular responses. A recent report (Hsieh *et al.*, 2007) described the preparation of a TiO_2 nanoparticle matrix capable of selectively binding phosphopeptides at pH 2.0 in 0.1% TFA and 50% acetonitrile to prevent nonspecific binding. In this case, elution was performed with 100 m*M* ammonium phosphate pH 8.5, which was compatible with detection of phosphorylated peptides by MALDI-MS analysis without further desalting.

The routine preparation of protein samples for MS analysis using HPLC techniques has driven vendors to provide their sorbents in an increasing array of column, cartridge, and capillary formats. This in turn has permitted the streamlining of sample preparation via the use of column switching procedures. In this approach a short precolumn is used to rapidly concentrate and desalt a protein prior to a subsequent, high-resolution separation via a longer column on the same HPLC system. As an example, PepMap microscale precolumns (Dionex, Sunnyvale, CA) can be used to rapidly concentrate and desalt protein samples over a packed bed length of only 5 mm. With a choice of resins available, the efficiency of analyte trapping is governed by the choice of packing material while the reduction in sample volume improves the resolution of the subsequent chromatography step. In this context, it is also worth comment that monolithic separation materials are being increasingly employed for protein concentration and desalting (Schley *et al.*, 2006). The effective pore size of up to 20 μm permits rapid convective transport of solutes through the monolithic matrix. This contrasts with the slower diffusion-mediated mass transport achieved within the submicron pores of bead-based sorbents and enables monoliths to support efficient separations at high flow rates with low backpressure, which is particularly useful for rapid desalting by column switching. Nevertheless, in a more direct approach, Mass-Prep columns (Waters, Milford, MA) have been used for in-line desalting of proteins on a hydrophobic matrix followed by direct injection into an ESI mass spectrometer (Wheat *et al.*, 2007). Furthermore, sample preparation for MALDI-MS analysis may be performed without the requirement for a microextraction device. Thus, a wide range of sorbents have been used for desalting proteins in batch mode prior to spotting onto a MALDI plate. A recently developed example is a ZnO-poly(methyl methacrylate) nanobead adsorbent capable of clearing saturating levels of NaCl (6.2 *M*) or NH_4HCO_3 (2.6 *M*) or 1 *M* urea from a protein sample prior to MALDI-MS (Shen *et al.*, 2008). Similar desalting has been achieved by applying a hydrophobic matrix directly to the wells of a MALDI plate (Jia *et al.*, 2007). In this case, the target protein was captured on the matrix and, following drying, desalting of the sample was achieved by applying a minimal on-plate wash which permitted MALDI-MS analysis without further purification.

1.5. Manufacturing-scale chromatographic applications

Similar to microscale extractions used in analytical applications, protein recovery at preparative-scale and manufacturing-scale frequently involves concentration and desalting. Column chromatography remains a central technique for the purification of proteins due to the high-selectivity achievable combined with the convenience and robustness of a packed resin bed to which the feed is delivered under flow. Commercial manufacturing of enzymes and biologic drugs commonly requires the processing of thousands of liters of a starting feedstock. To minimize operating complexity and processing time while controlling costs, it may be essential to concentrate the feed and remove the bulk of impurities and unwanted salts early in the downstream process. This can be achieved by the use of large resin beds of up to 2 m in column diameter for bind-and-elute chromatography that generates an eluate fraction smaller than the column load volume. Furthermore, the selectivity of the resin may be exploited to elute a purified product. A notable example is the use of protein A affinity resins for the production of antibody-based proteins expressed from mammalian cell culture. During affinity capture, which is mediated by a specific epitope present in the Fc region of the target molecule, the bulk of the media components, salts, and impurities flow through the column. Product recovery is typically achieved by elution at low pH and is generally associated with a 90% reduction in volume. The latest generation of Protein A affinity resins have been optimized for ligand density, bead structure, and mass transfer characteristics to maximize dynamic binding capacity while minimizing column cycling. This in turn minimizes intermediate volumes and effectively concentrates the stream at the first column step in the process. Examples of resins are MabSelect (GE Healthcare, Piscataway, NJ) and ProSep Ultra Plus (Millipore, Billerica, MA) for which binding capacities in the range of 20–50 g/l of resin can be achieved depending on the target molecule and its contact time with the resin during column loading. Conventional chromatography resins, including IEX and hydrophobic interaction chromatography (HIC) sorbents, have been similarly optimized for particle size, pore size, and ligand density to achieve highest capacity binding of molecules within a particular molecular weight range but with efficient elution. For example, ToyoPearl GigaCap S-650M (Tosoh Biosciences, Tokyo, Japan) is a cation-exchange resin that has been optimized for the capture of IgG1 molecules with a molecular weight of approximately 150 kDa at high dynamic binding capacity (>100 g/l of resin) while permitting recovery in a minimal elution volume, typically 2–3 column volumes. Finally, mixed mode sorbents have been developed that permit IEX chromatography to be performed with high-capacity binding at relatively high conductivity. This has been achieved through the combination of an IEX and hydrophobic moiety on the same ligand. An example is Capto MMC resin (GE Healthcare, Piscataway, NJ) which combines a cation exchange and hydrophobic group and binds bovine serum albumin with a capacity of 45 mg/ml of resin at a conductivity of 30 mS/cm. This property can be useful because, while optimal protein purification is generally achieved via the application of multiple orthogonal modes of chromatography performed as separate steps, it can be challenging to maintain the feed conductivity at a level compatible with a conventional IEX step without dilution. Therefore, although this type of sorbent may not concentrate the process stream directly, it may obviate the need for a reduction in conductivity by dilution during downstream processing. A further attraction of mixed mode resins is that they may provide unique selectivities which may enhance the overall purification process.

2. Electrophoresis

Concentration of proteins by gel electrophoresis is generally employed to enhance the sensitivity of a subsequent analytical technique. In one example, the sensitivity of Western blot analysis was enhanced by applying up to 250 μl of a protein sample over five consecutive loadings on a 1–2 cm stacking gel in a mini-gel system (Sheen and Ali-Khan, 2005). The approach permitted effective sample concentration with vertical band broadening as the only artifact. A similar concentration of a dilute sample prior to western blotting has been achieved via a funnel shaped loading well. Concentration of protein samples by stacking has been achieved by capillary electrophoresis (CE) (e.g., see Shihabi, 2002). Moreover, a recent report has described conditions for CE that permit simultaneous concentration and desalting of proteins. This is a notable advance, as desalting is normally required prior to CE because ionic salts cause band broadening through Joule heating and electrodispersion. In addition, a desirable stacking effect can be achieved when samples for CE are loaded at a lower ionic strength than the electrolyte, leading to protein enrichment and improved sensitivity in analytical applications. Desalting of proteins prior to CE can be achieved by standard RP extractions. However, while solid phase mini-beds that permit desalting in line with CE have been described, the approach may be technically challenging and generally the desalting step is performed as a separate operation. Thus, to overcome these drawbacks a novel technique termed capillary isoelectric trapping was developed to simultaneously concentrate and desalt protein samples during CE (Booker and Yeung, 2008). The approach utilizes a discontinuous buffer system to establish a stable pH boundary within the capillary. The stationary boundary traps zwitterionic proteins at their pI while removing contaminating ionic salts, which continue to migrate due to their constant charge. This technique facilitates the recovery of concentrated and desalted protein samples for MS or, following a switch to an alternative buffering system, permits the analysis of the concentrated protein by capillary zone electrophoresis. Finally, microfluidic devices are finding increasing utility for sample preparation prior to biochemical analyses, in part due to their ability to rapidly concentrate and desalt small protein samples via the application of an electric field (e.g., see Yu *et al.*, 2008).

3. Dialysis

Dialysis is the size-based separation of molecules in solution by selective diffusion through a semipermeable membrane. This technique is considered the most popular method employed for removal of low-molecular-weight solutes from large protein molecules. In particular, the nondenaturing desalting technique permits buffer exchange under benign or physiological conditions with minimum risk of impacting the target protein's properties.

The mechanism of dialysis has not changed, but the techniques and tools for employing dialysis have improved over the past decade. In particular, optimized techniques have allowed for higher throughput implementation (reduced processing times), increased flexibility for sample volume processing (10 μl → 100 ml), improved membrane morphology for reduced protein adsorption and loss, and enhanced convenience for sample handling.

Typically, the volume of the target buffer solution (dialysate) will be several orders of magnitude greater than the protein sample volume. The sample is transferred to a sealed compartment and exposed to dialysis membrane which functions as a semipermeable barrier. The membrane typically possesses a specific pore size, termed the molecular weight cutoff (MWCO) limit, and permits the free passage of molecules smaller than this in both directions while retaining large macromolecules such as proteins. The buffer concentration gradient across the membrane drives the diffusion of solutes from regions of high concentration to regions of low concentration (see Fig. 9.1). The diffusional transport or flux is described using Fick's law for diffusion:

$$J = -D\frac{\partial C}{\partial x}$$

where J is the diffusion flux in dimensions of [(amount of substance) $L^{-2}T^{-1}$], D is the diffusion coefficient or diffusivity in dimensions of [$L^2\ T^{-1}$], c is the concentration in dimensions of [(amount of substance) L^{-3}], and x is the position [L]. The diffusion coefficient is proportional to the squared velocity

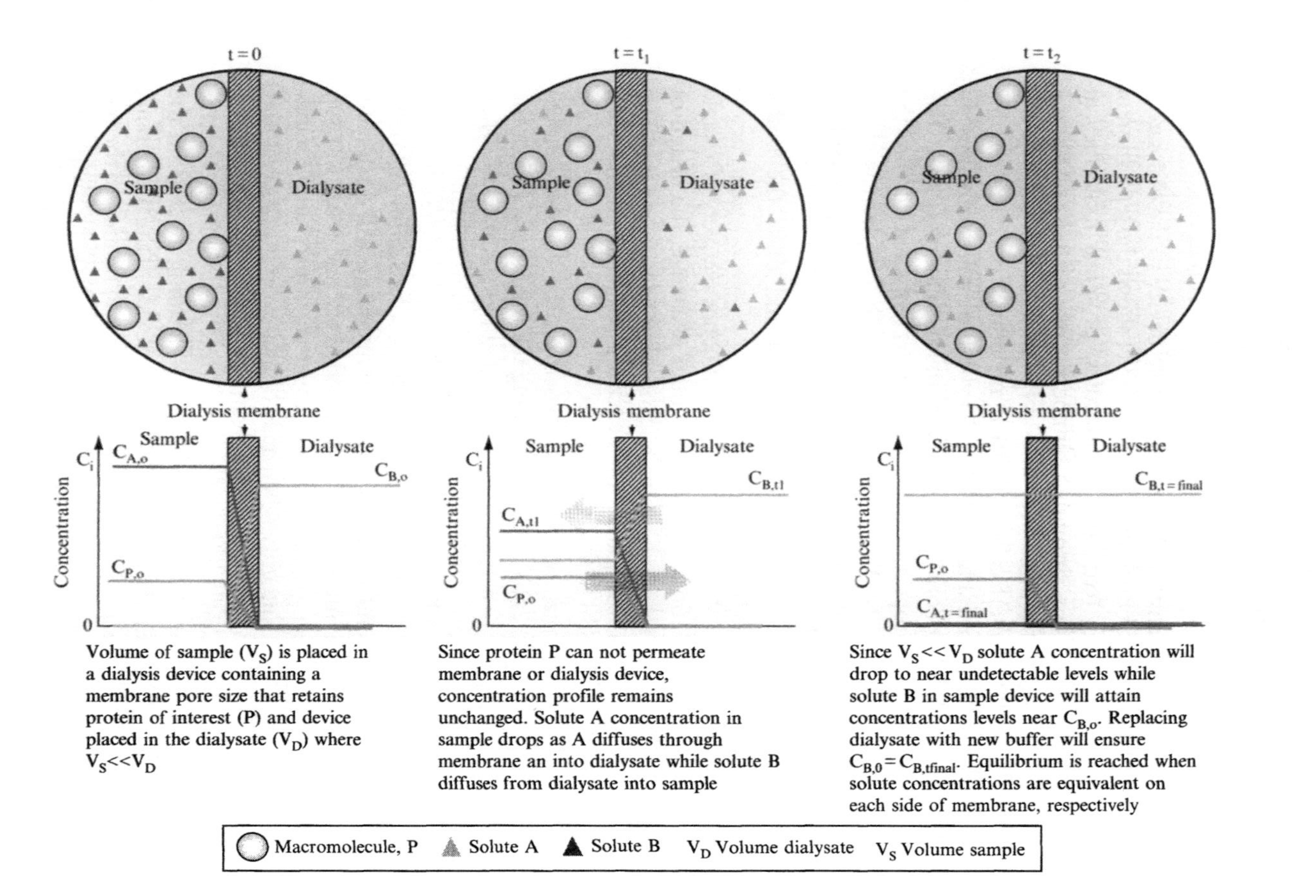

Figure 9.1 Schematic of the dialysis process. Top circular diagrams provide graphic representation of solute distributions within "sample" and "dialysate" pools throughout dialysis process. The bottom graphs show the respective concentration profiles from each pool. (See Color Insert.)

of the diffusing particles, which depends on the temperature, viscosity of the fluid, and the size of the particles according to the Stokes–Einstein relation (Cussler, 1997). Dialysis is most efficient when the dialysate volume is several orders of magnitude larger than the sample volume (see Fig. 9.1). This volume difference maintains a concentration differential across the membrane and ensures that solute being transferred from the sample device to the dialysate is diluted to near undetectable concentrations. Figure 9.1 shows the separation of small (e.g., salts) and large (e.g., macromolecules, proteins) molecules in solution by selective diffusion through a semipermeable dialysis membrane including the respective concentration profiles at various time points. Once the liquid–membrane–liquid interface is established, including concentration differences across the semipermeable membrane ($t = 0$), molecules will diffuse in the direction of decreasing concentration across the membrane ($t > 0$). The rate of diffusion is directly proportional to the diffusion coefficient "D" and the concentration gradient $\partial C/\partial x$. Assuming D is constant, $\partial C/\partial x$ decreases with time moving from left to right in Fig. 9.1, with the rate of dialysis decreasing as equilibrium is approached (Cussler, 1997). The starting dialysate can be refreshed to recreate the concentration differential and drive the buffer exchange to effective completion.

There are several aspects of dialysis which can influence recovery of a stable product. These include the time required for buffer exchange, the design of the dialysis system, and membrane chemistry and morphological properties. The time required to complete the required buffer exchange will depend on several factors. The inherent diffusion of molecules across a semipermeable membrane can be enhanced to minimize process time. Based on the equation shown previously, transport can be enhanced by increasing the membrane area or the solute flux. It may be easier to achieve a large ratio of sample to dialysate when sample volume is small. However, the impact of increased membrane area or increased time (for small membrane area) needs to be evaluated in the context of protein loss due to adsorption on the membrane. Since, solute diffusion coefficients are temperature dependent, an increase in temperature will drive faster molecular motion, increase intrinsic diffusion coefficients, and result in faster transport. However, increasing temperature can adversely impact sample stability. Therefore, since sample integrity is generally of highest importance, it may be necessary to predetermine the highest temperature compatible with stability.

The dialysis system design can also dictate the time required for effective dialysis and efficiency of product recovery. Solutes diffusing away from the sample permeate enter a boundary layer on the dialysate side of the membrane. The solute concentration in this boundary layer is higher in relation to the rest of the dialysate thus lessening the concentration gradient across the membrane and inhibiting diffusion. However, the boundary layer can be reduced by inducing convective flow or mixing on the dialysate side of the membrane.

Finally, the mechanical design of current dialysis membranes and cartridges allows for the processing of a wide range of sample volumes

(10 μl $\rightarrow$ 100 ml) with negligible product loss. The traditional dialysis tubing is difficult to prepare before it can be used. Pierce Biotech (Rockford, IL) offers SnakeSkin® dialysis tubing to simplify large sample dialysis. This technology comes in a pleated regenerated cellulose membrane format (from 3.5 to 10 kDa MWCO) allowing rapid preparation. In the case with small sample volume dialysis, Pierce offers compact Slide-A-Lyzer® Mini-dialysis Units (10–100 μl), and Cassettes (0.1–30 ml) to process limited small biological samples. These units are disposable cups made of polypropylene and regenerated cellulose allowing sample to be easily removed using standard lab pipettes. The new designs allow for improved recovery of valuable small sample volumes.

3.1. Concentration and affinity binding applications

Protein buffer-exchange applications have dominated the use of dialysis membranes. However, there is increased application of dialysis in the area of concentration and macromolecular affinity binding interactions.

The use of dialysis membranes for protein concentration is similar to aqueous two-phase extraction where reagents (e.g., polyethylene glycol, dextran, polymers) are used to generate a two-phase system with a dialysis membrane separating the phases (Waziri *et al.*, 2004). The system is designed such that protein can freely pass through the membrane and interact with the precipitating agent which can then be subsequently concentrated and product recovered.

The technique termed "equilibrium dialysis" has also been utilized for characterization of a candidate drug in serum binding assays, studying antigen–antibody interactions and evaluating low-affinity interactions that are undetectable using other methods. In equilibrium dialysis, a membrane is selected such that the desired protein or ligand can pass freely through but with the receptor molecule located only on one side and unable to permeate. As the protein diffuses across the membrane some of it will bind to the receptor and some will remain free in solution. Diffusion of the ligand across the membrane and binding to the receptor continues until equilibrium has been reached. The concentration of free ligand in the ligand chamber can then be used to determine the binding characteristics of the sample (Waters *et al.*, 2008)

4. Ultrafiltration

Ultrafiltration (UF) exploits a separation mechanism essentially equivalent to that employed for dialysis. Both techniques employ a semipermeable barrier or membrane to separate sample (containing desired product or solute) from sample-free solution. The membrane is designed to permit low-molecular-weight solutes (e.g., salts) to permeate while completely retaining the desired product. Ultrafiltration membranes are well suited for the concentration of biologics since they can operate at wide temperatures ranging from 2 to 26 °C and involve no phase changes or chemical additives thereby minimizing the extent of denaturation and/or degradation of labile biological products. The features that distinguish UF from dialysis are threefold: (1) application, (2) mode of operation, (3) membrane construction. First, dialysis is primarily used for solute or buffer exchange, whereas UF is employed for either buffer exchange or product concentration. In addition, dialysis operates via diffusional mass transfer driven by concentration differences across the semipermeable membrane, whereas UF works primarily by conventional mass transfer driven by pressure differences applied across the membrane. As a result of the different driving forces used for solute or solvent transfer, membranes employed for ultrafiltration are mechanically stronger than those for dialysis. Pressure-driven UF membranes are designed to withstand high pressure drops exceeding 75 psi.

In general, three modeling approaches have been used to describe the actual mechanism of solute transport via ultrafiltration (Denisov, 1994). A description of one transport mechanism by ultrafiltration is shown in Fig. 9.2. The pressure-driven fluid flow promotes convective transport of solutes and solvent toward the upstream surface of the membrane where the applied pressure drop is termed the transmembrane pressure (ΔP_{TM}). If the membrane is completely retentive to a given solute (e.g., protein), as in the case of concentration by UF, $C_{\mathrm{filtrate}} = 0$ and the retained solute will accumulate at the upstream surface of the membrane causing an increase concentration within a boundary layer thickness, δ at the membrane surface. This phenomenon is referred to as "concentration polarization." The accumulation of solutes at the membrane surface can affect solvent filtrate flux, J_{v}, by several mechanisms. First, the accumulated solute can generate an osmotic pressure drop across the membrane driving fluid from the filtrate to the feed or retentate side, thereby reducing the net rate of solvent transport. The flux can be described as shown below and in Fig. 9.2:

$$J_{\mathrm{v}} = L[\Delta P_{\mathrm{TM}} - \sigma_{\mathrm{o}}(\Pi_{\mathrm{W}} - \Pi_{\mathrm{F}})]$$

where L is the hydraulic permeability, σ_{o} is the osmotic reflection coefficient, and Π_{W} and Π_{F} are the solute osmotic pressures at the membrane wall and filtrate, respectively (Denisov, 1994). This model assumes the filtrate flux is a result of the difference between the applied pressure and induced osmotic pressure and that any membrane fouling or resistance to flow is negligible. The gel polarization model (Cheryan, 1986), however, assumes that as ΔP_{TM} the high concentration of retained solute at the

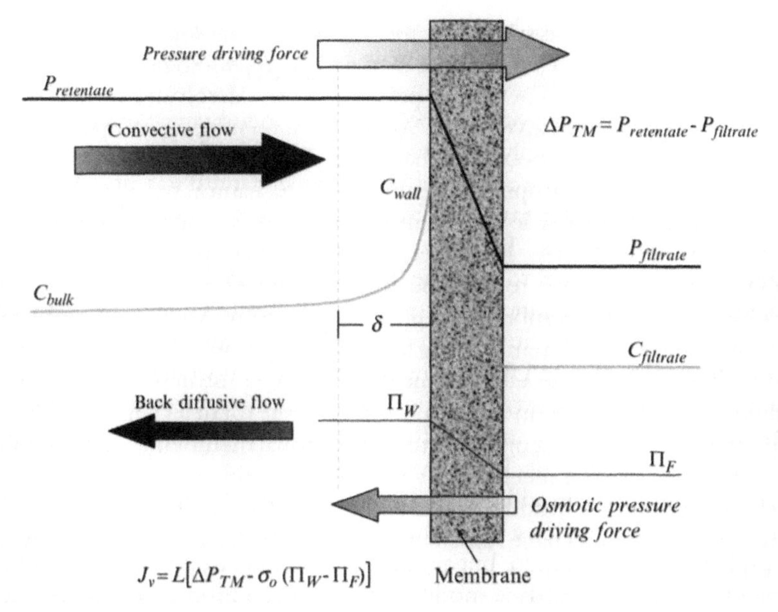

Figure 9.2 Schematic representation of convective and diffusive solute transport for ultrafiltration processes.

membrane surface forms a gel layer providing an additional hydraulic resistance to filtrate flow. The third modeling approach assumes solute at the membrane surface can irreversibly foul the membrane, thereby reducing its hydraulic permeability (Cheryan, 1986).

Overall, these models allow the user to better understand the impact of UF processing parameters on solute and solvent flux so that conditions can be optimized to improve membrane throughout. This is probably most important for developing and designing UF processes to concentrate expensive product such as protein therapeutics at large scale (>100 l) where processing productivity and recovery are critical. These factors are not as critical for ultrafiltration at the lab bench but are considered in designing UF systems for very small-scale preparations.

4.1. Ultrafiltration membranes

In general, most UF membranes consist of a strong substructure (e.g., Tyvek®) upon which a very thin skin polymeric layer is attached. This thin layer essentially provides the desired permselective membrane properties and dictates the flow resistance. The technologies employed to generate the membranes for ultrafiltration have not changed over the past few decades. However, there have been significant improvements in the control of membrane pore size distributions, membrane morphology, and membrane modifications techniques. The three main casting technologies are air, immersion, and melt casting. They differ primarily in the desolvation methods and equipment (Zeman and Zydney, 1996). Current membrane compositions can accommodate multiple polymers that are grafted, blended, or applied in the form of copolymers to improve chemical and mechanical stability. The cellulose-type polymers offer low protein binding that is required by many biotechnology applications. However, cellulosic-type membranes generally degrade upon exposure to alkaline hypochlorite cleaning solutions. New cellulose composite membranes (e.g., Ultracel® from Millipore Corp.) offer low fouling and low protein binding for excellent product retention, recovery, and higher yields. These Ultracel membranes are constructed of regenerated cellulose membrane cast on a microporous substrate for defect-free membranes with superior robustness compared to conventional membranes. The composite technology offers a mechanically robust design able to withstand extreme operating conditions. In addition, polysulfone (PS) and polyethersulfone (PES) polymeric membranes have sufficient alkaline resistance for applications where NaOH cleaning is required, but are more susceptible to fouling by biologics-containing solutions. Surface modification has been proposed to be the most versatile method to reduce fouling on these membrane types. Membrane modification is performed using negatively charged monomers (Rong *et al.*, 2006) or by hydrophilization of polymeric membranes with neutral polymers. Several key membrane vendors such as Millipore, Pall Corporation, Gelman Sciences, Sartorius, and Sterlitech provide a wide variety of modified polymeric and ceramic membranes for use in both small- and large-scale biologic applications.

4.2. Ultrafiltration devices

As mentioned previously, current ultrafiltration membrane and related devices can be used for solute or buffer exchange and concentration of the desired protein. Most small-scale UF membrane devices are designed to operate only in a dead-end mode whereas larger scale UF systems employ the crossflow mode of operation (see Fig. 9.3).

Dead-end devices employ a feed and permeate stream, each with the same flow rate. There have been a number of modules developed specifically for dead-end applications. The foremost approaches include:

1. *Vacuum filtration*: Membrane disk is placed between a support plate and an open glass or plastic funnel. The feed solution is placed in the funnel and drawn through the membrane into a collection container by application of a vacuum.
2. *Syringe-end filters*: UF membranes are bonded to plastic support plate and sealed into a small-disk shaped unit equipped with luer connectors that

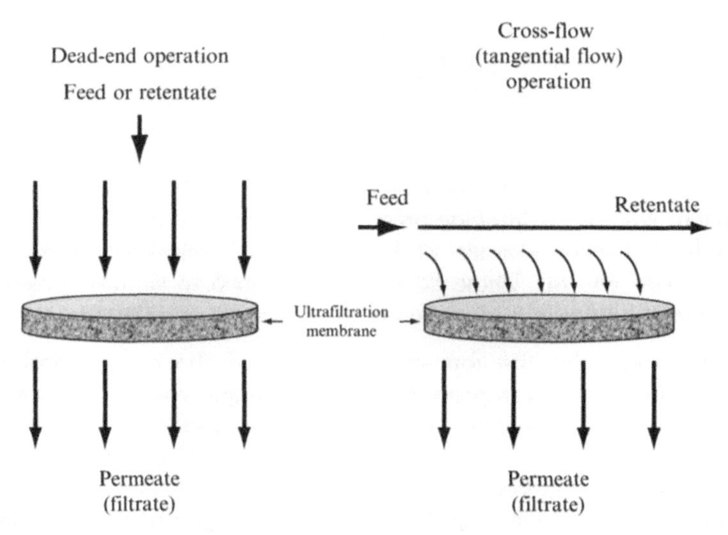

Figure 9.3 Schematic representation of dead-end and crossflow ultrafiltration operating modes (adapted from Zeman and Zydney, 1996).

can be attached to a syringe. The syringe is filled with sample and manually driven through the syringe filter to concentrate sample.

3. *Centrifugal filters*: A membrane disk is placed inside a small centrifuge tube and the permeate is collected in a space at the bottom of the tube beneath the membrane. The feed sample is driven across the UF filter by centrifugal force leaving concentrated sample above the membrane. Pall Corporation offers a wide range of centrifugal UF devices for processing samples < 1 ml up to 60 ml (Nanosep®, MicroSep™, MacroSep®, and JumboSep™) containing Omega™ modified PES membranes with MWCO ranging from 10 to 100 K. Millipore offers disposable Centriplus® centrifugal filter devices used for the concentration (100-fold) and desalting of 2–15 ml of sample through Amicon's low-adsorptive, hydrophilic, membranes. These devices are designed for use in most centrifuges that can accommodate 50 ml centrifuge tubes, with swinging-bucket or fixed-angle rotors. Finally, Millipore's Amicon Ultra-4 centrifuge filters are designed to process and recover extremely valuable biologics by accurately concentrating (80 to 100-fold) small sample volumes (1–4 ml).
4. *Multiwell filter plate*: UF membranes are incorporated into 96- or 384-multiwell filter plates for high-throughput concentration or buffer exchange of small samples (80–300 μl). Systems can be designed to contain membranes of various molecular weight cutoff (10–100 kDa) and provide greater than 90% recovery. The plates are constructed to eliminate lateral flow and mixing between wells and contain outlet tips and splash guards that ensure clean filtrate collection. Pall Corporation

offers the The AcroPrep™ filter plates (Pall Corporation, Meriden, CT) in 96- or 384-multiwell Filter Plate format.

5. *Stirred-cell device*: A flat sheet UF membrane disk is placed on an appropriate support plate and sealed in place at the bottom of a cylindrical housing. Fluid in the stirred device is agitated using a magnetic stirring bar suspended from the top of the housing with the device designed to operate with air to provide pressure driving force. Both Millipore and Sterlitech provide a variety of device sizes with sample volumes ranging from 3 to 400 ml. These devices are designed to simulate crossflow/ tangential filtration using different ultrafiltration membranes.

Even though the first four devices described above are probably the easiest to operate and can process very small sample volumes, they are not designed to provide precise control of concentration or concentration polarization on the membrane. In contrast, the stirred-cell device supports improved mass transport and concentration control due to its mixing capabilities and capability to perform buffer exchange and product concentration simultaneously. The process of buffer exchange using ultrafiltration is termed "diafiltration." In diafiltration, the exchange buffer is added to the retentate or feed during the filtration, with the UF membrane-permeating buffer species removed from the feed as the excess fluid is filtered through the membrane. In contrast to dialysis, in which differences in solute concentration across the membrane drive buffer exchange, the UF buffer-exchange process is faster due to convective flow of buffer across the membrane. This can be beneficial for processing proteins or biologics that are unstable and that require immediate buffer exchange. Diafiltration can be performed in one of two modes. In continuous diafiltration, the wash (exchange) buffer is added continuously throughout the filtration, with the system often designed so that the total feed volume remains constant during the process. In discontinuous diafiltration, the feed volume is first reduced by a predetermined amount using a standard ultrafiltration. The wash buffer is then added to the feed reservoir and the filtration continued. This process can be repeated to obtain desired final solute concentration and volume reduction.

4.3. Purification applications

The most recent advances in ultrafiltration membranes and systems technology have expanded the capability of UF from concentration and buffer-exchange applications to more complex biological separations. These new types of membrane-based "purification" applications have been driven by their low operating costs and ease of use compared to traditional bind-and-elute column chromatography. Recent work has shown that electrostatic interactions can be used to enhance traditional sized-based, membrane-based processes and the approach has been termed high-performance tangential-flow filtration (Sakena and Zydney, 1994; van Reis *et al.*, 1997). These membranes are designed to separate biological molecules of identical size but differing in isoelectric points (pI) by exploiting electrostatic interactions between the molecules and the membrane pores. In addition, there have been significant advances in the area of membrane chromatography (or affinity membrane chromatography) using membranes with larger pore distributions (microfiltration membranes) than ultrafiltration membranes. Membrane adsorbers have become accepted as an alternative to final polishing chromatography steps for removal of trace impurities (Ghosh, 2002). There are several commercial adsorbers available from Sartorius and Pall Corporation that serve as anion and cation exchangers which have designed to overcome the diffusion mass transfer limitations associated with bead-based chromatography. Typically, the charged ligand is immobilized in the membrane pores while convective flow delivers the solute molecules in close proximity to the ligand thus minimizing diffusional limitations. Nonetheless, the adoption of this technology has been slow because membrane chromatography has been limited by a lower binding capacity than that achieved by conventional chromatography resins even though the high-flux advantages provided by membrane adsorbers offer the potential for high productivity.

5. Lyophilization

Concentration of solutes in solution can be achieved thermodynamically by driving solvent into a gaseous head space (evaporation) or by boiling. Unfortunately, labile solutes such as protein molecules can be quickly degraded by the heat required to eliminate solvents. Since the boiling point of a solution is the temperature at which the vapor pressure of the solution equals the atmospheric pressure (headspace air pressure), solvents can boil by increasing solution temperature or by lowering the atmospheric pressure (e.g., applying vacuum). This latter technique is termed vacuum concentration. For the process of lyophilization the sample temperature is lowered to the point where the solution freezes and solvents are removed by sublimation. The freezing and vacuum step can be performed simultaneously when both heat and headspace gases are removed to concentrate the product of interest. The liquid protein product to be lyophilized typically contains the active biologic ingredient, a solvent system and several bulking and stabilizing agents. The bulking agents are used to provide support structure for the active molecule while the stabilizing agent will play a significant role in maintaining the activity of the target protein in liquid form prior to the lyophilization process and after sample reconstitution. Lyophilization generally consists of three major steps (Costantino and Pikal, 2004; Jennings, 1999; Przic *et al.*, 2004):

1. *Freezing*: The rate of freezing can affect final product quality. For example, slow freezing can cause ice crystals to form which enhances freeze-drying, but may denature labile proteins. Conversely, when sample freezing is performed rapidly small ice crystals form which can impede freeze-drying but protect protein stability.
2. *Primary drying*: During primary drying, greater than 80% of the solvent is directly converted from solid to vapor phase through sublimation. The residual solvent remains adsorbed on the product as moisture. Similar to freezing rates, drying rates can affect the structure and morphology of the lyophilized "cake" with rates designed to eliminate sample thaw.
3. *Secondary drying*: Residual solvent is desorbed during secondary drying to attain a moisture level low enough to inhibit microbial growth or chemical reactions, while still preserving the activity and integrity of the freeze-dried product. The product's physical properties determine the duration of secondary drying. For example, proteins generally require the presence of residual water to maintain structural integrity and biological activity. The requisite residual moisture is product dependent and must be determined empirically.

Figure 9.4 summarizes the key constituents and operations of a lyophilizer. Although lyophilizer design varies dependent on the operational scale, the key components and systems requirements do not change. Innovations over the last decade have improved the efficiency, cost-effectiveness, accuracy, and convenience of lyophilization equipment at all scales. These improvements have been augmented by computer-controlled automation. In particular, robotic sample loading/unloading systems have made the process more accurate by minimizing user handling. The application of enhanced real time in-process controls and monitoring such as SMART Freeze Dryer™ Temperature Monitoring technology (SP Industries, Warminister, PA) allows improved temperature control during critical freeze-drying cycles. Techniques such as nuclear magnetic resonance (NMR) weighing technology eliminates traditional sample weighing through noninvasive, noncontact check weigh technology.

In spite of these improvements, inherent limitations of the lyophilization process remain. One of the most difficult challenges is achieving the target moisture content in a specific product. While residual solute-bound water is removed from the product during secondary drying, overdrying can result in reduced product stability. Future freeze dry systems will incorporate systems for precise monitoring of water removal throughout a freeze-dry cycle using improved in-process mass flow meter technology.

Commensurate with the improved capabilities of current pilot and large-scale lyophilizers, an expansion of lab-scale systems has occurred

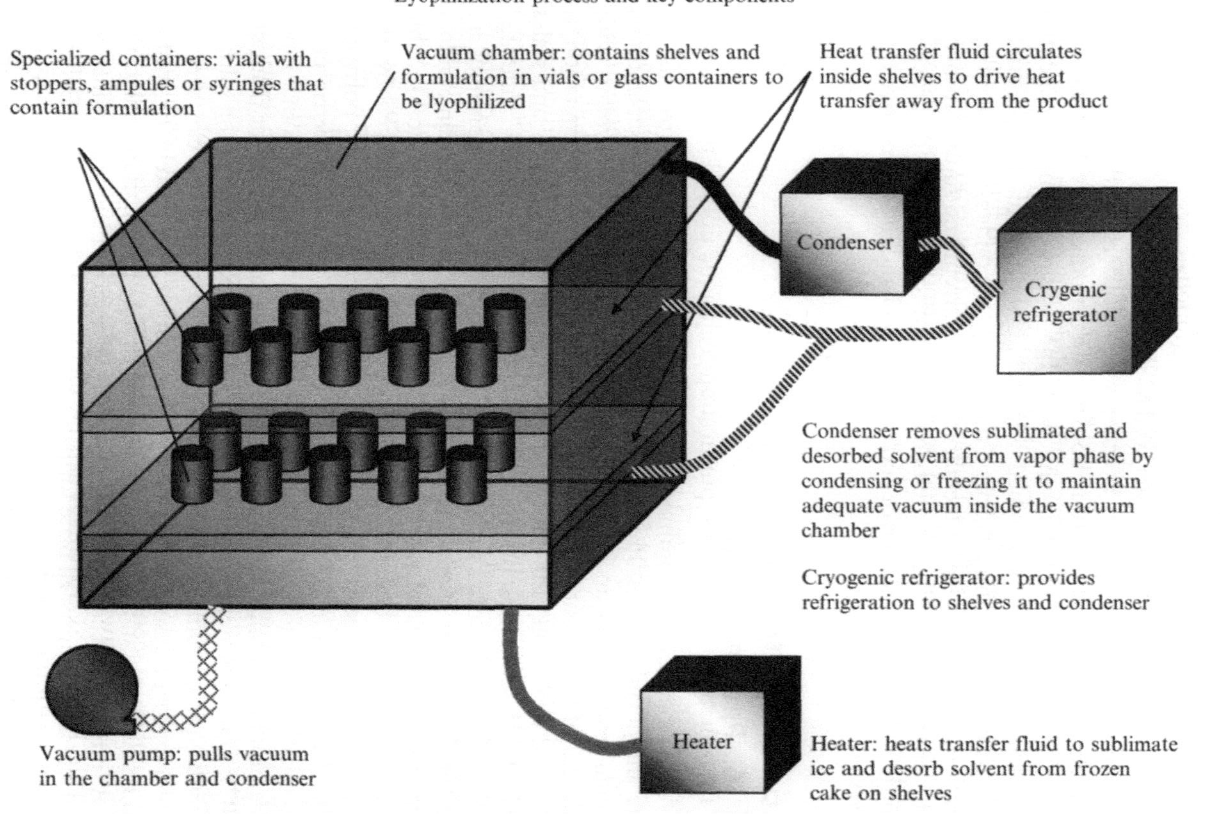

Lyophilization process and key components
Specialized containers: vials with stoppers, ampules or syringes that contain formulation
Vacuum chamber: contains shelves and formulation in vials or glass containers to be lyophilized
Heat transfer fluid circulates inside shelves to drive heat transfer away from the product
Condenser
Crygenic refrigerator
Condenser removes sublimated and desorbed solvent from vapor phase by condensing or freezing it to maintain adequate vacuum inside the vacuum chamber
Cryogenic refrigerator: provides refrigeration to shelves and condenser
Heater
Heater: heats transfer fluid to sublimate ice and desorb solvent from frozen cake on shelves
Vacuum pump: pulls vacuum in the chamber and condenser

Figure 9.4 Schematic of the lyophilization process and key system components.

permitting relatively small volumes to be processed. SP Industries, Labconco and Martin Christ corporations have specialized in the development of laboratory-scale systems. Novel benchtop systems provide a simple, economical device for freeze-drying via dry ice. For example, particular systems are equipped with a center well to accommodate dry ice and solvent and which can serve both as a water vapor collector and a convenient prefreezing bath. Flasks, serum bottles, and ampules may be frozen by dipping and rotating them in the well. In addition, compact benchtop freeze-dryers have been designed with rapid drying rate, an unrestricted vapor path for maximum efficiency, more efficient condensers, and multiple port manifold with valves. Overall, both the current and future lab-bench and commercial lyophilizers are being designed to include microprocessor control reducing operator error and minimizing handling to improve overall system performance.

6. Precipitation

Precipitation is a common purification technique utilized to concentrate and desalt proteins. Protein precipitation is the process of separating a protein from a solution as a solid by altering the protein solubility with addition of a reagent. A detailed discussion of protein precipitation is presented in Chapter 20. Precipitation is generally inexpensive and scales easily. In addition, most precipitations can often be performed in crude feed streams where impurities such as DNA, lipids, and contaminant proteins are present. There exist several methods or systems by which one can conduct precipitation. The challenge is to determine which method is most suitable to accomplish the desired goals. Precipitation may be induced by the addition of neutral salts, organic solvents, nonionic hydrophilic polymers, polyvalent metal ions, or by an acid or base to induce isoelectric point precipitation (Harrison *et al.*, 2003). The most important factors that influence protein solubility are structure, size, charge, and the solvent (Ladisch, 2007). Once a protein is precipitated, it can be separated from the solution by subsequent centrifugation or filtration. The precipitate can then be resolubilized in a smaller volume to concentrate the protein or washed and resolubilized in a different solution to change the solution conditions.

A number of commercially available filtration plates and tubes are available in 96-multiwell plate formats (Anachem, UK; Millipore, Bedford, MA; Pall, Meriden, CT) which can be useful in screening precipitation conditions with minimal resources. These plates can also enable precipitation techniques to be automated, further reducing resources. Crystallization screening kits, which contain ready-to-use solutions reflecting potential precipitation/crystallization conditions, can also be useful for exploring a broad range of buffers, pH, and precipitants (Hampton Research, Aliso Viejo, CA; Sigma-Aldrich, St Louis, MO).

Many precipitation methods, including precipitation with organic solvents and isoelectric precipitation can result in the denaturation of proteins or reduction of a protein's bioactivity. Although detrimental to the function of the protein, these precipitation methods are commonly used to eliminate components contained in biological fluids that can interfere with downstream applications and analyses. Several analytical techniques are routinely performed on denatured proteins and do not require a protein to be biologically active and others, such as SDS–PAGE, denature proteins as part of an analytical method. For these applications, precipitations that denature proteins are still appropriate. Some common precipitants utilized to concentrate proteins and remove interfering species for analytical analysis include ethanol, acetone, chloroform/methanol, and trichloroacetic acid.

Protein precipitation using acetone can be accomplished by adding a 4:1 ratio of cold acetone (– 20 °C) to a protein sample. The solution is then mixed, centrifuged, and the supernatant is discarded. This process may be repeated in order to achieve better removal of salts and other impurities. The precipitate is then dried and resolubilized in the desired solution. A similar procedure can be employed using a 9:1 volumetric ratio of cold ethanol (– 20 °C) to protein sample. An advantage of using ethanol is it is less flammable than acetone. Wessel and Flugge (1984) proposed a chloroform–methanol precipitation system with improved recoveries for dilute protein solutions.

TCA precipitation is considered a very efficient protocol for precipitating proteins from dilute solutions. An equal volume of 20% TCA is added to a protein sample and incubated on ice for 30 min. The solution is centrifuged at – 4 °C and decanted. The precipitate pellet is then washed with cold acetone (– 20 °C), centrifuged again at 4 °C, decanted, and dried. The precipitated pellet can then be dissolved in the desired buffer. Additional acetone washes of the precipitate or neutralization of the resolubilized sample may be necessary to reduce the acidity prior to further analysis. Sivaraman, *et al.* (1997) demonstrated that protein precipitation is optimized at a TCA concentration of roughly 15%.

In many cases, it is desirable to maintain the activity and structure of the protein after precipitation. This can be accomplished by neutral salt addition (salting out) or by the addition of nonionic hydrophilic polymer such as polyethylene glycol. Hofmeister (1887, 1890) first described the use of neutral salts to precipitate protein. The most commonly used salt is ammonium sulfate, as it is very water soluble and has no adverse effects upon the bioactivity of proteins. Ammonium sulfate precipitation is generally accomplished by adding a saturated ammonium sulfate solution (41.22 g/100 g water at 25 °C) slowly to a protein sample to a desired concentration. Protein solubilities differ greatly in ammonium sulfate so it may be beneficial to perform initial experiments to determine the ammonium sulfate concentration at which the protein of interest is precipitated from solution. This property may enable the ability to purify a protein from other proteins and contaminants through fractional precipitation. Hydrophilic polymers such as polyethylene glycol (PEG) can also be utilized to precipitate proteins while preserving bioactivity. Proteins are generally easier to precipitate using PEGs with molecular weights greater than 4000. Atha and Ingham (1981) demonstrate the precipitation of several proteins in the presence of PEG 4000 at concentrations of 3–30% (w/v).

7. Crystallization

Crystallization is another purification technique that can be used to concentrate and desalt proteins similarly to precipitation. Protein crystallization is a powerful separation technology because it simultaneously concentrates, purifies, and stabilizes the product (Etzel, 2006). Crystallization is a controlled precipitation from an aqueous solution with four main variables that control crystal morphology and recovery: protein concentration, precipitant concentration, pH, and temperature. It is similar to precipitation in that solid particles are formed from solution; however, precipitates have poorly defined morphology and are characterized by small particle size, whereas crystals are highly ordered with generally larger particle sizes. High-throughput screening systems are being developed in multiwell plate and smaller formats that, combined with robotic liquid handling and novel analysis methods, are enabling the rapid screening of broad ranges of conditions with nanoliter amounts (Brown *et al.*, 2003). In addition, applications for microfluidic chips have been developed for biomolecule crystallization (Teragenics, Watertown, MA). In practice, it is usually more difficult to crystallize than precipitate a protein and for purposes of concentrating and desalting a protein, precipitation is generally a preferable method. In addition, as crystallization processes are scaled up from the lab bench, mixing dynamics can significantly impact robustness.

REFERENCES

Atha, D. H., and Ingham, K. C. (1981). Mechanism of precipitation of proteins by polyethylene glycols. *J. Biol. Chem.* **256,** 12108–12117.

Booker, C. J., and Yeung, K. K. C. (2008). In-capillary protein enrichment and removal of nonbuffering salts using capillary electrophoresis with discontinuous buffers. *Anal. Chem.* **80,** 8598–8604.

Brown, J., Walter, T. S., Carter, L., Abrescia, N. G. A, Aricescu, A. R., Batuwangala, T. D., Bird, L. E., Brown, N., Chamberlain, P. P., and Davis, S. (2003). A procedure for setting up high-throughput nanolitre crystallization experiments. II. Crystallization results. *J. Appl. Crystallogr.* **36,** 315–318.

Cheryan, M. (1986). Ultrafiltration Handbook. Technomic, Lancaster, PA.

Costantino, H. R., and Pikal, M. J. (2004). Lyophilization of Biopharmaceuticals. AAPS Press.

Cussler, E. L. (1997). Diffusion Mass Transfer in Fluid Systems: Mass Transfer in Fluid Systems. 2nd edn. Cambridge University Press.

Denisov, G. A. (1994). Theory of concentration polarization in cross-flow ultrafiltration: Gel-layer model and osmotic–pressure model. *J. Membr. Sci.* **91,** 173–187.

Etzel, M. (2006). Bulk protein crystallization—Principles and methods. *In* "Process Scale Bioseparations for the Biopharmaceutical Industry" (A. Shukla, M. Etzel, and S. Gadam, eds.). Taylor & Francis, Boca Raton, FL.

Ghosh, R. (2002). Protein separation using membrane chromatography: Opportunities and challenges. *J. Chromatogr. A* **952,** 13–27.

Harrison, R. G., *et al.* (2003). Bioseparations Science and Engineering. Oxford University Press, New York, NY.

Hsieh, H.-C., Sheu, C., Shi, F. K., and Li, D.-T. (2007). Development of a titanium dioxide nanoparticle pipette-tip for the selective enrichment of phosphorylated peptides. *J. Chromatogr. A* **1165,** 128–135.

Hofmeister, F. (1887). Zur Lehre von der Wirkung der Sake. *Arch. Exp. Path. Pharm.* **24,** 247.

Hofmeister, F. (1890). Uber die Darstellung von krystallisirtem Eiralbumin und die Krystallisirbarkeit colloider Stoffe. *Z. Physiol. Chem.* **14,** 165.

Jennings, T. (1999). Lyophilization: Introduction and Basic Principles. Interpharm Press.

Jia, W., Wu, H., Lu, H., Li, N., Zhang, Y., Cai, R., and Yang, P. (2007). Rapid and automatic on-plate desalting protocol for MALDI-MS: Using imprinted hydrophobic polymer template. *Proteomics* **7,** 2497–2506.

Ladisch, M. (2007). Bioseparations Engineering: Principles, Practice, and Economics. John Wiley & Sons Inc.

Przic, D. S., Ruzic, D. S., Nenad, N. L., and Petrovic, S. D. (2004). Lyophilization—The process and industrial use. *Chem. Ind.* **58,** 552.

Rong, W., Zhansheng, L., and Fane, A. G. (2006). Development of a novel electrophoresis-UV grafting technique to modify PES UF membranes used for NOM removal. *J. Membr. Sci.* **273,** 47–57.

Sakena, S., and Zydney, A. L. (1994). Effect of solution pH and ionic strength on the separation of albumin from immunoglobulins (IgG) by selective filtration. *Biotechnol. Bioeng.* **43,** 960–968.

Schley, C., Swart, R., and Huber, C. G. (2006). Capillary scale monolithic trap column for desalting and preconcentration of peptides and proteins in one- and two-dimensional separations. *J. Chromatogr. A* **1136,** 210–220.

Shen, W., Xiong, H., Xu, Y., Cai, S., Lu, H., and Yang, P. (2008). ZnO-poly(methyl methacrylate) nanobeads for enriching and desalting low-abundant proteins followed by directly MALDI-TOF MS analysis. *Anal. Chem.* **80,** 6758–6763.

Sheen, H., and Ali-Khan, Z. (2005). Protein sample concentration by repeated loading onto SDS-PAGE. *Anal. Biochem.* **343,** 338–340.

Sivaraman, T., *et al.* (1997). The mechanism of 2, 2, 2-trichloroacetic acid-induced protein precipitation. *J. Protein Chem.* **16,** 291–297.

Shihabi, Z. K. (2002). Transient pseudo-isotachophoresis for sample concentration in capillary electrophoresis. *Electrophoresis* **23,** 1612–1617.

van Reis, R., Gadam, S., Frautschy, L. N., Orlando, S., Goodrich, E. M., Saksena, S., Kuriyel, R., Simpson, C. M., Pearl, S., and Zydney, A. L. (1997). High performance tangential flow filtration. *Biotechnol. Bioeng* **56,** 71–82.

Waters, N. J., *et al.* (2008). Validation of a rapid equilibrium dialysis approach for the measurement of plasma protein binding. *J. Pharm. Sci.* **97**(10), 4586–4595.

Waziri, S. M., Abu-Sharkh, B. F., and Ali, S. A. (2004). Protein partitioning in aqueous two-phase systems composed of a pH-responsive copolymer and poly(ethylene glycol). *Biotechnol. Prog.* **20,** 526–532.

Wessel, D., and Flugge, U. I. (1984). A method for the quantitative recovery of protein in dilute solution in the presence of detergents and lipids. *Anal. Biochem.* **138,** 141–143.

Wheat, T. E., Rainville, P. R., Gillece-Castro, B. L., Lu, Z., Cravello, L., and Mazzeo, J. R. (2007). Fast on-line desalting of proteins for determination of structural information using exact mass spectrometry. *Bioprocess. J.* **6,** 55–59.

Yu, H., Lu, Y., Zhou, Y.-G., Wang, F.-B., He, F.-Y., and Xia, X.-H. (2008). A disposable microfluidic device for rapid protein concentration and purification via direct-printing. *Lab Chip* **8,** 1496–1501.

Zeman, L., and Zydney, A. (1996). Microfiltration and Ultrafiltration. *Marcel Dekker Inc.*

CHAPTER TEN

Maintaining Protein Stability

Murray P. Deutscher

Contents

1. Causes of Protein Inactivation 121
2. General Handling Procedures 122
3. Concentration and Solvent Conditions 122
4. Stability Trials and Storage Conditions 123
5. Proteolysis and Protease Inhibitors 124
6. Loss of Activity 125

Abstract

Proteins are fragile molecules that often require great care during purification to ensure that they remain intact and fully active. Nowadays, many proteins are also purified in small amounts under denaturing conditions by various gel electrophoretic techniques, such that inactive proteins are obtained. But even here, it is usually advantageous to maintain the protein in an intact form. In the case of enzymes, and other proteins with assayable biological activities, maintenance of activity is generally of prime importance, both for following the protein during purification and for subsequent studies of function. This chapter will focus on the major points to keep in mind with regard to maintaining the stability of a protein during purification and storage. Various other chapters describe in detail stabilization procedures for specific biological systems and specific classes of proteins.

1. Causes of Protein Inactivation

Removal of proteins from the cellular environment subjects them to a variety of conditions and processes that can lead to loss of activity or alteration of structure. These include dilution, change in solution conditions, exposure to degradative enzymes, oxygen, heavy metals, and surfaces, and change in physical condition (e.g., freezing and thawing). Awareness that any one of these situations could affect the protein, and knowledge of

Department of Biochemistry and Molecular Biology, University of Miami School of Medicine, Miami, Florida, USA

Methods in Enzymology, Volume 463
ISSN 0076-6879, DOI: 10.1016/S0076-6879(09)63010-X

precautions that could be taken to minimize their effect, will go a long way toward ensuring a successful purification protocol. If the protein of interest is lost or inactivated during the course of any procedure, determination of the reason for this loss will often suggest a simple solution. Thus, if possible, examine whether the loss of activity is accompanied by loss of the protein or changes in its structure, or whether the protein remains, but is now inactive. Distinguishing among these different possibilities might indicate what type of process is behind the problem and, thus, what an appropriate solution might be.

2. General Handling Procedures

Obviously, to maintain the stability of a protein, avoid treatments that will denature it. Thus, protein solutions should generally not be stirred vigorously or vortexed since this may lead to oxidation or surface denaturation. Protein solutions should not be exposed to extremes of pH, high temperatures, organic solvents, or any other condition that might promote denaturation. Likewise, if one is storing a protein solution for extended periods of time in an unfrozen state, bacterial and fungal growth can become a problem. In these situations sterile solutions and antibacterial or antifungal agents may be necessary. Finally, it is best to make up all solutions that will come in contact with the protein with glass–distilled water, and to store the water in containers that do not have algal growth.

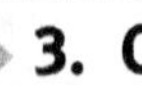

3. Concentration and Solvent Conditions

Extraction of proteins from cells inevitably leads to a change in their environment. Since proteins are generally stable *in vivo*, the theoretical goal is to try to reproduce the cellular milieu as closely as possible. This would mean very high protein concentrations, close to neutral pH, moderate ionic strengths, reducing conditions, etc. In practice, some of these conditions are compatible with protein purification, and some are not. As a first approximation, it is generally a good practice to keep the protein concentration high (>1 mg/ml). This would help to maintain protein complexes, possibly minimize the effects of deleterious contaminants and surface adsorption, and provide a general stabilizing environment for the protein of interest. It is relatively easy to maintain high protein concentrations early in a purification scheme, but this becomes more difficult as the protein is purified unless one resorts to concentration procedures after each step. Since these latter procedures often have their own problems, one may have to settle for more dilute solutions unless particular stability problems become obvious. It may be helpful to alternate purification steps between ones that concentrate proteins with ones that dilute them. For example, elution of proteins from a column to which they are bound using a batchwise procedure will tend to concentrate the eluted proteins, whereas gradient elution will tend to dilute them. Columns to which proteins bind will tend to concentrate, whereas gel filtration will dilute. By judicious arrangement of purification steps, one may be able to avoid extensive dilution.

The solution conditions are also extremely important. Although it is not possible to describe a universal stabilizing solvent applicable to every protein, the addition of certain components is generally helpful. These include a buffer, usually around neutrality, to avoid unnecessary pH changes. Careful attention should be given to the buffer anion since in many cases Cl^- may be harmful. EDTA is usually added at about 0.1 m*M* to chelate heavy metal ions that could interact with the protein or promote oxidation. A reducing agent such as 2-mercaptoethanol or dithiothreitol is often present to counteract oxidative effects, particularly of cysteine residues. The use of dithiothreitol at about 0.1–1 m*M* is preferred because it does not form mixed disulfides with proteins, as 2-mercaptoethanol does. Sufficient reducing agent should be present since it can oxidize relatively rapidly and lead to protein oxidation. In some cases, salts are also added to maintain a certain ionic strength, but only if they are compatible with the next purification or analytical step. Likewise, glycerol at 5–20% often helps to maintain stability and is compatible with most purification steps at these concentrations. Occasionally, low levels of a nonionic detergent are added to prevent aggregation or the adsorption of proteins to surfaces, such as glassware. Finally, it is good practice to include protease inhibitors, particularly at early steps.

4. Stability Trials and Storage Conditions

One of the most important studies that can be performed during the course of a new protein purification is a stability and storage study. What this means is that after every step of the purification procedure the stability and storage properties of the protein of interest should be determined. Although rapid purification of a protein is desirable, the situation will often arise, especially during a new purification, when it becomes necessary to keep a protein for some length of time prior to the next step. For this purpose you will need to know how stable it is under different storage conditions. The simplest way to test this is to take small portions of the protein solution, store them under a variety of conditions (e.g., in ice, frozen, at room temperature, with and without different stabilizing agents), and then assay the activity of the protein after different periods of time. Again, keep in mind what the next step in the purification procedure will be. Some storage conditions may be fine for stability, but not useful for further purification. A case in point is storage at − 20 °C in 50% glycerol (v/v). This is often a useful condition for maintaining stability, but terrible if one plans further purification. Sometimes, it may be necessary to use such a procedure (the glycerol could be removed by dialysis), but generally it should be a last resort.

A different situation arises when one has completed a purification procedure and wants to store the purified protein for long periods of time. Here, the primary concern is long-term stability, and many conditions that might be impractical during the course of purification could be used. These might include addition of high concentrations of glycerol, addition of stabilizing substrates, even addition of an extraneous protein such as serum albumin. The choice of storage condition depends on what is effective for stabilization, and for what the purified protein will be used. If one is primarily interested in studying enzyme activity, the presence of serum albumin may not matter. In contrast, one would not want an extraneous protein present if structural studies are planned. If one is unsure, the best course may be to store portions of the protein under a variety of conditions.

Related to the question of storage is the problem of freezing and thawing solutions of purified proteins. One way to avoid repeated cycles of freezing and thawing is to store the purified protein in small portions and to thaw individual samples once, as needed; alternatively, the protein may be stored under conditions in which it does not freeze, such as 50% glycerol at − 20 °C. If repeated freezing and thawing is necessary, it is best to use quick freezing and thawing procedures. During freezing, solutes are concentrated and the protein could be exposed to unusually harsh conditions. We routinely quick freeze protein solutions in dry ice–ethanol baths to avoid this problem. Likewise, in thawing protein solutions, this should be done rapidly with gentle mixing in lukewarm water until only a small amount of ice is present; the solution is then placed in ice or kept at room temperature during use. The final thawed solution should be mixed gently, or inverted if in a tube, to ensure even distribution of solutes.

5. Proteolysis and Protease Inhibitors

Proteolysis is a major problem for the purification of proteins. It is a particularly insidious problem because in many cases the protein of interest is only partially degraded and may retain biological activity. This can result in erroneous conclusions about the size and structure of the protein. Proteolysis can be a problem at any stage of a purification procedure. Total proteolytic activity is generally highest in the initial crude extract since purification will tend to eliminate these contaminating activities; however, there are also more proteins present in the extract that could act to protect the protein of

Table 10.1 Common protease inhibitors

Protease inhibitor	Protease class inhibited	Concentration used
PMSF (phenylmethylsulfonyl fluoride)	Serine proteases	0.1–1 m*M*
EDTA and EGTA	Metalloproteases	0.1–1 m*M*
Benzamidine	Serine proteases	~1 m*M*
Pepstatin A	Acid proteases	~1 μg/ml
Leupeptin	Thiol proteases	~1 μg/ml
Aprotinin	Serine proteases	~5 μg/ml
Antipain	Thiol proteases	~1 μg/ml

interest. As purification proceeds, even a small contamination with a protease could have a large effect because a larger fraction of the available protein substrate will be the one with which you are working.

How can you tell if proteolysis is a problem in your particular situation? The simplest test is to incubate the extract or partially purified protein at a moderate temperature (e.g., 30 °C), withdrawing portions at intervals, and assaying for biological activity. Although this method is not foolproof because there may be other reasons for loss of activity, most proteins will not be heat inactivated under these conditions. If activity is lost, the addition of protease inhibitors is recommended since even if proteins are kept at 0–4 °C throughout purification, some cleavages will occur unless the proteases are inactivated.

Cells contain a variety of different classes of proteases. Fortunately, protease inhibitors are available that can act on the various proteases. A list of some commonly used inhibitors is presented in Table 10.1. Protease inhibitors useful for particular systems or situations are described in other chapters of this volume. Probably the best approach in working with a new protein is to use a mixture of inhibitors that affect different classes of proteases. Such mixtures are commercially available. Once conditions for maintaining the protein of interest in a stable form are obtained, inhibitors can be removed one by one to determine which are really necessary. Some trial and error will be involved, as well as for deciding which inhibitors, if any, are needed as the purification proceeds. Note that protease inhibitors can be toxic and/or unstable under certain conditions. They should not be used without first learning their properties.

6. Loss of Activity

The most commonly heard lament during a protein purification is, "I've lost my activity." When this happens, a careful analysis of the situation is required to determine the cause. Most importantly, one should have a careful accounting of enzyme units to evaluate the extent of the activity loss. For many purification steps, percentage losses of as much as 50% are not unusual, but of course, these vary with each individual protein. Generally, purification methods that involve binding of a protein to a matrix, and which may require conformational changes during binding, have a greater effect on activity than a procedure such as gel filtration.

If activity is totally lost during a particular purification step, other possibilities need to be considered. In some cases proteins may bind very tightly to columns, and require more extreme procedures for elution. Depending on the type of chromatography (see Sections VI and VII of this volume), this may require increased ionic strength, use of a chaotropic salt (e.g., KBr), or inclusion of detergent or ethylene glycol in the elution buffer.

A second possibility is that more than one component is required for the activity of the protein, and these components (e.g., another subunit or a cofactor) are separated during the fractionation step. Thus, either component by itself would be inactive or all have to be present to observe activity. To test for this possibility, all the fractions from the previous step are mixed back together, and activity measured. In some cases, it may be necessary to concentrate the mixture back to the original volume in order to observe activity. If mixing of all the fractions results in the appearance of activity, one could then test fractions or groups of fractions in a pairwise fashion. Often it is found that a little activity remains and that the second component is needed for optimal activity. In this case, one of the required fractions is already known, and the other fractions can then be tested for their stimulating activity.

Sometimes activity may be lost between purification steps, such as during dialysis or concentration, or even during storage. In the former situations, one should again test for removal of a possible required component. The possibility also exists that the protein has stuck to the dialysis tubing or the concentration membrane. Here, washing the tubing or membrane with buffer containing some detergent may be helpful. Problems of stability during storage have been discussed above.

The most frustrating situation is if none of the above possibilities is the cause of the loss of activity. Under these circumstances the most likely explanation is actual inactivation of the protein due to denaturation, proteolysis, etc. If an independent measure for the protein is available (e.g., a Western blot), this can be shown directly. If not, an answer to the inactivation problem may require trial and error experiments to test various conditions. Sometimes, the best solution is simply to avoid that particular purification step.

Micropurification (purification of very small amounts of protein in the submilligram range) presents a special set of problems. In particular, one has to be especially mindful of the possibility of protein loss due to adsorption on surfaces. Small amounts of protein bind tightly to surfaces such as glass and many plastics, often on the order of 1 $\mu g/cm^2$. During micropurification procedures in which the amount of protein present is small to begin with, these adsorptive losses can represent a significant percentage of the protein present. Certainly, one should avoid containers made of polystyrene since this material (e.g., a microtiter dish) has a high protein binding capacity. In these situations, tubes should be carefully rinsed to ensure maximum recovery of protein. In addition, the presence of low concentrations of a nonionic detergent may help to prevent binding of protein to the surface. Careful accounting of protein throughout the procedure is important and may alert the investigator to the occurrence of unwanted protein loss.

SECTION THREE

RECOMBINANT PROTEIN EXPRESSION AND PURIFICATION

CHAPTER ELEVEN

Selecting an Appropriate Method for Expressing a Recombinant Protein

William H. Brondyk

Contents

1. Introduction 132
2. *Escherichia coli* 133
   2.1. *E. coli*: Temperature and molecular chaperones 133
   2.2. *E. coli*: Fusion partners 134
   2.3. *E. coli*: Disulfide bond formation 134
   2.4. *E. coli*: Posttranslational modifications 134
3. *Pichia pastoris* 135
4. Baculovirus/Insect Cells 136
5. Mammalian Cells 138
6. Protein Characteristics 139
   6.1. Protein characteristics: *E. coli* and codon usage 141
   6.2. Protein characteristics: Cytoplasmic proteins 141
   6.3. Protein characteristics: Secreted proteins 142
   6.4. Protein characteristics: Membrane proteins 142
   6.5. Protein characteristics: Toxic proteins 142
7. Recombinant Protein Applications 143
8. Conclusion 144
References 144

Abstract

Recombinant proteins are important tools for studying biological processes. Generating a recombinant protein requires the use of an expression system. Selection of an appropriate expression system is dependent on the characteristics and intended application of the recombinant protein and is essential to produce sufficient quantities of the protein. Over the last 30 years, there have been considerable advances in the technologies for expressing recombinant proteins. In this chapter the unique characteristics of four commonly used expression systems, *Escherichia coli*, *Pichia pastoris*, baculovirus/insect cell, and mammalian cells are described. The *E. coli* system is a rapid method for

Genzyme Corporation, Framingham, Massachusetts, USA

Methods in Enzymology, Volume 463
ISSN 0076-6879, DOI: 10.1016/S0076-6879(09)63011-1

expressing proteins but lacks many of the posttranslational modifications found in eukaryotes. The capacity of *E. coli* for protein folding and forming disulfide bonds is not sufficient for many recombinant proteins although there are a number of tools developed to overcome these limitations. In contrast to *E. coli*, the eukaryotic *P. pastoris*, baculovirus/insect cell, and mammalian systems promote good protein folding and many posttranslational modifications. How the characteristics and the downstream application of a recombinant protein can influence the choice of an expression system is then reviewed.

1. Introduction

Choosing an appropriate method for expressing a recombinant protein is a critical factor in obtaining the desired yields and quality of a recombinant protein in a timely fashion. Selecting a wrong expression host can result in the protein being misfolded or poorly expressed, lacking the necessary posttranslational modifications or containing inappropriate modifications. Factors to consider when selecting an expression system include the mass of the protein and number of disulfide bonds, type of posttranslational modifications desired on the expressed protein, and the destination of the expressed protein. The intended application of the purified recombinant protein is also critical in the decision-making process and the applications can be categorized into four broad areas: structural studies, *in vitro* activity assays, antigens for antibody generation, and *in vivo* studies. The purpose of this chapter is to help guide the investigator in the decision-making process for choosing an appropriate expression system. However, even with the described guidelines there are many circumstances when it is not obvious *a priori* which expression system is the best choice, and the use of multiple expressions systems must be attempted before an optimal system is identified.

Numerous expression systems are currently being used in academic and industrial settings. Some of these systems are too new and insufficiently tested to comment on their utility. In addition, some established systems for expressing recombinant proteins, such as transgenic animals, are too technically challenging, time consuming and prohibitively expensive to be a viable option for the average laboratory. For the purpose of this chapter, only *Escherichia coli*, *Pichia pastoris*, baculovirus/insect cell, and mammalian expression systems will be considered (for more detailed coverage of these systems, see Chapters 12–15). These four systems have straightforward protocols, are readily accessible either from colleagues or from research product companies (e.g., Invitrogen, EMD-Novagen, Stratagene, and Promega), and are relatively inexpensive for small-scale production. The characteristics and available options of these expression systems will be briefly reviewed with the focus on the differences between the systems. Strategies will then be presented to help guide the investigator in making the best choice for an expression system.

2. *Escherichia coli*

The bacteria *E. coli* was the first host used to express recombinant proteins and is still considered to be the workhorse in the field. Using the *E. coli* system offers a rapid and simple method for expressing recombinant proteins due to its short doubling time. Consequently, the assessment of recombinant gene expression in *E. coli* can take less than a week. The growth media for *E. coli* are inexpensive and there are relatively straightforward methods to scale-up bioproduction (see Chapter 12).

In *E. coli*, recombinant proteins are normally either directed to the cytoplasm or to the periplasm and, to a lesser extent, secreted. Proteins directed to the cytoplasm are the most efficiently expressed, giving yields of up to 30% of the biomass (Jana and Deb, 2005). However, the high expression of recombinant protein can often lead to the accumulation of aggregated, insoluble protein that forms inclusions bodies. Inclusion bodies have been observed not only with eukaryotic proteins but also to a lesser extent with overexpressed proteins from prokaryotes including *E. coli*. The rate of translation and folding in *E. coli* is almost 10-fold higher than that observed in eukaryotic cells, and this presumably contributes to the inclusion body formation of eukaryotic proteins (Andersson *et al.*, 1982; Goustin and Wilt, 1982). Inclusion bodies can be a significant hindrance in obtaining soluble, active protein in some situations. However, in some cases, inclusion bodies are advantageous because they are resistant to proteolysis, easy to concentrate by centrifugation, minimally contaminated with other proteins, and, with some effort, able to be refolded to form active, soluble proteins (see Chapter 17 for details).

2.1. *E. coli*: Temperature and molecular chaperones

Several methods have been described for maximizing the formation of soluble, properly folded proteins in the cytoplasm and minimizing inclusion body formation. The most straightforward method involves lowering the temperature to 15–30 °C during the expression period (Sahdev *et al.*, 2008). Presumably, the reduced temperature slows the rate of transcription, translation, and refolding, thereby allowing for proper folding (Vera *et al.*, 2006). In addition, lower temperature has been shown to decrease heat shock protease activity (Spiess *et al.*, 1999). Some investigators have coexpressed molecular chaperones in the cytoplasm along with the recombinant protein for promoting protein solubility (Young *et al.*, 2004). The utility of this approach appears to be quite protein-specific, and therefore needs to be tested individually for each recombinant protein of interest (Baneyx and Mujacic, 2004).

2.2. *E. coli*: Fusion partners

Alternatively, a method that promotes solubility with many proteins is to fuse the recombinant protein at either the N-terminus or C-terminus to a soluble fusion tag (Esposito and Chatterjee, 2006). Fusion partners that have been shown to increase solubility of recombinant proteins include glutathione-*S*-transferase (GST), thioredoxin, maltose-binding protein (MBP), small ubiquitin-modifier (SUMO), and N-utilization substrate (NusA). Both GST and MBP have the added advantage of also being an affinity purification tag. Unfortunately, no single tag appears to work for all recombinant proteins, and multiple fusion partners may need to be evaluated for promoting soluble expression. Fusion tags can be removed from the recombinant protein by several strategies, and a widespread approach involves adding a protease site between the fusion partner and the recombinant protein that can be cleaved with the specific protease. This approach must be carefully tested since removal of the fusion tag can, in some cases, render the recombinant protein insoluble (Waugh, 2005).

2.3. *E. coli*: Disulfide bond formation

E. coli is normally inefficient in promoting the correct formation of disulfide bonds when recombinant proteins are expressed in the cytoplasm; normally disulfide bond formation occurs only in the periplasm where it is catalyzed by the Dsb system (Andersen *et al.*, 1997; Bardwell, 1994). Consequently, if disulfide bond formation is needed, the recombinant protein can be directed to the periplasm via a cleavable signal peptide (e.g., pelB). However, a major disadvantage of periplasmic expression is the significant reduction in production yields. Through engineering of the *E. coli* genome, a more suitable environment for disulfide bond generation in the cytoplasm can be induced by disrupting the thioredoxin reductase (*trxB*) and glutathione reductase (*gor*) genes in the Dsb system which in turn enables thioredoxin and glutaredoxin to promote cytoplasmic reduction of cysteines (Bessette *et al.*, 1999). These engineered strains are commercially available through EMD-Novagen (Origami). If additional disulfide bond formation is still needed, the recombinant protein can be fused to thioredoxin, and the fusion protein expressed in a $trxB^{-}/gor^{-}$ *E. coli* strain (LaVallie *et al.*, 1993).

2.4. *E. coli*: Posttranslational modifications

Finally, it is important to recognize that *E. coli* has a limited capacity for posttranslational modifications compared to eukaryotic organisms. For example, *E. coli* does not support enzyme-mediated N-linked glycosylation, O-linked glycosylation, amidation, hydroxylation, myristoylation, palmitation, or sulfation.

3. *PICHIA PASTORIS*

Yeast is another traditional, powerful tool for expressing recombinant proteins and has been used successfully to express a multitude of proteins. Yeast has many of the advantageous features of *E. coli* such as a short doubling time and a readily manipulated genome, but also has the additional benefits of a eukaryote that includes improved folding and most posttranslational modifications. The first yeast routinely used for recombinant protein expression was *Saccharomyces cerevisiae* (Strausberg and Strausberg, 2001). However, in the last 15 years, *P. pastoris* has become the yeast of choice because it typically permits higher levels of recombinant protein expression than does *S. cerevisiae* (see Chapter 13). *P. pastoris* is a methyltropic yeast, and can use methanol as its only carbon source (Cregg *et al.*, 1985). The growth of *P. pastoris* in methanol-containing medium results in the dramatic transcriptional induction of the genes for alcohol oxidase (AOX) and dihydroxyacetone synthase (Cregg, 2007). After induction, these proteins comprise up to 30% of the *P. pastoris* biomass. Investigators have exploited this methanol-dependent gene induction by incorporating the strong, yet tightly regulated, promoter of the alcohol oxidase I (*AOX1*) gene into the majority of vectors for expressing recombinant proteins (Daly and Hearn, 2005). The *P. pastoris* expression vectors integrate in the genome whereas by contrast, *S. cerevisiae* vectors use the more unstable method of replicating episomally. The length of time to assess recombinant gene expression with the *P. pastoris* method is approximately 3–4 weeks which includes the transformation of yeast, screening the transformants for integration, and an expression time-course. An appealing feature of *P. pastoris* is the extremely high cell densities achievable under appropriate culture conditions (Cregg, 2007). Using inexpensive medium, the *P. pastoris* culture can reach 120 g/l of dry cell weight density. An important caveat is that the induction medium requires a low percentage of methanol. In large-scale cultures, the amount of methanol becomes a fire hazard requiring a new level of safety conditions.

P. pastoris has been used to obtain both intracellular and secreted recombinant proteins. Like other eukaryotes, it efficiently generates disulfide bonds and has successfully been used to express proteins containing many disulfide bonds. To facilitate secretion, the recombinant protein must be engineered to carry a signal sequence. The most commonly used signal sequence is the pre-pro sequence from *S. cerevisiae* α-mating factor (Daly and Hearn, 2005). Because *P. pastoris* secretes few endogenous proteins, purification of the recombinant protein from the medium is a relatively simple task. If proteolysis of the recombinant protein is a concern, expression can be completed using the *pep4* protease-deficient strain of *P. pastoris*

(Gleeson *et al.*, 1998). This strain has reduced vacuole peptidase A activity which is responsible for activation of carboxypeptidase Y and protease B1.

Yeast has the posttranslational capacity to add glycans at both specific asparagine residues (N-linked) and serine/threonine residues (O-linked). These glycan structures are substantially different from the modifications added by insect and mammalian cells. In *P. pastoris* the N-linked glycan is a high mannose type and usually contains 8–17 mannoses, which is quite different from *S. cerevisiae* structures that consist of approximately 50–150 mannose residues (Celik *et al.*, 2007; Gemmill and Trimble, 1999). Similar to insect and mammalian cells, the consensus sequence for N-linked glycans in yeast is Asn-Xaa-Ser/Thr. Two groups have completed extensive engineering to create *P. pastoris* strains that produce complex N-linked glycan structures comparable to those produced by mammalian cells (Hamilton and Gerngross, 2007). However, only the strains developed by Roland Contreras' group are available to investigators and must be licensed through Research Corporation Technologies. The O-linked structures in *P. pastoris* have not been studied comprehensively but are known to be formed by the addition of one to four mannose residues to serines/threonines (Goto, 2007; Trimble *et al.*, 2004). Several reports have indicated that expression of certain proteins in *P. pastoris* resulted in the addition O-linked glycans not observed when the protein was expressed endogenously in mammalian cells (Daly and Hearn, 2005).

4. Baculovirus/Insect Cells

Baculovirus-mediated expression in insect cells offers another useful tool for generating recombinant proteins (see Chapter 14). Baculovirus is a lytic, large (130 kb), double-stranded DNA virus, and the *Autographa californica* virus is the most commonly used baculovirus isolate for recombinant expression. Baculovirus is routinely amplified in insect cell lines derived from the fall armyworm *Spodoptera frugiperda* (*Sf*9, *Sf*21), and recombinant protein expression is completed either in the aforementioned lines or in a line derived from the cabbage looper *Trichoplusia ni* (High-Five) (Kost *et al.*, 2005). Originally, creating recombinant baculoviruses involved cotransfecting the gene of interest flanked by baculovirus sequence with baculovirus DNA into insect cells, and screening for rare homologous recombination events. Recombinants were identified by screening plaques with a modified morphology, and often additional rounds of plaque screening were required to ensure that the recombinant viral preparation was not contaminated with wild-type virus. This lengthy and laborious process for generating recombinant viruses has been largely replaced by using site-specific transposition (Bac-to-Bac or BaculoDirect, Invitrogen) or an improved homologous recombination method with an engineered baculovirus containing a lethal mutation in orf1629 (flashBAC from Oxford Expression Technologies or BacMagic from EMD-Novagen). Both of these approaches overcome the requirement to isolate plaques because the efficiency of recombination is 100%. Following one or two rounds of amplifying the recombinant baculovirus, the investigator can quantify the baculovirus concentration stock either by the plaque assay or by using the newer, more rapid real-time PCR or antibody-based assays (Hitchman *et al.*, 2007; Kwon *et al.*, 2002). The improvements in creating and quantifying recombinant baculoviruses have dramatically reduced the time for evaluating baculovirus expression to approximately 3 weeks, including a time-course study for optimizing expression.

The most common promoters used with baculovirus expression are the polH and p10 promoters, both of which induce a high level of expression in the very late phase of the baculovirus infection (Ikonomou *et al.*, 2003). During this phase, cells undergo cell death with the concomitant release of proteases, which can result in degradation of the expressed recombinant protein. To reduce proteolysis of the recombinant protein, promoters active in earlier phases of the lytic cycle such as the basic promoter have been used (Ikonomou *et al.*, 2003). Alternatively proteolytic activity can be minimized by using constructs deleted in the *chi*A and *v-cath* genes, which encode chitinase and a cathepsin protease, respectively (Monsma and Scott, 1997).

Baculovirus-mediated expression is routinely used to generate both cytoplasmic and secreted recombinant proteins. Efficient secretion generally requires the presence of a signal peptide. Both insect and mammalian signal sequences can promote entry into the insect cell secretory pathway. Insect cells were originally grown in serum-containing medium which complicated purification of the secreted proteins. Recent advances in media development permit the replacement of serum with protein hydrolysates derived from either animal tissues or plants, thereby greatly simplifying protein purification (Ikonomou *et al.*, 2003). However, the high cost of this specialized media can limit its use for large-scale bioproduction.

Insect cells efficiently generate disulfide bonds in recombinant proteins. They also produce the majority of the posttranslational modifications found in mammalian cells. However, the N-linked glycan structure formed in most insect cells is the predominantly fucosylated paucimannose structures ($Man_3GlcNAc_2$-*N*-Asn) (Harrison and Jarvis, 2006; Jenkins *et al.*, 1996). This finding has prompted the recent generation of insect cell lines that produce glycoproteins with the complex N-linked glycans normally found in mammalian cells (Harrison and Jarvis, 2006). A transgenic Sf-9 insect line expressing several glycosyltransferases is commercially available (Mimic cell line, Invitrogen) and produces N-linked glycans containing a biantennary, sialylated structure. There are only a few reports describing the O-linked glycans structures generated by insect cells (Chen *et al.*, 1991; Sugyiama *et al.*, 1993; Thomsen *et al.*, 2004).

5. Mammalian Cells

Mammalian expression methods have conventionally been considered to be the least efficient vehicle for expressing recombinant proteins. However, recent advances have significantly improved the expression levels from mammalian cell lines (see Chapter 15). For example, stably transfected Chinese hamster ovary (CHO) cells have been reported to express recombinant antibodies up to a level of a few grams per liter (Figueroa *et al.*, 2007; Wurm, 2004). While many cell lines and expression strategies have been tested, this chapter will focus on transient transfection in human embryonic kidney (HEK293) cells and stable transfection with CHO cells.

The HEK293 cell line was derived from human embryonic kidney cells transformed with adenovirus. HEK293 cells can be transiently transfected with a high efficiency (>80%) using certain cationic lipids, calcium phosphate, or polyethyleneimine as transfection reagents (Durocher *et al.*, 2002; Jordan *et al.*, 1996). For large-scale transient transfections (>100 ml), calcium phosphate or polyethyleneimine reagents are more cost-effective options when compared to cationic lipids (Baldi *et al.*, 2007). Transient transfections have been performed at even the bioreactor level but for most laboratories this scale is technically challenging (Girard *et al.*, 2002). The transient transfection method is relatively easy, and the evaluation for a given recombinant protein can be made in less than 2 weeks.

CHO cells are commonly used for mammalian expression when large quantities of recombinant protein are needed. For example, most therapeutic antibodies currently on the market are manufactured using this method. The standard method for stable CHO expression involves transfecting dihydrofolate reductase (DHFR)-deficient CHO cells with a DHFR selection cassette along with an expression cassette containing the gene of interest (Wurm, 2004). Dihydrofolate reductase converts dihydrofolate into tetrahydrofolate which is required for the *de novo* synthesis of purines, certain amino acids, and thymidylic acid. Methotrexate, which binds and inhibits DHFR, is used as a selection agent and only those cells that have integrated the DHFR selection cassette will survive. Sequentially increasing the concentration of methotrexate will result in amplification of the *DHFR* gene along with the linked gene of interest. Following at least one round of selection with the drug methotrexate, the stably transfected pools are subcloned using limiting dilution cloning into multiwell plates. Typically only a small percentage of the screened subclones will be expressing the recombinant gene at a high level since in the majority of the clones, the expression cassette has integrated into the heterochromatin region which is transcriptionally inactive. Unfortunately, the entire selection and screening process takes at least 2–3 months, making this the major drawback of the CHO method. However, recent high-throughput methods based on flow cytometry or automation have increased the ease in rapidly screening and selecting high expressing clones (Browne and Al-Rubeai, 2007). Another development has been to use specific *cis*-acting DNA elements flanking the recombinant gene cassette that confer active transcription to integration sites (Kwaks and Otte, 2006). Unfortunately, the majority of these DNA elements is owned by companies and must be licensed for use in the laboratory, and, even with the aforementioned advances in CHO expression, the timelines for generating a high expressing CHO clone have not changed considerably.

Mammalian expression systems are used primarily to generate secreted rather than intracellular recombinant proteins. Serum-free media have been developed for both the CHO and HEK293 cell lines, which simplifies the purification of secreted recombinant proteins. However, the cost of the media is quite high, making large-scale bioproduction rather costly.

Mammalian cells contain the most superior folding and disulfide bond formation when compared to other expression hosts. The N-linked and O-linked glycan structures formed by mammalian cells are extremely varied and are not only dependent on the protein but also on the mammalian cell type used as the expression host (Jenkins *et al.*, 1996). Furthermore, the cell culture conditions such as nutrient content, pH, temperature, oxygen levels and ammonia concentration can significantly affect the glycosylation profile (Butler, 2006). N-linked glycosylation can result in oligomannose, hybrid, and complex structures, and the structures all contain the $Man_3GlcNac_2$ core (Bhatia and Mukhopadhyay, 1998). The oligomannose glycans can have two to six additional mannoses and the mannoses can be phosphorylated or sulfated. The most common complex structures have two to four Gal β1,4-$GlcNac_2$ attached to the mannoses which result in bi-, tri-, and tetra-antennary branches. The branches can terminate with sialic acid, and fucose can also be attached to the structures. Hybrid structures contain features of both the oligomannose and complex structures. *O*-Glycosylation structures can be classified into eight types based on their core structures: *O*-GalNAc-type glycosylation, *O*-GlcNAc-type glycosylation, *O*-fucosylation, *O*-mannosylation, *O*-glucosylation, phosphoglycosylation, *O*-glycosaminoglycan-type glycosylation, and collagen-type glycosylation (Peter-Katalinic, 2005).

6. Protein Characteristics

When choosing an expression system, one can easily survey the literature to determine if the recombinant protein has previously been expressed in the past and then assess the success of the published strategy.

Table 11.1 Summary of expression methods

Expression systems	Advantages	Disadvantages
E. coli	Rapid expression method (days)	Limited capacity for posttranslational modifications
	Inexpensive bioproduction media and high density biomass	Difficult to produce some proteins in a soluble, properly folded state
	Simple process scale-up	
	Well characterized genetics	
P. pastoris	Moderately rapid expression method (weeks)	N-linked glycan structures different from mammalian forms
	Inexpensive bioproduction media and high density biomass	Enhanced safety precautions needed for large-scale bioproduction due to methanol in induction media
	Most posttranslational modifications and high folding capacity	
Baculovirus/ insect cell	Moderately rapid expression method (weeks)	N-linked glycan structures different from mammalian forms
	Most posttranslational modifications and high folding capacity	Low density biomass and expensive bioproduction media
		Difficult process scale-up
Mammalian—transient expression	Moderately rapid expression method (weeks)	Low density biomass and expensive bioproduction media
	All posttranslational modifications and high folding capacity	Difficult process scale-up
Mammalian—stable expression	All posttranslational modifications and high folding capacity	Lengthy expression method (months)
		Low density biomass and expensive bioproduction media
		Difficult process scale-up

In drawing from reports in the literature, it is important to consider whether the downstream application of the recombinant protein in the report is similar or compatible with your intended application. Lacking that information, reports about orthologs that have been expressed and described in detail can prove useful. The number of proteins being expressed has increased dramatically over the last decade which is in part due to the sequencing of many genomes, the development of high-throughput expression methods, and the large Protein Structure Initiative (PSI). This expression trend will likely continue, which will mean that eventually the amount of prior data will make the choice of an expression host a more straightforward decision.

6.1. Protein characteristics: *E. coli* and codon usage

In general, when considering the recombinant protein-host system options, prokaryotic proteins should be expressed with only *E. coli* and not with eukaryotic systems since usually the posttranslational modifications and improved folding found in eukaryotes are either not necessary or perhaps not desired for a prokaryotic protein. For eukaryotic proteins the circumstance is different since there is a preponderance of examples where eukaryotic proteins are successfully expressed in *E. coli* (Sahdev *et al.*, 2008). When using this strategy, an important consideration is that *E. coli*, like all organisms, has a bias for codon usage and the abundance of the tRNA pools in *E. coli* mirror this bias (Gustafsson *et al.*, 2004; Marin, 2008). Expressing a eukaryotic protein containing several codons that are rare in *E. coli* can be inefficient when the pool of corresponding tRNAs is limiting. The shortage of tRNAs can lead to frameshifts in translation, misincorporation of amino acids, or premature termination of translation. This problem is most evident when rare codons are grouped together at the N-terminus (Kane, 1995). However, the problem can be avoided by synthesizing a codon-optimized gene or by using engineered *E. coli* strains containing increased pools of rare tRNAs, which are available commercially (e.g., Rosetta strains, EMD-Novagen). In most cases, codon bias should also be corrected in those circumstances where the recombinant gene is being expressed in a phylogenetically distant organism.

6.2. Protein characteristics: Cytoplasmic proteins

For a cytoplasmic protein, the optimal choice of an expression system depends on the protein mass and the number of disulfide bonds in the protein. For proteins between 10 and 50 kDa and containing few disulfide bonds, *E. coli* is a good option for soluble protein expression (Dyson *et al.*, 2004). For larger proteins or those with many disulfide bonds, if soluble expression is desired then usually either baculovirus or yeast is the preferred choices. Successful expression of proteins smaller than 10 kDa, with few or no disulfide bonds, has been achieved through fusion with soluble tags and expression in *E. coli* (Esposito and Chatterjee, 2006). Alternatively, expression of these small proteins can be directed into the secretory pathway of *P. pastoris* (Daly and Hearn, 2005). However, in this pathway, care must be taken to monitor potential inadvertent glycosylation of the normally cytosolic protein when it is forced into the secretory pathway. This can be achieved by inspecting the sequence for the lack of the consensus N-linked glycosylation sites. Unfortunately, for O-linked glycosylation there are no consensus sequences, so the secreted recombinant protein must be analyzed to ensure the lack of O-linked glycans.

6.3. Protein characteristics: Secreted proteins

Any of the expression hosts can be used to produce secreted proteins. However, as described earlier, *E. coli* lacks most of the posttranslational capabilities found in eukaryotic hosts. Consequently, *E. coli* may be suboptimal for expressing secreted eukaryotic proteins but this is highly dependent on the downstream application.

6.4. Protein characteristics: Membrane proteins

Membrane proteins represent an extremely challenging class of proteins to express in large quantities. For some purposes, production of just the soluble, hydrophilic portion will suffice, in which case the membrane-spanning domain can be removed and the desired soluble portion can be expressed. There are no clear guidelines on choosing the best system to express intact membrane proteins (Sarramegna *et al.*, 2003). However, for most eukaryotic membrane proteins *E. coli* as an expression method is generally not a good option because of its limited capacity for folding and posttranslational modifications. By contrast, researchers have reported modest success with expressing G protein-coupled receptors using the baculovirus and yeast methods (Sarramegna *et al.*, 2003).

6.5. Protein characteristics: Toxic proteins

Recombinant proteins that are toxic to the expression host can be challenging to produce but this obstacle can usually be overcome. If the recombinant protein is toxic, it is often useful to determine whether the problem is host cell specific. If so, then the protein can be expressed in a more compatible expression host. Another option is to use a tightly regulated, inducible expression system such as those available for *E. coli* and *P. pastoris*. For example, several elaborate inducible expression systems have been developed for *E. coli* (Saida, 2007). In these systems, expression of the recombinant gene is regulated by an inducible promoter, transcription terminators, control of the plasmid copy number, or modification of the coding sequence of the recombinant gene. In the available *P. pastoris* system, the AOX1 promoter is tightly regulated by the combination of an induction mechanism as well as a repression/derepression method (Daly and Hearn, 2005). Alternatively, several studies have demonstrated that the baculovirus/insect system can be used to express toxic proteins (Aguiar *et al.*, 2006; Korth and Levings, 1993). Lastly, with mammalian systems the easiest option is to use transient expression. The several inducible systems compatible with CHO cells take a considerable amount of time to complete the necessary cell engineering and also with these systems it is difficult to obtain tight regulation of gene expression which is required to prevent cell death (Rossi and Blau, 1998).

7. Recombinant Protein Applications

Recombinant protein expression for structural studies requires proper folding, correct formation of disulfide bonds and homogeneity of the recombinant product. The intrinsic cellular capacity for folding and disulfide bond formation has been described for each of the expression systems. Potential sources of heterogeneity include phosphorylation, inefficient cleavage of the initiator methionine by methionine aminopeptidase, and glycosylation. Unfortunately, phosphorylation heterogeneity is often observed with recombinant kinases and has been reported for each of the expression hosts. In these cases, homogeneity can be attained by removing the phosphates on the recombinant protein through treatment with phosphatase. N-terminal methionine heterogeneity can occur when the recombinant protein is expressed in the cytoplasm. Prokaryotic and eukaryotic organisms contain methionine aminopeptidases, and the efficiency of methionine cleavage is influenced by the amino acid adjacent to the initiator methionine (Giglione *et al.*, 2004). In general the size of the side chain of the adjacent amino acid inversely affects the efficiency of methionine cleavage. However, in *E. coli* the high expression level of recombinant proteins can saturate the enzyme and alter the rules for efficient cleavage (Dong *et al.*, 1996). This heterogeneity can be avoided by carefully selecting the amino acid adjacent to the initiator methionine or by using a cleavable tag at the N-terminus. Glycosylation heterogeneity occurs with glycoproteins expressed in all eukaryotic organisms and normally less heterogeneity is observed with insect and yeast hosts (Jenkins *et al.*, 1996). Regardless of the expression host, the glycans are usually removed before attempting to crystallize the recombinant protein.

Normally, when generating recombinant proteins for use as immunizing antigens, any of the expression hosts can be used for this application. With glycoproteins it is not always clear whether the presence of glycans will alter the immunogenicity of the recombinant antigen (Bhatia and Mukhopadhyay, 1998; Prasad *et al.*, 1995).

Producing recombinant proteins suitable for *in vitro* activity studies as well as for *in vivo* experiments requires appropriate protein folding and disulfide bond formation. For glycoproteins, the presence of N-linked glycans along with the glycan structure can have a significant impact on both applications, and therefore must be considered when selecting a eukaryotic expression host. N-linked glycosylation has been shown to positively influence protein structure and increase protein stability (Bhatia and Mukhopadhyay, 1998). *In vitro*, the structure of the N-linked glycan on certain protein ligands has been demonstrated to affect the affinity for receptor binding and signal transduction. With recombinant immunoglobulins, the conserved N-linked glycan present in the Fc region influences *in vitro* effector activity (Presta, 2008). For instance, the presence of fucose on the N-linked glycan in human IgG1 reduces antibody-dependent cell cytotoxicity activity (Shinkawa *et al.*, 2003). *In vivo*, the N-linked glycan structure on proteins dramatically impacts metabolic clearance and biodistribution. For example, N-linked glycans that are not capped with sialic acid are cleared by hepatic receptors, including the asialoglycoprotein and mannose receptors (Weigel and Yik, 2002). The importance of O-linked glycosylation has not yet been defined. If the effect of glycosylating the recombinant protein is unknown then a safer strategy involves choosing an expression host similar to the source of the recombinant gene.

8. Conclusion

The *E. coli*, *P. pastoris*, baculovirus/insect cells and mammalian systems each have both advantages and disadvantages for expressing recombinant proteins (Table 11.1). Whether a given system will express a protein at a high level and generate a quality product is largely protein dependent. A careful evaluation of the characteristics of the recombinant protein along with the downstream application must be considered when selecting an expression method. Unfortunately, there will be circumstances when the expression choice will not be obvious and several expression hosts must be evaluated.

References

Aguiar, R. W., Martins, E. S., Valicente, F. H., Carneiro, N. P., Batista, A. C., Melatti, V. M., Monnerat, R. G., and Ribeiro, B. M. (2006). A recombinant truncated Cry1Ca protein is toxic to lepidopteran insects and forms large cuboidal crystals in insect cells. *Curr. Microbiol.* **53,** 287–292.

Andersen, C. L., Matthey-Dupraz, A., Missiakas, D., and Raina, S. (1997). A new *Escherichia coli* gene, dsbG, encodes a periplasmic protein involved in disulphide bond formation, required for recycling DsbA/DsbB and DsbC redox proteins. *Mol. Microbiol.* **26,** 121–132.

Andersson, D. I., Boham, K., Isaksson, L. A., and Kurland, C. G. (1982). Translation rates and misreading characteristics of rpsD mutants in *Escherichia coli. Mol. Gen. Genet.* **187,** 467–472.

Baldi, L., Hacker, D. L., Adam, M., and Wurm, F. M. (2007). Recombinant protein production by large-scale transient gene expression in mammalian cells: state of the art and future perspectives. *Biotechnol. Lett.* **29,** 677–684.

Baneyx, F., and Mujacic, M. (2004). Recombinant protein folding and misfolding in *Escherichia coli. Nat. Biotechnol.* **22,** 1399–1408.

Bardwell, J. C. (1994). Building bridges: Disulphide bond formation in the cell. *Mol. Microbiol.* **14,** 199–205.

Bessette, P. H., Aslund, F., Beckwith, J., and Georgiou, G. (1999). Efficient folding of proteins with multiple disulfide bonds in the *Escherichia coli* cytoplasm. *Proc. Natl. Acad. Sci. USA* **96,** 13703–13708.

Bhatia, P. K., and Mukhopadhyay, A. (1998). Protein glycosylation: Implications for *in vivo* functions and therapeutic applications. *Adv. Biochem. Eng. Biotechnol.* **64,** 155–201.

Browne, S. M., and Al-Rubeai, M. (2007). Selection methods for high-producing mammalian cell lines. *Trends Biotechnol.* **25,** 425–432.

Butler, M. (2006). Optimisation of the cellular metabolism of glycosylation for recombinant proteins produced by mammalian cell systems. *Cytotechnology* **50,** 57–76.

Celik, E., Calik, P., Halloran, S. M., and Oliver, S. G. (2007). Production of recombinant human erythropoietin from *Pichia pastoris* and its structural analysis. *J. Appl. Microbiol.* **103,** 2084–2094.

Chen, W., Shen, Q., and Bahl, O. P. (1991). Carbohydrate variant of the recombinant β-subunit of human choriogonadotropin expressed in baculovirus expression system. *J. Biol. Chem.* **266,** 4081–4087.

Cregg, J. M. (2007). Introduction: Distinctions between *Pichia pastoris* and other expression systems. *Methods Mol. Biol.* **389,** 1–10.

Cregg, J. M., Barringer, K. J., Hessler, A. Y., and Madden, K. R. (1985). *Pichia pastoris* as a host system for transformation. *Mol. Cell. Biol.* **5,** 3376–3385.

Daly, R., and Hearn, M. T. W. (2005). Expression of heterologous proteins in *Pichia pastoris*: A useful experimental tool in protein engineering and production. *J. Mol. Recognit.* **18,** 119–138.

Dong, M. S., Bell, L. C., Guo, Z., Phillips, D. R., Blair, I. A., and Guengerich, F. P. (1996). Identification of retained N-formylmethionine in bacterial recombinant mammalian cytochrome P450 proteins with the N-terminal sequence MALLLAVFL...: Roles of residues 3–5 in retention and membrane topology. *Biochemistry* **35,** 10031–10041.

Durocher, Y., Perret, S., and Kamen, A. (2002). High-level and high-throughput recombinant protein production by transient transfection of suspension-growing human 293-EBNA1 cells. *Nucleic Acids Res.* **30,** E9.

Dyson, M. R., Shadbolt, S. P., Vincent, K. J., Perera, R. L., and McCafferty, J. (2004). Production of soluble mammalian proteins in *Escherichia coli*: Identification of protein features that correlate with successful expression. *BMC Biotechnol.* **4,** 32–49.

Esposito, D., and Chatterjee, D. K. (2006). Enhancement of soluble protein expression through the use of fusion tags. *Curr. Opin. Biotechnol.* **17,** 353–358.

Figueroa, B., Ailor, E., Osborne, D., Hardwick, J. M., Reff, M., and Betenbaugh, M. J. (2007). Enhanced cell culture performance using inducible anti-apoptotic genes E1B-19 K and Aven in the production of a monoclonal antibody with Chinese Hamster Ovary cells. *Biotechnol. Bioeng.* **97,** 877–892.

Gemmill, T. R., and Trimble, R. B. (1999). Overview of *N*- and *O*-linked oligosaccharide structures found in various yeast species. *Biochim. Biophys. Acta* **1426,** 227–237.
Giglione, C., Boularot, A., and Meinnel, T. (2004). Review: Protein N-terminal methionine excision. *Cell. Mol. Life Sci.* **61,** 1455–1474.
Girard, P., Derousazi, M., Baumgartner, G., Bourgeois, M., Jordan, M., Jacko, B., and Wurm, F. M. (2002). 100-liter transient transfection. *Cytotechnology* **38,** 15–21.
Gleeson, M. A., White, C. E., Meininger, D. P., and Komives, E. A. (1998). Generation of protease-deficient strains and their use in heterologous protein expression. *Methods Mol. Biol.* **103,** 81–94.
Goto, M. (2007). Protein *O*-glycosylation in fungi: Diverse structures and multiple functions. *Biosci. Biotechnol. Biochem.* **71,** 1415–1427.
Goustin, A. S., and Wilt, F. H. (1982). Direct measurement of histone peptide elongation rate in cleaving sea urchin embryos. *Biochim. Biophys. Acta* **699,** 22–27.
Gustafsson, C., Govindarajan, S., and Minshull, J. (2004). Codon bias and heterologous protein expression. *Trends Biotechnol.* **22,** 346–353.
Hamilton, S. R., and Gerngross, T. U. (2007). Glycosylation engineering in yeast: The advent of fully humanized yeast. *Curr. Opin. Biotechnol.* **18,** 387–392.
Harrison, R. L., and Jarvis, D. L. (2006). Protein *N*-glycosylation in the baculovirus-insect cell expression system and engineering of insect cells to produce "mammalianized" recombinant glycoproteins. *Adv. Virus Res.* **68,** 159–191.
Hitchman, R. B., Siaterli, E. A., Nixon, C. P., and King, L. A. (2007). Quantitative real-time PCR for rapid and accurate titration of recombinant baculovirus particles. *Biotechnol. Bioeng.* **96,** 810–814.
Ikonomou, L., Schneider, Y., and Agathos, S. N. (2003). Insect cell culture for industrial production of recombinant proteins. *Appl. Microbiol. Biotechnol.* **62,** 1–20.
Jana, S., and Deb, J. K. (2005). Strategies for efficient production of heterologous proteins in *Escherichia coli*. *Appl. Microbiol. Biotechnol.* **67,** 289–298.
Jenkins, N., Parekh, R. B., and James, D. C. (1996). Getting the glycosylation right: implications for the biotechnology industry. *Nat. Biotechnol.* **14,** 975–981.
Jordan, M., Schallhorn, A., and Wurm, F. M. (1996). Transfecting mammalian cells: Optimization of critical parameters affecting calcium-phosphate precipitate formation. *Nucleic Acids Res.* **24,** 596–601.
Kane, J. F. (1995). Effects of rare codon clusters on high-level expression of heterologous proteins in *Escherichia coli*. *Curr. Opin. Biotechnol.* **6,** 494–500.
Korth, K. L., and Levings, C. S. (1993). Baculovirus expression of the maize mitochondrial protein URF13 confers insecticidal activity in cell cultures and larvae. *Proc. Natl. Acad. Sci. USA* **90,** 3388–3392.
Kost, T. A., Condreay, J. P., and Jarvis, D. L. (2005). Baculovirus as versatile vectors for protein expression in insect and mammalian cells. *Nat. Biotechnol.* **23,** 567–575.
Kwaks, T. H. J., and Otte, A. P. (2006). Employing epigenetics to augment the expression of therapeutic proteins in mammalian cells. *Trends Biotechnol.* **24,** 137–142.
Kwon, M. S., Dojima, T., Toriyama, M., and Park, E. Y. (2002). Development of an antibody-based assay for determination of baculovirus titers in 10 hours. *Biotechnol. Prog.* **18,** 647–651.
LaVallie, E. R., DiBlasio, E. A., Kovacic, S., Grant, K. L., Schendel, P. F., and McCoy, J. M. (1993). A thioredoxin gene fusion expression system that circumvents inclusion body formation in the *E. coli* cytoplasm. *Biotechnology (NY)* **11,** 187–193.
Marin, M. (2008). Folding at the rhythm of the rare codon beat. *Biotechnol. J.* **3,** 1047–1057.
Monsma, S. A., and Scott, M. (1997). BacVector-3000: An engineered baculovirus designed for greater protein stability. *Innovations* 16–19.
Peter-Katalinic, J. (2005). O-glycosylation of proteins. *Methods Enzymol.* **405,** 139–171.

Prasad, S. V., Mujtaba, S., Lee, V. H., and Dunbar, B. S. (1995). Immunogenicity enhancement of recombinant rabbit 55-kilodalton zona pellucida protein expressed using the baculovirus expression system. *Biol. Reprod.* **52,** 1167–1178.

Presta, L. G. (2008). Molecular engineering and design of therapeutic antibodies. *Curr. Opin. Immunol.* **20,** 460–470.

Rossi, F. M., and Blau, H. M. (1998). Recent advances in inducible gene expression systems. *Curr. Opin. Biotechnol.* **9,** 451–456.

Sahdev, S., Khattar, S. K., and Saini, K. S. (2008). Production of active eukaryotic proteins through bacterial expression systems: A review of existing biotechnology strategies. *Mol. Cell Biochem.* **307,** 249–264.

Saida, F. (2007). Overview on the expression of toxic gene products in *Escherichia coli*. *Curr. Protoc. Protein Sci.* 5.19.1–5.19.13.

Sarramegna, V., Talmont, F., Demange, P., and Milon, A. (2003). Heterologous expression of G-protein-coupled receptors: Comparison of expression systems from the standpoint of large-scale production and purification. *Cell. Mol. Life Sci.* **60,** 1529–1546.

Shinkawa, T., Nakamura, K., Yamane, N., Shoji-Hosaka, E., Kanda, Y., Sakurada, M., Uchida, K., Anazawa, H., Satoh, M., Yamasaki, M., Hanai, N., and Shitara, K. (2003). The absence of fucose but not the presence of galactose or bisecting *N*-acetylglucosamine of human IgG1 complex-type oligosaccharides shows the critical role of enhancing antibody-dependent cellular cytotoxicity. *J. Biol. Chem.* **278,** 3466–3473.

Spiess, C., Beil, A., and Ehrmann, M. (1999). A temperature-dependent switch from chaperone to protease in a widely conserved heat shock protein. *Cell* **97,** 339–347.

Strausberg, R. L., and Strausberg, S. L. (2001). Overview of protein expression in *Saccharomyces cerevisiae*. *Curr. Protoc. Protein Sci.* **Chapter 5,** Unit 5.6.

Sugyiama, K., Ahorn, H., Maurer-Fogy, I., and Voss, T. (1993). Expression of human interferon-α2 in Sf9 cells: Characterization of O-linked glycosylation and protein heterogeneities. *Eur. J. Biochem.* **217,** 921–927.

Thomsen, D. R., Post, L. E., and Elhammer, A. P. (2004). Structure of O-glycosidically linked oligosaccharides synthesized by the insect cell line Sf9. *J. Cell. Biochem.* **43,** 67–79.

Trimble, R. B., Lubowski, C., Hauer, C. R., Stack, R., McNaughton, L., Gemmill, T. R., and Kumar, S. A. (2004). Characterization of N- and O-linked glycosylation of recombinant human bile salt-stimulated lipase secreted by *Pichia pastoris*. *Glycobiology* **14,** 265–274.

Vera, A., Gonzalez-Montalban, N., Aris, A., and Villaverde, A. (2006). The conformational quality of insoluble recombinant proteins is enhanced at low growth temperatures. *Biotechnol. Bioeng.* **96,** 1101–1106.

Waugh, D. S. (2005). Making the most of affinity tags. *Trends Biotechnol.* **23,** 316–320.

Weigel, P. H., and Yik, J. H. N. (2002). Glycans as endocytosis signals: The cases of the asialoglycoprotein and hyaluronan/chondroitin sulfate receptors. *Biochim. Biophys. Acta* **1372,** 341–363.

Wurm, F. M. (2004). Production of recombinant protein therapeutics in cultivated mammalian cells. *Nature Biotechnol.* **22,** 1393–1398.

Young, J. C., Agashe, V. R., Siegers, K., and Hartl, P. U. (2004). Pathways of chaperone-mediated protein folding in the cytosol. *Nat. Rev. Mol. Cell Biol.* **5,** 781–791.

CHAPTER TWELVE

Bacterial Systems for Production of Heterologous Proteins

Sarah Zerbs, Ashley M. Frank, *and* Frank R. Collart

Contents

1. Introduction 150
2. Heterologous Protein Production Using *Escherichia coli* 150
3. Planning a Bacterial Expression Project 151
4. Evaluation of Project Requirements 152
5. Target Analysis 152
6. Cloning 153
7. Preparation of T4 DNA Polymerase-Treated DNA Fragments 155
8. Expression in the *E. coli* Cytoplasm 156
9. Expression of Cytoplasmic Targets in *E. coli* 157
    9.1. Analysis of protein expression and solubility 157
10. Analysis of Heterologous Protein Expression in *E. coli* 157
    10.1. Analysis of protein solubility 158
    10.2. Analysis of protein expression results 159
    10.3. Autoinduction method for protein expression 159
11. Small-Scale Expression Cultures in Autoinduction Media Protocol 160
12. Periplasmic Expression of Proteins 160
13. Expression of Periplasmic Targets in *E. coli* 161
14. Small-Scale Osmotic Shock Protocol 162
15. Alternative Bacterial Systems for Heterologous Protein Production 164
16. Alternative Vector and Induction Conditions 165
17. Production Scale 166
Acknowledgment 166
References 166

Abstract

Proteins are the working molecules of all biological systems and participate in a majority of cellular chemical reactions and biological processes. Knowledge of the properties and function of these molecules is central to an understanding of chemical and biological processes. In this context, purified proteins are a starting point for biophysical and biochemical characterization methods that can assist in

Biosciences Division, Argonne National Laboratory, Lemont, Illinois, USA

Methods in Enzymology, Volume 463
ISSN 0076-6879, DOI: 10.1016/S0076-6879(09)63012-3

the elucidation of function. The challenge for production of proteins at the scale and quality required for experimental, therapeutic and commercial applications has led to the development of a diverse set of methods for heterologous protein production. Bacterial expression systems are commonly used for protein production as these systems provide an economical route for protein production and require minimal technical expertise to establish a laboratory protein production system.

1. Introduction

Bacterial expression systems and *Escherichia coli* in particular are frequently used for production of heterologous proteins at laboratory and industrial process scales (Baneyx, 1999; Terpe, 2006). The widespread use of bacterial systems for protein production arises primarily from the nominal cost and minimal technical requirements for implementation in a laboratory scale environment. A variety of vector and host options are available and the short doubling time of most engineered strains enables rapid evaluation of experimental outcome and reduces the stringency of requirements for sterile technique and facilities. Many of these intrinsic advantages at the laboratory scale can be easily modified to accommodate automation of the process and implementation in a high throughput setting (Dieckman *et al.*, 2002). In most cases, the bacterial expression systems commonly used for benchtop scale processing can also be adapted to large-scale projects that require production and screening of hundreds of expression clones or large-scale production of selected proteins (Baneyx, 1999; Klock *et al.*, 2005; Terpe, 2006; Yokoyama, 2003).

In spite of these advantages, bacterial systems have a number of important limitations for expression of heterologous proteins that should be considered in the development of a protein production expression strategy. These limitations are especially apparent for eukaryotic proteins (Dyson *et al.*, 2004) or proteins that require the coexpression of maturation proteins (Londer *et al.*, 2008). This is not unexpected since the biological and chemical characteristics of the bacterial cellular compartments differ from eukaryotic organisms. In particular, accessory proteins such as chaperones, posttranslational modification proteins, or maturation proteins vary widely between eukaryotic and bacterial organisms. In some cases, these limitations can be circumvented by the use of genetically modified bacterial expression strains but the selection of an alternative expression system must often be considered.

2. Heterologous Protein Production Using *Escherichia coli*

There are an increasing number of bacterial systems available for the production of heterologous proteins. The selection of a particular system is influenced by the nature of the target protein, the experience of the user, and the intended use of the product. *E. coli* is the most commonly used bacterial system for production of heterologous proteins. Over a century of intensive study of *E. coli* has provided a great deal of information about regulatory mechanisms and the function of the host accessory proteins that may impact expression outcome. In addition, there is an extensive resource of methodological and technological materials in support of protein production in a laboratory or commercial setting. For many protein production projects, the availability of these resources and the minimal technical requirements make *E. coli* a host of first choice for preliminary protein expression screening. Consequently, we use this organism as a model to illustrate specific approaches and methods for protein production. The description of specific methods for protein production using *E. coli* will be followed by a discussion of alternative bacterial expression systems. Many of the core concepts and techniques described for protein production using *E. coli* are directly applicable to other bacterial systems.

3. Planning a Bacterial Expression Project

The successful implementation of a system for bacterial expression is dependent on a series of sequential steps that are illustrated in chronological order in Table 12.1. The initial stages of the process are not experimental but rely on defining a set of project requirements and analysis of the

Table 12.1 Scheme for heterologous protein production using a bacterial host

Stage	Impact on selection of expression system
Outline project requirements	Define requirements such as scale, functionality, resources, and intended application to select an appropriate expression system
Target analysis	Utilize coding region sequence features, available experimental data or historical data to define and optimize expression strategies
Cloning	Cloning options include vectors, selection strategy, fusion tags to enhance solubility or facilitate purification, inducer, or cellular targeting sequences
Selection of expression strain(s)	Rare codons, protease, accessory proteins
Characterization of the expression product	Expression, solubility, functionality

sequence features and biochemical characteristics of the expression target. These steps are critical to the ultimate success of the project and can contribute to a realistic assessment of project costs and likelihood for success. We present a brief review of these topics to provide continuity with the experimental components as some of these topics are addressed in greater detail in chapters included in this volume.

4. Evaluation of Project Requirements

The intended use of the protein product can have a major impact in the selection and prioritization of expression systems. Preservation of the native function and characteristics of the protein are essential for some applications (e.g., functional assignment or drug screening) but may be of less concern if the target is to be used as an immunogen. Other practical considerations include protein yield, time constraints, or production costs. A clear definition of the project requirements is vital to ensure the selection of an appropriate expression system and the strategy for cloning and production of the desired product (Dieckman *et al.*, 2006).

5. Target Analysis

The biochemical and biological attributes of the target are essential considerations for selection of an expression strategy and are primary predictors of expression and solubility outcomes. One set of attributes can be assembled by analysis of the primary sequence for prediction of secondary structure, biological localization (membrane, cytoplasmic, periplasmic, or extracellular), classification into families (fold and domains) and inference of biochemical properties (p*I*, disordered regions, ligands). Features such as a high probability for membrane spanning helices may dictate the use of a system designed for the expression of membrane proteins or the use of a domain cloning strategy for production of the soluble component of the protein. Use of the target sequence features to guide the selection of the expression host and vector construct will contribute to production of a mature and appropriately localized protein. Sequence information can be supplemented with experimental or historical data on the target protein or a homolog to provide further insight into expression system optimization. For example, proteins known to require a prosthetic group for proper function may necessitate selection of a specific host to ensure production of a fully functional protein. A case in point is the production of type c cytochromes which is facilitated by the use of genetically engineered strains of *E. coli* containing accessory proteins for covalent attachment of hemes to polypeptide chains of apocytochromes (Londer *et al.*, 2008). In practice, it is often difficult to predict the outcome of a protein expression trial in the absence of historical expression data for the target protein or a similar protein. A common experimental approach is to utilize several systems, beginning with the simplest and most cost effective method and invoke more complex systems for targets that fail.

6. Cloning

A variety of options are available for cloning the target sequence which can be generally categorized into serial and parallel systems. Serial systems, such as those that utilize restriction enzymes to generate compatible target and vector termini, are ubiquitous and provide several options to generate the target/vector compatible ends required to generate an expression construct. A disadvantage of this approach is the requirement of validating restriction enzyme cleavage strategies for each target and vector. There is growing interest in the use of parallel systems or universal cloning methods that enable easy transfer of the target to multiple vectors and expression systems regardless of the target sequence (Table 12.2). Examples of these systems include the Gateway (Esposito *et al.*, 2009), Infusion (Zhu *et al.*, 2007) or ligation independent cloning (LIC) methods (Aslanidis and de Jong, 1990; Haun *et al.*, 1992). An advantage of this approach is the ability to clone a target in multiple vectors and to simultaneously evaluate multiple expression strategies in a cost effective manner.

LIC is a cost effective method particularly suited for bacterial expression as the cloning reagents are nonproprietary and available from several commercial suppliers. In the LIC method, specific nucleotide sequences are appended to the PCR primers and allow any gene to be cloned regardless of DNA sequence. Compatible ends in both the vector and PCR fragments are generated by treatment with T4 DNA polymerase in the presence of a specific nucleoside triphosphate. The procedure generates complementary 10–15 bp overhangs in the vector and PCR fragment that anneal with sufficient strength to permit transformation without ligation. The process allows consistent design of PCR primers, is directional, and results in high cloning efficiency. Although commercial vectors are available, the approach has been used in several large-scale cloning projects which provide vector resources to individual investigators. The procedure described below was developed for use with the suite of vectors designed at the Midwest Center for Structural Genomics (Eschenfeldt *et al.*, 2009). The method is generally applicable to other LIC compatible systems provided adjustments are made for the specific sequences appended to the amplification primers.

Table 12.2 Commonly used *E. coli* expression hosts and vectors[a]

Strain	Feature	References or Suppliers
HB101 DH5α	Common strains for subcloning and preparation of plasmid DNA. Strains have reduced recombination and restriction capabilities that aid in plasmid stability and improved quality of plasmid DNA	Boyer and Roulland-Dussoix (1969) and Woodcock *et al.* (1989)
BL21(DE3)	Widely used expression strain which lacks the *lon* and *opmT* proteases and contains a copy of the T7 RNA polymerase gene under the control of the *lacUV5* promoter. These modifications enable stable expression of proteins using T7 promoter driven constructs	Studier and Moffatt (1986), multiple suppliers
C41 C43	Effective for expression of toxic and membrane proteins	Miroux and Walker (1996), Lucigen
ABLE strains	Strains enable copy number control of ColE1 derived plasmids	Stratagene
Origami	K-12 derivatives with mutations in the *trxB* and *gor* genes which enhance disulfide bond formation in the cytoplasm	Derman *et al.* (1993), EMD Biosciences
Vector		
pET series	Widely used expression systems for inducible expression of proteins using a T7 promoter construct	EMD Biosciences
pBAD series	Tightly regulated expression system with expression controlled by the presence of arabinose	Guzman *et al.* (1995)
His GST	Common purification fusion tags	Hengen (1995) and Smith and Johnson (1988)

Table 12.2 (*continued*)

Strain	Feature	References or Suppliers
MBP Nus	Common solubility enhancing fusion tags	De Marco *et al.* (2004) and Kapust and Waugh (1999)
Gateway vectors	A series of vectors that use a recombinational strategy to enable transfer of DNA fragments between different cloning vectors	Hartley *et al.* (2000), Invitrogen
Ligation independent cloning vectors	A series of vectors that use an anneal strategy to enable parallel cloning of coding region fragments	Aslanidis and de Jong, (1990) and Haun *et al.* (1992)
pET26b	Vector contains the *pelB* leader sequence that targets expression to the periplasmic space	Makrides (1996) and Matthey *et al.* (1999)

[a] The NIH funded PSI Materials Repository (http://www.hip.harvard.edu/PSIMR/index.htm) and Protein Expression Laboratory (http://web.ncifcrf.gov/atp/) maintain an extensive collection of expression strains and vectors that are available to the general public.

7. Preparation of T4 DNA Polymerase-Treated DNA Fragments

1. Append the appropriate LIC specific nucleotide sequences (e.g., forward primer: TACTTCCAATCCAATGCC, reverse primer with added stop: TTATCCACTTCCAATGTTA) to the target specific PCR primers.
2. The specified method is scaled to microwell plates (96 targets) but can be adjusted to any number of targets. Make LIC Reaction mix sufficient for one 96-well plate by combining the following reagents:

   a. 465 μL of 10 × T4 DNA polymerase buffer. T4 DNA polymerase reaction buffers supplied by most vendors can be substituted for the LIC reaction buffer described in this section. Our comparison of various common T4 DNA polymerase reaction buffers shows less than a 25% difference in the cloning efficiency of the final product.
   b. 465 μL of 25 m*M* dCTP, molecular biology grade (Promega cat. no. U1221).
   c. 228 μL of 100 m*M* dithiothreitol (DTT) solution (Novagen cat. no. 70099).
   d. 60 μL of water.
   e. 250 units of T4 DNA Polymerase (LIC quality, $\sim$2.5 units/μL, EMD Biosciences/Novagen).

3. Keep this mixture on ice and add the T4 DNA polymerase just before use. Pipette up and down several times to uniformly distribute the enzyme in the reaction mix.
4. Array 10.4 μL of the LIC reaction mix into a polypropylene 96-well plate.
5. Add 30 μL (40–100 ng) of purified PCR fragment to the LIC reaction mix and pipette up and down several times to mix. Incubate at room temperature for 30 min. Our studies of various fragment-to-vector ratios (Dieckman *et al.*, 2006) indicate a wide tolerance for variation in the amount of target DNA fragment on the annealing reaction.
6. Incubate on a heat block at 75 °C for 20 min to inactivate the T4 DNA polymerase.
7. In another 96-well plate, anneal 1–2 μL of the T4-Polymerase-treated PCR LIC fragments with 4 μL (20–50 ng) T4-Polymerase-treated LIC vector.
8. Incubate the annealing reaction 5–10 min at room temperature.
9. Use the entire annealing reaction to transform $\sim$ 50 μL competent *E. coli* cells and select for transformed colonies (Sambrook and Russell, 2001).

Following the heating process the LIC plates are stored in the refrigerator at 4 °C until needed. The preparation of LIC compatible vector is similar to the above procedure but uses the complementary base dGTP (Eschenfeldt *et al.*, 2009). The constructs can be validated by sequence analysis at this stage or the analysis can be performed after analysis for expression and solubility.

8. Expression in the *E. coli* Cytoplasm

The majority of vectors available for expression in *E. coli* are designed for cytoplasmic expression. These vectors have been engineered with an assortment of selectable markers, bacterial promoters, plasmid replication origins, localization signals, and fusion tags (Table 12.2). The protocol we describe is adapted for the T7 promoter (Novagen, pET vector series) which can overexpress a target at a level representative of the most abundant native proteins in *E. coli* (Studier and Moffatt, 1986). An *E. coli* strain that expresses the T7 RNA polymerase (e.g., BL21(DE3)) is required for protein expression using the pET family of vectors. Variants of these strains are available to coexpress rare tRNAs (Carstens, 2003), additional cofactors necessary for protein folding (Baneyx and Palumbo, 2003), or proteins that enable disulfide bond formation and promote protein folding activity in the cytoplasm (Prinz *et al.*, 1997). This protocol summarizes the process for use of IPTG in conjunction with an inducible system for production of the target protein.

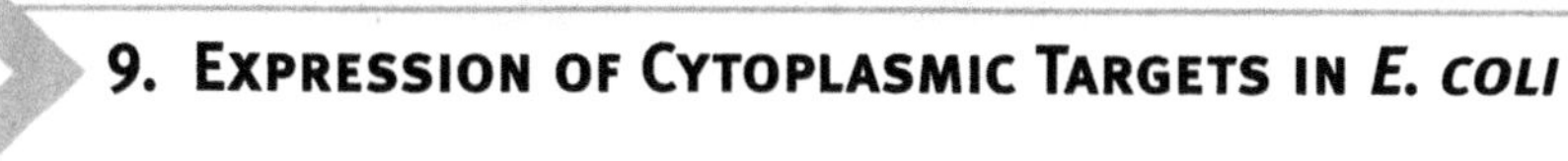

9. Expression of Cytoplasmic Targets in *E. coli*

1. Use a single colony to inoculate a 2-mL culture of media with appropriate antibiotic. Use a culture tube that will hold at least 5× the final culture volume for sufficient aeration.
2. Incubate samples at 37 °C, 250 rpm until the OD_{600} reaches 0.4–0.8. The culture should appear cloudy but not completely turbid. In general, this requires 3–4 h of growth with BL21(DE3) cells.
3. Add 20 μL 100 m*M* IPTG to each culture (1 m*M* final concentration). Return induced cultures to the incubator set at 37 °C, 250 rpm for 4 h. The cultures should be turbid after 4 h of induction.
4. Cultures should show significant expression after 3–8 h of induction at 37 °C. Remove sample and analyze as described below.

9.1. Analysis of protein expression and solubility

Validation of protein expression typically involves an assessment of expression and solubility of the target protein and a qualitative verification of the expected protein size. Most of the *E. coli* systems used for heterologous expression produce sufficient level of the target protein to enable assessment of expression/solubility outcome by denaturing gel electrophoresis (Fig. 12.1). The method is low cost, relatively easy, and returns results quickly. Proteins expressed at low levels or that may be marginally soluble may require a more sensitive detection method such as Western blotting.

10. Analysis of Heterologous Protein Expression in *E. coli*

1. Remove 200 μL of induced culture from each sample to a clean microcentrifuge tube. Centrifuge this portion at 14,000 rpm for 1 min. The cells should form a dense pellet with a clear supernatant. The remaining culture is used for assessment of target protein solubility described in Section 10.1.
2. Decant or aspirate spent media, being careful to retain the cell pellet. Add 50 μL of 2 × SDS loading dye and use repeated pipeting to resuspend cells.
3. Boil samples for 5 min and allow to cool slightly before loading on an SDS–PAGE gel. For a 17-well 8 × 10 cm gel stained with Coomassie, 5 μL of sample is usually sufficient.

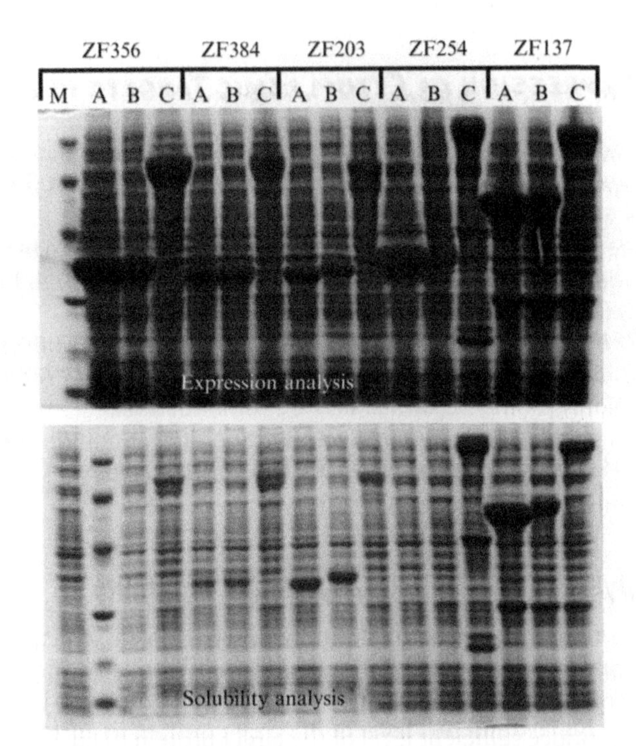

Figure 12.1 Expression and solubility for a set of zebrafish proteins. The top figure represents a Coomassie-stained gel displaying total expression products for proteins expressed in three different vector systems. The bottom figure is a Coomassie-stained gel with the soluble fractions of the same constructs. In the second gel, the order of ZF356 and the molecular weight marker is reversed from the top gel. Vector "A" is the pMCSG7 vector and produces an N-terminal fusion containing a TEV protease cleavable His tag. Vector "B" targets the protein to the periplasm and produces a N-terminal fusion similar to that described for Vector A. Vector "C" is pMCSG19 which contains a maltose binding protein (MBP) fusion sequence (Donnelly *et al.*, 2006). Protein accession numbers are as follows: ZF356; AAH56726.1, ZF384; AAH58296.1, ZF203; AAH46038.1, ZF254; AAH47843.1, ZF137; AAH67155.1.

10.1. Analysis of protein solubility

1. Pellet the remaining cell culture by centrifugation at 3500 rpm for 10 min. Carefully decant spent media and blot remaining liquid on a paper towel.
2. Freeze pellets at − 80 °C. Prepare enough lysis buffer (final concentration: 300 m*M* NaCl, 50 m*M* Na_2PO_4, protease inhibitor cocktail (Sigma), 120 kU/mL rLysozyme (Novagen) and 25 U/mL Benzonase (Novagen)), for all samples. Alternatively, Bugbuster (Novagen) or B-PER (Pierce) can be substituted for lysis buffer. Use volumes indicated for a 2-mL culture in reagent instructions.

3. Remove samples from freezer and thaw slightly. Add 180 μL of lysis buffer to each sample, cover tightly and vortex to completely resuspend pellet.
4. Return plate to -80 °C for 5 min, then remove and incubate at room temperature until the ice is completely melted. Repeat this freeze–thaw cycle once more. Samples may turn clear or they may remain cloudy.
5. Pellet cell debris at 3500 rpm for 10–15 min.
6. Remove a 50 μL portion from the top of each sample being careful not to remove any of the cell debris with the supernatant.
7. Add 60 μL of 2 $\times$ SDS loading dye to the supernatant and boil for 5 min. Analyze the samples by denaturing gel electrophoresis. For a 17-well 8 $\times$ 10 cm gel stained with Coomassie, 5 μL of sample is usually sufficient.
8. [Optional] To examine the insoluble fraction of the cell lysate, discard all of the remaining supernatant from the cell debris in step 7. Be careful not to disturb or remove the pellet.
9. Add 300 μL of 1 $\times$ SDS loading dye to the pellet and cover tightly. Vortex sample until entire pellet is resuspended.
10. Boil sample for 5 min and cool slightly before loading on an SDS–PAGE gel. For a 17-well 8 $\times$ 10 cm gel stained with Coomassie, 3 μL is usually enough to visualize expression.

10.2. Analysis of protein expression results

Use of the same culture for assessment of expression and solubility can compensate for the variance in expression level. Expression rates in *E. coli* typically exceed 80% but solubility can be a limiting factor and is target dependent (Fig. 12.1). Targets are scored as "no expression" or "insoluble" based on the absence of a detectable stained protein band of the correct molecular weight observed after SDS–PAGE analysis. Targets can be scored as positive for expression or solubility based on observation of a protein stained band of correct molecular weight. This is a qualitative assessment and additional validation such as sequence analysis and/or mass spectrometry is necessary to assure identity of the polypeptide. We use a relative ranking scale that compares staining of the target relative to the general intensity of native *E. coli* proteins. Target bands that are visible but with intensity level less than most of the *E. coli* proteins are scored as level 1 or low expression/solubility. Levels 2 and 3 (moderate and high expression/solubility) have staining intensity comparable to that of highly expressed *E. coli* proteins or more prominent than any *E. coli* protein, respectively.

10.3. Autoinduction method for protein expression

A disadvantage of induction by IPTG is the requirement to monitor bacterial growth to achieve optimal induction conditions. This attribute is especially apparent in experiments which survey expression in large clone sets. In this scenario, differences in growth rates will be observed and optimal induction conditions are unlikely to be achieved for all expression clones. Autoinduction systems provide an approach to circumvent these difficulties and simplify the expression protocol (Studier, 2005). Several systems have been described which provide high-level protein expression with pET and other IPTG-inducible bacterial expression systems (Blommel *et al.*, 2007).

11. Small-Scale Expression Cultures in Autoinduction Media Protocol

1. A transformation plate or a fresh streak of a frozen stock may be used for this experiment. The expression cell line must contain genes for *lac* permease (*lacY*) and ß-galactosidase (*lacZ*) as well as T7 RNA polymerase.
2. Prepare autoinduction media (e.g., Overnight Express Kit, Novagen) and add appropriate antibiotic to media.
3. Use a single colony to inoculate 2-mL cultures of autoinduction media. Use tubes that hold at least 5× the final volume for adequate aeration.
4. Incubate tubes at 37 °C and 250 rpm for at least 16 h. Alternatively, incubate cultures at 25 °C and 250 rpm for at least 20 h. The culture must reach stationary phase for robust induction of expression.

Expression and solubility outcome of individual targets is assessed as described in the previous section for IPTG-induced cultures. Expression yield of heterologous proteins is similar for autoinduced and IPTG-induced cultures but individual variation in target protein solubility may be observed (Fig. 12.2). In some studies, autoinduction produced 5–20 times as much target protein per volume of culture as conventional IPTG induction (Studier, 2005).

12. Periplasmic Expression of Proteins

Approximately 8–12% of the bacterial proteome is not destined for the cytoplasm. Heterologous expression strategies for this group include amplification of the nonsignal component of the coding region and processing these targets through the standard *E. coli* cloning and expression pipeline using a cytoplasmic or periplasmic targeting vector with outcomes similar to those observed for cytoplasmic proteins. This approach is also successful for some eukaryotic proteins containing disulfide bonds. Proteins are routed to the *E. coli* periplasmic space by addition of an appropriate targeting signal (e.g., the *E. coli* PelB signal sequence) to the N-terminus of the target

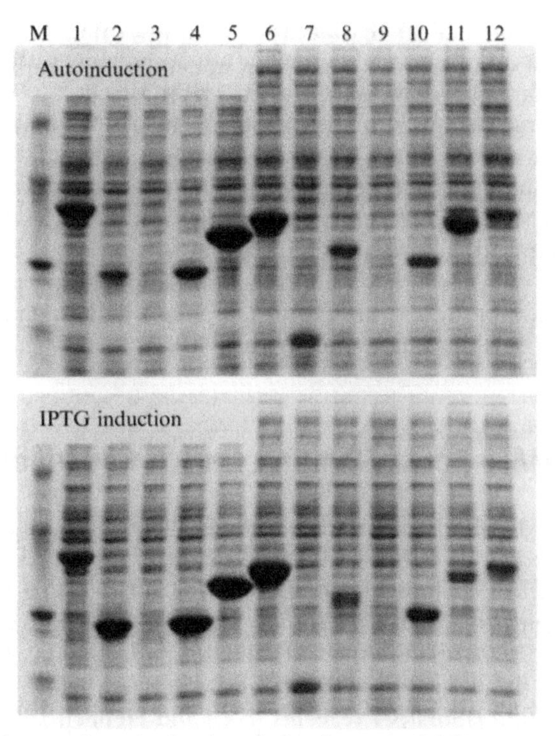

Figure 12.2 Coomassie-stained gel analysis of target solubility outcome for autoinduced and IPTG-induced cultures. Targets were amplified from *Shewanella oneidensis* genomic DNA and cloned into the cytoplasmic expression vector pMCSG7 (targets 1–6) or the periplasmic expression vector pBH31 SA (targets 7–12). Soluble fractions for analysis were prepared as described in the text. Targets are as follows: 1 = SO 3070; 2 = SO 2454; 3 = SO 2444; 4 = SO 1503; 5 = SO 1190; 6 = SO1560; 7 = SO 0809; 8 = SO 0837; 9 = SO 4048; 10 = SO 1503; 11 = SO 1190; 12 = SO 1560.

protein. As the periplasm accounts for 20–40% of the total volume of the cell, overall expression yields are typically lower when compared to cytoplasmic expression. This is apparent from a direct comparison of the expression levels for targets cloned into a cytoplasmic expression vector (Fig. 12.2, targets 4–6) and targets cloned into a periplasmic expression vector (Fig. 12.2, targets 10–12).

13. Expression of Periplasmic Targets in *E. coli*

1. Use a single colony to inoculate a 2-mL culture of media with appropriate antibiotic. Use a culture tube that will hold at least 5× the final culture volume for sufficient aeration.

2. Incubate samples at 30 °C, 250 rpm until the OD_{600} reaches 0.4–0.8. The culture should appear cloudy but not completely turbid. In general, this requires 3–5 h of growth with BL21(DE3) cells.
3. Drop the temperature of the incubator to 19 °C and incubate the cultures for at least 30 min to allow cells to equilibrate. The OD of the culture should continue to increase during the temperature shift.
4. Add 20 μL 100 m*M* IPTG to each culture (1 m*M* final concentration). Return induced cultures to the incubator set at 19 °C, 250 rpm overnight. The cultures should be turbid after the overnight induction.
5. Cultures should show some expression after 4 h and significant expression after 12–16 h of induction at 19 °C.

14. Small-Scale Osmotic Shock Protocol

If the localization of a target to the periplasm must be confirmed, an osmotic shock can be performed to analyze the contents of the periplasm separately from the cytoplasm. Many methods for releasing proteins from the outer compartment of bacteria are available. Addition of chloroform (Ames *et al.*, 1984) or Polymyxin B (Dixon and Chopra, 1986) releases proteins from the periplasm. The method described below is inexpensive and uses common laboratory reagents (Neu and Heppel, 1965; Nossal and Heppel, 1966). Lysozyme can be added to the osmotic shock SET (sucrose, EDTA, Tris–HCl) buffer to remove more of the outer cell wall and components (Birdsell and Cota-Robles, 1967; Malamy and Horecker, 1964); the spheroplasts made during this procedure are very prone to lysis and cytoplasmic proteins may contaminate the periplasmic fraction. For cells that lyse easily the addition of 0.5 m*M* $MgCl_2$ to the cold water shock can stabilize the spheroplasts (Neu and Chou, 1967). The method we describe works well for a majority of proteins targeted to the periplasm but may need to be optimized for some targets. Comparison of the amount of target protein in the SET and water fractions in Fig. 12.3 illustrates the variance obtained with individual proteins. However, the proportion of soluble heterologous proteins is enriched in the shock fractions compared to background host proteins (Fig. 12.3). This method can be scaled up as a first purification step to reduce background proteins and cell debris.

1. Use a 5-mL culture where protein expression has been induced for 4 h to overnight.
2. Pellet the induced culture at 3500 rpm for 5 min. Decant or aspirate spent media.
3. Wash cells twice with 5 mL of 10 m*M* Tris–HCl, pH 8.0. Centrifuge suspensions at 3500 rpm for 5 min, decant washes after pelleting cells. A vortexer may be used to resuspend the cell pellet.

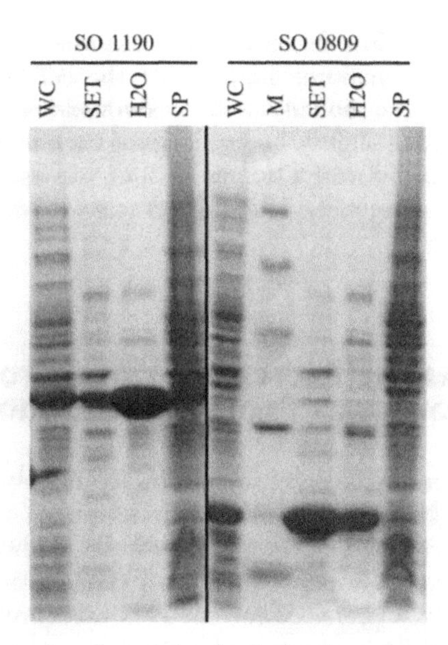

Figure 12.3 An example of osmotic shock fractions for targets from *Shewanella oneidensis* cloned into the periplasmic expression vector pBH31 SA. Fractions are labeled as follows: WC = whole cell expression; SET = SET buffer fraction; H2O = water shock fraction; SP = spheroplast fraction; M = molecular weight markers.

4. Resuspend cells in 1 mL 10 m*M* Tris–HCl and transfer to a 1.5 mL microcentrifuge tube. Pellet cells by centrifuging for 30 s at 14,000 rpm and aspirate all remaining Tris–HCl wash.
5. Resuspend pellet completely in 300 μL SET buffer (50 m*M* Tris–HCl, pH 7.5, 20% (w/v) sucrose, 1 m*M* EDTA). Incubate cells and buffer on ice for 10 min.
6. Centrifuge samples at 14,000 rpm for 30 s to 1 min. Some preps may require longer centrifugation times, as the cells do not pellet easily in SET buffer. Remove a 50 μL portion of supernatant from the top of the sample to avoid picking up cell debris. Discard remaining liquid, but retain the cell pellet.
7. Add 300 μL ice-cold sterile water to the pellet and quickly resuspend cells. Incubate cells in water on ice for 5 min, without shaking.
8. Centrifuge sample at 14,000 rpm for 20–30 s. Immediately remove a 50 μL portion from the top of the supernatant to avoid picking up debris.
9. Steps 6–9 can be repeated with remaining pellet to release additional material from the periplasm.
10. Add 50 μL 2 × SDS loading dye to the SET buffer fraction and H_2O shock fraction and mix well. Boil samples 5 min and cool slightly before loading on gel. For a 17-well 8 × 10 cm gel stained with Coomassie, 10 μL of each sample was usually sufficient to visualize bands.

11. [Optional] To examine the spheroplast fractions after the shock procedure remove all remaining liquid from the cell pellet. Add 500 μL 2 × SDS loading dye directly to the pellet and resuspend. Boil sample for 5 min and cool slightly before loading on the gel. For a 17-well, 8 × 10 cm gel stained with Coomassie, 3 μL was usually sufficient. This sample can be extremely viscous and may require additional SDS loading buffer.

15. Alternative Bacterial Systems for Heterologous Protein Production

A variety of alternative bacterial systems are available for heterologous protein expression. In many instances, the selection of a bacterial system is motivated by characteristics of the target proteins. Although *E. coli* can be utilized for the heterologous production of multiheme cytochromes, an alternative host is often preferred for complex cytochromes with multiple heme groups. *Shewanella oneidensis* is a Gram-negative bacterium often used for the production of this particular protein class. The genome of this organism encodes a large number of predicted cytochrome *c* genes and contains accessory proteins that promote correct processing of the apocytochromes into the mature protein (Takayama and Akutsu, 2007). A heterologous expression system that uses a genetically engineered *Pseudomonas fluorescens* strain (Pfenex, DOW Chemical Co.) has been developed for large-batch fermentation as well as high-throughput small-scale screening of target proteins. This modified Gram-negative organism is metabolically versatile and contains various microbial secretion systems that enable production in the cytoplasm, periplasm, or extracellular space.

Protein production in nonpathogenic Gram-positive *Bacillus* species provides an alternative to the Gram-negative host strains (Schmidt, 2004). These organisms contain a naturally efficient secretion system to direct expressed proteins into the culture supernatant often resulting in improved yield and ease of purification. The most frequently used species are modified *B. subtilis* strains which lack genes for both extracellular lipolytic enzymes and proteases, resulting in improved stability of heterologous proteins. *Bacillus megaterium* has several large plasmids and is known for the stable replication and maintenance of these plasmids. Expression strains typically have low intrinsic protease activity and a combination of features that result in stable high-yield protein expression.

Membrane proteins comprise a significant fraction of the proteins encoded by the genome but represent a challenge for protein production. In addition to *E. coli* (Neophytou *et al.*, 2007; Shaw and Miroux, 2003), several other membrane protein expression systems have been developed for production of integral membrane proteins. The Gram-positive lactic acid bacterium *Lactococcus lactis* grows to high densities and is suited for large-scale overproduction of membrane proteins (Kunji *et al.*, 2003). Several auxotrophic strains are available as well as an inducible expression system regulated by polycyclic peptide nisin. Functional screens for the characterization of the membrane protein can be performed with whole cells because ligands can act directly on the cytoplasmic membrane in which the membrane proteins are expressed.

The photosynthetic bacterium *Rhodobacter* has been engineered for heterologous production of full-length membrane proteins (Laible *et al.*, 2004). This system is unique in that it incorporates foreign membrane proteins into its own system of intracytoplasmic membranes by synthesizing new membrane coordinately with the expression of foreign target membrane proteins. These new membranes form as invaginations of the cytoplasmic membrane and allow integration of heterologous membrane proteins. One of the key strengths in using the *Rhodobacter* membrane expression system is that proteins produced are localized to the membranes nearly quantitatively in all cases. Membrane protein expression and purification is challenging and the functionality of proteins produced in heterologous expression systems must be carefully validated.

16. Alternative Vector and Induction Conditions

Aside from employing different bacterial hosts, the heterologous expression system may also be optimized by manipulating vector and induction conditions. One such method that has demonstrated recent success in protein expression and solubility is a cold-shock expression system using a pCold vector (Qing *et al.*, 2004). This system uses the knowledge of *E. coli* cell cold-shock response to inhibit endogenous protein production while enhancing target protein expression. It has been suggested that when the *cspA* mRNA is truncated and expressed under cold-shock conditions (37 °C dropped to 15 °C for 36 h), polysomes are occupied with translation of the truncated *cspA* gene and cannot adapt to ribosomal form III to translate non-cold-shock proteins (Jiang *et al.*, 1996). Consequently, only the genes in the *cspA* mRNA will be translated, while endogenous protein expression is inhibited. The pCold vectors include a cloning region after the *cspA* promoter for a target gene insert that is overexpressed upon 15 °C cold-shock induction and derepression with 1 m*M* IPTG. Due to the minimal level of expressed host background proteins, this technology often eliminates the need to purify the heterologously expressed protein, greatly cutting production costs.

17. Production Scale

The goal of most bacterial expression projects is to produce large amounts of soluble, correctly folded and active proteins. In general, the expression and solubility levels of a target protein decrease in larger cultures. However, targets that perform well in small-scale experiments are more likely to produce satisfactory results during scale-up (Moy *et al.*, 2004). A common approach to improve the final production yield is to select several of the most soluble constructs obtained from the pilot studies for evaluation in large-scale production processes. The second component of the strategy is optimization of growth and culture parameters of the large-scale process to improve the quality and yield of the final product. The ability to generate and screen multiple bacterial expression products in a timely manner contributes to the utility of bacterial expression systems for protein production.

Acknowledgment

The submitted manuscript has been created by UChicago Argonne, LLC, Operator of Argonne National Laboratory ("Argonne"). Argonne, a U.S. Department of Energy Office of Science laboratory, is operated under Contract No. DE-AC02-06CH11357. The U.S. Government retains for itself, and others acting on its behalf, a paid-up nonexclusive, irrevocable worldwide license in said article to reproduce, prepare derivative works, distribute copies to the public, and perform publicly and display publicly, by or on behalf of the Government.

References

Ames, G. F., Prody, C., and Kustu, S. (1984). Simple, rapid, and quantitative release of periplasmic proteins by chloroform. *J. Bacteriol.* **160,** 1181–1183.

Aslanidis, C., and de Jong, P. J. (1990). Ligation-independent cloning of PCR products (LIC-PCR). *Nucleic Acids Res.* **18,** 6069–6074.

Baneyx, F. (1999). Recombinant protein expression in Escherichia coli. *Curr. Opin. Biotechnol.* **10,** 411–421.

Baneyx, F., and Palumbo, J. L. (2003). Improving heterologous protein folding via molecular chaperone and foldase co-expression. *Methods Mol. Biol.* **205,** 171–197.

Birdsell, D. C., and Cota-Robles, E. H. (1967). Production and ultrastructure of lysozyme and ethylenediaminetetraacetate-lysozyme spheroplasts of Escherichia coli. *J. Bacteriol.* **93,** 427–437.

Boyer, H. W., and Roulland-Dussoix, D. (1969). A complementation analysis of the restriction and modification of DNA in Escherichia coli. *J. Mol. Biol.* **41,** 459–472.

Carstens, C. P. (2003). Use of tRNA-supplemented host strains for expression of heterologous genes in E. coli. *Methods Mol. Biol.* **205,** 225–233.

De Marco, V., Stier, G., Blandin, S., and de Marco, A. (2004). The solubility and stability of recombinant proteins are increased by their fusion to NusA. *Biochem. Biophys. Res. Commun.* **322,** 766–771.

Derman, A. I., Prinz, W. A., Belin, D., and Beckwith, J. (1993). Mutations that allow disulfide bond formation in the cytoplasm of *Escherichia coli*. *Science* **262,** 1744–1747.

Dieckman, L., Gu, M., Stols, L., Donnelly, M. I., and Collart, F. R. (2002). High throughput methods for gene cloning and expression. *Protein Expr. Purif.* **25,** 1–7.

Dieckman, L. J., Hanly, W. C., and Collart, F. R. (2006). Strategies for high-throughput gene cloning and expression. *Genet. Eng. (NY)* **27,** 179–190.

Dixon, R. A., and Chopra, I. (1986). Leakage of periplasmic proteins from *Escherichia coli* mediated by polymyxin B nonapeptide. *Antimicrob. Agents Chemother.* **29,** 781–788.

Donnelly, M. I., Zhou, M., Millard, C. S., Clancy, S., Stols, L., Eschenfeldt, W. H., Collart, F. R., and Joachimiak, A. (2006). An expression vector tailored for large-scale, high-throughput purification of recombinant proteins. *Protein Expr. Purif.* **47,** 446–454.

Dyson, M. R., Shadbolt, S. P., Vincent, K. J., Perera, R. L., and McCafferty, J. (2004). Production of soluble mammalian proteins in *Escherichia coli*: Identification of protein features that correlate with successful expression. *BMC Biotechnol.* **4,** 32.

Eschenfeldt, W. H., Lucy, S., Millard, C. S., Joachimiak, A., and Mark, I. D. (2009). A family of LIC vectors for high-throughput cloning and purification of proteins. *Methods Mol. Biol.* **498,** 105–115.

Esposito, D., Garvey, L. A., and Chakiath, C. S. (2009). Gateway cloning for protein expression. *Methods Mol. Biol.* **498,** 31–54.

Guzman, L. M., Belin, D., Carson, M. J., and Beckwith, J. (1995). Tight regulation, modulation, and high-level expression by vectors containing the arabinose PBAD promoter. *J. Bacteriol.* **177,** 4121–4130.

Hartley, J. L., Temple, G. F., and Brasch, M. A. (2000). DNA cloning using *in vitro* site-specific recombination. *Genome Res.* **10,** 1788–1795.

Haun, R. S., Serventi, I. M., and Moss, J. (1992). Rapid, reliable ligation-independent cloning of PCR products using modified plasmid vectors. *Biotechniques* **13,** 515–518.

Hengen, P. (1995). Purification of His-Tag fusion proteins from Escherichia coli. *Trends Biochem Sci.* **20,** 285–286.

Jiang, W., Fang, L., and Inouye, M. (1996). The role of the 5'-end untranslated region of the mRNA for CspA, the major cold-shock protein of Escherichia coli, in cold-shock adaptation. *J. Bacteriol.* **178,** 4919–4925.

Kapust, R. B., and Waugh, D. S. (1999). Escherichia coli maltose-binding protein is uncommonly effective at promoting the solubility of polypeptides to which it is fused. *Protein Sci.* **8,** 1668–1674.

Klock, H. E., White, A., Koesema, E., and Lesley, S. A. (2005). Methods and results for semi-automated cloning using integrated robotics. *J. Struct. Funct. Genomics* **6,** 89–94.

Kunji, E. R., Slotboom, D. J., and Poolman, B. (2003). *Lactococcus lactis* as host for overproduction of functional membrane proteins. *Biochim. Biophys. Acta* **1610,** 97–108.

Laible, P. D., Scott, H. N., Henry, L., and Hanson, D. K. (2004). Towards higher-throughput membrane protein production for structural genomics initiatives. *J. Struct. Funct. Genomics* **5,** 167–172.

Londer, Y. Y., Giuliani, S. E., Peppler, T., and Collart, F. R. (2008). Addressing *Shewanella oneidensis* "cytochromome": The first step towards high-throughput expression of cytochromes c. *Protein Expr. Purif.* **62,** 128–137.

Makrides, S. C. (1996). Strategies for achieving high-level expression of genes in *Escherichia coli*. *Microbiol. Rev.* **60,** 512–538.

Malamy, M. H., and Horecker, B. L. (1964). Release of alkaline phosphatase from cells of *Escherichia coli* upon lysozyme spheroplast formation. *Biochemistry* **3,** 1889–1893.

Matthey, B., Engert, A., Klimka, A., Diehl, V., and Barth, S. (1999). A new series of pET-derived vectors for high efficiency expression of *Pseudomonas* exotoxin-based fusion proteins. *Gene* **229,** 145–153.

Miroux, B., and Walker, J. E. (1996). Over-production of proteins in *Escherichia coli*: Mutant hosts that allow synthesis of some membrane proteins and globular proteins at high levels. *J. Mol. Biol.* **260,** 289–298.

Moy, S., Dieckman, L., Schiffer, M., Maltsev, N., Yu, G. X., and Collart, A. F. (2004). Genome-scale expression of proteins from *Bacillus subtilis*. *J. Struct. Funct. Genomics* **5,** 103–109.

Neophytou, I., Harvey, R., Lawrence, J., Marsh, P., Panaretou, B., and Barlow, D. (2007). Eukaryotic integral membrane protein expression utilizing the Escherichia coli glycerol-conducting channel protein (GlpF). *Appl. Microbiol. Biotechnol.* **77,** 375–381.

Neu, H. C., and Chou, J. (1967). Release of surface enzymes in Enterobacteriaceae by osmotic shock. *J. Bacteriol.* **94,** 1934–1945.

Neu, H. C., and Heppel, L. A. (1965). The release of enzymes from Escherichia coli by osmotic shock and during the formation of spheroplasts. *J. Biol. Chem.* **240,** 3685–3692.

Nossal, N. G., and Heppel, L. A. (1966). The release of enzymes by osmotic shock from Escherichia coli in exponential phase. *J. Biol. Chem.* **241,** 3055–3062.

Prinz, W. A., Aslund, F., Holmgren, A., and Beckwith, J. (1997). The role of the thioredoxin and glutaredoxin pathways in reducing protein disulfide bonds in the *Escherichia coli* cytoplasm. *J. Biol. Chem.* **272,** 15661–15667.

Qing, G., Ma, L. C., Khorchid, A., Swapna, G. V., Mal, T. K., Takayama, M. M., Xia, B., Phadtare, S., Ke, H., Acton, T., Montelione, G. T., Ikura, M., *et al.* (2004). Cold-shock induced high-yield protein production in Escherichia coli. *Nat. Biotechnol.* **22,** 877–882.

Sambrook, J., and Russell, D. W. (2001). Molecular cloning: A laboratory manual. Cold Spring Harbor Laboratory Press, Cold Spring Harbor, NY.

Schmidt, F. R. (2004). Recombinant expression systems in the pharmaceutical industry. *Appl. Microbiol. Biotechnol.* **65,** 363–372.

Shaw, A. Z., and Miroux, B. (2003). A general approach for heterologous membrane protein expression in Escherichia coli: the uncoupling protein, UCP1, as an example. *Methods Mol. Biol.* **228,** 23–35.

Smith, D. B., and Johnson, K. S. (1988). Single-step purification of polypeptides expressed in Escherichia coli as fusions with glutathione S-transferase. *Gene.* **67,** 31–40.

Studier, F. W. (2005). Protein production by auto-induction in high density shaking cultures. *Protein Expr. Purif.* **41,** 207–234.

Studier, F. W., and Moffatt, B. A. (1986). Use of bacteriophage T7 RNA polymerase to direct selective high-level expression of cloned genes. *J. Mol. Biol.* **189,** 113–130.

Takayama, Y., and Akutsu, H. (2007). Expression in periplasmic space of *Shewanella oneidensis*. *Protein Expr. Purif.* **56,** 80–84.

Terpe, K. (2006). Overview of bacterial expression systems for heterologous protein production: from molecular and biochemical fundamentals to commercial systems. *Appl. Microbiol. Biotechnol.* **72,** 211–222.

Woodcock, D. M., Crowther, P. J., Doherty, J., Jefferson, S., DeCruz, E., Noyer-Weidner, M., Smith, S. S., Michael, M. Z., and Graham, M. W. (1989). Quantitative evaluation of *Escherichia coli* host strains for tolerance to cytosine methylation in plasmid and phage recombinants. *Nucleic Acids Res.* **17,** 3469–3478.

Yokoyama, S. (2003). Protein expression systems for structural genomics and proteomics. *Curr. Opin. Chem. Biol.* **7,** 39–43.

Zhu, B., Cai, G., Hall, E. O., and Freeman, G. J. (2007). In-fusion assembly: Seamless engineering of multidomain fusion proteins, modular vectors, and mutations. *Biotechniques* **43,** 354–359.

CHAPTER THIRTEEN

Expression in the Yeast *Pichia pastoris*

James M. Cregg,*,† Ilya Tolstorukov,*,† Anasua Kusari,*
Jay Sunga,* Knut Madden,† *and* Thomas Chappell†

Contents

1. Introduction 170
2. Other Fungal Expression Systems 170
3. Culture Media and Microbial Manipulation Techniques 171
4. Genetic Strain Construction 172
   4.1. Mating, creating diploids 172
   4.2. Random spore analysis 173
5. Gene Preparation and Vector Selection 174
6. Transformation by Electroporation 176
7. DNA Preparation 176
8. Examination of Strains for Recombinant Protein Production 178
9. Assay Development—The Yeastern Blot 182
10. Posttranslational Modification of the Recombinant Protein (Proteinases and Glycosylation) 184
11. Selection for Multiple Copies of an Expression Cassette 185
References 187

Abstract

The yeast *Pichia pastoris* has become the premier example of yeast species used for the production of recombinant proteins. Advantages of this yeast for expression include tightly regulated and efficient promoters and a strong tendency for respiratory growth as opposed to fermentative growth. This chapter assumes the reader is proficient in molecular biology and details the more yeast specific procedures involved in utilizing the *P. pastoris* system for gene expression. Procedures to be found here include: strain construction by classical yeast genetics, the logic in selection of a vector and strain, preparation of electrocompetent yeast cells and transformation by electroporation, and the yeast colony western blot or Yeastern blot method for visualizing secreted proteins around yeast colonies.

* Keck Graduate Institute of Applied Life Sciences, Claremont, California, USA
† Biogrammatics, Inc., Carlsbad, California, USA

Methods in Enzymology, Volume 463
ISSN 0076-6879, DOI: 10.1016/S0076-6879(09)63013-5

1. Introduction

Relative to other expression systems, yeast got off to a slow start in the early 1980s, primarily due to poor results with several proteins in baker's yeast *Saccharomyces cerevisiae* (Romanos *et al.*, 1992). Promising results in shake flask culture, more often than not, led to disappointing yields when scaled up in fermentor cultures. In addition, yeast were not likely to be of value in producing human proteins containing N-linked carbohydrates as injectable pharmaceutical drugs, a major goal of many biotechnology and pharmaceutical companies, because their sugars were of a high mannose type in composition and configuration, quite different from that of humans. As a result, recombinant glycoproteins made in a yeast system have a typical fungal-like *N*-glycosylation pattern, which a human immune system recognizes as foreign and rejects. This results in the rapid clearance of yeast products from the blood and a strong immune response from a patient that can possibly result in death. Since the 1980s, these problems have been addressed and several yeasts have become productive alternative systems for recombinant protein production. As a eukaryotic microbial expression system, yeast are a good alternative for proteins for which expression in a bacterial system leads to the synthesis of improperly folded, and inactive protein aggregates or inclusion bodies.

This review will focus primarily on the most popular of these new yeast expression systems, *Pichia pastoris*. Only details of procedures that are specific or peculiar to expression in *P. pastoris* will be covered; for more common methods (e.g., agarose gel electrophoresis, immunoblotting, general recombinant DNA methodology, etc.), readers are referred to the many excellent books describing these methods including: (Sambrook *et al.*, 1989) or the series by (Ausubel *et al.*, 2001).

2. Other Fungal Expression Systems

In addition to *P. pastoris*, several other yeast species have been developed for expression; among them, *Hansenula polymorpha*, *Pichia methanolica*, *Kluyveromyces lactis*, *Arxula adeninivorans*, and *Yarrowia lipolytica* are the best known and developed (Gellissen, 2005). Many of the reasons for using one of these alternative yeast expression systems, as well as the methods needed to construct recombinant strains, are similar to each other and to *P. pastoris*.

Like *P. pastoris*, *H.* polymorpha and *P. methanolica* are methylotrophic yeast and virtually all of the advantages cited for *P. pastoris* are also true for these other species (Gellissen, 2002). In particular, all three have the potent promoter regulating the expression of the alcohol oxidase gene from their respective species available for controlling recombinant protein expression. However, both *H. polymorpha* and *P. methanolica* are different from *P. pastoris* in that the construction of expression strains is often somewhat more laborious. With *H. polymorpha*, this extra labor can be rewarded with strains harboring 100 or more copies of an expression cassette. In addition, *H. polymorpha* can readily grow at temperatures up to 50 °C providing the potential of significantly decreasing bioprocess times and thereby reducing costs.

K. lactis was the first yeast species after *S. cerevisiae* to be developed for recombinant protein expression and is well known for its use in production of rennin for cheese processing (van Ooyen *et al.*, 2006). An advantage of *K. lactis* for some purposes is that, like *S. cerevisiae*, it is officially on the generally recognized as safe ("GRAS") list of microorganisms.

A. adeninivorans is a dimorphic yeast with advantages for secreting recombinant proteins (Boer *et al.*, 2005). This yeast has both a vegetative yeast like growth state and a mycelial growth state in which the cells send out mycelia like filamentous fungi. During the mycelial growth phase, the synthesis of secreted proteins is enhanced relative to its vegetative growth phase. In addition, *A. adeninivorans* does not O-glycosylate secreted proteins in its mycelial growth phase. Finally, *A. adenininvorans* has a relatively high growth temperature of up to 48 °C and osmotolerance up to 3.4 *M* (10%) NaCl.

Y. lipolytica, like *A. adeninivorans* is a dimorphic yeast (Madzak *et al.*, 2005). Most recombinant genes are expressed in *Y. lipolytica* off the *XPR2* promoter which, in its native state, expresses the gene for alkaline extracellular protease. However, several other promoters, all of which are constitutive, are also available for the yeast.

Finally, certain filamentous fungi, such as *Neurospora crassa*, various *Aspergillus* species and *Sordaria macrospora*, have proved to be effective expression systems for certain recombinant products, particularly secreted proteins. However, the techniques for dealing with filamentous fungi are very different from yeast and will not be dealt with here. Readers interested in expression in filamentous fungi are referred to one of several excellent reviews on these systems (Heerikhuisen *et al.*, 2005; Kuck and Poggeler, 2005).

3. Culture Media and Microbial Manipulation Techniques

Techniques for culturing *P. pastoris* (and most other yeast species) at the bench level are identical to those used for *Escherichia coli* and *S. cerevisiae*. The most common rich medium for cultivation is YPD (1% yeast extract,

2% peptone, 2% dextrose) and defined medium is YNB (0.67% yeast nitrogen base with ammonium sulfate and without amino acids, 2% dextrose plus any amino acids or nucleotides required for growth at ~ 50 ug/ml each). Growth of *P. pastoris* on methanol requires that the dextrose is replaced with methanol to 0.5%. The recipe for mating/sporulation medium is 0.5% sodium acetate, 1% KCl 1% glucose. For Petri plates the media are prepared with 2% agar. Incubations are typically done at 30 °C. In liquid YPD, *P. pastoris* has a generation time of approximately 90 min and a generation time of approximately 3 h in defined medium. With methanol as sole carbon source and a defined culture medium, the generation time is around 5 h.

4. Genetic Strain Construction

Although a number of markers of various types exist for *P. pastoris*, the right combination for your purposes may not exist. Therefore, it may be necessary to construct a new host strain with the optimal set of genetic markers. The first step in strain construction is the mating and selection of diploid strains (Tolstorukov and Cregg, 2008). Because *P. pastoris* is functionally homothallic, the mating type of a strain is not a consideration in planning a genetic cross as cells of the same strain will also mate; However, the mating efficiency between *P. pastoris* cells is low. Therefore, it is essential that strains to be crossed contain complementary markers that allow for selective growth of crossed diploids, and against the growth of self-mated diploids and parental strains. Auxotrophic markers are generally most convenient for this purpose, but mutations in any gene that affect the growth or other phenotype of *P. pastoris* such as genes required for utilization of methanol or a nitrogen source (e.g., methylamine) can be used as well. Following are the steps needed to mate *P. pastoris*:

4.1. Mating, creating diploids

1. To begin a mating experiment, select a fresh colony from each strain to be mated from YPD plate (no more than one week old) and streak each across the length of an independent YPD plate.
2. After overnight incubation at 30 °C, transfer the cell streaks from both plates onto a single sterile velvet such that the streaks from one plate are perpendicular to those on the other.
3. Transfer the cross streaks from the velvet to a mating/sporulation plate and incubate for 2–3 days at room temperature to initiate mating.
4. After incubation, replica plate to an appropriate agar medium for the selection of complementing diploid cells. Diploid colonies will grow at

the junctions of the streaks after approximately 2–3 days of incubation at 30 °C. Diploid *P. pastoris* cells are approximately twice as large as haploid cells and easily distinguished by examination under a light microscope by their size and their high efficiency of sporulation.

5. Purify diploid colonies by streaking at least once for single colonies on diploid selection medium.
6. Diploid *P. pastoris* strains efficiently undergo meiosis and sporulation in response to nitrogen limitation. To initiate this phase of the life cycle, transfer freshly grown diploid colonies from a YPD or YNB plus glucose plate to a mating/sporulation plate either by replica-plating or with an inoculation loop; incubate the plate for 3–4 days at room temperature. Sporulation in all *Pichia* species correlates with accumulation of a red pigment in the ascus; therefore, sporulated diploid samples are easily distinguished by their tan color relative to the white color of haploid cultures. Diploids can also be distinguished by a high number of asci in the cell culture as observed by normal or phase-contrast microscopy.

4.2. Random spore analysis

P. pastoris spores are small and adhere to one another making tetrad dissection via micromanipulation difficult. Therefore, spore products are analyzed using a random spore analysis (RSA), as follows:

1. Transfer an inoculation loop full of sporulated *P. pastoris* cells (from Step 6 above) to a 1.5 ml microcentrifuge tube containing 0.25 ml of sterile water, vortex the mixture.
2. In a fume hood, add 0.25 ml of diethyl ether to the spore preparation and vortex thoroughly for approximately 5 min at room temperature. The ether treatment selectively kills vegetative cells remaining in spore preparations.
3. While still in the hood, centrifuge the cells for 2 min, remove the ether (upper phase) and resuspend the pellet in the remaining water. Spread 10 and 100 ul of the suspension onto a nonselective YPD plate.
4. After 4–5 days incubation at 30 °C, streak out single colonies onto a fresh YPD plate as a master plate for further analysis.
5. Replica plate the master plate onto a set of plates containing suitable diagnostic media. Alternatively/additionally, the initial plates with 100–600 colonies can also be replica plated. For example, spore products from a *his4* and *arg4* cross would be analyzed on YNB plus glucose supplemented with:

   a. No amino acids
   b. Arginine
   c. Histidine
   d. Arginine and Histidine

6. Compare the phenotype of individual colonies on each of the diagnostic plates to identify strains with the desired phenotype(s).
7. Select several colonies that appear to have the appropriate phenotype and streak for single colonies onto a nonselective medium plate such as YPD; then, retest a single colony from each streak on the same set of diagnostic plates. This step is important since *P. pastoris* spores adhere tightly to one another and colonies resulting from spore germination frequently contain cells derived from more than one spore. Another consequence of spore clumping is that markers appear not to segregate 1:1 but to be biased toward the dominant or wild-type phenotype. For example, in a*his4* × *arg4* cross as described above, more His^+Arg^+ spore products will be apparent than the 25% expected in the population and His^-Arg^- spore products will appear to be underrepresented.

5. Gene Preparation and Vector Selection

The first consideration in the selection of a *P. pastoris* expression vector is whether you intend to secrete a protein product or produce it intracellularly. A general rule of thumb is to produce a recombinant protein in the same way it is expressed in its native host: if a protein is produced intracellularly by its native host, one should also produce it intracellularly in the yeast host; if the protein is secreted from its native host, secrete it from the yeast system. Although there have been exceptions to this general rule, it is generally best to follow it since the intracellular and secretory environments are very different from each other and synthesis of a protein in the wrong compartment may result in a protein that is improperly folded and inactive. A number of vectors have been constructed for *P. pastoris* expression; a list and detailed discussion of these vectors can be found in (Lin-Cereghino and Lin Cereghino, 2008) and at http://www.biogrammatics.com (Fig. 13.1).

To clone a gene into a *P. pastoris* expression vector, an available template can be amplified with appropriate primers by the polymerase chain reaction (PCR), or the gene can be synthesized. *De novo* synthesis can facilitate cases where DNA optimization or gene modification is required. In any case, suitable restriction sites at the termini can be generated to facilitate the cloning. For example, an *Eco*RI site can be added to the 5′ (ATG-containing) end of the gene and a *Not*I site to the 3′ end of the gene, to facilitate cloning into the pPICZ series of vectors (Invitrogen, Carlsbad, CA), as long as sites for these enzymes do not exist within the sequence of the gene. For *P. pastoris* expression vectors from Biogrammatics, Inc., an appropriate type IIS restriction enzyme site and "seamless" cloning sequences can be added to flank a gene for cloning into an optimal expression context.

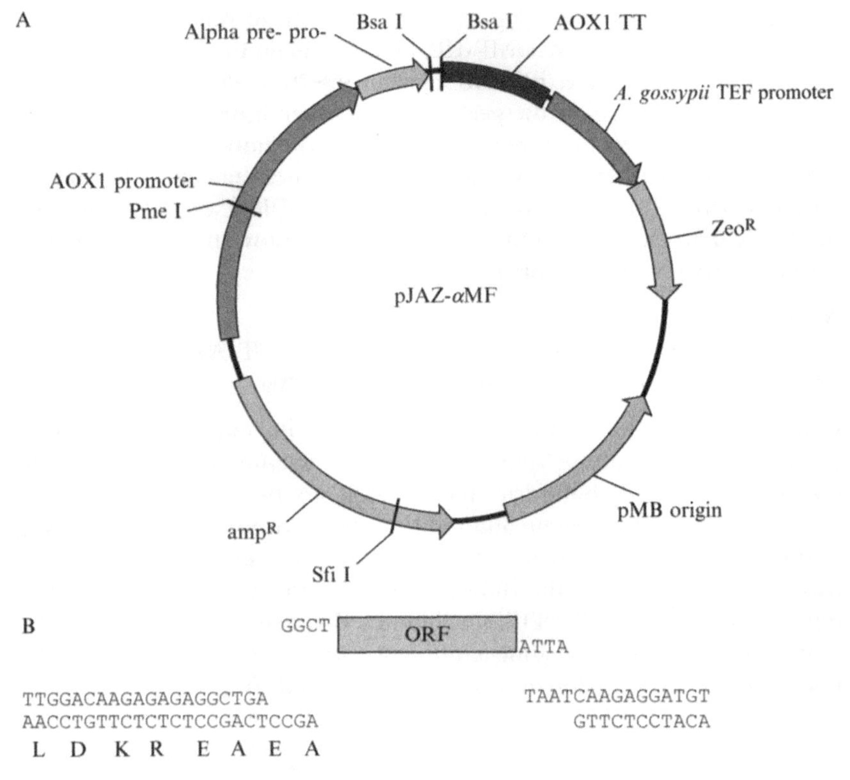

Figure 13.1 Map of Biogrammatics expression vector, pJAZ-αMF.

Cloning a gene whose product is to be secreted can be a little trickier. Extracellular, or secreted expression using a genes native secretion signal sequence will follow the same procedures as above for intracellular expression; however, several options exist when using a foreign signal sequence, such as the *S. cerevisiae* alpha mating factor secretion signal sequence (αMF), encoded in the expression vector. In this case, the subcloning procedure must place the gene of interest in frame with the αMF codons. This αMF secretion signal is very commonly utilized for secretion because it has proved to be very good at secreting recombinant proteins of many types. Although the recombinant protein may be successfully secreted using the αMF signal, proper processing for the αMF at the NH_2-terminus of the desired protein may not occur and modifications of the αMF signal, or the use of an alternative secretion signal sequence, may ultimately be necessary to obtain a properly processed protein. In this regard, the αMF signal sequence, may include two Glu–Ala repeats at the junction between the signal peptide and the NH_2-terminus of the mature protein of interest. The Biogrammatics vector, pJAZ-αMF, is designed for making a construction with these Glu–Ala repeats (Fig. 13.1). Another set of Alternative Biogrammatics vectors such as pJAZ-aMF-KR do not contain the Glu–Ala repeats in the cloning site. To utilize the αMF signal in a pPICZ alpha vector one must add sequences to the 5′ of the gene such that when ligated to the portion of αMF present in the vector it results in the reconstruction of a functional αMF signal sequence (either with or without the sequences encoding the Glu–Ala repeats in αMF). Typically, *Xho*I is used to cut the pPICZ vector which cuts it inside the αMF signal encoding sequences. To restore these sequences, an oligonucleotide with the sequence:

*Xho*I
TCT CTC GAG AAA AGA GAG GCT GAA GCT-ABC-DEF....
Ser Leu Glu Lys Arg Glu Ala Glu Ala

is synthesized where "ABC DEF..." denotes the nucleotide sequences encoding the first amino acids of the mature recombinant protein. This oligonucleotide will hybridize appropriately to the 5′ end of the gene encoding the mature protein and result in the incorporation of the missing portion of the αMF sequence. Similarly, the "seamless" cloning scheme used to clone genes into the Biogrammatics vectors utilize type IIS restriction sites to join the ABC DEF nucleotides of a gene of interest to the last Alanine in the αMF by creating a four base "sticky" end comprising the last nucleotide in a Glu codon and an entire Ala codon (Fig. 13.1).

6. Transformation by Electroporation

At least four different procedures to introduce foreign plasmid DNA into *P. pastoris* have been developed using: spheroplast-generation, LiCl, polyethelene glycol1000 and electroporation. The electroporation procedure is most commonly used and therefore a modified version of that described by (Becker and Guarante, 1991) will be outlined in detail. For the other procedures, readers are referred to either of the volumes of Methods in Molecular Biology: *Pichia* Protocols (Cregg, 2008; Higgins and Cregg, 1998).

7. DNA Preparation

For all transformation methods, linear plasmid DNA is most commonly transformed into *P. pastoris* for integration into the yeast genome. The DNA sequence at the ends of the linear plasmid DNA stimulate integration by a single crossover recombination event into the locus shared by vector and host genome. Therefore, linearization of an expression plasmid is performed in

Pichia DNA in the plasmid, such as in the promoter (i.e., a *Pme*I site in the AOXI promoter). The final vector, prepared in *E. coli*, is cut with a restriction enzyme that linearizes the vector, and then the DNA is purified and concentrated to at least 100 ng/ul in water prior to transformation. At this point, the vector is ready for transformation into *P. pastoris*.

Procedure for preparation of electrocompetent cells (Lin-Cereghino *et al.*, 2005).

Prepare the following (all solutions should be autoclaved except for the DTT and HEPES solutions, which should be filter sterilized):

1. 500 ml liquid YPD media in a 2.8 l Fernbach culture shaking flask.
2. H_2O (1 l).
3. 1 *M* sorbitol (100 ml).
4. Appropriate selective agar plates.
5. 1 *M* DTT (2.5 ml).
6. BEDS solution (9 ml): 10 m*M* Bicine–NaOH, pH 8.3, 3% ethylene glycol, 5% DMSO (dimethyl sulfoxide), 1 *M* sorbitol and 0.1 *M* DTT.
7. 1 *M* HEPES buffer, pH 8.0 (50 ml).
8. Sterile 250 ml centrifuge tubes.
9. Sterile electroporation cuvettes.
10. Electroporation instrument: BTX Electro Cell Manipulator 600 (BTX, San Diego, CA); Bio-Rad Gene Pulser (Bio-Rad, Hercules, CA); Electroporator II (Invitrogen, San Diego, CA). Parameters for electroporation with different instruments vary with the instrument. Be sure to check instructions for each type of instrument. (Becker and Guarante, 1991; Grey and Brendel, 1995; Pichia Expression Kit Instruction Manual; Stowers *et al.*, 1995).

Protocol:

1. Inoculate 10 ml YPD media with a single fresh *P. pastoris* colony of the strain to be transformed from an agar plate and grow overnight shaking at 30 °C.
2. Use the overnight culture to inoculate a 500 ml YPD culture in a 2.8 l Fernbach culture flask to a starting OD_{600} of 0.01 and grow to an OD_{600} of 1.0 (~ 12 h).
3. Harvest the cells by centrifugation at 2000*g* at 4 °C, discard the supernatant and resuspend the cells in 100 ml of fresh YPD medium plus HEPES (pH 8.0, 200 m*M*) in a sterile 250 ml centrifuge tube.
4. Add 2.5 ml of 1 *M* DTT and gently mix.
5. Incubate at 30 °C for 15 min with slow rotating.
6. Add 150 ml cold water to the culture and wash by centrifugation at 4 °C with an additional 250 ml of cold water. At this stage and from here on, keep the cells ice cold and do not vortex the cells to resuspend them (slow pipetting is best).

7. Wash cells a final time in 20 ml of cold 1 *M* sorbitol, then resuspend in 0.5 ml of cold 1 *M* sorbitol (final volume including cells will be 1.0–1.5 mls).
8. Use these cells directly without freezing to achieve the most transformants.
9. To freeze competent cells, distribute in 40 ul aliquots to sterile 1.5 ml minicentrifuge tubes and place the tubes in a − 70 °C freezer.

Electroporation procedure:

1. Add up to 1 ug of linearized plasmid DNA sample in no more than 5 ul of water to a tube containing 40 ul of frozen or fresh competent cells and then transfer the entire mixture to a 2 mm gap electroporation cuvette held on ice.
2. Pulse cells according to the parameters suggested for yeast by the manufacturer of the specific electroporation instrument (Table 13.1).
3. Immediately add 0.5 ml of cold 1 *M* sorbitol and 0.5 ml of cold YPD, then transfer the entire cuvette contents to a sterile 1.5–2.0 ml minicentrifuge tube.
4. Incubate for 3.5–4 h at 30 °C with slow shaking (100 rpm).
5. Spread aliquots onto selective agar plates and incubate for 2–4 days.
6. To avoid mixed colonies, pick and restreak transformants on selective medium at least once before proceeding with further analysis.

Rapid preparation of electrocompetent *P. pastoris*:

1. Grow a 5 ml culture of *P. pastoris* in YPD overnight, shaking at 30 °C.
2. Dilute the overnight culture to an OD_{600} of 0.15–0.20 in 50 ml YPD medium in a flask large enough to provide good aeration.
3. Grow to an OD_{600} of 0.8–1.0 at 30 °C with shaking (4–5 h).
4. Centrifuge cells at 500*g* for 5 min at room temperature and discard supernatant.
5. Resuspend cells in 9 ml of ice-cold BEDS solution supplemented with DTT.
6. Incubate the cell suspension for 5 min, shaking at 30 °C.
7. Centrifuge cells at 500*g* for 5 min at room temperature and resuspend in 1 ml of BEDS (without DTT).
8. Perform electroporation as described above, immediately or freeze cells in small aliquots at − 80 °C.

8. Examination of Strains for Recombinant Protein Production

Yeast expression systems have been successful at generating large quantities of recombinant proteins. For example, production levels of between 1 and 10 g/l of culture supernatant have been secreted from

Table 13.1 Parameters for electroporation using selected instruments

Instrument	Cuvette gap (mm)	Sample volume (ul)	Charging voltage (V)	Capacitance (uF)	Resistance (Ω)	Field strength (kV/cm)	Pulse length (~ ms)	References
ECM600 (BTX)	2	40	1500	Out	129	7500	5	Becker and Guarante (1991)
Electroporator II (Invitrogen)	2	80	1500	50	200	7500	10	Pichia Manual
Gene-Pulser (Bio-Rad)	2	40	1500	25	200	7500	5	Grey and Brendel (1995)
Cell-Porator (BRL)	1.5	20	480	10	Low	2670	NS	Lorow-Murray and Jesse (1991) and Stowers *et al.* (1995)

P. pastoris strains. However, one should not expect to see a band on a coomassie-stained gel in initial expression experiments. Thus, it is imperative that by the time a recombinant strain is ready to begin expression studies, one or more sensitive assays for the detection of the targeted protein is in place. Assays can be based on enzyme activity, an epitope tag fused to ones product or an antibody against the desired gene product; however, the point is: serious efforts to develop sensitive detection methods should begin at the same time or even prior to expression studies. Good assays facilitate strain development as illustrated in the following examples.

The most convenient way to detect foreign protein expression in yeast is via a plate activity assay. Plate assays allow one to crudely quantify and compare productivity of 100s to 1000s of transformants directly on diagnostic plates at a single glance using replica-plating or other techniques.

An example of a plate activity assay is shown in Fig. 13.2 for secretion of the enzyme phytase. This enzyme degrades phytate which results in the clearing of a zone around the expressing colony on agar plates. The size of the zone roughly correlates to the amount of phytase being secreted. As a second example, the expression of bacterial β-lactamase as an intracellular protein is shown in *P. pastoris* (Fig. 13.3). The colonies, which are typically yellow in color, turn pink to purple with the expression of β-lactamase. The intensity of the purple color roughly indicates the amount of enzyme expressed and, in this case, the number of copies of the expression cassette in

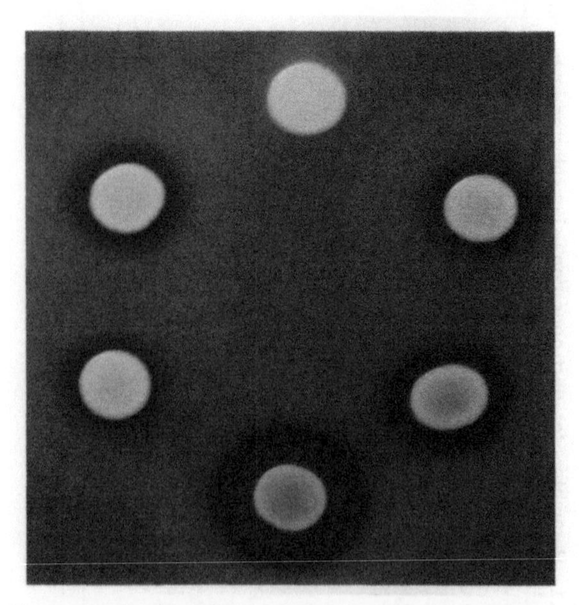

Figure 13.2 Plate assay for detection of phytase constitutively expressed and secreted by selected *P. pastoris* strains. Top spot: negative control; remaining spots show five transformants secreting various amounts of the enzyme.

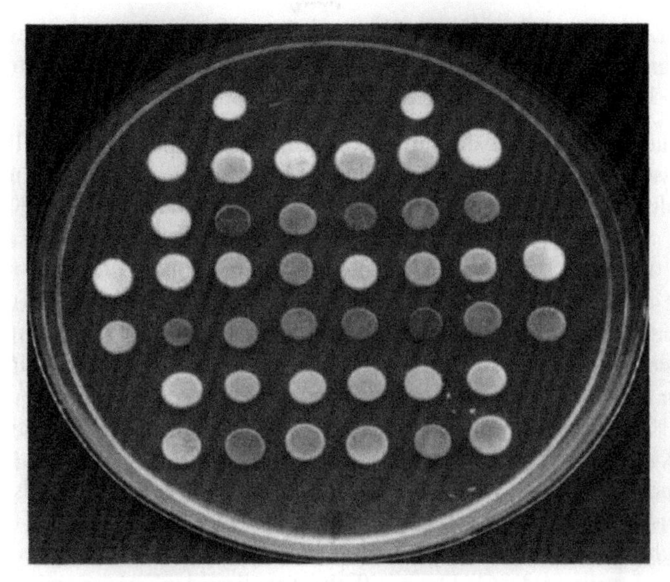

Figure 13.3 Expression of intracellular β-lactamase in *P. pastoris*. The two spots at the top of the plate are non-β-lactamase expressing negative control strains. (See Color Insert.)

each strain. Finally, with good quality polyclonal or monoclonal antibodies against a tag region or the recombinant protein itself, a plate antibody assay or "Yeastern Blot" is possible, as described below.

Once colonies of transformed cells have been selected, single cell purified and collected onto a master plate, samples of multiple strains are examined for expression of the foreign protein. Before analysis of expression, one can screen for the presence of the recombinant gene in transformants by PCR. Simply prepare genomic DNA in a cell-free extract by glass bead disruption as described below and utilize it as template in a PCR reaction with primers that are complementary to a portion of the recombinant gene. Different methods of detection for the expression of proteins in *P. pastoris* can be applied depending on what kind of promoter is used for expression (inducible or constitutive), and what kind of vector is used (intracellular or secretory):

A. If the gene of interest is expressed constitutively, inoculate a sample colony into YPD medium, grow the culture for 2–3 days with good aeration and analyze the proteins in samples taken periodically during this time by any available detection method.
B. For methanol-induced expression, a colony should be grown for 17–24 h in YPD and then transferred to a fresh methanol-containing medium for induction. The induction can be performed in 15-ml tubes

with 2 ml medium containing 0.5% methanol with good aeration at a starting OD_{600} of about 10. For intracellular proteins an induction time of 6–12 h in methanol medium is sufficient.

Intracellularly expressed samples require the preparation of cell extracts for analysis. Culture samples can be harvested after 10–12 h and prepared for cell breakage by glass bead disruption. Harvest approximately 10–50 OD_{600} units of cells and resuspend in 150 ul of a breakage buffer. Add an equal volume of sterile acid washed 0.45 mm diameter glass beads. Vortex the mixture on the highest setting for 1 min, then place the sample on ice for 1 min; repeat this process at least five times. Alternatively, load samples into a vortexer with a head made to hold multiple microfuge tubes. Place the vortexer in a 4 °C cold box or cold room and vortex on high for approximately 10 min. Examine cultures for cell breakage under a microscope, 80–90% of the cells should be disrupted. After disruption, draw off the cell debris and buffer into a fresh microcentrifuge tube. Rinse the beads with an additional 100 ul portion of buffer by a brief vortexing and transfer the wash to the tube with the rest of the cell debris. Centrifuge the samples on high speed for 5 min at 4 °C. Draw off the top liquid phase containing your protein and transfer to a fresh microcentrifuge tube. This is your crude protein sample ready for SDS–PAGE, western blotting or enzyme assay.

Secreted proteins build up in the medium much more slowly and require at least 2 days to reach high levels. Allow the cultures to incubate with shaking for 2–5 days. Add fresh methanol to a final concentration of 0.5% every 12 h and collect 50–100 ul supernatant samples during this induction period. The supernatant samples are ready for SDS–PAGE, western blotting or enzyme assay and can be stored at – 20 °C.

9. Assay Development—The Yeastern Blot

There is much in common in different antibody assays. One useful and yeast-specific antibody assay is the yeast colony western blot, sometimes referred to as the "Yeastern" blot for secreted proteins (Fig. 13.4). For standard western blot assay procedures, you are referred to (Sambrook *et al.*, 1989).

The Yeastern blot is a convenient means of qualitatively screening large numbers of yeast colonies directly on plates for expression of a recombinant protein. However, readers should be aware that the correlation between the size of the "halo" surrounding a colony and the amount of recombinant product is not always linear and all results with this method should be confirmed with a standard western blot or other assay. The procedure for Yeastern blotting is as follows:

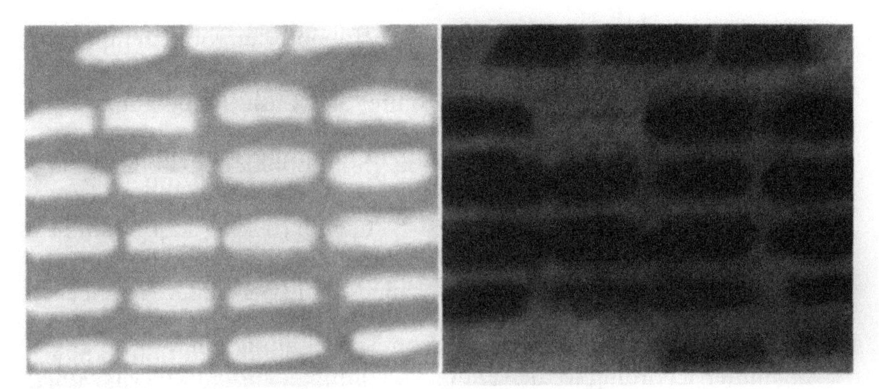

Figure 13.4 "Yeastern Blot" of *P. pastoris* colonies secreting both heavy and light chains of IgG antibodies. Negative controls of *P. pastoris* strain that does not secrete antibody shown on second row second streak from the left and at first two positions from the left on bottom row.

1. Transfer freshly grown yeast colonies from the surface of an agar Petri plate onto a sterile piece of Whatman No. 1 filter paper by a standard replica-plating method. The filter paper should be cut to a size that exactly fits within the plate.
2. Place the filter with yeast cells onto the surface of a fresh plate containing an appropriate induction medium and incubate the plate for 1–2 days.
3. Prepare a piece of nitrocellulose membrane the same size and dimensions as the filter paper by soaking the membrane for 5 min or more in 15 ml of transfer buffer (25 m*M* Tris base, pH 8.5, 0.2 *M* glycine, 20% methanol). Soak two additional pieces of cut filter paper in transfer buffer.
4. Prepare a sandwich of the papers as described below:

   a. Place one piece of the filter paper on the anode platform of a western blotter. (It is essential to remove all bubbles between membrane layers. This can be done by rolling a pipette over their surface.).
   b. Place the soaked nitrocellulose filter on top of the filter paper.
   c. Place the filter paper with replica-plated yeast cells on top of the nitrocellulose paper.
   d. Place the second piece of soaked filter paper on top of the filter with cells.
   e. Finally, place the cathode plate on top of the sandwich.

5. Transfer proteins to the nitrocellulose membrane with a constant current (1–4 mA/cm^2) for 1 h.
6. Remove the nitrocellulose membrane and wash it for 5 min in 15 ml TBS buffer (50 m*M* Tris–HCl, pH 7.6, 150 m*M* NaCl). Replace the TBS buffer with 15 ml of TBST (TBST is prepared by adding Tween-20 to 0.05% to TBS) containing 1–5% bovine serum albumin (BSA).

Place filter in blocking buffer and rock gently at room temperature for 1 h.

7. Transfer filter to 15 ml of TBST buffer and wash by rocking at room temperature for 5 min.
8. Prepare primary antibody dilution in 15 ml of blocking buffer according to the vendor's recommendations (typically, 0.5 ug/ml).
9. Incubate membrane overnight at 2–8 °C with rocking action (can be as short as 1–3 h at room temperature depending on antibody).
10. Wash membrane in at least four changes of TBST buffer (15 ml each, 5 min/wash) then briefly rinse in fresh TBST.
11. Prepare an enzyme-conjugated secondary antibody dilution in blocking buffer according to the vendor's recommendations (typically 3 ul in 15 ml of blocking buffer).
12. Incubate membrane in secondary antibody solution for 1 h, rocking at room temperature.
13. Repeat Step 10.
14. Treat membrane in the dark room with visualization reagents according to the vendor's recommendations (e.g., Pierce ECL Western Blotting Substrate), remove excess liquid by blotting with filter paper and place the filter in a plastic protector and expose to X-ray film for anywhere from 30 s to several minutes. The intensity and size of the signal around each colony approximately reflects the secretion level of the recombinant product by the different colonies/strains.

10. Posttranslational Modification of the Recombinant Protein (Proteinases and Glycosylation)

Posttranslational problems that affect the quality of recombinant proteins expressed in *P. pastoris* include proteolysis and glycosylation.

A proteinase deficient strain of *P. pastoris* can be tested if initial results by SDS–PAGE analysis (coomassie staining or western blots) suggest proteolytic degradation of the protein. Signs of proteolysis include low recombinant protein levels and active or immunoreactive products that are smaller than the full-length product. Degraded protein can also run as a "smear" after PAGE, running from approximately the correct size of the product to smaller sizes. The Pichia strain SMD1168 (*pep4 his4*) is deleted for much of the *PEP4* gene (Gleeson *et al.*, 1998). The PEP4 gene product is responsible for activating many of the proteases in the vacuole *of P. pastoris* which enter the vacuole as inactive zymogens and are activated there by the *PEP4* product. Although secreted recombinant proteins do not go to the vacuole, they can contact proteases in the culture medium from the lysis of a small portion of the cells. Due to the extremely high culture densities used with *P. pastoris*, the concentration of vacuolar proteases in the medium can be considerable. To utilize a *PEP4* strain, one must transform one of these strains (i.e., SMD1168 (*pep4 his4*) or SMD1168H (*pep4*)) with the expression vector or create deletions in the PEP4 gene in a strain already expressing a protein of interest. To determine if PEP4 deficient strains benefit the expression of the protein, the recombinant protein should be examined from wild type and the PEP4 deficient strains after an induction performed in parallel. Note: *PEP4* deficient strains of *P. pastoris* are not as robust as wild-type strains. In particular, *PEP4* deficient strains die more quickly when stored on plates, do not transform as efficiently as wild-type strains, grow more slowly in culture, and are more difficult to induce on methanol. Furthermore, *PEP4* deficient strains are difficult to mate with other non-*PEP4 P. pastoris* strains.

If a recombinant protein expressed *in P. pastoris* is larger than expected and somewhat heterogenous in size by SDS–PAGE analysis, it may be glycosylated. One should first examine the amino acid sequence of the protein for potential glycosylation sites: the signal for addition of N-linked oligosaccharides is ASN-X-Ser/Thr, and O-linked sugars can be added to the OH-group of any Ser or Thr residue. Glycosylation can be confirmed by deglycoslyating suspect protein with PNGase F and examining the product by SDS–PAGE. A good protocol for deglycoslation of proteins can be found at: http://www.neb.com/nebecomm/products/productP0704.asp. If this treatment reduces the apparent molecular size of your protein resulting in a more homogenous product, the protein is almost certainly glycosylated.

11. Selection for Multiple Copies of an Expression Cassette

Perhaps the most productive means of increasing the per cell amount of a recombinant protein using the *P. pastoris* system is by increasing the number of copies of the expression cassette in a strain (Brierley, 1998; Thill *et al.*, 1990). Two general approaches have been developed to create multi-copy expression strains in *P. pastoris*. The first approach involves constructing a vector with multiple head-to-tail copies of an expression cassette (Brierley, 1998). The key to generating this construction is a vector that has an expression cassette flanked by restriction sites that have complementary termini (e.g., *Bam*HI–*Bgl*II, *Sal*I–*Xho*I combinations). The process of repeated cleavage and reinsertion results in the generation of a series of vectors that contain increasing numbers of expression cassettes. A particular advantage to this approach, especially in the production of human pharmaceuticals, is that the precise number of expression cassettes is known and can be recovered for direct verification by DNA sequencing.

The second approach utilizes expression vectors that contain a drug resistance gene as the selectable marker, and selection for strains resistant to higher levels of the drug (Scorer *et al.*, 1994). Drug resistant genes used in *P. pastoris* include the bacterial *Kan*^{R}, *Zeo*^{R}, and *Bsd*^{R} genes, as well as, the *P. pastoris FLD1* gene (Shen *et al.*, 1998). Each of these genes supports significant enrichment for strains with increased copy number of the expression vector with higher levels of drug resistance. For example, selection of zeocin resistant *P. pastoris* transformants is performed on plates with 1–2 mg/ml zeocin instead of the standard 100–200 ug/ml zeocin. However, no matter which drug is used, the vector copy number will still vary greatly. After selection most transformants still contain only a single vector copy, even if they are resistant to high drug levels. Thus, 50–100 independent transformants selected at the high drug concentration should be analyzed for copy number and expression level to identify better expressing strains. By this approach, strains carrying up to 30 copies of an expression cassette have been isolated (Scorer *et al.*, 1994). Importantly, once isolated, multicopy strains are stable with standard microbial handling procedures and do not require continued drug selection on plates or in liquid medium (i.e., maintain a stock of a given strain selected on medium with the drug stored frozen at −80 °C, then use a working stock kept on plate of noninducing medium for a limited number of experiments or a single production run).

One drawback of this selection procedure is the difficulty in obtaining enough clones to screen for multiple expression cassettes. First, the number of transformants resistant to high levels of drug is very low, often 0.1–1% of the number on low levels of drug (100 ug/ml for zeocin). Therefore, unless ones transformation efficiency is at its peak, there may not be any transformants resistant to the highest drug levels. Furthermore, 50–100 transformants are needed to screen for a multicopy strain since only about 1–5% of the transformants resistant to high levels of the drug are due to added copies of the resistance gene and most are resistant due to other unknown factors.

In part due to the difficulty in obtaining multicopy strains by direct selection a new method for obtaining strains with high copy numbers of vector and elevated recombinant protein expression levels was developed (Sunga *et al.*, 2008). Briefly, *P. pastoris* transformants selected on a low level of drug and containing only one or a few copies of the vector are subsequently subjected to higher drug levels to obtain strains with higher numbers of copies. Simply streak transformants on agar plates containing higher and higher levels of zeocin. For example, if the original transformant was selected on 100 ug/ml of zeocin, streak the strain on plates containing 500 ug/ml of the drug. Collect colonies that are resistant to the higher level of drug as individual strains and confirm that one or more have elevated recombinant protein expression levels and are resistant due to a higher number of vector copies by PCR. Once the strain is confirmed as a better expressing/multicopy strain, the process can be repeated at an even higher concentration of zeocin (2 mg/ml). Again, high resistance strains are collected and examined for expression and the number of vector copies. This iterative process has been termed Posttransformational vector amplification (PTVA) and results in strains containing multiple head-to-tail copies of the entire vector integrated at a single locus in the genome. An analysis of PTVA-selected clones indicates 40% showed a three- to fivefold increase in vector copy number. So-called "jackpot" clones, with greater than 10 copies of the expression vector, represented 5–6% of selected clones and in some cases had a proportional increase in recombinant protein production.

Although the molecular details of the process(s) by which these amplification events occur are not well understood, key observations about the process have been made. First, the amplification process appears to occur naturally in a small percentage of cells in virtually any vector containing strain. Second, the PTVA process leads to a considerable increase in copy number of the entire vector and not just portions of the vector such as the resistance gene. This is clearly important if a uniform recombinant product is desired. Third, Southern blot data demonstrated that all the copies are inserted into the *P. pastoris* genome in the same location as the original copy and in a head-to-tail configuration (Sunga *et al.*, 2008).

Finally, the PTVA method works with other drug resistant selectable markers in *P. pastoris* and not just with zeocin vectors. Thus, this amplification process seems to be a general response to high drug levels in this yeast species. Given that most yeast species have similar homologous recombination systems, this technique should work in other yeast species as well.

REFERENCES

Ausubel, E. M., Brent, R., Kingston, E., Moore, D. D., Seidman, J. G., Smith, J. A., and Struhl, K. (eds.) (2001). *In* "Current Protocols in Molecular Biology" John Wiley and Sons, Inc., New York.

Becker, D. M., and Guarante, L. (1991). High-efficiency transformation of yeast by electroporation. *Methods Enzymol.* **194,** 182–187.

Boer, E., Gellisson, G., and Kunze, G. (2005). Arxula adeninivorans. *In* "Production of Recombinant Proteins" (G. Gellisson, ed.), pp. 89–110. Wiley-VCH Verlag GmbH and Co. KGaA, Weinheim, Germany. Chapter 5.

Brierley, R. A. (1998). Secretion of recombinant human insulin-like growth factor I (IGF-I). *In* "Methods in Molecular Biology Vol. 103 Pichia Protocols" (D. R. Higgins and J. M. Cregg, eds.), pp. 149–178. Humana Press, Totowa, NJ.

Cregg, J. M. (2008). DNA-mediated transformation. *In* "Methods in Molecular Biology: Pichia Protocols" (J. M. Cregg ed.), 2nd ed. pp. 27–42. Humana Press, Totowa, NJ. Chapter 3.

Gellissen, G. (ed.) (2002). *In* "*Hansenula polymorpha*: Biology and Applications" Wiley-VCH Verlag GmbH and Co. KGaA, Weinheim, Germany.
Gellissen, G. (ed.) (2005). *In* "Production of Recombinant Proteins" Wiley-VCH Verlag GmbH and Co. KGaA, Weinheim, Germany.
Gleeson, M. A. G., White, C. E., Meininger, D. P., and Komives, E. A. (1998). Generation of protease-deficient strains and their use in heterologous protein expression. *In* "Methods in Molecular Biology, Vol. 103, Pichia Protocols" (D. Higgins and J. M. Cregg, eds.), pp. 81–94. Humana Press, Totowa, NJ.
Grey, M., and Brendel, M. F. (1995). Ten-minute eletrotransformation of *Saccharomyces cerevisiae. In* "Methods in Molecular Biology, Vol. 47: Electroporation Protocols for Microorganisms" (J. A. Nickoloff, ed.), pp. 269–272. Humana Press, Totowa, NJ.
Heerikhuisen, M., van den Hondel, C., and Punt, P. (2005). Aspergillus sojae. *In* "Production of Recombinant Proteins" (G. Gellissen, ed.), pp. 191–214. Wiley-VCH Verlag GmbH and Co. KGaA, Weinheim, Germany. Chapter 9.
Higgins, D. R., and Cregg, J. M. (eds.) (1998). Methods in Molecular Biology, Vol. 103, Pichia Protocols, Humana Press, Totowa, NJ.
Kuck, U., and Poggeler, S. (2005). Sordaria macrospora. *In* "Production of Recombinant Proteins" (G. Gellissen, ed.), pp. 215–232. Wiley-VCH Verlag GmbH and Co. KGaA, Weinheim, Germany. Chapter 10.
Lin-Cereghino, J., and Lin Cereghino, G. P. (2008). Vectors and strains for expression. *In* "Methods in Molecular Biology: Pichia Protocols" (J. M. Cregg, ed.), pp. 111–126. Humana Press, Totowa, NJ. Chapter 2.
Lin-Cereghino, J., Wong, W. W., Xiong, S., Giang, W., Luong, L. T., Vu, J., Johnson, S. D., and Lin-Cereghino, G. P. (2005). Condensed protocol for competent cell preparation and transformation of the methylotrophic yeast *Pichia pastoris. Biotechniques* **38,** 44–48.
Lorow-Murray, D., and Jesse, J. (1991). High efficiency transformation of *Saccharomyces cerevisiae* by electroporation. *Focus* **13,** 65–68.
Madzak, C., Nicaud, J.-M., and Gaillardin, C. (2005). Yarrowia lipolytica. *In* "Production of Recombinant Proteins" (G. Gellissen, ed.), pp. 163–190. Wiley-VCH Verlag GmbH and Co. KGaA, Weinheim, Germany. Chapter 8.
Pichia Expression Kit Instruction Manual p. 63. Version E, Invitrogen, San Diego, CA.
Romanos, M. A., Scorer, C. A., and Clare, J. J. (1992). Foreign gene expression in yeast: A review. *Yeast* **8,** 423–488.
Sambrook, J., Fritsch, E. F., and Maniatis, T. (1989). Molecular Cloning: A Laboratory Manual. 2nd ed. Cold Spring Harbor Laboratory Press, Cold Spring Harbor, NY.
Scorer, C. A., Clare, J. J., McCombie, W. R., Romanos, M. A., and Sreekrishna, K. (1994). Rapid selection using G4118 of high copy number transformants in *Pichia pastoris* for high-level foreign gene expression. *Bio/Technology* **12,** 181–184.
Shen, S., Sulter, G., Jeffries, T. W., and Cregg, J. M. (1998). A strong regulated promoter for controlled expression of foreign genes in the yeast *Pichia pastoris. Gene* **216,** 93–102.
Stowers, L., Gautsch, J., Dana, R., and Hoekstra, M. F. (1995). Yeast transformation and preparation of frozen spheroplasts for electroporation. *In* "Methods in Molecular Biology, Vol 47: Electroporation Protocols for Microorganisms" (J. A. Nickoloff, ed.), pp. 261–267. Humana Press, Totowa, NJ.
Sunga, A. J., Tolstorukov, I., and Cregg, J. M. (2008). Post transformational vector amplification in the yeast *Pichia pastoris. FEMS Yeast Res.* **8,** 870–876.
Thill, G. P., Davis, G. R., Stillman, C., Holtz, G., Brierley, R., Engel, M., Buckholtz, R., Kinney, J., Provow, S., Vedvick, T., and Siegel, R. S. (1990). Positive and negative effects of multi-copy integrated expression vectors on protein expression in *Pichia pastoris. In* "Proceedings of the 6th International Symposium on Genetics of Microorganisms"

(H. Heslot, J. Davies, J. Florent, L. Bobichon, G. Durand, and L. Penasse, eds.) Vol. II, pp. 477–490. Societe Francaise de Microbiologie, Paris.

Tolstorukov, I., and Cregg, J. M. (2008). Classical Genetics. *In* "Methods in Molecular Biology: Pichia Protocols" (J. M. Cregg, ed.), 2nd ed. pp. 189–202. Humana Press, Totowa, NJ. Chapter 14.

van Ooyen, A. J., Dekker, P., Huang, M., Olsthoorn, M. M., Jacobs, D. I., Colussi, P. A., and Taron, C. H. (2006). Heterologous protein production in the yeast *Kluyveromyces lactis*. *FEMS Yeast Res.* **6,** 381–392.

CHAPTER FOURTEEN

Baculovirus–Insect Cell Expression Systems

Donald L. Jarvis

Contents

1. Introduction 192
2. A Brief Overview of Baculovirus Biology and Molecular Biology 193
3. Baculovirus Expression Vectors 195
4. Baculovirus Expression Vector Technology—The Early Years 196
5. Baculovirus Expression Vector Technology—Improved 198
6. Baculovirus Transfer Plasmid Modifications 198
7. Parental Baculovirus Genome Modifications 200
8. The Other Half of the Baculovirus–Insect Cell System 210
9. A New Generation of Insect Cell Hosts for Baculovirus Expression Vectors 212
10. Basic Baculovirus Protocols 214
   10.1. Insect cell maintenance 215
   10.2. Isolation of baculovirus genomic DNA 215
   10.3. Baculovirus plaque assays 216
References 218

Abstract

In the early 1980s, the first-published reports of baculovirus-mediated foreign gene expression stimulated great interest in the use of baculovirus–insect cell systems for recombinant protein production. Initially, this system appeared to be the first that would be able to provide the high production levels associated with bacterial systems and the eukaryotic protein processing capabilities associated with mammalian systems. Experience and an increased understanding of basic insect cell biology have shown that these early expectations were not completely realistic. Nevertheless, baculovirus–insect cell expression systems have the capacity to produce many recombinant proteins at high levels and they also provide significant eukaryotic protein processing capabilities. Furthermore, important technological advances over the past 20 years have improved upon

Department of Molecular Biology, University of Wyoming, Laramie, Wyoming, USA

Methods in Enzymology, Volume 463
ISSN 0076-6879, DOI: 10.1016/S0076-6879(09)63014-7

the original methods developed for the isolation of baculovirus expression vectors, which were inefficient, required at least some specialized expertise and, therefore, induced some frustration among those who used the original baculovirus–insect cell expression system. Today, virtually any investigator with basic molecular biology training can relatively quickly and efficiently isolate a recombinant baculovirus vector and use it to produce their favorite protein in an insect cell culture. This chapter will begin with background information on the basic baculovirus–insect cell expression system and will then focus on recent developments that have greatly facilitated the ability of an average investigator to take advantage of its attributes.

1. Introduction

It seems that nearly every book chapter or review focusing on the baculovirus–insect cell expression system begins with a statement emphasizing the popularity of this system and/or noting its widespread use for recombinant protein production. While this introduction has become increasingly redundant, it also has become increasingly accurate over the past 20 years. An advanced PubMed search using the terms "baculovirus" and "expression" and "vector" at the time of this writing yielded over 2000 hits. While this is a significant number, it vastly underestimates the actual number of published studies involving recombinant protein production using the baculovirus–insect cell expression system. Moreover, it does not include the large number of studies performed behind closed doors in the biotechnology industry, which are clearly evidenced by the number of oral presentations given by industrial scientists at various baculovirus–insect cell technology conferences held over the past two decades.

The previous edition of this Guide to Protein Purification (1990) included a comprehensive description of the technical details involved in using the baculovirus–insect cell expression system, as it existed at that time. This exercise will not be repeated here and the reader should refer to the previous edition of this book and other sources given below for those details. In this edition, I will begin by providing the background information needed for the reader to understand the basic principles underlying the original baculovirus–insect cell expression system. I will then focus on the new tools and approaches developed for the isolation of recombinant baculoviruses since the publication of the previous edition of this book, as these have greatly facilitated the ability of virtually any biomedical investigator to utilize the baculovirus–insect cell system, relative to the state of the art in 1990.

2. A Brief Overview of Baculovirus Biology and Molecular Biology

Baculoviridae is a large family of viruses with relatively large, double-stranded, circular DNA genomes (see Miller, 1997 for a comprehensive review). The natural hosts for these viruses are arthropods, mainly insects, and most baculoviruses have a very narrow host range, which is usually restricted to just one insect species. The most extensively studied and exploited member of the Baculoviridae is *Autographa californica* multicapsid nucleopolyhedrovirus, or Ac*M*NPV. Compared to most baculoviruses, this isolate has a relatively broad host range, as it can productively infect around 25 different lepidopteran insects and can infect most tissue types within an individual target insect species. Ac*M*NPV was the isolate used to develop the original recombinant baculovirus vectors (Pennock *et al.*, 1984; Smith *et al.*, 1983b) and its genome is still used as the backbone for the production of most baculovirus vectors today. Thus, while they can be produced using other types of baculoviruses, most generic references to "baculovirus vectors" actually refer more specifically to recombinant variants of Ac*M*NPV. Because this is a relatively unimportant detail for most baculovirus–insect cell expression system users, I will slip into this rather loose language in the latter parts of this chapter.

With respect to the development of the original baculovirus–insect cell expression system (see Summers, 2006 for a detailed history), one of the most important features of Ac*M*NPV and other baculoviruses was the ability of these viruses to produce "polyhedra" or "occlusion bodies" during productive viral infections. Polyhedra are large particles that appear in the nuclei of Ac*M*NPV-infected insect cells near the end of the infectious cycle. At late times after infection, host cell nuclei are typically packed with 2–3 dozen of these complex particles, which consist, in part, of progeny virions embedded within a protective, paracrystalline array. Importantly, this paracrystalline array is composed of a single, virus-encoded protein called polyhedrin. Thus, it is obvious that Ac*M*NPV and other baculoviruses must produce extremely large amounts of polyhedrin protein in order to fulfill their need to produce large numbers of polyhedra. In fact, polyhedrin is the major protein in Ac*M*NPV-infected cells near the end of the infectious cycle, typically representing over half of the total protein found in these cells at that time.

The ability of baculoviruses to produce large amounts of polyhedrin is important because it was one of the fundamental features that spurred the development of these viruses as vectors for foreign protein production. The basic notion was that the ability of these viruses to produce extremely large amounts of polyhedrin could be harnessed to produce large amounts of other proteins of far greater interest to the biomedical research community. This notion was extended by the finding that the polyhedrin protein is not required for the replication of baculoviruses in cultured insect cells (Smith *et al.*, 1983a). Thus, it was theoretically possible to create helper-independent recombinant baculoviruses in which the viral DNA sequence encoding polyhedrin was replaced by a foreign DNA sequence encoding a protein of interest. This theory was reduced to practice when the first recombinant baculoviruses were produced by homologous recombination between the polyhedrin region in the Ac*M*NPV genome and "transfer" plasmids containing a foreign gene placed under the control of the polyhedrin promoter (Pennock *et al.*, 1984; Smith *et al.*, 1983b).

The original recombinant baculoviruses were designed to express foreign genes under the control of the polyhedrin promoter because the strength of this promoter is ultimately responsible for the ability of Ac*M*NPV and other baculoviruses to produce such large amounts of the polyhedrin protein. However, it is relevant that baculoviruses actually have three or four distinct classes of genes, which are expressed in a temporally regulated, sequential fashion. The first to be expressed are the early genes, which can be subclassified as immediate-early and delayed-early genes (reviewed in Friesen, 1997). The salient feature of the early genes is that they have host-like promoters, which can be recognized and transcribed by host transcriptional machinery. With some important caveats that will not be discussed here, these promoters can function in the absence of other baculoviral factors to induce viral gene expression at the very beginning of the infectious cycle. For this reason, baculovirus early promoters have been used to drive constitutive foreign gene expression in uninfected lepidopteran insect cells (reviewed in Douris *et al.*, 2006; Harrison and Jarvis, 2007a; Jarvis and Guarino, 1995) and this is an important facet of efforts to develop transgenic insect cell lines as improved hosts for baculovirus-mediated foreign protein production (reviewed in Harrison and Jarvis, 2006, 2007b; Shi and Jarvis, 2007). The next class of genes to be expressed during baculovirus infection is the late genes, which is a set of genes expressed after the onset of viral DNA replication (reviewed in Lu and Miller, 1997; Passarelli and Guarino, 2007). The salient feature of these genes is that they have virus-specific promoters, which can be recognized and transcribed only by virus-encoded transcriptional machinery. Thus, these promoters can function only in the context of baculovirus infection and are activated at later times of infection, once a virus-specific transcription complex has been produced. The last class of genes to be expressed during baculovirus infection is the very late genes, which are expressed after the onset of viral DNA replication, like the late genes, but later, closer to the end of the infectious cycle. The very late genes encode proteins such as polyhedrin that are involved in the production of polyhedra. Like the late genes, the very late genes have virus-specific promoters that can be recognized and transcribed only by a virus-encoded transcription complex and, therefore, can function only in the context of baculovirus infection. In fact, the very late promoter-specific transcription complex includes the same proteins as the late promoter-specific complex plus one or more additional proteins (McLachlin and Miller, 1994; Mistretta and Guarino, 2005). Late and very late promoter elements also are quite similar, but the very late elements include an additional downstream "burst" sequence, which leads to the extremely high levels of transcription ("hypertranscription") observed during the very late phase of baculovirus infection (Ooi *et al.*, 1989).

3. Baculovirus Expression Vectors

The classic baculovirus expression vector, in its simplest form, is a recombinant baculovirus whose genome contains a foreign nucleic acid sequence, almost always a cDNA, encoding a protein of interest under the transcriptional control of the polyhedrin promoter. The chimeric gene consisting of the polyhedrin promoter and foreign protein coding sequence is found in the polyhedrin locus of the viral genome in place of the nonessential, wild-type polyhedrin gene. The recombinant virus can be used to infect cultured insect cells or larvae (caterpillars) in the laboratory and this leads to high-level transcription of the foreign cDNA during the very late phase of infection. The resulting mRNA then can be translated to produce the protein of interest. The polyhedrin promoter seems to invariably provide extremely high levels of foreign gene expression at the transcriptional level and this leads, in many cases, to high levels of foreign protein production, as originally anticipated from the ability of baculoviruses to produce large amounts of the polyhedrin protein. Indeed, the potential for high-level recombinant protein production is one of the major advantages of the baculovirus–insect cell system, with "high-level" defined rather loosely here as $\geq$100 mg of recombinant protein per liter of infected insect cell culture, or $\sim$4 g of cells at the usual density of 1×10^6 cells/mL. In addition, high-level production of recombinant proteins in baculovirus-infected insect cells is rarely associated with inclusion body formation, which is commonly observed in bacterial systems. However, anyone who works in the area of recombinant protein production knows that the actual production and solubility levels achieved in any system are highly dependent upon the specific protein under investigation. Thus, a more useful generalization that has arisen from 25 years of collective experience with baculovirus–insect cell systems is that these systems are much more likely to produce high levels of foreign nuclear and cytoplasmic proteins than secretory pathway proteins. The latter are typically produced at lower levels than the former, often at the level of single to tens of mg/L (Jarvis, 1997).

Another major advantage typically associated with the baculovirus–insect cell system is its eukaryotic protein processing capabilities, which include the ability to provide protein modifications such as phosphorylation and glycosylation, among many others. Again, however, this claim comes with a caveat, as it is now well recognized that the protein processing pathways of the lepidopteran insect cell hosts for baculovirus vectors are not identical to those of higher eukaryotes. An additional complication is that baculovirus infection can have an adverse impact on host protein processing functions (Azzouz *et al.*, 2000; Jarvis and Summers, 1989). Obviously, these are important considerations for anyone interested in producing recombinant proteins with eukaryotic modifications, particularly those known to directly or indirectly influence their functions.

Finally, it should be noted that the baculovirus–insect cell system has proved to be particularly useful for the production of multiprotein subunit complexes (reviewed in Berger *et al.*, 2004; Kost *et al.*, 2005). The power of this system for this increasingly important application is exemplified by its ability to produce virus-like particles composed of multiple virion components, which are outstanding vaccine candidates. For example the baculovirus–insect cell system has been used to produce virus-like particles consisting of multiple proteins from polio, bluetongue, adeno-associated, hepatitis C, and papilloma viruses, among others. This has been accomplished by using multiple recombinant baculoviruses, each encoding individual proteins, or single recombinant viruses encoding multiple proteins to infect insect cell cultures.

4. Baculovirus Expression Vector Technology—The Early Years

The first recombinant baculovirus vectors designed to express chimeric genes consisting of the polyhedrin promoter and a foreign coding sequence, as described in the preceding section, were produced using a basic homologous recombination approach. The methodological details of this approach were described in the last edition of this volume (Bradley, 1990) and also are available in primary papers and excellent technical manuals from the original contributors (O'Reilly *et al.*, 1992; Summers and Smith, 1987). Thus, this exercise will not be repeated here, as indicated above. However, for background purposes it is important to briefly note that this general method involved (1) construction of a bacterial "transfer" plasmid containing the chimeric gene flanked by sequences derived from the polyhedrin region of the viral genome (Fig. 14.1) and (2) cotransfection of cultured insect cells with a mixture of this transfer plasmid DNA and genomic DNA extracted from purified preparations of wild-type AcMNPV (Fig. 14.2).

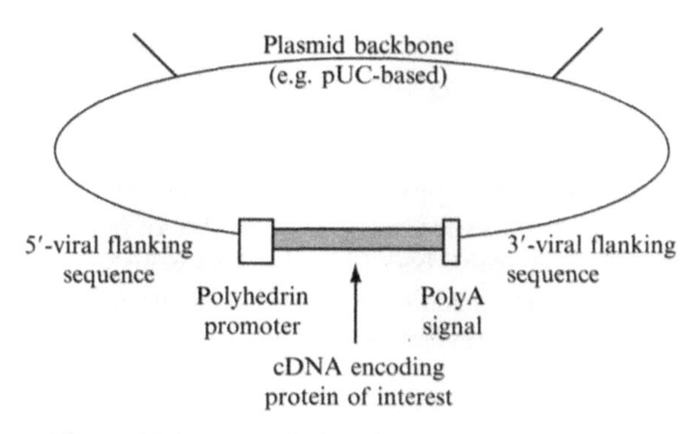

Figure 14.1 A simple baculovirus transfer plasmid.

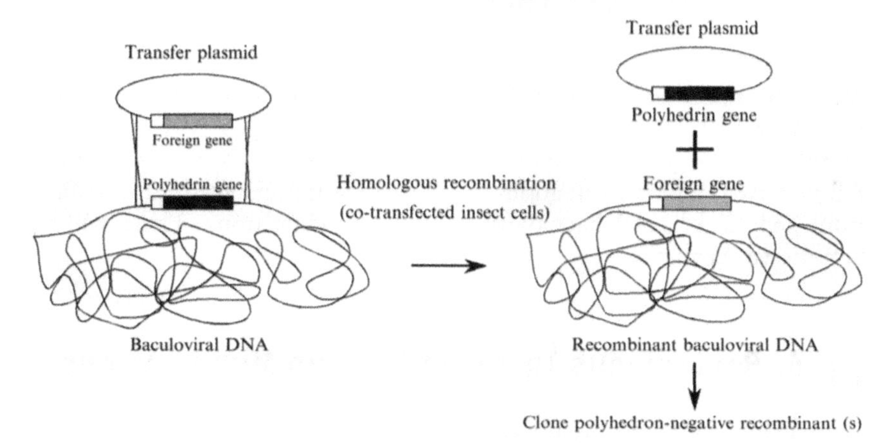

Figure 14.2 Producing a baculovirus expression vector by homologous recombination.

Homologous recombination between the sequences flanking the chimeric gene of interest in the transfer plasmid and the sequences upstream and downstream of the polyhedrin gene in the wild-type AcMNPV genome produced recombinant viral DNA molecules in these cotransfected insect cells. A double crossover recombination event was necessary to simultaneously knockout the polyhedrin gene and knock-in the chimeric gene encoding the protein of interest. Of course, this was a relatively rare event with an estimated frequency of $\sim 0.1\%$ (Smith *et al.*, 1983a). Thus, it was necessary to separate the small minority of recombinant virus progeny from the vast majority of parental viral progeny produced by the cotransfected insect cells by cloning. This was easily accomplished by baculoviral plaque assays, but then one had to be able to distinguish recombinant viral plaques from the much larger background of plaques derived from the parental virus. Initially, this was accomplished using a simple visual screen, as the parental viral progeny produced polyhedron-positive plaques while the recombinant viral progeny, which lacked a functional polyhedrin gene, produced polyhedron-negative plaques. An investigator with a trained eye could visually identify polyhedron-negative plaques by examining the assay plate under a dissecting microscope. However, the trained eye was a key to success and the inability of many investigators to recognize polyhedron-negative baculoviral plaques was a serious problem that constrained the use of the baculovirus–insect cell system as a recombinant protein production tool for several years.

5. Baculovirus Expression Vector Technology—Improved

Starting around 1990, various investigators began to address the technical constraints associated with the isolation of baculovirus vectors and to improve the system in other ways by developing a wide variety of modifications of the basic approach described above. These modifications fell into two different categories, including (1) transfer plasmid modifications and (2) parental baculovirus genome modifications. These will be considered separately.

6. Baculovirus Transfer Plasmid Modifications

Transfer plasmid modifications were designed to serve two different purposes. The most important purpose was to facilitate the identification of recombinant baculovirus plaques by visual screening, which had been a difficult task for the reasons described above. One general type of modification that fell within this category was to incorporate chimeric marker genes under the control of baculovirus promoters encoding products such as the *E. coli* β-galactosidase protein, which could be far more easily recognized by visual screening of plaque assays than the polyhedron-negative plaque phenotype (Vialard *et al.*, 1990). The introduction of a marker gene into transfer plasmids was a clever development that served its intended purpose, but it also was important to be aware of a potential trap associated with this approach (O'Reilly *et al.*, 1992). While the introduction of a marker gene could signal a double crossover homologous recombination event, which was the desired outcome, it also could indicate a single crossover recombination event between the transfer plasmid and viral DNA. Single crossover recombination was a far more frequent event that produced recombinant viral genomes containing the entire transfer plasmid, including the bacterial replicon, at somewhat random loci. These recombinants were genetically unstable and the newly acquired foreign gene typically would be lost within a round or two of viral replication. Despite this major limitation, transfer plasmids that incorporated marker genes into the recombinant baculovirus genome were widely used and particularly useful in the hands of investigators aware of the potential single crossover recombination trap. These investigators would simply use incorporation of the marker gene as a prescreen and then perform additional screening to map the genomic position of a foreign gene and confirm that a double crossover recombination event was responsible for transfer of the gene of interest to the recombinant baculovirus vector.

The second purpose served by transfer plasmid modifications was to facilitate recombinant protein expression and purification in the baculovirus–insect cell system. While the specific, individual modifications that were undertaken are far too numerous to discuss here, three general types of transfer plasmid modifications that fell within this category included the addition of sequences encoding secretory signal peptides, the addition of sequences encoding amino- or carboxy-terminal purification tags (e.g., 6× HIS), replacement of the polyhedrin promoter with alternate baculovirus promoters, and replacement of the polyhedrin promoter with multiple promoter elements, which could simultaneously drive the expression of multiple recombinant proteins during baculovirus infection.

A nearly comprehensive list of baculovirus transfer plasmids with the two general types of modifications discussed above, together with short descriptions of their specific, functional features was recently published and this list is a good source of information on this topic (Possee and King, 2007). However, one type of modification that was not included and deserves further consideration is the "immediate-early" transfer plasmid, in which the polyhedrin promoter is replaced by a promoter from a baculovirus immediate-early gene, such as the AcMNPV *ie1* gene (Jarvis *et al.*, 1996). The idea of using promoters from earlier classes of baculovirus genes, such as the *ie1* gene, might seem counterintuitive because these promoters are weaker than the polyhedrin promoter. However, there is evidence to suggest that higher quality products can be obtained using recombinant baculoviruses that express foreign genes under the control of earlier classes of promoters, despite their inability to drive equally high levels of transcription (Chazenbalk and Rapoport, 1995; Hill-Perkins and Possee, 1990; Jarvis *et al.*, 1996; Murphy *et al.*, 1990; Rankl *et al.*, 1994; Sridhar *et al.*, 1993). In fact, this approach can be particularly useful for secretory pathway proteins, which are often produced at relatively low levels when their genes are expressed under polyhedrin control, as mentioned above. In this circumstance, an investigator can potentially gain the advantage of initiating foreign gene expression at earlier times of infection, thereby avoiding the potentially adverse effects of baculovirus infection on host protein processing pathways, without sacrificing high production levels.

7. Parental Baculovirus Genome Modifications

Like transfer plasmid modifications, parental baculovirus genome modifications have been designed to serve a variety of different purposes. Initially, the most important purpose was to find a way to overcome the low efficiency of recombinant baculovirus vector production and isolation, which was a major problem with the original baculovirus–insect cell system. It was clear that the heart of this problem was the extremely low efficiency of homologous recombination between the transfer plasmid and parental baculovirus DNA that occurred in cotransfected insect cells. Ultimately, various investigators developed both indirect and direct solutions to this problem.

The first major step toward a solution came when Kitts and Possee developed the first baculovirus with a linearizable DNA genome (Kitts *et al.*, 1990). The key feature of this viral DNA genome, which was derived from a recombinant baculovirus produced using the original approach, was that it had a novel sequence in the polyhedrin locus that contributed a unique *Bsu*36I site (Fig. 14.3). Thus, this genome could be linearized with *Bsu*36I and used as the parental viral DNA to be mixed with a transfer plasmid and used to cotransfect insect cells. The most important feature of this approach was that the linearized parental viral DNA molecule could not replicate. Thus, linearization vastly reduced the number of parental progeny produced by the cotransfected insect cells. On the other hand, the linearized viral DNA could still undergo homologous recombination with the transfer plasmid, which would recircularize the viral DNA molecule and restore its ability to replicate. Together, these factors increased the efficiency of recombinant baculovirus vector production by cotransfected insect cells from ~0.1 to ~10–20% and produced a quantum leap in the simplification of baculovirus–insect cell technology. This basic approach was improved when Kitts and Possee created a recombinant baculovirus known as BakPAK6, which could be gapped with *Bsu*36I (Kitts and Possee, 1993). The BakPAK6 genome included an *E. coli lacZ* gene in the polyhedrin locus, which contributed one *Bsu*36I site, and it also had two additional *Bsu*36I sites that had been introduced into upstream and downstream viral genes (Fig. 14.4). Importantly, *Bsu*36I digestion deleted a portion of *orf*1629, which is an essential viral gene located just downstream of polyhedrin that encodes a viral nucleocapsid phosphoprotein (Vialard and Richardson, 1993). With BakPAK6 as the parental viral DNA, the average efficiency of recombinant baculovirus vector production by cotransfected insect cells increased to about 95%. The success of this latter development spurred commercialization and widespread dissemination of several "linearized" baculoviral DNAs as starting materials for recombinant

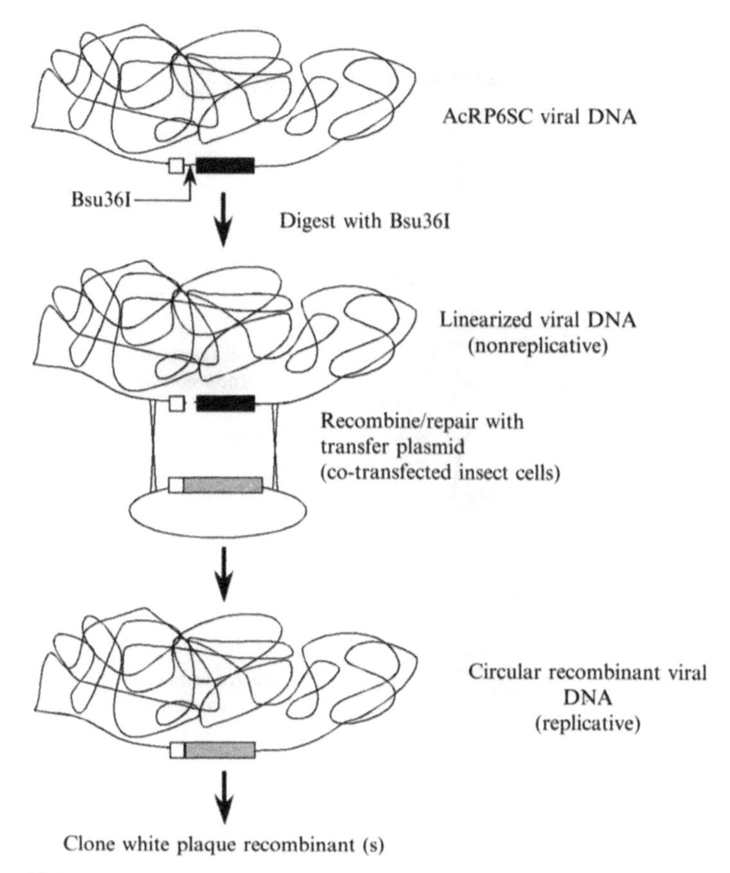

Figure 14.3 **Producing a baculovirus expression vector by homologous recombination with a linearized parental viral genome.**

baculovirus vector production. Among others, these included the original BakPAK6™ viral DNA, which was commercialized by ClonTech; analogous, prelinearized viral DNAs called BaculoGold™ (Pharmingen/BD Biosciences), BacVector® (Novagen), and DiamondBac™ (Sigma-Aldrich); and a slightly modified, linearized viral DNA/transfer plasmid system known as Bac-N-Blue™ (Invitrogen), in which the *E. coli lacZ* marker in the viral genome is regenerated upon recombination.

At this point in this chapter, it is important to emphasize that the development of these viral DNA backbones effectively solved the major problem that had been associated with the production of baculovirus expression vectors at the time of publication of the first edition of this book. These linearizable baculovirus DNAs were widely marketed by various companies and were quickly recognized and adopted by the

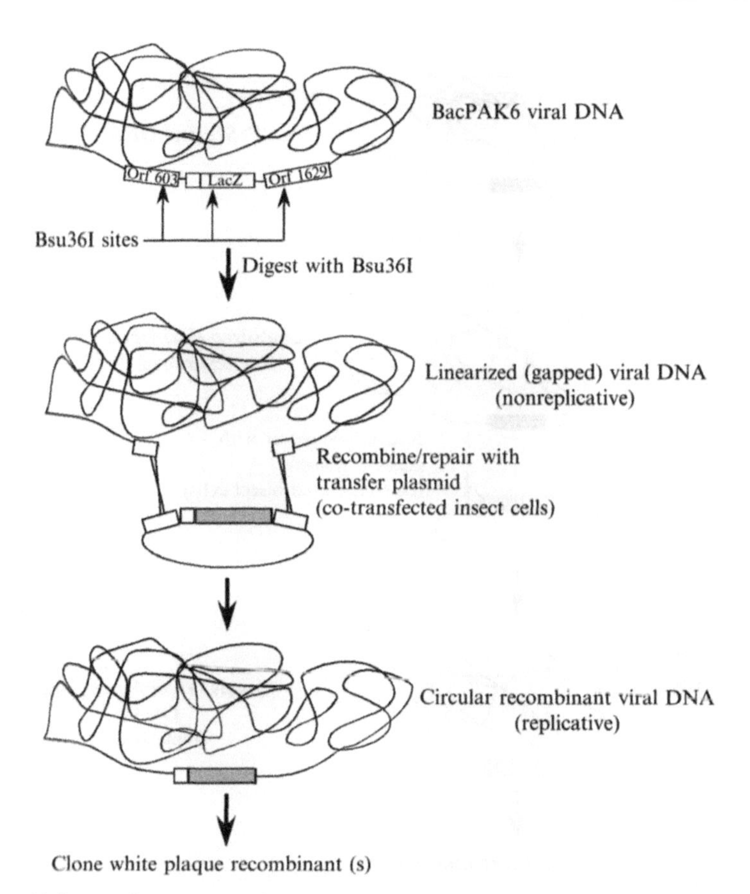

Figure 14.4 Producing a baculovirus expression vector by homologous recombination with a linearized/gapped parental viral genome.

biomedical research community as improved tools for recombinant baculovirus production. This was important because these and other emerging tools and approaches (see below) were responsible for making baculovirus expression vector technology far more accessible to the average laboratory investigator with a basic background in molecular biology. Ultimately, this led to more widespread use and increased general recognition of the baculovirus–insect cell system as an excellent tool for recombinant protein production.

In parallel with the development of linearizable baculovirus genomes, Luckow and his colleagues developed a totally different approach for the isolation of baculovirus expression vectors, which relied upon the process of genetic transposition, rather than homologous recombination (Luckow *et al.*, 1993). This new approach allowed these investigators to address the low homologous recombination efficiency problem directly by using a

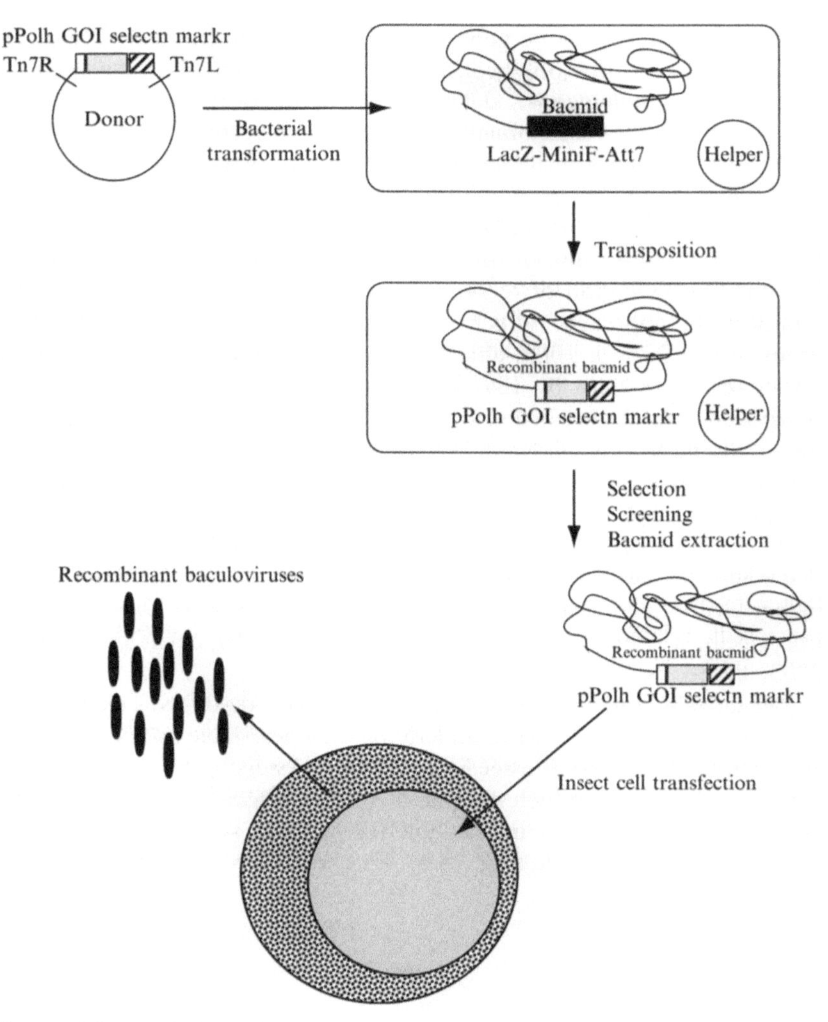

Figure 14.5 **Producing a baculovirus expression vector by transposition between a bacmid and a transfer plasmid.**

different molecular mechanism to produce recombinant baculovirus DNAs. A key element of this approach was the creation of a new strain of *E. coli* containing an autonomously replicating plasmid, or bacmid, which included a cloned copy of the entire baculovirus genome (Fig. 14.5). The polyhedrin region of the bacmid included an *E. coli lacZ* gene and a "mini-Att Tn7" site, which is an attachment site used in the transposition process. The new *E. coli* strain also included a helper plasmid encoding the transposase. Upon transformation with a baculovirus transfer plasmid containing a polyhedrin-promoted gene of interest flanked by the left and right ends of Tn7, the chimeric gene could be efficiently transposed from the transfer plasmid to the polyhedrin locus of the bacmid. This transposition event would simultaneously delete the *lacZ* gene in the bacmid, allowing bacteria containing recombinant bacmids to be detected by standard blue–white screening on a selective bacterial medium. Thus, one could simply isolate the recombinant bacmid DNA encoding the gene of interest from an *E. coli* clone with a white colony phenotype and use this DNA to transfect insect cells. Transfection of the DNA would initiate a viral infection and lead to the production of recombinant baculovirus progeny. This *in vivo* transposition approach, which can be used to produce recombinant baculovirus DNAs at 100% efficiency, was initially commercialized as the Bac-to-Bac^{TM} system by Life Technologies and is now available from Invitrogen. The bacteriocentric nature of the Bac-to-Bac^{TM} system was important because it allowed investigators who were less familiar with virological and more familiar with basic bacteriological and molecular biological methods to work within their comfort zones. One disadvantage of this system, however, is the fact that it yields recombinant baculoviruses that retain the bacterial replicon, which appears to be associated with a higher level of genetically instability upon serial passage of these viruses in insect cells, relative to baculoviruses lacking this element (Pijlman *et al.*, 2003; and see below).

Most recently, the Possee and King groups have described a clever cross-hybridization of the basic ideas underlying the linearizable baculoviral DNA and bacmid approaches (Possee *et al.*, 2008). In essence, they have created a new type of bacmid, which consists of a recombinant baculoviral genome with a bacterial replicon in the polyhedrin locus and a deletion in the downstream *orf*1629 gene (Fig. 14.6). Due to the presence of the replicon

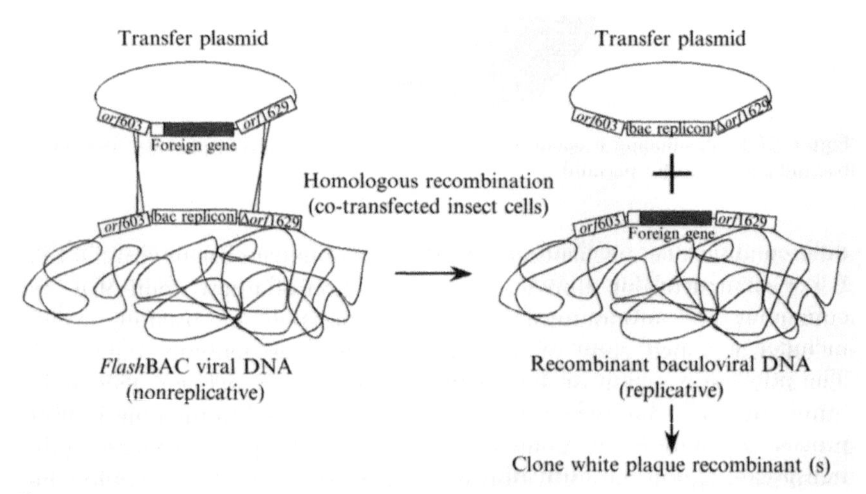

Figure 14.6 Producing a baculovirus expression vector using the *flash*BAC approach.

and the *orf*1629 deletion, this bacmid can replicate in *E. coli*, but not in insect cells. Therefore, an *E. coli* strain carrying this bacmid provides a convenient source of replication defective parental baculovirus genomic DNA for recombinant viral vector production. One can simply isolate the viral DNA from *E. coli*, mix it with a transfer plasmid, and then use the mixture to cotransfect insect cells. Homologous recombination between the viral and transfer plasmid DNAs will simultaneously restore *orf*1629, knockout the bacterial replicon in the polyhedrin locus, and knock-in the gene of interest at this same site. While this approach does not improve the efficiency of homologous recombination, it can provide an extremely high efficiency of recombinant baculovirus production by the cotransfected insect cells because the parental viral DNA is defective and cannot replicate on its own. This approach has been commercialized as *flash*BAC™ by Oxford Expression Technologies. One feature of the *flash*BAC™ system that warrants additional emphasis is that the bacterial replicon is deleted from the bacmid upon homologous recombination with the transfer plasmid, as mentioned above. This is advantageous because it eliminates potential problems with the genetic stability of baculovirus expression vectors produced using conventional bacmids, which retain the bacterial replicon within their genomes (Pijlman *et al.*, 2003). Another feature that warrants additional emphasis is that the bacmid used in the *flash*BAC™ system consists of a defective baculovirus genome that cannot replicate in insect cells. Clearly, this means that homologous recombination between the defective parental viral genome and the transfer plasmid is required to produce helper-independent baculovirus progeny in this system. However, it is important to recognize that this does not necessarily mean that the cotransfected insect cells will produce only recombinant baculovirus progeny. These cells also can produce progeny derived from the defective parental viral genome as a result of genetic complementation. Specifically, the recombinant virus produced in the cotransfected insect cells could provide the *orf*1629 product as a helper function needed to package defective viral DNAs in *trans*. Thus, while Oxford Expression Systems' commercial literature indicates that it is not necessary, baculovirus vectors produced using the *flash*BAC™ system should be subjected to at least one round of plaque purification. Without this step, primary recombinant virus stocks produced by cotransfected insect cells are likely to be contaminated with defective, parental viruses that could interfere with downstream vector replication and foreign gene expression.

Arguably, the simplest way to produce baculovirus vectors is to use a cross-hybridization of the basic ideas underlying Gateway® technology (Hartley, 2003; Walhout *et al.*, 2000) and the linearized parental viral DNA approach, which involves *in vitro*, rather than *in vivo* recombination between a Gateway® entry plasmid encoding a protein of interest and the linearized parental viral DNA. In this system, the parental viral DNA is a prelinearized baculovirus genome in which the polyhedrin coding sequence is replaced by an *E. coli LacZ* gene and a herpes simplex virus thymidine kinase gene flanked by bacteriophage l site-specific recombination sites (attR1 and attR2; Fig. 14.7). The *LacZ* gene contributes a blue plaque phenotype and the thymidine kinase gene contributes a marker that confers negative selection against the parental viral DNA in insect cells treated with certain nucleoside analogs, such as gancyclovir. One can mix this prelinearized viral DNA with an entry plasmid encoding a protein of interest flanked by lambda virus site-specific recombination sites (attL1 and attL2), add a purified recombinase (LR Clonase™; Invitrogen), and this enzyme will mediate site-specific recombination between the att sites in the entry plasmid and parental viral DNA. It is important to note that this *in vitro* reaction provides a high, but not quantitative efficiency of recombination and therefore yields a mixture of parental and recombinant baculovirus DNAs. This mixture is subsequently transfected into insect cells, which are cultured in the presence of gancyclovir to select against replication of the parental viral DNAs, as noted above. One round of selection gives rise to mostly recombinant baculovirus vector progeny, which can be resolved from a relatively low level of parental progeny by performing a plaque assay. Putative recombinants can be easily identified by their white-plaque phenotypes in the presence of a chromogenic substrate for β-galactosidase. The development of this *in vitro* approach for the production of recombinant baculovirus vectors was initiated by Franke and colleagues at Life Technologies and

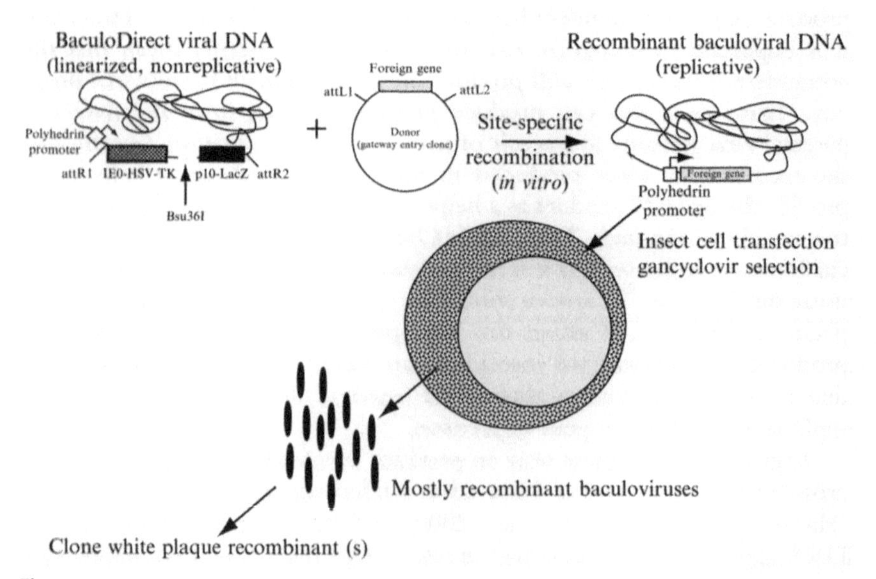

Figure 14.7 Producing a baculovirus expression vector using the BaculoDirect approach.

subsequently commercialized as the BaculoDirect™ system by Harwood and colleagues at Invitrogen. This system is extremely rapid, efficient, and simple to use. However, it is important to note that the BaculoDirect™ system, like the *flash*BAC™ system, is visibly marketed as one that "can be used to clone and express genes from baculovirus without plaque purification or selection in bacteria." Again, this is not good advice because insect cells transfected with the *in vitro* recombination reaction will produce both recombinant and parental viral progeny, as mentioned above, despite the application of negative selection pressure against the parental virus by the thymidine kinase gene product and gancyclovir. Thus, investigators using the BaculoDirect™ system should ignore the manufacturer's claims to the contrary and perform at least one round of plaque purification to resolve the recombinant baculovirus vector of interest from contaminating parental viral progeny. In fact, those who follow this advice and incorporate a chromogenic β-galactosidase substrate into the agarose overlay used for the plaque purification process will typically observe at least some blue plaques, which will indicate the presence of parental viruses and confirm their wise decision to plaque purify their recombinant baculovirus vector. This process is facilitated by the commercial BaculoDirect™ instruction manual (Invitrogen), which, in fairness, includes protocols for baculovirus plaque purification in the presence of a chromogenic β-galactosidase substrate.

In context of the *in vitro* approach described above, it should be briefly noted that there are several published examples of direct *in vitro* recombinant DNA approaches that involve digestion of modified baculovirus genomic DNAs followed by direct ligation of the products with foreign DNA fragments. One example of this type of approach involved the introduction of two *Bsu*36I sites downstream of the polyhedrin promoter in the AcMNPV genome (Lu and Miller, 1996). These sites had slightly different recognition sequences and, upon digestion with *Bsu*36I, produced single-stranded 5′-TTA-3′ overhanging sequences that could be converted to 5′-TT-3′ with dTTP and the Klenow fragment of *E. coli* DNA polymerase I. This created a linear baculoviral DNA molecule that could be ligated with any DNA fragment that had been digested with *Eco*RI to generate single-stranded 5′-AATT-3′ overhanging sequences and treated with Klenow and dATP to convert those ends to 5′-AA-3′. The focused purpose in developing this approach was to facilitate the production of baculovirus-based cDNA expression libraries. The other two-published examples of direct cloning approaches both involved the introduction of homing endonuclease sites into the baculovirus genome. One involved the introduction of a unique I-*Sce*-I site into the AcMNPV genome to produce a recombinant designated Ac-Omega (Ernst *et al.*, 1994). Genomic DNA isolated from this virus could be digested with I-*Sce*-I and then ligated with a DNA fragment produced using the same enzyme. The other example, designated the Homingbac system, involved the introduction of a unique I-*Ceu*-I site into the genomes of several different baculoviruses (Lihoradova *et al.*, 2007). Genomic DNA isolated from these viruses could be linearized with I-*Ceu*-I and ligated with a DNA fragment containing compatible overhangs produced using *Bst*XI. At this time, none of the three direct cloning vectors described in this paragraph have been commercialized, nor have they been widely used in published studies. Thus, it is difficult to evaluate the relative success of these direct cloning approaches or the general utility of these parental baculovirus backbones for the production of baculovirus expression vectors.

In addition to solving the early technical problems associated with the production of recombinant baculovirus vectors, the purpose of some parental baculovirus genome modifications was to enhance foreign protein production by the recombinant vectors. The basic approach was to delete baculovirus genes that were (1) known to be nonessential for baculovirus replication in cultured insect cells, provisionally defined as "accessory" genes and (2) thought to interfere in some way with foreign protein production and/or to degrade the foreign protein of interest. The first baculoviral DNA of this type was a linearized viral DNA developed by Bishop and his colleagues and commercialized as BacVector®-2000 by Novagen. In addition to polyhedrin, this viral DNA lacked five other undisclosed baculoviral accessory genes. The impact of these deletions on foreign protein production was and is unclear, as no studies comparing the performance of a matched pair of recombinant baculoviruses produced using viral DNAs with and without these deletions have been published.

Subsequently, two additional viral genes, one encoding a chitinase (Hawtin *et al.*, 1995) and the other a cathepsin-like protease (Slack *et al.*, 1995), were deleted from BacVector®-2000 to produce an AcMNPV-based parental viral DNA called BacVector®-3000 (Novagen). Other AcMNPV-based viral DNAs lacking the viral chitinase and cathepsin-like protease genes include a commercial bacmid called BestBac (Expression Systems) and a noncommercial modified bacmid, AcBacDCC (Kaba *et al.*, 2004). The bacmid used in the *flash*BAC™ system also lacks a functional viral chitinase gene. The impact of the viral chitinase and cathepsin-like protease gene deletions in these different parental viral DNAs on foreign gene expression by their recombinant baculoviral vector offspring is not totally clear, but at least some information is available. One study showed that there was less degradation of a foreign glycoprotein produced by AcBacDCC, as compared to a matched baculovirus vector without the viral chitinase and cathepsin-like protease deletions (Kaba *et al.*, 2004). This is consistent with the seemingly obvious expectation that deleting a viral protease gene would have a generally positive impact on recombinant protein and secretory pathway protein production. However, more work needs to be done to determine if the deletion of this particular protease gene broadly impacts foreign protein production in the baculovirus–insect cell system. The potential impact of deleting the viral chitinase gene is less obvious. It has been suggested that the viral chitinase gene product, which is a resident ER protein (Thomas *et al.*, 1998), interferes with the production of secretory pathway proteins by contributing to the saturation of host protein translocation machinery (Kaba *et al.*, 2004). If this is true, then deletion of the viral chitinase gene might be expected to increase secretory pathway protein production levels by eliminating production of this product. However, there are no published studies documenting the impact of deleting the viral chitinase gene, alone, in an AcMNPV-based vector such as *flash*BAC™, and the study documenting the reduced degradation of a foreign glycoprotein produced by AcBacDCC is complicated by the fact that this vector lacks both the viral chitinase and cathepsin-like protease genes. On the other hand, one publication has documented the impact of deleting the viral chitinase gene, cathepsin-like protease gene, or both, on the production of a foreign protein by recombinant silkworm baculoviruses (*Bombyx mori* nucleopolyhedrovirus; BmNPV; Lee *et al.*, 2006). In this system, recombinant BmNPV vectors encoding an insect cellulase produced higher levels of both the foreign protein and its enzymatic activity in silkworms when either the chitinase or the cathepsin-like protease genes were deleted. Furthermore, a vector lacking both of these viral genes produced the highest levels of cellulase protein and enzyme activity. One could speculate that these results indicate that deleting the viral chitinase gene from AcMNPV-based vectors would have similar benefits. However, this speculation would be weakened by the fact that silkworms, not insect cell lines, were used as the hosts in this study.

DiamondBac™ is another example of a parental baculovirus DNA that has a deletion in an accessory gene to potentially enhance recombinant protein production in the baculovirus–insect cell system. As mentioned above, DiamondBac™ is a commercial, prelinearized baculovirus DNA analogous to BakPAK6, but this viral DNA also lacks a functional *p10* gene, which has been implicated in host cell lysis (Williams *et al.*, 1989). Accordingly, the manufacturer's commercial literature suggests that cells infected with a recombinant baculovirus vector produced using this parental virus retain higher cell viabilities throughout the course of infection, which contributes to higher recombinant protein production levels (Sigma-Aldrich, 2008). Another interesting and potentially useful feature of DiamondBac™ is that the *p10* gene in this baculoviral genome has been replaced by a protein disulfide isomerase (PDI) gene, which encodes a chaperone that drives disulfide bonding and contributes to protein folding. Published evidence suggests that coexpressing PDI in the baculovirus–insect cell system increases the solubility and secretion of recombinant IgG (Hsu *et al.*, 1996). The commercial literature from Sigma-Aldrich states that the *p10* gene deletion and PDI gene insertion provide "up to a 10-fold increase in overall protein production for many recombinant proteins" (Sigma-Aldrich, 2008). However, there are no published studies comparing foreign protein production levels by matched pairs of recombinant baculovirus vectors with and without the *p10* gene deletion and/or the PDI insertion. Thus, in the final analysis, directed studies with appropriately matched pairs of AcMNPV-based vectors will be needed to directly determine the general impact of deleting nonessential viral genes, such as the viral chitinase, cathepsin-like protease, and *p10* genes, on foreign protein production in the baculovirus–insect cell system.

In addition to the DiamondBac™ parental baculovirus DNA, other baculovirus vectors with gene insertions encoding heterologous protein processing enzymes designed to enhance foreign protein production have been described in the literature. One is a recombinant baculovirus that encodes a polydnavirus *vankyrin* gene under the control of the *p10* promoter (Fath-Goodin *et al.*, 2006). It has been found that the expression of certain *vankyrin* gene products prolongs the viability of baculovirus-infected Sf9 cells and this, in turn, can increase the amounts of a foreign protein produced by coexpression under the control of the polyhedrin promoter. Other types of baculovirus vectors that fall within this genre include those encoding heterologous glycosyltransferases (Jarvis *et al.*, 2001; Tomiya *et al.*, 2003) or enzymes involved in CMP-sialic acid biosynthesis (Hill *et al.*, 2006) under the control of baculovirus *ie1* promoters. The use of the immediate-early promoter in these vectors allows the newly introduced processing functions to be expressed early in the infection cycle, well before expression of the foreign gene encoding the protein of interest. This allows for functional extension of the endogenous insect cell protein *N*-glycosylation pathway and the establishment of a pathway capable of producing glycoproteins with humanized carbohydrate side chains prior to the onset of production of the glycoprotein of interest.

A transfer plasmid that can be used to simultaneously introduce a polydnavirus vankyrin gene and a foreign gene encoding a protein of interest into the baculovirus genome is commercially available from Paratechs. Similarly, parental baculoviral DNAs encoding higher eukaryotic *N*-glycosylation functions are likely to be commercially available for recombinant baculovirus vector isolation in the near future. These tools will facilitate a more global analysis of the capabilities of these types of vectors in increasing expression levels, on the one hand, or producing foreign glycoproteins with humanized carbohydrate side chains, on the other, as neither has been broadly assessed with a wide array of different foreign protein/glycoprotein products.

8. The Other Half of the Baculovirus–Insect Cell System

Considering the focus on baculovirus expression vectors to this point in this chapter, it might be necessary to remind the reader that the baculovirus–insect cell system is a binary system that consists of two essential components. The first one is, of course, the baculovirus expression vector, which is an insect virus whose obvious function is to deliver a foreign gene(s) encoding a protein(s) of interest to the host. In addition, the baculovirus expression vector typically serves another function, which is to induce the production of the transcriptional complex needed to transcribe the foreign gene of interest under the control of a late or very late baculovirus promoter. The second component, which has been given almost no attention so far, is the host, which is typically a lepidopteran insect cell line, but alternatively can be a lepidopteran insect.

The Order Lepidoptera includes the moths and butterflies, which are the hosts for many viruses in the Family Baculoviridae, including AcMNPV. Grace described the first-established lepidopteran insect cell cultures in 1962 (Grace, 1962) and as of today, over 250 insect cell lines have been described (Lynn, 2007). However, we can focus on just two lepidopteran insect cell lines, which are the two most commonly used hosts for baculovirus expression vectors. One is Sf9, a cell line described by Smith and Summers in 1987 (Summers and Smith, 1987) as a subclone of IPLB-SF-21, which is a line that had been isolated from pupal ovarian tissue of *Spodoptera frugiperda* (fall armyworm) in 1977 (Vaughn *et al.*, 1977). Sf9 cells can be obtained from several different companies, including Invitrogen and Novagen, or the American Type Culture Collection (ATCC). The other is High Five^{TM}, a cell line originally isolated from adult ovarian tissue of *Trichoplusia ni* (cabbage looper) and described as BTI Tn 5B-1 by Granados' and Wood's groups in 1992 (Davis *et al.*, 1992; Wickham *et al.*, 1992) and then commercialized by Invitrogen as High Five^{TM}.

One interesting and useful feature of Sf9 and High Five^{TM} cells is that they can grow in either adherent or suspension formats. Thus, it is convenient to perform lab-scale protein production experiments by infecting a few million Sf9 or High Five^{TM} cells as adherent cultures in plates or flasks. Adherent cultures of Sf21 or Sf9 cells also are routinely used to plaque purify and quantify recombinant baculovirus expression vectors. On the other hand, both Sf9 and High Five^{TM} cells can be scaled up to varying degrees in spinner flasks, shake flasks, or in airlift, stirred tank (Elias *et al.*, 2007), or Wave (Kadwell and Hardwicke, 2007) bioreactors to produce larger amounts of recombinant proteins.

The conditions and methods used to culture insect cells are quite different from those used to culture mammalian cells, which are usually more familiar to most investigators. For example, the optimal temperature for insect cell culture is 28 °C, rather than 37 °C. In addition, both Sf9 and High Five^{TM} cells are loosely adherent and neither EDTA nor trypsin is required for their subculture. It is not necessary to have a CO_2 incubator to grow insect cells because insect cell culture media are buffered with phosphate, rather than carbonate. Sf9 and High Five^{TM} cells both can be cultured in either growth media supplemented with serum or in serum-free media and both types of growth media are now widely available. In fact, since the publication of the first edition of this book in 1990, many different insect cell culture media have become available from many different manufacturers, including Expression Systems, Invitrogen (GIBCO), HyClone, and Sigma-Aldrich, among others. However, it is important to note that Sf9 and High Five™ cells tend to become adapted to a specific growth medium. Thus, the abrupt transition of an Sf9 or High Five™ culture from a growth medium supplemented with serum to a serum-free medium can result in loss of the culture. A much higher rate of success can be achieved by slowly weaning the cells out of serum containing and into serum-free medium. For example, one can increase the proportion of serum-free medium from 0 to 25, 50%, 75, and 100% over four consecutive passages. Even with this relatively slow type of weaning process, these cells typically undergo a temporary arrest or grow very slowly when first placed into the 100% serum-free medium. Thus, it is important to be patient, as the cells will typically recover after a lag of a day or two and begin growing normally again in the new medium. This type of weaning process is not only useful for the transition from a serum containing to a serum-free medium, but is also sometimes needed to move insect cell cultures from one specific type of serum-free medium into another.

9. A New Generation of Insect Cell Hosts for Baculovirus Expression Vectors

Technology for the genetic transformation of lepidopteran insect cells was first described in 1990 (Jarvis *et al.*, 1990). At that time, the main reason for developing this technology was to enable the creation of stably transformed insect cell lines that would constitutively produce a recombinant protein of interest in the absence of baculovirus infection. However, the vectors and methods used to create these "producer" cell lines also set the stage for the creation of transgenic insect cell lines with new traits that would improve their ability to support recombinant baculovirus-mediated foreign protein production. The first example of this type of host cell improvement was directed toward an issue raised earlier in this chapter, which is that the baculovirus–insect cell system has eukaryotic protein processing capabilities, but these are not always identical to those of higher eukaryotes. One of the best-recognized examples of this limitation is the ability of both Sf9 and High Five™ cells to *N*-glycosylate newly synthesized proteins, but their inability to process *N*-linked carbohydrate side chains (*N*-glycans) to the same extent as mammalian cells (reviewed in Harrison and Jarvis, 2006, 2007b; Shi and Jarvis, 2007). The first cell-based approach used to begin to address this problem was to genetically transform

Sf9 cells with a bovine β1,4-galactosyltransferase cDNA placed under the control of the Ac*M*NPV *ie1* promoter (Hollister *et al.*, 1998). The addition of this cDNA, which encodes an *N*-glycan processing enzyme that is not present in Sf9 cells, was designed to extend the endogenous processing pathway and permit the transgenic cell line to produce *N*-glycoproteins with more extensively processed, partially "humanized" *N*-glycans. Indeed, the resulting cell line, designated Sfβ4GalT, encoded and constitutively expressed this mammalian gene and, unlike the parental cell line, was able to produce foreign glycoproteins with partially humanized *N*-glycans. However, this was only a first step, as further humanization of the Sf9 cell protein *N*-glycosylation pathway of Sf9 cells, which involved knocking-in several additional mammalian functions, was required. In general, this was accomplished by constructing *ie1*-driven mammalian genes encoding these additional functions and using the resulting constructs to produce additional transgenic Sf9 cell subclones (Aumiller *et al.*, 2003; Hollister and Jarvis, 2001; Hollister *et al.*, 2002; Seo *et al.*, 2001). Each of these new insect cell lines retained the ability to support baculovirus replication and baculovirus-mediated foreign gene expression and encoded and constitutively expressed the mammalian genes with which they had been transformed. Finally, each of these transgenic insect cell lines also had more extensive *N*-glycan processing capabilities than the parental Sf9 cells and could produce recombinant glycoproteins with increasingly humanized *N*-glycans. An analogous transgenic subclone of High Five™ cells was produced using *ie1*-based transgenes encoding two different mammalian glycosyltransferases (Breitbach and Jarvis, 2001). One Sf9-based transgenic insect cell line originally designated SfSWT-1, which has an extended *N*-glycosylation pathway and can produce humanized recombinant glycoproteins when cultured in the presence of serum (Hollister *et al.*, 2002), has been commercialized by Invitrogen under the trade name "MIMIC™." More advanced transgenic insect cell lines with a larger repertoire of mammalian genes, which are capable of producing recombinant glycoproteins with terminally sialylated *N*-glycans when cultured in serum-free media (Aumiller *et al.*, 2003), are likely to be commercially available in the near future.

Another transgenic lepidopteran insect cell line that holds promise as an improved host for baculovirus expression vectors is an Sf9 cell derivative transformed with a polydnavirus *vankyrin* gene under the control of an immediate-early baculovirus promoter (Fath-Goodin *et al.*, 2006). As discussed above, baculovirus-mediated expression of this polydnavirus *vankyrin* gene prolonged the viability of baculovirus-infected Sf9 cells and increased the production of a foreign protein coexpressed under the control of the *polyhedrin* promoter. By analogy, a transgenic Sf9 cell derivative engineered to constitutively express this *vankryin* gene had a longer life span after and produced higher levels of yellow fluorescence protein when infected with a recombinant baculovirus expression vector encoding this product

(Fath-Goodin *et al.*, 2006). Three different "vankyrin-enhanced" (VE) Sf9 derivatives are now commercially available from ParaTechs.

The availability of MIMIC^{TM} and VE cells and the future availability of other transgenic insect cell lines of these types will permit a more global analysis of their abilities to produce foreign glycoproteins with humanized carbohydrate side chains or to increase recombinant protein expression levels, respectively. This is important because neither one of these two different types of "improved" hosts for baculovirus expression vectors has been broadly assessed using a wide variety of different foreign protein/glycoprotein products. In fact, there is one published report of the inability of MIMIC^{TM} (SfSWT-1) cells to sialylate one specific foreign glycoprotein, equine lutenizing hormone/chorionic gondotropin (Legardinier *et al.*, 2005). These two equine glycopeptide hormones are encoded by the same gene and, therefore, have identical amino acid sequences. From a careful analysis of the *N*-glycosylation profiles of the native products (Smith *et al.*, 1993), one would expect the recombinant hormone produced by MIMIC^{TM} (SfSWT-1) cells to have terminal alpha 2,3-linked sialic acid residues. Thus, the results of Legardinier and coworkers most likely reflect the inability of MIMIC^{TM} (SfSWT-1) cells to sialylate equine lutenizing hormone/chorionic gonadotropin in alpha 2,3-linked fashion, which is a capability that had not been assessed before. In fact, we recently found that MIMIC^{TM} (SfSWT-1) cells have no alpha 2,3-sialyltransferase activity, despite the fact that they encode and express a murine alpha 2,3-sialyltransferase (data not shown). Thus, it is now clear that MIMIC^{TM} (SfSWT-1) cells do not have the universal capacity to sialylate every foreign *N*-glycoprotein of interest and efforts are underway to produce a new cell line that can provide both alpha 2,6- and alpha 2,3-sialylation. However, even a new transgenic insect cell line with alpha 2,3-sialyltransferase activity might not have the capacity to sialylate every recombinant glycoprotein of interest. Furthermore, it will not be surprising if this limitation also applies to VE cells or any other type of transgenic insect cell line engineered to serve as an improved host for baculovirus expression vectors.

10. Basic Baculovirus Protocols

Chapter 10 in the first edition of this book contained a significant number of detailed protocols describing the construction of baculovirus transfer plasmids, insect cell culture, the production, isolation and characterization of baculovirus expression vectors, and methods for the analysis of recombinant protein production in the baculovirus–insect cell system (Bradley, 1990), which are still relevant today. In addition, a large number of other books and book chapters containing detailed protocols describing these technical aspects of the baculovirus–insect cell system have been published before and since that time (King and Possee, 1992; Murhammer, 2007; O'Reilly *et al.*, 1992; Richardson, 1995; Summers and Smith, 1987). Finally, each manufacturer of the new baculovirus transfer plasmids and genome backbones described in this chapter, which were created to facilitate the production and isolation of recombinant baculovirus expression vectors, provides detailed maps, sequences, and stepwise protocols for the use of their materials, which are freely available at their Web sites. Accordingly, I have chosen to share a relatively small, but essential set of time-tested protocols used in my lab, which were originally taken from the original baculovirus users "manuals" by Summers and Smith (1987) and O'Reilly *et al.* (1992) and, in some cases, modified to fit our needs. The reader is referred to the sources given above for more direct and detailed information on any of the specific approaches, transfer plasmids, and parental baculovirus vectors described in this chapter, which need not be reiterated here.

10.1. Insect cell maintenance

We routinely maintain 50 mL Sf9 cell cultures in 125 mL DeLong shake flasks (Bellco) in either serum-containing or serum-free media. We use TNM-FH medium supplemented with 10% (v/v) fetal bovine serum (HyClone) and 0.1% pluronic F68 (Sigma-Aldrich) for the former and we use ESF 921 (Expression Systems) for the latter. We do not heat-inactivate the fetal bovine serum nor do we use cell culture antibiotics in our maintenance cultures, as they can interfere with the creation of transgenic subclones by our standard coselection approach (Harrison and Jarvis, 2007a; Jarvis and Guarino, 1995). We split all of our maintenance cultures each Monday, Wednesday, and Friday of the week, using seeding densities for Sf9 cells of 0.5×10^6 cells/mL on Mondays and Wednesdays and 0.3×10^6 cells/mL on Fridays. We measure cell densities using a standard trypan blue dye exclusion procedure with a hemocytometer (Murhammer, 2007). Cell viabilities should remain very high in shake flask cultures and Sf9 cell doubling times should be around 24 h. For larger numbers of cells, the culture can be scaled up to 100 mL in 250 mL DeLong flasks, 200 mL in 500 mL DeLong flasks, or 800 mL in 2.8 L Fernbach flasks (Bellco). The seeding densities used for transgenic insect cell lines are variable and typically higher than those used for Sf9 cells. In our recent experience, transgenic lines seeded at the same densities used for Sf9 cells tend to grow slowly and/or undergo a temporary arrest before starting to grow at normal rates again.

10.2. Isolation of baculovirus genomic DNA

Unless you are in a position to constantly purchase prepurified viral DNA from commercial sources, the isolation of adequately purified baculovirus genomic DNA is a critical aspect of producing recombinant baculovirus expression vectors. We prepare viral DNA from budded virus progeny isolated using a differential centrifugation protocol. Sf9 cells are infected with an appropriate parental baculovirus using a low input multiplicity (e.g., 0.01 plaque-forming units/cell) to avoid the generation of defective interfering particles (Kool *et al.*, 1991). The cells are then cultured for 3–5 days in TNM-FH medium supplemented with fetal bovine serum and pluronic F68, as described above, and monitored under a phase-contrast microscope for signs of cytopathology. Once clear signs of infection are observed, the infected cells are aseptically transferred to disposable centrifuge tubes and then centrifuged at ~1000*g* for 15 min. The clear supernatants are transferred to ultracentrifuge tubes (e.g., 33 mL per Beckman SW28 tube) and then carefully underlaid with a solution of 25% (w/v) sucrose in 5 m*M* NaCl and 10 m*M* EDTA (e.g., 3 mL per Beckman SW28 tube). The tubes are ultracentrifuged for 75 min at 4 °C at ~100,000*g* (e.g., 28,000 rpm for a Beckman SW28 rotor) and the supernatant is carefully decanted away from the resulting budded virus pellet. The pellet is then gently resuspended using a blue tip with the narrow end cut off to produce a larger bore diameter and a disruption buffer (0.1 mL for each mL of starting infected cell culture medium) consisting of 10 m*M* Tris, pH 7.6, 10 m*M* EDTA, and 0.25% (w/v) sodium dodecyl sulfate. Proteinase K (10 mg/mL in water) is then added to a final concentration of 500 ug/mL and the preparation is incubated at 37 °C with occasional swirling until it clears, which typically requires 4 h to overnight. If the solution does not clear, one should add more disruption buffer and proteinase K. The solution is then gently extracted with phenol, then phenol–chloroform and the viral DNA is ethanol precipitated in the presence of 0.3 *M* sodium acetate. Finally, the DNA pellet is gently resuspended in 10 uL TE for each mL of starting culture medium. The concentration of the resuspended viral DNA can be estimated using spectrophotometry, but we find this is highly unreliable. Thus, we typically refine the estimate by digesting 10 ug of the viral DNA with *Eco*RI or *Hin*dIII, running the digests on an agarose gel against commercial DNA size standards of known concentrations, and staining the gel with ethidium bromide or another DNA stain. It should be noted that this protocol is essentially the same as the one described in the Baculovirus Expression Vectors manual by O'Reilly and coworkers (O'Reilly *et al.*, 1992).

10.3. Baculovirus plaque assays

Plaque assays are a critical aspect of virology, in general, and are the single best way to ensure that a recombinant baculovirus is initially isolated in a clonal form. Two-independent steps that need to be taken in preparation for baculovirus plaque assays include (1) preseeding the indicator cells into culture plates and (2) making serial 10-fold dilutions of the virus stock. We seed the cells first from a log-phase shake flask culture of Sf9 cells in TNM-FH containing 10% fetal bovine serum and 0.1% pluronic F68, as described above. The requisite number of cells is gently centrifuged out of the old growth medium, resuspended in fresh TNM-FH with no additives, and then seeded into 6-well cell culture plates (Corning) at a density of 0.75×10^6 cells/well. The cells are then allowed to settle and attach to the plastic at 28 °C for about an hour. The cells will adhere more tightly to the plastic when seeded in the absence than when seeded in the presence of serum. While the cells are attaching, the serial 10-fold dilutions of the virus stock are performed, working under the assumption that a typical baculovirus stock will have a plaque assay titer of $\sim 1 \times 10^7$ pfu/mL. We use TNM-FH supplemented with fetal bovine serum and pluronic F68 as the diluent, with 0.1 mL of virus and 9.9 mL dilution blanks for 100-fold dilutions and 0.5 mL of virus and 4.5 mL dilution blanks for 10-fold dilutions. The dilution blanks are set up and the dilutions performed under aseptic conditions and each dilution is mixed very well with a vortex mixer between each transfer. After the dilutions have been performed and the cells have attached to the culture plates, the cells are inoculated using two wells for each appropriately diluted virus. The medium is removed from the wells and replaced with 2 mL/well of the diluted virus. This is a critical step in the assay because the cells dry out rather quickly and those along the edges will die if you move too slowly. Thus, it is best to drain the media from only two wells at a time and immediately inoculate those wells with one virus dilution in duplicate before moving on to another pair of wells and a new dilution. We typically plate the virus at dilutions of 10^{-5}, 10^{-6}, and 10^{-7} to try to obtain reasonable numbers of plaques for counting. The virus is then allowed to attach to the cells for ~ 1 h at 28 °C and the overlay medium is prepared during the viral adsorption period. We prepare 125 mL bottles containing 50 mL of 2% (w/v) Seaplaque® low melting temperature agarose (Cambrex) in water, then autoclave the bottles and allow the agarose to solidify. These bottles of solid agarose then can be microwaved to remelt the agarose, cooled to ~ 60 °C, and the liquid agarose can be mixed with an equal volume of 2× Grace's medium prewarmed to ~ 30 °C (you will need 3 mL per well). If a blue–white screen is being used to help identify putative recombinant viruses, the chromogenic substrate should be added to the overlay at this time. The overlay is then swirled to obtain an even mixture and cooled to about 40–42 °C. Finally, the infected cells are removed from the incubator, the viral inocula are completely removed with pipettes, and 3 mL of overlay are very gentle dribbled down the edge of the well. Again, draining two wells at a time and overlaying those wells with agarose before moving on to another pair of wells will prevent the cells from drying out and dying. After the overlay hardens for ~ 15 min, the plates are incubate in sealed plastic baggie, upside down, for 7–10 days at 28 °C. A dissecting microscope is used to visualize plaques for careful counting and the plaque counts are used to calculate the titer of the original virus stock, taking the dilution factor and plating volume into account.

REFERENCES

Aumiller, J. J., Hollister, J. R., and Jarvis, D. L. (2003). A transgenic lepidopteran insect cell line engineered to produce CMP-sialic acid and sialoglycoproteins. *Glycobiology* **13,** 497–507.

Azzouz, N., Kedees, M. H., Gerold, P., Becker, S., Dubremetz, J. F., Klenk, H. D., and Eckert, V. R. T. S. (2000). An early step of glycosylphosphatidyl-inositol anchor biosynthesis is abolished in lepidopteran insect cells following baculovirus infection. *Glycobiology* **10,** 177–183.

Berger, I., Fitzgerald, D. J., and Richmond, T. J. (2004). Baculovirus expression system for heterologous multiprotein complexes. *Nat. Biotechnol.* **22,** 1583–1587.

Bradley, M. K. (1990). Overexpression of proteins in eukaryotes. *Methods Enzymol.* **182,** 112–132.

Breitbach, K., and Jarvis, D. L. (2001). Improved glycosylation of a foreign protein by Tn-5B1–4 cells engineered to express mammalian glycosyltransferases. *Biotechnol. Bioeng.* **74,** 230–239.

Chazenbalk, G. D., and Rapoport, B. (1995). Expression of the extracellular domain of the thyrotropin receptor in the baculovirus system using a promoter active earlier than the polyhedrin promoter Implications for the expression of functional highly glycosylated proteins. *J. Biol. Chem.* **270,** 1543–1549.

Davis, T. R., Trotter, K. M., Granados, R. R., and Wood, H. A. (1992). Baculovirus expression of alkaline phosphatase as a reporter gene for evaluation of production, glycosylation and secretion. *Nat. Biotechnol.* **10,** 1148–1150.

Douris, V., Swevers, L., Labropoulou, V., Andronopoulou, E., Georgoussi, Z., and Iatrou, K. (2006). Stably transformed insect cell lines: Tools for expression of secreted and membrane-anchored proteins and high-throughput screening platforms for drug and insecticide discovery. *Adv. Virus Res.* **68,** 113–156.

Elias, C. B., Jardin, B., and Kamen, A. (2007). Recombinant protein production in large-scale agitated bioreactors using the baculovirus expression vector system. *Methods Mol. Biol.* **388,** 225–246.

Ernst, W. J., Grabherr, R. M., and Katinger, H. W. (1994). Direct cloning into the Autographa californica nuclear polyhedrosis virus for generation of recombinant baculoviruses. *Nucleic Acids Res.* **22,** 2855–2856.

Fath-Goodin, A., Kroemer, J., Martin, S., Reeves, K., and Webb, B. A. (2006). Polydnavirus genes that enhance the baculovirus expression vector system. *Adv. Virus Res.* **68,** 75–90.

Friesen, P. D. (1997). Regulation of baculovirus early gene expression. *In* "The Baculoviruses" (L. K. Miller, ed.), pp. 141–170. Plenum Press, New York.

Grace, T. D. C. (1962). Establishment of four strains of cells from insect tissues grown *in vitro*. *Nature* **195,** 788–789.

Harrison, R. L., and Jarvis, D. L. (2006). Protein *N*-glycosylation in the baculovirus-insect cell expression system and efforts to engineer insect cells to produce "mammalianized" recombinant glycoproteins. *Adv. Virus Res.* **68,** 159–191.

Harrison, R. L., and Jarvis, D. L. (2007a). Transforming lepidopteran insect cells for continuous recombinant protein expression. *Methods Mol. Biol.* **388,** 299–316.

Harrison, R. L., and Jarvis, D. L. (2007b). Transforming lepidopteran insect cells for improved protein processing. *Methods Mol. Biol.* **388,** 341–356.

Strausberg, R. L., and Strausberg, S. L. (2003). Overview of protein expression in Saccharomyces cerevisiae. *Curr. Protoc. Protein Sci.* **Chapter 5,** Unit 5.17.

Hawtin, R. E., Arnold, K., Ayres, M. D., Zanotto, P. M., Howard, S. C., Gooday, G. W., Chappell, L. H., Kitts, P. A., King, L. A., and Possee, R. D. (1995). Identification and preliminary characterization of a chitinase gene in the *Autographa californica* nuclear polyhedrosis virus genome. *Virology* **212,** 673–685.

Hill, D. R., Aumiller, J. J., Shi, X., and Jarvis, D. L. (2006). Isolation and analysis of a baculovirus vector that supports recombinant glycoprotein sialylation by SfSWT-1 cells cultured in serum-free medium. *Biotechnol. Bioeng.* **95,** 37–47.

Hill-Perkins, M. S., and Possee, R. D. (1990). A baculovirus expression vector derived from the basic protein promoter of Autographa californica nuclear polyhedrosis virus. *J. Gen. Virol.* **71,** 971–976.

Hollister, J., and Jarvis, D. L. (2001). Engineering lepidopteran insect cells for sialoglycoprotein production by genetic transformation with mammalian β1, 4-galactosyltransferase and alpha 2, 6-sialyltransferase genes. *Glycobiology* **11,** 1–9.

Hollister, J. R., Grabenhorst, E., Nimtz, M., Conradt, H. O., and Jarvis, D. L. (2002). Engineering the protein N-glycosylation pathway in insect cells for production of biantennary, complex N-glycans. *Biochemistry* **41,** 15093–15104.

Hollister, J. R., Shaper, J. H., and Jarvis, D. L. (1998). Stable expression of mammalian beta 1, 4-galactosyltransferase extends the N-glycosylation pathway in insect cells. *Glycobiology* **8,** 473–480.

Hsu, T. A., Watson, S., Eiden, J. J., and Betenbaugh, M. J. (1996). Rescue of immunoglobulins from insolubility is facilitated by PDI in the baculovirus expression system. *Prot. Exp. Purif.* **7,** 281–288.

Jarvis, D. L. (1997). Baculovirus expression vectors. *In* "The Baculoviruses" (L. K. Miller, ed.), pp. 389–431. Plenum Press, New York.

Jarvis, D. L., Fleming, J. A., Kovacs, G. R., Summers, M. D., and Guarino, L. A. (1990). Use of early baculovirus promoters for continuous expression and efficient processing of foreign gene products in stably transformed lepidopteran cells. *Nat. Biotechnol.* **8,** 950–955.

Jarvis, D. L., and Guarino, L. A. (1995). Continuous foreign gene expression in transformed lepidopteran insect cells. *In* "Baculovirus Expression Protocols" (C. D. Richardson, ed.), Vol. 39, pp. 187–202. Humana Press, Clifton, NJ.

Jarvis, D. L., Howe, D., and Aumiller, J. J. (2001). Novel baculovirus expression vectors that provide sialylation of recombinant glycoproteins in lepidopteran insect cells. *J. Virol.* **75,** 6223–6227.

Jarvis, D. L., and Summers, M. D. (1989). Glycosylation and secretion of human tissue plasminogen activator in recombinant baculovirus-infected insect cells. *Mol. Cell. Biol.* **9,** 214–223.

Jarvis, D. L., Weinkauf, C., and Guarino, L. A. (1996). Immediate early baculovirus vectors for foreign gene expression in transformed or infected insect cells. *Prot. Exp. Purif.* **8,** 191–203.

Kaba, S. A., Salcedo, A. M., Wafula, P. O., Vlak, J. M., and van Oers, M. M. (2004). Development of a chitinase and v-cathepsin negative bacmid for improved integrity of secreted recombinant proteins. *J. Virol. Methods* **122,** 113–118.

Kadwell, S. H., and Hardwicke, P. I. (2007). Production of baculovirus-expressed recombinant proteins in wave bioreactors. *Methods Mol. Biol.* **388,** 247–266.

King, L. A., and Possee, R. D. (1992). The baculovirus system: A laboratory guide. Chapman and Hall, London.

Kitts, P. A., Ayres, M. D., and Possee, R. D. (1990). Linearization of baculovirus DNA enhances the recovery of recombinant virus expression vectors. *Nucleic Acids Res.* **18,** 5667–5672.

Kitts, P. A., and Possee, R. D. (1993). A method for producing recombinant baculovirus expression vectors at high frequency. *Biotechniques* **14,** 810–817.

Kool, M., Voncken, J. W., van Lier, F. L., Tramper, J., and Vlak, J. M. (1991). Detection and analysis of Autographa californica nuclear polyhedrosis virus mutants with defective interfering properties. *Virology* **183,** 739–746.

Kost, T. A., Condreay, J. P., and Jarvis, D. L. (2005). Baculovirus as versatile vectors for protein expression in insect and mammalian cells. *Nat. Biotechnol.* **23,** 567–575.

Lee, K. S., Je, Y. H., Woo, S. D., Sohn, H. D., and Jin, B. R. (2006). Production of a cellulase in silkworm larvae using a recombinant Bombyx mori nucleopolyhedrovirus lacking the virus-encoded chitinase and cathepsin genes. *Biotechnol. Lett.* **28,** 645–650.

Legardinier, S., Klett, D., Poirier, J. C., Combarnous, Y., and Cahoreau, C. (2005). Mammalian-like nonsialyl complex-type N-glycosylation of equine gonadotropins in Mimic insect cells. *Glycobiology* **15,** 776–790.

Lihoradova, O. A., Ogay, I. D., Abdukarimov, A. A., Azimova Sh, S., Lynn, D. E., and Slack, J. M. (2007). The Homingbac baculovirus cloning system: An alternative way to introduce foreign DNA into baculovirus genomes. *J. Virol. Methods* **140,** 59–65.

Lu, A., and Miller, L. K. (1996). Generation of recombinant baculoviruses by direct cloning. *Biotechniques* **21,** 63–68.

Lu, A., and Miller, L. K. (1997). Regulation of baculovirus late and very late gene expression. *In* "The Baculoviruses" (L. K. Miller, ed.), pp. 193–216. Plenum Press, New York.

Luckow, V. A., Lee, S. C., Barry, G. F., and Olins, P. O. (1993). Efficient generation of infectious recombinant baculoviruses by site-specific transposon-mediated insertion of foreign genes into a baculovirus genome propagated in *Escherichia coli*. *J. Virol.* **67,** 4566–4579.

Lynn, D. E. (2007). Available lepidopteran insect cell lines. *Methods Mol. Biol.* **388,** 117–138.

McLachlin, J. R., and Miller, L. K. (1994). Identification and characterization of vlf-1, a baculovirus gene involved in very late gene expression. *J. Virol.* **68,** 7746–7756.

Miller, L. K. (1997). The Baculoviruses. Plenum Press, New York.

Mistretta, T. A., and Guarino, L. A. (2005). Transcriptional activity of baculovirus very late factor 1. *J. Virol.* **79,** 1958–1960.

Murhammer, D. W. (2007). Baculovirus and Insect Cell Expression Protocols. Humana Press, Totowa, NJ.

Murphy, C. I., Lennick, M., Lehar, S. M., Beltz, G. A., and Young, E. (1990). Temporal expression of HIV-1 envelope proteins in baculovirus-infected insect cells: Implications for glycosylation and CD4 binding. *Gen. Anal. Tech. Appl.* **7,** 160–171.

Ooi, B. G., Rankin, C., and Miller, L. K. (1989). Downstream sequences augment transcription from the essential initiation site of a baculovirus polyhedrin gene. *J. Mol. Biol.* **210,** 721–736.

O'Reilly, D. R., Miller, L. K., and Luckow, V. A. (1992). Baculovirus Expression Vectors. W. H. Freeman and Company, New York.

Passarelli, A. L., and Guarino, L. A. (2007). Baculovirus late and very late gene regulation. *Curr. Drug Targets* **8,** 1103–1115.

Pennock, G. D., Shoemaker, C., and Miller, L. K. (1984). Strong and regulated expression of *Escherichia coli* beta-galactosidase in insect cells with a baculovirus vector. *Mol. Cell. Biol.* **4,** 399–406.

Pijlman, G. P., van Schijndel, J. E., and Vlak, J. M. (2003). Spontaneous excision of BAC vector sequences from bacmid-derived baculovirus expression vectors upon passage in insect cells. *J. Gen. Virol.* **84,** 2669–2678.

Possee, R. D., Hitchman, R. B., Richards, K. S., Mann, S. G., Siaterli, E., Nixon, C. P., Irving, H., Assenberg, R., Alderton, D., Owens, R. J., and King, L. A. (2008). Generation of baculovirus vectors for the high-throughput production of proteins in insect cells. *Biotechnol. Bioeng.* **101,** 1115–1122.

Possee, R. D., and King, L. A. (2007). Baculovirus transfer vectors. *Meth. Mol. Biol.* **388,** 55–76.

Rankl, N. B., Rice, J. W., Gurganus, T. M., Barbee, J. L., and Burns, D. J. (1994). The production of an active protein kinase C-delta in insect cells is greatly enhanced by the use of the basic protein promoter. *Prot. Exp. Purif.* **5,** 346–356.

Richardson, C. D. (1995). Baculovirus Expression Protocols. Humana Press, Totowa, NJ.

Seo, N. S., Hollister, J. R., and Jarvis, D. L. (2001). Mammalian glycosyltransferase expression allows sialoglycoprotein production by baculovirus-infected insect cells. *Prot. Exp. Purif.* **22,** 234–241.

Shi, X., and Jarvis, D. L. (2007). Protein N-glycosylation in the baculovirus-insect cell system. *Curr. Drug Targets* **8,** 1116–1125.

Sigma-Aldrich, D. L. (2008). D6192 DiamondBac^{TM} Baculovirus DNA.

Slack, J. M., Kuzio, J., and Faulkner, P. (1995). Characterization of v-cath, a cathepsin L-like proteinase expressed by the baculovirus *Autographa californica* multiple nuclear polyhedrosis virus. *J. Gen. Virol.* **76,** 1091–1098.

Smith, P. L., Bousfield, G. R., Kumar, S., Fiete, D., and Baenziger, J. U. (1993). Equine lutropin and chorionic gonadotropin bear oligosaccharides terminating with SO4–4-GalNAc and Sia alpha 2, 3Gal, respectively. *J. Biol. Chem.* **268,** 795–802.

Smith, G. E., Fraser, M. J., and Summers, M. D. (1983a). Molecular engineering of the Autographa californica nuclear polyhedrosis virus genome: Deletion mutations within the polyhedrin gene. *J. Virol.* **46,** 584–593.

Smith, G. E., Summers, M. D., and Fraser, M. J. (1983b). Production of human beta interferon in insect cells infected with a baculovirus expression vector. *Mol. Cell. Biol.* **3,** 2156–2165.

Sridhar, P., Panda, A. K., Pal, R., Talwar, G. P., and Hasnain, S. E. (1993). Temporal nature of the promoter and not relative strength determines the expression of an extensively processed protein in a baculovirus system. *FEBS Lett.* **315,** 282–286.

Summers, M. D. (2006). Milestones leading to the genetic engineering of baculoviruses as expression vector systems and viral pesticides. *Adv. Virus Res.* **68,** 3–73.

Summers, M. D., and Smith, G. E. (1987). A manual of methods for baculovirus vectors and insect cell culture procedures. *Tx. Agric. Exp. Stn. Bull.* **No. 1555.**

Thomas, C. J., Brown, H. L., Hawes, C. R., Lee, B. Y., Min, M. K., King, L. A., and Possee, R. D. (1998). Localization of a baculovirus-induced chitinase in the insect cell endoplasmic reticulum. *J. Virol.* **72,** 10207–10212.

Tomiya, N., Howe, D., Aumiller, J. J., Pathak, M., Park, J., Palter, K., Jarvis, D. L., Betenbaugh, M. J., and Lee, Y. C. (2003). Complex-type biantennary N-glycans of recombinant human transferrin from Trichoplusia ni insect cells expressing mammalian β1, 4-galactosyltransferase and β1, 2-N-acetylglucosaminyltransferase II. *Glycobiology* **13,** 23–34.

Vaughn, J. L., Goodwin, R. H., Thompkins, G. J., and McCawley, P. (1977). The establishment of two insect cell lines from the insect *Spodoptera frugiperda* (Lepidoptera: Noctuidae). *In Vitro* **13,** 213–217.

Vialard, J., Lalumiere, M., Vernet, T., Briedis, D., Alkhatib, G., Henning, D., Levin, D., and Richardson, C. (1990). Synthesis of the membrane fusion and hemagglutinin proteins of measles virus, using a novel baculovirus vector containing the beta-galactosidase gene. *J. Virol.* **64,** 37–50.

Vialard, J. E., and Richardson, C. D. (1993). The 1, 629-nucleotide open reading frame located downstream of the Autographa californica nuclear polyhedrosis virus polyhedrin gene encodes a nucleocapsid-associated phosphoprotein. *J. Virol.* **67,** 5859–5866.

Walhout, A. J., Temple, G. F., Brasch, M. A., Hartley, J. L., Lorson, M. A., van den Heuvel, S., and Vidal, M. (2000). GATEWAY recombinational cloning: Application to the cloning of large numbers of open reading frames or ORFeomes. *Methods Enzymol.* **328,** 575–592.

Wickham, T. J., Davis, T., Granados, R. R., Shuler, M. L., and Wood, H. A. (1992). Screening of insect cell lines for the production of recombinant proteins and infectious virus in the baculovirus expression system. *Biotechnol. Progr.* **8,** 391–396.

Williams, G. V., Rohel, D. Z., Kuzio, J., and Faulkner, P. (1989). A cytopathological investigation of Autographa californica nuclear polyhedrosis virus p10 gene functions using insetion/deletion mutants. *J. Gen. Virol.* **70,** 187–202.

CHAPTER FIFTEEN

Recombinant Protein Production by Transient Gene Transfer into Mammalian Cells

Sabine Geisse* *and* Cornelia Fux†

Contents

1. Introduction 224
2. HEK293 and CHO Cell Lines Commonly Used in TGE Approaches 224
3. Expression Vectors for HEK293 and CHO Cells 226
4. Cultivation of HEK293 Cells and CHO Cell Lines in Suspension 228
5. Transfection Methods 228
   5.1. Small-scale transfection by lipofection in six-well plates 229
   5.2. Medium scale transfection in shake flasks with polyethylenimine as transfer reagent 231
   5.3. Large-scale transient transfection in the Wave™ Bioreactor with polyethylenimine as transfer reagent 232
6. Conclusions 234
Acknowledgments 234
References 235

Abstract

The timely availability of recombinant proteins in sufficient quantity and of validated quality is of utmost importance in driving drug discovery and the development of low molecular weight compounds, as well as for biotherapeutics. Transient gene expression (TGE) in mammalian cells has emerged as a promising technology for protein generation over the past decade as TGE meets all the prerequisites with respect to quantity and quality of the product as well as cost-effectiveness and speed of the process. Optimized protocols have been developed for both HEK293 and CHO cell lines which allow protein production at any desired scale up to >100 l and in milligram to gram quantities. Along with an overview on current scientific and technological knowledge, detailed protocols for expression of recombinant proteins on small, medium, and large scale are discussed in the following chapter.

* Novartis Institutes for BioMedical Research, Department of NBC/PPA, Basel, Switzerland
† Novartis Pharma AG, Department of NBx/PSD: Scale up, Basel, Switzerland

Methods in Enzymology, Volume 463
ISSN 0076-6879, DOI: 10.1016/S0076-6879(09)63015-9

1. Introduction

Recombinant proteins including antibodies are essential tools for drug discovery in the early stage and used in a multitude of applications. While for production of therapeutic proteins criteria such as growth behavior, clonality, and stability, as well as the productivity of the producer cell line are of major importance, recombinant protein production for research purposes is mainly driven by the cost-effectiveness, simplicity, and speed of the process in conjunction with adequate yields of the product.

Over the past decade efficient transient transfection protocols have reportedly been developed, which meet and even exceed the above-mentioned prerequisites. Transient production of proteins nowadays is predominantly done in HEK293-derived cell lines, yet strong efforts are ongoing to develop similarly efficient protocols for CHO cells. This bears the advantage of maintaining the same host cell background and thus product characteristics throughout the entire process of therapeutic protein development in R&D. Detailed summaries of current scientific knowledge with respect to TGE can be found in some recently published reviews (Baldi *et al.*, 2007; Geisse, 2009; Geisse *et al.*, 2005; Pham *et al.*, 2006).

2. HEK293 and CHO Cell Lines Commonly Used in TGE Approaches

Several descendants of the original HEK293 cell line established by Graham and coworkers in 1977 (Graham *et al.*, 1977; Shaw *et al.*, 2002) are currently in use in large-scale transient expression approaches, as summarized in Table 15.1. While 293 Freestyle cells are a derivative of a 293 wild-type strain selected for good growth performance and high expression levels (Zhang *et al.*, 2009), most of the other candidates used in TGE trials are actually engineered cell lines expressing, for instance, the EBNA-1 gene from Epstein–Barr virus (EBV). The EBNA-1 protein acts in conjunction with the origin of replication from EBV, *oriP*, encoded on the expression plasmids; upon transfection, the plasmids remain in the nucleus as extrachromosomal entities—episomes—and are segregated onto the daughter cells once per cell cycle (Kishida *et al.*, 2008; Lindner and Sugden, 2007). In addition, the EBNA-1 gene contains a nuclear localization signal (NLS) sequence as well as a transcriptional enhancer element which both foster nuclear import and enhanced transcription rates (Dean *et al.*, 1999, 2005; Laengle-Rouault *et al.*, 1998).

Similarly, HEK293 T cells feature an integrated SV40 large T antigen copy, which upon interaction with the SV40 origin of replication (SV40*ori*)

Table 15.1 Overview of HEK293 cell lines and expression systems for transient expression

Expression system	Features of cell line	Cultivation medium[a]	Used in conjunction with plasmid vectors	References	Comments
HEK.EBNA (HE, 293-EBNA)	EBNA-1 transformed HEK293 cell line, originally marketed by Invitrogen	Suspension: for example, ExCell 293 (SAFC Biosciences), Freestyle 293 (Invitrogen)	pCEP4 (Invitrogen), pEAK8, pcDNA 3.1, pTT (NRC Canada)	Many, citations, for example in Baldi *et al.* (2007) and Pham *et al.* (2006)	Most commonly used cell line in literature, but has not been marketed by Invitrogen for several years
HEK.EBNA (HE, 293-EBNA)	EBNA-1 transformed HEK293 cell line	Adherent: DMEM + 10% FCS + 400 μg/ml G418, suspension not done	Same as above	ATCC/Stanford University	ATCC CRL-10852
293-SFE (293SF-3F6, NRC)	Suspension adapted EBNA-1 transformed 293 cell line	Suspension: Hybridoma serum-free medium + 1% BCS	pTT plasmid series	Pham *et al.* (2005)	System available upon license from NRC
HEK293 T	SV40 T-antigen transformed 293 cell line	Adherent: DMEM + 10% FCS, suspension not routinely done	pCMV/myc/ER (Invitrogen) + derivatives	Li *et al.* (2007) and Pham *et al.* (2005)	ATCC CRL-11268, also available in many labs
293 Freestyle (293-F)	HEK293 wild type	Suspension: Freestyle medium (Invitrogen)	pcDNA 3.1, pCMV SPORT	Zhang *et al.* (2008) Manual on Web site of Invitrogen	Cell line has been carefully selected for growth in Freestyle medium and for 293 fectin reagent
HKB-11 (Hybrid of Kidney and B-cell)	Fusion of 293 cell with B-Cell lymphoma cell line	Suspension: Bayer proprietary medium	pTAT/TAR vector (Bayer)	Cho *et al.* (2001)	System available upon license from Bayer Healthcare

[a] According to our experience, all of these cell lines can be cultivated adherently in Dulbecco's Modified Eagle medium (DMEM) or in a 1:2 mixture of DMEM/Ham's F12 medium, supplemented with 10% fetal calf serum.

supplied on the expression vector supports extrachromosomal maintenance of plasmids. Furthermore, the SV40 enhancer element as part of the SV40*ori* is capable of binding not only to SV40 large T antigen, but also to newly synthesized cytosolic transcription factors such as AP-1 and NFκB, which thus serve to "cargo" the plasmids into the nucleus (Dean *et al.*, 1999, 2005; Graessmann *et al.*, 1989).

The HKB-11 cell line results from a fusion between a Burkitt lymphoma-derived cell clone, 2B8 and 293 S cells (Cho *et al.*, 2001, 2002), facilitating single cell suspension growth while maintaining high transfection rates and recombinant product yields.

All of the six HEK293-derived cell lines described in Table 15.1 were extensively tested in parallel for growth behavior and "transfectability" in our lab. For proteins, such as antigens, HEK293-6E, and HKB-11 cells gave in most instances the best results, while for expression of antibodies, transfection into HEK293 T cells frequently resulted in higher expression levels (Geisse, 2009). In summary, it is certainly worthwhile to try parallel expression in several HEK293-derived cell lines in order to find the optimal candidate for a given gene to be expressed.

CHO cell lines used for transient expression approaches are descendants of the CHO K1 line or the dhfr-negative variants DG44 and DUK X B11 (Gottesman, 1987; Kao and Puck, 1968; Urlaub and Chasin, 1980; Urlaub *et al.*, 1983). Suspension-adapted CHO-S cells, a CHO K1 derivative, along with a suitable serum-free medium and a lipofection reagent were recently introduced as "Freestyle Max™ system" by Invitrogen (Zhang *et al.*, 2009). Genetically engineered CHO sublines are as yet infrequently used, even though a beneficial effect on protein expression exerted via the inclusion of viral elements derived from bovine papillomavirus, polyoma virus, and EBV on the expression vector could be demonstrated (Kivimäe *et al.*, 2001; Kunaparaju *et al.*, 2005; Silla *et al.*, 2005).

3. Expression Vectors for HEK293 and CHO Cells

To take advantage of the multiple beneficial functions of the EBV-or SV40-derived viral elements described above suitable expression vectors for transient expression in HEK293 cells should harbor the *oriP* or *SV40ori* sequences, respectively. The human cytomegalovirus-derived immediate early promoter is most commonly chosen to drive transgene expression (Lee *et al.*, 2007; Makrides, 1999; Van Craenenbroeck *et al.*, 2000), but also other promoters of viral origin, such as the mouse CMV promoter or a mouse/human hybrid promoter (Meissner *et al.*, 2001; Xia *et al.*, 2006) or of nonviral origin (e.g., the human elongation factor 1α [EF1alpha] promoter

(Backliwal *et al.*, 2008a; Pham *et al.*, 2006)) have been described as well. In brief, a striking advantage of a single promoter/cell line combination could so far not be demonstrated. Moreover, genetic elements active at the posttranscriptional level, such as an intron splice element or the Woodchuck hepatitis virus regulatory element (WPRE) impact significantly on protein yields by enhancement of mRNA stability, increased nuclear export rates of mRNA and increased translation rates (Backliwal *et al.*, 2008a; Klein *et al.*, 2006; Le Hir *et al.*, 2003; Matsumoto *et al.*, 1998). More details on the design of various expression plasmids and their performance can be found in the public domain (Backliwal *et al.*, 2008a; Durocher *et al.*, 2002; Meissner *et al.*, 2001; Pham *et al.*, 2006; Xia *et al.*, 2006) (Fig. 15.1).

It should, however, be mentioned that despite all benefits to gene expression by specifically tailored expression plasmids featuring additional enhancing elements, a large sized expression vector impacts negatively the stability and yield in plasmid preparation. We have observed a three- to fivefold reduction in plasmid yields when the size of the empty backbone vector was doubled or tripled. As quite large quantities of plasmid DNA need to be prepared for large-scale transient expression approaches and the size of genes/cDNAs to be expressed appears to be steadily increasing, this point should be taken into consideration in the design of backbone expression vectors for TGE approaches.

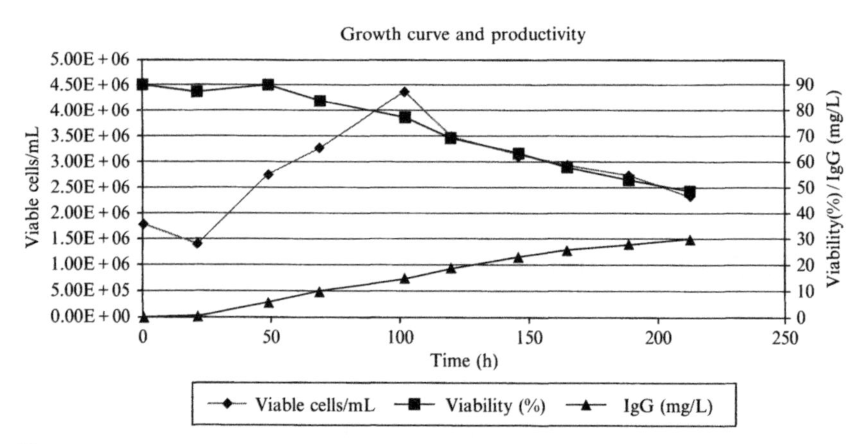

Figure 15.1 Production of a human IgG2 antibody by PEI-mediated transient transfection into HEK293 T cells cultivated in M11V3 medium (Novartis proprietary) in the Wave™ bioreactor over a period of 213 h. Heavy and light chain genes were encoded on the same plasmid backbone. Symbols denote viable cell count (diamonds), viability in % (squares) and antibody titer (triangles) determined by Protein A HPLC.

4. Cultivation of HEK293 Cells and CHO Cell Lines in Suspension

Transient transfections into these two most frequently used cell lines are in most instances performed in suspension culture, even though for both lines the option to use adherent cell cultures—either by cultivation in, for example, classical DMEM-based media containing fetal calf serum, by supplementation of serum-free media formulations with serum, or by cell-surface attachment mediated by coating of tissue culture flasks with, for example, poly-D-lysine prior to transfection—is retained.

CHO and HEK cell lines are readily adaptable to serum-free conditions and a large variety of specially designed media formulations (serum-free, protein-free, or chemically defined) are commercially available. Adaptation to serum-free suspension conditions is done by gradually "weaning" the cells from the serum by dilution; simultaneously, the cells are transferred from stationary culture in tissue culture flasks to agitated cultivation in Erlenmeyer shake flasks, spinner flasks, or roller bottles.

The suspension adapted cells should exhibit good growth behavior and high cell viability. Thus, cell counting and viability determination, done either manually using a Neubauer hemocytometer or by employing cell counting devices such as the Cedex™ (Innovatis AG, Bielefeld, FRG) or the Vi-Cell™ (Beckman Coulter, Krefeld, FRG), and dilution of cells twice per week is mandatory. In particular, at the day of transfection the cells should be in logarithmic growth phase; thus, a dilution of the culture 1 or 2 days prior to gene transfer is important. We keep our backup cultures at cell densities between $3–4 \times 10^5$ cells/ml for HEK293 cells and at approx. 2×10^5 cells/ml for CHO cells in Erlenmeyer shake flasks of variable size (Corning by Fisher Scientific, Wohlen, CH) in a shaker incubator (Kuehner, Climo-Shaker ISF1-X, Kuehner AG, Birsfelden, CH) at 100 rpm/37 °C and 5% CO_2.

5. Transfection Methods

For lipofection a wealth of formulations with individual, mostly proprietary compositions are commercially available, and doubtlessly many of them are highly efficacious in introducing the plasmid DNA into the cells. Their drawback is the cost-of-goods—transient transfection beyond the scale of several hundred milliliters is economically not feasible when using these reagents. Large-scale transient transfection approaches call for transfection reagents readily available in bulk quantities with little batch-to-batch variability to ensure consistency of the process. Calcium–phosphate-mediated transfection as well as polyethylenimine (PEI)-based plasmid complexation and uptake ofDNA into the cells are the only two commonly applied methods which meet these requirements, even though some other options have been described, such as Chitosan and derivatives thereof, and 14DEA2 as transfer reagents (Dang and Leong, 2006; Jiang and Sharfstein, 2008; Kusumoto *et al.*, 2006).

The details of calcium–phosphate-mediated transfection have been extensively studied and described (Batard *et al.*, 2001; Jordan and Wurm, 2004; Jordan *et al.*, 1996, 1998; Meissner *et al.*, 2001), and similarly, the mechanisms of action of PEI-mediated gene transfer have attracted much research efforts. An in-depth mechanistic overview on PEI-based transfection would exceed the scope of this chapter, but can be easily retrieved from the literature (Bertschinger *et al.*, 2006; Eliyahu *et al.*, 2005; Lungwitz *et al.*, 2005; Payne, 2007; Vicennati *et al.*, 2008).

In a nutshell, two individual aspects are essential to mention in this context. For both methods, calcium–phosphate-mediated transfection and PEI-mediated transfection, the compatibility of the cell cultivation medium with the transfection reagent is of key importance. It has been shown that the composition of some commonly used, commercially available culture media may hamper or even preclude an efficient uptake of plasmid DNA (Bertschinger *et al.*, 2006; Schlaeger and Christensen, 1999; Sun *et al.*, 2006), necessitating either a medium exchange prior to transfection or a biphasic expression strategy using two different media during the transfection and the production phases.

Furthermore, an efficiency of 90–100% of transduced cells, as observed with lipofection reagents, has not been achieved when using calcium–phosphate or PEI as transfer reagents. At best, a transfection rate of 60% is realistically obtained. However, for the sake of cost-effectiveness and scalability of the process, a reduction in transfection efficiency and yields appears to be acceptable. In a very recent publication by Chenuet and coworkers (Chenuet *et al.*, 2008) comparative analyses of calcium–phosphate- and PEI-mediated transfection demonstrated a higher plasmid uptake for PEI-mediated transfection into CHO DG44 cells; yet, for stable recombinant cell line generation the uptake and integration of plasmid copies resulting in high specific productivities appeared to be more favorable when calcium–phosphate-mediated transfection was applied.

5.1. Small-scale transfection by lipofection in six-well plates

Prior to embarking on transfection approaches on large scale, it is desirable to assess the integrity of the expression vector and the functionality of the protein. To do so, possibly also in combination with multiconstruct screening, a rapid, small-scale approach using lipofection as reagent for gene transfer is the method of choice. In the following, a protocol is described which is applicable to HEK293 as well as CHO cells. The expression vector we use is a composite plasmid derived from commercially available plasmids and features elements such as the CMV promoter, an intron splice element, and the *oriP* for extrachromosomal replication in HEK293–EBNA-1 positive cells. The inserted cDNA fragments are in most instances equipped with a protein detection tag which can be used for analytical purposes as well as for protein purification via affinity chromatography on Protein A column, for example:

1. At the day of transfection harvest the appropriate amount of cells (calculate 1×10^6 cells per well of a 6-well plate) and spin them down at 216*g* (1000 rpm) for 5 min.
2. Resuspend the cell pellet in 2 ml of medium/well; transfer the cell suspension to a 6-well plate and distribute the cells into the wells.
3. Place the plate into a cell culture incubator at 37 °C/5% CO_2.
4. For preparation of the lipoplexes, transfer 100 μl of OPTI-MEM medium (Invitrogen) into a sterile polypropylene tube.
5. Add 2 μg of supercoiled recombinant plasmid DNA (ideally at 1 μg/μl concentration) into this tube.
6. Add 7 μl of FuGENE® HD reagent (Roche Diagnostics, Indianapolis, IN) to the same tube and tap the tube gently for a few seconds for mixing.
7. Incubate the mixture for 15 min at room temperature.
8. Add the mix dropwise directly to one well of cells of the 6-well plate.
9. Incubate the 6-well plate at 37 °C/5% CO_2 in a tissue culture incubator for 72 h.
10. Harvest the culture supernatant (in case of secreted proteins) and/or the cells (for membrane-bound or intracellular expression) and perform the desired analytical experiments (e.g., ELISA, Western Blot, HPLC affinity chromatography, surface plasmon resonance, etc.).

Notes:

(i). We have achieved excellent results when using the FuGENE® HD reagent for transfection, but other lipofection reagents from other vendors, which we did not try, may work equally well. Care should be taken, however, to optimize process parameters such as the DNA to reagent ratio as well as the amount of cells in an appropriate volume of medium.

(ii). Harvest of samples beyond 72 h of incubation/production is possible; however, due to the relatively high cell density in a small volume nutrient depletion, cell death and potentially proteolytic damage to the protein may occur. Thus, if protein analysis following a prolonged production phase is desired, the medium scale protocol described below appears to be the better choice.

5.2. Medium scale transfection in shake flasks with polyethylenimine as transfer reagent

1. At the day of transfection harvest 1.25×10^7 cells and pellet them by centrifugation at 216*g* (1000 rpm) for 5 min.
2. Resuspend the cells in 9 ml of fresh culture medium and transfer the suspension into a 125 ml disposable shake flask. A total medium replacement at this stage enhances transfection efficiency.
3. Polyplex formation: Take two sterile 15 ml polypropylene tubes and add into each 1.75 ml of fresh culture medium.

Note: The original protocol for PEI transfection by Boussif and coworkers (Boussif *et al.*, 1995) describes the polyplex formation in a 150 m*M* NaCl solution. This protocol has been applied frequently; however, more recently DNA complexation carried out in RPMI 1640 medium (Invitrogen) for HEK293 cells or ProCHO5 medium (Lonza, Vervier, Belgium) for CHO cells has been shown to lead to successful transfection as well (Backliwal *et al.*, 2008c; Bertschinger *et al.*, 2006). We routinely use our cultivation medium M11V3 for complexing of the plasmid DNA without any adverse effects.

1. Into one of the tubes add 50 μg of polyethylenimine from a sterile stock solution: 25 kDa linear PEI (Polysciences, Warrington PA) is dissolved in water at a concentration of 1 mg/ml, and the pH is adjusted to 7.0.

Note: The PEI stock solution should be sterile, filtered, aliquotted, and stored at − 80 °C until use. A transfection reagent based on linear PEI is also commercially available under the trade name jetPEI™ from Polyplus-transfection, New York, NY.

2. Add 25 μg of supercoiled, recombinant plasmid DNA into the other tube.
3. Tap both tubes a few times gently and incubate them for 15 min at room temperature.
4. Pipet the diluted PEI-solution into the diluted DNA solution (*not the inverse!*), mix gently and incubate the mix for 15 min at room temperature. The DNA:PEI ratio employed corresponds to 1:2 (μg:μg).

Notes:

(i). If the complexes mature in 150 m*M* NaCl solution, a maturation time beyond 10 min does not lead to efficient release of DNA from the particles. If, however, culture medium is used for maturation, the duration of incubation does not seem to have any effect (Backliwal *et al.*, 2008b; Bertschinger *et al.*, 2006).

(ii). The DNA:PEI ratio is of critical importance for the success of these experiments. Commonly used are ratios of 1:2–1:3 (μg:μg), but for

CHO cells a ratio of 1:4–1:5 has been determined as optimal (Chenuet *et al.*, 2008). At high cell densities of 20×10^6 cells/ml, an *a priori* complex formation does not even seem to be necessary—the plasmid DNA can be added directly to the culture, followed by addition of the PEI solution (Backliwal *et al.*, 2008b).

5. Add the polyplexes formed in 3.5 ml to the 9 ml of suspension culture in the shake flask and incubate the flask on an orbital shaker platform at 37 °C/5% CO_2/100 rpm. The cell density at transfection corresponds to 1×10^6 cells/ml.
6. After 4 h of incubation, add 12.5 ml of culture medium to the shake flask and continue the incubation for 72 h (or more). The cell density at the beginning of the production phase is then 5×10^5 cells/ml.
7. After 72 h (or at a later time-point; 5–6 days posttransfection appears to be the maximum, otherwise severe nutrient limitations will occur) harvest the cells and/or the cell supernatant for analyses of recombinant protein expression.

5.3. Large-scale transient transfection in the Wave™ Bioreactor with polyethylenimine as transfer reagent

The disposable Wave™ bioreactor has gained widespread acceptance in recent years due its ease of maintenance and handling versus classical stirred tank reactors. Most commonly used are working volumes of 10 and 20 l, even though reactors of 100 l working volume are commercially available, and a prototype of 300 l working volume is under developmental construction. Whether to use the Wave™ bioreactor for transfection cultures below 10 l is a question of economical feasibility—we prefer to use shake flask cultures for smaller production runs (Fernbach flasks, Corning). Reportedly, also other reactors or shaking devices (Muller *et al.*, 2005; Pham *et al.*, 2006; Stettler *et al.*, 2007) have been successfully used for transiently transfected cultures.

The following protocol describes production of an antibody at a scale of 10 l using HEK293 T cells as host and M11V3 serum-free culture medium (Novartis proprietary). Both chains of the antibody were cloned into the same backbone expression vector.

1. A 20 l Wave™ bag (Sartorius Stedim Biotech, Göttingen, Germany) is mounted onto a Wave™ platform (Wave-Bioreactor SPS50) and linked to a DASGIP gas-mixing module (DASGIP, Juelich, Germany). Subsequently, the bag is inoculated with 4 l HEK293T cell culture at a cell density of approximately 1.8×10^6 viable cells/ml.
2. The following process parameters and conditions are applied: gas flow: 20 l/h; gas mix consisting of 21–25% O_2/0% CO_2; temperature 37 °C, pH: 6.8–7.4, rocking speed: 10 rpm, rocking angle: 7°.

3. 10 mg plasmid DNA (1 mg/ml) are mixed with M11V3 medium to a final volume of 500 ml and incubated for 10 min at room temperature. Afterward, the diluted DNA is sterile filtered through a 0.22 μm GP EXPRESS PLUS membrane (Millipore, Billerica, MA).
4. 30 ml PEI solution (1 mg/ml) are mixed with 500 ml M11V3 medium and incubated for 10 min at room temperature. Afterwards, the diluted PEI solution is sterile filtered through a 0.22 μm GP EXPRESS PLUS membrane.

Note: The PEI stock solution should be sterile, filtered, aliquotted, and stored at −80 °C until use.

5. Next, the PEI-solution is added to the diluted DNA and the mix is incubated for 15 min at room temperature for polyplex formation to occur.
6. The DNA–PEI–M11V3 mix is then aseptically added to the cells in the Wave^{TM} bag to achieve a final volume of 5 l. Incubation is continued for 5–6 h applying the following parameters: gas-flow: 25 l/h, gas mix: 25% O_2/0% CO_2, temperature: 37 °C, pH: 6.8–7.4, rocking speed: 10 rpm, angle: 7°.
7. Subsequently, 5 l M11V3 medium supplemented with 100 ml RX1 combined feed (feed solution consisting of amino acids, glucose and glutamine; custom-made by Irvine Scientific, Santa Ana, CA) are added aseptically to the cells in the Wave^{TM} bag. During the production phase the following process parameters are applied: gas flow: 30–40 l/h, gas mix: 30–40% O_2/0% CO_2, temperature: 37 °C, pH: 6.8–7.4 (adjusted with bicarbonate solution pH 9.2, Sigma Aldrich, Buchs, Switzerland), air saturation: 60–100%, rocking speed: 14–18 rpm, angle: 7°.
8. The transfected cells are cultivated for 10 days in the Wave^{TM} bioreactor for antibody production.

Note: For production of recombinant proteins the production phase until harvest is usually between 5 and 7 days, but may be even shorter, if the desired protein is sensitive toward proteolytic degradation. Production runs of antibody molecules may be extended to 10 days.

9. Each day a sample is taken and the cell density and viability is determined using a Vi-Cell cell counter device (Beckman Coulter). Nutrient status, pH, and air saturation are measured using the Bioprofile 400 Analyzer (Labor-Systeme Flückiger, Switzerland). The IgG concentration is assessed by Protein A HPLC determination (Fig. 15.1).
10. After 10 days, the cells are aseptically harvested and the cell removal is performed by cross-flow filtration (Fresenius Filter PlasmaFlux, 0.2 μm). Afterward, the cell free supernatant is concentrated 10-fold by means of hollow fiber filtration (Hemoflow F 10 HPS, Fresenius, Stans, Switzerland, 10 kDa cutoff).

11. The concentrate is subjected to protein purification by Protein A affinity chromatography and size exclusion chromatography.

A variety of modifications to these standard protocols described above have been tried and published, some of which are listed in brief below:

a. Enhancement of expression rates by "temperature shift" to 30–32 °C during the production phase when using CHO cells (Backliwal *et al.*, 2008a; Galbraith *et al.*, 2006).
b. Cell cycle arrest in G2/M by treatment with the microtubule depolymerizing agent nocodazole (Tait *et al.*, 2004).
c. Treatment with inhibitors of histone deacetylase and DNA methyl transferase (e.g., sodium butyrate, valproic acid) in CHO and HEK293 cells (Backliwal *et al.*, 2008c).
d. Genetic engineering of cell lines to prevent apoptosis or to overcome inhibitions exerted by the unfolded protein response (UPR) (Backliwal *et al.*, 2008c; Majors *et al.*, 2007; Tigges and Fussenegger, 2006).

6. Conclusions

The generation of recombinant proteins by means of transient transfection technologies has reached an impressive state in terms of method development and yields during the past several years, as reflected in the wealth of data published on the subject. As yet, it is probably premature to envisage the replacement of the tedious process of cell line development for production of biotherapeutics by large-scale transient approaches, even though remarkable antibody titers of >1 g/l have been reported already (Backliwal *et al.*, 2008a). Our own observations based on >100 transient expression trials indicate a huge heterogeneity in expression yields depending on the gene(s) expressed, ranging from 1 to >170 mg/l for proteins; antibody titers amount to 20–40 mg/l on average with frequent outliers to both sides. There is certainly a need for further technical advancement with respect to construct design, cultivation conditions, and possibly cell line engineering, but these efforts are more than justified by the overall impressive success of the approach.

Acknowledgments

We thank our collaborators Agnès Patoux, Thomas Cremer, Mirjam Buchs, Sibylle Bossart, Stefan Dalcher, and Rainer Uhrhahn for the development of the protocols and the generation of the experimental data. Special thanks go to Prof. Bertram Opalka, Uniklinikum Essen-Duesburg, FRG for critically reviewing the manuscript.

REFERENCES

Backliwal, G., Hildinger, M., Chenuet, S., Wulhfard, S., De Jesus, M., and Wurm, F. M. (2008a). Rational vector design and multi-pathway modulation of HEK 293E cells yield recombinant antibody titers exceeding 1 g/l by transient transfection under serum-free conditions. *Nucleic Acids Res.* **36,** e96.

Backliwal, G., Hildinger, M., Hasija, V., and Wurm, F. M. (2008b). High-density transfection with HEK-293 cells allows doubling of transient titers and removes need for a priori DNA complex formation with PEI. *Biotechnol. Bioeng.* **99,** 721–727.

Backliwal, G., Hildinger, M., Kuettel, I., Delegrange, F., Hacker, D. L., and Wurm, F. M. (2008c). Valproic acid: A viable alternative to sodium butyrate for enhancing protein expression in mammalian cell cultures. *Biotechnol. Bioeng.* **101,** 182–189.

Baldi, L., Hacker, D. L., Adam, M., and Wurm, F. M. (2007). Recombinant protein production by large-scale transient gene expression in mammalian cells: State of the art and future perspectives. *Biotechnol. Lett.* **29,** 677–684.

Batard, P., Jordan, M., and Wurm, F. (2001). Transfer of high copy number plasmid into mammalian cells by calcium phosphate transfection. *Gene* **270,** 61–68.

Bertschinger, M., Backliwal, G., Schertenleib, A., Jordan, M., Hacker, D. L., and Wurm, F. M. (2006). Disassembly of polyethylenimine-DNA particles *in vitro*: Implications for polyethylenimine-mediated DNA delivery. *J. Control Release* **116,** 96–104.

Boussif, O., Lezoualc'h, F., Zanta, M. A., Mergny, M. D., Scherman, D., Demeneix, B., and Behr, J. P. (1995). A versatile vector for gene and oligonucleotide transfer into cells in culture and *in vivo*: Polyethylenimine. *Proc. Natl. Acad. Sci. USA* **92,** 7297–7301.

Chenuet, S., Martinet, D., Besuchet-Schmutz, N., Wicht, M., Jaccard, N., Bon, A. C., Derouazi, M., Hacker, D. L., Beckmann, J. S., and Wurm, F. M. (2008). Calcium phosphate transfection generates mammalian recombinant cell lines with higher specific productivity than polyfection. *Biotechnol. Bioeng.* **101,** 937–945.

Cho, M.-S., Yee, H., Brown, C., Jeang, K.-T., and Cahn, S. (2001). An oriP expression vector containing the HIV Tat/TAR transactivation axis produces high levels of protein expression in mammalian cells. *Cytotechnology* **37,** 23–30.

Cho, M. S., Yee, H., and Chan, S. (2002). Establishment of a human somatic hybrid cell line for recombinant protein production. *J. Biomed. Sci.* **9,** 631–638.

Dang, J. M., and Leong, K. W. (2006). Natural polymers for gene delivery and tissue engineering. *Adv. Drug Deliv. Rev.* **58,** 487–499.

Dean, D., Dean, B., Muller, S., and Smith, L. (1999). Sequence requirements for plasmid nuclear import. *Exp. Cell Res.* **253,** 713–722.

Dean, D., Strong, D., and Zimmer, W. (2005). Nuclear entry of nonviral vectors. *Gene Ther.* **12,** 881–890.

Durocher, Y., Perret, S., and Kamen, A. (2002). High-level and high-throughput recombinant protein production by transient transfection of suspension-growing human 293-EBNA1 cells. *Nucleic Acids Res.* **30,** E9.

Eliyahu, H., Barenholz, Y., and Domb, A. J. (2005). Polymers for DNA delivery. *Molecules* **10,** 34–64.

Galbraith, D. J., Tait, A. S., Racher, A. J., Birch, J. R., and James, D. C. (2006). Control of culture environment for improved polyethylenimine-mediated transient production of recombinant monoclonal antibodies by CHO cells. *Biotechnol. Prog.* **22,** 753–762.

Geisse, S. (2009). Reflections on more than 10 years of TGE approaches. *Protein Expr. Purif.* **64,** 99–107.

Geisse, S., Jordan, M., and Wurm, F. (2005). Large-scale transient expression of therapeutic proteins in mammalian cells. *In* "Therapeutic Proteins", (C. Smales and D. James, eds.), Vol. 308, pp. 87–98. Humana Press, Totowa, NJ.

Gottesman, M. M. (1987). Chinese hamster ovary cells. *In* "Methods in Enzymology", (M. Gottesman, ed.), Vol. 151, pp. 3–8. Academic Press Inc, San Diego.

Graessmann, M., Menne, J., Liebler, M., Graeber, I., and Graessmann, A. (1989). Helper activity for gene expression, a novel function of the SV40 enhancer. *Nucleic Acids Res.* **17,** 6603–6612.

Graham, F. L., Smiley, J., Russell, W. C., and Nairn, R. (1977). Characteristics of a human cell line transformed by DNA from human Adenovirus Type 5. *J. Gen. Virol.* **36,** 59–72.

Jiang, Z., and Sharfstein, S. T. (2008). Sodium butyrate stimulates monoclonal antibody over-expression in CHO cells by improving gene accessibility. *Biotechnol. Bioeng.* **100,** 189–194.

Jordan, M., Koehne, C., and Wurm, F. (1998). Calcium-phosphate mediated DNA transfer into HEK-293 cells in suspension: Control of physicochemical parameters allows transfection in stirred media. *Cytotechnology* **26,** 39–47.

Jordan, M., Schallhorn, A., and Wurm, F. M. (1996). Transfecting mammalian cells: Optimization of critical parameters affecting calcium-phosphate precipitate formation. *Nucleic Acids Res.* **24,** 596–601.

Jordan, M., and Wurm, F. (2004). Transfection of adherent and suspended cells by calcium phosphate. *Methods* **33,** 136–143.

Kao, F.-T., and Puck, T. T. (1968). Genetics of somatic mammalian cells. VII. Induction and isolation of nutritional mutants in Chinese hamster cells. *Proc. Natl. Acad. Sci.* **60,** 1275–1281.

Kishida, T., Asada, H., Kubo, K., Sato, Y. T., Shin-Ya, M., Imanishi, J., Yoshikawa, K., and Mazda, O. (2008). Pleiotrophic functions of Epstein-Barr virus nuclear antigen-1 (EBNA-1) and oriP differentially contribute to the efficacy of transfection/expression of exogenous gene in mammalian cells. *J. Biotechnol.* **133,** 201–207.

Kivimäe, S., Allikas, A., Kurg, R., and Ustav, M. (2001). Replication of a chimeric origin containing elements from Epstein-Barr virus ori P and bovine papillomavirus minimal origin. *Virus Res.* **75,** 1–11.

Klein, R., Ruttkowski, B., Knapp, E., Salmons, B., Gunzburg, W. H., and Hohenadl, C. (2006). WPRE-mediated enhancement of gene expression is promoter and cell line specific. *Gene* **372,** 153–161.

Kunaparaju, R., Liao, M., and Sunstrom, N.-A. (2005). *Epi-CHO*, an episomal expression system for recombinant protein production in CHO cells. *Biotechnol. Bioeng.* **91,** 670–677.

Kusumoto, K., Akao, T., Mizuki, E., and Nakamura, O. (2006). Gene transfer effects on various cationic amphiphiles in CHO cells. *Cytotechnology* **51,** 57–66.

Laengle-Rouault, F., Patzel, V., Benavente, A., Taillez, M., Silvestre, N., Bompard, A., Sczakiel, G., Jacobs, E., and Rittner, K. (1998). Up to 100-fold increase of apparent gene expression in the presence of Epstein-Barr virus *oriP* sequences and EBNA1: Implications of the nuclear import of plasmids. *J. Virol.* **72,** 6181–6185.

Lee, J., Lau, J., Chong, G., and Chao, S. H. (2007). Cell-specific effects of human cytomegalovirus unique region on recombinant protein expression. *Biotechnol. Lett.* **29,** 1797–1802.

Le Hir, H., Nott, A., and Moore, M. J. (2003). How introns influence and enhance eukaryotic gene expression. *Trends Biochem. Sci.* **28,** 215–220.

Li, J., Menzel, C., Meier, D., Zhang, C., Dubel, S., and Jostock, T. (2007). A comparative study of different vector designs for the mammalian expression of recombinant IgG antibodies. *J. Immunol. Methods* **318,** 113–124.

Lindner, S., and Sugden, B. (2007). The plasmid replicon of Epstein-Barr virus: Mechanistic insights into efficient, licensed, extrachromosomal replication in human cells. *Plasmid* **58,** 1–12.

Lungwitz, U., Breunig, M., Blunk, T., and Gopferich, A. (2005). Polyethylenimine-based non-viral gene delivery systems. *Eur. J. Pharm. Biopharm.* **60,** 247–266.

Majors, B. S., Betenbaugh, M. J., and Chiang, G. G. (2007). Links between metabolism and apoptosis in mammalian cells: applications for anti-apoptosis engineering. *Metab. Eng.* **9,** 317–326.

Makrides, S. C. (1999). Components of vectors for gene transfer and expression in mammalian cells. *Protein Expr. Purif.* **17,** 183–202.

Matsumoto, K., Wassarman, K. M., and Wolffe, A. P. (1998). Nuclear history of a pre-mRNA determines the translational activity of cytoplasmic mRNA. *EMBO J.* **17,** 2107–2121.

Meissner, P., Pick, H., Kulangara, A., Chatellard, P., Friedrich, K., and Wurm, F. M. (2001). Transient gene expression: Recombinant protein production with suspension-adapted HEK293-EBNA cells. *Biotechnol. Bioeng.* **75,** 197–203.

Muller, N., Girard, P., Hacker, D. L., Jordan, M., and Wurm, F. M. (2005). Orbital shaker technology for the cultivation of mammalian cells in suspension. *Biotechnol. Bioeng.* **89,** 400–406.

Payne, C. K. (2007). Imaging gene delivery with fluorescence microscopy. *Nanomed* **2,** 847–860.

Pham, P. L., Kamen, A., and Durocher, Y. (2006). Large-scale transfection of mammalian cells for the fast production of recombinant protein. *Mol. Biotechnol.* **34,** 225–237.

Pham, P. L., Perret, S., Cass, B., Carpentier, E., St-Laurent, G., Bisson, L., Kamen, A., and Durocher, Y. (2005). Transient gene expression in HEK293 cells: Peptone addition posttransfection improves recombinant protein synthesis. *Biotechnol. Bioeng.* **90,** 332–344.

Schlaeger, E.-J., and Christensen, K. (1999). Transient gene expression in mammalian cells grown in serum-free suspension culture. *Cytotechnology* **30,** 71–83.

Shaw, G., Morse, S., Ararat, M., and Graham, F. L. (2002). Preferential transformation of human neuronal cells by human adenoviruses and the origin of HEK 293 cells. *FASEB J.* **16,** 869–871.

Silla, T., Hääl, I., Geimanen, J., Janikson, K., Abroi, A., Ustav, E., and Ustav, M. (2005). Episomal maintenance of plasmids with hybrid origins in mouse cells. *J. Virol.* **79,** 15277–15288.

Stettler, M., Zhang, X., Hacker, D. L., de Jesus, M., and Wurm, F. M. (2007). Novel orbital shake bioreactors for transient production of CHO derived IgGs. *Biotechnol. Prog.* **23,** 1340–1346.

Sun, X., Goh, P. E., Wong, K. T., Mori, T., and Yap, M. G. (2006). Enhancement of transient gene expression by fed-batch culture of HEK 293 EBNA1 cells in suspension. *Biotechnol. Lett.* **28,** 843–848.

Tait, A. S., Brown, C. J., Galbraith, D. J., Hines, M. J., Hoare, M., Birch, J. R., and James, D. C. (2004). Transient production of recombinant proteins by Chinese hamster ovary cells using polyethyleneimine/DNA complexes in combination with microtubule disrupting anti-mitotic agents. *Biotechnol. Bioeng.* **88,** 707–721.

Tigges, M., and Fussenegger, M. (2006). Xbp1-based engineering of secretory capacity enhances the productivity of Chinese hamster ovary cells. *Metab. Eng.* **8,** 264–272.

Urlaub, G., and Chasin, L. A. (1980). Isolation of Chinese hamster cell mutants deficient in dihydrofolate reductase activity. *Proc. Natl. Acad. Sci.* **77,** 4216–4220.

Urlaub, G., Käs, E., Carothers, A. M., and Chasin, L. A. (1983). Deletion of the diploid dihydrofolate reductase locus from cultured mammalian cells. *Cell* **33,** 405–412.

Van Craenenbroeck, K., Vanhoenacker, P., and Haegeman, G. (2000). Episomal vectors for gene expression in mammalian cells. *Eur. J. Biochem.* **267,** 5665–5678.

Vicennati, P., Giuliano, A., Ortaggi, G., and Masotti, A. (2008). Polyethylenimine in medicinal chemistry. *Curr. Med. Chem.* **15,** 2826–2839.

Xia, W., Bringmann, P., McClary, J., Jones, P. P., Manzana, W., Zhu, Y., Wang, S., Liu, Y., Harvey, S., Madlansacay, M. R., McLean, K., Rosser, M. P., *et al.* (2006). High levels of protein expression using different mammalian CMV promoters in several cell lines. *Protein Expr. Purif.* **45,** 115–124.

Zhang, J., Liu, X., Bell, A., To, R., Baral, T. N., Azizi, A., Li, J., Cass, B., and Durocher, Y. (2009). Transient expression and purification of chimeric heavy chain antibodies. *Protein Expr. Purif.* **65,** 77–82.

CHAPTER SIXTEEN

Tagging for Protein Expression

Arun Malhotra

Contents

1. Introduction 240
2. Some Considerations When Designing a Tagged Protein 241
   2.1. Affinity and/or solubility? 241
   2.2. Which tag(s) to use? 243
   2.3. Tandem tags? 244
   2.4. N- or C-terminal? 244
   2.5. Cleavage sites to remove tags 245
3. Protein Affinity Tags 245
   3.1. His-tag 245
   3.2. GST tag 246
   3.3. Other purification tags 247
4. Solubility Tags 249
   4.1. MBP tag 249
   4.2. Trx tag 250
   4.3. NusA tag 250
   4.4. Other solubility tags 250
5. Removal of Tags 251
6. Conclusions 253
Acknowledgment 254
References 254

Abstract

Tags are frequently used in the expression of recombinant proteins to improve solubility and for affinity purification. A large number of tags have been developed for protein production and researchers face a profusion of choices when designing expression constructs. Here, we survey common affinity and solubility tags, and offer some guidance on their selection and use.

Department of Biochemistry and Molecular Biology, University of Miami Miller School of Medicine, Miami, Florida, USA

Methods in Enzymology, Volume 463
ISSN 0076-6879, DOI: 10.1016/S0076-6879(09)63016-0

1. Introduction

Most proteins are made using recombinant techniques in expression systems. Additional residues or tags can be engineered on either the N- or C-terminal end of the protein of interest during the cloning step. These tags, which can range in size from just a few residues to full-length proteins or domains, can be used to improve protein production or to confer new properties that can be used for characterization and study of the target protein. The term "fusion protein" is also often used instead of the term "tag," and fusion sometimes refers to the simpler end-to-end joining of two proteins while tags are typically shorter and include linker regions; here, we will use these terms interchangeably.

The focus of this chapter is to survey some of the common tags that are used for improving the production of proteins, and highlight the advantages and pitfalls of their use. Given the wild profusion of tags, often in different combinations and with different cleavage sites, this review is by no means exhaustive and the reader is encouraged to look at constructs available from many commercial sources as well as repository sources such as Addgene (http://www.addgene.org) or ATCC (http://www.atcc.org). Detailed protocols related to the use of many individual tags have been described previously in *Methods in Enzymology* (e.g., Volumes 326 and 327), and in specialized journals in this field such as *Protein Expression and Purification* and the newer *Microbial Cell Factories*. Handbooks from suppliers of vectors for most expression systems (Amersham/GE Healthcare Inc., Clontech Inc., Invitrogen Inc., New England Biolabs Inc., Novagen/EMD Biosystems Inc., Roche Inc., Sigma-Aldrich Inc., and others) are another rich source of protocols.

Tags used to improve the production of recombinant proteins can be roughly divided into purification and solubility tags. The former are used along with affinity binding to allow rapid and efficient purification of proteins, while the latter refer to tags that enhance the proper folding and solubility of a protein. Tables 16.1 and 16.2 list some of the common purification and solubility tags that are used for protein expression. Table 16.3 summarizes common endoproteases used to remove tags and recover the target protein of interest.

While such tags are quite useful, other tags can be fused to proteins for a broad range of applications—labeling for imaging and localization studies, protein detection and quantification, protein–protein interaction studies, subcellular localization or transduction, and many others. It is important to keep in mind some of these additional capabilities that can be engineered into a protein as recombinant constructs are being designed, since multiple tags can be added together in different combinations (Fig. 16.1).

Table 16.1 Common affinity tags

Tag	Size	Affinity matrix
His-tag	6–10 His	Immobilized metal ions—Ni, Co, Cu, Zn
GST (glutathione-*S*-transferase)	211 aa	Glutathione resin
FLAG-tag	8 aa (DYKDDDDK) (22 aa for 3xFLAG)	Anti-FLAG mAB
Strep-II-tag	8 aa (WSHPQFEK)	Strep-Tactin (modified streptavidin)
Protein A (staphylococcal Protein A)	280 aa	Immobilized IgG
MBP (maltose-binding protein)	396 aa	Cross-linked amylose
CBP (calmodulin-binding protein)	26 aa	Immobilized calmodulin
CBD (chitin-binding domain)	51 aa	Chitin
HaloTag	~300 aa	Chloroalkane

Table 16.2 Common solubility tags

Tag	Size	Protein
MBP	396 aa	Maltose-binding protein
NusA	495 aa	N-utilization substance
Trx	109 aa	Thioredoxin
SUMO	~100 aa	Small ubiquitin modifier
GB1	56 aa	IgG domain B1 of streptococcus Protein G
SET/SEP	<20 aa	Hydrophilic solubility enhancing peptide sequences
HaloTag	~300 aa	Mutated dehalogenase

2. Some Considerations When Designing a Tagged Protein

2.1. Affinity and/or solubility?

Two challenges in the production of heterologous proteins in *Escherichia coli*, the workhorse of protein expression systems, are poor or low expression, and the misfolding of the expressed protein into insoluble aggregates called

Table 16.3 Common endoproteases used to remove protein tags

Protease	Cleavage site
Enterokinase	DDDDK↑
Factor Xa	IEGR↑
Thrombin	LVPR↑GS
PreScission (human rhinovirus 3C protease)	LEVLFQ↑GP
TEV (tobacco etch virus protease)	ENLYFQ↑G
TVMV (tobacco vein mottling virus protease)	ETVRFQG↑S
SUMO protease (catalytic core of Ulp1)	Recognizes SUMO tertiary structure and cleaves at the C-terminal end of the conserved Gly–Gly sequence in SUMO

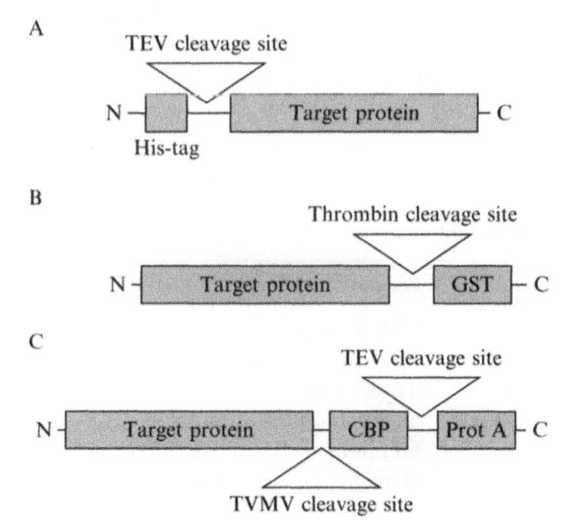

Figure 16.1 Example schematics of single and tandem tagged proteins. (A) N-terminally His-tagged protein; (B) C-terminally GST tagged protein; and (C) target protein with a removable (using TVMV proteases) C-terminal tandem affinity tag (TAP tag).

inclusion bodies. Expression problems caused by weak promoters, poor translation initiation, or the presence of rare codons, can be alleviated by introducing corrective sequences in the gene when assembling overexpression plasmids and the use of *E. coli* strains supplemented with rare tRNAs. Attaching a highly translated native gene as a fusion on the N-terminal end of the heterologous target protein is another approach to improving yield, and has also the added benefit of increasing the solubility of the target protein; this is the basis of solubility tags (Table 16.2).

Affinity tags, on the other hand, are crucial during protein purification, and allow the use of a variety of strategies to bind the target protein on an affinity matrix (Table 16.1). Some protein tags can function in both affinity and solubility roles—for example, the glutathione-*S*-transferase (GST) tag improves solubility of some proteins, while solubility tags such as maltose-binding protein (MBP) can be used for affinity purification of the target protein.

2.2. Which tag(s) to use?

Protein affinity and solubility tags vary widely in their size, and thus in the metabolic load that they impose on the host cell. For example, an MBP tag (43 kDa) on a 100-residue target protein will require the expression of 5 mg of fusion protein for every mg of the target protein produced. Affinity tags also vary in the cost of purification, since different affinity media have different expenses, both for the resin itself, and for operating costs (ease of regeneration and reuse, elution agent, binding capacity, etc.). Lichty *et al.* (2005) did a comparison of eight affinity tags looking at many different aspects such as purity, yield, and cost, and many similar studies have been carried out. Several recent reviews provide good overviews (Terpe, 2003), and list advantages and disadvantages of affinity tags (Arnau *et al.*, 2006b; Waugh, 2005) and solubility tags (Esposito and Chatterjee, 2006).

Apart from the cost, the choice between different affinity tags often depends on finding purification buffer conditions suitable for the target protein. For proteins susceptible to oxidation or proteolytic damage, the His-tag may not be very suitable since immobilized metal affinity chromatographic (IMAC) media cannot tolerate reducing agents or EDTA. Similarly, care has to be taken when IMAC media is used with target proteins that are sensitive to metal ions. Conversely, for target proteins requiring denaturing conditions or refolding, the His-tag and IMAC purification is an excellent choice.

Expression levels can dictate the choice of tags in some cases—solubility tags such as MBP, Trx, NusA, as well as GST have strong translational initiation signals and can drive expression levels higher which is useful for structural studies. On the other hand, when low expression levels are desirable, such as when studying complexes or physiological interactions, more stringent epitope tags or tandem tagging may be more appropriate.

A major source of comparative studies and innovation in protein tags is coming from ongoing structural genomics efforts, since protein production is a big roadblock for these projects (Pédelacq *et al.*, 2002; Structural Genomics Consortium *et al.*, 2008; Yokoyama, 2003). The reader is encouraged to take advantage of the consensus "what to try first" strategies that are emerging from these efforts (Structural Genomics Consortium *et al.*, 2008).

Many high-throughput studies have looked more specifically at the effectiveness of individual solubility tags (e.g., Cabrita *et al.*, 2006; Kataeva *et al.*, 2005), and reporter systems have been developed to monitor the solubility and proper folding of proteins (Listwan *et al.*, 2009; Liu *et al.*, 2006; Waldo *et al.*, 1999). Directed evolution of proteins can also be used for improving protein expression and solubility (Roodveldt *et al.*, 2005; Waldo, 2003). Use of tags for production of transmembrane proteins is another area of active development (Cunningham and Deber, 2007).

2.3. Tandem tags?

Multiple tags can be attached on target proteins, allowing for improved purification, expression, or tracking. The tandem affinity purification (TAP) tag was originally developed to allow for two purification steps—first on Protein A IgG beads, followed by cleavage of the Protein A tag and subsequent purification on calmodulin coated beads (Fig. 16.1; Puig *et al.*, 2001). The use of tandem tags allows TAP-tagged proteins to be detected and purified in native conditions even when expressed at very low levels. TAP tagging has proved very useful for studying protein complexes (Bauer and Kuster, 2003), and many other variations of the TAP tags have been designed (Collins and Choudhary, 2008).

Solubility tags such as Trx or NusA can also be linked with affinity tags such as His-tags for efficient purification of the fusion protein. Other tags such as S-tags or FLAG-tags can be added to allow protein detection when low levels of protein expression are expected and less specific antibodies (such as those against His-tags) may not be adequate. Affinity tags can also be attached at both ends of a target protein (Mueller *et al.*, 2003).

2.4. N- or C-terminal?

Tags can be placed at either the N- or C-terminus of a target protein. One advantage of placing a tag on the N-terminal end is that the construct can take advantage of efficient translation initiation sites on the tag. Solubility tags based on highly expressing proteins such as MBP, Trx, and NusA are also more efficient at solubilizing target proteins when positioned at the N-terminal end (Sachdev and Chirgwin, 1998), though recent high-throughput studies have shown than the MBP tag is still quite effective when positioned at the C-terminal end (Dyson *et al.*, 2004). Another advantage of placing a tag on the N-terminal site is that the tag can be removed more cleanly, since most endoproteases cut at or near the C-terminus of their recognition sites.

While placing a tag, care should be taken to preserve the positioning of any signal sequences or modification sites. Sequences at termini of the fusion protein should be examined for effects on the stability of the final construct, especially at the N-terminal end, which should be inspected for the host cell's N-end rule degradation signals (Bachmair *et al.*, 1986; Wang *et al.*, 2008). It is also useful to examine the sequence of the tagged protein for any inadvertently created interaction or cleavage sites using motif databases such as PROSITE (Hulo *et al.*, 2008).

2.5. Cleavage sites to remove tags

Tags can interfere with the structure and function of the target protein, and provision must be made to remove tags after the expression and purifications steps. Multiple cleavage sites can be engineered into the expression construct to remove individual tags at different stages of purification (Fig. 16.1). Table 16.3 lists some of the endoproteases used for tag removal, and these are discussed in this chapter.

3. Protein Affinity Tags

3.1. His-tag

The His-tag (also called 6xHis-tag) is one of the simplest and most widely used purification tags, with six or more consecutive histidine residues. These residues readily coordinate with transition metal ions such as Ni^{2+} or Co^{2+} immobilized on beads or a resin for purification. IMAC is the preferred choice as a first step during the purification of His-tagged proteins, though small batch reactions or spin columns with IMAC beads can be used for expression tests or small-scale preparations. Metal ions are immobilized using linkages such as Ni(II)-nitrilotriacetic acid (Ni-NTA) or Co^{2+}-carboxymethyl-aspartate on resins and beads available from many commercial sources. IMAC media typically has high binding capacities (5–40 mg of His-tagged protein/ml of media), is relatively low cost, and can easily be sanitized. For nickel binding media, the metal ion can often be stripped (using buffers with EDTA) and recharged for multiple use cycles. Some cobalt based resins (such as Talon, Clontech Inc.) use proprietary linkages that are more durable and cannot be recharged; such resins can be reused three and four times, but offer the advantage of being more specific for polyhistidine tags and almost no metal leakage during protein elution. IMAC can also be used under denaturing conditions, since the His-tag does not need a specific protein conformation for metal binding; indeed, binding to IMAC resins is stronger under denaturing conditions as the His-tag becomes more exposed.

His-tags bind the immobilized metal via the histidine imidazole ring, and tagged protein bound on IMAC media can be easily eluted using elution buffers with imidazole (100–250 m*M*) or low pH (4.5–6). Since some endogenous proteins also exhibit weak binding to IMAC media (Bolanos-Garcia and Davies, 2006), a low level of imidazole (5–20 m*M*) should be included in the loading and wash buffers to minimize nonspecific binding.

While the mild elution conditions for IMAC is one of the positive aspects of His-tags, care has to be taken to avoid EDTA (or EGTA) in any of the buffers. Cell extracts loaded on IMAC columns should not contain any EDTA, and only EDTA-free protease inhibitor cocktails should be used for sample preparation. TRIS salts weakly chelate metal ions as well, and the use of TRIS buffers should be minimized (50 m*M* or less). Most IMAC media is also very sensitive to reducing agents such as DTT or DTE, and low levels of β-mercaptoethanol (<10 m*M*) should be used instead.

The small size of the His-tag minimizes interference with the folding and structure of the target protein, and this tag can be positioned at either the N- or C-terminal ends (Fig. 16.1A). While the tag can be removed by introducing a protease cleavage site, there are many examples of proteins that have been crystallized with intact His-tags and little or no impact on the structure of the tagged protein (Carson *et al.*, 2007). In some cases, the His-tag can actually assist in crystal formation (e.g., Smits *et al.*, 2008). The His-tag can also be used with commercially available His-tag-specific antibodies for protein detection.

There are several variations on the standard 6xHis sequence used in His-tags (Terpe, 2003). These tags intersperse multiple histidines among other residues to lower the charge or improve stability, such as the HAT-tag and 6xHN tag which also bind better to the Co^{2+}-Talon resin (Clontech Inc.), and the MAT-tag (Sigma-Aldrich Inc.).

3.2. GST tag

GST is an abundantly expressed 26 kDa eukaryotic protein, and GST cloned from *Schistosoma japonicum* was shown to promote solubility and expression as an N-terminal fusion (Smith and Johnson, 1988). When positioned at the C-terminal end (Fig. 16.1B), GST is less efficient at improving protein solubility but still functions well as an affinity tag. GST binds to resin immobilized glutathione, and this property is used for affinity purification of GST tagged proteins. After the fusion protein is bound to the resin, it can be eluted under rather mild conditions using free reduced glutathione (10–40 m*M*) at neutral pH. Resins used for GST fusion purification, such as Glutathione-Sepharose beads, are relatively cheap, have high binding capacity (5–10 mg of GST/ml of resin), and can be regenerated and reused multiple times.

The GST tag has to be properly folded to bind glutathione, and thus the fusion protein needs to be soluble and in nondenaturing conditions for efficient purification. GST fusion proteins are often expressed at high levels, and thus protein solubility has to be monitored carefully. For some proteins, GST can act as a solubility tag (e.g., Kim and Lee, 2008). GST can be detected by a colorimetric assay with the substrate 1-chloro-2,4 dinitrobenzene (CDNB) (Habig *et al.*, 1974); this assay can also be used to monitor proper folding and accessibility of GST in the fusion protein when binding to glutathione resin is suboptimal. Commercial anti-GST antibodies are also available for detecting this tag.

The large size of the GST tag increases potential for degradation by proteases, and GST fusion proteins need to be purified quickly to minimize sample loss. Unlike for His-tagged proteins, EDTA can be used in buffers during sample preparation to reduce proteolytic damage. Care should also be taken to use reducing conditions since GST has four solvent exposed cysteines that can be involved in oxidative aggregation (Kaplan *et al.*, 1997). GST forms a homodimer in solution, which makes it a poor choice for tagging oligomeric proteins.

The kinetics of GST binding to glutathione and its elution are relatively slow, and so GST fusion proteins need to be loaded and eluted from GST columns at slow flow rates. The monitoring of protein during elution also has to be done carefully since glutathione absorbs strongly at 280 nm.

3.3. Other purification tags

3.3.1. Epitope tags

A number of short amino acid (aa) sequences, recognized by commercially available antibodies, can be used as tags for detection and purification of proteins. Epitope tagging is widely used in molecular biology as a general method for tracking recombinant proteins (Fritze and Anderson, 2000), and has the advantage of high specificity and the use of short tags that minimizes deleterious effects on the structure and function of the target protein. These tags are mostly placed at the N- or C-termini, but can also be placed within a target protein (in loops or between structural domains in a solvent exposed region). Epitope tags are usually sequences absent in the host cell, making the detection of the target protein straightforward. For purification, however, epitope tag binding media (typically, monoclonal antibodies immobilized on chromatographic resins) is expensive and less suitable for large-scale preparations than other affinity media. Short peptides corresponding to the tag, low pH, or other approaches (e.g., chelation of calcium required for tag binding, or use of salt and polyol) can be used to elute the target protein, but some of these procedures tend to be more harsh than other affinity purification methods (however, see Chapter 28 in this volume).

The FLAG-tag is a short, eight-residue (DYKDDDDK) hydrophilic peptide tag that can be used for detection and purification of target proteins (Einhauer and Jungbauer, 2001; Prickett *et al.*, 1989). Apart from availability of several high specificity anti-FLAG mABs, the FLAG-tag has the advantage of incorporating the enterokinase cleavage site (DDDK) that allows the complete tag to be cleanly removed after purification. Another variant of the FLAG-tag is the 3xFLAG-tag (Sigma-Aldrich, Inc.) that is made up of three tandem repeats of FLAG-like sequences (Hernan *et al.*, 2000). Other commonly used epitope tags are the HA-tag and the c-Myc tag (Fritze and Anderson, 2000).

3.3.2. The S-tag

S-tag, reviewed by Raines *et al.* (2000), takes advantage of the tight association between the N-terminal S-peptide (residues 1–20) and the S-protein (residues 21–124) in RNase S (Richards and Vithayarhil, 1959). The S-tag system (Novagen Inc.) uses the N-terminal 15 residues of the S-peptide (Potts *et al.*, 1963), with S-protein immobilized on agarose beads for target protein purification. S-tag vectors typically encode a site-specific protease cleavage site, and elution of the target protein requires cleavage of the tag or harsher denaturing conditions that disrupt the S-tag S-protein interaction.

3.3.3. STREP-II-tag

STREP-II-tag (WSHPQFEK) is a tag that takes advantage of the strong and specific interaction between biotin and streptavidin (Schmidt and Skerra, 1994). This peptide tag binds in the biotin pocket of streptavidin; Strep-Tactin, a recombinant form of streptavidin optimized to bind the Strep-II-tag, is used in affinity media. Bound target protein can be eluted with low levels (2.5 m*M*) of desthiobiotin, a biotin analog that competes for binding to Strep-Tactin in a reversible manner (Skerra and Schmidt, 2000). The elution conditions are gentle, and buffers can include high levels (up to 50 m*M*) of reducing agents such as DTT or β-mercaptoethanol, as well as chelating agents such as EDTA, which makes this an excellent purification technique for proteins that are very sensitive to oxidation. Strep-Tactin chromatographic resins can regenerated and reused a few times. While there is a low background of naturally biotinylated proteins in most cell extracts, these usually do not interfere in the purification; if necessary, avidin can be added to clear biotinylated host proteins (Schmidt and Skerra, 2007). There has also been some work on using streptavidin/avidin in fusion tags (Sano and Cantor, 2000), but they are difficult to use for affinity purification because the very strong interaction between these proteins is difficult to disrupt.

3.3.4. CBP-tag

The calmodulin-binding peptide (CBP) is a short 26-residue sequence derived from the C-terminus of skeletal muscle myosin light chain kinase that binds specifically to calmodulin (reviewed in Terpe, 2003). Calmodulin (CaM) immobilized on chromatographic media (such as CaM-Sepharose) can be used for affinity purification of target proteins tagged with CBP. Though calmodulin binds very strongly with CBP (nanomolar affinity), this interaction is dependent on calcium and the bound protein can be eluted in a single step using gentle buffers containing a calcium-chelating agent, such as EGTA. Another application of the CBP-tag is for ^{32}P isotopic labeling of the fusion protein, since the tag includes a protein kinase A target sequence (Vaillancourt *et al.*, 2000). The CBP-tag cannot be used for expression in eukaryotic systems, since many endogenous proteins in eukaryotes also interact with calmodulin.

4. Solubility Tags

Production of well-folded, soluble proteins is a major bottleneck in recombinant protein expression, especially when heterologous eukaryotic proteins are expressed in bacterial cells. Several soluble proteins are used as tags to improve folding of the target protein. These tags should be used in conjunction with other approaches to improve protein folding such as lowering temperature after protein induction or coexpression of chaperones (Baneyx and Mujacic, 2004; de Marco *et al.*, 2007; Sahdev *et al.*, 2008). It is also useful to screen multiple solubility tags (Peleg and Unger, 2008). For recalcitrant proteins, protein purification using denaturing conditions and refolding can be tried (Cabrita and Bottomley, 2004; Jungbauer and Kaar, 2007; Qoronfleh *et al.*, 2007; see Chapter 17 in this volume).

4.1. MBP tag

MBP is a solubility enhancing tag (di Guan *et al.*, 1988) that can also be used for effective affinity purification, since it binds specifically to maltose or amylose. MBP is a large 43 kDa secreted *E. coli* protein that can be expressed at very high levels, and helps keep proteins fused at its C-terminal end soluble (Kapust and Waugh, 1999). More recent surveys have shown than the MBP tag is also effective when placed on the C-terminal end of target proteins (Dyson *et al.*, 2004). The large size of this tag puts a heavy metabolic load on the host cell, but in high-throughput tests, MBP ranks as one of the best tags for making soluble fusions (Dyson *et al.*, 2004; Kataeva *et al.*, 2005). About a quarter of proteins expressed as MBP fusions, however, remain insoluble or are prone to aggregation when the MBP tag is removed. In our work on the expression of human calcitonin gene-related peptide-receptor component protein (CGRP-RCP) in *E. coli*, for example, we observed protein aggregation when the N-terminal MBP tag was removed by enterokinase cleavage even after successful expression and purification (Tolun *et al.*, 2007). Each solubility tag also has different effects, and several tags may need to be tried for recalcitrant proteins. In the work on CGRP-RCP, a thioredoxin tag was not effective and the fusion protein remained insoluble (Tolun *et al.*, 2007). Nallamsetty and Waugh (2006) have suggested that solubility tags such as MBP and NusA act as passive partners in the folding of target proteins, and that solubility of aggregation-prone target proteins after removal of the tag appears to depend on the characteristics of the target protein rather than the tag used.

Commercial vectors for tagging MBP are available for both cytoplasmic and periplasmic expression of target proteins, with a variety of cleavage sites (New England Biolabs Inc.). Cross-linked amylose resin is used to bind MBP tagged proteins, and the bound fusion protein can be easily eluted by adding 10 m*M* maltose to the wash buffers. This allows for an easy one-step purification of MBP tagged proteins under very gentle conditions; however, amylose affinity purification cannot be carried out with reducing agents or under denaturing conditions. Amylose resins are degraded somewhat by amylase activity in crude extracts, especially from cells grown in rich LB media, but this can be minimized by including glucose in the media (0.2%). Amylose resins can be regenerated and reused several times.

4.2. Trx tag

Thioredoxin (Trx) is a thermostable, 12-kDa intracellular *E. coli* protein that is easily overexpressed and soluble even when overexpressed up to 40% of the total cellular protein (LaVallie *et al.*, 1993), and is very useful as a tag in avoiding inclusion body formation in recombinant protein production (LaVallie *et al.*, 2000). Tests by Dyson *et al.* (2004) indicate that the Trx tag is more effective when placed on the N-terminal end of the target protein.

Thioredoxin accumulates at cytoplasmic membrane adhesion sites (Bayer *et al.*, 1987), which allows Trx fusion proteins to be released by simple osmotic shock or freeze/thaw treatments, providing a simple initial purification step. Additional affinity tags such as His-tags are typically attached to the Trx tag for further purification steps.

4.3. NusA tag

NusA is a large (495 aa) N-utilizing substance A transcription antitermination factor that was chosen as a potential tag as it ranked among the most soluble *E. coli* proteins as a fusion for heterologous protein expression, based on a statistical solubility model (Davis *et al.*, 1999; De Marco *et al.*, 2004). In some large-scale screening tests (Busso *et al.*, 2005; Kataeva *et al.*, 2005), NusA has performed as well as or better than MBP as a solubility tag, but the results vary for individual proteins. NusA tags are used in conjunction with other affinity tags such as His-tags (de Marco, 2006).

4.4. Other solubility tags

A number of smaller solubility enhancement tags (SETs) or solubility enhancement peptide (SEP) tags have been developed, which use highly acidic sequences to enhance solubility of some target proteins (Kato *et al.*, 2007;

Zhang *et al.*, 2004). Other small tags are the GB1-tag (56 residues) which is based on the IgG binding B1 domain of the streptococcal Protein G (Cheng and Patel, 2004; Zhou *et al.*, 2001), as well as the IgG binding domain of Protein A (ZZ domain, 116 residues; Inouye and Sahara, 2009; Rondahl *et al.*, 1992). These small tags are especially useful in protein preparations for NMR studies (Kato *et al.*, 2007), and some progress has been made in the creation of NMR-invisible solubility tags using protein ligation methods (Durst *et al.*, 2008; Kobashigawa *et al.*, 2009).

Small ubiquitin-like modifier (SUMO) protein (~11 kDa) has been shown to significantly improve protein stability and solubility as an N-terminal fusion (Marblestone *et al.*, 2006). The tag can be removed after purification using SUMO protease (catalytic domain of Ulp1) that recognizes the SUMO structure (Lee *et al.*, 2008; Panavas *et al.*, 2009). The SUMO-fusion system has also been adapted for use in insect cells and other eukaryotic expression systems (Liu *et al.*, 2008).

HaloTag is a recently created modular tagging system that uses a 34-kDa modified haloalkane dehalogenase protein that can bind a variety of synthetic ligands (HaloTag ligands; Promega Inc.). These ligands are comprised of a constant reactive linker that binds covalently to the HaloTag, and a variable reporter end that can impart a variety of useful properties to the fusion protein. Thus, a single tag can be used for subcellular imaging within live cells, cell labeling and sorting, affinity purification, or even immobilization on solid supports (Los *et al.*, 2008). Ohana *et al.* (2009) have used a version of this tag (HaloTag7) for affinity purification, using the chloroalkane linker attached to agarose beads. Since the HaloTag binds covalently to this linker in a very specific nonreversible manner, even poorly expressed proteins can be bound efficiently to the chloroalkane resin. The target protein can then be eluted off the resin using tobacco etch virus (TEV) protease that cuts at a cleavage site engineered between the HaloTag and the target protein. Surprisingly, the monomeric compact HaloTag was seen to dramatically improve fusion protein solubility when tested on a panel of difficult-to-express recombinant human proteins being expressed in *E. coli* (Ohana *et al.*, 2009). In these tests, the HaloTag fared significantly better than MBP, and appears to function as a bona fide solubility tag.

5. Removal of Tags

After a protein tag has been used for solubility enhancement or affinity purification, it is often useful to remove it for biological and functional studies since the tag can potentially interfere with the proper functioning of the target protein. This is especially true for large tags such GST or MBP, though there are some examples where the fusion protein was more amenable for crystallization (Smyth *et al.*, 2003).

Most commercial expression vectors that are used to add tags on target proteins also include cleavage sites with specific sequences that allow the tag to be removed using recombinant endoproteases. After the initial affinity purification step, the sample can be treated with the endoprotease to cleave off the tag, which can subsequently be separated from the target protein by passing the sample back on the affinity column and collecting the flow-through. The recombinant endoprotease usually also comes with an affinity tag, allowing for its easy removal after the cleavage reaction.

Some commonly used endoproteases are listed in Table 16.3. Enterokinase and Factor Xa are useful for removal of N-terminal tags since they cut at the C-terminal end of their recognition sequence, allowing for the complete tag and recognition sequence to be removed. However, both these enzymes are somewhat promiscuous and can cleave at secondary sites, often at other basic residues. Similar secondary cleavage sites have also been seen for thrombin, another protease used to remove tags (Jenny *et al.*, 2003; Liew *et al.*, 2005). One advantage with proteases such as thrombin, especially for large-scale protein production, is that it is cheaper and more efficient than other more specific proteases.

The PreScission protease is a more specific protease with a longer and stricter recognition sequence. This is the protease 3C from human rhinovirus-14 (3Cpro) with a GST tag, which makes it especially useful for removing GST tags. Another very specific and popular protease is the TEV protease (Kapust *et al.*, 2001). The TEV protease prefers a Gly after the cleavage site (Table 16.3), but can tolerate other residues with only a modest decrease in activity, which allows it to cleave N-terminal tags with no additional residues remaining on the target protein in many cases (Kapust *et al.*, 2002). The TEV protease is easy to produce in-house in large quantities and is usually expressed with a His-tag for convenient purification and removal after the cleavage reaction (Tropea *et al.*, 2009). Many forms of TEV proteases, with other affinity tags or enhanced for higher activity and stability, are sold commercially.

A variant of the TEV protease is the tobacco vein mottling virus (TVMV) protease that recognizes a different sequence (Nallamsetty *et al.*, 2004), and can be used to design separate cleavage sites between different tags (Fig. 16.1C).

Exoproteases can also be used to remove N-terminal tags, such as the TAGzyme system (Arnau *et al.*, 2006a, available commercially from Qiagen Inc.). This approach uses dipeptide aminopeptidase I (DAPase) to sequentially chew up the N-terminal tag until a dipeptide "stop point" in the sequence is reached. Additional variations of this system have been designed to work with a variety of sequences (Arnau *et al.*, 2006b, 2008).

Complete removal of C-terminal tags is more problematic, since most endoproteases cut toward the C-terminal end of their recognition sequence.

If complete removal of a tag is necessary, more specialized cleavage sites can be designed that take advantage of structure-based recognition (such as with the SUMO protease; Malakhov *et al.*, 2004) or an autocatalytic protein self-splicing element (Inteins; Saleh and Perler, 2006). Both of these cleavage systems have been coupled with affinity purification tags—the SUMO-fusion system (Butt *et al.*, 2005; Lee *et al.*, 2008), and intein-chitin-binding domain (CBD; also commercialized as the IMPACT system by New England Biolabs Inc.; Chong *et al.*, 1997) or the intein-polyhydroxybutyrate-binding (PHB) and similar protein purification systems (Gillies *et al.*, 2008).

In some cases, tags can also be cleaved chemically. Though chemical cleavage uses cheap reagents and can be very efficient, these reactions require harsh solvents and denaturing conditions, and are used typically for preparation of small peptides. Cyanogen bromide (CNBr) cleaves at methionines, and can be used when the fusion protein can be designed to have a unique methionine between the tag and the target peptide (Döbeli *et al.*, 1998; Fairlie *et al.*, 2002). Another chemical cleavage agent, hydroxylamine, cleaves the peptide bond between Asn and Gly (Hu *et al.*, 2008).

Practical considerations: While designing a specific cleavage site between a tag and a target protein is relatively straightforward, efficient cleavage of the tag is not always possible or easily predictable. Each construct has to be experimentally tested both for cleavage efficiency, and for any secondary cleavages that may occur when promiscuous proteases are used. Often, the level of the protease and the duration of incubation have to be optimized. Maximizing the efficiency of cleavage is especially important for oligomeric proteins where tags have to be removed for each monomer for productive yields of the target protein (Kenig *et al.*, 2006). The cleavage sequence also has to be sterically accessible to the protease and relatively unstructured; poor cleavage can sometimes be alleviated by introducing a spacer or linker of a few residues between the recognition site and the target protein, or by using sequences around the cleavage site that are unlikely to form secondary structures. For proteases that lack an affinity tag, on-column cleavage (where the protease is injected onto the affinity column with the fusion protein bound) is often more efficient than batch reactions.

6. Conclusions

A large variety of protein tags are available for facilitating the soluble expression and purification of recombinant proteins. However, even with this wide arsenal of tags, structural genomics protein production facilities are getting soluble purified proteins with success rates of less than 50% (Structural Genomics Consortium *et al.*, 2008). While many good strategies have emerged from these large-scale studies, given the diversity in how proteins fold and their distinct biochemical characteristics, there is no common set of solubility or affinity tags that works for all proteins. Rather, the choice of tags largely depends on the protein being expressed and the task at hand. Herein, we have surveyed the most effective protein expression tags and issues related to their use.

ACKNOWLEDGMENT

The author is supported in part by a grant (R01-GM69972) from the National Institutes of Health.

REFERENCES

Arnau, J., Lauritzen, C., and Pedersen, J. (2006a). Cloning strategy, production and purification of proteins with exopeptidase-cleavable His-tags. *Nat. Protoc.* **1,** 2326–2333.

Arnau, J., Lauritzen, C., Petersen, G. E., and Pedersen, J. (2006b). Current strategies for the use of affinity tags and tag removal for the purification of recombinant proteins. *Protein Expr. Purif.* **48,** 1–13.

Arnau, J., Lauritzen, C., Petersen, G. E., and Pedersen, J. (2008). The use of TAGZyme for the efficient removal of N-terminal His-tags. *Methods Mol. Biol.* **421,** 229–243.

Bachmair, A., Finley, D., and Varshavsky, A. (1986). *In vivo* half-life of a protein is a function of its amino-terminal residue. *Science* **234,** 179–186.

Baneyx, F., and Mujacic, M. (2004). Recombinant protein folding and misfolding in *Escherichia coli. Nat. Biotechnol.* **22,** 1399–1408.

Bauer, A., and Kuster, B. (2003). Affinity purification-mass spectrometry. Powerful tools for the characterization of protein complexes. *Eur. J. Biochem.* **270,** 570–578.

Bayer, M. E., Bayer, M. H., Lunn, C. A., and Pigiet, V. (1987). Association of thioredoxin with the inner membrane and adhesion sites in *Escherichia coli. J. Bacteriol.* **169,** 2659–2666.

Bolanos-Garcia, V. M., and Davies, O. R. (2006). Structural analysis and classification of native proteins from *E. coli* commonly co-purified by immobilised metal affinity chromatography. *Biochim. Biophys. Acta* **1760,** 1304–1313.

Busso, D., Delagoutte-Busso, B., and Moras, D. (2005). Construction of a set Gateway-based destination vectors for high-throughput cloning and expression screening in *Escherichia coli. Anal. Biochem.* **343,** 313–321.

Butt, T. R., Edavettal, S. C., Hall, J. P., and Mattern, M. R. (2005). SUMO fusion technology for difficult-to-express proteins. *Protein Expr. Purif.* **43,** 1–9.

Cabrita, L. D., and Bottomley, S. P. (2004). Protein expression and refolding—A practical guide to getting the most out of inclusion bodies. *Biotechnol. Annu. Rev.* **10,** 31–50.

Cabrita, L. D., Dai, W., and Bottomley, S. P. (2006). A family of *E. coli* expression vectors for laboratory scale and high throughput soluble protein production. *BMC Biotechnol.* **6,** 12.

Carson, M., Johnson, D. H., McDonald, H., Brouillette, C., and Delucas, L. J. (2007). His-tag impact on structure. *Acta Crystallogr. D Biol. Crystallogr.* **63,** 295–301.

Cheng, Y., and Patel, D. J. (2004). An efficient system for small protein expression and refolding. *Biochem. Biophys. Res. Commun.* **317,** 401–405.

Chong, S., Mersha, F. B., Comb, D. G., Scott, M. E., Landry, D., Vence, L. M., Perler, F. B., Benner, J., Kucera, R. B., Hirvonen, C. A., *et al.* (1997). Single-column purification of free recombinant proteins using a self-cleavable affinity tag derived from a protein splicing element. *Gene* **192,** 271–281.

Collins, M. O., and Choudhary, J. S. (2008). Mapping multiprotein complexes by affinity purification and mass spectrometry. *Curr. Opin. Biotechnol.* **19,** 324–330.

Cunningham, F., and Deber, C. M. (2007). Optimizing synthesis and expression of transmembrane peptides and proteins. *Methods* **41,** 370–380.

Davis, G. D., Elisee, C., Newham, D. M., and Harrison, R. G. (1999). New fusion protein systems designed to give soluble expression in *Escherichia coli. Biotechnol. Bioeng.* **65,** 382–388.

de Marco, A. (2006). Two-step metal affinity purification of double-tagged (NusA-His6) fusion proteins. *Nat. Protoc.* **1,** 1538–1543.

De Marco, V., Stier, G., Blandin, S., and de Marco, A. (2004). The solubility and stability of recombinant proteins are increased by their fusion to NusA. *Biochem. Biophys. Res. Commun.* **322,** 766–771.

de Marco, A., Deuerling, E., Mogk, A., Tomoyasu, T., and Bukau, B. (2007). Chaperone-based procedure to increase yields of soluble recombinant proteins produced in *E. coli. BMC Biotechnol.* **7,** 32.

di Guan, C., Li, P., Riggs, P. D., and Inouye, H. (1988). Vectors that facilitate the expression and purification of foreign peptides in *Escherichia coli* by fusion to maltose-binding protein. *Gene* **67,** 21–30.

Döbeli, H., Andres, H., Breyer, N., Draeger, N., Sizmann, D., Zuber, M. T., Weinert, B., and Wipf, B. (1998). Recombinant fusion proteins for the industrial production of disulfide bridge containing peptides: Purification, oxidation without concatemer formation, and selective cleavage. *Protein Expr. Purif.* **12,** 404–414.

Durst, F. G., Ou, H. D., Löhr, F., Dötsch, V., and Straub, W. E. (2008). The better tag remains unseen. *J. Am. Chem. Soc.* **130,** 14932–14933.

Dyson, M. R., Shadbolt, S. P., Vincent, K. J., Perera, R. L., and McCafferty, J. (2004). Production of soluble mammalian proteins in *Escherichia coli*: Identification of protein features that correlate with successful expression. *BMC Biotechnol.* **4,** 32.

Einhauer, A., and Jungbauer, A. (2001). The FLAG peptide, a versatile fusion tag for the purification of recombinant proteins. *J. Biochem. Biophys. Methods* **49,** 455–465.

Esposito, D., and Chatterjee, D. K. (2006). Enhancement of soluble protein expression through the use of fusion tags. *Curr. Opin. Biotechnol.* **17,** 353–358.

Fairlie, W. D., Uboldi, A. D., De Souza, D. P., Hemmings, G. J., Nicola, N. A., and Baca, M. (2002). A fusion protein system for the recombinant production of short disulfide-containing peptides. *Protein Expr. Purif.* **26,** 171–178.

Fritze, C. E., and Anderson, T. R. (2000). Epitope tagging: General method for tracking recombinant proteins. *Methods Enzymol.* **327,** 3–16.

Gillies, A. R., Hsii, J. F., Oak, S., and Wood, D. W. (2008). Rapid cloning and purification of proteins: Gateway vectors for protein purification by self-cleaving tags. *Biotechnol. Bioeng.* **101,** 229–240.

Habig, W. H., Pabst, M. J., and Jakoby, W. B. (1974). Glutathione S-transferases. The first enzymatic step in mercapturic acid formation. *J. Biol. Chem.* **249,** 7130–7139.

Hernan, R., Heuermann, K., and Brizzard, B. (2000). Multiple epitope tagging of expressed proteins for enhanced detection. *Biotechniques* **28,** 789–793.

Hu, J., Qin, H., Sharma, M., Cross, T. A., and Gao, F. P. (2008). Chemical cleavage of fusion proteins for high-level production of transmembrane peptides and protein domains containing conserved methionines. *Biochim. Biophys. Acta* **1778,** 1060–1066.

Hulo, N., Bairoch, A., Bulliard, V., Cerutti, L., Cuche, B. A., de Castro, E., Lachaize, C., Langendijk-Genevaux, P. S., and Sigrist, C. J. (2008). The 20 years of PROSITE. *Nucleic Acids Res.* **36,** D245–D249.

Inouye, S., and Sahara, Y. (2009). Expression and purification of the calcium binding photoprotein mitrocomin using ZZ-domain as a soluble partner in *E. coli* cells. *Protein Expr. Purif.* **66,** 52–57.

Jenny, R. J., Mann, K. G., and Lundblad, R. L. (2003). A critical review of the methods for cleavage of fusion proteins with thrombin and factor Xa. *Protein Expr. Purif.* **31,** 1–11.

Jungbauer, A., and Kaar, W. (2007). Current status of technical protein refolding. *J. Biotechnol.* **128,** 587–596.

Kaplan, W., Hüsler, P., Klump, H., Erhardt, J., Sluis-Cremer, N., and Dirr, H. (1997). Conformational stability of pGEX-expressed *Schistosoma japonicum* glutathione S-transferase: A detoxification enzyme and fusion-protein affinity tag. *Protein Sci.* **6,** 399–406.

Kapust, R. B., and Waugh, D. S. (1999). *Escherichia coli* maltose-binding protein is uncommonly effective at promoting the solubility of polypeptides to which it is fused. *Protein Sci.* **8,** 1668–1674.

Kapust, R. B., Tözsér, J., Fox, J. D., Anderson, D. E., Cherry, S., Copeland, T. D., and Waugh, D. S. (2001). Tobacco etch virus protease: Mechanism of autolysis and rational design of stable mutants with wild-type catalytic proficiency. *Protein Eng.* **14,** 993–1000.

Kapust, R. B., Tözsér, J., Copeland, T. D., and Waugh, D. S. (2002). The P1′ specificity of tobacco etch virus protease. *Biochem. Biophys. Res. Commun.* **294,** 949–955.

Kataeva, I., Chang, J., Xu, H., Luan, C. H., Zhou, J., Uversky, V. N., Lin, D., Horanyi, P., Liu, Z. J., Ljungdahl, L. G., *et al.* (2005). Improving solubility of *Shewanella oneidensis* MR-1 and *Clostridium thermocellum* JW-20 proteins expressed into *Escherichia coli. J. Proteome Res.* **4,** 1942–1951.

Kato, A., Maki, K., Ebina, T., Kuwajima, K., Soda, K., and Kuroda, Y. (2007). Mutational analysis of protein solubility enhancement using short peptide tags. *Biopolymers* **85,** 12–18.

Kenig, M., Peternel, S., Gaberc-Porekar, V., and Menart, V. (2006). Influence of the protein oligomericity on final yield after affinity tag removal in purification of recombinant proteins. *J. Chromatogr. A* **1101,** 293–306.

Kim, S., and Lee, S. B. (2008). Soluble expression of archaeal proteins in *Escherichia coli* by using fusion-partners. *Protein Expr. Purif.* **62,** 116–119.

Kobashigawa, Y., Kumeta, H., Ogura, K., and Inagaki, F. (2009). Attachment of an NMR-invisible solubility enhancement tag using a sortase-mediated protein ligation method. *J. Biomol. NMR* **43,** 145–150.

LaVallie, E. R., DiBlasio, E. A., Kovacic, S., Grant, K. L., Schendel, P. F., and McCoy, J. M. (1993). A thioredoxin gene fusion expression system that circumvents inclusion body formation in the *E. coli* cytoplasm. *Biotechnology (NY)* **11,** 187–193.

LaVallie, E. R., Lu, Z., Diblasio-Smith, E. A., Collins-Racie, L. A., and McCoy, J. M. (2000). Thioredoxin as a fusion partner for production of soluble recombinant proteins in *Escherichia coli. Methods Enzymol.* **326,** 322–340.

Lee, C. D., Sun, H. C., Hu, S. M., Chiu, C. F., Homhuan, A., Liang, S. M., Leng, C. H., and Wang, T. F. (2008). An improved SUMO fusion protein system for effective production of native proteins. *Protein Sci.* **17,** 1241–1248.

Lichty, J. J., Malecki, J. L., Agnew, H. D., Michelson-Horowitz, D. J., and Tan, S. (2005). Comparison of affinity tags for protein purification. *Protein Expr. Purif.* **41,** 98–105.

Liew, O. W., Ching Chong, J. P., Yandle, T. G., and Brennan, S. O. (2005). Preparation of recombinant thioredoxin fused N-terminal proCNP: Analysis of enterokinase cleavage products reveals new enterokinase cleavage sites. *Protein Expr. Purif.* **41,** 332–340.

Listwan, P., Terwilliger, T. C., and Waldo, G. S. (2009). Automated, high-throughput platform for protein solubility screening using a split-GFP system. *J. Struct. Funct. Genomics* **10,** 47–55.

Liu, J. W., Boucher, Y., Stokes, H. W., and Ollis, D. L. (2006). Improving protein solubility: The use of the *Escherichia coli* dihydrofolate reductase gene as a fusion reporter. *Protein Expr. Purif.* **47,** 258–263.

Liu, L., Spurrier, J., Butt, T. R., and Strickler, J. E. (2008). Enhanced protein expression in the baculovirus/insect cell system using engineered SUMO fusions. *Protein Expr. Purif.* **62,** 21–28.

Los, G. V., Encell, L. P., McDougall, M. G., Hartzell, D. D., Karassina, N., Zimprich, C., Wood, M. G., Learish, R., Ohana, R. F., Urh, M., *et al.* (2008). HaloTag: A novel protein labeling technology for cell imaging and protein analysis. *ACS Chem. Biol.* **3,** 373–382.

Malakhov, M. P., Mattern, M. R., Malakhova, O. A., Drinker, M., Weeks, S. D., and Butt, T. R. (2004). SUMO fusions and SUMO-specific protease for efficient expression and purification of proteins. *J. Struct. Funct. Genomics* **5,** 75–86.

Marblestone, J. G., Edavettal, S. C., Lim, Y., Lim, P., Zuo, X., and Butt, T. R. (2006). Comparison of SUMO fusion technology with traditional gene fusion systems: Enhanced expression and solubility with SUMO. *Protein Sci.* **15,** 182–189.

Mueller, U., Büssow, K., Diehl, A., Bartl, F. J., Niesen, F. H., Nyarsik, L., and Heinemann, U. (2003). Rapid purification and crystal structure analysis of a small protein carrying two terminal affinity tags. *J. Struct. Funct. Genomics* **4,** 217–225.

Nallamsetty, S., and Waugh, D. S. (2006). Solubility-enhancing proteins MBP and NusA play a passive role in the folding of their fusion partners. *Protein Expr. Purif.* **45,** 175–182.

Nallamsetty, S., Kapust, R. B., Tözsér, J., Cherry, S., Tropea, J. E., Copeland, T. D., and Waugh, D. S. (2004). Efficient site-specific processing of fusion proteins by tobacco vein mottling virus protease *in vivo* and *in vitro*. *Protein Expr. Purif.* **38,** 108–115.

Ohana, R. F., Encell, L. P., Zhao, K., Simpson, D., Slater, M. R., Urh, M., and Wood, K. V. (2009). HaloTag7: A genetically engineered tag that enhances bacterial expression of soluble proteins and improves protein purification. *Protein Expr. Purif.* **68,** 110–120.

Panavas, T., Sanders, C., and Butt, T. R. (2009). SUMO fusion technology for enhanced protein production in prokaryotic and eukaryotic expression systems. *Methods Mol. Biol.* **497,** 303–317.

Pédelacq, J. D., Piltch, E., Liong, E. C., Berendzen, J., Kim, C. Y., Rho, B. S., Park, M. S., Terwilliger, T. C., and Waldo, G. S. (2002). Engineering soluble proteins for structural genomics. *Nat. Biotechnol.* **20,** 927–932.

Peleg, Y., and Unger, T. (2008). Application of high-throughput methodologies to the expression of recombinant proteins in *E. coli*. *Methods Mol. Biol.* **426,** 197–208.

Potts, J. T., Young, D. M., and Anfinsen, C. B. (1963). Reconstitution of fully active RNase S by carboxypeptidase-degraded RNase S-peptide. *J. Biol. Chem.* **238,** 2593–2594.

Prickett, K. S., Amberg, D. C., and Hopp, T. P. (1989). A calcium-dependent antibody for identification and purification of recombinant proteins. *Biotechniques* **7,** 580–589.

Puig, O., Caspary, F., Rigaut, G., Rutz, B., Bouveret, E., Bragado-Nilsson, E., Wilm, M., and Séraphin, B. (2001). The tandem affinity purification (TAP) method: A general procedure of protein complex purification. *Methods* **24,** 218–229.

Qoronfleh, M. W., Hesterberg, L. K., and Seefeldt, M. B. (2007). Confronting high-throughput protein refolding using high pressure and solution screens. *Protein Expr. Purif.* **55,** 209–224.

Raines, R. T., McCormick, M., Van Oosbree, T. R., and Mierendorf, R. C. (2000). The S.Tag fusion system for protein purification. *Methods Enzymol.* **326,** 362–376.

Richards, F. M., and Vithayarhil, P. J. (1959). The preparation of subtilisin-modified ribonuclease and the separation of the peptide and protein components. *J. Biol. Chem.* **234,** 1459–1465.

Rondahl, H., Nilsson, B., and Holmgren, E. (1992). Fusions to the 5′ end of a gene encoding a two-domain analogue of staphylococcal protein A. *J. Biotechnol.* **25,** 269–287.

Roodveldt, C., Aharoni, A., and Tawfik, D. S. (2005). Directed evolution of proteins for heterologous expression and stability. *Curr. Opin. Struct. Biol.* **15,** 50–56.

Sachdev, D., and Chirgwin, J. M. (1998). Order of fusions between bacterial and mammalian proteins can determine solubility in *Escherichia coli*. *Biochem. Biophys. Res. Commun.* **244,** 933–937.
Sahdev, S., Khattar, S. K., and Saini, K. S. (2008). Production of active eukaryotic proteins through bacterial expression systems: A review of the existing biotechnology strategies. *Mol. Cell. Biochem.* **307,** 249–264.
Saleh, L., and Perler, F. B. (2006). Protein splicing in cis and in trans. *Chem. Rec.* **6,** 183–193.
Sano, T., and Cantor, C. R. (2000). Streptavidin-containing chimeric proteins: Design and production. *Methods Enzymol.* **326,** 305–311.
Schmidt, T. G., and Skerra, A. (1994). One-step affinity purification of bacterially produced proteins by means of the "Strep tag" and immobilized recombinant core streptavidin. *J. Chromatogr. A* **676,** 337–345.
Schmidt, T. G., and Skerra, A. (2007). The Strep-tag system for one-step purification and high-affinity detection or capturing of proteins. *Nat. Protoc.* **2,** 1528–1535.
Skerra, A., and Schmidt, T. G. (2000). Use of the Strep-tag and streptavidin for detection and purification of recombinant proteins. *Methods Enzymol.* **326,** 271–304.
Smith, D. B., and Johnson, K. S. (1988). Single-step purification of polypeptides expressed in *Escherichia coli* as fusions with glutathione S-transferase. *Gene* **67,** 31–40.
Smits, S. H., Mueller, A., Grieshaber, M. K., and Schmitt, L. (2008). Coenzyme- and His-tag-induced crystallization of octopine dehydrogenase. *Acta Crystallogr. Sect. F Struct. Biol. Cryst. Commun.* **64,** 836–839.
Smyth, D. R., Mrozkiewicz, M. K., McGrath, W. J., Listwan, P., and Kobe, B. (2003). Crystal structures of fusion proteins with large-affinity tags. *Protein Sci.* **12,** 1313–1322.
Structural Genomics Consortium, China Structural Genomics Consortium, Northeast Structural Genomics Consortium, Gräslund, S., Nordlund, P., Weigelt, J., Hallberg, B. M., Bray, J., Gileadi, O., Knapp, S., *et al.* (2008). Protein production and purification. *Nat. Methods* **5,** 135–146.
Terpe, K. (2003). Overview of tag protein fusions: From molecular and biochemical fundamentals to commercial systems. *Appl. Microbiol. Biotechnol.* **60,** 523–533.
Tolun, A. A., Dickerson, I. M., and Malhotra, A. (2007). Overexpression and purification of human calcitonin gene-related peptide-receptor component protein in *Escherichia coli*. *Protein Expr. Purif.* **52,** 167–174.
Tropea, J. E., Cherry, S., and Waugh, D. S. (2009). Expression and purification of soluble His(6)-tagged TEV protease. *Methods Mol. Biol.* **498,** 297–307.
Vaillancourt, P., Zheng, C. F., Hoang, D. Q., and Breister, L. (2000). Affinity purification of recombinant proteins fused to calmodulin or to calmodulin-binding peptides. *Methods Enzymol.* **326,** 340–362.
Waldo, G. S. (2003). Genetic screens and directed evolution for protein solubility. *Curr. Opin. Chem. Biol.* **7,** 33–38.
Waldo, G. S., Standish, B. M., Berendzen, J., and Terwilliger, T. C. (1999). Rapid protein-folding assay using green fluorescent protein. *Nat. Biotechnol.* **17,** 691–695.
Wang, K. H., Oakes, E. S., Sauer, R. T., and Baker, T. A. (2008). Tuning the strength of a bacterial N-end rule degradation signal. *J. Biol. Chem.* **283,** 24600–24607.
Waugh, D. S. (2005). Making the most of affinity tags. *Trends Biotechnol.* **23,** 316–320.
Yokoyama, S. (2003). Protein expression systems for structural genomics and proteomics. *Curr. Opin. Chem. Biol.* **7,** 39–43.
Zhang, Y. B., Howitt, J., McCorkle, S., Lawrence, P., Springer, K., and Freimuth, P. (2004). Protein aggregation during overexpression limited by peptide extensions with large net negative charge. *Protein Expr. Purif.* **36,** 207–216.
Zhou, P., Lugovskoy, A. A., and Wagner, G. (2001). A solubility-enhancement tag (SET) for NMR studies of poorly behaving proteins. *J. Biomol. NMR* **20,** 11–14.

CHAPTER SEVENTEEN

Refolding Solubilized Inclusion Body Proteins

Richard R. Burgess

Contents

1. Introduction 260
2. General Refolding Consideration 262
3. General Procedures 262
4. General Protocol 263
5. Comments on this General Procedure 264
   5.1. Overexpression of recombinant proteins in *E. coli* 264
   5.2. Washing IBs 265
   5.3. Solubilizing IBs 265
   5.4. Refolding 267
   5.5. High-resolution ion-exchange chromatography 268
   5.6. Reoxidation to form correct disulfide bonds 269
   5.7. Characterization 270
6. Performing a Protein Refolding Test Screen 271
   6.1. Systematic refolding screens 271
   6.2. Variables in refolding 271
   6.3. A practical, inexpensive rational approach 274
7. Other Refolding Procedures 275
8. Refolding Database: Refold 277
9. Strategies to Increase Proportion of Soluble Protein 277
10. Conclusion 279
References 279

Abstract

The vast majority of protein purification is now done with cloned, recombinant proteins expressed in a suitable host. The predominant host is *Escherichia coli*. Many, if not most, expressed proteins are found in an insoluble form called an inclusion body (IB). Since the target protein is often relatively pure in a washed IB, the challenge is not so much to purify the target, but rather to solubilize an IB and refold the protein into its native structure, regaining full biological

McArdle Laboratory for Cancer Research, University of Wisconsin-Madison, Madison, Wisconsin, USA

Methods in Enzymology, Volume 463
ISSN 0076-6879, DOI: 10.1016/S0076-6879(09)63017-2

activity. While many of the operations of this process are quite general (expression, cell disruption, IB isolation and washing, and IB solubilization), the precise conditions that give efficient refolding differ for each protein. This chapter describes the main techniques and strategies for achieving successful refolding.

1. Introduction

When proteins are expressed at high levels in *Escherichia coli* and other expression hosts, it is common to find that most of the protein is not soluble, but in an insoluble form called an inclusion body (IB). While the exact mechanism for IB formation is not fully understood and may vary with different proteins and expression conditions, it is generally thought that target proteins are being made faster than they can fold into the native structure. If a protein is partially folded or misfolded, it will generally have hydrophobic or "sticky" regions exposed that can interact with other similar proteins and form aggregates. These aggregates form IBs that are dense, refractile bodies on the order of 0.2–0.5 μm in diameter. Once in an IB, the protein is usually protected from proteolytic attack and is the predominant protein. IBs can vary from being mostly native protein, relatively easily solubilized under mild condition to being misfolded, effectively irreversibly insoluble material that requires high concentrations of denaturants to be solubilized (see Bowden *et al.*, 1991; Ventura and Villaverde, 2006). The latter category is by far the most common and thus, the major barrier to obtaining native, active material is to find conditions under which the denatured protein can be refolded efficiently. *This refolding process and strategies to achieve it are the focus of this chapter.*

Anfinsen, in the early 1960s (see Anfinsen, 1973), observed that small protein molecules could fold spontaneously to the native state. This observation has led to a thermodynamic hypothesis: the native state has the lowest free energy. While the thermodynamic hypothesis has become a paradigm of protein folding, a simple and general model for describing how an unfolded protein molecule finds its native state is still not available. One cannot yet use computers to accurately predict protein structure from sequence.

Levinthal (1968) pointed out the paradox that an unfolded protein molecule does not have the time needed to search all possible conformations, yet it reaches the single native state rapidly. Consider a (hypothetical) small protein with 100 residues. Levinthal calculated that, if each residue can assume only three different conformations (it is probably many more than this), the total number of structures would be 3^{100}, which is equal to 5×10^{47}. If it takes only 10^{-13} s to convert one structure into another, the total search time would be 5×10^{47} times10^{-13} s, which is equal to 5×10^{34} s or 1.6×10^{27} years! Clearly, it would take much too long for even a small protein to fold properly by randomly trying out all possible conformations. This paradox has shaped many studies on the theory of refolding (e.g., see Clark, 2004; Karplus, 1997).

A great many experimental studies and theoretical analyses have been carried out to better understand the folding process. The consensus is that when a denatured protein is no longer in denaturing conditions, some secondary structure (α-helix and β-sheet) forms, a hydrophobic collapse occurs to form a "molten globule" state with some secondary structure, but nonnative tertiary structures, the globular states go through multiple intermediates, and finally reach the lowest energy state, the native conformation. Solution conditions can have significant effects on the folding pathway of a given protein. If you think of the refolding energy funnel as a deep, fog-filled crater in the top of a volcanic mountain, then the rim is the denatured state and the very bottom of the crater is the native state (the lowest energy state). If you roll a ball down into the crater from one point on the rim of the crater, then it may be able to roll all the way to the bottom. From another point on the rim, the ball may become trapped in a local minimum and not reach the bottom. In many ways the challenge of finding conditions for efficient refolding is like trying to guess where on the rim to start rolling the ball.

The rate of protein refolding is rapid, with folding halftimes on the order of seconds for proteins lacking prolines and on the order of minutes for proteins with prolines (see Nall, 1994). The amino acid proline can consist of two isomers, a cis and trans configuration, and the halftime of this chemical isomerization is on the order of 1 min. In a native protein, each proline is in a particular configuration, but in a denatured protein an equilibrium mixture of the two isomers forms consisting of about 70% trans and 30% cis isomer. Therefore, as a protein refolds, those prolines that are in the incorrect conformation cannot form the native structure and must wait for an isomerization event to occur. Thus, proline isomerization becomes rate limiting for correct protein folding. The enzyme, protein proline isomerase (PPI), can speed up this isomerization rate.

Even though recombinant proteins were first prepared by refolding solubilized IBs in the late 1970s and early 1980s, and many improvements have been made in refolding procedures and techniques, refolding of any given protein still presents a significant challenge. During the last 10 years the protein structure initiative (PSI), through its numerous Structural Genomics Centers, has cloned and expressed over 110,000 proteins (as of December 2008), but were only able to purify about 29,000 (about 26%). On the order of 50% of Eubacterial and Archaeal proteins and only about 10% of Eukaryal proteins could be expressed as soluble proteins in *E. coli* (Graslund *et al.*, 2008). The proteins that were insoluble were generally abandoned rather than trying to find effective refolding conditions. I believe that many of these insoluble proteins could have been salvaged or rescued by finding suitable refolding conditions as described below.

Many excellent reviews report approaches to protein refolding (Baneyx and Mujacic, 2004; Cabrita and Bottomley, 2004; Clark, 2001; Jungbauer and Kaar, 2006; Middelberg, 2002; Panda, 2003; Quronfleh *et al.*, 2007; Singh and Panda, 2005; Swietnicki, 2006; Thatcher and Hitchcock, 1994; Tsumoto *et al.*, 2003; Vallejo and Rinas, 2004). *I will try to present the kind of advice below that I would give to colleagues embarking on a project that involves protein refolding.*

2. General Refolding Consideration

The main problem observed in many refolding attempts is that upon dilution or dialysis of the solubilized IB to decrease the concentration of the denaturant to a nondenaturing level, you get major precipitation, especially if the protein concentration is high during refolding. In a way this is likely to be similar to the process in the cell that leads to formation of IBs in the first place. You want to go from denatured to partially folded intermediate to native structure, and bias the reaction away from aggregation of the sticky folding intermediate and precipitation. Aggregation is dependent on collision of sticky folding intermediates. At low protein concentrations the probability of collision is diminished. *The best general strategy in refolding is to refold at the lowest protein concentration that is feasible.* In screening refolding conditions, one searches for conditions where little aggregation or turbidity is observed. But even when there is no obvious precipitation (the dilute protein in refolding solution is not turbid), there can be significant "soluble multimer." Soluble multimers are generally from 2- to 20-mers that are soluble and not large enough to cause much light scattering (turbidity). Usually these soluble multimers are not active and tend to bind tightly to ion exchange resins.

3. General Procedures

Given that it is very often the case that protein expressed in *E. coli* and other hosts is found in IBs, the following steps are almost always needed to obtain active, native protein:

1. Cell growth and protein overexpression
2. Cell lysis, isolating and washing IBs
3. Solubilizing IBs
4. Identifying suitable refolding conditions
5. Protein refolding
6. Reoxidation to form disulfide bonds where necessary
7. High-resolution ion-exchange chromatography
8. Characterization of final material to determine if refolding has been successful

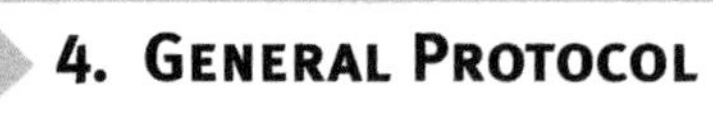

4. General Protocol

Below is a typical procedure that works well for many proteins. This protocol is based on and adapted from the protocol first developed by Nguyen *et al.* (1993) and used in the Cold Spring Harbor Protein Purification and Characterization Course section on Purification of Insoluble Recombinant Proteins (Burgess and Knuth, 1996). Other similar procedures are likely to give similar results, but this is the procedure used almost exclusively in my laboratory. The critical steps will be discussed in the following steps.

1. *E. coli* BL21(DE3)pLysS containing the target gene cloned into a pET expression vector (Studier *et al.*, 1990) is grown in 1 L of LB in a 2-L flask at 37 °C with shaking until the $A_{600\ \text{nm}}$ reaches about 0.6.
2. IPTG is added to 0.5–1 m*M* to induce expression of the T7 RNA polymerase that then transcribes the target gene.
3. Three to four hours after induction, the cells are harvested by centrifugation at 15,000 rpm for 15 min, the cells are resuspended in a small volume of the culture supernatant, transferred into a preweighed, 40-ml Oak Ridge tube, and centrifuged to pellet the cells. The wet weight of the cell pellet is noted and the cells are stored frozen at − 80 °C until use. It is common to get 1.5–2.0 g wet weight *E. coli* cells per liter using this procedure.
4. Cells are thawed, resuspended in 30 ml of a lysis buffer (50 m*M* Tris–HCl, pH 7.9, 0.1 m*M* EDTA, 5% glycerol, 0.1 m*M* DTT, 0.1 *M* NaCl), and then sonicated at 60% power for 3–4 intervals of 20 s each with about 1 min on ice to cool between each interval.
5. Purified Triton X-100 (from a 10%, w/v stock) is added to 1% (w/v) to dissolve membranes and solubilize membrane proteins. The lysate is incubated on ice for 10 min and then centrifuged at 15,000 rpm for 15 min to pellet the IB and remove the soluble supernatant.
6. The IB is resuspended in 30 ml of lysis buffer with 1% Triton X-100, incubated on ice for 10 min, and then centrifuged for 15,000 rpm for 15 min.
7. The drained pellet is resuspended in 30 ml of lysis buffer without Triton X-100 to remove the Triton X-100 and recentrifuged as above. The pellet is called the washed IB fraction and is usually over 90% pure.
8. The washed IB pellet is resuspended in a suitable denaturant, incubated, to allow denaturation and solubilization, and then centrifuged at 15,000 rpm for 15 min to remove any residual insoluble material. We routinely use the lysis buffer above (lacking the NaCl if diluting from GuHCl) containing either 6 *M* guanidine hydrochloride (GuHCl), 8 *M* urea, or 0.3% Sarkosyl (*n*-lauroyl sarcosinate) for IB solubilization.

9. More recently, we have carried out a relatively simple refolding test (see below) to identify a suitable refolding buffer.
10. Adjust the protein concentration in the denatured sample to about 1 mg/ml.
11. Dilute the denatured protein about 15–60 fold to dilute the denaturant concentration to a point where the protein can refold. We usually either flash dilute or slowly drip dilute the denatured IB into refolding buffer in a beaker with vigorous stirring to cause very rapid mixing. The dilution process is performed at room temperature and after mixing is complete, we let the solution stand for 1–2 h to allow the refolding process to complete and to allow any aggregated material to form and flocculate.
12. The refolded material is filtered through a low protein binding, 0.22 μm membrane filter (e.g., Stericup-GV 0.22 μm, 500 ml, Millipore #SCGVU05RE) to remove any particulate material (this can be preceded by a centrifugation step if there is significant precipitated material that would otherwise immediately clog the filter).
13. Pump the filtered solution onto a 10–15-ml ion-exchange column as fast as the pressure constraints of the column and system will allow (hopefully, at least 10 ml/min). Monitor the absorbance at 280, 260, and 320 nm, if possible. The absorbance at 320 nm is a measure of light scattering and for some proteins can indicate a peak that is composed of multimers.
14. Wash the column with 5–10 column volumes of Buffer A (50 m*M* Tris–HCl, pH 7.9, 5% glycerol, 0.1 m*M* EDTA, 0.1 m*M* DTT) plus 0.1 *M* NaCl and then elute with a 10-column volume, linear salt gradient in the same buffer from 0.1 to 1.0 *M* NaCl at a flow rate of 5 ml/min, collecting 3–4 ml fractions.
15. Analyze the $A_{280\ nm}$ peaks by SDS–PAGE to determine purity of the various fractions. If an enzyme assay is available, assay fractions to determine those with the highest specific activity.
16. Pooled peak fractions are dialyzed against lysis buffer containing 50% glycerol and stored at − 20 or − 80 °C.
17. Characterize the pooled peak material, if possible, by specific activity compared to a standard sample of the protein purified by a more conventional procedure that does not involve refolding.

5. Comments on this General Procedure

5.1. Overexpression of recombinant proteins in *E. coli*

The issue of achieving a high level of expression is a topic somewhat separate from the refolding issue and will not be discussed further here (see Chapter 12 in this volume). Many systems will allow expression levels of the target protein of 10–40% of the total cell protein (Makrides, 1996; Murby *et al.*, 1996; Sorensen and Mortensen, 2005; Studier *et al.*, 1990) (see Section 9). Usually one carries out a test to see at what temperature to grow the cells, how much inducer to use, and how long after induction you should harvest the cells. After induction of the high-level expression of the target protein, the cells are usually centrifuged, and frozen at − 80 °C until use. There are anecdotal reports that IBs from cells stored for weeks in the frozen state are harder to refold, but in my experience, we have gotten excellent refolding results from cells stored frozen for months. Unfortunately, to my knowledge, no systematic study has addressed this point.

5.2. Washing IBs

We originally used 2% sodium deoxycholate to wash IBs (Burgess and Knuth, 1996; Nguyen *et al.*, 1993), but later found that 1% Triton X-100 worked better and was much easier to work with. It is highly recommended that one use what I call "gourmet" Triton X-100 (Pierce, Surfact-Amps X-100, 10% (w/v) solution) that has been purified (to remove peroxides sometimes found in old, yellowish bottles of Triton X-100) and stored under nitrogen in sealed ampoules. It is possible to test various salts and detergents to see which are particularly effective at washing out protein impurities from an IB, but not solubilizing a given target protein. Sometimes washes with 1–2 *M* urea can be used. In general, it is desirable to remove as much membrane material, DNA, and other proteins as possible from an IB before solubilization.

5.3. Solubilizing IBs

As mentioned earlier, some IBs are nearly native and can be solubilized by mild conditions such as nonionic detergent or even 0.5 *M* NaCl (Vera *et al.*, 2006). In most cases, this will not work and much stronger denaturing conditions must be used.

The solubilizing agent/denaturant is prepared in a buffer, much like the lysis buffer above, to control pH, chelate heavy metals, and maintain a reducing environment. Usually one resuspends the IB with the solubilizing agent, lets it incubate for 30–60 min, and then centrifuges out any insoluble material. Since the material is already at least partially denatured in the IB, this solubilization can be performed at room temperature and in some cases it is even necessary to heat the solution to achieve full solubilization. For example, when native green fluorescent protein (GFP) is adjusted to 6 *M* GuHCl it retains its fluorescence at room temperature, but at 75 °C it denatures and loses it fluorescence within 1 min (R. Burgess and N. Thompson, unpublished results). A detailed discussion of solubilization is found in Marston and Hartley (1990).

Some comments on the use of several solubilizing agents follow:

1. *GuHCl.* This is perhaps the most commonly used solubilizing agent, usually used at 6 *M* in a compatible buffer. Most proteins will rapidly denature in this strong chaotropic agent, but in many cases it may help to incubate at higher temperatures to achieve complete denaturation. Once solubilization occurs, it is often possible to dilute to 3 *M* GuHCl without a problem, since the transition from denatured to native state often occurs in the 1–2 *M* GuHCl range (Pace, 1986). One usually dilutes to about 0.1 *M* GuHCl so that the salt concentration is low enough to allow the refolded target protein to bind to a suitable ion-exchange column (see below).
2. *Urea.* 8 *M* urea is a common solubilizing agent, and is especially useful if one wants to further purify the target protein in the denatured state (Knuth and Burgess, 1987). The 8 *M* urea is generally not as effective as 6 *M* GuHCl in solubilizing protein and causing complete denaturation. One must be aware of the possibility of carbamylation of proteins by the cyanate always present in urea solutions (see Chapter 44 in this volume for advice on how to minimize cyanate in urea).
3. *Sarkosyl or SDS.* Sarkosyl has been found to be a quite effective solubilizing agent that allows refolding at higher protein concentrations (see Burgess, 1996). Sarkosyl is a strong anionic detergent, much like SDS, but it seems to bind to proteins more weakly and dissociate more easily than SDS. Usually an IB pellet can be solubilized in 0.3% Sarkosyl, but you must remember that you cannot solubilize more than about an equal weight of protein per weight of Sarkosyl. For example, if you have a large washed IB that contains 150 mg of target protein, it will not be solubilized completely by 20 ml of 0.3% Sarkosyl (60 mg of Sarkosyl). You will have to add at least 50 ml of 0.3% or 30 ml of 0.5% Sarkosyl.

We have noticed that if there is much Triton X-100 present in an IB, it will take more Sarkosyl to solubilize the target protein. This is almost certainly due to the fact that Triton X-100 forms large micelles and absorbs some of the Sarkosyl to form mixed micelles that are not effective at solubilization.

If one dilutes 0.3% Sarkosyl-solubilized target protein to about 0.01% Sarkosyl, most of the Sarkosyl will dissociate from the protein and the protein will refold. The residual detergent seems to bind to the hydrophobic regions (the "sticky" sites) on partially refolded protein and prevent aggregation. It is effectively acting as a chemical chaperone. If your target protein is able to bind to a cation-exchange column such as a POROS HS, then the diluted solution can be filtered and pumped on a 10-ml column, washed with 10–20 column volumes of 0.1 *M* NaCl in buffer, and eluted with a salt gradient (see Section 5.5). The target protein will bind, but the free Sarkosyl will flow-through and any residual Sarkosyl bound to the protein will dissociate during the long wash. Experimental analysis of the eluted target protein indicates essentially no (<1 molecule detergent per 10 molecules of protein) residual Sarkosyl in the protein (R. Burgess, unpublished). Do not use MonoS for this purpose since Sarkosyl binds to the column and can damage the column and contaminate your eluted protein.

SDS can also be used in a somewhat similar fashion, but it is important to use SDS that is free from higher alkane lengths such as C14, C16, etc. Refolding experiments using GFP denatured in SDS work quite well (R. Burgess N. Thompson, and R. Chumanov, unpublished), so SDS should be considered a viable solubilizing agent.

The successful use of the cationic detergent cetyltrimethylammonium chloride (CTAC) in solubilization and refolding has also been reported (Puri *et al.*, 1992). Even distilled water at very low salt has been reported to achieve solubilization (Song, 2009).

5.4. Refolding

Assuming you have carried out refolding trials as described below, then one basically dilutes out the denaturant by diluting the solubilized protein into a suitable refolding buffer. However, there are three ways to dilute during refolding.

1. *Reverse dilution*: Add refolding buffer to denatured protein with mixing between each addition. This method has been successful in several cases (e.g., Gribskov and Burgess, 1983). However, on reflection, it is clear that this method of dilution results in the highest concentration of protein at the critical time when the protein is starting to refold. For example, if the solubilized protein concentration is 1 mg/ml in 6 *M* GuHCl, then if one adds 2–5 volumes of refolding buffer (dilutes to 1–2 *M* GuHCl), one is likely to go through the refolding transition between protein concentrations of 330–160 μg/ml.
2. *Flash dilution*: Add denatured protein to refolding buffer quickly. For example, add 10 ml of solubilized protein in 6 *M* GuHCl at one time with rapid mixing to 590 ml of a suitable refolding buffer to achieve a 60-fold dilution. The protein concentration will be 16 μg/ml during the refolding, much lower than with reverse dilution and less likely to result in aggregation.
3. *Drip dilution*: Add denatured protein to refolding buffer very slowly, drop-by-drop, over period of 1 h. In theory, this should be the optimal method, since protein is always refolding at minimal protein concentration (see Singh and Panda, 2005; Vallejo and Rinas, 2004). For example, if you add 0.1 ml of solubilized protein as above into 590 ml refolding buffer, then the protein concentration will only be 0.16 μg/ml. As you add small incremental amounts of the sample, the protein will always be

low and when the last has been added, the protein concentration will be 16 μg/ml. However, if the denatured protein is added over a period of 1 h, and significant protein refolding to native state is achieved with a refolding half-life of 1–3 min, then the concentration of partially refolded sticky protein will always be quite low and the native protein will not be available to aggregate. While adding solubilized protein over 1 h period is tedious when added drop-by-drop, one can set up a peristaltic pump at a very slow flow rate to deliver the denatured protein over 1 h period. This is more reproducible and causes considerably less graduate student burnout.

Let the protein sit after dilution for 30–60 min and then filter as described in the generic protocol. If this is not done, then as the protein solution is loaded on the column (see below) very small particulate material is likely to clog up the column, causing the pressure to rise, either aborting the run or requiring one to decrease the flow rate to such low levels that it takes many hours to load the column.

5.5. High-resolution ion-exchange chromatography

After refolding and filtering, a final, high-resolution ion-exchange column step accomplishes five important jobs:

1. *It concentrates protein* (e.g., from 600 to 4–8 ml).
2. *It removes denaturant* (in the flow-through).
3. *It removes minor impurities* (in flow-through or binding weaker or stronger than the target protein to the column). If a high-resolution anion-exchange column is used, like POROS HQ or MonoQ, any DNA contamination will bind tightly and elute at 0.6–0.9 *M* NaCl. If the presence of *E. coli* lipopolysaccharide (LPS) in the final product is particularly undesirable, one can wash the column with isopropanol to greatly reduce LPS before beginning the salt gradient elution.
4. *It selects for a homogeneous, active monomer.* If one has a sample of the native protein, purified by nonrefolding methods, then one can see at what salt it elutes from this same ion-exchange column. If a major peak is eluted at the same salt from the refolded material, then one has some confidence that the material is properly refolded because misfolded molecules are likely (but not necessarily) to elute slightly differently if the column is capable of high resolution.
5. *It removes soluble protein multimers.* Soluble multimers are very often found in refolded material (see Section 5.7). If you have chosen the refolding solution well, they may be minimal; if not, they may dominate. These soluble *N*-mers have *N* times as much charge as the monomer. They tend to bind tighter to the ion-exchange column and elute later due to their higher charge. Removal of multimers is

an extremely important step if the protein is being used for crystallization for structure determination, since heterodisperse material is much less likely to crystallize.

In my experience, in most cases the protein eluting in the first major peak from the column will be high quality, pure material with full biological activity.

5.6. Reoxidation to form correct disulfide bonds

If your protein has no cysteines this is not an issue. However, most proteins contain cysteines, and many have disulfide bonds as important parts of their three-dimensional structure. Since the cytoplasm of *E. coli* is very reducing, most internal proteins are in the reduced state. If you add a reducing agent like 0.1–1 m*M* DTT to the lysis buffer, you can usually keep the protein reduced and prevent unwanted or incorrect disulfide bonds from forming during the early steps above. However, once you refold your target protein, you must at some stage allow the reoxidation of the protein to form disulfide bonds if they can form. Proteins that have cysteines, but that do not normally contain disulfide bonds have cysteines that are not positioned in the precise geometric configuration needed to form a disulfide bond (see Anthony *et al.*, 2002). Therefore, even though there are often a very large number of possible incorrect disulfide bonds, they do not form readily. The best strategy is to have a redox buffer present during protein refolding. Such a buffer (see below) has a mixture of reducing and oxidizing agents that allow disulfide shuffling to occur (see Gilbert, 1994). Disulfide shuffling consists of allowing disulfide bonds to be formed and be reduced repeatedly. If an incorrect bond forms in a misfolded protein and cannot be reduced, the protein is frozen into an incorrect conformation and will not reach the native state. If that bond is reduced, the protein can continue to move between intermediate folding states until it reaches the correct structure. If the correct bond now forms, it stabilizes the final native protein. Even if it is occasionally reduced, the protein is in a stable state and will reform the correct bond upon reoxidation. For this reason, it is often successful merely to have reducing agent in the solubilized protein but not in the refolding buffer. After refolding, the protein is allowed to undergo slow air oxidation (exposed to air for a day or two) and often achieves the correct disulfide bonded structure. For more on reoxidation see Vallejo and Rinas (2004) and Kersteen and Raines (2003).

Some typical methods for forming disulfide bonds are

1. *Air oxidation*: expose to air, without reducing agent for several days.
2. *Redox buffer*: for example, reduced/oxidized glutathione = 10/1 (3 m*M* GSH/0.3 m*M* GSSG). Various redox pairs are used including reduced

and oxidized cysteine, dithiothreitol, and glutathione. The molar ratio of reduced to oxidized form is sometimes varied to achieve optimal disulfide reoxidation.

3. *Protein disulfide isomerase (PDI)*: this is an enzyme that catalyzes disulfide shuffling (see review by Kersteen and Raines, 2003). One can also use small molecule PDI mimics (see Woycechowsky *et al.*, 1999).

5.7. Characterization

One of the most common problems I see when reviewing papers involving refolding is that one has no idea if the material at the end of the procedure is correctly refolded, monodisperse, or has full biological activity. It is common to see an enzyme or biological assay, but often without any standard. You see phrases like "since the final product has activity it is shown that it is correctly folded and fully active." Only a comparison with a known, fully active standard will allow an estimation of the percent of the final product that is active. Without this comparison you know there is activity, but do not know if 100%, 10%, 1%, 0.1%, or 0.01% of the refolded material is active. Such a determination of specific activity of the final product is essential in a refolding procedure, if possible.

Another important characterization is to determine the size of the refolded material. Is it monomeric (monodisperse) or does it contain large amounts of soluble multimer (heterodisperse)? Crystallographers routinely use dynamic light scattering (DSL) to determine if purified protein, refolded or not, is monodisperse. Another method, developed by Mark Knuth at GNF in La Jolla, CA is to analyze the final product by analytical size exclusion chromatography (ANSEC). We have found this to be very useful since very small amounts of protein can be analyzed. Typically, we run 20–50 μg of protein on a 12-ml Shodex KW-802.5 column at 0.5–1.0 ml/min in a buffer containing 0.25 M NaCl, monitoring absorbance at 280 and 215 nm. When the column is calibrated with suitable MW markers, one can quickly determine if most of the protein is in a single peak (monodisperse) of the size expected for the monomer (this is good), or if much of the protein elutes earlier indicating that it is much larger (heterodisperse) due to formation of multimers (this is bad). Use of this procedure is an important part of a thorough set of refolding test trials (see below).

One cannot use circular dichroism (CD) or Western blot analysis to determine activity or monodispersity. There are plenty of examples of proteins whose CD spectra are very similar or identical to native standard material and which react just fine in a Western blot analysis that are not monodisperse or active, respectively.

6. Performing a Protein Refolding Test Screen

6.1. Systematic refolding screens

In the general procedure and discussion earlier, it was assumed that you knew a suitable refolding solution for your protein. This, however, is the most essential and the most difficult part in designing an efficient refolding protocol. In early refolding papers, usually a single refolding solution was chosen and it either worked or did not work. The ones that worked were published and the unsuccessful attempts were often abandoned. As a result, the early successes were usually proteins that refolded readily under many different conditions. In recent years, it has become increasingly clear that many proteins can only be refolded under very specific conditions. The challenge has become how to screen the very many possible conditions to find one that is effective in promoting refolding. An important advance was the use of fractional factorial approaches to systematically determine the effects of many different variables (Armstrong *et al.*, 1999; Cowieson *et al.*, 2006; Quronfleh *et al.*, 2007; Trésaugues *et al.*, 2004; Vincentelli *et al.*, 2004; Willis *et al.*, 2005).

Several commercial products have been developed to aid the researcher in identifying suitable refolding conditions. Much more information on these refolding screens and related protocols are available at the company web sites.

AthenaES QuickFold™ (15-solution kit); (http://www.athenaes.com/QuickFoldProteinRefolding Kit).
EMD/Novagen's iFOLD1™, iFOLD2™, and iFOLD3™ (96-solution kit); (http://www.novagen.com).
Pierce Biotechnology's ProMatrix™ (96-solution kit components); (http://www.fishersci.com).

It is quite likely that this list will grow and that the screens will evolve as more experience is gained in the use of this type of screen and new refolding aids are identified. It should be noted that nothing prevents a researcher who cannot afford to buy these rather expensive kits from designing a set of test solutions of their own and customizing the refolding screen to fit their special needs. *The key concept here is the systematic, parallel screening of multiple refolding conditions.*

6.2. Variables in refolding

There are many solution variables (such as pH, temperature, salt concentration, redox environment, and the presence of divalent ions) and additives that have been reported to improve the efficiency of refolding of solubilized proteins from IBs. Such variables and additives have been taken into account in designing the commercial refolding screens mentioned above and serve as examples for those wishing to design their own refolding screen. These variables are discussed below. Other reviews on protein refolding go into greater detail on some of these (Armstrong *et al.*, 1999; Cowieson *et al.*, 2006; Middelberg, 2002; Quronfleh *et al.*, 2007; Schein, 1991; Singh and Panda, 2005; Trésaugues *et al.*, 2004; Vallejo and Rinas, 2004; Vincentelli *et al.*, 2004; Willis *et al.*, 2005).

pH: Most refolding is done in the range of pH 5–9 and, in our experience, most proteins refold best at about pH 8–8.5. In general it is a good idea to use a pH that is at least one pH unit away from the isoelectric point or pI (the pH at which the protein has zero net charge and is most prone to precipitation).

Temperature: So far no obvious trend has developed as to any generally optimal temperature to use during refolding. Most people carry out the refolding process at near room temperature. This is a low enough to prevent thermal damage to the protein and high enough to increase the thermal motion of the molecules that is likely important in melting out transient misfolded conformations and reaching the native state. One could argue that higher temperatures will strengthen the hydrophobic interactions that likely lead to aggregation, but then again it should also strengthen the hydrophobic interactions needed to bury hydrophobic residues in the interior of a native protein.

A very interesting paper by Xie and Wetlaufer (1996) reported the refolding of carbonic anhydrase II at 4 mg/ml in 1 *M* GuHCl and 50 m*M* Tris sulfate, pH 7.5 at different temperatures. It appears that refolding at low temperatures (4–12 °C) for 120 min followed by a "temperature-leap" to 36 °C for 30 min gave excellent recovery of activity (>90%). They argue that hydrophobic interaction is diminished at the low temperature, minimizing aggregation and allowing slow conversion to an intermediate that is not active, or aggregation prone. Upon temperature-leap from 4 to 36 °C, this intermediate converts to a highly active, native form.

Salt concentration: Some salt is probably desirable to cause "salting in" to increase solubility of the native protein (see Chapter 20 in this volume). Often 50–100 m*M* salt is used. In the case of a 60-fold dilution of 6 *M* GuHCl, the final concentration of GuHCl is 0.1 *M*. Additional salt is often avoided to keep the salt low enough to allow protein binding to an appropriate ion-exchange column.

Redox agents: The use of redox buffers to aid in correct disulfide bond formation is discussed in the section on reoxidation above. In some of the commercial kits there are solutions with no reducing agent, with reducing agent and redox buffers with several ratios of reducing to oxidizing agents.

Divalent cations: This variable has not been thoroughly explored, but it is clear that native proteins often have divalent cations such as Mg^{2+} or Zn^{2+} as part of their structure. Obviously, anything that will stabilize the native structure will bias refolding toward native and away from aggregated protein.

Arginine and other amino acids: There is an extensive literature on the use of arginine to promote refolding of solubilized protein and how it functions to diminish aggregation (Arakawa *et al.*, 2007; Baynes *et al.*, 2005; Das *et al.*, 2007; Dong *et al.*, 2004; Reddy *et al.*, 2005). It appears that arginine decreases aggregation by slowing the rate of protein–protein interactions. Das *et al.* argue that this is because arginine forms supramolecular assemblies in solution. One of its drawbacks is that it is usually used and is most effective at concentrations of 0.5–1.0 *M*. At these concentrations, it interferes with subsequent chromatography on Ni^{2+}-chelate affinity columns and will prevent the binding of most protein to an ion-exchange column without further dilution or dialysis. The natural osmoprotectant proline has also been found to be effective in some cases at increasing solubility both *in vitro* and *in vivo* (Ignatova and Gierasch, 2006).

Glycerol and sugars (sucrose, mannitol, sorbitol, and trehalose): We have found glycerol to be an excellent refolding additive in many cases, usually used in the 5–30% range. One extreme method by Shimamoto *et al.* (1998) involves solubilization of IB protein to 2–10 mg/ml by 6 *M* GuHCl, addition of glycerol to 50% (v/v), dialysis into 75% (v/v) glycerol to remove denaturant, and then rapid dilution to lower glycerol to 5–10%. Several sugars in the 0.5–1.5 *M* range have been seen to promote successful refolding (Bowden and Georgiou, 1988).

Other chemical additives: A large variety of papers have reported the use of materials such as polyethylene glycol (PEG) (Cleland *et al.*, 1992), cyclodextrin (Rozema and Gellman, 1996), and various nondetergent sulfobetaines (NDSBs) (Expert-Bezancon *et al.*, 2003) to increase refolding for certain proteins. The effects of certain ionic liquids (i.e., organic salts with melting points below 100 °C) such as *N′*-alkyl-*N*-methylimidazolium chlorides as refolding additives have been reported (Lange *et al.*, 2005). The ethyl and propyl derivatives, at concentrations of 0.5–1.0 *M*, showed effects in increasing activity and refolding yields comparable to L-arginine.

Detergents: In theory, detergents should be helpful in preventing aggregation during refolding. At low concentrations they bind weakly to exposed hydrophobic or sticky regions and mask them, preventing aggregation. As their concentration decreases they dissociate and allow reformation of native structure. This is thought to be why Sarkosyl works as well as it does. At high concentration it is a denaturant, but at low concentration it acts as an artificial chaperone and promotes refolding without aggregation.

Chaotropic agents (denaturants at high concentrations): Since aggregation is due to interactions among partially folded intermediates, it has sometimes been useful to have a nondenaturing amount of a chaotrope in the refolding solution to dissociate aggregates, but not the more stable, native structures. 1 *M* urea and 0.5 *M* GuHCl have been used.

Target protein-specific additives: The same argument given above for the potential benefit of certain divalent cations can be used to suggest that the presence of appropriate substrates, or cofactors, or essential heme groups, for example, would stabilize native structure and improve refolding yield.

6.3. A practical, inexpensive rational approach

With all the refolding screens, you still have to obtain some sort of a readout that indicates which condition works best. The best situation is when your target enzyme is a known enzyme and can be easily assayed. You merely make your dilutions from solubilized protein into the various refolding solutions, wait some time for refolding (often several hours), and then assay a portion of the dilute solution for enzyme activity.

It is even easier if you want to examine the effects of conditions on refolding of a fluorescent protein such as GFP since the fluorescence is restored when the GFP refolds correctly and can very conveniently be measured during and after the refolding with an appropriate fluorescence plate reader. This makes an excellent exercise for teaching protein refolding and is used at the Protein Course at Cold Spring Harbor (R. Burgess, N. Thompson, R. Chumanov, unpublished results).

However, most proteins being studied these days are either not enzymes or have a very difficult biological assay. How do you know which refolding condition worked best? The following protocol has been used with great success (R. Chumanov and R. Burgess, unpublished results). This procedure is quite general and takes advantage of the turbidity of solutions where protein aggregation has occurred (see Trésaugues *et al.*, 2004; Vincentelli *et al.*, 2004).

1. Prepare washed IBs solubilized in several different chaotropic agents as listed above and centrifuged to remove insoluble material. Protein concentration should be 3–5 mg/ml.
2. Prepare a set of test solutions based on variables and additives that you think might be effective.
3. Pipette 10 μl of solubilized protein into the appropriate number of wells of a 96-well plate microtiter dish (Linbro, flat bottomed, polystyrene).
4. Carry out a 20-fold flash dilution by quickly adding 190 μl of the test refolding solution and mixing the solution up and down several times in the pipette at room temperature.
5. Wait 15–60 min and then read the absorbance of the plate at 320 nm. The solution is not absorbing the light, but rather the protein aggregates scatter light and decrease the amount of transmitted light measured. This works well because light scattering increases as you go to lower wavelengths, proteins do not absorb at 320 nm, and 320 nm is about a low as you can go in wavelength before absorbance of the plate gets too great.

At 320 nm, the absorbance of the empty well due to the plastic is about 0.2 and can be subtracted from the apparent absorbance readings due to solution turbidity to give a corrected turbidity.

6. The results of the absorbance readings with the plate absorbance subtracted can be graphed versus the solution number and one can see which solutions give the lowest turbidity. Corrected turbidity values range from 0.0 to about 0.4. We often find several conditions that give low- or zero-corrected turbidity. One often gets somewhat different results with protein denatured using different solubilizing agents such as 6 *M* GuHCl, 8 *M* urea, or 0.3% Sarkosyl.
7. Choose the best 2–3 conditions that are compatible with loading directly (after filtration) onto an analytical size exclusion column (ANSEC, see Section 5.7). To be really useful for large-scale refolding, it must also be compatible with loading onto an appropriate ion-exchange column at a salt concentration of about 0.1 m NaCl. The solution (about 200 μl) can be removed from the corresponding well, filtered through a syringe filter or centrifuged in a microfuge tube and loaded onto a 12-ml ANSEC column. *A refolding condition that gives low turbidity and mostly monomeric material on the ANSEC column is the condition that is likely to be useful in the large-scale refolding of the target protein.*

7. Other Refolding Procedures

Although the main focus of this chapter has been on refolding of solubilized IB protein by dilution into a suitable refolding solution, there are several other significant approaches to protein refolding that need to be described briefly.

1. *On-column refolding.* This is a large and active area of research and application. The basic principle is that denatured proteins when bound to a column resin or residing within the pores of a gel filtration resin are tethered or sequestered and if they experience a gradual decrease in the concentration of denaturant, they will refold, but be less likely to aggregate. This idea was first proposed by Tom Creighton in the late 1980s and is now widely used in various column modes such as ion exchange, affinity, metal chelate affinity, hydrophobic interaction, and gel filtration chromatography. For example, one can load an IB protein solubilized with 8 *M* urea onto an ion-exchange column, wash away some impurities with 8 *M* urea, wash with a gradually decreasing urea gradient, and elute with a linear salt gradient. Another approach is to preload a gel filtration column with a reverse gradient of denaturant. The top of the column is 8 *M* urea, the gradient decreases linearly to 0 *M* urea

in the middle of the column, and the bottom half of the column is 0 *M* urea. One then loads the denatured protein and denaturant is applied at a slow flow rate to the column. The protein will move down the column faster than the buffer (because it is partially or completely excluded from the bead) and move gradually into lower and lower concentrations of denaturant until it has refolded and elutes from the column. The difficulty here is to decide on the steepness and volume of the reverse gradient. The various protocols and applications are summarized nicely in several recent reviews and articles (Jungbauer and Kaar, 2006; Jungbauer *et al.*, 2004; Oganesyan *et al.*, 2005; Swietnicki, 2006; Veldkamp *et al.*, 2007). While this on-column refolding method seems ideal, it is often the case that as the denaturant concentration is decreased, the protein precipitates on the column and is lost. This is usually due to the tendency of many researchers to overload their column or to load from one end that becomes saturated with protein. In these cases, the fact that the protein is tethered to the resin does not help because the bound proteins are close enough together on the column that they can still aggregate during refolding. One general procedure that sometimes helps is to bind the protein in denaturant to the column in batch (obviously not applicable to size exclusion chromatography) at well below saturating amounts, then load the resin into a column, and finally carry out the subsequent washing and elution steps in column.

2. *Use of folding catalysts such as protein and chemical chaperones, PPI, PDI.* Another nice approach is to covalently attach various enzymes and chaperones to a gel filtration column resin and then refold in column. The enzymes that have been immobilized, either individually or perhaps better in combinations are: PPI (catalyze proline isomerization), PDI (promote disulfide shuffling), DsbA and DsbC to catalyze disulfide formation, and the chaperones GroEL/GroES to aid in refolding (Baneyx and Mujacic, 2004; Jungbauer *et al.*, 2004; Paul *et al.*, 2007; Swietnicki, 2006).
3. *High-pressure refolding.* A number of papers, summarized by Quronfleh *et al.* (2007) report the successful solubilization of IBs by exposure to high hydrostatic pressure (1.5–2.5 kbar). It is claimed that exposure of IBs to high pressure within a certain range will disaggregate the aggregated protein but not denature it. The process of pressurizing, holding at high pressure, and slowly decreasing the pressure in a variety of buffers has been reported to result in high recovery of soluble and active enzymes. While this method has been commercialized by Barofold (Boulder, CO) using a specialized pressure cells (PreEMP^{TM} high-pressure chamber), the equipment is not widely available. In some cases, while the IB is solubilized, the protein is largely in the form of soluble multimers. Like the solubilization with denaturants and refolding discussed for most of this chapter, one must find the right conditions for optimal recovery of active protein by screening multiple conditions.

4. *Alkaline pH shift to solubilize and refold.* Singh and Panda (2005) argue that proteins in IBs are largely folded and can more effectively be refolded if not subjected to high concentrations of GuHCl or urea. They state that IB protein (recombinant human growth hormone) was solubilized efficiently at pH 12.5 in 2 *M* urea and a concentration of 2 mg/ml, but retained significant secondary structure. The solubilized protein was then diluted into 2 *M* urea, pH 8 with a 40% recovery of activity.

8. Refolding Database: Refold

A database of protein refolding information is available at http://refold.med.monash.edu.au.

This database, developed by S. Bottomley and his colleagues at Monash University, contains over 1000 proteins that have been successfully refolded and presents protocols and statistics on the frequency of use of various refolding techniques, disruption methods, fusion proteins, and preparation prior to refolding (Chow *et al.*, 2006). For example, almost 85% of the examples utilize some form of dilution or dialysis to remove denaturant, while about 12% utilize on-column refolding.

9. Strategies to Increase Proportion of Soluble Protein

The standard method for determining the proportion of soluble and insoluble material is to prepare a cell lysate and then centrifuge to sediment insoluble material (unbroken cells, cell debris, and IBs). The total (crude lysate), soluble (supernatant), and insoluble (pellet) fractions are analyzed by SDS–PAGE and the intensity of the target protein band in the three fractions is compared by stain intensity or by Western blot analysis. If the vast majority of the target protein is in the soluble fraction, then you say that the protein is soluble. If the majority is in the pellet, then you say the protein is insoluble. Often target protein is in both soluble and insoluble fractions. While this is often an accurate estimate of the proportion of the protein that is soluble, I have seen two cases where the results can be misleading. The first is when cell breakage is incomplete and the SDS gel pattern of the pellet looks much like that of the lysate. This suggests that more effective lysis conditions be used (see Chapter 18 in this volume). Sometimes one sees significant amounts of target protein in the soluble fraction due to soluble multimers or to ineffective pelleting of IBs. If sonication is too vigorous, sometimes the IBs are broken up into smaller particles that may not be completely sedimented. This can be remedied by longer, harder centrifugations. Also sedimentation of IBs can be incomplete if the viscosity of the lysate is high due to the presence of too much large DNA. Treatment of the lysate with a nuclease such as Benzonase can decrease this viscosity.

Even though I believe it is possible to find effective refolding conditions for most expressed proteins, many researchers strongly prefer to decrease IB formation and increase the proportion of soluble product if possible. In many expression systems so much protein is produced that even if only 20% of the protein is soluble there can be quite significant amounts of protein available for purification by more standard methods. Some of the strategies for increasing the amount of soluble product are listed below (see Schein, 1989; Chapters 11 and 12 in this volume).

1. *Induce overproduction at lower temperatures, for example, 20 °C.* It was observed in the late 1980s (Schein, 1989) that one could increase the proportion of soluble material by inducing the target protein overexpression in *E. coli* grown below the normal growth temperature of 37 °C. Temperatures of 20–25 °C are commonly used, although cell growth is quite slow at the lower temperature. The basic rationale seems to be that at the lower temperature, transcription and translation are slower, so that proteins have more time to refold into native structures and the internal concentration of the partially folded, sticky intermediate is lower, resulting in less aggregation, and IB formation (see Vera *et al.*, 2006).
2. *Add 0.4 M sucrose to the culture medium.* This approach has been reported by Bowden and Georgiou (1988). This phenomenon may be due to the high osmolyte concentration inducing the osmotic shock response, which increases the internal level of glutamate, proline, and trehalose, thus providing conditions more favorable for refolding.
3. *Coexpression of molecular chaperones.* Since molecular chaperones such as DnaK/J and GroEL/ES are known to suppress misfolding and aggregation in the cell, one can increase the levels of these proteins by introducing a separate plasmid that contains inducible genes for these chaperones.
4. *Subject cells to brief heat shock at 42 °C, then shift to 20 °C, and induce.* This is an elegant approach that avoids the need to coexpress chaperones as above, but rather takes advantage of the natural heat shock response where DnaK/J and GroEL/ES levels increase. A reasonable procedure is to grow cells at 35 °C to $A_{600\ nm}$ of about 0.5, shift the cells rapidly to 42 °C for 15–20 min to induce the heat shock response, decrease the temperature to 20–25 °C, and then induce expression of target protein for 4–8 h.
5. *Coexpression of several subunits of a complex.* It has been observed that if two proteins (say protein X and Y) that form a stable heterodimer are expressed in *E. coli* individually they are mostly insoluble, but if they are coexpressed they refold, form their normal dimer (XY) and are soluble. Vectors for such coexpression are marketed by EMD/Novagen as the Duet vectors.

6. *Fuse to easily refolded protein*, for example, a "solubility tag" such as NusA, Trx, MBP, GST, SUMO, and HaloTag (see Chapter 16 in this volume).
7. *Form disulfide bonds in the E. coli cytoplasm.* The environment in the *E. coli* cytoplasm is reducing, preventing most disulfide bond formation. Mutations in the glutathione reductase (*gor* gene) and thioredoxin reductase (*trxB* gene) enhance formation of disulfide bonds in the *E. coli* cytoplasm (Bessette *et al.*, 1999). Strains AD494 (DE3) and BL21trxB(DE3) are *trxB* deficient. Novagen Origami strains are *trxB* and *gor* deficient.

10. Conclusion

Recombinant proteins overexpressed in *E. coli* often form insoluble IBs in the cell. The IBs are easy to purify and solubilize, but finding suitable conditions for efficient refolding of the solubilized protein is sometimes difficult. The best general strategy seems to be to screen many different refolding conditions in parallel to find ones that give the highest recovery of enzyme activity or the lowest turbidity due to aggregation and that show the least amount of soluble multimers. Many additives may prove useful in refolding, but the surest way to prevent aggregation and precipitation upon refolding is to refold at low protein concentration.

References

Anfinsen, C. B. (1973). Principles that govern the folding of protein chains. *Science* **181,** 223–230.

Anthony, L. C., Dombkowski, A. A., and Burgess, R. R. (2002). Using disulfide bond engineering to study conformational changes in the β'260–309 region of *E. coli* RNA polymerase β' during σ70 binding. *J. Bacteriol.* **184,** 2634–2641.

Arakawa, T., Ejima, D., Tsumoto, K., Obeyama, N., Tanaka, Y., Kita, Y., and Timasheff, S. N. (2007). Suppression of protein interactions by arginine: A proposed mechanism of the arginine effects. *Biophys. Chem.* **127,** 1–8.

Armstrong, N., De Lencastre, A., and Gouaux, E. (1999). A new protein folding screen: Application to the ligand binding domains of a glutamate and kainite receptor and to lysozyme and carbonic anhydrase. *Protein Sci.* **8,** 1475–1483.

Baneyx, F., and Mujacic, M. (2004). Review. Recombinant protein folding and misfolding in *E. coli. Nat. Biotechnol.* **22,** 1399–1408.

Baynes, B. M., Wang, D. I. C., and Trout, B. L. (2005). Role of arginine in the stabilization of proteins against aggregation. *Biochemistry* **44,** 4919–4925.

Bessette, P. H., Åslund, F., Beckwith, J., and Georgiou, G. (1999). Efficient folding of proteins with multiple disulfide bonds in the *E. coli* cytoplasm. *Proc. Natl. Acad. Sci. USA* **96,** 13703–13708.

Bowden, G. A., and Georgiou, G. (1988). The effect of sugars on β-lactamase aggregation in *E. coli. Biotech. Prog.* **4,** 97–101.

Bowden, G. A., Paredes, A. M., and Georgiou, G. (1991). Structure and morphology of protein inclusion bodies in *E. coli. Biotechnology (NY)* **9,** 725–730.

Burgess, R. R. (1996). Purification of overproduced *E. coli* RNA polymerase sigma factors by solubilizing inclusion bodies and refolding from Sarkosyl. *Meth. Enzymol.* **273,** 145–149.

Burgess, R. R., and Knuth, M. (1996). Purification of a recombinant protein overproduced in *E. coli*. *In* "Strategies for Protein Purification and Characterization: A Laboratory Manual", (D. Marshak, J. Kadonaga, R. Burgess, M. Knuth, W. Brennan Jr., and S.-H. Lin, eds.), pp. 205–274. Cold Spring Harbor Press, New York.

Cabrita, L. D., and Bottomley, S. P. (2004). Protein expression and refolding—A practical guide to getting the most out of inclusion bodies. *Biotechnol. Annu. Rev.* **10,** 31–54.

Chow, M. K., Amin, A. A., Fulton, K. F., Whisstock, J. C., Buckle, A. M., and Bottomley, S. P. (2006). REFOLD: An analytical database of protein refolding methods. *Protein Expr. Purif.* **46,** 166–171.

Clark, E. D. (2001). Protein refolding for industrial processes. *Curr. Opin. Biotechnol.* **12,** 202–207.

Clark, P. L. (2004). Protein folding in the cell: Reshaping the folding funnel. *Trends Biochem. Sci.* **29,** 527–534.

Cleland, J. L., Builder, S. E., Swartz, J. R., Winkler, M., Chang, J. Y., and Wang, D. I. (1992). Polyethylene glycol enhanced protein refolding. *Biotechnology* **10,** 1013–1019.

Cowieson, N. P., Wensely, B., Listwan, P., Hume, D. A., Kobe, B., and Martin, J. L. (2006). An automatable screen for the rapid identification of proteins amenable to refolding. *Proteomics* **6,** 1750–1757.

Das, U., Hariprasad, G., Ethayathulla, A. S., Manral, P., Das, T. K., Pasha, S., Mann, A., Ganguli, M., Verma, A. K., Bhat, R., Chandrayan, S. K., Ahmed, S., Sharma, S., Kaur, P., Singh, T. P., and Srinivasan, A. (2007). Inhibition of protein aggregation: Supramolecular assemblies of arginine hold the key. *PLoS ONE* **2,** e1176.

Dong, X.-Y., Huang, Y., and Sun, Y. (2004). Refolding kinetics of denatured-reduced lysozyme in the presence of folding aids. *J. Biotechnol.* **114,** 135–142.

Expert-Bezancon, N., Rabilloud, T., Vuillard, L., and Goldberg, M. E. (2003). Physico-chemical features of non-detergent sulfobetaines active as protein folding helpers. *Biophys. Chem.* **100,** 469–479.

Gilbert, H. F. (1994). The formation of native disulfide bonds. *In* "Mechanisms of Protein Folding", (R. H. Pain, ed.), pp. 104–136. IRL Press, Oxford.

Graslund, S., *et al.* (2008). Protein production and purification. *Nat. Methods* **5,** 135–146.

Gribskov, M., and Burgess, R. R. (1983). Overexpression and purification of the sigma70 subunit of *E. coli* RNA polymerase. *Gene* **26,** 109–118.

Ignatova, Z., and Gierasch, L. M. (2006). Inhibition of protein aggregation *in vitro* and in vivo by a natural osmoprotectant. *Proc. Natl. Acad. Sci. USA* **103,** 13357–13361.

Jungbauer, A., and Kaar, W. (2006). Review. Current status of technical protein refolding. *J. Biotechnol.* **128,** 587–596.

Jungbauer, A., Kaar, W., and Schlegl, R. (2004). Folding and refolding of proteins in chromatographic beds. *Curr. Opin. Biotechnol.* **15,** 487–494.

Karplus, M. (1997). The Levinthal paradox: Yesterday and today. *Fold. Des.* **2,** S69–S75.

Kersteen, E. A., and Raines, R. T. (2003). Minireview. Catalysis of protein folding by protein disulfide isomerase and small-molecule mimics. *Antioxid Redox Signal.* **5,** 413–424.

Knuth, M. W., and Burgess, R. R. (1987). Purification in the denatured state. *In* "Protein Purification: Micro to Macro", (R. R. Burgess, ed.), pp. 297–305. Alan R. Liss, New York.

Lange, C., Patil, G., and Rudolph, R. (2005). Ionic liquids as refolding additives: N'-alkyl and N'-(ω-hydroxyalkyl) N-imidazolium chlorides. *Protein Sci.* **14,** 2693–2701.

Levinthal, C. (1968). Are there pathways for protein folding? *J. Chim. Phys.* **65,** 44–45.

Makrides, S. C. (1996). Strategies for achieving high-level expression of genes in *E. coli*. *Microbiol. Rev.* **60,** 512–538.
Marston, F. A. O., and Hartley, D. L. (1990). Solubilization of protein aggregates. *Meth. Enzymol.* **182,** 264–276.
Middelberg, A. P. J. (2002). Preparative protein refolding. *Trend Biotechnol.* **20,** 437–443.
Murby, M., Uhlen, M., and Stahl, S. (1996). Review. Upstream strategies to minimize proteolytic degradation upon recombinant production in *E. coli*. *Protein Expr. Purif.* **7,** 129–136.
Nall, B. (1994). Proline isomerization as a rate limiting step. *In* "Mechanisms of Protein Folding", (R. H. Pain, ed.), pp. 80–103. IRL Press, Oxford.
Nguyen, L., Jensen, D. B., and Burgess, R. R. (1993). Overproduction and purification of sigma32, the *E. coli* heat-shock transcription factor. *Protein Expr. Purif.* **4,** 425–433.
Oganesyan, N., Kim, S.-H., and Kim, R. (2005). On-column protein refolding for crystallization. *J. Struct. Funct. Genomics* **6,** 177–182.
Pace, C. N. (1986). Determination and analysis of urea and guanidine hydrochloride denaturation curves. *Meth. Enzymol.* **131,** 266–280.
Panda, A. K. (2003). Bioprocessing of therapeutic proteins from the inclusion bodies of *E. coli*. *Adv. Biochem. Eng. Biotechnol.* **85,** 43–93.
Paul, S., Punam, S., and Chaudhuri, T. K. (2007). Chaperone-assisted refolding of *E. coli* maltodextrin glucosidase. *FEBS J.* **274,** 6000–6010.
Puri, N. K., Crivelli, E., Cardamone, M., Fiddes, R., Bertolini, J., Ninham, B., and Brandon, M. R. (1992). Solubilization of growth hormone and other recombinant proteins from *E. coli* inclusion bodies by using a cationic surfactant. *Biochem. J.* **285,** 871–879.
Quronfleh, M. W., Hesterberg, L. K., and Seefeldt, M. B. (2007). Review. Confronting high-throughput protein refolding using high pressure and solution screens. *Protein Expr. Purif.* **55,** 209–224.
Reddy, R. C. K., Lilie, H., Rudolph, R., and Lange, C. (2005). L-Arginine increases the solubility of unfolded species of hen egg white lysozyme. *Protein Sci.* **14,** 929–935.
Rozema, D., and Gellman, S. H. (1996). Artificial chaperone-assisted refolding of carbonic anhydrase B. *J. Biol. Chem.* **271,** 3478–3487.
Schein, C. H. (1989). Production of soluble recombinant proteins in bacteria. *Biotechnology* **7,** 1141–1147.
Schein, C. H. (1991). Optimizing protein refolding to the native state in bacteria. *Curr. Opin. Biotechnol.* **2,** 746–750.
Shimamoto, N., Kasciukovich, T., Nagai, H., and Hayward, R. S. (1998). Efficient solubilization of proteins overproduced as inclusion bodies by use of an extreme concentration of glycerol. *J. Tech. Tips Online* **23,** t01576.
Singh, S. M., and Panda, A. K. (2005). Review. Solubilization and refolding of bacterial inclusion body protein. *J. Biosci. Bioeng.* **99,** 303–310.
Song, J. (2009). Insight into insoluble proteins with pure water. *FEBS Lett.* **583,** 953–959.
Sorensen, H. P., and Mortensen, K. K. (2005). Review. Advanced genetic strategies for recombinant protein expression in *E. coli*. *J. Biotechnol.* **115,** 113–128.
Studier, F. W., Rosenberg, A. H., Dunn, J. J., and Dubendorff, J. W. (1990). Use of T7 RNA polymerase to direct expression of cloned genes. *Meth. Enzymol.* **185,** 60–89.
Swietnicki, W. (2006). Folding aggregated proteins into functionally active forms. *Curr. Opin. Biotechnol.* **17,** 367–372.
Thatcher, D. R., and Hitchcock, A. (1994). Protein folding in biotechnology. *In* "Mechanisms of Protein Folding", (R. H. Pain, ed.), pp. 229–261. IRL Press, Oxford.
Trésaugues, L., Collinet, B., Minard, P., Henkes, G., Aufrère, R., Blondeau, K., Liger, D., Zhou, C.-Z., Janin, J., and van Tilbeurgh, H. (2004). Refolding strategies from inclusion bodies in a structural genomics project. *J. Struct. Funct. Genomics* **5,** 195–204.

Tsumoto, K., Ejima, D., Kumagai, I., and Arakawa, T. (2003). Practical considerations in refolding proteins from inclusion bodies. *Protein Expr. Purif.* **28,** 1–8.
Vallejo, L. F., and Rinas, U. (2004). Review. Strategies for the recovery of active proteins through refolding of bacterial inclusion body proteins. *Microb. Cell Fact.* **3,** 11–22.
Veldkamp, C. T., Person, F. C., Hayes, P. L., Mattmiller, J. E., Haugner, J. C. III, de la Cruz, N., and Volkman, B. F. (2007). On-column refolding of recombinant chemokines for NMR studies and biological assays. *Protein Expr. Purif.* **52,** 202–209.
Ventura, S., and Villaverde, A. (2006). Protein quality in bacterial inclusion bodies. *Trends Biotechnol.* **24,** 179–185.
Vera, A., Gozalez-Montalban, N., Aris, A., and Villaverde, A. (2006). The conformation quality of insoluble recombinant proteins is enhanced at low growth temperatures. *Biotechnol. Bioeng.* **96,** 1101–1106.
Vincentelli, R., Canaan, S., Campanacci, V., Valencia, S., Maurin, D., Frassinetti, F., Scappucini-Calvo, L., Bourne, Y., Cambillau, C., and Bignon, C. (2004). High-throughput automated refolding screening of inclusion bodies. *Protein Sci.* **13,** 2782–2792.
Willis, M. S., Hogan, J. K., Prabhakar, P., Liu, X., Tsai, K., Wei, Y., and Fox, T. (2005). Investigation of protein refolding using a fractional factorial screen: A study of reagent effects and interactions. *Protein Sci.* **14,** 1818–1826.
Woycechowsky, K. J., Wittrup, K. D., and Raines, R. T. (1999). A small-molecule catalyst of protein refolding *in vitro* and *in vivo*. *Chem. Biol.* **6,** 871–879.
Xie, Y., and Wetlaufer, D. B. (1996). Control of aggregation in protein refolding: The temperature-leap tactic. *Protein Sci.* **5,** 517–523.

SECTION FOUR

PREPARATION OF EXTRACTS AND SUBCELLULAR FRACTIONATION

CHAPTER EIGHTEEN

Advances in Preparation of Biological Extracts for Protein Purification

Anthony C. Grabski

Contents

1. Introduction 286
2. Chemical and Enzymatic Cell Disruption 287
3. Mechanical Cell Disruption 290
4. Concluding Remarks 293
5. Procedures, Reagents, and Tips for Cell Disruption 293
   5.1. Buffer composition 293
   5.2. Cell disruption buffers 296
   5.3. Microscale protocols for *E. coli* cell lysis 296
   5.4. Gram-scale mechanical disruption of *E. coli* 299
References 301

Abstract

There are a variety of reliable methods for cellular disintegration and extraction of proteins ranging from enzymatic digestion and osmotic shock to ultrasonication, and pressure disruption. Each method has inherent advantages and disadvantages. Generally vigorous mechanical treatments reduce extract viscosity but can result in the inactivation of labile proteins by heat or oxidation, while gentle treatments may not release the target protein from the cells, and resulting extracts are extremely viscous. Depending on the cell type selected as the source for target protein expression, cellular extracts contain large amounts of nucleic acid, ribosomal material, lipids, dispersed cell wall polysaccharide, carbohydrates, chitin, small molecules, and thousands of unwanted proteins. Isolation and recovery of a single protein from this complex mixture of macromolecules presents considerable challenges. The first and possibly most important of these challenges is generation of a cellular extract that can be efficiently manipulated in downstream processes without inactivation or degradation of labile protein targets. Cell disruption techniques must rapidly and efficiently lyse cells to extract proteins with minimal proteolysis or oxidation while reducing extract viscosity caused by cell debris and genomic DNA contamination.

Department of Research and Development, Semba Biosciences, Inc., Madison, Wisconsin, USA

Methods in Enzymology, Volume 463
ISSN 0076-6879, DOI: 10.1016/S0076-6879(09)63018-4

Advanced bioprocessing equipment and reagents have been developed over the past twenty years to complement established disruption procedures and accomplish these tasks with even greater success. This chapter will summarize these advances and describe detailed protocols for some of the most popular methods for protein extraction.

1. Introduction

Early protein chemists worked primarily with small extracellular proteins that were stable, plentiful, and easily isolated without cell disruption. Intracellular proteins expressed at physiological levels might constitute <0.001% of the total cellular protein and were difficult to extract and recover without proteolysis or loss of enzymatic activity. Modern recombinant protein production through genetic engineering of *Escherichia coli* can result in expression of the target protein at over 40% of the cellular total. Recombinant yeasts such as *Saccharomyces cerevisiae, Pichia pastoris*, and *Kluvyeromyces lactis* are capable of expressing complex proteins requiring posttranslational modifications (Spencer and Spencer, 1997). Recovery of the intracellular recombinant proteins from these hosts requires effective cell disruption and extraction techniques. Well established protocols have been reviewed and detailed procedures described for laboratory and process scale disruption of various cell types by both mechanical and nonmechanical methods (Andrews and Asenjo, 1987; Cull and McHenry, 1990; Dignam, 1990; Engler, 1985; Gegenheimer, 1990; Harrison, 1991; Hopkins, 1991; Jazwinski, 1990; Middelberg, 1995). Radical changes to standard disruption equipment and procedures had not occurred up to 1995 (Foster, 1995). However, the evolution of structural and functional proteomics demanded new reagents and automated methods to streamline the process steps converting gene sequences to purified proteins (Grabski *et al.*, 2002).

Proteomics and structural genomics efforts require cell extraction methods that allow screening of multiple host vector combinations for hundreds of proteins in parallel (Stevens *et al.*, 2001). The best combinations for correctly folded high-level expression identified in these multiparallel screens are scaled up in order to produce milligram quantities of purified protein for functional and structural analysis (Dieckman *et al.*, 2002; Lesley, 2001; Stevens and Wilson, 2001; Zhu *et al.*, 2001). These high-throughput (HT) protein expression and purification strategies have fostered development and optimization of reagents and instrumentation that allow efficient cell lysis and protein extraction at both micro- and macroscales.

2. Chemical and Enzymatic Cell Disruption

Micro-scale cell disruption is often accomplished through chemical and enzymatic methods or combinations of the two. Lysis methods such as sonication and French press are not easily applied to cell pellets harvested from 5 ml of culture or less, and excess heat and oxidation are common problems. A significant advancement for simplification of microscale cell lysis and has been the development of detergent-based reagents such as B-Per® (Chu *et al.*, 1998) and BugBuster® (Grabski *et al.*, 1999). These reagents do not require expensive equipment, are very fast and easy to use, and are the most practical and effective means to HT cell extract preparation. Highly active enzymes and enzyme mixtures have also been developed to improve lysis efficiencies and decrease extract viscosity by completely digesting genomic DNA. Used alone and in combinations these commercially available lysis reagents and enzymes (Table 18.1) efficiently disrupt cells and extract proteins from bacteria, yeasts, plants, insect cells, and higher eukaryotes. Detergent-based methods for extraction and enrichment of proteins and subcelluar organelles from eukaryotic cells and tissues are covered extensively in chapter 19 by Michelsen and von Hagen of this volume.

Another advancement in chemical cell disruption tailored for HT robotic processing of samples, is the development of reagents for extraction of recombinant proteins from *E. coli* without cell harvest from the culture media (Grabski *et al.*, 2001, 2003; Stevens and Kobs, 2004). Traditional protein purification methods require an initial cell harvest step, which concentrates cell mass and removes spent media components (Burgess, 1987; Deutscher, 1990; Scopes, 1994). This step, as well as any mechanical lysis step is difficult to automate and scale down for parallel extraction and purification of small amounts of proteins simultaneously. Concentrated, detergent-based reagents (Table 18.1) such as PopCulture™, FastBreak™, and B-PER® Direct, combined with high-activity lysozyme and nuclease overcome these bioprocessing hurdles to automation. The advantages of these reagents for HT extraction and purification are eliminating the need for multiple centrifugation steps to separate cells from culture media and clarify the crude extract, and eliminating the need for mechanical disruption (Nguyen *et al.*, 2004). These innovations allow direct affinity adsorption of target proteins from the total culture extract, and the entire cell growth, extraction, and purification can be done in a single tube or well.

The active ingredients in many of the bacterial lysis reagents are nonionic or zwitterionic detergents functioning to disrupt cell membrane and cell wall structures weakening the cells for rupture by osmotic shock,

Table 18.1 Commercial cell disruption reagents and enzymes

Vendor	Reagents	Cell types	Vendor web site
EMD Chemicals-Novagen	BugBuster®, YeastBuster™, CytoBuster™ PopCulture® rLysozyme™, Benzonase®, Lysonase™ M-Pek, S-Pek	Bacteria, yeast, insect, mammalian	www.emdbiosciences.com
Epicentre	EasyLyse™, OmniCleave™, ReadyLyse™	Bacteria	www.epicentre.com
Promega	FastBreak™	Bacteria	www.promega.com
Roche	Complete™ Lysis (B, M, Y)	Bacteria, mammalian, yeast	www.roche-applied-science.com
Semba Biosciences, Inc.	Recombinant Lysozyme, Benzonase®, Liquisonic™	Bacteria	www.sembabio.com
Sigma Aldrich	CelLytic™ (B, IB, M, MEM, MT, NuClear™, P, PN, and Y)	Bacteria, mammalian, plant, yeast	www.sigmaaldrich.com
Thermo Fisher-Pierce	POP-PERS® (B-, I-, M-, NE-,P-, and Y-Per®)	Bacteria, insect, mammalian, plant, yeast	www.thermofisher.com

freeze-thaw, or enzymatic attack by phage or chicken egg white lysozyme. Gram positive bacteria are easily disrupted by lysozyme treatment alone, with a few exceptions (Cull and McHenry, 1990). Gram negative bacteria are difficult-to-disrupt with lysozyme alone because their outer lipid bilayer must be permeabilized first to expose the peptidoglycan cell wall to lysozyme digestion. Tris buffer plus EDTA liberates about 50% of the polyanionic lipopolysaccharide in the bilayer (Lieve, 1974), but EDTA interferes with the most popular downstream purification step for recombinant proteins, immobilized metal affinity chromatography (IMAC). The detergent lysis reagents are extremely effective for outer membrane permeabilization without IMAC interference. However, detergent-based reagent cell disruption has several disadvantages. Solubilities for some proteins extracted with detergents may be enhanced by detergent–protein interactions, and protein solubilities will change depending on the detergent–protein ratio. The solubility of a particular protein extracted by these reagents may not correlate with the protein's solubility as extracted by scale-up mechanical methods. Detergents and detergent impurities can interfere with downstream processing and structural characterization (Swiderek *et al.*, 1997), and fractionation methods that depend on hydrophobic interaction should be avoided (Marshak *et al.*, 1996). Despite these drawbacks, detergent-based lysis reagents are widely used in modern proteomics. The exact concentrations and types of detergents employed in these lysis reagents are proprietary and the cellular architecture to be disrupted dictates the chemical properties and concentration of detergent required for the task. Several references from this chapter (Neugebauer, 1990; Chu and Mallia, 2001; Eshaghi *et al.*, 2005; Kashino, 2003) on applications of detergents for protein extraction and solubilization provide a wealth of information allowing home brewed extraction reagents to be formulated with reasonable success.

Yeasts are more difficult-to-disrupt than bacteria. Their thick complex cell wall can compose up to 25% of the cells' dry weight. Typical components are glucans, cellulose, mannoproteins, and chitin interconnected by covalent, disulfide, hydrogen, and hydrophobic bonds. Yeast cells should be harvested in late log or early stationary phase, for more efficient disruption. Yeasts cultured longer into stationary phase develop thicker cell walls and multiple budding scars or aggregates in the case of budding yeast (Catley, 1988). Protease inhibitors should be included in lysis buffers or added to commercial lysis reagents and protease deficient strains used for expression. Yeast lysis with reagents such as Y-Per® and YeastBuster™ (Table 18.1) and lysis by controlled manipulation of genes involved in cell wall biogenesis have provided new tools to complement previously developed mechanical and enzymatic methods (Drott *et al.*, 2002).

Enzymatic treatment of microbial cells for lysis and viscosity reduction has several distinct advantages over mechanical and chemical methods. The lytic enzymes are highly specific for targeted cell wall components, they are gentle and do not generate shear, high temperature or oxidative damage, and they are simple to use requiring no specialized equipment for processing. Enzyme treatments of cells and extracts are also combined with mechanical disruption methods to increase the selectivity of product release, to increase the rate and yield of extraction, to minimize product damage, and to reduce viscosity for downstream processes (Andrews and Asenjo, 1987; Grabski *et al.*, 1999). Improved bioprocessing enzymes and enzyme mixtures (Table 18.1) include high-specific activity phage lysozymes such as rLysozyme™, Ready-Lyse™, and Recombinant Lysozyme, nonspecific nucleases like Benzonase® and OmniCleave™, and the lysozyme-nuclease cocktails Liquisonic™, Lysonase™, and EasyLyse™. Recombinant Lysozyme, Ready-Lyse™ and rLysozyme™ phage lysozymes have specific activities over 200-fold greater than that of chicken egg white lysozyme. The nuclease of *Serratia marcesens*, available as a highly purified recombinant enzyme Benzonase®, is one of the most promiscuous nucleases known. The enzyme cleaves all forms of DNA and RNA (Meiss *et al.*, 1995; Nestle and Roberts, 1969). Compared to bovine pancreatic DNase I, Benzonase attacks substrates more evenly making it an excellent choice for reducing extract viscosity caused by nucleic acids. Zymolase® and Lyticase glucanases are enzymes useful for yeast protoplasting or yeast cell lysis.

Potential problems with enzymatic treatment limit its use especially for industrial cell disruption. These problems include the added lytic enzyme and impurities in the enzyme preparation that complicate downstream purification, degradation of the recovered product during lysis, temperature, and pH optima for the lytic enzyme may be incompatible with the target protein, and the expense and limited availability of most lytic enzymes prohibits their use at an industrial scale (Hopkins, 1991).

3. Mechanical Cell Disruption

Sonication and high-pressure disruption have been effectively applied for disruption of microorganisms, plants, and animal cells. These methods have been used for decades and the equipment and mechanisms involved described in detail (Harrison, 1991; Hopkins, 1991; Middelberg, 1995). Sonication is based on the shear forces created by high frequency ultrasonic vibrations generated by resonation (15–25 kHz) of a tuned probe or horn. The sonic pressure waves created cause the collapse of formed microbubbles and their implosion generates shock waves with sufficient energy to disrupt cell walls, and reduce viscosity by shearing nucleic acids.

High-pressure homogenizers and pressure extruders operate by forcing a pressurized cell suspension through a narrow orifice valve. The valve may be a simple restricting needle valve as in the French Press or a more complex design with a combined valve seat and impact ring found in the APV Manton-Gaulin homogenizer. The mechanism for disruption is a combined pressure drop and shear at the nozzle in pressure extrusion. Numerous mechanisms for cell disruption with pressure homogenizers have been proposed including turbulence, cavitation, viscous shear, and impingement (Kleinig and Middelberg, 1998; Pandolfe, 1999). Universal acceptance for a single mechanism has not been reached. However, cavitation and impingement have been reported as the major forces responsible for disruption (Shirgaonkar *et al.*, 1998; SPX Corp., 2009).

Cell disruption using glass beads to grind the cells in suspension, also known as bead milling or bead homogenization, is a frequently used procedure at both laboratory and production scale. Bead milling can be accomplished with laboratory equipment as simple as a magnetic stirrer, vortex mixer, or blender and with commercial equipment specialized for the process in the form of high speed mills, agitators, and mixers. The cell disintegration is dependent on cell concentration, bead size and composition, ratio of beads to suspension, processing duration, and the forces applied (Ramanan *et al.*, 2008). The method is effective with difficult-to-disrupt cells like yeasts, spores, and microalgae. It is used for bacteria, plant and animal cells and is a preferred method for large-scale disruption of fungi (Hopkins, 1991). The mechanism and models of the bead milling process have been reviewed (Harrison, 1991; Middelberg, 1995), but basically the cells are crushed, ground, and torn apart by abrasive contact and shear forces created between the beads, cells, and reaction chamber itself.

The specialized equipment necessary for mechanical cell disruption techniques has undergone functional innovation and new instruments have been developed. The majority of mechanical methods are not easily applied to cell pellets harvested from 5 ml of culture or less, and excess heat and oxidation are common problems with mechanical disruption. However, 96-well sonicator heads and microplate horns are available (Misonix, Inc. Farmingdale, NY; www.misonix.com) and for cases where traces of detergent in the purified protein preparation might interfere with biochemical characterization or crystallography, this HT physical method is a viable option. The SonicMan high-throughput sonication system (MatriCal, Inc. Spokane, WA; www.matrical.com) is a stand alone or integrated unit for 96-, 384-, and 1536-well plate sonications. This touch screen controlled microplate sonicator uses disposable gasketed pin lids to prevent well-to-well cross-contamination and has a plate shuttle allowing direct integration into robotics platforms. Another new sonicator available from BioSpec Products is the cordless handheld Sonozap Ultrasonic Homogenizer. The 1/8 in. diameter auto-tuned probe is ideal for small samples of 0.3–5.0 ml. Pressure Biosciences, Inc. (www.pressurebiosciences.com) has developed the Barocycler^{TM}, bench top, and PCT Shredder, hand held, sample preparation systems. These instruments are capable of rapid, high-pressure (up to 35 kpsi) cycling using specialized PULSE™ tubes. The 1.2–1.5 ml tubes contain a ram and pored lysis disk that facilitate the hydrostatic pressure induced lysis of plant, animal, insect, or microbial cells (Garrett *et al.*, 2002). The FastPrep (MP Biomedicals, Irvine, CA; www.mpbio.com), Geno/Grinder (SPEX Certiprep, Inc., Metuchen, NJ; www.spexcsp.com) MagNA Lyser (Roche Diagnostics, Penzberg, Germany; www.roche.com), Mikro-Dismembrator (Sartorius Stedim Biotech, Aubagne, France; www.sartorius-stedim.com), Mini-Bead Beater (BioSpec Products, Bartlesville, OK; http://www.biospec.com), and Retsch Mixer Mill (Retsch GmbH, Haan, Germany; www.retsch.com) are all shaking-type bead mills that process multiple milliliter size samples. These devices are not truly HT but are capable of high-yield disruption results from recalcitrant cells in 1–5 min.

The Microfluidics Microfluidizer (Newton, MA; www.microfluidicscorp.com) differs from other high-pressure homogenizers. The instrument pressurizes and accelerates split streams of the cell suspension by gas driven pumping of the liquid through fixed-geometry micro-channels. The two high-velocity streams directly impinge on each other creating very high shear forces and pressure drop as the suspension exits the device. Microfluidizer models range from the M-110P capable of processing small sample volumes of 25 ml and ideal for laboratory use to the M-700 biopharmaceutical process models with up to 900 l/h throughput, complete process control monitoring, CIP, and validatable under 21 CFR to cGMP. Constant Systems Ltd. (Low March Daventry, Northants, England; www.constantsystems.com) has designed hydraulically operated disruption instruments that function similar to the original French press by exerting the disruptive forces of pressure extrusion but under controlled, contained and repeatable conditions. The Constant Systems instruments are available in single shot bench (1–20 ml) to continuous process scale (405–565 ml/min.) models. Useful features include touch screen monitoring and control, cooling jacket, and CIP/SIP processes. Avestin, Inc. (www.avestin.com) offers high-pressure EmulsiFlex homogenizers capable of cell disruption from 1.0–1000 l/h. The homogenizers generate disrupting pressure from 500 to 30,000 psi using either gas or electrically driven piston pumps. They can be equipped with heat exchangers for cooling cell extracts and are SIP sterilizable for GMP manufacturing. The BioNeb cell disruption systems of Glas-Col, LLC (Terre Haute, IN; www.glascol.com) employ nebulizing gas pressures from 10 to 250 psi to break open cells without heat generation. Laminar flow in the nebulization channel creates cell disrupting shear forces. The magnitude of the force depends on the pressure applied, type of gas (argon $<$ nitrogen $<$ helium) and viscosity of the liquid. The smaller the nebulized droplets, the higher the viscosity, and the greater the gas pressure applied, the greater the shear force created (Surzycki *et al.*, 1996).

4. Concluding Remarks

The cell disruption methods, reagents, and instrumentation described in this chapter are useful and effective if appropriately modified for different cell types and scales, but the question of which method is superior for each application has certainly been asked. Studies have been conducted comparing methods of microbial disruption and protein quantitation in resulting extracts (Benov and Al-Ibraheem, 2002; De Mey *et al.*, 2008; Guerlava *et al.*, 1998; Ho *et al.*, 2008). The mechanical methods of high-pressure homogenization and bead milling are favorite methods at large scale because of their effectiveness and ability to rapidly handle various process volumes at low cost. Sonication, chemical reagents including detergents, enzymatic treatments, freeze-thaw, and enzymatic plus chemical or physical method combinations are also very effective and are used more frequently at laboratory and especially microscale. The uniqueness of proteins and differences in host cell structure force a high degree of empiricism in selection and optimization of cell disruption and extract preparation techniques. However, the tools and methods for success in biological extract preparation are available and have kept pace with the demands of modern structural and functional proteomics.

5. Procedures, Reagents, and Tips for Cell Disruption

The following strategies and guidelines serve as starting points to produce quality cellular extracts suitable for most downstream purification and analysis processes. Although the focus is on disrupting *E. coli* at micro- and macrolaboratory scale, many of the techniques can be applied to disrupting and isolating intracellular proteins from other sources and at larger scales. Comprehensive references for protein purification provide additional information on some of the guidelines for protein extract preparation outlined here as well as on many important aspects of protein purification and characterization (Burgess, 1987; Deutscher, 1990; Harris and Angal, 1989; Hopkins, 1991; Marshak *et al.*, 1996; Scopes, 1994).

5.1. Buffer composition

The components and volume of cell disruption buffers are critical not only for efficient disruption but will also affect subsequent purification steps and the target protein's stability and recovery after it has been released from the cells. Each protein extracted is an individual and ideally would have an extraction and purification buffer tailored specifically for its own biochemical requirements and intended direction through the purification pipeline. In most cases a more generic extraction buffer can be used with good results if a few basic criteria are met. These criteria are pH, ionic strength, additives to prevent degradation and improve stability, and buffer to cell paste ratio. A very good reference for maintenance of active enzymes through buffer composition and other means is Scopes (1994; "Protein Purification"). Technical information on pH and buffers including tables for preparation of pH 1–13 buffers, buffer properties, and the influence of salt, temperature, and dilution values can be found by Dawson *et al.* (1986).

The pH selected for the extraction buffer should be at least one pH unit above or below the protein's isoelectric point. This pH difference from the pI will prevent isoelectric precipitation by maintaining a positive or negative charge on the protein and also facilitate ion-exchange purification as a purification step. The buffer ionic strength should be 20–50 m*M* with a pK_a within 0.5 unit of the desired pH in order to maintain buffering capacity with minimal conductivity increase. The ionic strength inside the cytoplasm of a typical cell is 150–200 m*M* with very high concentrations of charged biomolecules for ionic protein interaction. The lysis buffer should contain at least 50–100 m*M* NaCl. Increased ionic strength of the extraction buffer will reduce these ionic interactions and precipitating losses from charged particulates that would adsorb protein and be removed by centrifugation or filtration steps. Finally, buffer components to prevent degradation and improve stability should be added as necessary. These include protease inhibitors (Table 18.2), reducing agents such as dithiothreitol (DTT), tris (2-carboxyethyl)phosphine (TCEP), or tris(hydroxypropyl)phosphine (THP), divalent cations, cofactors, and kosmotropes like glycerol, sorbitol, or trehalose. The water soluble odorless phosphines TCEP and THP are more stable and effective for the maintenance of reduced protein disulfide bonds than the more commonly used thiol reductants 2-mercaptoethanol and DTT (Cline *et al.*, 2004; Getz *et al.*, 1999; Han and Han, 1994). Nonionic or zwitterionic detergents may be added to increase solubility of hydrophobic proteins. Reducing agents, protease inhibitors, and detergents may interfere with some purification and detection methods and assays. The potential interference of these buffer components must be considered when selecting the chemicals and concentrations employed. A very versatile buffer (Buffer B) for bacterial cell disruption is 50 m*M* Tris/HCl or sodium phosphate pH 7.5–8.0, 50 m*M* NaCl, 5% glycerol, 0.5 m*M* EDTA, and 0.5 m*M* DTT. Buffer formulations recommended for yeast disruption (Buffer Y) and insect cell disruption (Buffer I) are given below.

The volume of buffer used for resuspension of cell pellets must be at least three times the volume of the original pellet for effective disruption and good recovery of the liquid fraction after removal of insoluble cell

Table 18.2 Protease inhibitors

Inhibitor	Proteases inhibited	Stock solution	Effective concentration
Aprotinin	Serine	10 mg/ml PBS	0.6–2.0 μg/ml
Benzamidine	Serine	50 mg/ml water	0.5–4.0 m*M*
E-64 (L- transepoxysuccinyl-leucylamido-[4-guanidino]butane)	Cysteine	5.0 mg/ml water	1.0–10 μ*M*
EDTA (ethylenediaminetetraacetic acid)	Metallo	1.9 g/10 ml water	1.0–10 m*M*
Leupeptin	Serine/cysteine	50 mg/11 ml water	10–100 μ*M*
PMSF (phenylmethylsulfonyl fluoride)	Serine	120 mg/ml water	0.1–1.0 m*M*
AEBSF (4-[2-aminoethyl] benzenesulfonyl fluoride hydrochloride)			
AEBSF is less toxic than PMSF			
Pepstatin A	Aspartic	5 mg/7.3 ml (1.0 m*M*) methanol or DMSO	1.0–10 μ*M*
Complete tablets[a]	Broad spectrum	Tablets	1X of multiple inhibitors
Inhibitor cocktails I–VII[b]	Broad spectrum	100X	1X of multiple inhibitors
Inhibitor mixes	Broad spectrum	100X	1X of multiple inhibitors
B, FY, G, HP, M, P[c]			

[a] www.roche.com
[b] www.emdchemicals.com
[c] www.serva.de

debris and precipitated material by centrifugation or filtration. Since approximately 50% of the volume of the insoluble residue will be trapped liquid, at least three volumes are required for greater than 85% liquid recovery. Proteins are more stable at high concentration, but highly concentrated extracts are difficult to process and aggregation can occur. Although a 3:1 ratio of disruption buffer to cell paste may be used to produce a more concentrated extract, 5–10 volumes of buffer are preferred and will yield more soluble protein and a less viscous extract.

5.2. Cell disruption buffers

Buffer B: 50 m*M* Tris/HCl (pH 8.0), 50 m*M* NaCl, 5% glycerol, 0.5 m*M* EDTA, 0.5 m*M* DTT or 1.0 m*M* THP, 0.5 m*M* benzamidine, 1.0 m*M* AEBSF.

Buffer I: 50 m*M* sodium phosphate (pH 7.5), 150 m*M* NaCl, 5% glycerol, 0.03% Brij 35, 0.5 m*M* EDTA, 0.5 m*M* DTT or 1.0 m*M* THP, 1.0% Triton X-100, 0.5 m*M* benzamidine, 1.0 μM pepstatin A, 1.0 m*M* AEBSF. Alternatively, to avoid nuclear lysis in insect cells, substitute 1.0% NP-40 for Triton X-100.

Buffer Y: 50 m*M* Tris/HCl (pH 8.0), 150 m*M* LiCl, 10% ethylene glycol, 0.5 m*M* EDTA, 2.0 m*M* THP, 0.1% Triton X-100, 0.5 m*M* benzamidine, 1.0 μM E-64, 1.0 μM pepstatin A, 1.0 m*M* AEBSF.

Note: Protease inhibitors may be omitted from or supplemented to (Table 18.2) these buffers depending on the sensitivity of the protein target to proteolysis, and the degree or class of inhibition required. The ionic strength of buffers I and Y may need to be reduced or appropriate dilution of the extract performed if the initial protein fractionation step is ion-exchange chromatography.

5.3. Microscale protocols for *E. coli* cell lysis

5.3.1. Detergent-based reagent lysis

Lysis reagent recipe for 10 ml solution: 10 ml B-Per or BugBuster, 20 μl Lysonase Bioprocessing Reagent. Additives such as EDTA, protease inhibitors, 5–10% glycerol, and reducing agents may be included in the lysis reagent as required to limit proteolysis and improve target protein solubility and stability. Downstream purification and analysis requirements must be considered when selecting lysis buffer components and their concentrations.

1. Culture and express the target proteins using 1.0 ml medium in 2 ml × 96 or 5 ml medium in 10 ml 24-deep well plates with air permeable sealing film or BugStopper^{TM} cap mats.
2. Pellet the cells by centrifugation.
3. Remove and discard the spent medium by aspiration.

4. Resuspend the cell pellets in 100–200 μl of lysis reagent by pipetting.
5. Mix the reaction on a shaking platform 10 min.
6. Remove 10 μl of the mixture for total cell extract analysis.
7. Centrifuge the mixture 5 min and remove a 10-μl sample of the supernatant for analysis of soluble proteins. The remaining clarified supernatant is ready for other fractionation or analysis procedures.
8. The amount of target protein present in the insoluble fraction may be estimated indirectly by differential comparison of the total cell extract in step 6 and the soluble protein in step 7. The samples are typically compared by electrophoresis, ELISA, or enzymatic assay. Direct analysis of the insoluble fraction can be done electrophoretically by aspiration of the soluble supernatant, solubilization of the pelleted fraction in sodium dodecyl sulfate polyacrylamide gel electrophoresis (SDS-PAGE) sample buffer, and SDS-PAGE.

Note: Procedures and tips for preparation of protein samples for SDS-PAGE are available in Grabski and Burgess (2001). The article provides details for SDS-PAGE sample buffer recipes, protein sample preparation, and gel loading recommendations, and gives alternatives for analysis of difficult samples.

5.3.2. Detergent-based reagent whole culture in-media cell lysis

Lysis reagent recipe for 10 ml solution: 10 ml PopCulture, FastBreak, or B-Per Direct, 200 μl Lysonase Bioprocessing Reagent. The following protocol can be performed on multiple plates using robotic liquid handling platforms, on single plates by multichannel pipette, and on individual cultures. Multiwell filtration plates are available from several manufacturers for separation of affinity resin from culture extract and processing solutions. Separation of magnetic particles is accomplished using specialized magnetic pin plates or platforms.

1. Culture and express the target proteins using 1.0 ml medium in 2 ml × 96 or 5.0 ml medium in 10 ml x 24-deep well plates with air permeable sealing film or BugStopper^{TM} cap mats.
2. Add 0.1 culture volume lysis reagent.
3. Pipette to mix the reaction and continue to mix on a shaking platform for 10 min.
4. Remove 50 μl of the mixture for total cell extract analysis. The sample can be analyzed by electrophoresis, ELISA or enzymatic assay. Direct SDS-PAGE analysis and Coomassie staining requires high-expression levels and maximum sample loading or sample concentration by precipitation to detect dilute protein. Screening soluble expression levels is easily performed at this step using specialized filter plates (Pall, Millipore, 3 *M*) capable of separating protein aggregates and inclusion bodies from soluble proteins.

5. Add equilibrated affinity capture resin or magnetic affinity particles directly to the crude cell lysate. Typically 50–100 μl/ml (50% particle slurry in capture buffer) of the original culture volume is added to the lysate depending on resin capacity for the recombinant protein and target protein expression level.
6. Wash the capture resin with 10–30 resin volumes of wash buffer to remove contaminants.
7. Elute the target protein with three resin volumes elute buffer.

5.3.3. Freeze-thaw plus enzymatic lysis

Lysis reagent recipe for 10 ml solution: 10 ml lysis buffer at 50–100 m*M* concentration (glycine, sodium acetate, Tris–HCl, sodium phosphate, or HEPES depending on pH desired), 50–300 m*M* NaCl, and 30 μl Lysonase Bioprocessing Reagent. Additives such as EDTA, detergents, protease inhibitors, 5–10% glycerol, and reducing agents may be included in the lysis buffer as required to limit proteolysis and improve target protein solubility and stability. Downstream purification and analysis requirements must be considered when selecting lysis buffer components and their concentrations.

Freeze-thaw is dependent on the density of the suspension, number of freeze-thaw cycles, and rate of freezing, but has been shown to be less effective for protein release from *E. coli* than bead vortexing, French press, or sonication (Benov and Al-Ibraheem, 2002). However, when combined with enzymatic lysis by lysozyme it is an effective and gentle method of protein release for labile proteins and does not require specialized equipment. The slow freeze-thaw cycle ruptures the cell membrane exposing the cell wall to enzymatic digestion.

1. Culture and express the target proteins using 1.0 ml medium in 2 ml × 96 or 5 ml medium in 10 ml × 24-deep well plates with air permeable sealing film or BugStopper™ cap mats.
2. Pellet the cells by centrifugation.
3. Remove and discard the spent medium by aspiration.
4. Freeze the cell pellets completely at − 20 °C.
5. Resuspend the cell pellets in 100–200 μl of lysis reagent by pipetting.
6. Mix the reaction on a shaking platform 10 min at room temperature.
7. Freeze the suspension again at − 20 °C.
8. Thaw at room temperature, mix, and remove 10 μl of the mixture for total cell extract analysis.
9. Centrifuge the mixture 5 min and remove a 10-μl sample of the supernatant for analysis of soluble proteins. The remaining clarified supernatant is ready for other fractionation or analysis procedures.
10. The amount of target protein present in the insoluble fraction may be estimated indirectly by differential comparison of the total cell extract

in step 8 and the soluble protein in step 9. The samples are typically compared by electrophoresis, ELISA or enzymatic assay. Direct analysis of the insoluble fraction can be done electrophoretically by aspiration of the soluble supernatant, solubilization of the pelleted fraction in SDS-PAGE sample buffer, and SDS-PAGE.

5.4. Gram-scale mechanical disruption of *E. coli*

5.4.1. Sonication

The sonication procedure below is convenient for cell pellets of 2–50 g.

1. Pellet the cells by centrifugation in tared centrifuge bottles or tubes at 9000×*g* for 15 min.
2. Decant and discard the spent medium and drain residual medium by inverting for a few minutes on paper towels. Record the cell pellet weights.
3. Freeze the cell pellets completely at − 20 °C and process the cells within 1 week for best results. Longer term storage of cell paste should be at − 80 °C to minimize proteolytic degradation. Fresh unfrozen cells may be used, but enzymatic pretreatment with lysozyme will be less effective because the cell membrane is intact.
4. Completely resuspend the cell pellets in 7–10 ml/g 4 °C lysis buffer using a hand held homogenizer at low speed. Avoid vigorous mixing and foaming to prevent oxidation.
5. Optional: add chicken egg (0.2 mg/ml) or recombinant lysozyme (60 KU or 0.2 μl/ml) and nuclease such as DNase (10 μg/ml) or Benzonase® (1.0 μl/ml).
6. Sonicate the suspension in a glass beaker on ice using 60–90 s bursts. Avoid excessive heating and aeration by adjustment of instrument settings, probe immersion depth, and duration of bursts. *Caution*: Do not allow the sonicator tip to contact the glass beaker during operation or the glass may break. For 100 ml cell suspension, four bursts at a setting of eight and 70% duty cycle using a Branson Model S-450 sonifier (Branson Sonic Power Co., Danbury, CT; www.bransonultrasonics.com) and 1/2-in. diameter horn is usually sufficient. Cool the extract with stirring on ice for several minutes between each burst. If desired, save a 50-μl sample of the lysate at this point (prior to centrifugation) for analysis of total cellular protein.
7. Centrifuge the lysate at 15,000×*g* for 15 min at 4 °C. Carefully transfer the supernatant to another container without disturbing the pellet. The soluble fraction may be further clarified by filtration with a low protein binding filter depending on the tolerance for particulates in the next fractionation step. If using syringe filters, it is convenient to attach a 0.8-μ*M* filter on top and a 0.45-μ*M* filter on the bottom in tandem.

The larger pore size filter in-line first prevents rapid fouling of the smaller pore size filter. This final clarified supernatant represents the soluble cell protein fraction and is ready for chromatography or other downstream fractionation or analysis procedures.

5.4.2. High-pressure homogenization

The batch, high-pressure homogenization procedure below is convenient for cell pellets of 50–500 g with an APV Gaulin homogenizer (APV Fluid Handling, Delevan, WI; www.apv.com), Microfluidizer, or Constant Systems hydraulic disrupter. Cell disruption of greater than 500 g of cells should be performed in multiple batches or using continuous disruption equipment described above.

1. Follow steps 3 and 4 in the sonication procedure above to resuspend the fresh or frozen cell paste, but resuspend the cell paste using a Waring or similar blender (Waring Products, Torrington, CT; www.waringproducts.com) at slow speed. Do not blend excessively and keep the extract cold at all times. The resuspension volume should be less than 10 ml/g depending on the amount of buffer removed to rinse the disruption instrument after processing. Remove and set aside 50–100 ml of the lysis buffer from the calculated resuspension volume prior to blending. This buffer will be used for rinsing the instrument after sample processing. The rinse solution contains residual extract from the processing equipment's dead volume.
2. If the Gaulin press is used, two process passes at 10,000 psi gives excellent disruption of *E. coli*. Additional process passes or continuous recycling may be necessary for difficult-to-disrupt cells such as yeast. Follow the manufacturer's operating instructions if the Microfludizer or hydraulic disrupter is used. Significant heating of the sample will occur during processing especially with smaller volumes. Chill the lysate to 5–10 °C between passes by immersion of the collection vessel in a dry ice ethanol bath. Minimize freezing of the lysate to the sides and bottom of the collection vessel by continuous stirring and periodic removal of the vessel from the cooling bath.
3. Centrifuge the lysate at 15,000×*g* for 30 min at 4 °C. Carefully decant and filter the supernatant using Miracloth (EMD Biosciences-Calbiochem, Gibbstown, NJ; www.emdbiosciences.com) into another container without disturbing the pellet. The soluble fraction may be further clarified by filtration with a smaller pore size low protein binding filter depending on the tolerance for particulates in the next fractionation step. This final clarified supernatant represents the soluble cell protein fraction and is ready for chromatography or other downstream fractionation or analysis procedures.

REFERENCES

Andrews, B. A., and Asenjo, J. A. (1987). Enzymatic lysis and disruption of microbial cells. *TIBTECH* **5,** 273–277.

Benov, L., and Al-Ibraheem, J. (2002). Disrupting *Escherichia coli*: A comparison of methods. *J. Biochem. Mol. Biol.* **35,** 428–431.

Burgess, R. R. (1987). *In* "Protein Purification: Micro to Macro", (R. R. Burgess, ed.), Alan R. Liss, New York.

Catley, B. J. (1988). Isolation and analysis of cell walls. *In* "Yeast: A Practical Approach", (I. Campbell and J. H. Duffus, eds.), pp. 163–183. IRL Press, Washington.

Chu, R., and Mallia, K. (2001). Method for recovery of proteins prepared by recombinant DNA procedures. U. S. Patent 6,174,704.

Chu, R., Mallia, K., Brennan, T., and Klenk, D. (1998). Recombinant protein extraction from *E. coli* using B-PER® Bacterial Protein Extraction Reagent. *Previews* **2,** 12–13.

Cline, D. J., Redding, S. E., Brohawn, S. G., Psathas, J. N., Schneider, J. P., and Thorpe, C. (2004). New water-soluble phosphines as reductants of peptide and protein disulfide bonds: Reactivity and membrane permeability. *Biochemistry* **43,** 15195–15203.

Cull, M., and McHenry, C. S. (1990). Preparation of extracts from prokaryotes. *In* "Guide to Protein Purification", (M. P. Deutscher, ed.), *Meth. Enzymol.* **182,** pp. 147–153.

Dawson, R. M. C., Elliott, D. C., Elliott, W. H., and Jones, K. M. (1986). pH, buffers, and physiological media. *In* "Data for Biochemical Research", pp. 417–448. Oxford University Press, Inc., New York.

De Mey, M., Lequeux, G. J., Maertens, J., De Muynck, C. I., Soetaert, W. K., and Vandamme, E. J. (2008). Comparison of protein quantification and extraction methods suitable for *E. coli* cultures. *Biologicals* **36,** 198–202.

Guide to protein purification. Deutscher, M. P. (ed.) (1990). *Meth. Enzymol.* **182,** pp. 147–203.

Dieckman, L., Gu, M., Stols, L., Donnelly, M. I., and Collart, F. R. (2002). High-throughput methods for gene cloning and expression. *Protein Expr. Purif.* **25,** 1–7.

Dignam, J. D. (1990). Preparation of extracts from higher eukaryotes. *In* "Guide to Protein Purification", (M. P. Deutscher, ed.), *Meth. Enzymol.* **182,** pp. 194–203.

[1]Drott, D., Bahairi, S., and Grabski, A. (2002). Yeast Buster protein extraction reagent for fast, efficient extraction of proteins from yeast. *inNovations* **15,** 14–16.

Engler, C. R. (1985). Disruption of microbial cells. *In* "Comprehensive Biotechnology", (C. L. Cooney and A. E. Humphrey, eds.), Vol. 2, pp. 305–324. Pergamon Press, Toronto.

Eshaghi, S., Hedren, M., Abdel Nasser, M. I., Hammarberg, T., Thornell, A., and Norlund, P. (2005). An efficient strategy for high-throughput expression screening of recombinant integral membrane proteins. *Protein Sci.* **14,** 676–683.

Foster, D. (1995). Optimizing recombinant protein recovery through improvements in cell-disruption technologies. *Curr. Opin. Biotechnol.* **6,** 523–526.

Garrett, P. E., Tao, F., Lawrence, N., Ji, J., Schumacher, R. T., and Manak, M. M. (2002). Tired of the same old grind in the new genomics and proteomics era? *TARGETS* **5,** 156–162.

Gegenheimer, P. (1990). Preparation of extracts from plants. *In* "Guide to Protein Purification", (M. P. Deutscher, ed.), *Meth. Enzymol.* **182,** pp. 174–193.

Getz, E. B., Xiao, M., Chakrabarty, T., Cooke, R., and Selvin, P. R. (1999). A comparison between the sulfhydryl reductants tris(2-carboxyethyl) phosphine and dithiothreitol for use in protein biochemistry. *Anal. Biochem.* **273,** 73–80.

[1] References available online: www.emdbiosciences.com.

[1]Grabski, A., and Burgess, R. R. (2001). Preparation of samples for SDS-polyacrylamide gel electrophoresis: Procedures and tips. *inNovations* **13,** 10–12.

[1]Grabski, A., McCormick, M., and Mierendorf, R. (1999). BugBuster™ and Benzonase®: The clear solutions to simple efficient extraction of *E. coli* proteins. *inNovations* **10,** 17–19.

[1]Grabski, A., Drott, D., Handley, M., Mehler, M., and Novy, R. (2001). Extraction and purification of proteins from *E. coli* without harvesting cells. *inNovations* **13,** 1–4.

[1]Grabski, A., Mehler, M., Drott, D., and Van Dinther, J. (2002). Automated purification of recombinant proteins in a 96-well format with RoboPop™ Kits and robotic sample processing. *inNovations* **14,** 2–5.

[1]Grabski, A., Mehler, M., and Luedke, R. (2003). Automated methods for solubility screening and recombinant protein purification. *Amer. Biotechnol. Lab.* **35,** 34–38.

Guerlava, P., Izac, V., and Tholozan, J.-L. (1998). Comparison of different methods of cell lysis and protein measurements in *Clostridium perfringens*: Application to the cell volume determination. *Curr. Microbiol.* **36,** 131–135.

Han, J. C., and Han, G. Y. (1994). A procedure for quantitative determination of tris (2-carboxyethyl)phosphine, an odorless reducing agent more stable and effective than dithiothreitol. *Anal. Biochem.* **220,** 5–10.

Harris, E. L. V., and Angal, S. (eds.) (1989). *In* "Protein Purification Methods, a Practical Approach", IRL Press, Oxford.

Harrison, S. T. L. (1991). Bacterial cell disruption: A key unit operation in the recovery of intracellular products. *Biotechnol. Adv.* **9,** 217–240.

Ho, C. W., Tan, W. S., Yap, W. B., Ling, T. C., and Tey, B. T. (2008). Comparative evaluation of different cell disruption methods for the release of recombinant hepatitis B core antigen from *E. coli. Biotechnol. Bioproc. Eng.* **13,** 577–583.

Hopkins, T. R. (1991). Physical and chemical cell disruption for the recovery of intracellular proteins. *In* "Purification and Analysis of Recombinant Proteins", (R. Seetharam and S. K. Sharma, eds.), pp. 57–83. Marcel Dekker, Inc., New York.

Jazwinski, S. M. (1990). Preparation of extracts from yeast. *In* "Guide to Protein Purification", (M. P. Deutscher, ed.), *Meth. Enzymol.* **182,** pp. 154–174.

Kashino, Y. (2003). Separation methods in the analysis of protein membrane complexes. *J. Chromatgr. B* **797,** 191–216.

Kleinig, A. R., and Middelberg, A. P. J. (1998). On the mechanism of microbial cell disruption in high-pressure homogenization. *Chem. Eng. Sci.* **53,** 891–898.

Lesley, S. A. (2001). High-throughput proteomics: Protein expression and purification in the postgenomic world. *Protein Expr. Purif.* **22,** 159–164.

Lieve, L. (1974). The barrier function of the gram-negative envelope. *Ann. N.Y. Acad. Sci.* **235,** 109–129.

Marshak, D. R., Kadonaga, J. T., Burgess, R. R., Knuth, M. W., Brennan, W. A., and Lin, S.-H. (1996). Solubilization and purification of the rat liver insulin receptor. *In* "Strategies for Protein Purification and Characterization: A Laboratory Manual", pp. 275–336. Cold Spring Harbor Laboratory Press, New York.

Meiss, G., Friedhoff, P., Hahn, M., Gimadutdinow, O., and Pingoud, A. (1995). Sequence preferences in cleavage of dsDNA and ssDNA by extracellular *Serratia marcescens* endonuclease. *Biochemistry* **34,** 11979–11988.

Middelberg, A. P. J. (1995). Process-scale disruption of microorganisms. *Biotechnol. Adv.* **13,** 491–551.

Nestle, M., and Roberts, W. K. (1969). An extracellular nuclease from *Serratia marcescens. J. Biol. Chem.* **244,** 5219–5225.

Neugebauer, J. M. (1990). Detergents: An overview. *In* "Guide to Protein Purification", (M. P. Deutscher, ed.), *Meth. Enzymol.* **182,** pp. 239–253.

Nguyen, H., Martinez, B., Oganesyan, N., and Kim, R. (2004). An automated small-scale protein expression and purification screening provides beneficial information for protein production. *J. Struct. Funct. Genom.* **5,** 23–27.

Pandolfe, W. D. (1999). Homogenizers. *In* "Wiley encyclopedia of food science and technology", (F. J. Francis, ed.), pp. 1289–1294. John Wiley and Sons, Inc., New York.
Ramanan, R. N., Ling, T. C., and Ariff, A. B. (2008). The performance of a glass bead shaking technique for the disruption of *E. coli* cells. *Biotechnol. Bioproc. Eng.* **13,** 613–623.
Scopes, R. K. (1994). Protein Purification: Principles and Practice. 3rd edn. Springer Verlag, New York.
Shirgaonkar, I. Z., Lothe, R. R., and Pandit, A. B. (1998). Comments on the mechanism of microbial cell disruption in high-pressure and high-speed devices. *Biotechnol. Prog.* **14,** 657–660.
Spencer, J. F. T., and Spencer, D. M. (1997). Yeasts and the life of man: Part II: Genetics and molecular biology of industrial yeasts and processes. *In* "Yeasts in Natural and Artificial Habitats", (J. F. T. Spencer and D. M. Spencer, eds.), pp. 243–263. Springer-Verlag, Berlin.
SPX Corp. (2009). *In* "Cell Disruption by Homogenization", pp. 1–18; Pub. #3006-01-06-2008-US.
Stevens, J., and Kobs, G. (2004). Taking the spin out of cell lysis. *Promega Notes.* **86,** 23–24.
Stevens, R. C., and Wilson, I. A. (2001). Industrializing structural biology. *Science* **293,** 519–520.
Stevens, R. C., Yokoyama, S., and Wilson, I. A. (2001). Global efforts in structural genomics. *Science* **294,** 89–92.
Surzycki, S., Togasaki, R. K., and Kityama, M. (1996). Process and apparatus for fragmenting biomaterials. *U. S. Patent* **5**(506), 100.
Swiderek, K. M., Alpert, A. J., Heckendorf, A., Nugent, K., and Patterson, S. D. (1997). Structural analysis of proteins and peptides in the presence of detergents: Tricks of the trade. *ABRF News Article Methodol.* **8,** 17–25.
Zhu, H., Bilgin, M., Bangham, R., Hall, D., Casamayor, A., Bertone, P., Lan, N., Jansen, R., Bidlingmaier, S., Houfek, T., Mitchell, T., Miller, P., *et al.* (2001). Global analysis of protein activities using proteome chips. *Science* **293,** 2101–2105.

CHAPTER NINETEEN

Isolation of Subcellular Organelles and Structures

Uwe Michelsen *and* Jörg von Hagen

Contents

1. Introduction 306
2. Extraction and Prefractionation of Subproteomes 308
   2.1. Density velocity centrifugation 308
   2.2. Density gradient centrifugation 311
   2.3. Mitochondria enrichment 311
   2.4. Endomembrane enrichment (LOPIT) 314
   2.5. Differential detergent fractionation 316
   2.6. Lipid raft enrichment (detergent-resistant fractionation) 320
   2.7. Nuclei and histone enrichment 322
   2.8. Purification of core histones 324
   2.9. Nuclear extract enrichment 325
   2.10. Immunoaffinity reagents for organelle enrichment 326
References 327

Abstract

One of the major challenges in functional proteomics is the separation of complex protein mixtures to allow detection of low abundance proteins and provide for reliable quantitative and qualitative analysis of proteins impacted by environmental parameters. Prerequisites for the success of such analyses are standardized and reproducible operating procedures for sample preparation prior to protein separation. Due to the complexity of total proteomes, especially of eukaryotic proteomes, and the divergence of protein properties, it is often beneficial to prepare standardized partial proteomes of a given organism to maximize the coverage of the proteome and to increase the chance to visualize low abundance proteins and make them accessible for subsequent analysis. In this chapter we will describe with detailed recipes procedures for the enrichment and isolation of the currently most investigated organelles and subcellular compartments in mammalian cells using classical centrifugation techniques to more sophisticated immunoaffinity-based procedures.

Merck KGaA, Darmstadt, Germany

Methods in Enzymology, Volume 463
ISSN 0076-6879, DOI: 10.1016/S0076-6879(09)63019-6

Abbreviations

2D	Two-dimensional
DDF	Differential detergent fractionation
DRF	Detergent-resistant fractions
DRM	Detergent-resistant membranes
EMSA	electro mobility shift assay
g	Gravity unit
GPCR	G-Protein coupled receptor
HM	Homogenization medium
IEF	Isoelectric focussing
iTRAQ	Isotope tags for relative and absolute quantification
LC	Liquid chromatography
LOPIT	Localization of organelle proteins by isotope tagging
M	Molar
β-MCD	β-Methylcyclodextrin
MS	Mass spectrometry
MudPIT	Multidimensional protein identification technology
PBS	Phosphate buffered saline
PMSF	Phenylmethylsulfonyl fluoride
RAM	Restricted access material
RP	Reversed phase
SCX	Strong cation exchange
SDS	Sodium dodecyl sulfate

1. Introduction

In proteomics research, one essential step among enrichment techniques is subcellular fractionation, which is of special importance for analysis of intracellular organelles and multiprotein complexes. To reduce sample complexity, subcellular fractionation is a flexible and adjustable approach and is most efficiently combined with high-resolution 2D gel/mass spectrometry analysis as well as with gel-independent techniques.

Isolating distinct subcellular compartments by fractionation has been among the standard strategies established in biochemistry-oriented laboratories for decades. Determination of marker enzyme activities was to assay the efficiency of the subcellular fractionation and a major goal was the identification of individual new proteins. At that time, the power of protein identification techniques was very limited, the characterization of the protein subsets specific to subcellular compartments was time-consuming, of limited sensitivity, or even impossible.

By increasing their degree of complexity, organisms acquire a broader repertoire of options to meet environmental challenges. Increased complexity is realized in two ways: by cell differentiation and by compartmentalization within cells. Subsets of cells of a given organism serve different purpose and own distinct properties, for example, neurons, germ cells, or epithelial cells. In addition certain cell functions are compartmentalized such as storage of genetic material, degradation of proteins, or the generation of the cell fuel.

The identification of proteins at the subcellular level is therefore pivotal for understanding of cellular function. There are proteins that are associated with subcellular structures only in certain physiological states, but localized elsewhere in the cell in other states. Protein translocation between different compartments, cycling of proteins between the cell surface and intracellular pools or shuttling between nucleoplasm and cytoplasm is very common in mammalian cells. To be able to monitor the subcellular distribution of gene products, the classic proteome analysis approach must be refocused on a subcellular level. Such an approach fits very well to the way in which cells are organized.

Finally one should note that many current proteome analysis projects are aimed at the comparative analysis of tissue samples. Tissue samples are more complex than samples from cultured cells as any tissue contains many different cell types and contains structural material like connective tissue that may not be the target of the analysis. This fact implies an even higher challenge to sample preparation procedures to avoid artifacts. An overview of the proteomics workflow for all types of samples is presented in Fig. 19.1 highlighting the steps for the enrichment of organelles and subproteomes.

Based on the scheme shown in Fig. 19.1, the following chapter is subdivided into three sections summarizing the most important and prominent sample preparation techniques used for enrichment and fractionation.

It is important to mention at this point, that based on the various downstream separation techniques no general workflow is established. All the diverse analysis techniques require a specific sample preparation to assure the compatibility of the remaining sample with the selected technique. More detailed information how to specifically prepare samples and various sample types is found in "Proteomics Sample Preparation" (von Hagen, 2008). The described procedures are enrichment-based techniques and do not claim to exclusively purify organelle proteomes. Moreover, the outlined procedures are a starting point and need to be optimized depending on species, tissue, and cell type.

In contrast, techniques based on chromatographic principles like multidimensional protein identification technology (MudPit) (Washburn *et al.*, 2001), PF2D (Billecke *et al.*, 2006), or FFE (Weber *et al.*, 2004) are

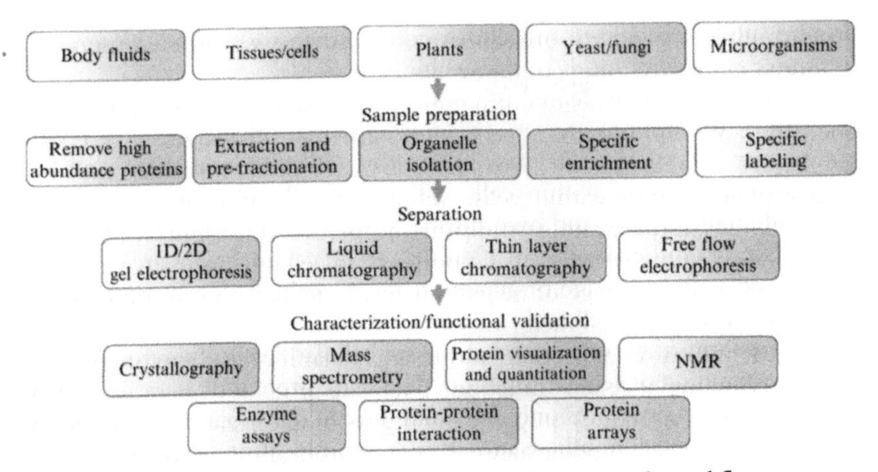

Figure 19.1 Schematic representation of the proteomic workflow.

commonly used to reduce proteome complexicity and do rely on preenriched organelles. These separation techniques are briefly described and protocols can be found elsewhere.

2. Extraction and Prefractionation of Subproteomes

2.1. Density velocity centrifugation

2.1.1. Introduction

Centrifugation is a common sample preparation technique that accomplishes separation based on the density and size differences in a mixture of components. In the absence of Brownian motion and thermal mixing, a centrifuge is often not required. If samples are diverse with regard to density or shape and given enough time, sample components eventually will settle at the bottom of a tube at 1*g*. The process can be sped up using a centrifuge. The density velocity centrifugation takes advantage of the difference in the terminal velocities of different particles as determined by Stoke's law:

$$V_t = \frac{2R^2(\rho_S - \rho)a}{(9\mu)}$$

where V_t is the terminal velocity of the biomolecule, μ is the viscosity of the medium, R is the radius of the particle, ρ_s is the density of the particle, a is the centrifugal acceleration, and ρ is the density of the medium.

The terminal velocity is a function of the particle radius and density. Thus, on the average, bigger and heavier particles migrate through the medium faster and settle at the bottom of a centrifuge tube in a shorter time. Because a typical mixture of cell homogenate contains organelles of varying sizes and densities, as well as shapes, they can be separated according to the sedimentation speed. The decrease in the sedimentation time in a centrifuge over gravitational settling is mainly due to the considerable increase in the variable a in Stoke's equation (Table 19.1).

Each fraction obtained through differential centrifugation contains quite a few different types of organelles which have similar sedimentation velocities. Because sedimentation velocity is a combination of both the size and the density, the fraction can be further separated based on density alone irrespective of the sizes. This can be accomplished by a process known as density gradient centrifugation (Fig. 19.2).

Procedure:

1. Centrifuge the cellular homogenate for 2 min at 100*g* to remove unbroken cells.
2. Centrifuge the supernatant at 600*g* for 10 min to sediment nuclei.
3. Centrifuge the supernatant at 15,000*g* for 5 min to deposit mitochondria, chloroplasts, lysosomes, and peroxisomes.
4. Ultracentrifuge at 100,000*g* for 60 min for sedimentation of the plasma membrane, fragments of the endoplasmic reticulum, and large polyribosomes.
5. The cytosol remains unpelleted after centrifugation at 300,000*g* for 2 h.

Table 19.1 Density and dimensions of cellular organelles

Organelle/compartment	Diameter (μm)	Density (g/cm^3)
Nuclei	3–12	1.4
Mitochondria	0.5–2	1.15
Golgi apparatus	–	1.08
Lysosomes	0.5–0.8	1.2
Peroxisomes	0.5–0.8	1.25
Plasma membranes	0.05–3	1.15
smooth endoplasmic reticulum	0.05–3	1.16
Nucleic acid	0.03	1.7
Ribosomes	0.03	1.6
soluble proteins	0.001–0.002	1.3

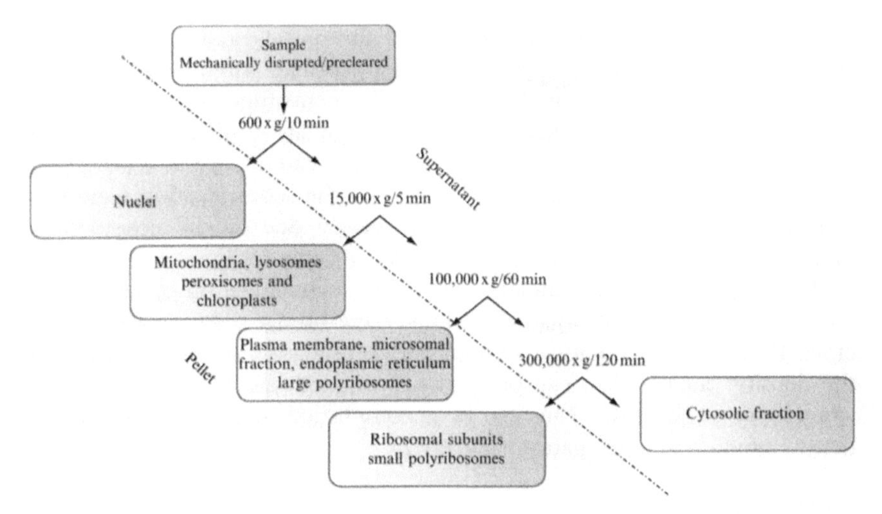

Figure 19.2 Cell fractionation by differential velocity centrifugation of a precleared mechanical disrupted cell homogenate.

2.1.2. Further reduction in proteome complexity

2.1.2.1. Free-flow electrophoresis Free-flow electrophoresis (FFE) is a highly versatile technique for the separation of a wide variety of charged or chargeable analytes such as low-molecular-weight organic compounds, peptides, proteins, protein complexes, membranes, organelles, and whole cells. A detailed protocol is outlined in Current Protocols in Proteins Sciences (Weber *et al.*, 2004). This technology is used mainly for the purification of prepurified organelle extracts as obtained by the procedure described above and thus represents solely an additional separation technology.

2.1.2.2. Multidimensional protein identification technology MudPIT is based on a gel-free two-dimensional high-performance liquid chromatography (HPLC) approach initially developed by John Yates III lab (Washburn *et al.*, 2001). A complex protein mixture is denatured and digested with proteases and separated in a capillary packed with a strong cation exchange (SCX) resin back-to-back with a reversed phase (RP) material. The column is interfaced with a quaternary HPLC system coupled directly to a tandem mass spectrometer.

Moreover other orthogonal chromatographic resins are described to achieve efficient reduction in proteome complexity like restricted access materials (RAM) (Willemsen *et al.*, 2004) or the PF2 D system (Billecke *et al.*, 2006) using IEF columns. This technology can be used to reduce organelle proteome complexity obtained from prepurified organelles from density velocity centrifugation or density gradient centrifugation described in Section 2.2.

2.2. Density gradient centrifugation

2.2.1. Introduction

A density gradient is simply established by placing sugar crystals at the bottom of a test tube (Fisher and Cline, 1963). The sugar dissolves in the solution and diffuses slowly toward the top which is a time-consuming process and not very reproducible. A more practical approach is to place layer after layer of sucrose solutions of decreasing concentrations in a test tube, resulting in the heaviest layer at the bottom and the lightest layer on top. Finally, the cell fraction is placed on top of the sucrose layers. If the density of a given particle is higher than that of the surrounding solution it will sink until the density of the surrounding solution is exactly the same as that of the particle. Using a centrifuge will accelerate this process. Unlike the differential centrifugation technique described above, centrifugation time does not matter, as long as the whole system is allowed to come to quasiequilibrium.

This chapter will give a more detailed view on the classical density gradient technique with regard to the isolation of mitochondria which are currently in the focus of apoptosis research. For the enrichment of other organelles please refer to the previous chapter in *Methods in Enzymology* (Storrie and Madden, 1990). This section gives detailed protocols for the isolation of lysosomes, postnuclear supernatants, mitochondria, nuclei as well as assays to determine marker enzyme activity and organelle intactness (Fig. 19.3).

2.3. Mitochondria enrichment

Proteomic analyses of particular mitochondrial proteins and their complexes have provided important information about mitochondrial function (Edelman *et al.*, 1966). Percoll gradients are generally used in protocols for the purification of mitochondrial fractions after tissue homogenization and differential centrifugation of a cell homogenate.

One drawback is the removal of Percoll which is necessary before further analyses. To circumvent difficulties in Percoll removal from the isolated mitochondrial fraction, some workers used sucrose in the gradient buffer as described in detail below. One should note that this again may cause problems because disaccharides are able to enter the mitochondrial matrix through membrane porin channels and thus might interfere with the physiological interpretation of results obtained.

The described sucrose gradient separation procedure is a protein subfractionation method optimized for mitochondria. It can be used to reduce sample complexity and resolves a sample into at least 10 fractions. The sucrose gradient is designed for an initial sample volume of up to 0.5 mL at 5 mg/mL protein. The sample should be solubilized in a nonionic detergent. It has been determined that at protein concentration of 5 mg/mL

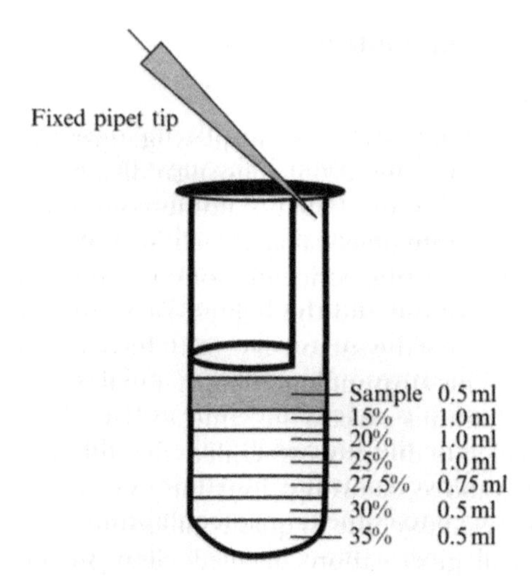

Figure 19.3 Sucrose gradient setup: 15–35% sucrose density gradient. The gradient is prepared by layering less dense sucrose solutions upon one another; therefore, the first solution applied is the 35% sucrose solution. Firstly, a Beckman polyallomer tube is held upright in a tube stand. Next a tip is placed held steady by a clamp stand and the end of the tip is allowed to make contact with the inside wall of the tube as shown above.

mitochondria are completely solubilized by 20 m*M* *n*-dodecyl-*β*-D-maltopyranoside (1% w/v lauryl maltoside). By this treatment membranes are disrupted while the previously membrane embedded multisubunit complexes remain intact, a step necessary for the procedure described below:

Material and equipment:

- Sucrose solutions 15, 20, 25, 27.5, 30, and 35% (w/v).
- Protease inhibitor cocktail.
- Phosphate buffered saline solution (PBS): 1.4 m*M* KH_2PO_4, 8 m*M* Na_2HPO_4, 140 m*M* NaCl, 2.7 m*M* KCl, pH 7.3.
- 200 m*M* *n*-dodecyl-*β*-D-maltopyranoside (10% w/v lauryl maltoside).
- Swing-out compatible ultracentrifuge and rotor.
- Polyallomer centrifuge tubes.

Homogenize and lyse cells as described above. Preferentially mechanical disruption methods (Goldberg, 2008) should be used to avoid addition of detergents which might result in leakage of the mitochondria and other cellular compartments as well.

General procedure:

- Wash cells twice in PBS.
- Resuspend cells in PBS (1 mL/10^9 cells) including phosphatase and protease inhibitors. Shear the cells with >20 strokes using a Potter–Elvehjem or Dounce homogenizer (Check by viability stain).

- Alternatively use five and more freeze thaw cycles, five or more sonications (5 pulses of 5 s), French press, high pressure homogenization (2 cycles of 500–1000 bar) or a bead mill according to the protocol provided by the various suppliers. For the following application preclear the lysate by a centrifugation step of 5 min at 500*g* at 4 °C.

Procedure:

1. A mitochondrial membrane suspension at max. 5 mg/mL protein in PBS is adjusted with lauryl maltoside to a final concentration of 1%.
2. Mix well and incubate on ice for 30 min.
3. Centrifuge at 70,000*g* for 30 min.
4. The supernatant is collected and the pellet discarded.
5. Add a protease inhibitor cocktail and keep the sample on ice until centrifugation is performed.
6. Once the sucrose gradient is poured, discrete layers of sucrose should be visible. Having applied the sample to the top of the gradient the tube should be handled carefully.
7. The polyallomer tubes should be centrifuged in a swinging bucket type rotor at 130,000*g* for 16 h at 4 °C with the lowest acceleration and deceleration profile.
8. Carefully remove 500 μL fractions from top to the bottom and analyze the fractions. The fractions can now be stored at − 80 °C.

2.3.1. Other density gradient formulations

For larger amounts of proteins, multiple gradients can be prepared or scaling-up to the gradients is also possible. For example, 40 mg of mitochondria can be resolved by a sucrose gradient of 40 mL total volume.

1. Combine as follows 55 (5 mL), 50 (5 mL), 35 (10 mL), 30 (10 mL), 25% (2 mL) sucrose solution.
2. Add on top 8 mL mitochondria supernatant (solubilized in 1% lauryl maltoside centrifuged 30 min at 72,000*g*).
3. Centrifuge at 115,000*g* for 17 h at 4 °C with the lowest acceleration and deceleration profile.

Besides the classical sucrose density medium, different media from various suppliers are also used routinely to isolate subcellular structures and organelles. The most prominent materials to mention are Ficoll (Kurokawa *et al.*, 1965), Percoll (Pertoft *et al.*, 1977), Nycodenz (Rickwood *et al.*, 1982), and Optiprep (van Veldhoven *et al.*, 1996). For detailed protocols, please refer to the corresponding manuals provided by the different suppliers.

2.4. Endomembrane enrichment (LOPIT)

2.4.1. Introduction

Identifying the localization of a novel protein is an important step toward assigning function and will also improve our understanding of cell compartments. To allow confident protein localization organelle preparations must be free of contaminants. For some organelles, such as mitochondria, it is relatively easy to produce a largely homogeneous preparation of the organelle. However, Golgi apparatus and endoplasmic reticulum of the endomembrane system are difficult to separate from one another. Furthermore, proteins associated with this class of organelle traffic through the endomembrane system to their final destination or cycle between the compartments. So it is a tough job to discriminate between genuine organelle residents and contaminants.

Several techniques have emerged which enable assignment of proteins to organelles. Here we discuss the sample preparation necessary for the use with one of these methods—localization of organelle proteins by isotope tagging (LOPIT).

The LOPIT workflow involves partial separation of organelles using equilibrium density gradient centrifugation. The relative abundance of proteins amongst the fractions of the gradient is compared with the profiles of known organelle markers. The relative abundance is measured by a stable isotope tagging procedure and tandem mass spectrometry (MS/MS) whereby primary amine groups present on tryptic peptides are derivatized with iTRAQ reagents (Ross *et al.*, 2004). The LOPIT technology is applicable to the study of protein localization in many sample types including cultured cells, tissues, and whole organisms.

The procedure starts with the collection of crude membrane extracts and partial separation of organelles through a dense medium. The degree of organelle separation is monitored using one-dimensional SDS–PAGE and western blotting employing specific antibodies. Alternatively, enzyme assays measuring the activity characteristic for a particular organelle can be utilized. This has been described in detail by Storrie and Madden (1990) in Methods in Enzymology (Table 19.2).

The gradient fractions enriched in specific organelles are selected for quantitative labeling by tagging tryptic peptides with the iTRAQ reagents. The labeled mix of peptides is separated by SCX chromatography, and further separated by reverse phase chromatography coupled to tandem mass spectrometer (LC-MS/MS). Finally peptides are identified and quantified using fragmentation data generated by MS/MS and analyzed with statistical approaches. The protocol described here is based on the work of Sherrier *et al.*, (1999) and can be optimized for membranes from various sources (Ford *et al.*, 1994; Graham *et al.*, 1994). A typical preparation would involve the homogenization of 60 g of tissue in an equal volume of homogenization

Table 19.2 Enzyme and substrate markers for different cellular organelles

Organelle	Enzyme or substance
Nuclei	DNA, RNA, Parp
Mitochondria	Cytochrome *c* oxidase, succinate dehydrogenase
Plasma membrane	Alkaline phosphodiesterase I, GPCRs
Lysosome	β-Galactosidase, β-hexosaminidase, acid-phosphatase
Endosome	Peroxidase
Peroxisome	Catalase
Golgi apparatus	α-Mannosidase II, galactosyl transferase
Smooth endoplasmic reticulum	Glucose-6-phosphatase
Rough endoplasmic reticulum	RNA
Cytosol	Lactate dehydrogenase
Cytoskeleton	Vimentin, keratin

medium using a Polytron homogenizer for two 7 s pulses. The amount of medium used and the time of homogenization are crucial to minimize organelle disruption during this stage of the preparation. The amount of starting material should yield at a minimum 100 μg of membrane protein in each gradient fraction which will be labeled with iTRAQ tags. The first part of the protocol describes membrane fractionation using equilibrium density gradient centrifugation. Utilizing a 16% iodixanol gradient enables the separation of the Golgi apparatus, endoplasmic reticulum, vacuoles, plasma membrane, and mitochondria/plastids

After harvesting the fractions in step 8 it is important to check the degree of membrane separation by performing western blot analysis of gradient fractions with antibodies specific for different organelle marker proteins. Selected fractions can then be used in the following wash steps. The high pH of the carbonate buffer disrupts electrostatic interactions between peripheral proteins and the membrane. In addition, closed vesicles are transformed into membrane sheets and soluble proteins are released. The generated pellets can be used directly for the iTRAQ labeling reaction which is described in Sadowski *et al.* (2006) and Shadforth *et al.* (2005).

Material and equipment:

- *Homogenization buffer*: 250 m*M* sucrose. 10 m*M* HEPES-NaOH (pH 7.4), 1 m*M* EDTA (pH 8.0), 1 m*M* DTT.
- *Optiprep stock solution*: 60% solution of iodixanol in water.

- *System buffer*: 60 m*M* HEPES-NaOH (pH 7.4), 6 m*M* EDTA (pH 8.0), 6 m*M* DTT.
- *Iodixanol solution*: Prepared by diluting five parts of Optiprep stock solution with one part of system buffer.

Note: All solution must be prepared freshly before use and kept at 4 °C.

Procedure:

1. Preclear the homogenate from intact cells, nuclei, and cell wall fragments by centrifuging for 5 min at 2200*g* at 4 °C.
2. Decant supernatants into clean centrifuge tubes and recentrifuge as above.
3. Aliquot the supernatants into a total of six centrifuge tubes and underlay each supernatant with 6 mL of 18% iodixanol cushion solution.
4. Centrifuge at 100,000*g* at 4 °C for 2 h using an ultracentrifuge.
5. Collect the crude membranes from the interphases (typically 6–8 mL)
6. Dilute membranes with iodixanol gradient solution and adjust to 16% iodixanol using iodixanol solution.
7. Decant the adjusted solution into two polycarbonate tubes and centrifuge at 350,000*g* at 4 °C for 3 h in an ultracentrifuge with the slowest deceleration rate selected.
8. Harvest twenty 0.5 mL fractions from the top of each gradient into polycarbonate tubes.
9. Add 800 μL of 162.5 m*M* Na_2CO_3 to 0.5 mL of each selected density fraction to give a final concentration of 100 m*M* of Na_2CO_3.
10. Incubate for 30 min on ice and centrifuge for 25 min at 100,000*g* at 4 °C using an ultracentrifuge and discard the supernatants.
11. Wash the pellets with 1 mL of deionized water at 4 °C, centrifuge for 10 min like above and discard the supernatants.

2.5. Differential detergent fractionation

2.5.1. Introduction

To solubilize proteins, especially proteins with extended hydrophobic domains the uses of amphipathic molecules like detergents are essential. Typical detergents contain both polar and hydrophobic groups. They contain a polar head at the end of a long hydrophobic chain or tail. In contrast to purely polar or nonpolar molecules, detergent exhibit unique properties in water. Their polar groups form hydrogen bonds with water molecules, while the hydrocarbon chains aggregate due to hydrophobic interactions. These properties allow detergents to be soluble in water. In aqueous solutions, they form organized spherical structures called micelles. Because of their amphipathic nature, detergents are able to solubilize hydrophobic compounds in water. The following procedure describes the

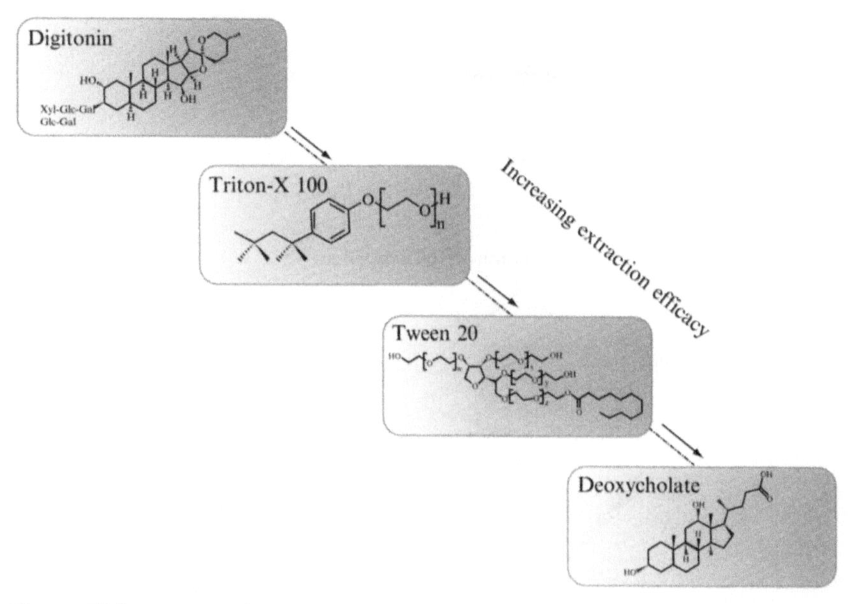

Figure 19.4 Commonly used detergents with increased extraction efficacy from Digitonin to sodium Deoxycholate.

use of detergents with increased extraction efficacy properties which result in the enrichment of four distinct subproteom fractions (Ramsby *et al.*, 1994) (Fig. 19.4).

A typical sequential extraction method based on detergents enables simple fractionation of proteins in their native state according to their subcellular localization, yielding four subproteomes enriched in (Fig. 19.4A) cytosolic; (B) membrane and membrane organelle-localized; (C) soluble and DNA-associated nuclear, and (D) cytoskeletal proteins.

Due to the complexity of total proteomes and the divergence of protein properties, it is beneficial to prepare standardized partial proteomes of a given organism to make them accessible for subsequent analysis. Such analyses often require that the proteins of interest remain in a nondenatured state and their subcellular localization can be determined. Changes in the topology of proteins, that is, spatial rearrangements to another subcellular compartment, are important cellular events and crucial for biological processes such as signal transduction or apoptosis. This cannot be achieved by selective purification of cellular organelles. Furthermore, the homogenization techniques employed, require usually relatively large amounts of starting material, and are generally more efficient with tissues than with tissue culture cells (Abdolzade-Bavil *et al.*, 2004).

The protocol described enables simple fractionation of the total proteome into four distinct subcellular fractions. Various related protocols in kit

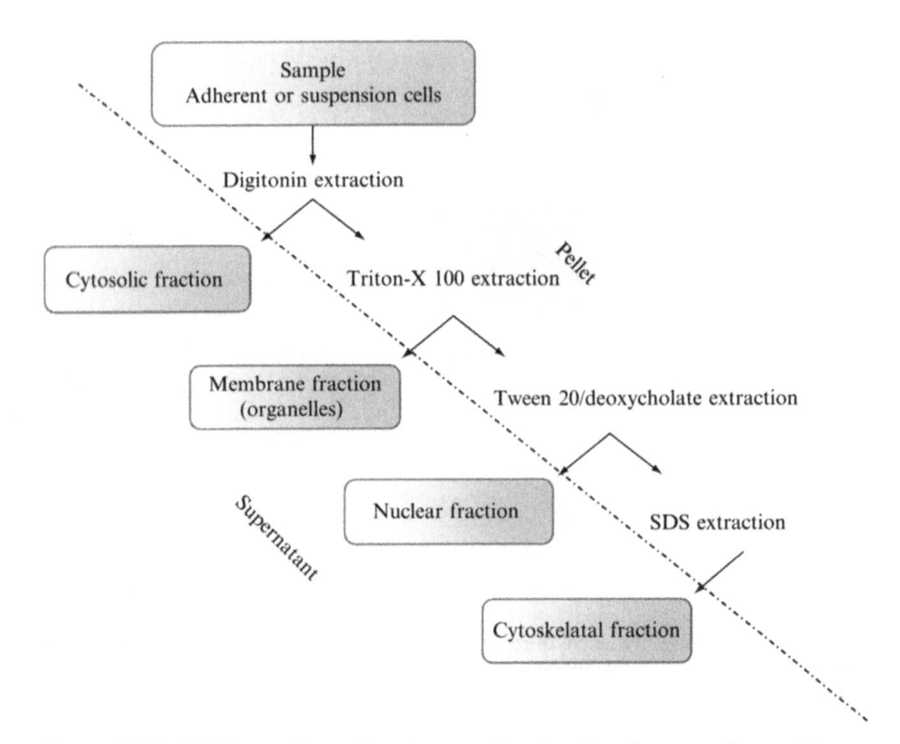

Figure 19.5 *DDF procedure*: **Nonultracentrifugation-based organelle enrichment technique.**

format have recently become commercially available by various suppliers (e.g., ProteoExtract kits from Calbiochem, San Diego, USA) (Fig. 19.5).

Reagents and equipment:

- 100 m*M* EDTA: dissolve 3.36 g EDTA in 100 mL water. Store at room temperature.
- 100 m*M* phenylmethylsulfonyl fluoride (PMSF): dissolve 174 mg PMSF in 10 mL isopropanol.
- Piperazine-*N*,*N*′-bis(2-ethanesulfonic acid) (PIPES) stock buffer (10×): Filter through 0.45 μm sterile filter. Store dark at 4 °C.
- *Cytosolic protein extraction buffer* (0.015% digitonin, pH 6.8 at 4 °C): dissolve by heating 18.75 mg digitonin (Calbiochem) in 4 mL 10× stock buffer in a small flask with a stir bar. Add 1 mL PMSF. Combine with remaining reagents: 6 mL 10× stock buffer and 5 mL EDTA. Add water to 100 mL.
- *Membrane protein extraction buffer* (0.5% Triton X-100, pH 7.4 at 4 °C): combine 10 mL 10× buffer, 0.1 mL PMSF, 3 mL EDTA, and 5 mL freshly prepared 10% Triton X-100. Add water to 100 mL.

- *Nuclear protein extraction buffer* (1% Tween/0.5% DOC, pH 7.4 at 4 °C): separately dissolve 0.5 g DOC in 2.5 mL 10× buffer and 1 mL Tween-40 in 2.5 mL 10× buffer (heat to dissolve if necessary). Combine and add 5 mL 10× buffer and 1 mL PMSF. Add water to 100 mL.
- *Cytoskeleton solubilization buffer* (5% sodium dodecyl sulfate, 10 m*M* sodium phosphate, pH 7.4): for nonreducing buffer, solubilize 0.5 g SDS in 5 mL 20 m*M* sodium phosphate buffer, pH 7.4. Add water to 10 mL. For denaturing buffer, add 1 mL β-mercaptoethanol. Adjust water appropriately.

Procedure:

1. Add ice-cold *cytosolic protein extraction buffer* to washed cell pellets (5 volumes/g wet weight, gently resuspend by swirling) or monolayers (1 mL/T25 flask).
2. Incubate cells on ice with gentle agitation (platform mixer) until 95–100% of cells are permeabilized (5–10 min) as assessed by trypan blue exclusion.
3. For suspension-cultured cells, centrifuge the extraction mixture at 500*g* and remove supernatant.
4. For cell monolayers, tilt the culture flask and remove extract (*cytosolic protein-enriched fraction*) with a pipet and store at − 70 °C.
5. Carefully resuspend insoluble pellet in ice-cold *membrane protein extraction buffer* in a volume equivalent to that used for cytoplasma protein extraction (5 volumes relative to starting wet weight) to obtain a homogeneous suspension.
6. For monolayer cultures, add 1 mL *membrane protein extraction buffer* per T25 flask equivalent (approx. 5×10^6 cells).
7. Incubate on ice with gentle agitation for 30 min.
8. Remove the extract (*membrane and organellar protein-enriched fraction*) by centrifuging the suspensions for 10 min at 5000*g* or decanting monolayers.
9. Aliquot, and store at − 70 °C.
10. Resuspend the insoluble pellets from suspension cultures in *nuclear protein extraction buffer* at one-half the volume used for the membrane protein extraction.
11. Remove the extract (*nuclear protein-enriched fraction*) by pelleting the detergent-resistant residue 10 min at 6000*g* at 4 °C.
12. Extract cell monolayers with 0.5–1 mL *nuclear protein extraction buffer* per T25 flask equivalent.
13. Aliquot, and store at − 70 °C.
14. The detergent-resistant pellet is washed in ice-cold PBS (pH 7.4, 1.2 m*M* PMSF) by resuspension (Teflon/glass homogenizer) and centrifugation 20 min at 4 °C and 12,000*g* to mechanically shear DNA or use 1000 units Benzonase (EMD/Novagen) to enzymatically degrade nucleic acids.

15. Pellets from suspension cultures are washed once with − 20 °C of 90% acetone, lyophilized, and weights determined in tared centrifuge tubes. Samples are stored at − 70 °C.
16. Monolayers are rinsed *in situ* with PBS, and the detergent-resistant residue (*Cytoskeleton*) is suspended directly into 0.5–1 mL nondenaturing *cytoskeleton solubilization buffer* without β-mercaptoethanol. Store at − 70 °C.

2.6. Lipid raft enrichment (detergent-resistant fractionation)

2.6.1. Introduction

A lipid raft is a cholesterol-enriched microdomain in cell membranes. Many functions have been attributed to rafts, from cholesterol transport, endocytosis, and signal transduction. The analysis of membrane signaling complexes in the natural environment requires a tailor made procedure depending on the complex of interest. Based on the various sample types no general protocol can be outlined. The protocol given is a good starting point. For further optimization, Table 19.3 gives a brief summary of mild detergents commonly used to isolate membrane proteins and lipid rafts.

One drawback of the use of mild detergents is that classes of pharmacological relevant proteins which are anchored in the membrane by various hydrophobic transmembrane domains are resistant to efficient extraction. Besides the usage of sodium carbonate and OptiPrep also a magnetic-bead immunoisolation approach might be used to isolate subpopulations of rafts enriched for different markers such as caveolin-1, flotillin (Shah and Sehgal, 2007). Recently, another nondetergent-based extraction procedure claims to efficiently extract membrane proteins with several passing transmembrane domains (7TMs or GPCRs; Nakamura *et al.*, 2008) and this opens an in-depth analysis of GPCR transmembrane signaling complexes in a two-step procedure using a mild detergent to enrich lipid rafts first and after enrichment extract the entire components of such complexes using the novel chemistry approach as mentioned above or by diluting the small molecule based chemistry and use it for both steps to enrich and extract lipid rafts and the membrane signaling complexes.

Material and equipment:

- TNE-buffer (25 m*M* Tris–HCl pH 7.5, 50 m*M* NaCl, 5 m*M* EDTA) containing 1%.
- Triton X-100.
- Protease and phosphatase inhibitors (Calbiochem).
- Nycodenz.
- Swing-out compatible ultracentrifuge and rotor.
- Polyallomer centrifuge tubes.

Table 19.3 Commonly used detergents for the solubilization of lipid rafts, membrane protein complexes and transmembrane proteins

Name	Abbr.	Type[a]	MW	cmc (mM)[b]	cmc (% w/v)[c]	Percentage for solubilization (%)
n-Dodecyl-β-D-maltopyranoside	DDM	N	510.6	0.17	0.0087	1
n-Undecyl-β-D-maltopyranoside	UDM	N	496.6	0.59	0.029	1
n-Decyl-β-D-maltopyranoside	DM	N	482.6	1.8	0.087	1
Cymal 7	Cymal 7	N	522.5	0.19	0.0099	1
Cymal 6	Cymal 6	N	508.5	0.56	0.028	2
n-Octyl-β-D-glucopyranoside	nOG	N	292.4	18	0.53	2
N,N-Dimethyldodecylamine *N*-oxide	LDAO	Z	229.4	1–2	0.023	1
3[(3-Cholamidopropyl) dimethylammonio]propane sulfonic acid	CHAPS	Z	614.9	8	0.49	1
Triton X-100	TX-100	N	647	0.23	0.015	1 (v/v)

[a] The type of detergents: N = Nonionic, Z = Zwitterionic.

[b] The cmc is the critical micelle concentration and depends on temperature and solution conditions. For solubilizing and purification of membrane proteins one has to work always above the cmc.

[c] Percentage (w/v) of detergent as used in the detergent screen.

Procedure:

1. Plate 2.0–3.0 × 10^7 culture cells (adherent or suspension) in culture flasks.
2. Wash the cells twice with prewarmed PBS.
3. Where required treat the cells either with β-MCD (10 m*M*) for 45 min and/or with cholesterol for 1 h in the serum free medium at 37 °C.
4. After treatment or 72 h of infection stimulate the cells where required
5. Lyse the cells in cold 750 μL of TNE-buffer containing 1% Triton X-100 supplemented with protease and phosphatase inhibitors for 30 min on ice.
6. Mix the lysate with 1500 μL of 70% Nycodenz, dissolved in TNE-buffer.
7. Load this mix in the bottom of 4 mL of polyallomer ultracentrifuge tube.
8. Overlay this bottom loaded mix with 25, 21.5, 18, 15, and 8% Nycodenz, prepared in TNE-buffer.
9. Spin the tube at 200,000*g* for 4 h at 4 °C using a swing-out rotor.
10. Fractions each of 350 μL from top to the bottom of the tube were analyzed by western blotting.
11. Probe the blot to confirm caveolin-1 positive (fractions 3–6) or detergent-resistant membrane fractions.

2.7. Nuclei and histone enrichment

2.7.1. Introduction

Understanding mechanisms of regulation of transcription by signal transduction events and other processes that occur on DNA in vitro analysis of chromatin is often important. The basic unit of chromatin is the nucleosome core, containing two copies each of the core histones H2A, H2B, H3, and H4 to form a histone octamer that wraps 145 bp of DNA in a left-handed superhelix. *In vivo*, chromatin is associated with linker histones (such as H1), which facilitate the ordered packing of nucleosomes. The particle with the linker histone is properly termed the nucleosome (or chromatosome), while the linker histone-free particle is the nucleosome core. The procedure described below is divided in two parts: preparation of washed nuclei and isolation of histones.

2.7.2. Preparation of a washed nuclear pellet

The isolation of nucleosomes and free histones requires clean nuclei as starting material. In this protocol, nuclei are released from cultured cells by homogenization and nonionic detergent NP-40. Several washes with a buffer containing detergent are then used to remove membranes and yield a pellet containing relatively clean nuclei. Subsequent extraction with sodium chloride removes most loosely bound proteins from chromatin. However, the NP-40 washes result in extracts less active than nuclear extracts prepared by other means. If both nuclear extract and chromatin are desired, the nuclear extract should be prepared as described below and the nuclear pellet from this procedure can be used to prepare chromatin (Schnitzler, 2001).

Materials and equipment:

- 2 *M* NaCl.
- Nuclear Pellet Buffer2 (NPB)/0.6 *M* KCl/10% (v/v) glycerol.
- Dounce homogenizer with type B pestle.
- Nuclear Pellet Buffer (NPB).
  - 20 m*M* HEPES, pH 7.5.
  - 3 m*M* $MgCl_2$.
  - 0.2 m*M* EGTA.
- Store up to several weeks at 4 °C. Immediately before use add:
  - 3 m*M* 2-mercaptoethanol.
  - 0.5 m*M* PMSF.
  - 1 μM pepstatin A.
  - 1 μM leupeptin.
- Leupeptin.
  - Prepare a 1 m*M* aqueous stock solution and store up to 1 year at − 20 °C.
- Lysis buffer (LB).
  - 20 m*M* HEPES, pH 7.5.
  - 0.25 *M* sucrose.
  - 3 m*M* $MgCl_2$.
  - 0.5% (v/v) Nonidet P-40 (NP-40).
- Store up to several weeks at 4 °C. Immediately before use add:
  - 3 m*M* 2-mercaptoethanol.
  - 0.5 m*M* PMSF.
  - 1 μM pepstatin A.
  - 1 μM leupeptin.

Note: Unless otherwise indicated, keep all solutions and materials on ice or at 4 °C.

Procedure:

1. 3×10^9 cells (HeLa) were washed twice in 1 L PBS.
2. Resuspend pellet in 40 mL *lysis buffer.*
3. Lyse the cells with ~15 strokes of a type B pestle Dounce homogenizer. Monitor lysis by light microscopy.
4. Centrifuge 15 min at 3000*g*, 4 °C.
5. Repeat resuspension and pelleting twice with lysis buffer and once with *NPB.*

6. Resuspend nuclei in 2 pellet vol *NPB* and measure the total volume of the suspension.
7. While gently stirring, add dropwise 1 total vol of *NPB2*.
8. Continue gentle stirring for 10 min at 4 °C.
9. Pellet nuclei 30 min at 17,500*g*, 4 °C.
10. Nuclear pellets can be frozen in liquid nitrogen or dry ice and kept for more than a year at − 80 °C.

2.8. Purification of core histones

2.8.1. Introduction

The four core histones can be readily purified by binding chromatin fragments to hydroxylapatite resin. Hydroxylapatite binds DNA very strongly (Simon and Felsenfeld, 1979). Histone H1 is eluted with 0.6 *M* NaCl and core histones are eluted from the resin-bound DNA at 2.5 *M* NaCl (Zagariya *et al.*, 1993).

Materials and equipment:

- Nuclear pellet (see above).
- HP buffer (sec below) with and without 2.5 *M* NaCl .
- BioGel HTP powder (Bio-Rad), with adsorption capacity 0.6 mg DNA per g dry powder.
- 2 × 15-cm column and accessories.
- Centriprep-10 concentrators (Amicon; optional).
- Additional reagents and equipment to determine protein concentration and SDS–PAGE.
- Hydroxylapatite (HP) buffer.
  50 m*M* sodium phosphate, pH 6.8, 0.6 *M* NaCl.
  Store up to several weeks at 4 °C.
  Immediately before use add: 1 m*M* 2-mercaptoethanol, 0.5 m*M* PMSF.

Procedure:

1. Resuspend ~ 2 mL nuclear pellet containing ~ 6 mg DNA in 25 mL HP buffer.
2. Stir gently for 10 min at 4 °C. To avoid proteolysis add 1 μM pepstatin A and 1 μM leupeptin to HP buffer.
3. Add 10 g dry BioGel HTP powder (hydroxyl apatite).
4. Add HP buffer to prepare 1:1 slurry. Pour this into a 2 × 15-cm column and collect the eluent.
5. Wash resin with 300 mL HP buffer (30 mL/h). The eluent from step 3 and the first column volume of wash contain H1.
6. Elute the core histones with HP buffer containing 2.5 *M* NaCl, collecting 8-mL fractions.
7. Analyze the purity of the core histones by SDS–PAGE.

8. If necessary, concentrate the core histones up to 10 mg/mL using Centriprep-10 concentrators.
9. Divide into aliquots, freeze on dry ice, and store up to 4 years at − 80 °C.

2.9. Nuclear extract enrichment

2.9.1. Introduction

The interaction of proteins with DNA is central to the control of many cellular processes including DNA replication, recombination and repair, transcription, and viral assembly. One technique to studying gene regulation and determining protein–DNA interactions is the electrophoretic mobility shift assay (EMSA) also referred as gel retardation analysis.

Material and equipment:

- *Nuclear Buffer 1 (NB1)*: 25 m*M* HEPES; pH 7.9, 5 m*M* KCl, 0.5 m*M* $MgCl_2$, 1 m*M* DTT, 100 m*M* PMSF and ad. 100 mL.
- *Nuclear Buffer 2 (NB2)*: 100 mL Buffer 1 + 1 mL NP-40 (1% IGEPAL).
- *Nuclear Buffer 3 (NB3)*: 1:1 mixture of buffer 1 + 2.
- *Nuclear Buffer 4 (NB4)*: 25 m*M* HEPES, pH 7.9; 350 m*M* NaCl.
- leupeptin, PMSF, aprotinin, pepstatin, and DTT (Calbiochem).
- IGEPAL CA-630 (Sigma).

Note: All components (buffers, protease inhibitors, etc) must be kept on ice during the procedure

Procedure:

The following protocol is optimized for about 10^7 cells.

1. Discard media and add 1 mL of ice-cold PBS.
2. Scrape cells into PBS, transfer to eppendorf tube and spin at 10,000*g* for 5 min.
3. Discard supernatant and resuspend pellet in 200 μL of *NB1* (scale-up or down proportionally).
4. Add 200 μL of *NB2* and rotate at 4 °C for 15 min.
5. Spin at 2500*g* for 1 min.
6. Transfer supernatant to new eppendorf tube (*cytoplasmic protein*-enriched fraction).
7. Add 100 μL of *NB3* and gently mix. Transfer the wash to the cytoplasmic protein preparation from step 6. A short spin may be necessary if the pellet was disturbed.
8. Add 500 μL of *NB4* and rotate for 1 h at 4 °C.
9. Spin at max speed for 10 min. Supernatant after this spin is the *nuclear protein* preparation.

2.10. Immunoaffinity reagents for organelle enrichment

2.10.1. Introduction

Cell fractions often contain more than one type of organelle even after differential and equilibrium density gradient centrifugation. The usage of antibodies is described since 1975 (Thompson and Miller, 1975) for diverse organelles and depends heavily on the quality of the antibody used. Besides antibodies also the labeling of organelles like plasma membranes with biotin and the isolation via streptavidin is a common procedure (Zeheb and Orr, 1986). Preenriched density gradient fractions can be further purified by immunological techniques, using monoclonal antibodies for various organelle-specific membrane proteins. One example is the purification of a particular class of cellular vesicles whose outer surface is coated with the protein clathrin. An antibody to clathrin, bound to protein A or G (depending on antibody subtype and species) coupled to a solid phase like agarose, polymeric resins or preferably to magnetic particles which allows for an orthogonal separation, not only gravity, which might interfere with large organelles in crude extracts. The target specific antibody selectively binds to the vesicles in a crude preparation of membranes, and the whole antibody complex can then be isolated by low-speed centrifugation or magnetic separation.

All cells contain a dozen or more different types of small membrane-limited vesicles of about the same size (50–100 nm in diameter) and density. Because of their similar size and density, these vesicles are difficult to separate from one another by centrifugation techniques. Immunological techniques are particularly useful for purifying specific classes of such vesicles.

Material and equipment:

- Protein A/G coupled solid phase in HEPES, pH 7.0, containing 1% BSA, 100 m*M* sodium chloride, and 0.01% sodium azide.
- Monoclonal or polyclonal AB.
- Wash buffer I (WB1).
  - 20 m*M* HEPES, pH 7.0
  - 1% BSA
- Wash buffer II (WB2):
  - 20 m*M* HEPES, pH 7.0
  - 100 m*M* sodium chloride
- Elution buffer (EB):
  - 0.1% SDS, 50 m*M* Tris, pH 6.5
  - 100 m*M* DTT and 10% glycerol

Procedure:

Prepare a precleared cell lysate as described above:

1. Incubate the lysate, antibody (according to the suppliers information), Protein A/G and solid phase (30 μL 1:1 slurry solution) overnight at 4 °C with gentle agitation.

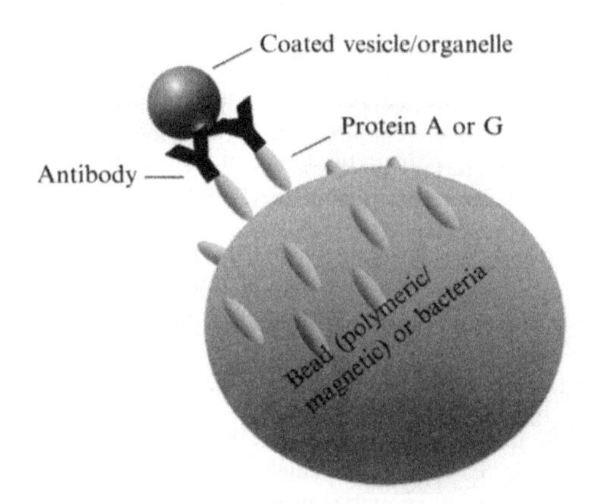

Figure 19.6 Scheme of affinity organelle and vesicle isolation.

2. Centrifuge at 500*g* for 5 min at 4 °C.
3. Wash the pellet three times with 500 μL Wash buffer I.
4. Wash the pellet twice with 500 μL Wash buffer II.
5. Elute the coated vesicles/organelles with 30 μL elution buffer or under mild extraction conditions if required.
6. Centrifuge at 5000*g* for 1 min. The supernatant contains the enriched clathrin coated vesicles (Fig. 19.6).

REFERENCES

Abdolzade-Bavil, A., Hayes, S., Goretzki, L., Kröger, M., Anders, J., and Hendriks, R. (2004). Convenient and versatile subcellular extraction procedure, that facilitates classical protein expression profiling and functional protein analysis. *Proteomics* **4,** 1397–1405.

Billecke, C., Malik, I., Movsisyan, A., Sulghani, S., Sharif, A., Mikkelsen, T., Farrell, N. P., and Bögler, O. (2006). Analysis of glioma cell platinum response by metacomparison of two-dimensional chromatographic proteome profiles. *Mol. Cell Proteomics* **5,** 35–42.

Edelman, M., Epstein, H. T., and Schiff, J. A. (1966). Isolation and characterization of DNA from the mitochondrial fraction of Euglena. *J. Mol. Biol.* **17,** 463–469.

Fisher, W. D., and Cline, G. B. (1963). A density gradient for the isolation of metabolically active thymus nuclei. *Biochim. Biophys. Acta* **68,** 640–642.

Ford, T., Graham, J., and Rickwood, D. (1994). Iodixanol: A nonionic iso-osmotic centrifugation medium for the formation of self-generated gradients. *Anal. Biochem.* **220,** 360–36612.

Goldberg, S. (2008). Mechanical/physical methods of cell disruption and tissue homogenization. *Methods Mol. Biol.* **424,** 3–22.

Graham, J., Ford, T., and Rickwood, D. (1994). The preparation of subcellular organelles from mouse liver in self-generated gradients of iodixanol. *Anal. Biochem.* **220,** 367–373.

Kurokawa, M., Kato, M., and Sakamoto, T. (1965). Distribution of sodium-plus-potassium-stimulated adenosine-triphosphatase activity in isolated nerve-ending particles. *Biochem. J.* **97,** 833–844.

Nakamura, E., Kozaki, K., Tsuda, H., Suzuki, E., Pimkhaokham, A., Yamamoto, G., Irie, T., Tachikawa, T., Amagasa, T., Inazawa, J., and Imoto, I. (2008). Frequent silencing of a putative tumor suppressor gene melatonin receptor 1 A (MTNR1A) in oral squamous-cell carcinoma. *Cancer Sci.* **99,** 1390–1400.

Pertoft, H., Rubin, K., Kjellén, L., Laurent, T. C., and Klingeborn, B. (1977). The viability of cells grown or centrifuged in a new density gradient medium. *Percoll. Exp. Cell Res.* **110,** 449–457.

Ramsby, M. L., Makowski, G. S., and Khairallah, E. A. (1994). Differential detergent fractionation of isolated hepatocytes: Biochemical, immunochemical and two-dimensional gel electrophoresis characterization of cytoskeletal and noncytoskeletal compartments. *Electrophoresis* **15,** 265–277.

Rickwood, D., Ford, T., and Graham, J. (1982). Nycodenz: A new nonionic iodinated gradient medium. *Anal Biochem.* **123,** 23–31.

Ross, P. L., Huang, Y. L. N., Marchese, J. N., Williamson, B., Parker, K., Hattan, S., Khainovski, N., Pillai, S., Dey, S., Daniels, S., *et al.* (2004). Multiplexed protein quantitation in Saccharomyces cerevisiae using amine-reactive isobaric tagging reagents. *Mol. Cell. Proteomics* **3,** 1154–1169.

Sadowski, P. G., Dunkley, T. P., Shadforth, I. P., Dupree, P., Bessant, C., Griffin, J. L., and Lilley, K. S. (2006). Quantitative proteomic approach to study subcellular localization of membrane proteins. *Nat. Protoc.* **1,** 1778–1789.

Schnitzler, G. R. (2001). Isolation of histones and nucleosome cores from mammalian cells. *Curr. Protoc. Mol. Biol.* Chapter 21, Unit 21.5.

Shadforth, I., Dunkley, T., Lilley, K., and Bessant, C. (2005). i-Tracker: For quantitative proteomic using iTRAQ. *BMC Genomics* **6,** 145.

Shah, M. B., and Sehgal, P. B. (2007). Nondetergent isolation of rafts. *Methods Mol. Biol.* **398,** 21–28.

Sherrier, D. J., Prime, T. A., and Dupree, P. (1999). Glycosylphosphatidylinositol-anchored cell-surface proteins from Arabidopsis. *Electrophoresis* **20,** 2027–2035.

Simon, R. H., and Felsenfeld, G. (1979). A new procedure for purifying histone pairs H2A + H2B and H3 + H4 from chromatin using hydroxylapatite. *Nucleic Acids Res.* **6,** 689–696.

Storrie, B., and Madden, E. A. (1990). Isolation of subcellular organelles. *Methods Enzymol.* **182,** 203–225.

Thompson, E. B., and Miller, J. V. (1975). Enrichment of polysomes synthesizing a specific protein by use of affinity chromatography. *Methods Enzymol.* **40,** 266–273.

Van Veldhoven, P. P., Baumgart, E., and Mannaerts, G. P. (1996). Iodixanol (Optiprep), an improved density gradient medium for the iso-osmotic isolation of rat liver peroxisomes. *Anal. Biochem.* **237,** 17–23.

von Hagen, J. (2008). Proteomics Sample Preparation. New York: Wiley VCH, New York.

Washburn, M. P., Wolters, D., and Yates, J. R. (2001). Large-scale analysis of the yeast proteome by multidimensional protein identification technology. *Nat. Biotechnol.* **19,** 242–247.

Weber, P. J., Weber, G., and Eckerskorn, C. (2004). Isolation of organelles and prefractionation of protein extracts using free-flow electrophoresis. *Curr. Protoc. Protein Sci.* Chapter 22, Unit 22.5.

Willemsen, O., Machtejevas, E., and Unger, K. K. (2004). Enrichment of proteinaceous materials on a strong cation-exchange diol silica restricted access material: protein-protein displacement and interaction effects. *J. Chromatogr. A.* **1025,** 209–216.

Zagariya, A., Khrapunov, S., and Zacharias, W. (1993). Rapid method for the fractionation of nuclear proteins and their complexes by batch elution from hydroxyapatite. *J. Chromatogr.* **648,** 275–278.

Zeheb, R., and Orr, G. A. (1986). Use of avidin-iminobiotin complexes for purifying plasma membrane proteins. *Methods Enzymol.* **122,** 87–94.

SECTION FIVE

PURIFICATION PROCEDURES: BULK METHODS

CHAPTER TWENTY

Protein Precipitation Techniques

Richard R. Burgess

Contents

1. Introduction	332
2. Ammonium Sulfate Precipitation	332
2.1. Principles	332
2.2. Basic procedure	334
2.3. Doing an ammonium sulfate precipitation test	336
2.4. Comments/problems/solutions	337
3. Polyethyleneimine Precipitation	337
3.1. Principles	337
3.2. Different modes of use of PEI	338
3.3. Basic procedure for Strategy C	338
3.4. Doing an PEI precipitation test	340
3.5. An example of using PEI to precipitate a basic protein bound to DNA	340
4. Other Methods	341
4.1. Ethanol and acetone precipitation	341
4.2. Isoelectric precipitation	341
4.3. Thermal precipitation	341
4.4. Polyethylene glycol (nonionic polymer) precipitation	341
5. General Procedures When Fractionating Proteins by Precipitation	341
References	342

Abstract

After cell lysis, the most often used second step in a protein purification procedure is some sort of a rapid, bulk precipitation step. This is commonly accomplished by altering the solvent conditions and taking advantage of the changes in solubility of your protein of interest relative to those of many of the other proteins and macromolecules in a cell extract. This chapter will focus on the two most widely used precipitation methods: (1) ammonium sulfate precipitation and (2) polyethyleneimine (PEI) precipitation. These two methods work through entirely different principles, but each can achieve significant enrichment of target protein if optimized and applied carefully.

McArdle Laboratory for Cancer Research, University of Wisconsin–Madison, Madison, Wisconsin, USA

Methods in Enzymology, Volume 463
ISSN 0076-6879, DOI: 10.1016/S0076-6879(09)63020-2

1. Introduction

For both laboratory scale and larger scale protein fractionation, there is a need for a quick, bulk precipitation to remove much of cellular protein and other components. It is especially important to remove proteases as early in the procedure as possible to avoid protein degradation. This precipitation must be rapid, gentle, scalable, and relatively inexpensive. In addition to the fractionation, it can also achieve a significant concentration of the enriched protein. While many different precipitation methods have been used over the last hundred years, ammonium sulfate (AS) has remained the most widely used and polyethyleneimine (PEI) has increased in popularity, especially for acidic proteins. These two methods will be discussed in detail, followed briefly by several other precipitation methods and some general advice on handling precipitates and obtaining maximal purification from your precipitation step. An extensive and very useful general overview of various types of protein precipitation procedures can be found in Scopes (1994). An excellent review of AS and organic solvent (ethanol and acetone) precipitation can be found in Englard and Seifter (1990).

2. Ammonium Sulfate Precipitation

2.1. Principles

While several salts can be used as precipitants, AS has several properties that make it the most useful. It is very stabilizing to protein structure, very soluble, relatively inexpensive, pure material is readily available, and the density of a saturated solution (4.1 *M*) at 25 °C ($\rho = 1.235$ g/cm^3) is not as high as another salting-out agent, potassium phosphate (3 *M*, $\rho = 1.33$ g/cm^3).

Figure 20.1 shows a typical protein solubility curve where the log of the protein solubility is plotted as a function of AS concentration. The main features of this curve are a region at low salt where the solubility increases (called "salting in"), and then a region where the log solubility decreases linearly with increasing AS concentration (called "salting out"). The latter part of the curve can be described by the equation $\log_{10}S = \beta - K_s(\Gamma/2)$ where S in the solubility of the protein in mg/ml of solvent, $\Gamma/2$ is the ionic strength, and β and K_s are constants characteristic of the protein in question. K_s is a measure of the slope of the line and β is the log of the solubility if the salting-out curve is extrapolated to zero ionic strength. In general, most proteins have similar K_s values but vary considerably in their β value.

Suppose that the curve in Fig. 20.1 is valid for your protein and that the concentration of your protein in a cell extract is 1 mg/ml. The upper

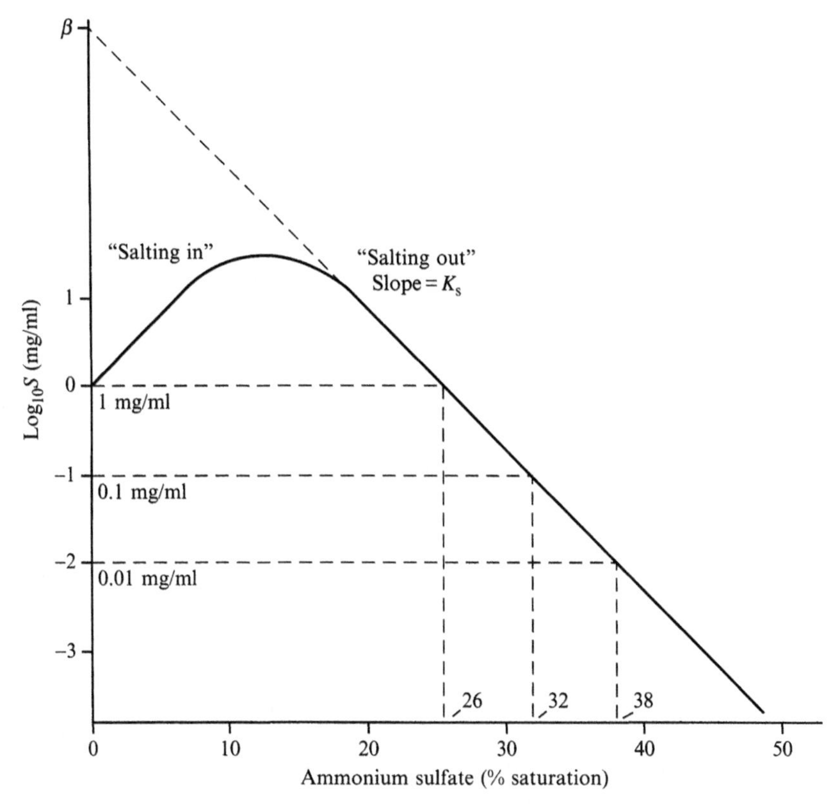

Figure 20.1 Ammonium sulfate solubility curve for a hypothetical protein. This represents the log solubility of a hypothetical protein as a function of percent saturation of ammonium sulfate. The "salting-out" line follows the relationship $\log S = \beta - K_s(\Gamma/2)$ as described in the text, where $\Gamma/2$ is the ionic strength, which here is given as ammonium sulfate percent saturation.

horizontal dotted line intercepts the solubility line at $\log S = 0$ ($S = 1$ mg/ml) and at an AS percent saturation of 26%. This means that if you add AS to 26% saturation, all of your protein would be soluble. Now if you increased the AS to 32% saturation (the middle horizontal dotted line), the $\log S$ would be -1 ($S = 0.1$ mg/ml) so 90% of your protein would become insoluble and precipitate out. For this extract, an excellent strategy would be to make a 26–32% saturated AS cut: add AS to 26%, spin out insoluble material, and then make the supernatant 32% saturated and collect what precipitates, which would contain 90% of your protein. You would remove those proteins and cell components that precipitate at 26% saturation and those that fail to precipitate at 32% saturation.

It is instructive to consider what would happen if you diluted the extract 10-fold with buffer. Now the initial concentration of your protein in the extract is 0.1 mg/ml or log $S = -1$. You can add AS to 32% saturation and your protein will not precipitate. To achieve 90% precipitation of your protein, you would have to increase the AS to about 38% saturation (bottom horizontal dotted line) or carry out a 32–38% saturated AS cut. You would end up having to use more than 10 times as much AS with the diluted extract to obtain your protein. This illustrates how important it is to specify the concentration of your extract.

You do not usually have a curve like that shown in Fig. 20.1 for your protein of interest so you have to determine the appropriate AS concentrations experimentally as described below.

2.2. Basic procedure

While there are numerous variations on AS precipitation, the most common ones are to add solid AS to a protein extract to give a certain percent saturation. Adding an amount of solid AS based on Table 20.1 is convenient, reproducible, and practical.

1. Generally one determines a lower percent saturation at which the protein of interest just does not precipitate and a higher percent saturation that gives $>90\%$ precipitation as described in the section below.
2. Add solid AS to reach the lower value. Take care to add the AS slowly with rapid stirring so that the local concentration does not "overshoot" the target value. Some people carefully grind the solid AS with a mortar and pestle to a fine powder that dissolves rapidly. Once the AS is completely dissolved, allow the precipitation to continue for about 30 min. This is a compromise between waiting several hours as precipitation slowly approaches equilibrium and the desire to move along with the purification and not to introduce long delays in the procedure. Generally, one carries out all operations in an ice bucket or cold room.
3. Centrifuge at about 10,000 $\times\, g$ for about 10 min in a precooled rotor to pellet the material that is insoluble.
4. Carefully pour off the supernatant and determine its volume. Determine the grams of AS from Table 20.1 to go from the lower desired percent saturation to the final higher percent saturation. Again add the AS slowly with rapid mixing to avoid high local concentrations and let the solution sit for 30 min to allow precipitation to occur.
5. Centrifuge as above. Let the pellet drain for about 1 min to remove as much as possible of the supernatant. If you have carried out the test precipitation carefully, the pellet will contain 90% or more of your target protein. This protein can be dissolved in an appropriate buffer and after either dialysis, desalting, or dilution used in the next step of the purification.

Table 20.1 Final concentration of ammonium sulfate: Percentage saturation at 0 °C[a]

Initial concentration of ammonium sulfate (percentage saturation at 0 °C)	Percentage saturation at 0 °C																
	20	25	30	35	40	45	50	55	60	65	70	75	80	85	90	95	100
	Solid ammonium sulfate (g) to be added to 1 l of solution																
0	106	134	164	194	226	258	291	326	361	398	436	476	516	559	603	650	697
5	79	108	137	166	197	229	262	296	331	368	405	444	484	526	570	615	662
10	53	81	109	139	169	200	233	266	301	337	374	412	452	493	536	581	627
15	26	54	82	111	141	172	204	237	271	306	343	381	420	460	503	547	592
20	0	27	55	83	113	143	175	207	241	276	312	349	387	427	469	512	557
25		0	27	56	84	115	146	179	211	245	280	317	355	395	436	478	522
30			0	28	56	86	117	148	181	214	249	285	323	362	402	445	488
35				0	28	57	87	118	151	184	218	254	291	329	369	410	453
40					0	29	58	89	120	153	187	222	258	296	335	376	418
45						0	29	59	90	123	156	190	226	263	302	342	383
50							0	30	60	92	125	159	194	230	268	308	348
55								0	30	61	93	127	161	197	235	273	313
60									0	31	62	95	129	164	201	239	279
65										0	31	63	97	132	168	205	244
70											0	32	65	99	134	171	209
75												0	32	66	101	137	174
80													0	33	67	103	139
85														0	34	68	105
90															0	34	70
95																0	35
100																	0

[a] Reprinted from Englard and Seifter (1990), which was adapted from Dawson *et al.* (1969).

2.3. Doing an ammonium sulfate precipitation test

Generally one can precipitate 90% of a given protein with a 10% increase in AS saturation so one should restrict the range of the "AS cut" to no more than 10% (the proteins that are just soluble at 30% saturation but precipitate at 40% saturation are referred to as the 30–40% AS cut).

Figure 20.2 illustrates a method to determine the optimal AS precipitation conditions using only two centrifugation steps. Basically you place a volume of cell extract, for example, 10 ml in each of five tubes. You add with mixing amounts of solid AS to give 20%, 30%, 40%, 50%, and 60% saturation based on Table 20.1, let sit 30 min to allow precipitation, and then centrifuge to pellet the insoluble material. The pellets represent the 20%, 30%, 40%, 50%, and 60% saturated AS pellets. The volumes of the corresponding supernatants are determined and again solid AS is added to raise each to a 10% higher level of saturation. Again you mix, allow 30 min to precipitate, and then spin. The five pellets are the 20–30%, 30–40%, 40–50%, 50–60%, and 60–70% AS cuts. Each of these is dissolved in buffer and assayed for enzyme activity and total protein and perhaps subjected to SDS gel analysis. Most of the activity should be in one of the cuts, but if, for example, half is in 30–40% cut and half is in the 40–50% cut, then perhaps a 35–45% cut would be optimal. While this test may seem onerous, it is really

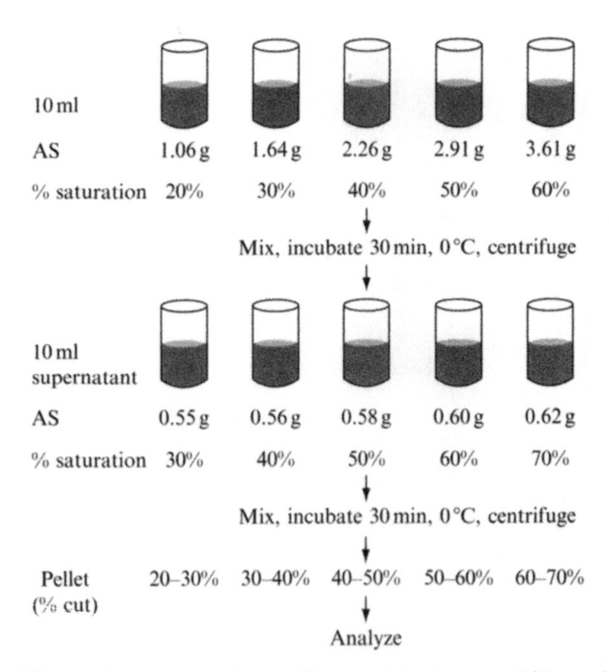

Figure 20.2 How to do an ammonium sulfate precipitation test. This test is carried out as described in the text and is self-explanatory.

quite an efficient way to determine the optimal conditions that will result in higher enrichment in this important step.

2.4. Comments/problems/solutions

1. Pellet is not solid. If the AS pellet is not firm after centrifugation it will be difficult to cleanly pour off the supernatant. One solution is simply to centrifuge 50% longer such that the precipitate that sedimented to the bottom of the tube has more time to compact. Another reason for a loose pellet is the presence of DNA that increases viscosity and slows sedimentation rates. If viscosity is a problem, it can be reduced by sonicating longer to break cells and shear DNA into shorter pieces. Another approach is to treat with the recombinant nuclease Benzonase (EMD/Novagen).
2. Pellet floats in high concentrations of AS. Since the density of very high concentrations of AS approach that of protein aggregates, the AS precipitate might float rather than sediment to the bottom of the tube during centrifugation. This can be a problem especially when your protein contains lipid or if there are nonionic detergents around that bind to the protein and decrease its density.
3. Published protocols are often hard to follow. Many published procedures fail to indicate the AS saturation convention (saturated at 0, 20, or 25 °C) or the protein concentration of the extract. As cautioned above, the amount of AS needed to achieve a given precipitation is dependent on the protein concentration.
4. You must interrupt an AS precipitation procedure. If you must stop the procedure, leave the protein as an AS precipitate. Proteins are remarkably stable in AS, either as a suspension of precipitated protein or as a pellet.

3. Polyethyleneimine Precipitation

3.1. Principles

PEI, whose trade name is Polymin P, is a basic cationic polymer made by BASF in large quantities for use in the textile and paper industry. PEI is a product of polymerization of ethyleneimine to yield a basic polymer with the structure: $CH_3CH_2N{-}({-}CH_2CH_2{-}NH{-})_n{-}CH_2CH_2NH_2$. Typically, n equals 700–2000 to give a molecular weight range of 30,000–90,000 Da. Since the pK_a value of the imino group is 10–11, PEI is a positively charged molecule in solutions of neutral pH. The use of PEI in protein fractionation originated at Boehringer Mannheim and was published by Zillig *et al.* (1970). More extensive examples of its application in protein purification and several reviews have been published (Burgess, 1991; Burgess and Jendrisak, 1975; Jendrisak, 1987; Jendrisak and Burgess, 1975).

PEI can be thought of as similar to soluble DEAE cellulose. It binds to negatively charged macromolecules such as nucleic acid and acidic proteins and forms a network of PEI and bound acidic molecules that rapidly precipitates. The binding is stoichiometric. A heavy precipitate rapidly forms that can be pelleted by centrifugation for 5 min at 5000 rpm. Whether an acidic protein binds to PEI depends on the salt concentration. At low salt (0.1 *M* NaCl), a mildly acidic protein will bind and precipitate, but at intermediate salt (0.4 *M* NaCl), it will elute from the polymer and become soluble. A highly acidic protein will bind at low salt, not be solubilized at intermediate salt, and will be eluted at high salt (0.9 *M* NaCl). It should be noted that when a protein is eluted from a PEI pellet both the protein and the PEI become soluble. Thus it is necessary to remove the PEI from the protein before returning to low salt (see below).

3.2. Different modes of use of PEI

There are three different strategies for the use of PEI precipitation.

Strategy A: Precipitate with PEI at high salt (1 *M* NaCl). This precipitates the nucleic acids and leaves almost all protein in the supernatant.

Strategy B (for neutral or basic proteins): Precipitate with PEI at 0.1 *M* NaCl to remove nucleic acids and acidic proteins. This leaves protein of interest in the supernatant.

Strategy C (for acidic proteins such as E. coli RNA polymerase): This protocol will be presented in detail below and is based on Burgess and Jendrisak (1975) as refined by Burgess and Knuth (1996).

3.3. Basic procedure for Strategy C

1. Prepare a 10% (v/v) [5% (w/v)] PEI stock solution. PEI comes as viscous liquid that is 50% (w/v). (We use PEI from MP Biochemicals, $M_w = 50{,}000$–100,000, but one can also use material from other sources, for example, Sigma and Aldrich, $M_w = 750{,}000$). Ten milliliter is diluted to 70 ml with ddH_2O and concentrated HCl (3.8–4.0 ml) is added until the pH reaches 7.9. The volume is made up to a final volume of 100 ml with ddH_2O. This stock solution is stable at cold room or room temperature indefinitely. It should be noted that some companies sell PEI solutions that have been diluted one to one with ddH_2O to reduce viscosity and aid in dispensing. This material is only 25% (w/v).
2. Break *E. coli* cells (about 3 g wet weight cell pellet) by sonication in 30 ml buffer containing 50 m*M* Tris–HCl, pH 7.9, 5% glycerol, 0.1 m*M*

EDTA, 0.1 m*M* DTT, and 0.15 *M* NaCl. Centrifuge out cell debris at 15,000 rpm for 15 min. All operations are carried out on ice.

3. Based on a PEI precipitation test for your system (see below) add 10% (v/v) PEI, pH 7.9 to a final concentration of, for example, 0.3% (v/v) and mix well for 5 min to allow formation of a dense white precipitate.
4. Centrifuge at 5000 rpm for 5 min. *Note: Do not centrifuge too hard or the pellet will be harder to resuspend.* Decant the 0.3% PEI supernatant and save for later analysis. Let the pellet drain for 1–2 min to get rid of as much of the supernatant as possible.
5. Thoroughly resuspend the 0.3% PEI pellet in 30 ml of the above buffer containing 0.4 *M* NaCl. If available, the Tissue Tearor homogenizer (BioSpec Products, Inc. Cat # 985370-07) works very well to finely resuspend the pellet and effectively washes out proteins physically trapped in the pellet, and elutes out mildly acidic protein that are weakly bound to the PEI in the pellet. Let sit 5 min and then centrifuge at 5000 rpm for 5 min and decant the 0.4 *M* NaCl wash.
6. Resuspend thoroughly the 0.4 *M* NaCl pellet in 30 ml of buffer above but containing 0.9 *M* NaCl. This elutes more acidic proteins (like *E. coli* RNA polymerase), but leaves the nucleic acids in the pellet (it takes about 1.6 *M* NaCl to elute the nucleic acids). Let sit about 5 min, mix, and centrifuge at 15,000 rpm for 10 min.
7. To the 0.9 *M* NaCl eluate, add solid AS to 60% saturation (add 3.61 g per 10 ml of eluate). Mix well and let precipitation occur for at least 30 min. Centrifuge at 15,000 rpm for 10 min. Let pellet drain for 5 min. The pellet contains the AS precipitated protein, but almost all of the PEI remains in the supernatant. While traces of PEI are trapped in the pellet, they usually do not interfere with subsequent operations. If it is necessary to more completely remove these PEI traces, one can resuspend the pellet in buffer containing 60% saturated AS and recentrifuge.

This procedure typically gives a sixfold purification of RNA polymerase from other proteins, greater than 90% recovery, and a nearly complete removal of nucleic acids in 1–2 h.

3.3.1. Additional comments

1. It is important to emphasize that it is necessary to remove PEI from the 0.9 *M* NaCl eluate. If you merely dilute to low salt or dialyze to low salt, the proteins will again bind PEI and reprecipitate.
2. We have found that the PEI precipitation can occur even in the presence of 1% Triton X-100.
3. In contrast to AS precipitation, when you dilute an extract 10-fold with buffer, you use the same total amount of PEI (e.g., if you found that 0.3% PEI gave good precipitation of your protein with a normal extract,

you would only have to add 0.03% PEI to achieve the same precipitation with the 10-fold diluted extract, but of course there would be 10 times the volume of extract). This reflects the fact that PEI binds tightly to the acidic components and essentially titrates them.

3.4. Doing an PEI precipitation test

Since one cannot predict how much PEI to add to precipitate a given acidic protein nor how much salt it will take to elute the protein from the PEI pellet, one is advised to carry out simple PEI precipitation and elution tests (Burgess and Jendrisak, 1975; Burgess and Knuth, 1996). Basically this involves taking about six 200-μl samples into six microfuge tubes and adding 10% (v/v) PEI to final concentrations of 0%, 0.1%, 0.2%, 0.3%, 0.4%, and 0.5% (v/v). Mix well and microfuge for 1 min at high speed. Analyze the supernatants by enzyme activity or by SDS polyacrylamide gel electrophoresis and, if necessary, by Western blot to determine the minimum amount of PEI needed to precipitate all of the target protein. Let us say that it takes 0.3%. Now prepare a set of six microfuge tubes which each contain a small 0.3% PEI pellet prepared as above and add 200 μl of buffer containing 0, 0.2, 0.4, 0.6, 0.8, and 1.0 *M* NaCl. Resuspend well. Let sit for 15 min, microcentrifuge, and analyze supernatant for your protein as above. The highest salt concentration that does not elute any of your protein is used as a wash and the salt concentration where all of your protein is eluted is used as the eluent.

3.5. An example of using PEI to precipitate a basic protein bound to DNA

Recently (Duellman and Burgess, 2008), in trying to purify the very basic protein Epstein–Barr virus nuclear antigen 1 (EBNA1) expressed in *E. coli*, we carried out a PEI precipitation test at 0.1 *M* NaCl to see if we could precipitate the nucleic acid and acidic proteins and leave the basic EBNA1 in the supernatant (this is Strategy B). To our surprise, the EBNA1 precipitated as we added some PEI (0.15%), but then failed to precipitate when we added more PEI (0.4%). It appears that the EBNA1 was bound to DNA and precipitated out with the DNA. However, at higher PEI concentrations the PEI preferentially bound the DNA and displaced the EBNA1. We found that precipitating with 0.15% PEI, washing with 0.3 *M* NaCl, and then eluting out of the PEI pellet with 0.8 *M* NaCl gave both a good enrichment for EBNA1 and rapid removal of nucleic acids.

4. Other Methods

This chapter has focused on AS and PEI precipitation. Other methods for protein precipitation mentioned very briefly below have been well described in numerous publications (see, Englard and Seifter, 1990; Ingham, 1990; Scopes, 1994).

4.1. Ethanol and acetone precipitation

Precipitation with organic solvents, such as ethanol and acetone, has been in use for well over a hundred years, but is probably best known for its use in fractionating human serum in the classic work of Cohen and Edsall. Care must be taken to carry out precipitations at very cold temperatures to avoid protein denaturation.

4.2. Isoelectric precipitation

Proteins are less soluble at their isoelectric point where they have zero net charge and can most easily approach each other with minimal charge repulsion. Since proteins are also less soluble at very low ionic strength, isoelectric precipitation is usually done at very low or no salt.

4.3. Thermal precipitation

In this method, cell extracts are heated to a temperature at which many proteins denature and precipitate, where the protein of interest is more stable and stays soluble. This approach is particularly useful in purifying enzymes from thermophiles expressed in *E. coli* where the extract is heated to a high enough temperature, often to denature and precipitate almost all *E. coli* protein, leaving the heat stable enzyme in solution.

4.4. Polyethylene glycol (nonionic polymer) precipitation

This subject has been reviewed extensively by Ingham (1990).

5. General Procedures When Fractionating Proteins by Precipitation

1. Thorough resuspension of precipitated protein pellets is important during washing or elution. While a pellet may seem quite solid, there is a very significant amount of supernatant trapped in a pellet and adhering

to the walls of the centrifuge tube. As mentioned earlier, it is wise to let the pellet drain well to remove as much of the supernatant as possible. If the pellet is large compared to the total amount of supernatant then it is recommended that one resuspend the pellet in 10 volumes of an appropriate buffer to remove supernatant proteins trapped in the pellet. For example, with a 40% saturated AS precipitate, one can wash by resuspending the pellet in 40% saturated AS and recentrifuging. For washing pellets of PEI precipitated material, washing is very useful and the use of a homogenizer like a Tissue Tearor is recommended to break up the precipitate into a very fine suspension. If resuspension is not thorough, then material that you want to wash out is not efficiently removed and the fractionation is less effective.

2. Try to avoid frothing when mixing. Whipping air into a protein solution can promote oxidation of proteins and also cause protein denaturation at the air–water interface.

REFERENCES

Burgess, R. R. (1991). The use of polyethyleneimine in the purification of DNA binding proteins. *Meth. Enzymol.* **208,** 3–10.

Burgess, R. R., and Jendrisak, J. J. (1975). A procedure for the rapid, large-scale purification of *E. coli* DNA-dependent RNA polymerase involving Polymin P precipitation and DNA-cellulose chromatography. *Biochemistry* **14,** 4634–4638.

Burgess, R. R., and Knuth, M. W. (1996). Purification of a recombinant protein overproduced in *E. coli*. *In* "Strategies for Protein Purification and Characterization: A Laboratory Manual", (D. Marshak, J. Kadonaga, R. Burgess, M. Knuth, W. Brennan, Jr., and S.-H. Lin, eds.), pp. 219–274. Cold Spring Harbor Press, Cold Spring Harbor, NY.

Dawson, R. M. C., Elliot, D. C., Elliot, W. H., and Jones, K. M. (1969). *In* "Data for Biochemical Research." 2nd edn., p. 616. Oxford University Press, Oxford.

Duellman, S. J., and Burgess, R. R. (2008). Large-scale Epstein-Barr virus EBNA1 protein purification. *Protein Expr. Purif.* **63,** 128–133.

Englard, S., and Seifter, S. (1990). Precipitation techniques. *Meth. Enzymol.* **182,** 287–300.

Ingham, K. C. (1990). Precipitation of proteins with polyethylene glycol. *Meth. Enzymol.* **182,** 301–306.

Jendrisak, J. J. (1987). The use of PEI in protein purification. *In* "Protein Purification: Micro to Macro", (R. Burgess, ed.), pp. 75–97. A.R. Liss, New York.

Jendrisak, J. J., and Burgess, R. R. (1975). A new method for the large-scale purification of wheat germ DNA-dependent RNA polymerase II. *Biochemistry* **14,** 4639–4645.

Scopes, R. K. (1994). Protein Purification, Principles and Practice. 3rd edn. pp. 7–101. Springer-Verlag, New York.

Zillig, W., Zechel, K., and Halbwachs, H. J. (1970). A new method of large scale preparation of highly purified DNA-dependent RNA-polymerase from *E. coli*. *Hoppe Seylers Z. Physiol. Chem.* **351,** 221–224.

CHAPTER TWENTY-ONE

Affi-Gel Blue for Nucleic Acid Removal and Early Enrichment of Nucleotide Binding Proteins

Murray P. Deutscher

Contents

1. A Representative Protocol 344
Reference 345

Abstract

Passage of an extract or supernatant fraction through a column of Affi-Gel Blue and batchwise elution can be a rapid and effective early procedure for removal of nucleic acid, concentration of the sample and purification of nucleotide binding proteins.

Important points to be considered for the early steps of a protein purification procedure are: (1) how to deal with the large volume and large amounts of material; (2) how to remove nucleic acids; and (3) how to enrich, at least to a small degree, the protein of interest, being mindful as well of points 1 and 2.

In the initial steps of a protein purification, it is usual to be working with large volumes of fairly concentrated extract. This can be quite cumbersome, and as a consequence, one generally will use a bulk procedure such as precipitation to partially fractionate the extract and also to reduce the volume of material as the precipitate can be redissolved in a smaller volume. Common precipitants for these purposes include ammonium sulfate, ethanol or acetone, polyethylene glycol, and polyethyleneimine (see Chapter 20).

The nucleic acids present in crude extracts can cause a variety of problems and need to be removed early in the purification procedure. High-molecular-weight DNA can lead to elevated viscosity making samples difficult to work with. DNA and RNA also can interfere with assays,

Department of Biochemistry and Molecular Biology, University of Miami School of Medicine, Miami, Florida, USA

Methods in Enzymology, Volume 463
ISSN 0076-6879, DOI: 10.1016/S0076-6879(09)63021-4

particularly with enzymes that act on nucleic acids. As a consequence, assays of extracts may dramatically underestimate enzyme activity. Most importantly, nucleic acids may hamper subsequent purification steps. In as much as DNA and RNA are polyanions, they can compete with chromatographic resins for protein binding. For all of these reasons, early removal of nucleic acids is beneficial, and various protocols have been used for this purpose. Among them are treatment with nucleases, precipitation with certain reagents or binding to anion exchange resins.

An ideal early purification step would be one that could simultaneously concentrate the sample, remove nucleic acids and afford some enrichment of the desired protein as well. We have found that the resin, Affi-Gel Blue, is particularly useful in this regard. Over the years, Affi-Gel Blue has been used as an affinity chromatography resin in the purification of hundreds of proteins, primarily later in a purification procedure (see Bio-Rad Bulletin 1107 for references, available at biorad.com). In those situations, the partially purified protein of interest is bound to the resin and eluted with a salt gradient or with ligand specific to the protein. For the purposes of the discussion here, we focus on the use of Affi-Gel Blue very early in a purification protocol, generally in a bulk fashion, to accomplish the goals stated above.

Affi-Gel Blue is a cross-linked agarose bead to which the dye Cibacron Blue F3GA has been covalently attached. Conjugates of this dye to other supports also are commercially available, but have not been examined with regard to the procedure described here. The Cibacron Blue dye has ionic, hydrophobic, and aromatic character, and consequently, displays affinity for many proteins. One large class of enzymes for which binding to the dye appears particularly strong are those that have nucleotide cofactors or which act on nucleotide-containing substrates, including DNA and RNA. Thus, the procedure to be described has proven to be extremely useful for enzymes of this group (e.g., Deutscher and Marlor, 1985).

1. A Representative Protocol

The sample to be loaded may be a low-speed (S10) or high-speed (S100) supernatant fraction. The latter is preferable as there will be no interference by ribosomes or microsomes. The size of the column to be used will be determined by the amount of sample to be loaded. A useful rule of thumb is to use 1 ml of packed gel (50–100 mesh) per 100 mg of protein in the sample. According to the manufacturer (Bio-Rad, Bulletin 1107), 1 ml of packed Affi-Gel Blue binds 11 mg of albumin. Generally, only 5–10% of total protein binds to the columns under the conditions described, so the suggested ratio of protein to gel should be sufficient. However, this should be determined for each extract used by specific assay of the flow-through material to ensure that the protein of interest is completely retained on the column. If not, the ratio should be adjusted accordingly. A column with a large diameter is preferable to allow rapid flow rates.

We have found that a column buffer of 10 m*M* Tris–Cl, pH 7.5, containing 0.1 *M* KCl and appropriate stabilizing agents to be a good starting point. The sample should also be adjusted to 0.1 *M* KCl. As noted, under these conditions, less than 10% of the total protein is retained by the column. In addition, nucleic acid is not bound and is washed through the column. After the sample is added, the column continues to be washed with the same solution until the A_{280} of the eluate has ceased decreasing. The protein of interest and most of the bound proteins can be eluted batchwise in the same buffer solution containing 1 *M* KCl. Usually two column volumes will be sufficient to elute most of the protein, although some trailing may be observed. This general procedure will suffice for most situations, and will result in removal of most nucleic acid, concentration of the sample and a 5- to 10-fold purification of the protein of interest.

However, this protocol can be modified to increase the amount of purification. One can use an intermediate elution step with a KCl concentration that removes as much extraneous protein as possible, but not the protein of interest. The desired protein can then be eluted with the lowest concentration of KCl necessary, leaving other extraneous proteins still bound. This enhanced protocol will require some trial and error elutions, but 20-fold purification of the desired protein is often attainable. In some cases, the protein of interest may be bound extremely tightly and requires stronger elution conditions. We have found 1 *M* KBr to be a good choice, but the eluate should be dialyzed rapidly because Br^- can lead to denaturation of some proteins. Alternatively, increasing the KCl concentration to 2 *M* may also be successful.

The Affi-Gel Blue column can be reused, but the column should be regenerated with several column volumes of 2 *M* guanidine HCl or 1.5 *M* sodium thiocyanate followed by equilibration with the starting buffer. Columns should be stored in the cold with 0.1% sodium azide. Under these conditions, columns and resin are stable for years.

REFERENCE

Deutscher, M. P., and Marlor, C. W. (1985). Purification and characterization of *Escherichia coli* RNase T. *J. Biol. Chem.* **260,** 7067–7071.

SECTION SIX

PURIFICATION PROCEDURES: CHROMATOGRAPHIC METHODS

CHAPTER TWENTY-TWO

Ion-Exchange Chromatography

Alois Jungbauer *and* Rainer Hahn

Contents

1. Introduction 349
2. Principle 351
3. Stationary Phases 353
4. Binding Conditions 355
5. Elution Conditions 361
6. Operation of Ion-Exchange Columns 363
7. Example: Separation of Complex Protein Mixture 366
   7.1. Pretreatment of milk 366
   7.2. Chromatography conditions 366
8. Example: High-Resolution Separation with a Monolithic Column 367
   8.1. Mn peroxidase production 367
   8.2. Chromatography conditions 367
References 370

Abstract

Ion-exchange chromatography is the most popular chromatographic method for separation of proteins. It is a versatile and generic tool and is suited for discovery of proteins, high-resolution purification, and industrial production of proteins. Separation conditions are within physiological range of salt and pH and in the most cases a native protein can be obtained. In this chapter, the guidance will be provided for binding and elution conditions and selection of stationary phases.

1. Introduction

Proteins, polynucleotides, and other biomacromolecules expose charged moieties at the surface and thus they can interact with ion exchangers. Ion-exchange chromatography is a versatile and generic tool for protein and plasmid separation. It is frequently used for analytical and

Department of Biotechnology, University of Natural Resources and Applied Life Sciences, Vienna, Austria

Methods in Enzymology, Volume 463
ISSN 0076-6879, DOI: 10.1016/S0076-6879(09)63022-6

preparative purposes. From the early application to the current state of the art of ion-exchange chromatography, the method substantially improved over the years. In the early days of protein separation using ion exchangers, only extremely low flow rates were applicable due to the soft nature of the chromatography medium. Often the columns have been operated under gravity flow and linear flow velocities of only 1–2 cm/h were obtained. This led to purification cycles lasting several days. As a consequence, protein purification had to be performed in the cold room to maintain the biological activity and to prevent microbial growth. Currently, ion-exchange chromatography is operated at linear flow velocities of up to 500 cm/h. This is two orders of magnitude faster than two decades ago. Whenever the stability of a protein solution allows, ion-exchange chromatography is performed at room temperature. A substantial step toward improvement and acceleration of the purification cycles was achieved by introduction of FPLC in the early 1980s (Jungbauer, 1993). Since then, ion-exchangers have been constantly improved (Jungbauer, 2005). The latest jump in development was introduction of monoliths (Jungbauer and Hahn, 2008). With this chromatography media it is possible to separate proteins in less than 5 min (Jungbauer and Hahn, 2004).

The popularity of the methods is based on the high resolution that can be achieved with ion-exchange chromatography. A dilute protein or polynucleotide solution can be rapidly concentrated and simultaneously purified. Usually simple salt buffers are sufficient and concentrations are used in a concentration range, where the so-called salting-in effect on proteins is observed. This is the range where protein becomes more soluble with increasing salt concentration. In contrast, this is not always possible with hydrophobic interaction chromatography. With hydrophobic interaction chromatography often salt concentrations are required which are in the salting-out regime.

One drawback is frequently associated with ion-exchange chromatography—this is incompatibility with mass spectrometry—especially in case of ionization mode. The ionization of proteins and peptides is severely perturbed by ions.

We can define four general cases, where ion-exchange chromatography is applied for protein and biomolecular purification and separation.

1. *De novo purification*: Here only limited information on the molecular structure of the compounds is available. Often only a biological activity of a protein solution or a cell extract is known. The task is to identify the compounds responsible for the biological activity. In this case, intuition is requested. Only limited rules are available and by observation of how the biological activity behave on an ion-exchange column will guide to selection of buffers and materials. Here, the work is really exploratory and often several methods are interconnected to identify a protein.

The most difficult situation is the presence of protein complex. Often the complexes are destroyed during purification and the biological activity is lost or if not lost the biophysical properties of a complex is different than a single protein.

2. *Purification of a protein with known pI, size, and primary structure*: Here, it is possible to develop purification strategies based on rational rules. The most popular one is that a protein binds at pH above pI to anion exchangers and below pI to cation exchangers. From primary sequence, the pI can be easily calculated. This design rule works as long as the protein did not undergo posttranslationial modification such as glycosylation, phosphorylation, etc. Then, the surface charge is substantially altered and the rule becomes obsolete. It has also been observed that certain proteins bind also at pH identical to pI. This has been explained that the surface charge of a protein is not homogenously distributed and the binding occurs in an oriented manner. Of course, if the protein is part of a multiprotein complex, then all predictions are useless (see Chapter 3 in this volume).
3. *Preparative and industrial separation of proteins*: In this case, a purification protocol already exists and the focus is on productivity, throughput or scale-up. In such a scenario, attention is paid on maximizing/optimizing dynamic binding capacity, geometry of the column, and the residence time. Often gradient elution is replaced by step gradients, since in an industrial environment sophisticated gradient mixing is not available. Here the pressure drop is also an important parameter.
4. *High resolution separation of protein variants and/or isoforms*: Proteins undergo posttranslational modifications, chemical modifications such as deamidation or they are artificially modified such as PEGylation. Proteins with the same primary structure have different surface structure and thus different retention behavior on ion exchangers. This can be exploited for separation into the individual isoforms or variants. Proteins differing in a single charge can be separated by ion exchangers. This high-resolution purification became very important since the different isoforms may also differ in biological activity. Ion-exchange chromatography is also often used for separation of intact from truncated forms.

2. Principle

The simplest explanation of ion-exchange chromatography of proteins is based on the attraction of oppositely charged molecules. The protein carries surface charges depending on pI and pH of the environment. The ion exchanger is composed of a base matrix usually in the form of porous beads to provide enough surface area for adsorption. On this base matrix, a charged ligand, either positively or negatively charged, is immobilized. To improve capacity, instead of a small charged ligand, a charged polymer has also been grafted onto the matrix. Irrespectively of the molecular size of the charged ligand, we distinguish between cation and anion exchangers. Cation exchangers are negatively charged and anion exchangers are positively charged. Above its pI, the protein is negatively charged and binds to an anion exchanger, below its pI, it is positively charged and binds to cation exchangers. Earlier we have already discussed the limitations of this simple rule. The ion exchanger itself behaves like an acid or base and the disproportionation of the charges depends on the pH. Strong ion exchangers behave like a strong acid or base and do not change the charge within a wide range of pH, whereas weak ion exchangers do. This property can also be exploited to gain selectivity or by applying pH gradients for elution.

The binding of a charged species to an ion exchanger can be described by the mass action law which has been postulated by Boardman and Partridge (1955). Fred Regnier and his group have extended this work (Kopaciewicz *et al.*, 1983). The retention of a protein and the salt concentration has been used to determine the number of binding sites interacting with the surface. The binding of a protein is achieved at low and elution at high salt concentration (Fig. 22.1). The elution window for isocratic elution is very low, thus linear salt gradients are very popular.

K is the distribution coefficient of the protein between mobile and stationary phase. K is inversely proportional to salt concentration and power of interacting sites (z); Eq. (22.1):

$$\log K = \log\left(K_e q_0^z\right) - z \log(\mathrm{Na}^+), \tag{22.1}$$

where K_e is the affinity of the protein to the stationary phase and q_0 is the ion-exchange capacity. Steve Cramer and his group have further refined the relationship and have also considered the shielded charges at the ion-exchange surface by the adsorbed protein (Brooks and Cramer, 1992).

In respect to how many binding sites are involved in binding to a surface, a different behavior of oligo-, polynucleotides and proteins have been observed (Yamamoto *et al.*, 2007). Especially for small oligonucleotides, the number of binding sites correlated with the number of moieties. Exceeding a certain size the number of binding sites does not increase linearly and levels off. Plasmids bind with many binding sites thus it is possible to bind them even at salt concentration of up to 800 mM. Proteins usually bind with only a few charges.

In proteins, the number of binding sites compared to the charged moieties on the surface is rather small. Only up to 10 binding sites have been observed (Fig. 22.2).

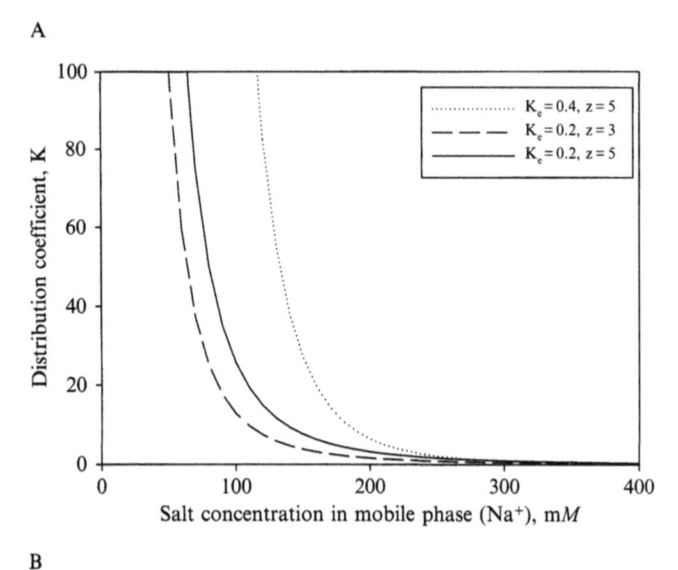

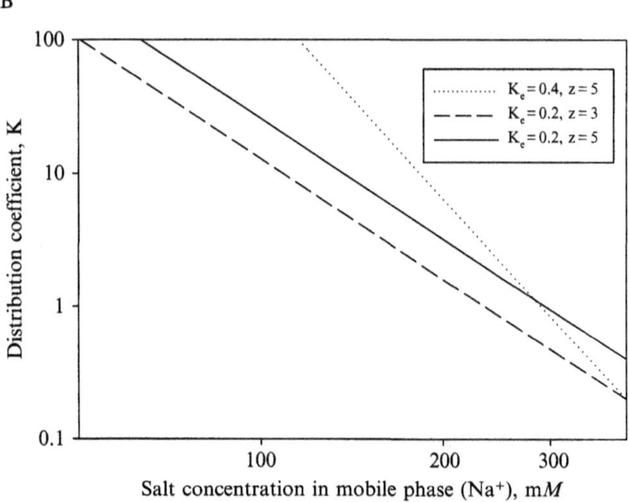

Figure 22.1 Relationship between protein retention and salt concentration in ion-exchange chromatography: (A) linear plot and (B) log–log relationship.

3. Stationary Phases

The evolution of ion-exchange media paralleled the development of stationary phases, which took place over the last several decades. The driving force for development of new media has been the need for better

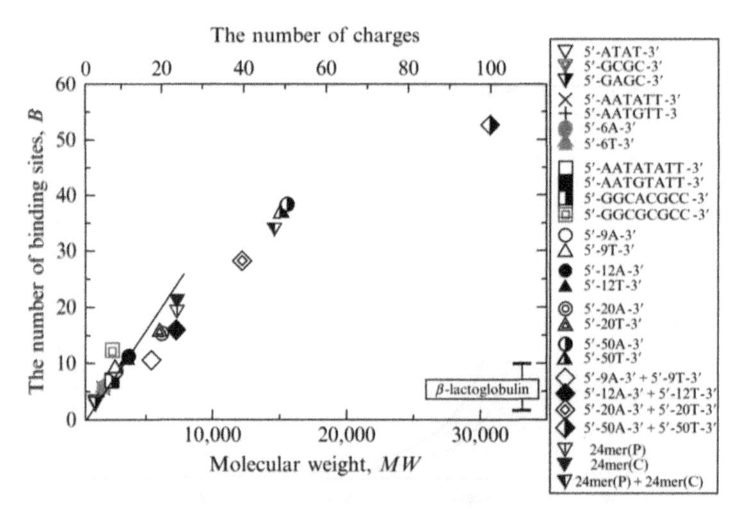

Figure 22.2 Relationship between the number of binding sites and the molecular mass or number of charges of polyA and polyT DNAs. As comparison, the number of binding sites of β-lactoglobulin is shown. Figure from Yamamoto *et al.* (2007).

mechanical stability, reduced tendency to unspecific adsorption, higher binding capacity, and accelerated mass transfer. From an engineering point of view, these features are to some extent incompatible with one another (Mueller, 2005). A strategy for the development of new materials must balance contradictive physical characteristics on the chromatographer's wish list: Large particles allow a low-pressure drop but slow mass transfer kinetics due to long diffusional pathways through the porous network of the bead. Large pores enable fast mass transfer but reduce the available surface area and thus the equilibrium capacity. Mechanical stability also suffers from oversized pores. The chemical composition of the stationary phase (natural or synthetic polymers, inorganic materials) places intrinsic limits for the preparation of porous spherical materials. Traditionally, stationary phases common in chromatography of biomolecules are typically composed of a bead-shaped matrix comprising liquid filled pores. A variety of materials has been used in the design of chromatography matrices (Janson and Ryden, 1998). Among these the most common are polysaccharides (cellulose, dextran, and agarose), synthetic organic polymers (polyacrylamide, polymethacrylate, polystyrene), and inorganic materials (silica, hydroxyapatite). To produce a mechanically stable and functional matrix, the materials are chemically cross-linked and provided with a functional ligand. The physical and chemical conditions of the raw materials present in the production (solvents, concentration of base materials and cross-linkers, temperature, etc.) determine the properties of the stationary phase. Particle sizes of such media range from 2 μm for analytical purposes up to about 200 μm for low-pressure preparative applications. Pore sizes are in the range of 10–100 nm. In many cases, chromatographic media exhibit a typical size distribution of both particles and pores. The mean surface area for media used for preparative applications is in the range of 5–100 m^2/ml of gel. A major improvement to the early, single material-based porous media has been the invention of composite media which combine the properties of two different base materials. Different types of preparations are available (Mueller, 2005). One example is to fill polymers in pores of a rigid matrix with subsequent cross-linking. These types of media are distinguished by a solid (or surface-type) diffusion mechanism of proteins though the hydrogel which is formed inside the pore space. Another, more frequently used approach is the attachment of polymers to the surface by grafting techniques. Extremely high binding capacities and accelerated mass transfer have been shown for such media although the exact mechanism(s) for this behavior is not clear. These media also exhibit a certain salt tolerance and thus direct application of fermentation broths or feed stocks with moderate salt concentrations is possible. Other developments worth mentioning include the introduction of perfusion media (Afeyan *et al.*, 1990) and monoliths (Tennikova *et al.*, 1990). In perfusion chromatography, the pressure drop at high velocity causes a convective fluid flow in large, permeable pores, while capacity is provided by smaller pores with high surface areas. Monoliths are polymerized as single blocks, where the transport of fluid is solely based on convective flow through the channels. The entire mobile phase is driven through the whole volume of the matrix. Components to be separated are conveyed to the active groups located on the surface of flow-through pores, which form an interconnected network. A selection of cation exchangers for the separation of proteins is given in Table 22.1. The choice is much larger, but would exceed the scope of this manuscript. Typical types with different transport mechanism have been selected to cover the spectrum of available material.

Stationary phases for analytical applications are not discussed here. Equivalent to the cation exchangers, one can also find anion exchangers with the same base matrix but different ligands. Several suppliers also offer the same base matrix but different particle diameter and/or ligands.

4. Binding Conditions

According to the general principles of ion-exchange equilibria, it is clear that loading of proteins should be preferably at very low salt concentration (Fig. 22.1). Depending of the ligand density of the ion exchanger, the salt concentration must be adjusted. Usually a concentration of 10–100 m*M* buffer concentration is recommended. This corresponds to a conductivity of 1–4 mS/cm. Regular feedstock from bacterial culture or

Table 22.1 Selected cation exchangers suited for preparative protein separation

Name of stationary phase	Supplier	Base matrix	Mean particle diameter (μm)	Mass transport mechanism	Typical application and features
SP Sepharose fast flow	GE Healthcare	Cross-linked agarose	90	Pore diffusion	Preparative protein capture
Capto S	GE Healthcare	Highly cross-linked agarose with flexible dextran surface extender	90	Unknown mechanisms (accelerated transport)	Preparative protein capture; salt tolerant, very high capacity
UNOsphere rapid S	Bio-Rad	Polyacrylamide interpenetrating network	80	Parallel diffusion (pore and surface diffusion)	Preparative protein capture; very high capacity
S-HyperD *M*	Pall	Ceramic shell filled with polyacryalamide soft gel	80	Surface diffusion	Preparative protein capture; very high capacity
CIM SO_3	BIA Separations	Polymethacrylate monolith	1–2	Convective mass transport	Ultrafast analytical separations; capture of large biomolecules
Fractogel EMD SO_3^- *M*	Merck	Polymethacrylate with polyacrylamide surface extender	65	Unknown mechanisms (accelerated transport)	Preparative protein capture, salt tolerant, very high capacity

POROS HS	Applied Biosystems	Polystyrene–divinylbenzene with through pores	50	Pore diffusion, convective mass transport at high flow rates	Preparative protein purification, improved resolution
Source 30 S	GE Healthcare	Polystyrene–divinylbenzene monobeads with hydrophilic coating	30	Pore diffusion	Preparative protein purification, improved resolution
Toyopearl SP-650 *M*	Tosoh Bioscience	Polymethacrylate	65	Pore diffusion	Preparative protein capture
GigaCap S	Tosoh Bioscience	Polymethacrylate with flexible polymeric surface extender	65	Unknown mechanisms (accelerated transport)	Preparative protein capture; salt tolerant; very high capacity

cell culture has a conductivity ranging from 15 to 40 mS/cm. Thus the protein solution has to be either diluted or desalted. Dilution is a simple method, but only suitable for small scale. The more appropriate methods are dialysis, diafiltration, or desalting by size exclusion chromatography. For laboratory scale quantities size exclusion chromatography is recommended. This can be effected either by simple columns in centrifuge tube, prepacked, or self-packed columns. The maximal loading volume is 1/3 of the total column volume. Limiting factor is the viscosity of the protein solution. Due to viscous fingering, band spreading increases and the load must be reduced.

Ion exchangers with grafted surfaces can be loaded with a feedstock of higher conductivity (Necina *et al.*, 1998). Sometimes it is possible to load a clarified feedstock directly. This is definitely only possible with feedstock of low conductivity. The grafted layer binds a lot of ions and the equilibrium is shifted to conditions still allowing protein adsorption. A compromise must be found between binding capacity and efforts of preconditioning the feedstock. As trade off of a simple process by direct loading, a lower capacity must be taken into account (Fig. 22.3). With modern ion changers a lower capacity can be often accepted because they can be run at higher velocity.

The selection of the buffering species depends on the isoelectric point of the protein. Above the isoelectric point, the protein is negatively charged and binds to anion exchangers, below positively charged and binds to cation exchangers. The actual selection depends on the impurity profile and the stability of the protein. A protein with a high isoelectric point is preferably purified with a cation exchanger. Such proteins are definitely easier to purify. In Fig. 22.4, a histogram of isoelectric points of proteins is shown. It is clear that slightly acidic proteins are more difficult to separate, since the chance of impurities present in the feedstock with similar pI is very high.

The second rule for binding is that a buffer should be selected with a pK_a value close to the selected pH, in extreme maximum one pH unit below or above.

The third rule is that the buffering species such as acetate or Tris should not bind to the ion exchanger. So do not use, for example, acetate with anion exchangers and Tris with cation exchangers. Binding of the buffering species causes unstable pH and nonreproducible conditions.

The fourth rule is that the protein is exposed to a different pH when it is close to the ion-exchanger surface. In cation exchangers, the pH can be up to one pH unit lower due to binding of hydronium ions, whereas in anion exchangers the pH can be one unit higher due to binding of hydroxyl ions. This has to be obeyed especially when working with labile proteins.

When the pI and other information on the protein are not available, then the binding conditions must be searched on small scale, either in a batch or in a column mode. For the batch mode, the ion exchangers are filled in small test tubes or in microtiter plates. The adsorbent is equilibrated with the appropriate buffer of varying pH or/and salt concentration and

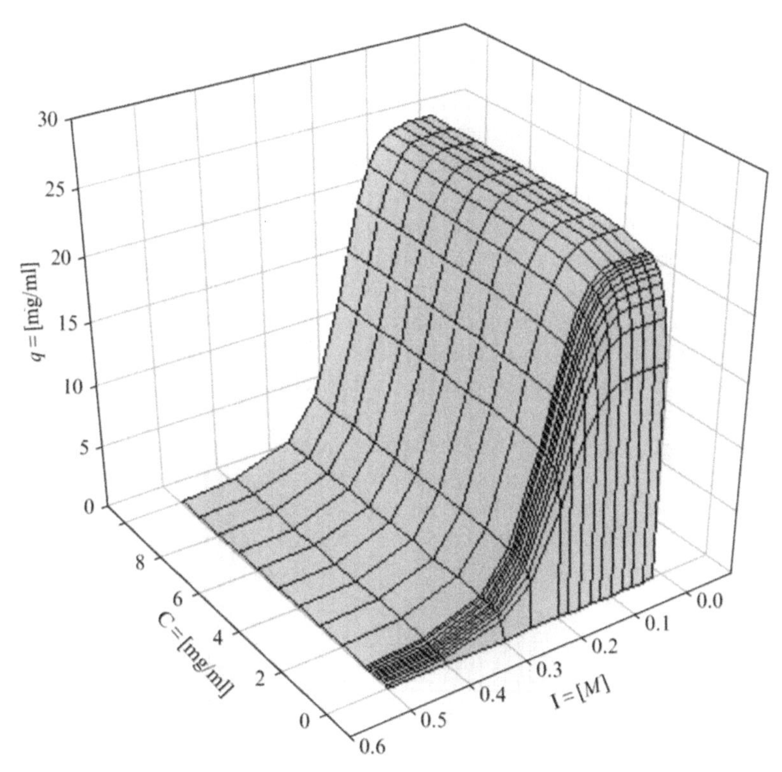

Figure 22.3 Relationship among protein binding capacity, protein concentration, and salt concentration in a protein solution; I is the ionic strength of the loading buffer, C is the protein concentration in the mobile phase in equilibrium, and q is the concentration of protein in the stationary phase.

then the protein solution is added. The supernatant is tested for the protein and by addition of salt the protein is eluted from the column and it is again tested in the supernatant. The advantage of microtiter plates are the multiple experiments which can be run in parallel. Separation of liquid and solid phase can be easily achieved in a centrifuge. Currently, chromatography manufacturers also provide ready-to-use microtiter plates.

Several manufacturers have also designed small-scale, prepacked columns, which can be operated by a syringe or a laboratory robot. With these tiny columns it is easy to optimize binding conditions without wasting precious proteins. Optimization of ion exchangers is very material consuming, since the capacity is extremely high. These small columns such as MediaScout MiniColumns from Atoll (Weingarten, Germany) are of 5 mm internal diameter and up to 10 mm bed height (Fig. 22.5). They contain each up to 0.2 ml of the chromatography adsorbent and are prepacked to defined density, corrected for each individual resin material, in order to replicate conditions in larger columns.

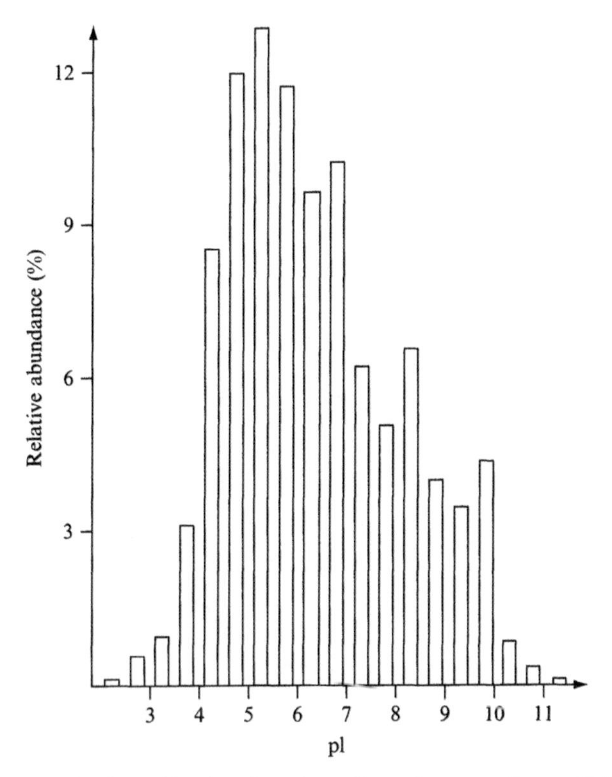

Figure 22.4 Distribution of isoelectric points of proteins according to G. Righetti 13.

Figure 22.5 Robotic applications of prepacked minicolumns for operation with a positive displacement pipette and columns situated at frame equivalent to microtiter plates, reproduced by kind permission of Atoll (Weinheim, Germany).

In several steps, the different solutions are applied. Flow is controlled by a positive displacement pipette. The applied volume varies from μl to ml range depending on the type of pipette used.

5. Elution Conditions

Elution of a bound protein can be effected by four modes:

1. Linear salt gradient
2. Step salt gradient
3. pH gradient
4. Displacement development.

Sometime combination between pH gradient and salt is used, but this is difficult to optimize and robust conditions often are not achieved. Linear salt and pH gradients are preferred for high-resolution separation, while step salt gradients are preferred for concentration of proteins. Displacement chromatography is in principle suited for separation of related species, but due to the more complicated optimization, is less used in general.

The retention window of proteins is very small; at low salt concentration proteins are tightly bound and then with increasing salt concentration not retained at all; compare (Fig. 22.1) the relationship between k' and salt concentration in the mobile phase. So in ion-exchange chromatography linear gradients and even more complex segmented gradients are used.

The resolution R_s

$$R_s = 2\frac{t_A - t_B}{W_A + W_B} \tag{22.2}$$

with the retention time of two different compounds (t_A, t_B) and the peak width at the base (W_A, W_B) depends on the slope of the gradient β. Where β is defined as

$$\beta = \frac{C_M - C_M^0}{t_G} \tag{22.3}$$

with C_M^0 and C_M as the start and final salt concentration of the buffer, and t_G the gradient time. The retention of a compound depends in a complex manner on the normalized gradient slope (γ). A detailed discussion would exceed the scope of this chapter. A detailed analysis can be found in the book published by Yamamoto *et al.* (1988):

$$\gamma = \frac{\beta L}{v} = \frac{\beta V_0}{Q}, \tag{22.4}$$

L is the length of the column and v is the interstitial velocity, V_0 the void volume, and Q is the flow rate.

Resolution is increased with the normalized gradients (γ). From Eq. (22.4) one can infer three possibilities to increase resolution.

1. Increasing the gradient length
2. Increasing the column length/volume
3. Decreasing the interstitial velocity/flow rate.

All three parameters can be varied at once. Equation (22.4) also helps to transfer a successful elution condition to a larger column. The same resolution is obtained when γ is kept constant.

The resolutions in these cases are 5.00 and 4.97 min, respectively, with a difference ion scale by a factor of 7 (Fig. 22.6).

As a consequence of the interrelation between protein retention and γ, it has been observed that the peak salt concentration at shallower gradients is lower compared with steeper gradients. So the protein does not always elute at the same salt concentration. Resolution depends on γ in a reciprocal manner.

$$R_s \propto \frac{1}{\gamma}\sqrt{\frac{L}{\text{HETP}}} \tag{22.5}$$

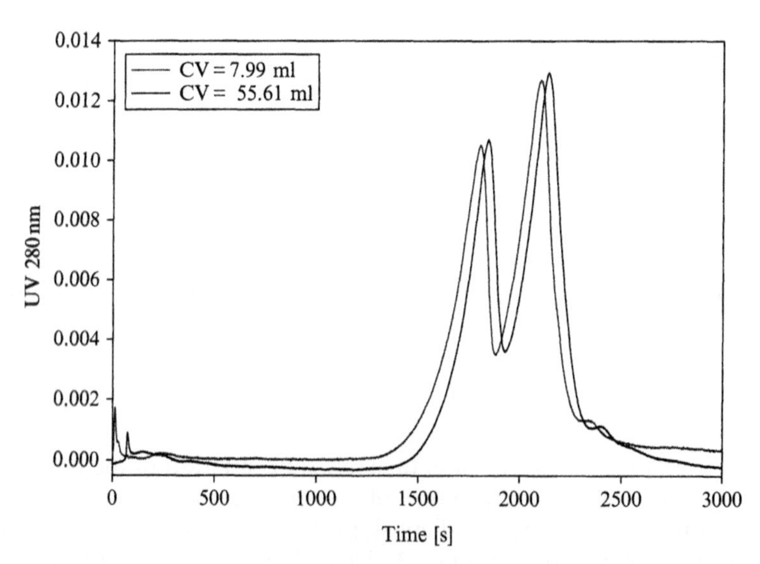

Figure 22.6 Example of a α-lactoglobulin isoform separation on Source30Q (GE Healthcare, Uppsala, Sweden). The flow rate, gradient length, and load are scaled to the column volume thus consequently γ was kept constant. The flow rate is 30 CV/h. Column dimensions are: 1×10 and 2×17.7 cm. Reproduced by kind permission of Jorgen Mollerup. (See Color Insert.)

With a shallower gradient (low γ) resolution increases. Resolution can also be influenced by the particle size. This is reflected by the term $\sqrt{L/\mathrm{HETP}}$ which is equivalent to $\sqrt{N}$, with N the number of theoretical plates, for protein chromatography being a mass transfer limited process. In this case for well-packed columns, HETP is proportional to the velocity and the square of the particle diameter. So with reduction of the particle diameter the resolution can be substantially improved. Once nonporous material is used then HETP is only proportional to the diameter not to the square of the diameter of the particle. Thus, materials for ultrahigh resolution have particle diameters below 5 μm. They are manufactured as nonporous particles.

Peak volume also increases with the length of the gradient. A step gradient can be considered as a very steep linear gradient. In practice, a step never can be generated. So in a step gradient the protein elutes at lower salt concentration compared to a linear gradient.

6. Operation of Ion-Exchange Columns

Here some general rules are given on how to operate an ion-exchange column. Several dedicated steps are required to ensure a stable operation with an ion exchanger. These steps are

1. Loading with salt
2. Equilibration
3. Loading of protein
4. Washout of unbound material
5. Elution
6. Regeneration
7. (Sanitization)

General conditions are listed in Table 22.2.

The loading with salt is crucial since the charged ligands must be entirely saturated with counter ions. This is usually done with a buffer of a molarity between 10 and 100 mM supplemented with 1 M NaCl. Less frequently KCl is used. Bivalent ions should be avoided. A least one column volume of buffer should be prepared in order to saturate the column. Further optimization needs an on- or off-line conductivity probe.

Equilibration is performed with the buffer which is suited for protein binding. pH, buffer, salt, and molarity depend on the individual conditions. Very rough guidance for selection of the appropriate buffers is given in the section on binding conditions. For this purpose, prepare at least 10 column volumes. The large volume for equilibration is required to get a stable pH. Since the different ions are migrating differently through the column, the

Table 22.2 Overview of operation conditions for ion-exchange chromatography of proteins

Step	Molarity of buffer	Buffer volume
Loading with salt	1 *M* and higher	At least 1 CV
Equilibration	10–100 m*M*	Up to 10 CV
Loading of protein	Similar to equilibration buffer	Depends on dynamic binding capacity
Washout of unbound material	Equilibration buffer with eventually supplements	At least 1 CV
Elution		
Linear gradient	0–500 m*M* salt	10 CV
Step gradient	Between 150 and 500 m*M* salt	1 CV
Linear pH gradient	Below 50 m*M*	
Induced pH gradient	10 m*M* and lower	
Regeneration	High salt concentration 1 *M*	1 CV
Sanitization		At least 1 h residence time

counter ion is more retained than the species with the same charge. Thus a wave with different ions is traveling through the column. To maintain electroneutrality, the pH changes. A detailed explanation of this phenomenon can be found at Pabst and Carta (2007). Especially for weak ion-exchange ligands, a larger volume is required to reach the actual pH.

Loading of the protein solution should be performed at the same pH and conductivity as the equilibration buffer. Often this is not possible; especially the composition of the ions in a crude feedstock and the equilibration buffer may be completely different. Again as a consequence, a pH transition will be observed during loading. When the protein solution is desalted by size exclusion chromatography, dialysis, or diafiltration, the same ion composition can be obtained leading to most robust loading conditions.

After loading the protein solution, the unbound material is washed out. A first try is always the equilibration buffer. To obtain a higher purity in the eluate, the wash procedure may consist of several steps; for a tightly bound protein of interest, washing with equilibration buffer with salt added is often used. A lot of protocols exist and it seems very protein specific which compounds are added to a washing buffers. These may be detergents, urea, sugars, ethylene glycol, etc.

Elution can be effected by a linear or step salt gradient or by displacement development. It is best to start with a linear salt gradient. The gradient volume should be about 10 column volumes. The two buffers for the preparation of the gradient are in the simplest case the equilibration buffer (Buffer A) and the buffer used for loading with counterions (Buffer B). In most cases, the second half of the gradient is not required, since most proteins elute from a conventional ion exchanger in the range between 150 and 500 n*M* salt. So gradients to maximal 50% B are often sufficient. The step gradient can also be generated from Buffers A and B. It should be taken into consideration that an ideal step never can be achieved in a column. First, due to retention of salt and the dead volume in HPLC often denoted as dwell time and secondly, due to the washing-out characteristics of a buffer mixer only a sigmoidal-shaped curve will be obtained (Kaltenbrunner and Jungbauer, 1997). According to the selection of the salt concentration in the step elution, the protein will migrate with the salt front or travel behind the salt front. Yamamoto *et al.* (1988) named this behavior type I or II elution. When a concentrated eluate is washed then salt concentration must be selected for type I elution. This is a salt concentration somewhat higher than the corresponding peak salt concentration in linear gradient elution. For the type I elution less than one column volume buffer is required.

Elution by pH gradient is more difficult to optimize. So the best start is to find a condition according to the pI of the protein where the protein does not bind to the ion exchanger. This is below the pI for an anion exchanger and above the pI for a cation exchanger. So pH has to be decreased or raised. Keep in mind that the ion exchanger behaves like an acid or a base and the pH gradient produced follows a titration curve. It is extremely difficult to produce a linear pH gradient. This can be done by reacting boronate buffers with diols such as mannitol (Kaltenbrunner *et al.*, 1993). The fourth possibility is the elution with induced pH gradients. When working with buffer of low molarity in the range of 10 m*M*, the application of salt induces a pH gradient due to the differential migration of the different salt species. Such an induced pH gradient can be used for high resolution and additional for focusing of the protein peak (Pabst *et al.*, 2008a,b). In this case also, several column volumes of buffer are required.

For regeneration of the column, the optimal condition is to use the same buffer as for loading with salt. This is only possible when a relatively clean feedstock is processed. Then it is possible to combine two steps into one. Especially when lipids, endotoxins, and other sticky compounds are loaded onto an ion-exchange column, then the regeneration with a caustic agent, preferably up to 1 *M* NaOH, is recommended. The maximum concentration of NaOH depends on the resistance of the stationary phase. One column volume is sufficient but for very tough impurities the column should be soaked in NaOH. So after pumping one column volume over the column, flow is stopped and NaOH is left several hours in the column.

The NaOH treatment also has some sanitization effects. It will definitely reduce the bacterial count in a column, but will not sterilize a column.

It is extremely difficult to sterilize a chromatography column (Jungbauer and Lettner, 1994; Jungbauer *et al.*, 1994). So many dead-end holes are in a packed column and the sanitization agent can only diffuse into these holes. Oxidizing agents are often not compatible with the stationary phase and the equipment. Smaller columns can be autoclaved and the connected to the system under sterile conditions. Usually it is not necessary to sterilize a chromatography column.

Storage of columns should be done after careful cleaning with regeneration puffers, sanitization when necessary, and under conditions where the column is fully saturated with counter ions. After saturation, the column can be washed with buffer and stored, for instance in 20% ethanol. For laboratory use, it is also recommended to add bacteriocidal compounds to the storage solution; NaN_3 is often used.

7. Example: Separation of Complex Protein Mixture

At typical example for ion-exchange chromatography of a complex protein mixture is the separation of proteins from sweet whey. This has some industrial relevance but may also serve as an inexpensive test system for evaluation of ion exchangers. We have actually used this protein mixture for comparison of materials of different manufactures of ion-exchange adsorbents (Hahn *et al.*, 1998).

7.1. Pretreatment of milk

Milk was centrifuged at 4420×*g* at room temperature for 30 min for delipidation. The pH of the skimmed milk was adjusted to 4.7 by slow addition of 5 *M* HCl. After casein precipitation, the solution was stirred for further 30 min to complete precipitation. Casein was removed by centrifugation at 17,700*g* and 4 °C for 30 min. The obtained whey was diluted with distilled water until a conductivity of 2.7 mS/cm was obtained. The pH was readjusted to 4.7 since pH shifted during dilution. Prior to chromatography, the whey was filtered through a 0.45-μm Millipak-60 filter (Millipore, Bedford, MA, USA).

7.2. Chromatography conditions

The cation exchange resins were packed into HR 10 and HR 16 columns (GE Healthcare, Uppsala, Sweden). Column dimensions were 30/10 and 35/16 mm height × I.D., respectively. The buffers were prepared from citric acid and the pH was adjusted with 1.0 N NaOH. The cation exchangers were regenerated with 1 *M* NaCl in 20 m*M* citric acid and equilibrated with 20 m*M* citric acid, pH 4.7, 2.7 mS/cm. After loading and washing with equilibration buffer, the bound proteins were eluted with sequential gradients with increasing NaCl concentration.

The chromatograms are shown in Fig. 22.7. The flow through and the eluates were collected and then further analyzed by analytical size exclusion chromatography and SDS–PAGE.

The composition of the flow through at beginning and ending of the loading step is shown in Fig. 22.8.

Although cation exchangers are used, the selectivity is different. It is difficult to predict the selectivity for the various materials.

8. Example: High-Resolution Separation with a Monolithic Column

An impressive example for high-resolution separation using monolithic columns is the separation of manganese peroxidase (MP) using an anion-exchanger monolithic column (Podgornik and Podgornik, 2004). The separation is completed in less than 3 min.

8.1. Mn peroxidase production

Phanerochaete chrysosporium MZKI B-223 (ATCC 24725) was grown in a nitrogen-limited medium in agitated 500 ml Erlenmeyer flasks containing 100 ml of the growth medium at 32 °C. The growth medium contained 1 m*M* Mn(II) favoring MnP production only. Two hundred and fifty milliliters of *P. chrysosporium* growth medium was harvested after 4 days of cultivation, when the highest MnP activities were recorded. The growth medium filtrate was frozen overnight and rethawed to remove mucilaginous polysaccharides by centrifugation.

8.2. Chromatography conditions

MnP isoenzymes were isolated using CIM disk monolithic column containing 0.34 ml CIM QA (quaternary amine) disk (BIA Separations, Ljubljana, Slovenia). Experiments were performed on a gradient HPLC system built with two pumps, an injection valve with a 100-μl stainless steel sample loop, a variable wavelength monitor with a 10-mm optical path set to 280 or 409 nm and with a 10-μl volume flow cell, connected by means of 0.25 mm I.D. PEEK capillary tubes. Isoenzymes were separated with a linear gradient of sodium acetate of different concentrations (from 10 m*M* to 1 *M*) and pH values (from 4 to 6), at a flow rate of 4 ml/min (Fig. 22.9).

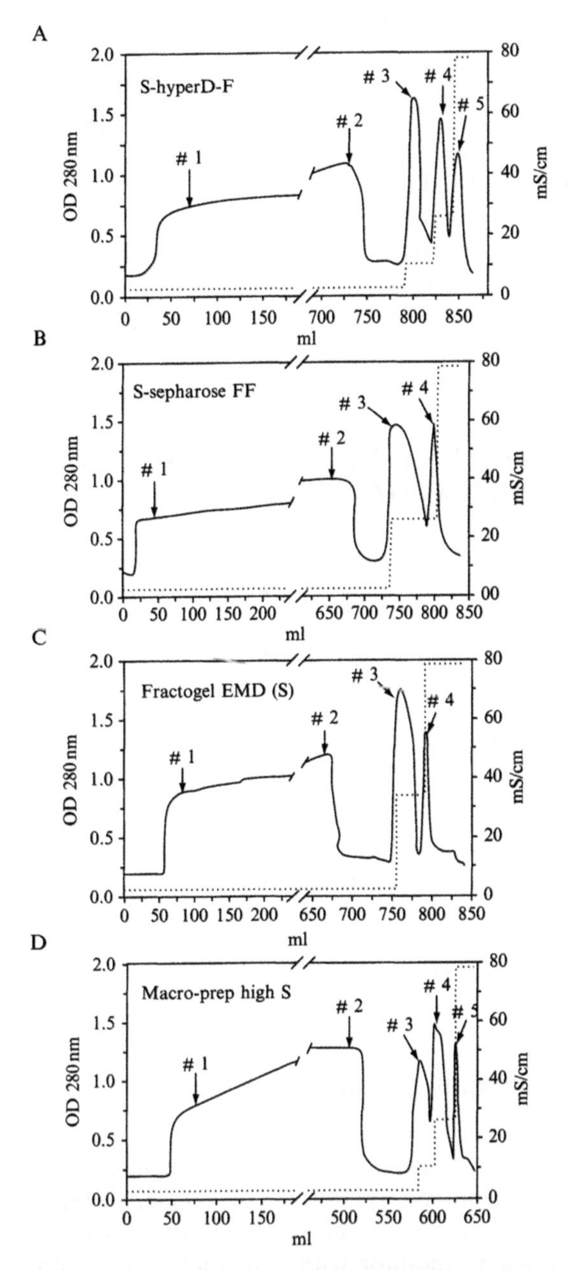

Figure 22.7 Purification of bovine whey proteins by cation-exchange chromatography. Clarified whey was applied to four different cation-exchange columns (3.531.6 cm), equilibrated with 20 m*M* citric acid, pH 4.7. The flow rate was 3.3 ml/min (200 cm/h). Elution of bound material was carried out with sequential NaCl gradients. Unbound material and eluted fractions were characterized by analytical size-exclusion chromatography (Fig. 22.8). (A) S-HyperD-F, (B) S-Sepharose FF, (C) Fractogel EMD SO S, and (D) Macro-Prep High-S Support (···, theoretical salt gradient; —, UV absorbance at 280 nm), according to Hahn *et al.* (1998).

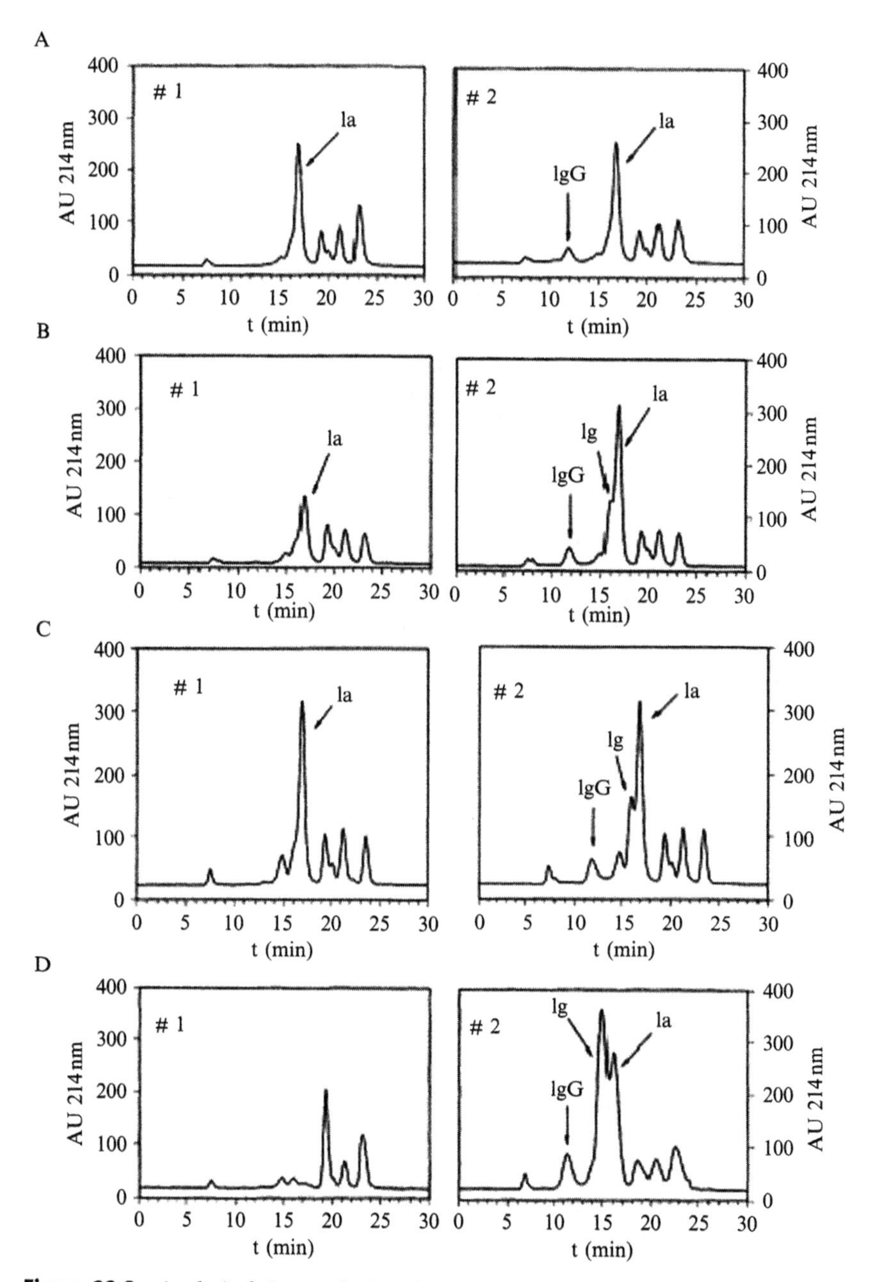

Figure 22.8 Analytical size-exclusion chromatography on Superdex 200 of unbound material (flow through) when bovine whey was loaded onto cation-exchange sorbents. Fractions of the flow through, as indicated in Fig. 22.1, were analyzed. (A) S-HyperD-F 1 and 2, (B) S-Sepharose FF 1 and 2, (C) Fractogel EMD SO S 1 and 2, and (D) Macro-Prep High-S Support 1 and 2; lg, lactoglobulin; la, lactalbumin.

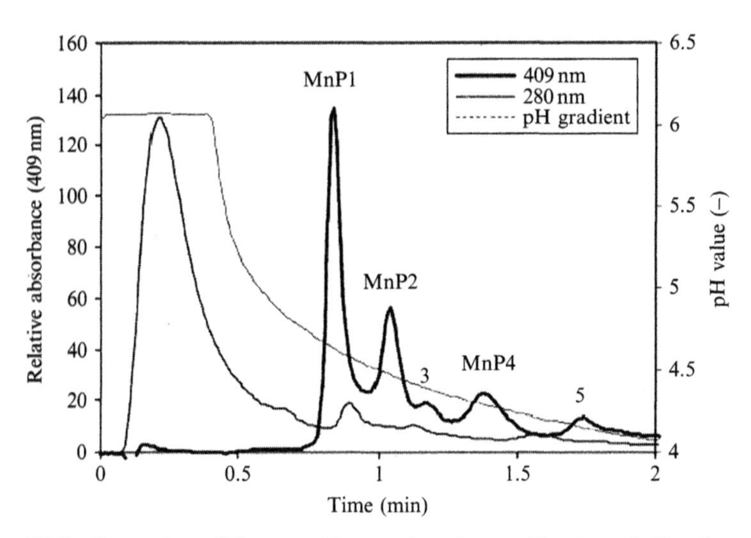

Figure 22.9 Separation of Mn-peroxidase variants by combination of pH and concentration gradient. Stationary phase: CIM QA disk monolithic column; mobile phase: buffer A: 10 m*M* sodium acetate, pH 6; buffer B: 20 m*M* sodium acetate, pH 4; gradient: 0–100% of buffer B in 100 s; sample: 100 μl of desalted protein solution; flow rate: 4 ml/min; detection: UV at 280 nm, Vis at 409 nm, and on-line pH meter [pH gradient?].

This very fast method allows separation of isoforms and two other forms in less than 2 min. The method is suited for preparative and analytical applications.

REFERENCES

Afeyan, N. B., Gordon, N. F., Mazsaroff, I., Varady, L., Fulton, S. P., Yang, Y. B., and Regnier, F. E. (1990). Flow-through particles for the high-performance liquid chromatographic separation of biomolecules: Perfusion chromatography. *J. Chromatogr.* **519,** 1–29.

Boardman, N. K., and Partridge, S. M. (1955). Separation of neutral proteins on ion-exchange resins. *Biochem. J.* **59,** 543–552.

Brooks, C. A., and Cramer, S. M. (1992). Steric mass-action ion exchange: Displacement profiles and induced salt gradients. *AIChE J.* **38,** 1969–1978.

Hahn, R., Schulz, P. M., Schaupp, C., and Jungbauer, A. (1998). Bovine whey fractionation based on cation-exchange chromatography. *J. Chromatogr. A* **795,** 277–287.

Janson, J.-C., and Ryden, L. (1998). Protein Purification: Principles, High Resolution Methods, and Applications 2nd edn. Wiley-VCH, New York.

Jungbauer, A. (1993). Preparative chromatography of biomolecules. *J. Chromatogr.* **639,** 3–16.

Jungbauer, A. (2005). Chromatographic media for bioseparation. *J. Chromatogr. A* **1065,** 3–12.

Jungbauer, A., and Hahn, R. (2004). Monoliths for fast bioseparation and bioconversion and their applications in biotechnology. *J. Sep.. Sci.* **27,** 767–778.

Jungbauer, A., and Hahn, R. (2008). Polymethacrylate monoliths for preparative and industrial separation of biomolecular assemblies. *J. Chromatogr. A* **1184,** 62–79.

Jungbauer, A., and Lettner, H. (1994). Chemical disinfection of chromatographic resins, Part 1: Preliminary studies and microbial kinetics. *BioPharm* **7,** 46–56.

Jungbauer, A., Lettner, H., Gurrier, L., and Boschetti, E. (1994). Chemical sanitization in process chromatography, Part 2: *In situ* treatment of packed columns and long-term stability of resins. *BioPharm* **7,** 37–42.

Kaltenbrunner, O., and Jungbauer, A. (1997). Simple model for buffer blending in ion-exchange chromatography. *J. Chromatogr. A* **769,** 37–48.

Kaltenbrunner, O., Tauer, C., Brunner, J., and Jungbauer, A. (1993). Isoprotein analysis by ion exchange chromatography using a linear pH gradient combined with a salt gradient. *J. Chromatogr.* **639,** 41–49.

Kopaciewicz, W., Rounds, M. A., Fausnaugh, J., and Regnier, F. E. (1983). Retention model for high-performance ion-exchange chromatography. *J. Chromatogr. A* **266,** 3–21.

Mueller, E. (2005). Properties and characterization of high capacity resins for biochromatography. *Chem. Eng. Technol.* **28,** 1295–1305.

Necina, R., Amatschek, K., and Jungbauer, A. (1998). Capture of human monoclonal antibodies from cell culture supernatant by ion exchange media exhibiting high charge density. *Biotechnol. Bioeng.* **60,** 689–698.

Pabst, T. M., and Carta, G. (2007). pH transitions in cation exchange chromatographic columns containing weak acid groups. *J. Chromatogr. A* **1142,** 19–31.

Pabst, T. M., Carta, G., Ramasubramanyan, N., Hunter, A. K., Mensah, P., and Gustafson, M. E. (2008a). Separation of protein charge variants with induced pH gradients using anion exchange chromatographic columns. *Biotechnol. Prog.* **24,** 1096–1106.

Pabst, T. M., Antos, D., Carta, G., Ramasubramanyan, N., and Hunter, A. K. (2008b). Protein separations with induced pH gradients using cation-exchange chromatographic columns containing weak acid groups. *J. Chromatogr. A* **1181,** 83–94.

Podgornik, H., and Podgornik, A. (2004). Separation of manganese peroxidase isoenzymes on strong anion-exchange monolithic column using pH-salt gradient. *J. Chromatogr. B Analyt. Technol. Biomed. Life Sci.* **799,** 343–347.

Tennikova, T. B., Belenkii, B. G., and Svec, F. (1990). High-performance membrane chromatography. A novel method of protein separation. *J. Liq. Chromatogr.* **13,** 63–70.

Yamamoto, S., Nakanishi, K., and Matsuno, R. (1988). Ion-exchange chromatography of proteins Marcel Dekker, New York.

Yamamoto, S., Nakamura, M., Tarmann, C., and Jungbauer, A. (2007). Retention studies of DNA on anion-exchange monolith chromatography: Binding site and elution behavior. *J. Chromatogr. A* **1144,** 155–169.

CHAPTER TWENTY-THREE

GEL FILTRATION[1]

Earle Stellwagen

Contents

1. Principle 373
2. Practice 374
   2.1. Matrices 376
   2.2. Sample preparation 380
   2.3. Chromatographic solvents 380
   2.4. Preliminary screening 380
   2.5. Chromatography using conventional matrix 382
   2.6. Scaling upward 383
   2.7. Trouble shooting 384
   2.8. Further information 385

Among the chromatographic techniques employed for protein purification, gel filtration is unique in that fractionation is based on the relative size of protein molecules. In contrast to conventional filtration, none of the proteins is retained by a gel-filtration column. This feature is at once both the strength and weakness of gel filtration; a strength because the function of fragile proteins is not damaged by binding to a chromatographic support, and a weakness because the absence of such binding limits the resolution of the chromatography.

1. PRINCIPLE

Gel filtration is performed using porous beads as the chromatographic support. A column constructed from such beads will have two measurable liquid volumes, the external volume, consisting of the liquid between the beads, and the internal volume, consisting of the liquid within the pores of the beads. Large molecules will equilibrate only with the external volume while small molecules will equilibrate with both the external and internal volumes. A mixture of proteins is applied in a discrete volume or zone at the top of a gel-filtration column and allowed to percolate through the column.

Department of Biochemistry, University of Iowa, Iowa City, Iowa

[1] Reprinted from *Methods in Enzymology*, Volume 182 (Academic Press, 1990)

Methods in Enzymology, Volume 463
ISSN 0076-6879, DOI: 10.1016/S0076-6879(09)63023-8

The large protein molecules are excluded from the internal volume and therefore emerge first from the column while the smaller protein molecules, which can access the internal volume, emerge later.

The dimensions important to gel filtration are the diameter of the pores that access the internal volume and the hydrodynamic diameter of the protein molecules. The latter is defined as the diameter of the spherical volume created by a protein as it rapidly tumbles in solution. Proteins whose hydrodynamic diameter is small relative to the average pore diameter of the beads will access all of the internal volume and are described as being included in the gel matrix. Proteins whose hydrodynamic diameter is comparable to the average pore diameter will access some but not all of the internal volume and are described as being fractionally excluded. Proteins whose hydrodynamic diameter is large relative to the average pore diameter will be unable to access the internal volume and are described as being excluded.

This conceptualization has led to the gradual renaming of gel filtration as size-exclusion chromatography. The order of elution of a mixture of proteins from a size-exclusion column will then be the inverse of their hydrodynamic diameters. If all the proteins in a mixture are known, or can be assumed to have the same shape, then the order of elution will be the inverse of their molecular weights. This discussion will treat protein dimensions in terms of molecular weight since common usage assumes that protein mixtures contain only globular proteins. However, the reader should bear in mind that hydrodynamic volume is the operative protein dimension and that an asymmetrical protein will appear to elute with an abnormally high-molecular weight compared with globular proteins of similar molecular weight.

2. Practice

An elution profile obtained by size-exclusion chromatography is illustrated in Fig. 23.1A. Zero elution volume is defined as the entry of the sample into the chromatographic support. The elution volume for the excluded component is designated V_0 for the void volume, which represents the volume external to the beads. The elution volume for the included component is designated V_t for the total volume, which represents the sum of the external volume and the internal volume within the beads. Elution volumes intermediate between these values are designated V_e. A partition coefficient, designated K_{av}, relating these values is given in Eq. (1):

$$K_{av} = \frac{V_e - V_0}{V_t - V_0} \tag{1}$$

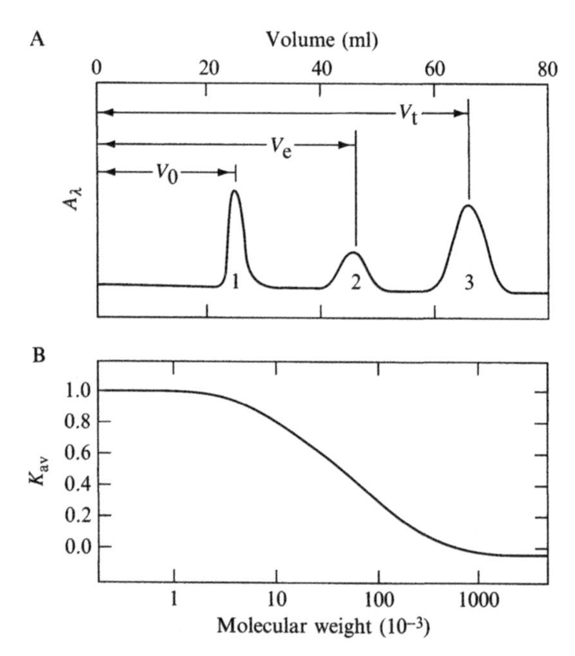

Figure 23.1 Chromatographic performance of a size-exclusion matrix. Part (A) illustrates a relatively simple elution profile. The ordinate represents concentration expressed as spectral absorbance at some fixed wavelength, λ, and the abscissa represents effluent volume subsequent to the application of the sample into the column. If the effluent flow rate is constant then the abscissa could be expressed in time. Component 1 is excluded from the matrix and its elution position is denoted as V_0. Component 2 is partially excluded and its elution position is denoted as V_e. Component 3 is included and its elution position is denoted as V_t. The assignment of a component to an elution position is established by application of each component individually to the column. Part (B) illustrates the sigmoidal dependence of the partition coefficient K_{av} as defined in Eq. (1) on the logarithm of the molecular weight of a series of components having the same shape.

A semilogarithmic plot of the dependence of the partition coefficient on molecular weight is illustrated in Fig. 23.1B. The separation of proteins based on molecular weight will be greatest in the central linear region of this sigmoidal relationship, spanning K_{av} values between 0.2 and 0.8. This span is described as the fractionation range of a size-exclusion matrix.

The steeper the slope of the sigmoidal relationship in the fractionation range the greater the resolving power of a matrix. Accordingly, the best separation among proteins having similar molecular weights will be achieved using a matrix with a narrow fractionation range.

Fewer than 10 proteins can be resolved from one another in the effluent from any size-exclusion column. This relatively low resolution occurs because none of the proteins is retained by the column during chromatography and because nonideal flow occurs around the beads. Accordingly, prospects for a significant enhancement in purification (-fold) by size-exclusion chromatography are most promising if the desired protein has a molecular weight either considerably larger or smaller than that of the majority of proteins in a mixture. Since this will generally not be the case, an investigator can anticipate only a modest enhancement in purification (-fold). Accordingly, it is wise to perform size-exclusion chromatography relatively late in a purification procedure when the numbers of other proteins are small and when the preceding step has fractionated the protein mixture on the basis of a completely different property. For example, pooled fractions obtained from ion-exchange chromatography will likely contain a mixture of proteins, each having about the same net charge but a range of molecular weights.

2.1. Matrices

The properties of some conventional and high-performance size-exclusion matrices are given in Tables 23.1–23.4. It should be noted that suppliers use a variety of terms and abbreviations to index these products in their catalogs, including gel-filtration chromatography (GFC), gel-permeation chromatography (GPC), and size-exclusion chromatography (SEC).

The conventional matrices are distinguished by their relative economy and slow flow rates. These matrices are available in bulk, requiring an investigator to pour columns of any desired dimensions to accommodate the volume of the sample to be chromatographed. The flow rates normally used for chromatography are obtained by multiplying the linear flow rate listed in Table 23.2 by the cross-sectional area of the column in cm^2 to yield the flow rate in mm/h. A column can be packed with a flow rate approximately five times that used during chromatography.

The high-performance matrices are distinguished by their convenience, rapid flow rates, and expense. These matrices are usually purchased as poured columns which are attached to an existent high-performance chromatograph available to the investigator. The smaller analytical columns, about 8×300 mm, are normally loaded with not more than a few milligrams of protein and operated at a flow rate of about 1 ml/min. The larger preparative columns generally contain beads having a diameter of 30 μm. The approximately 20×300 mm columns can be loaded with between 10 and 100 mg of protein and can be operated at a flow rate of about 5 ml/min while the very large columns can be loaded with up to 2 g of protein and be operated at a flow rate of up to 30 ml/min.

Table 23.1 Matrix parameters

Stability						
Name	Supplier	Chemistry[a]	Form supplied	pH	Temperature (°C)	Bead diameter (μm)
Conventional						
BioGel A	Bio-Rad	AG	Suspension	4–13	1–30	40–300[b]
BioGel P	Bio-Rad	PA	Powder	2–10	1–80	40–30[b]
Sephacryl HR	Pharmacia	DX	Suspension	2–13	1–100	25–75
Sephadex G	Pharmacia	DX/PA	Powder	2–10	1–100	20–300[b]
Sepharose	Pharmacia	AG	Suspension	4–10	1–40	45–200[b]
Ultrogel A	IBF	AG	Suspension	3–10	2–36	60–140
Ultrogel AcA	IBF	AG/PA	Suspension	3–10	2–36	60–140
High performance						
Protein Pak	Waters	S	Packed column	2–8	1–90	10
Shodex	Showa Denko	S	Packed column	3–7.5	10–45	9
Superose	Pharmacia	AG	Packed column	1–14	4–40	10–13
SynChropak	SynChrom	S	Packed column	2–7	1–60	5–10
TSK-SW	Toyo-Soda	S	Packed column	3–7.5	1–45	10–13
Zorbax	DuPont	S	Packed column	3–8.5	1–100	4–6

[a] The following symbols are used to denote the chemical nature of the matrix: AG, cross-linked agarose; PA, cross-linked polyacrylamide; DX, cross-linked dextran; DX/DA, copolymer of allyl dextran and bisacrylamide; AG/PA, mixture of agarose and polyacrylamide; S, bonded silica.

[b] Individual matrices have narrower ranges.

Table 23.2 Powdered matrix parameters

Hydration time (h)						
Name	Code	Fractionation range[a] (kDa)	Swollen volume (ml/g)	20 °C	90 °C	Linear flow[a,b] (cm/h)
BioGel	P-60	3–60	14	4	1	5
	P-100	5–100	15	4	1	5
	P-200	30–200	29	4	1	4
	P-300	60–400	36	4	1	3
Sephadex	G-50	2–30	10	3	1	5
	G-100	4–150	18	72	5	5
	G-150	5–300	25	72	5	3
	G-200	5–600	35	72	5	2

[a] The values listed are for beads of a medium mesh size.
[b] The linear flow indicated is appropriate for moderately high-resolution chromatography. The volume flow in mm/h is obtained by multiplying the linear flow by the cross-sectional area of a column in cm^2.

Table 23.3 Suspended matrix parameters

Name	Code	Fractionation range (kDa)	Linear flow[a] (cm/h)
BioGel	A-0.5m	<10–500	18
	A-1.5m	10–1500	18
	A-5m	10–5000	18
Sephacryl	S-200 HR	5–600	15
	S-300 HR	10–1500	15
	S-400 HR	20–8000	15
Sepharose	CL-6B	10–4000	18
Ultrogel	A6	25–2400	5
	A4	55–9000	4
	AcA 54	5–70	4.5
	AcA 44	10–130	4.5
	AcA 34	20–350	4
	AcA 22	100–1200	2.5

[a] The linear flow indicated is appropriate for moderately high-resolution chromatography. The volume flow in mm/h is obtained by multiplying the linear flow by the column cross-sectional area in cm^2.

Table 23.4 Packed column matrix parameters

Name	Code	Pore diameter (Å)	Diameter × length (mm)	Fractionation range (kDa)
Protein Pak	60	50	7.8 × 300	1–20
	125	125		2–80
	300	300		10–500
Shodex	WS 802.5	150	8 × 300[a]	4–150
	WS 803	300		10–700
	WS804	500		10–2000
Superose	12	–	10 × 300	1–300
	6	–		5–5000
SynChropak	60	60	7.8 × 300[b]	5–30
	100	100		5–130
	300	300		15–800
	500	500		30–2000
TSK	G2000SW	125	7.5 × 300[c]	5–60
	G3000SW	250		1–300
	G4000SW	500		5–1000
Zorbax	GF-250	150	9.4 × 250	4–500
	GF-450	300		5–900

[a] Also 8 × 500 and 20 × 300 (mm).
[b] Also 21.5 × 250 (mm).
[c] Also 21.5 × 600 and 55 × 600 (mm).

2.2. Sample preparation

The sample should not have a protein concentration in excess of about 50 mg/ml and should be clarified by centrifugation, if necessary, in order to prevent particulate matter from slowing the flow rate of the column. The solvent for the protein sample is of little consequence since the protein will advance ahead of the application solvent during chromatography.

2.3. Chromatographic solvents

The solvents used to flow through the column have wide latitude, subject only to the pH and temperature constraints listed in Table 23.1. However, the ionic strength of the chromatographic solvent should be at least 0.2 *M* to minimize the binding of proteins to the matrix by electrostatic or by van der Waals interactions. Most proteins are inherently stable at room temperature and require only low temperatures in order to reduce the rate of peptide hydrolysis catalyzed by any proteolytic enzymes present in the protein sample. However, proteolysis becomes an increasing problem during purification as the desired protein becomes the more abundant substrate for the proteases. In some cases, rather expensive proteolytic inhibitors or effectors need be present in the chromatographic solvent in order to maintain the function of the desired protein. Some economy can be realized by equilibration with only one column volume of the solvent containing the expensive component(s) prior to application of the sample, since the sample advances into the column solvent during chromatography. The solvent following the sample application need not contain the expensive component(s).

Columns poured in glass cylinders should be equilibrated with a simple solvent, such as 0.1 *M* NaCl containing about 0.02% sodium azide, to prevent the growth of microorganisms. Methanol is the preferred storage solvent for columns poured in stainless steel cylinders in order to avoid the corrosion accelerated by the continued presence of salt solutions.

2.4. Preliminary screening

In order to optimize the purification (-fold) achieved by size-exclusion chromatography, it is necessary to use a matrix which will best resolve the desired protein from the remaining proteins. Accordingly, a preliminary screening is useful to estimate the molecular weight of the desired protein and the molecular weights over which the remaining proteins are distributed. The elements needed for screening in addition to a protein sample include a size-exclusion column, a fraction collector, an assay for total protein, an assay for the desired protein, and a molecular weight calibration mixture. The assay for total protein can either be ultraviolet absorbance or a colorimetric procedure (see Chapter 6 by Stoscheck, Vol. 182). A sufficient concentration of the sample must be applied to the column so that the function of the desired protein can be measured with confidence in the eluate fractions. It should be anticipated that the concentration of the desired protein will be diluted at least an order of magnitude by the chromatography.

Molecular weight calibration mixtures, often termed gel-filtration standards, can be purchased from several suppliers, including Bio-Rad Laboratories (Richmond, CA), Pharmacia LKB Biotechnology (Piscataway, NJ), and Sigma Chemical (St. Louis, MO). These calibration mixtures contain several identified proteins of known molecular weight as well as components to establish V_0 and V_t. Alternatively, an investigator can customize a calibration mixture using purified components. Blue dextran and DNA restriction fragments are frequently used to determine V_0. It is important not to use a small aromatic or heterocyclic compound to determine V_t since such molecules are particularly prone to reversible adsorption by size-exclusion chromatographic matrices.

If a high-performance size-exclusion analytical column and chromatograph is available, the screening is both rapid and simple. The column used for screening should have a broad fractionation range. A guard column should be placed in front of the analytical column to retain any particulate material which has escaped notice. A protein sample containing a minimal volume appropriate for analysis of the desired protein in the column effluent should be injected. The effluent should be monitored for protein concentration using an absorbance flow detector set either at the more sensitive 225 nm, if the solvent absorbance permits, or at the less sensitive 280 nm. Effluent fractions should be collected and analyzed for the total protein, if a flow absorbance detector is not available, and for the desired protein. Finally, a gel-filtration standard should be injected into the column and the effluent monitored again at the same wavelength. Comparison of the elution profile for the gel-filtration standard with the profiles for the total protein and the desired protein in the sample should facilitate selection of a matrix that will optimize the purification (-fold) achievable by size-exclusion chromatography.

If a high-performance analytical column is not available, then the screening must be done with a conventional matrix having a broad fractionation range. It is likely that the matrix selected will have to be poured into a column. Instructions for pouring a column using a conventional matrix are detailed below. Again, a conventional matrix that can optimize the purification (-fold) obtained by size-exclusion chromatography can be selected from the screening based on the elution profiles obtained for the gel-filtration standard and for the total protein and the desired protein in the sample.

2.5. Chromatography using conventional matrix

The volume of a conventional matrix used for protein purification should be 30–100 times the volume of the sample to be fractionated. The amount of matrix required to form the column is suspended in the chromatographic solvent and brought to the temperature at which chromatography will be performed. The volume of the suspension should be no more than twice the volume of the column to be made. Fine particles should be removed by gently swirling the suspension and the supernatant removed by suction after about 90% of the beads have settled. Finally, the suspension should be placed under negative pressure to reduce the volume of dissolved air. A filter flask and a laboratory aspirator are useful for this purpose.

If the matrix is supplied as a dry powder, the matrix should be swollen in the chromatographic solvent prior to removal of the fine particles. The matrix may be swollen at either ambient temperature or at 100 °C, depending upon the time available to the investigator. As shown in Table 23.2, swelling of a matrix proceeds much faster at 100 °C without damage to the matrix.

The chromatographic column should be made in a glass or transparent plastic cylinder of either commercial design or laboratory improvisation. The ratio of the length of the cylinder to its diameter may vary from 20 to 100. When improvising, elements of the following procedure can be used. The bottom of the column can be formed from a rubber stopper containing a short length of a thick-walled capillary tube positioned flush with the narrow end of the stopper. The cylinder is oriented vertically and clamped securely in the location in which the chromatography will be performed. The stopper is inserted into the bottom end of the cylinder. A short length of flexible tubing is attached to the protruding glass tube and a clamping device attached to the tubing to control the liquid flow through the cylinder. A nylon or Teflon mesh is placed inside the cylinder and pushed to the bottom to fit snugly against the stopper. The clamp is closed and the cylinder filled with the matrix suspension. The excess suspension is placed in a vessel with a bottom exit and stopcock, such as a separatory funnel, and the exit attached to the top of the cylinder with a length of flexible tubing and a one-hole stopper containing a short length of glass tubing. This assembled apparatus should be airtight between the surface of the excess suspension in the separatory funnel and the flexible tubing extending from the bottom of the cylinder. The flow rate is controlled by the height of the separatory funnel relative to the column. The column can be packed using a flow rate about five times greater than that listed in Table 23.2. Once the desired column height is packed, the clamp and stopcock are closed, the excess matrix suspension removed, and some chromatographic solvent passed through the column using the separatory funnel as the reservoir. A pool of solvent several centimeters in height should be continuously maintained at the top of the column to buffer the impact of the chromatographic solvent as it enters the cylinder so as not to disturb the packing at the top of the column. The packed column should never be allowed to run dry, as it will produce channeling within the column which will severely perturb protein resolution.

To apply a sample to the column, the stopcock should be closed, the stopper at the top of the cylinder removed, and the solvent pooled above the column drained through the column until the solvent just dips below the top of the packed column. The clamp is then closed and the sample or standard solution added carefully to minimally disturb the packing at the top of the column. The clamp is then opened and the sample solution allowed to enter the column until it just dips below the top of the column. The clamp is then closed and a small amount of chromatographic solvent added with minimal disturbance to the packing at the top of the column. This solvent is then admitted to the column, the clamp again closed, and more chromatographic solvent added to the column to form a pool of desired height. A supply of chromatographic solvent is placed in the separatory funnel and connected by an air-tight seal to the top of the cylinder with the flexible tubing. The height of the separatory funnel is then adjusted to achieve and maintain the desired flow rate.

The absorbance of the column effluent can be continuously monitored at a desired wavelength using a flow monitor. It is important that the tubing at the bottom of the column and the flow optical cell in the monitor have a small diameter to prevent convective mixing of the liquid emerging from the column. It is also important that a flexible tubing be used which does not contribute ultraviolet-absorbing material to the chromatography solvent. Alternatively, the column effluent can be directed to a fraction collector and the fractions assayed for both total protein and desired protein. A drop counter is ideal for this purpose.

2.6. Scaling upward

Size-exclusion chromatography using conventional matrices can be easily scaled upward by increasing the volume of the column appropriate to the volume of the sample to be fractionated. Very large sample volumes may be best handled with repetitive chromatography as opposed to construction of columns of monumental dimensions. Semipreparative and preparative scale high-performance columns are available as indicated in Table 23.4 and some suppliers will provide bulk material for packing by the investigator. Although these larger high-performance columns can be quite expensive it should be remembered that they represent a considerable saving in investigator time and that the investment can be amortized over many different uses.

2.7. Trouble shooting

2.7.1. Poor resolution

This is a common lament because size-exclusion chromatography has an inherent low resolution. Nonetheless, changes in some operational parameters may improve resolution. First, since flow rate and resolution are inversely related, decreasing the flow rate may improve the resolution. Second, use of a bead size having a smaller diameter should improve resolution. Third, use of a matrix having a narrower fractionation range may be helpful.

2.7.2. Low flow rate

This usually results from plugging of the filters or the matrix with particulate material in the samples. The column should be first washed by reverse flow with a solubilization agent such as a nonionic or ionic detergent, a protein denaturant such as urea or guanidinium hydrochloride, an organic solvent such as methanol or, within the stability of the matrix, brief exposure to a strong acid or base. If this does not succeed for a conventional matrix, then the column should be disassembled, the individual components cleaned, and the column repacked. If this does not succeed for a high-performance matrix, either the column may be sent to Phenomenex or another supplier for cleaning and repacking at a fee or the column may be simply replaced. Laboratories which have facilities for repacking columns at pressure can clean and repack high-performance columns themselves.

2.7.3. Skewed peaks

A primary cause is poor sample application. For a conventional column, the quality of sample application can be observed by placing an inert colored component in the protein sample such as blue dextran or potassium dichromate. If the sample has an irregular appearance in the column it will likely generate an asymmetrical peak in the elution profile. For a high-performance column, the injector can be disassembled and cleaned. Tailing of peaks generally results from adsorption of proteins to the matrix. This situation can be improved by using a more potent lyotropic salt, such as sodium perchlorate instead of sodium chloride, as the principal ionic component in the chromatographic solvent. In the case of a high-performance column, tailing may indicate the loss of the coating on the silica beads, a situation requiring replacement of the column. Skewed peaks may also result from a reversible equilibrium between different states of polymerization of the protein. For example, hemoglobin can exhibit a dynamic equilibrium between the di- and tetrameric forms of the protein. Since polymerization involves a change in molecular weight, the matrix will favor dissociation while chemical equilibrium will favor association. These opposing forces can result in the appearances of a skewed peak characteristic for a dynamic exchange. Changes in the pH, temperature, or chemical composition of the chromatographic solvent may shift the chemical equilibrium such that only one polymeric form is significantly populated.

2.7.4. Disappearance of desired protein

This may occur for at least two reasons. The desired protein may be moderately adsorbed to the column so that its elution occurs after V_t in a very broad peak that is difficult to distinguish from noise in the baseline. If this is the case, a protein solubilization agent such as a nonionic detergent or a modest concentration of a protein denaturant should be added to the chromatographic solvent. A second possibility involves the dissociation of a functional protein complex into discrete proteins of different molecular weight in which none of the dissociated proteins retains the function. Mixing aliquots from different fractions should facilitate complexation of the component proteins and restoration of the function.

2.8. Further information

Virtually all the suppliers of size-exclusion matrices and customized chromatographic columns have prepared detailed instructions regarding the use of their products. These instructions are quite helpful and generally free of charge.

CHAPTER TWENTY-FOUR

Protein Chromatography on Hydroxyapatite Columns

Larry J. Cummings, Mark A. Snyder, *and* Kimberly Brisack

Contents

1. Introduction 388
2. Mechanisms 389
3. Chemical Characteristics 392
    3.1. Cationic and anionic modification of HA surfaces 392
    3.2. Metal adsorption 393
    3.3. HA solubility 394
4. Purification Protocol Development 396
5. Packing Laboratory-Scale Columns 397
6. Process-Scale Column Packing 399
7. Applications 401
References 402

Abstract

The introduction of spherical forms of hydroxyapatite has enabled protein scientists to separate and purify proteins multiple times with the same packed column. Biopharmaceutical companies have driven single column applications of complex samples to simpler samples obtained from upstream column purification steps on affinity, ion exchange or hydrophobic interaction columns. Multiple column purification permits higher protein loads to spherical forms of hydroxyapatite and improved reduction in host cell protein, aggregates, endotoxin, and DNA from recombinant proteins. Adsorption and desorption mechanisms covering the multimodal properties of hydroxyapatite are discussed. The chemical interactions of hydroxyapatite surface with common ions, metals, and phosphate species affect column lifetimes. Adsorbed hydroxonium ions from low ionic strength buffers are noted by a shift in effluent pH during column equilibration. Hydroxonium ion desorption is observed by acidic shifts in the column effluent with the magnitude and duration surprisingly extreme. Buffering reagents with high buffering capacity reduce both the magnitude and duration of the acidic shift. Column packing methods for the

Bio-Rad Laboratories, Inc., Hercules, California, USA

Methods in Enzymology, Volume 463
ISSN 0076-6879, DOI: 10.1016/S0076-6879(09)63024-X

robust spherical particles as well as the microcrystalline hydroxyapatite particles are reviewed. Applications covering extracted proteins and recombinant protein purification, especially monoclonal antibodies, with multiple chromatography media in concert with hydroxyapatite are reviewed.

1. Introduction

The use of hydroxyapatite, a hydroxylated calcium phosphate material, in columns for protein chromatography has increased substantially between 1991 and 2009 primarily due its use for recombinant protein purification. The methods for using hydroxyapatite (HA) follow those discussed by Tiselius *et al.* (1956) and reviewed by Gorbunoff (1985). Although microcrystalline HA is still manufactured in production-scale quantities, it is generally used in batch mode operations rather than in packed columns because the crystals are mechanically unstable in applications requiring multiple column cycles. Suppliers of HA developed manufacturing processes to make spherical particles that allow multiple column cycles. One process forms microcrystalline HA in agarose gel particles, HA Ultrogel® sorbent as described in French patent 2231422, British patent 1468592, Swiss patent 597893, and German patent 2426501. Another described by Thompson and Miles (1973) forms large spheroids, approximately 50–600 μm in diameter utilizing microcrystalline HA and a ceramic hardening process. The spheroidal ceramic particles were commercialized in the early 1970s by BDH Ltd. The adsorption capacity for protein on agarose–HA is similar if not lower than the microcrystalline HA. The elution profile of bovine serum albumin is similar between the spheroidal ceramic particles and microcrystalline HA. A granular, irregularly shaped form of porous ceramic HA is produced by Clarkson Chromatography. The most commonly used porous spherical ceramic HA for downstream processing of recombinant proteins (CHT™ Ceramic Hydroxyapatite, Bio-Rad Laboratories). The spherical particles are manufactured by spray drying nanoparticles to obtain porous particles in 20, 40, or 80 μm sizes. The particles are mechanically sized to obtain a narrow size distribution. Mechanical stability and porosity is imparted to the sized particles by sintering them at high temperatures. Packed beds of these particles have 12–20 times the surface area of microcrystalline HA particles as discussed by Kadoya *et al.* (1986) and Ichitsuka *et al.* (1992). The increase in surface area results in increased protein adsorption to the commercialized porous CHT™ Types I and II. Type I adsorbs greater than 25 mg of lysozyme per dry gram of resin from a 10-m*M* phosphate buffer, pH 6.8 and Type II between 12 and 19 mg/g due to its lower surface area. This relationship was confirmed by Zhu *et al.* (2009) studying protein adsorption to porous and nonporous bioceramics.

The newer, robust forms of hydroxyapatite such as agarose–HA and ceramic HA are used to manufacture production-scale quantities of recombinant proteins. Large columns ranging from 20 to 600 l are used for many column cycles and thus exposed to very large quantities of buffers prepared from USP grade reagents and water for injection. These larger scale processes revealed some new challenges concerning HA such as adsorption of metals and excursions in effluent pH. The remainder of this chapter covers a review of the mechanism of protein adsorption and desorption, a discussion of hydroxyapatite chemical characteristics, a description of purification protocol development, lab-scale and process-scale column preparation, and a brief review of recent protein chromatography applications utilizing HA.

2. Mechanisms

The mechanism of protein adsorption and desorption to and from HA surfaces has been reviewed periodically since 1971 (Bernardi, 1971; Gorbunoff, 1990). A recent contribution cites (Kandori *et al.*, 2004) earlier mechanisms in explaining that acidic proteins bind through C (calcium)-sites while basic proteins bind through P (phosphate)-sites. C- and P-sites were previously described by Kawasaki *et al.* (1986) and later further discussed by Gagnon (1996) with an emphasis on the importance of C-sites in the purification of monoclonal antibodies; monoclonal antibody binding occurs in part through calcium coordination complexes with carboxyl clusters in the antibody. The mechanisms of HA–protein interactions were simplified to the following description. Amino groups are adsorbed to P-sites but repelled by C-sites. The situation is opposite and more complex for carboxyl groups. Although amine binding to P-sites and the initial attraction of carboxyl groups to C-sites are electrostatic, the binding of carboxyl groups to C-sites involves formation of much stronger chelation, coordination bonds than anion exchange. Phosphoryl groups on proteins and other solutes interact more strongly with C-sites than do carboxyl groups as discussed by Kawasaki (1991).

Further work has involved bone–protein interactions by Dong *et al.* (2007) to study the mechanistic role of zero net charge $-OH$ and $-NH_2$ groups in protein for the surface of HA. Steered molecular dynamic simulation demonstrated the strong interaction between the HA surface and protein $-OH$ and $-NH_2$ groups of bone morphogenetic proteins (growth factors for bone formation) through a water-bridged H-bond. Their observations were confirmed by Shen *et al.* (2008) using a fibronectin fragment (FN-III_{10}). These and similar investigations support the C- and P-site model. Hydroxyapatite surfaces have been proposed to interact differentially with hexahistidine tagged proteins and histidine-rich regions of immunoglobulin G (IgG) (Ng *et al.*, 2007). In this case, calcium–histidine interaction appears to be insignificant based on the coelution of hexahistidine tagged and untagged fusion protein. The adsorption mechanism for proteins remain consistent with those described previously by Gorbunoff (1990) and has been advanced by the addition of the water-bridged H-bond mechanism.

Some proteins do not adsorb to HA if the loading buffer contains phosphate. Good's buffers developed by Ferguson *et al.* (1980) are frequently used under these circumstances. For example, serine hydroxymethyl transferase from *Escherichia coli* has been purified by Schirch *et al.* (1985) to homogeneity by applying the anion-exchange fraction in 20 m*M* *N*,*N*-bis(2-hydroxyethyl)-2-aminoethanesulfonic acid (BES), pH 7.0 to a column of HA equilibrated with the loading buffer followed by elution with a shallow linear phosphate gradient (Schirch *et al.*, 1985).

Protein desorption is conducted by reversing P-site interactions, the two types of C-site interactions and/or the H-bond. Linear and step gradients of phosphate buffer are the most often used desorption agent. In addition, Freitag and Breier (1995) noted that proteins retained by electrostatic interaction of their net positive charge with negative net charge on the HA surface are desorbed by anions added to the phosphate buffer or by cations to desorb those retained through their net negative charge with the positive net charge on the HA surface. Anions such as Cl^-, F^-, ClO_4^-, SCN^-, and phosphate have been discussed by Gorbunoff (1984a,b) and Gorbunoff and Timasheff (1984) in series of publications. Fluoride ion was used by Schlatterer *et al.* (2006) to desorb prostaglandin D synthase from a column of ceramic hydroxyapatite. The enzyme was adsorbed from bovine cerebrospinal fluid followed by sequential desorption with 150 and 375 m*M* NaF using a linear gradient to 750 m*M* NaF in 10 m*M* sodium phosphate buffer, pH 6.25.

Desorption is also highly influenced by buffer pH. According to the study by Ogawa and Hiraide (1996), proteins desorbed more quickly at lower phosphate concentration above pH 7 than below pH 7 and the elution sequence was similar for Types I and II porous ceramic HA but desorption of proteins from Type I required stronger solutions of sodium phosphate buffer. Similarly, buffer pH also affects protein desorption from porous ceramic fluoroapatite for the pH range 5.5–6.8. Table 24.1 compares the elution time in minutes of several proteins in this range obtained on ceramic HA Type I and ceramic fluoroapatite using 1.1 cm × 15 cm packed columns. The columns were packed using 0.2 *M* phosphate buffer, pH 10 and equilibrated with 5 m*M* sodium phosphate at pH 5.5, 6.0, or 6.8. The elution order of the proteins is not dependant solely on the pI of the proteins. With some proteins desorption appears to depend on their structure and carboxyl group content as explained by Gorbunoff (1984a,b).

Table 24.1 Elution time in minutes of proteins on ceramic hydroxyapatite Type I (CHT I) and ceramic fluoroapatite (CFT)

Protein	pI	CHT I pH 6.8	CFT pH 6.8	CHT I pH 6.0	CFT pH 6.0	CHT I pH 5.5	CFT pH 5.5
α-Lactalbumin	4.5	24.3	22.3	27.4	23.7	33.2	29.2
Transferrin	4.7	24.2	23.7	30.8	31.8	46.6	50.0
Bovine serum albumin	4.8	24.3	22.3	29.7	24.2	43.0	39.3
Ovalbumin	5.0	22.0	22.0	24.3	23.8	30.7	27.3
Carbonic anhydrase	5.3	24.3	22.6	29.1	27.6	43.0	39.8
Catalase	5.4	23.5	22.0	29.7	24.1	35.1	35.7
Conalbumin	6.8	32.6	33.7	38.7	37.9	46.4	46.1
Myoglobin	7.0	30.3	29.4	34.8	33.1	46.5	44.7
Ribonuclease a	9.7	30.8	31.4	34.1	35.9	40.5	42.5
α-Chymotrypsinogen	9.8	34.9	31.1	37.9	37.2	51.3	50.0
Lysozyme	10.5	31.6	30.7	34.9	34.9	41.8	42.1
Cytochrome *c*	10.6	45.0	43.4	49.9	49.2	63.1	60.6

Proteins were applied in 5 m*M* phosphate application buffer. The columns were rinsed at 1 ml/min for 10-min with application buffer then eluted with 50-min linear gradient to 0.4 *M* sodium phosphate at pH 5.5, 6.0, or 6.8, respectively. The retention time of each protein excludes the 10-min rinse time with 5 m*M* phosphate application buffer.

3. Chemical Characteristics

3.1. Cationic and anionic modification of HA surfaces

Calcium and magnesium cations change the surface of HA through the formation of a phosphate–Ca or Mg bridge. Although initially commented upon over 20 years ago, Gorbunoff (1984a,b) and Gagnon *et al.* (2009) have recently used Ca-modified ceramic HA surfaces to improve the purification of F_{ab} from Mab and F_c antibody components. These workers showed that Ca-modified ceramic HA is restored to its native form after treatment with 20 column volumes of dilute phosphate buffer.

HA also has a high affinity for pyrophosphate (PPi) and polyphosphates as explained by Krane and Glimcher (1962). PPi is a contaminant in anhydrous salts of mono- and disodium phosphate and may be a contaminant in its potassium salt counter parts. There are advantages and disadvantages to PPi in phosphate buffers relative to the adsorption and desorption of proteins to the HA. Figure 24.1 shows the separation of bovine serum albumin, ovalbumin, α-chymotrypsinogen A, and cytochrome *c* with

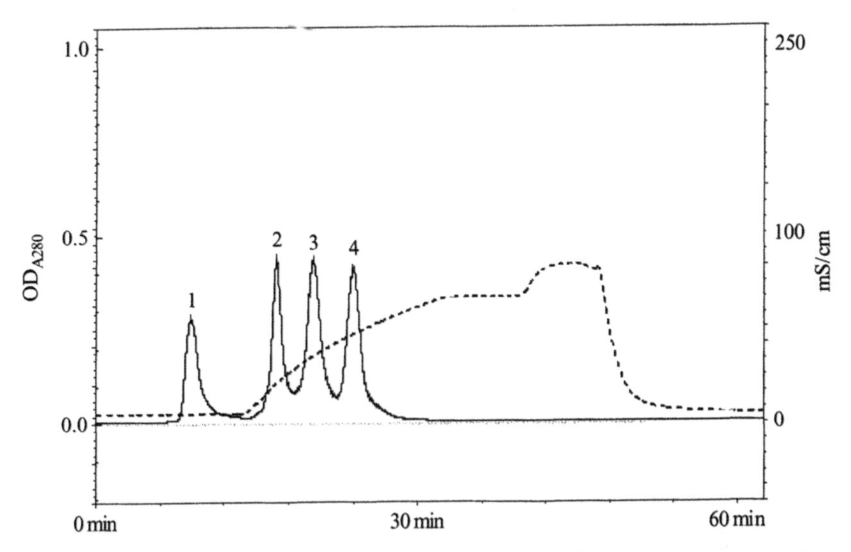

Figure 24.1 Separation of protein mixture using phosphate buffers prepared from anhydrous phosphate salts: (1) bovine serum albumin, (2) ovalbumin, (3) α-chymotrypsinogen, and (4) cytochrome *c*. Column: 10 mm × 40 mm. Flow rate: 2.5 ml/min. Temperature: 20 °C. Program: sample inject, 5-min isocratic A; 20-min 0–80% linear gradient to B; 10-min isocratic 80% B; 10-min isocratic 100% B; 15-min isocratic A. Buffer A: 10 m*M* sodium phosphate, pH 6.8. Buffer B: 400 m*M* sodium phosphate, pH 6.8. Reagents: sodium phosphate monobasic anhydrous and sodium phosphate dibasic anhydrous. Dashed line indicates the relative conductivity of the column effluent.

phosphate buffers prepared from anhydrous sodium phosphate. Figure 24.2 shows the separation of these proteins when applied and desorbed with sodium phosphate buffer prepared from dibasic sodium phosphate hepta hydrate and monobasic sodium phosphate monohydrate. The column used with sodium phosphate buffer prepared from anhydrous sodium phosphate required four program cycles in order to diminish the effects of the PPi (see Program: in Fig. 24.1). As mentioned previously, PPi may be used advantageously in the chromatography of proteins on HA particularly for the exclusion of otherwise weakly adsorbed proteins. Care must be taken, however, to ensure that the levels of higher-order phosphates are kept the same between all experiments to avoid artifacts.

3.2. Metal adsorption

Columns used multiple times can discolor due to the accumulation of metals. The causes of discoloration are found in studies conducted by dental scientists relative to caries in teeth, medical scientists relative to bone

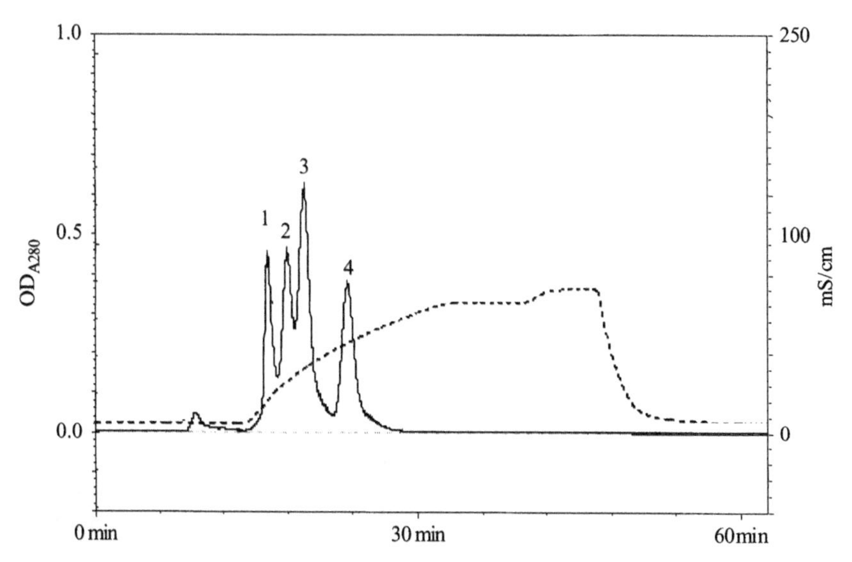

Figure 24.2 Separation of protein mixture using phosphate buffers prepared from hydrated phosphate salts: (1) bovine serum albumin, (2) ovalbumin, (3) α-chymotrypsinogen, and (4) cytochrome *c*. Column: 10 mm × 40 mm. Flow rate: 2.5 ml/min. Temperature: 20 °C. Program: sample inject, 5-min isocratic A; 20-min 0–80% linear gradient to B; 10-min isocratic 80% B; 10-min isocratic 100% B; 15-min isocratic A. Buffer A: 10 m*M* sodium phosphate, pH 6.8. Buffer B: 400 m*M* sodium phosphate, pH 6.8. Reagents: sodium phosphate monobasic monohydrate and sodium phosphate dibasic, heptahydrate. Dashed line indicates the relative conductivity of the column effluent.

integrity and tissue fluids for artificial bone or cadaver implants, and material scientists examining soil remediation and metals absorption.

According to Shepard *et al.* (2000), a column of ceramic HA discolored after several recombinant protein purification cycles. The column of ceramic HA was the third column in the process following cation and anion exchange. They concluded that the metals that discolored the ceramic HA came from various sources: process equipment, process reagents, process water, and fermentation nutrients. However, the discoloration did not impair the purification of their recombinant protein. Agronomists explain the preferential absorption of multivalent metals to HA (Ma *et al.*, 1994) and show that HA has a high capacity for divalent heavy metals. Many of the described metals were observed in the discolored ceramic HA column, especially Fe, Al, Zn, and Mn.

3.3. HA solubility

Voids in the column headspace or the formation of channels can occur with HA resulting from the massive amounts of buffer applied to the column. The solubility constant for HA is 2.4×10^{-59} which is equivalent to about 4 ppm. Phosphate ion and basic pH suppress the solubility. However, adsorbing protein to HA from very dilute phosphate buffer, 1 m*M* (pH 6.8), in unbuffered alkali metal solutions, 1 m*M* NaCl or KCl or in zwitterionic buffers is common. Some acidic proteins absorb only in water as determined by Schirch *et al.* (1985). The low to unbuffered loading conditions reduce the control over the surface pH of the HA as explained by Scopes (1993). Some researchers (Schroder *et al.*, 2003) determined that cobuffers such as 4-morpholinepropanesulfonic acid (MES) could be used to permit phosphate elution while maintaining the pH of the loading and elution buffers. However, HA solubility is a risk even with phosphate suppression. For example, one column was equilibrated with 20 m*M* sodium phosphate, pH 6.5, loaded with protein applied in the equilibration buffer then sequentially eluted with increased sodium chloride buffered with 20 m*M* sodium phosphate. In this case, voids formed and the column displayed increasing pressure from cycle to cycle. The voids and increased pressure were caused by chemical damage to the HA largely due to the liberation of hydroxonium ion that accumulated on the surface of the HA. Figure 24.3 shows the pH profile of the column effluent and the zones where hydroxonium ion adsorbed (A and B) and desorbed (C and D).

Harding *et al.* (2005) discuss the adsorption and desorption of hydroxonium ion as a function of the zero point charge of HA. The results of these research (Skartsila and Spanos, 2007) show the relationship of zero point charge, pH, hydroxonium adsorption, desorption, and calcium content. The zero point charge determined by Skartsila and Spanos differs, 7.3 ± 0.1 and 6.5 ± 0.2. These research contributions explain the

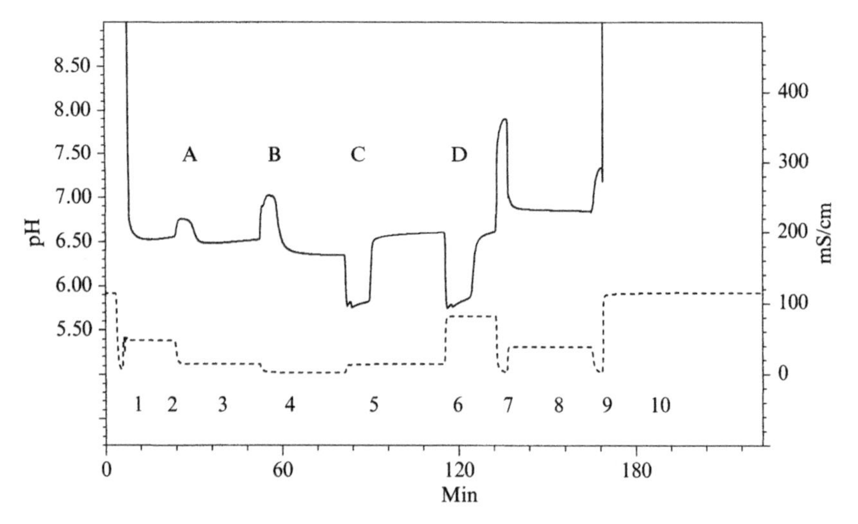

Fig. 24.3 Effluent pH changes relative to buffer sequence. Column 11 mm × 230 mm. Flow rate: 7.0 ml/min. Temperature: 20 °C. Program: (1) 3-min 20 m*M* sodium phosphate, pH 6.5 to rinse 0.1 *M* NaOH from the column; (2) 17-min pH adjustment with 500 m*M* sodium phosphate/0.1 *M* NaCl, pH 6.5; (3) 29-min 20 m*M* sodium phosphate/low concentration NaCl, pH 6.5; (4) 29-min 20 m*M* sodium phosphate, pH 6.5; (5) 34-min 20 m*M* sodium phosphate/low concentration NaCl, pH 6.5; (6) 18-min 20 m*M* sodium phosphate/low concentration NaCl, pH 6.5; (7) 3-min 20 m*M* sodium phosphate to rinse calcium ion from the column; (8) 29-min regeneration with 500 m*M* sodium phosphate/0.1 *M* NaCl, pH 6.5; (9) 3-min 20 m*M* sodium phosphate to rinse concentrated phosphate from the column; (10) 29-min sanitization with 0.5 N NaOH. Points A and B indicate the change to a more basic effluent than the pH 6.5 input buffer. Points C and D indicate the change to more acidic effluent than the pH 6.5 input buffer. Dashed line indicates the relative conductivity of the column effluent. The solid line indicates the pH of the column effluent.

differences between input buffer pH and deviations in the pH of column effluent. It is reasonable to assume that changing the pH of the input buffers or increasing their buffer capacity could reduce the intensity of the pH shift and perhaps the duration of the shift. However, as noted previously, some proteins do not adsorb to HA when the phosphate concentration is too high. In some cases, even 5 m*M* phosphate prevents protein adsorption as noted by these process development engineers (McCue *et al.*, 2007). Alternative buffering agents were used to preserve the integrity of the target protein while allowing adsorption to HA. This was best accomplished without any phosphate in the equilibration and load buffers; however, the packed column failed after a few cycles. Studies showed that as little as 2 m*M* phosphate increased the column life without compromising the purity of the target protein. Phosphate in combination with MES or 3-(*N*-morpholino) propanesulfonic acid (MOPS) and minor amounts of calcium ion was

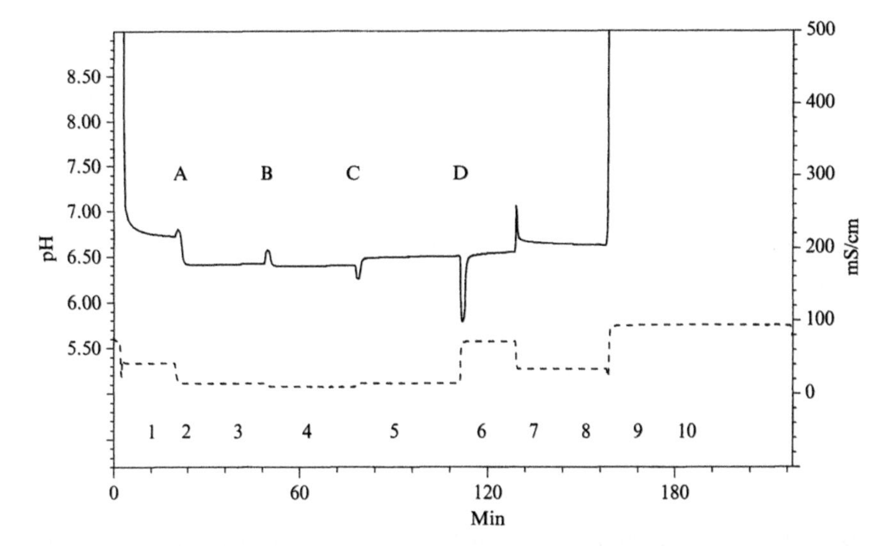

Figure 24.4 Effluent pH changes relative to buffer sequence. The buffer sequence is the same as in Fig. 24.3 except that buffers 1, 3, 4, and 5 contain 75 m*M* MES. Dashed line indicates the relative conductivity of the column effluent. The solid line indicates the pH of the column effluent.

sufficient to suppress the solubility of HA. pH shifts were not explicitly mentioned by these development engineers (McCue *et al.*, 2007). Figure 24.4 shows the benefits of MES as a cobuffer in terms of minimizing pH shifts and durations of the shift observed in Fig. 24.3. Calcium ion was measured in the high sodium chloride fraction for the two elution buffers. It was 36 ppm for the 0.02 *M* sodium phosphate buffered solution and 5 ppm for the 75 m*M* MES/0.02 *M* phosphate buffered solution.

4. Purification Protocol Development

Most proteins are adsorbed to HA low ionic strength buffers with phosphate at concentrations as low as 1 m*M*. Adsorbed proteins are often eluted with higher concentrations of phosphate but also with NaCl or KCl (salts). It is uncommon to see other types of desorbing agents for used for these purposes.

If the protein sample contains a high amount of salt, dilute it sufficiently so that it adsorbs to the HA but not below 1 m*M* in phosphate. A sufficient concentration of an alternative buffer should be incorporated for pH control. Elute the column with a linear salt gradient to determine if the protein desorbs. If retained then repeat the experiment using a linear phosphate gradient as eluant. Test the protein for purity. Determine the concentration of the salt or phosphate necessary to elute the protein and apply it as a desorption step in the purification protocol. Compare the purity of the step desorbed protein to the gradient desorbed protein. Make adjustment to the desorption protocol if necessary. Desorb other bound proteins and biological components from the column by rinsing with 3 column volumes of phosphate-containing, salt-free, low ionic strength (LIS) buffer followed by 3–5 column volumes of 0.5 *M* phosphate buffer, pH 7, 3 column volumes LIS and 3–5 column volumes of 0.5–1 *M* NaOH. Prepare the column for subsequent column cycles by rinsing with 3 column volumes LIS, 3 column volumes of 0.5 *M* phosphate buffer at the pH of the equilibration buffer then with 3 column volumes load buffer.

5. Packing Laboratory-Scale Columns

Packing protocols for laboratory-scale HA columns differ depending on the source of HA. To pack HA Ultrogel sorbent, use a column with a height/diameter ratio between 1 and 6 with diameters ranging from 1.6 to 5.0 cm. Prepare the sorbent for packing by mildly agitating the product container by inverting it or by stirring the contents with a plastic paddle. Transfer a suitable volume of slurry to a vacuum flask with a volume capacity about twice the planned packed volume. Dilute the slurry with about 40% additional water. Mix gently, then connect a vacuum pump to remove dissolved gas. Mildly stir the degassed sorbent to obtain a homogeneous suspension. Pour it into the column in one continuous motion minimizing entrapping air into the sorbent stream. Settle the suspension for 5–10 min until a 1-cm clear supernatant layer is visible at the top of the column. Insert the inlet adapter into the column, connect it to a liquid chromatography then flow pack with water or equilibration buffer at between 55 and 250 cm/h depending on the desired packed bed height; 55 cm/h for a 20-cm height bed, 250 cm/h for a 5-cm height packed bed. Adjust the inlet adapter so that it contacts the surface of the packed bed excluding any trapped air while lowering the adapter. If air accidently intrudes the packed bed, the column requires repacking.

CHT is a spherical porous ceramic media. Suitable columns include Millipore Vantage™ L and Waters AP with variable height adapters with pressure ratings of at least 3 bar. Calculate the weight of CHT required for the packed column from the packed bed dimensions and the density of CHT, 0.63 g/ml. For example, a 1.1-cm ID × 20.0 cm height packed bed has a volume of 19 ml or 12 g of CHT. Suspend the CHT in 3.5 volumes of packing buffer (0.2 *M* sodium phosphate dibasic heptahydrate, pH 9–10) in a suitably size beaker using a plastic spatula. Alternative packing solutions should be at least 0.1 *M* in ionic strength and contain at least 20 m*M* buffer component that buffer in the range of pH 6.8–10. Allow the suspension to settle to allow occluded air to separate from the particles. Attach a packing extension tube to the column so that all of the suspended ceramic HA can be dispensed in a single step. Pour packing buffer into the column/extension assembly to a height of 1–2 cm. Suspend the particles in the beaker with the plastic spatula and pour it into the column/extension assembly. Rinse any remaining ceramic HA into the column/extension assembly with packing buffer. Open the column outlet and allow the column to pack by gravity flow to constant height. Close the column outlet, remove the extension tube then insert the inlet adapter and lower to just touch the surface of the gravity packed ceramic HA. Connect the column to a liquid chromatography, open the column outlet, and equilibrate with 3 column volumes of packing buffer at 250 cm/h to flow pack the ceramic HA. Adjust the adapter so that it just touches the surface of the flow packed ceramic HA. If air accidently intrudes the bed, it is usually eliminated during column conditioning steps. Condition the column by equilibrating with 3 column volumes of equilibration buffer, 3 column volumes of 0.5–1.0 *M* NaOH, 3 column volumes of 0.4–0.5 *M* sodium phosphate, pH 6.5–7.2 then 3 column volumes equilibration buffer. Equilibration buffers are generally low ionic strength buffers such as 2–20 m*M* phosphate with Good's cobuffers added when the phosphate concentration is 2–10 m*M*.

Microcrystalline hydroxyapatite is available from various companies. Bio-Gel® HTP and HT require preparation before packing them into a column. Decant the supernatant from the HT then add an equal volume of packing buffer. Suspend the HT by rotating the container then pour approximately twice the volume of the suspension for the desired packed volume it into a beaker. Settle the suspension for 30 min then decant the supernatant which may contain fines. For HTP, dispense 1 g for each 2.5 ml of desired packed volume. Add the powder to a beaker containing two to six times the desired packed volume while mixing slowly with a plastic stirring rod. Settle the suspension for 30 min then decant the supernatant which may contain fines. Add packing buffer equal to the desired packed volume to make a 50% (v/v) suspension. Gently suspend the HT or HTP with a plastic stirring rod. Pour the suspension into the column using a funnel or column extension to contain the total volume of the suspension. Wait for 10 min then open the column outlet and pack the column by gravity flow. Do not allow air to intrude into the packed bed. Close the column outlet when approximately 2 cm of buffer remains above the packed bed, remove the extension tube or funnel then insert the inlet adapter and lower to eject air from the adapter inlet line. Continue lowering until the adapter just touches the surface of the gravity packed bed. If using DNA Grade Bio-Gel HTP, prepare it and pack it similarly to HTP. The maximum flow rate for HT and HTP is 100 and 40 cm/h for the DNA

Table 24.2 Hydroxyapatite sources

Source	Product name	Particle size(s)
Bio-Rad Laboratories	Bio-Gel HT	Microcrystalline
	Bio-Gel HTP	Microcrystalline
	DNA-Grade Bio-Gel HTP	Microcrystalline
	CHT ceramic Type I	20, 40, and 80 μm
	CHT ceramic Type II	20, 40, and 80 μm
Clarkson Chromatography	Ceramic hydroxyapatite	53–124 μm
	Hypatite C	Microcrystalline
Pall Corporation	HA Ultrogel	60–80 μm, microcrystalline in agarose gel particles
Sigma-Aldrich	Hydroxyapatite, Type I	Microcrystalline
GFS Chemicals	Calcium Hydroxyapatite	Microcrystalline
Spectrum Chemicals	Calcium Hydroxyapatite	Microcrystalline

Microcrystalline sizes of hydroxyapatite are not described. The product specification describes the maximum flow rate under gravity flow where the maximum column height is 100 mm and the hydrostatic head is 200 mm. All of the microcrystalline sizes are produced by the method of Tiselius (Tiselius *et al.*, 1956).

Grade. The preparation and packing instructions for calbiochem hydroxylapatite fast flow and high resolution are similar to HTP and DNA Grade HTP, respectively. Clarkson's ceramic hydroxyapatite and Hypatite C should be prepared and packed similarly to HTP and HT.

Suppliers of HA do not often list their entire HA product line. Table 24.2 lists current suppliers of HA and ceramic HA.

6. Process-Scale Column Packing

Well-packed large-scale columns, in which the beds are homogeneous and continuous from top to bottom, exhibit the best chromatographic performance. Several methods exist for packing columns with CHT that depend on the type of column and equipment used. Review the relevant instruction manuals for columns, media transfer devices, and media packing devices prior to packing the columns.

The maximum packed bed height of open columns such as EasyPack™ (Bio-Rad Laboratories), BPG™ (GE Healthcare), and Moduline 2™ (Millipore) cannot be greater than 50% of the height between the surfaces of the media retention plates (frits or nets). For example, if the distance is 54 cm, the maximum packed height is 27 cm. Calculate the volume of the packed column. For each liter of packed bed volume, use 630 g of dry powder and 1.79 l of packing buffer to prepare a 50% (v/v) slurry. With the column outlet closed, dispense the packing buffer to the column followed by the dry powder. Agitate the CHT-buffer mixture with a plastic paddle to hydrate the powder and blend the components to a homogenous slurry. Reverse the direction of agitation to minimize the motion of the slurry. Assemble the top adapter following the manufacturer's instruction and insert into the column tube. Wait 5 min to allow for a resin free zone then lower the adapter allowing air to vent through the top process inlet and to purge the adapter flow distributor and inlet line with packing buffer. Flow pack at 200–300 cm/h with 2 column volumes of packing buffer. Once the bed is consolidated, lower the adapter, leaving a headspace of 1–5 mm between the media retention plate and the top of the packed bed. Do not lower the adapter into the packed bed to avoid irreversible damage to the CHT particles.

Closed columns such as the InPlace™ (Bio-Rad Laboratories) and BioProcess™ LPLC (GE Healthcare) columns are packed with externally prepared slurries. As with open columns, the maximum packed bed height cannot be greater than 50% of the height between the surfaces of the media retention plates (frits). Calculate the volume of the packed column. For each liter of packed bed volume, use 630 g of dry powder and 1.79 l of packing buffer to prepare a 50% (v/v) slurry. Dispense the packing buffer to the media slurry tank followed by the dry powder. Agitate the CHT-buffer mixture to with a low-shear hydrofoil impeller (Type A3) to hydrate the powder and blend the components to a homogenous slurry. Transfer all of the slurry to the column using a media transfer device. Apply up flow at 50 cm/h to expel air bubbles from the transferred slurry. Stop the flow and wait 5 min to allow a resin free zone to form. Lower the adapter, allowing air to vent through the top process inlet and to purge the adapter flow distributor and inlet line with packing buffer. Axially, flow pack at 200–300 cm/h to consolidate the packed bed. Continue lowering until the adapter media retention plate is 1–5 mm above the surface of the packed bed. When packing stainless steel InPlace or LPLC columns axial flow pack until the adapter reaches 2 cm above the target height of the column then lower at 10 cm/h until the adapter sensor signals contact with the top of the bed.

Process-scale columns utilizing microcrystalline HA is not achievable. HA Ultrogel is utilized in shallow bed columns with large diameters. Packing this type of column and media requires assistance from the column and media suppliers.

The evaluation of packed process columns is often required by end users per the recommendation of regulatory agencies. The United States Food and Drug Administration and its counterparts in Europe, Canada, Japan, and Asia issued guidelines for evaluating packed columns. Chromatography media suppliers have issued simple methods for lab-scale and process-scale columns but do not necessarily correlate the results between the two. However, innovators (Teeters and Quiñones-García, 2005) within the biopharmaceutical companies have been resourceful and proposed residence time distribution (RTD) to track the condition of a packed column during the column's lifetime.

7. Applications

Phosphate gradient or step elution remains a dominant desorbing strategy in recombinant purification protocols. Recombinant human catalase expressed in *Pichia pastoris* was purified to 95% in a three-step process—ammonium sulfate precipitation, anion-exchange chromatography, and hydroxyapatite chromatography. The recombinant catalase was desorbed from the HA column with a linear gradient of 0.05–0.3 *M* phosphate, pH 6.8. Shi *et al.* (2007) obtained a 60% yield of the catalase that was secreted into the culture medium. Numerous proteins purified by chromatography on HA columns utilize phosphate linear or step gradients as described by Hsieh *et al.* (2003), Stránská *et al.* (2007), Luellau *et al.* (1998), and Nuss *et al.* (2008).

Clones of cDNA for human and two *E. coli* pyridoxal kinase genes were used by di Salvo *et al.* (2004) to express pyridoxal kinases in *E. coli*. The ammonium sulfate precipitated cell lysate pellet containing the enzyme was dissolved with phosphate buffer then purified using three types of chromatography media—hydrophobic interaction gel, anion-exchange resin, and ceramic HA. The HA purification step for the human pyridoxal kinase utilized a 20-m*M* sodium *N*,*N*-bis(2-hydroxyethyl)-2-aminoethanesulfonic acid (BES), pH 7.3 adsorption buffer and a linear gradient to 100 m*M* potassium phosphate, pH 7.3 to desorb the enzyme. Methyl jasmonate hydrolyzing esterase was purified from cell cultures of *Lycopersicon esculentum* by Stuhlfelder *et al.* (2002) using anion-exchange chromatography, gel-filtration, and chromatography on ceramic HA. The HA step utilized a 20-m*M* potassium phosphate, 20-m*M* β-mercaptoethanol, 0.3-m*M* calcium chloride, pH 6.8 adsorption buffer and a linear gradient to 500 m*M* potassium phosphate, 20 m*M* β-mercaptoethanol, pH 6.8 to desorb the esterase. Recombinant erythropoietin (rEPO) is manufactured with a multistep chromatography process using affinity, hydrophobic interaction, HA, and anion-exchange chromatography as explained by these inventors in a recent application for a patent (Schumann *et al.*, 2007). The ceramic HA step utilized a 20-m*M* Tris, 5-m*M* calcium chloride, 250-m*M* sodium chloride, 9% isopropanol, pH 6.9 adsorption buffer. The rEPO is desorbed with 10 m*M* Tris, 0.5 m*M* calcium chloride, 10 m*M* potassium phosphate, pH 6.8. In another purification process for rEPO, the inventors (Kawasaki *et al.*, 2008) use HA Ultrogel sorbent equilibrated with 0.05 *M* Tris buffer (pH 7.5) containing 1 *M* sodium chloride and 2 m*M* calcium chloride to adsorb the protein. The rEPO desorbed with 20 m*M* phosphate buffer solution (pH 7.5) containing 0.005% (w/v) of polysorbate 80.

Immunoglobulins have been isolated and purified utilizing multiple chromatography steps. For example, human immunoglobulin A (IgA) from Cohn fraction III of human serum was further purified by Leibl *et al.* (1996) used heparin-affinity chromatography followed by fractionation with ceramic HA. The ceramic HA was equilibrated with 10 m*M* phosphate, 137 m*M* NaCl, pH 7.4, and the IgA heparin fraction applied. The adsorbed IgA monomer was desorbed with 15.3 m*M* phosphate, 287 m*M* NaCl, pH 6.8 then further purified by anion exchange, size exclusion, and affinity chromatography to remove IgG from the IgA component. In another example, monoclonal IgG (Mab) was obtained from mouse ascites fluid and isolated by hydrophobic charge induction chromatography. Guerrier *et al.* (2001) applied the Mab fraction to HA gel in 10 m*M* sodium phosphate, pH 8 then desorbed it with 0.5 *M* potassium chloride, 10 m*M* sodium phosphate, pH 6.8. HA was used as a polishing step by Sinacola and Robinson (2002) to remove aggregates of single-chain antibodies (scFv) and inactive monomeric scFv from active scFv. The active scFv did not adsorb to HA equilibrated with 200 m*M* NaCl, 1 m*M* ethylenediaminetetraacetic acid buffered with 100 m*M* Tris–HCl, pH 8.3.

Mab expressed in high titers often contain clinically significant amounts of immunoglobulin aggregates. The selectivity of the monomer and polymer aggregates is often the same for HA surfaces in phosphate desorption systems. Sodium chloride was used to purify monomer from aggregates on HA (Sun, 2003). The details are given in US patent application 20050107594. Aggregates, endotoxin and DNA were desorbed with 0.5 *M* sodium phosphate buffer, pH 6.8. A significant improvement in aggregate removal with phosphate desorption was demonstrated by Gagnon (2008) with buffers supplemented with polyethylene glycol (PEG). With higher concentrations of PEG in the adsorption buffer, monomer was not adsorbed but aggregates, endotoxin and DNA were.

REFERENCES

Bernardi, G. (1971). Chromatography of proteins on hydroxyapatite. *Methods Enzymol.* **22,** 325–339.

di Salvo, M. L., Hunt, S., and Schirch, V. (2004). Expression, purification, and kinetic constants for human and *Escherichia coli* pyridoxal kinases. *Protein Expr. Purif.* **36,** 300–306.

Dong, X., Pan, H.-H., Wang, Q., and Wu, T. (2007). Understanding adsorption–desorption dynamics of BMP-2 on hydroxyapatite (001) surface. *Biophys. J.* **93,** 750–759.

Ferguson, W. J., Bell, D. H., Braunschweiger, K. I., Braunschweiger, W. R., Good, N. E., Jarvis, N. P., McCormick, J. J., Smith, J. R., and Wasmann, C. C. (1980). Hydrogen ion buffers for biological research. *Anal. Biochem.* **104,** 300–310.

Freitag, R., and Breier, J. (1995). Displacement chromatography in biotechnological downstream process. *J. Chromatogr. A.* **691,** 101–112.

French patent 2231422, British patent 1468592, Swiss patent 597893, and German patent 2426501.

Gagnon, P. (1996). Purification tools for monoclonal antibodies. p. 87. Validated Biosystems Inc., Tucson, AZ.

Gagnon, P. (2008). Improved antibody aggregate removal by hydroxyapatite chromatography in the presence of polyethylene glycol. *J. Immunol. Methods* **336,** 222–228.

Gagnon, P., Cheung, C.-W., and Yazaki, P. (2009). Reverse calcium affinity purification of Fab with calcium derivatized hydroxyapatite. *J. Immunol. Methods* **342,** 115–118.

Gorbunoff, M. J. (1984a). The interaction of proteins with hydroxyapatite: I. Role of protein charge and structure. *Anal. Biochem.* **136,** 425–432.

Gorbunoff, M. J. (1984b). The interaction of proteins with hydroxyapatite: II. Role of acidic and basic groups. *Anal. Biochem.* **136,** 433–439.

Gorbunoff, M. J. (1985). Protein chromatography on hydroxyapatite columns. *Methods Enzymol.* **117,** 370–380.

Gorbunoff, M. J. (1990). Protein chromatography on hydroxyapatite columns. *Methods Enzymol.* **182,** 329–339.

Gorbunoff, M. J., and Timasheff, S. (1984). The interaction of proteins with hydroxyapatite: III. Mechanism. *Anal. Biochem.* **136,** 440–445.

Guerrier, L., Boschetti, E., and Flayeux, I. (2001). A dual-mode approach to the selective separation of antibodies and their fragments. *J. Chromatogr. B* **755,** 37–46.

Harding, I. S., Hing, K. A., and Rashid, N. (2005). Surface charge and the effect of excess calcium ions on the hydroxyapatite surface. *Biomaterials* **26,** 6818–6826.

Hsieh, H.-Y., Calcutt, M. J., Chapman, L. F., Mitra, M., and Smith, D. S. (2003). Purification and characterization of a recombinant alpha-*N*-acetylgalactosaminidase from *Clostridium perfringens*. *Protein Expr. Purif.* **32,** 309–316.

Ichitsuka, T., Kawamura, K., Ogawa, T., Sumita, M., and Yokoo, A. (1992). Packing material for liquid chromatography *US Patent 5039408.*

Kadoya, T., Ebihara, M., Ishikawa, T., Isobe, T., Kobayashi, A., Kuwahara, H., Ogawa, T., Okuyama, T., and Sumita, M. (1986). A new spherical hydroxyapatite for high performance liquid chromatography of proteins. *J. Liq. Chromatogr.* **9,** 3543–3557.

Kandori, K., Ishikawa, T., and Miyagawa, K. (2004). Adsorption of immunogamma globulin onto various synthetic calcium hydroxyapatite particles. *J. Colloid Interface Sci.* **273,** 406–413.

Kawasaki, T. (1991). Hydroxyapatite as a liquid chromatographic packing. *J. Chromatogr.* **544,** 147–184.

Kawasaki, T., Ikeda, K., Kuboki, Y., and Takahashi, S. (1986). Further study of hydroxyapatite high-performance liquid chromatography using both proteins and nucleic acids, and a new technique to increase chromatographic efficiency. *Eur. J. Biochem.* **155,** 249–257.

Kawasaki, A., Kirihara, S., and Mukai, K. (2008). Method for production of erythropoietin *WO/2008/068879.*

Krane, S. M., and Glimcher, M. J. (1962). Transphosphorylation from nucleoside di and triphosphates by apatite crystals. *J. Biol. Chem.* **237,** 2991–2998.

Leibl, H., Eibl, M. M., Mannhalter, J. W., Tomasits, R., and Wolf, H. (1996). Method for the isolation of biologically active monomeric immunoglobulin A from plasma fraction. *J. Chromatogr. B* **678,** 173–180.

Luellau, E., Freitag, R., Vogt, S., and von Stockar, U. (1998). Development of a downstream process for isolation and separation of immunoglobulin monomers, dimers, and polymers from cell culture supernatant. *J. Chromatogr. A* **796,** 165–175.

Ma, Q. Y., Logan, T. J., and Traina, S. J. (1994). Effects of aqueous Al, Cd, Cu, Fe(II), Ni, and Zn on Pb immobilization by hydroxyapatite. *Environ. Sci. Tech.* **28,** 1219–1228.

McCue, J. T., Cecchini, D., Dolinski, E., and Hawkins, K. (2007). Use of an alternative scale-down approach to predict and extend hydroxyapatite column lifetimes. *J. Chromatogr. A* **1165,** 78–85.

Ng, P. K., Gagnon, P., and He, J. (2007). Mechanistic model for adsorption of immunoglobulin on hydroxyapatite. *J. Chromatogr. A* **1142,** 13–18.

Nuss, J. E., Choksi, K. B., DeFord, J. H., and Papaconstantinou, J. (2008). Decreased enzyme activities of chaperones PDI and BiP in aged mouse livers. *Biochem. Biophys. Res. Commun.* **365,** 355–361.

Ogawa, T., and Hiraide, T. (1996). Effect of pH on gradient elution of different proteins on two types of Macro-Prep ceramic hydroxyapatite. *Am. Lab.* **28,** 31–34.

Schirch, V., Angelaccio, S., Hopkins, S., and Villar, E. (1985). Serine hydroxymethyltransferase from *Escherichia coli*: Purification and properties. *J. Bacteriol* **163,** 1–7.

Schlatterer, J. C., Baeker, R., Kehler, W., Klose, J., Schlatterer, B., and Schlatterer, K. (2006). Purification of prostaglandin D synthase by ceramic and size exclusion chromatography. *Prostaglandins Other Lipid Mediat.* **81,** 80–89.

Schroder, E., Jonsson, T., and Poole, L. (2003). Hydroxyapatite chromatography: Altering the phosphate-dependent elution profile of protein as a function of pH. *Anal. Biochem.* **313,** 176–178.

Schumann, C., Hesse, J.-O., and Mack, M. (2007). Method for purifying erythropoietin *US Patent Application Publication, US2007/0293420 A1.*

Scopes, R. K. (1993). Protein Purification Principles and Practice. 3 edn. pp. 172–175. Springer, New York.

Shen, J. W., Pan, H.-H., Wang, Q., and Wu, T. (2008). Molecular simulation of protein adsorption and desorption on hydroxyapatite surfaces. *Biomaterials* **29,** 513–532.

Shepard, S. R., Brickman-Stone, C., Koch, G., and Schrimsher, J. L. (2000). Discoloration of ceramic hydroxyapatite used for protein chromatography. *J. Chromatogr. A* **891,** 93–98.

Shi, X.-L., Feng, M.-Q., Shi, J.-H., Shi, J., Zhong, J., and Zhou, P. (2007). High-level expression and purification of recombinant human catalase in *Pichia pastoris*. *Protein Expr. Purif.* **54,** 24–29.

Sinacola, J. R., and Robinson, A. S. (2002). Rapid refolding and polishing of single-chain antibodies from *Escherichia coli* inclusion bodies. *Protein Expr. Purif.* **26,** 301–308.

Skartsila, K., and Spanos, N. (2007). Surface characterization of hydroxyapatite: Potentiometric titrations coupled with solubility measurements. *J. Colloid Interface Sci.* **308,** 405–412.

Stránská, J., Chmelík, J., Peč, P., Popa, I., Řehulka, P., Šebela, M., and Tarkowski, P. (2007). Inhibition of plant amine oxidases by a novel series of diamine derivatives. *Biochimie* **89,** 135–144.

Stuhlfelder, C., Lottspeich, F., and Mueller, M. J. (2002). Purification and partial amino acid sequences of an esterase from tomato. *Photochemistry* **60,** 233–240.

Sun, S. (2003). Removal of high molecular weight aggregates from an antibody preparation using ceramic hydroxyapatite chromatography. *In* "Third International Hydroxyapatite Conference", Lisbon, Portugal.

Teeters, M. A., and Quiñones-García, I. (2005). Evaluating and monitoring the packing behavior of process-scale chromatogragphy columns. *J. Chromatogr. A* **1069,** 53–64.

Thompson, A. R., and Miles, B. J. (1973). New materials, especially for chromatography. *Methodol. Dev. Biochem. Prep. Tech.* **2,** 95–101.

Tiselius, A., Hjertén, S., and Levin, O. (1956). Protein chromatography on calcium phosphate columns. *Arch. Biochem. Biophys.* **65,** 132–155.

Zhu, X. D., Fan, H. S., Li, D. X., Luxbacher, T., Xiao, Y. M., Zhang, H. J., and Zhang, X. D. (2009). Effect of surface structure on protein adsorption to biphasic calcium-phosphate ceramics in vitro and in vivo. *Acta Biomater.* **5,** 1311–1318.

CHAPTER TWENTY-FIVE

Theory and Use of Hydrophobic Interaction Chromatography in Protein Purification Applications

Justin T. McCue

Contents

1. Theory 406
2. Latest Technology in HIC Adsorbents 408
3. Procedures for Use of HIC Adsorbents 409
   3.1. Introduction 409
   3.2. Choice of adsorbent 409
   3.3. Feed/load preparation 410
   3.4. Adsorbent preparation 410
   3.5. Product elution 410
   3.6. Gradient elution 412
   3.7. Stepwise (isocratic) elution 412
   3.8. Adsorbent regeneration and sanitization 413
References 413

Abstract

Hydrophobic interaction chromatography (HIC) is a valuable tool used in protein purification applications. HIC is used in the purification of proteins over a broad range of scales—in both analytical and preparatory scale applications. HIC is used to remove various impurities that may be present in the solution, including undesirable product-related impurities. In particular, HIC is often employed to remove product aggregate species, which possess different hydrophobic properties than the target monomer species and can often be effectively removed using HIC. In this chapter, we provide a description of the basic theory of HIC and how it is used to purify proteins in aqueous-based solutions. Following the theoretical background, the latest in HIC adsorbent technology is described, including a list of commonly used and commercially available adsorbents. The basic procedures for using HIC adsorbents are described next, in order to provide the reader with useful starting points to apply HIC in protein purification applications.

Biogen Idec Corporation, Cambridge, Massachusetts, USA

Methods in Enzymology, Volume 463
ISSN 0076-6879, DOI: 10.1016/S0076-6879(09)63025-1

1. Theory

Hydrophobic proteins will self-associate, or interact, when dissolved in an aqueous solution. This self-association forms the basis for a variety of biological interactions, such as protein folding, protein–substrate interactions, and transport of proteins across cellular membranes (Janson and Rydén, 1997). Hydrophobic interaction chromatography (HIC) is used in both analytical and preparatory scale protein purification applications. HIC exploits hydrophobic regions present in macromolecules that bind to hydrophobic ligands on chromatography adsorbents. The interaction occurs in an environment which favors hydrophobic interactions, such as an aqueous solution with a high salt concentration.

By itself, water (a polar solvent) is a poor solvent for nonpolar molecules. Under such an environment, proteins will self-associate, or aggregate, in order to achieve a state of lowest thermodynamic energy. Prior to self-association, water molecules form highly ordered structures around each individual macromolecule (Fig. 25.1A). The self-association of nonpolar molecules (such as proteins) in the polar solvent is driven by a net increase in entropy of the environment. During the aggregation process, the overall surface area of hydrophobic sites of the protein exposed to the polar solvent

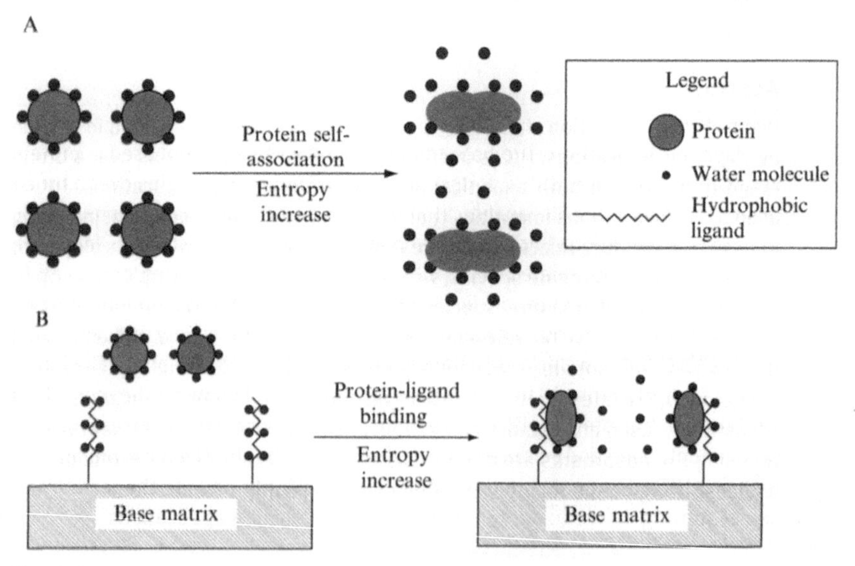

Figure 25.1 Schematic diagram showing hydrophobic interaction between proteins in an aqueous solution (A) and between proteins and a hydrophobic ligand on an HIC adsorbent (B).

is decreased, which results in a less structured (higher entropy) condition, which is the favored thermodynamic state.

This same concept is responsible for the interaction (association) between hydrophobic ligands attached to an adsorbent and the proteins of interest (Fig. 25.1B). Association, or hydrophobic interaction, between the protein and the hydrophobic ligand is driven primarily by an increase in the overall entropy (compared with the condition when no interaction is occurring between the protein and the adsorbent).

The polarity of the solvent can be controlled through the addition of salts or organic solvents, which can strengthen or weaken hydrophobic interactions between the HIC adsorbent and the protein. The influence of ions on hydrophobic interaction follows the well-known Hofmeister series (Hofmeister, 1988). Anions which promote hydrophobic interaction the greatest are listed (in decreasing strength of interaction) from left to right (Påhlman *et al.*, 1977):

$$PO_4^{3-} > SO_4^{2-} > CH_3COO^- > Cl^- > Br^- > NO_3^- > CLO_4^- > I^- > SCN^-$$

Ions which promote hydrophobic interactions are called lyotropes, while those which disrupt (weaken) hydrophobic interactions are called chaotropes. In the above series, phosphate ions promote the strongest hydrophobic interaction, while thiocyanate ions disrupt hydrophobic interactions.

For cations, the Hofmeister series consists of the following (listed in order of decreasing lyotropic strength):

$$NH_4^+ > Rb^+ > K^+ > Na^+ > Cs^+ > Li^+ > Mg^{2+} > Ca^{2+} > Ba^{2+}$$

Two of the most common Lyotropic salts used to promote hydrophobic interaction in aqueous solution are ammonium sulfate and sodium chloride. These salts are commonly employed when using HIC for protein purification.

In addition to salts, organic solvents can also be used to alter the strength of hydrophobic interactions (Fausnaugh and Regnier, 1986; Melander and Horvath, 1977). Organic solvents commonly used to weaken, or disrupt hydrophobic interactions include glycols, acetonitrile and alcohols. The organic solvents alter the polarity of the mobile phase, thereby weakening potential interactions that may occur. They may be added to the solution during the elution process, in order to disrupt hydrophobic interactions and elute the strongly bound protein of interest.

Protein hydrophobicity is a complex function of several properties, which include the amino acid sequence, as well as protein tertiary and quaternary structure in a given solution (Ben-Naim, 1980; Tanford, 1980). Hydrophobicity scales have been created for particular amino acids, which are based upon the solubility in water and organic solvents

(Jones, 1975; Nozaki and Tanford, 1971; Tanford, 1962; Zimmerman *et al.*, 1968). Empirical hydrophobic scales for proteins have also been created (Chotia, 1976; Krigbaum and Komoriya, 1979; Manavalan and Ponnuswamy, 1978; Rose *et al.*, 1985; Wertz and Scheraga, 1978) which are based upon the fraction of amino acids exposed on the protein surface, as well as the degree of amino acid hydrophobicity. The ability to predict the hydrophobicity of complex proteins has been only semiquantitative to date, and experiments are usually required to accurately understand protein hydrophobicity in a given aqueous solution.

2. Latest Technology in HIC Adsorbents

HIC adsorbents consist of a base matrix which is coupled to a hydrophobic ligand. The base matrix, which typically consists of porous beads with diameters ranging from 5 to 200 μm, provides high surface area for ligand attachment and protein binding. Common base matrices include agarose, methacrylate, polystyrene–divinylbenzene and silica (Table 25.1). For analytical applications, the bead size of the adsorbent is in the lower range (5–20 μm). Small beads are used in order to maximize resolution when performing analytical separations. For preparatory scale applications, larger bead sizes are usually required ($\geq$20 μm). Larger bead sizes are required for preparatory scale columns due to pressure drop limitations associated with the column hardware.

HIC adsorbents containing hydrophobic ligands with various degrees of hydrophobicity are available. The ligands consist of alkyl or aryl chains. As a

Table 25.1 Properties of commercially available HIC adsorbents

Base matrix	Available ligand types	Adsorbent manufacturers
Cross-linked agarose	Butyl Octyl Phenyl	GE Healthcare
Polystyrene–divinylbenzene	Phenyl	GE Healthcare Applied Biosystems
Methacrylate	Butyl Ether Phenyl Hexyl	 TosoHaas EM Industries
Silica	Propyl Diol Pentyl	J. T. Baker Synchrom Supelco YMC

general rule, the strength of hydrophobic binding of the ligand will increase with the length of the organic chain. Several of the most common ligands include butyl, octyl, and phenyl, which are linked to the base bead support through several different coupling approaches (Hjertén *et al.*, 1974; Ulbrich *et al.*, 1964). Aromatic ligands, such as phenyl, can also interact with the adsorbed compounds through so-called "π–π interactions," which can further strengthen the hydrophobic interaction (Porath and Larsson, 1978).

The hydrophobic interaction strength of the ligand can also be influenced by the ligand loading (ligand density) on the base matrix. The strength of interaction can increase with higher ligand densities. In order to have reproducible performance, manufacturers of HIC adsorbents must often produce adsorbents with narrow ranges of ligand density to ensure consistent performance from lot to lot.

3. Procedures for Use of HIC Adsorbents

3.1. Introduction

The application of HIC adsorbents for use in the purification of protein compounds is fairly straightforward. The purification process is composed of a series of subsequent steps, which are described in this section.

3.2. Choice of adsorbent

Since the hydrophobic properties of a given protein are often unknown, several HIC adsorbents may need to be screened prior to selection of a final adsorbent. The screening process is used to determine how strongly compounds bind to the adsorbent, and determine which adsorbents are viable candidates to purify the protein. Adsorbents with a broad range of hydrophobic binding properties should be included in the initial screening studies to determine the hydrophobicity of the compound, as well as which adsorbents effectively purify the protein.

Adsorbents may be available in the form of prepacked columns from the adsorbent manufacturer or may have to be packed by the user. In the case when columns need to be packed, the packing instructions from the manufacturer should be followed. Height equivalent to the theoretical plate (HETP) measurements can be used to verify that the column is packed correctly. Once packed, the adsorbent manufacturer often provides recommended operating ranges for use, which include allowable flow rate ranges, column conditioning, and column cleaning procedures. Acceptable operating conditions will vary for different adsorbents, and the vendor documentation should be consulted prior to use.

3.3. Feed/load preparation

Prior to column loading, the salt concentration of the protein mixture (which will be purified using the HIC adsorbent) must be increased to a level in which the target protein binds to the adsorbent using a high salt buffer. Proteins may precipitate in high salt solutions, so the compound solubility in a given salt solution should be evaluated prior to its selection. The influence of buffer pH on compound binding to HIC adsorbents has no general trend, but can influence the strength of interaction (Hjertén *et al.*, 1974). A buffer pH should be chosen in which the protein and the adsorbent are stable (e.g., avoid the use of extreme pH solutions). During the protein load adjustment step, the salt concentration may range from approximately 0.5 to 2.0 *M*, and will be increased high enough to ensure the protein binds effectively to the adsorbent. The salt concentration required to bind the protein to the adsorbent will depend greatly on the choice of salt, as described in Section 3.2. Selection of the appropriate salt concentration in which the protein binds to the adsorbent will require experimental work in most cases.

3.4. Adsorbent preparation

Prior to loading the protein feed, the column should first be equilibrated in a high salt buffer solution which possesses a similar composition (salt concentration) and pH as the feed solution to ensure the protein will bind tightly to the adsorbent. This step is referred to as the equilibration step.

Following the equilibration step, the adjusted feed (which contains the protein of interest) is loaded onto the HIC column at an appropriate velocity. After the protein-containing solution is loaded onto the column, the column is often washed with the equilibration buffer prior to product elution. Additional wash steps may be implemented prior to the elution step to remove undesirable impurity species. The wash steps may contain a salt concentration at an intermediate salt concentration—less than the load step, but greater than the elution step.

3.5. Product elution

After being bound to the adsorbent, the desired protein must be eluted and then collected in the column effluent. In many cases, the elution process is used to separate, or resolve, unwanted species from the desired protein. The unwanted species may bind less tightly to the adsorbent, and will be eluted prior to the product. In other cases, undesirable species bind more tightly to the adsorbent and will remain bound after the product is eluted. This is usually the case when HIC is used to separate protein aggregate species, which bind more tightly to the adsorbent than the desired protein monomer species. During the elution step, a portion (or fraction) of the eluate may contain highly purified product, while fractions before and after contain higher levels of undesirable impurities. A schematic of an elution process (during a gradient elution) is shown in Fig. 25.2. Figure 25.2 illustrates that the column effluent collected during the elution step may need to be fractionated in order to achieve acceptable product purity when using HIC. The gradient elution process is described in more detail in Section 3.6.

The elution process can be done using either a stepwise (isocratic) or a gradient approach. The four most common methods (listed from most common to least common) used to elute the bound protein include the following:

1. *Decrease in the salt concentration (relative to the binding conditions).* A decrease in the salt concentration will decrease the strength of hydrophobic interaction between the protein and the ligand, and the protein will be desorbed and eluted from the column.
2. *Addition of organic solvents.* Addition of an organic solvent (such as ethylene or propylene glycol) changes the solvent polarity, which disrupts the hydrophobic interaction.
3. *Increase in the salt concentration (using a chaotropic salt).* Addition of a chaotropic salt will disrupt the hydrophobic interaction.
4. *Detergent addition.* Detergents are used as protein displacers, and have been used mainly for the purification of membrane proteins when using HIC (Janson and Rydén, 1997).

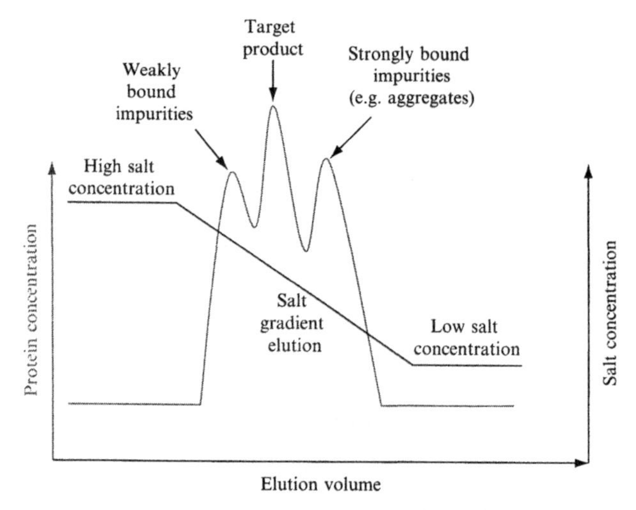

Figure 25.2 Schematic chromatogram showing a gradient elution of a protein mixture using hydrophobic interaction chromatography. In the diagram, the salt concentration is linearly decreased (from high salt to low salt), which results in elution of both impurities and the target protein.

This most common approach used to elute proteins from HIC adsorbents is by lowering the salt concentration during the elution step. This should be the first method that is attempted when using HIC for purification of a new protein compound. The other approaches described above have the disadvantage that an additional component (such as a chaotropic salt or an organic solvent) needs to be added, which may impact protein stability. However, such agents may be required in order to effectively elute a strongly bound protein species from the adsorbent. Each protein must be evaluated case by case to determine which elution method is appropriate. The HIC adsorbent used in the purification may also influence which elution method is effective.

3.6. Gradient elution

Gradient elutions are an extremely effective method useful for screening different HIC adsorbents in protein purification. During the gradient elution process, the salt concentration is decreased gradually (in a linear fashion) from a high salt concentration to a low salt concentration over a defined volume. During the initial screening of a bound compound on an adsorbent, the salt concentration may be decreased to as low as 0 m*M* to determine the salt concentration when the product elutes. As a starting point, a typical gradient elution process is performed over 10 column volumes, during which fractions are collected and evaluated for product purity. The gradient in salt concentration may be decreased (performed over a larger volume) in order to improve protein resolution (Yamamoto *et al.*, 1988).

In the event that the protein remains bound to the adsorbent following the gradient elution process, this may indicate that either a weaker lyotropic salt should be selected to bind the protein to the adsorbent or that a stronger elution condition is required to elute the protein. Stronger elution solutions may include the use of an organic solvent. For preparatory scale applications, nonflammable organic solvents (such as propylene or ethylene glycol) are often selected. Organic solvents in analytical scale applications may include such solvents as acetonitrile and alcohols. Alternatively, an adsorbent with weaker hydrophobic binding strength may need to be selected to decrease the strength of hydrophobic interaction.

3.7. Stepwise (isocratic) elution

After identifying the appropriate adsorbent and salt concentration to effectively elute the protein of interest, an isocratic elution can be used if desired. An advantage of using isocratic elution is its simplicity—it requires a simple switch in the inlet buffer (from a high to a low salt concentration). Use of an isocratic elution is a preferable approach to simplify the equipment requirements, as gradient elution requires multiple pumps and additional process control to generate a linear change in the buffer salt concentration.

3.8. Adsorbent regeneration and sanitization

HIC adsorbents are reusable for multiple cycles and have a relatively long lifetime before having to be replaced. However, adsorbents must be cleaned and regenerated between uses in order to ensure reproducible performance over many cycles. The adsorbent manufacturers' provide regeneration procedures for the adsorbents, which should be consulted prior to use. In general, the cleaning procedures depend upon the stability of the base matrix and the hydrophobic ligand. For strongly bound proteins, 6 *M* guanidine hydrochloride is often recommended. If detergents have been used during the process, ethanol or methanol can be used to as part of the regeneration procedure (GE Healthcare, 2006). For sanitization, a caustic solution (1.0 *M* NaOH) can be used for most of the adsorbents (with the exception of silica). The manufacturer should also provide information on the appropriate storage conditions. Storage solution should be selected that prevents microbial growth, but does not impact ligand or base matrix stability.

REFERENCES

Ben-Naim, A. (1980). Hydrophobic Interactions. Plenum Press, New York.

Chotia, C. (1976). Surface of monomelic proteins. *J. Mol. Biol.* **105,** 112.

Fausnaugh, J. L., and Regnier, F. E. (1986). Solute and mobile phase contributions to retention in hydrophobic interaction chromatography of proteins. *J. Chromatogr.* **359,** 131–146.

GE Healthcare (2006). Data File No. 18-1127-63 AC.

Hjertén, S., Rosengren, J., and Påhlman, S. (1974). Hydrophobic interaction chromatography. *J. Chromatogr.* **101,** 281–288.

Hofmeister, F. (1988). On regularities in the albumin precipitation reactions with salts and their relationship to physiological behavior. *Arch. Exp. Pathol. Pharmakol.* **24,** 247–260.

Janson, J.-C., and Rydén, L. (eds.) (1997). Protein Purification: Principles, High-Resolution Methods, and Applications, 2nd edn., p. 284. Wiley-VCH, New York.

Jones, D. D. (1975). Amino acid properties and side-chain orientation in proteins. *J. Theor. Biol.* **50,** 167–183.

Krigbaum, W. R., and Komoriya, A. (1979). Local interactions as a structure determinant for protein molecules. *Biochim. Biophys. Acta* **576,** 204–248.

Manavalan, P., and Ponnuswamy, P. K. (1978). Hydrophobic character of amino acid residues in globular proteins. *Nature* **275,** 673–674.

Melander, W., and Horvath, C. (1977). Salt effects on hydrophobic interactions in precipitation and chromatography of proteins: An interpretation of the lyotropic series. *Arch. Biochem. Biophys.* **183,** 200–215.

Nozaki, Y., and Tanford, C. (1971). The solubility of amino acids and two glycine peptides in aqueous ethanol and dioxane solutions. Establishment of a hydrophobicity scale. *J. Biol. Chem.* **246,** 2211–2217.

Påhlman, S., Rosengren, J., and Hjerten, S. (1977). Hydrophobic interaction chromatography on uncharged Sepharose derivatives. *J. Chromatogr.* **131,** 99–108.

Porath, J., and Larsson, B. (1978). Charge-transfer and water-mediated chromatography. I. Electron-acceptor ligands on cross-linked dextran. *J. Chromatogr.* **155,** 47–68.

Rose, G. D., Geselowitz, A. R., Lesser, G. J., Lee, R. H., and Zehfus, M. H. (1985). Hydrophobicity of amino acid residues in globular proteins. *Science* **229,** 834–838.

Tanford, C. (1962). Contribution of hydrophobic interactions to the stability of the globular conformation of proteins. *J. Am. Chem. Soc.* **84,** 4240–4247.

Tanford, C. (1980). *In* The Hydrophobic Effect 2nd edn. Wiley, New York.

Ulbrich, V., Makes, J., and Jurecek, M. (1964). Identification of giycidyl ethers. Bis(phenyl-) and bis(a-naphthylurethans) of glycerol a-alkyl (aryl)ethers. *Collect. Czech. Chem. Commun.* **29,** 1466–1475.

Wertz, D. H., and Scheraga, H. A. (1978). Influence of water on protein structure. *Macromolecules* **11,** 9–15.

Yamamoto, S., Nakanishi, K., and Matsuno, R. (1988). Ion-Exchange Chromatography of Proteins Mercel Dekkar, New York.

Zimmerman, J.M, Eliezer, N., and Simha, R. (1968). The characterization of amino acid sequences in proteins by statistical methods. *J. Theor. Biol.* **21**(2), 170–201.

SECTION SEVEN

PURIFICATION PROCEDURES: AFFINITY METHODS

CHAPTER TWENTY-SIX

Affinity Chromatography: General Methods

Marjeta Urh, Dan Simpson, *and* Kate Zhao

Contents

1. Introduction 418
2. Selection of Affinity Matrix 419
   2.1. General features of the support material 419
   2.2. Selectivity 420
   2.3. Stability 422
   2.4. Magnetic affinity beads 422
3. Selection of Ligands 423
   3.1. General considerations for ligands design and selection 423
   3.2. Characterization of immobilized ligand 424
   3.3. Affinity matrices carrying specific ligands 425
   3.4. Immunoglobulin binding proteins 425
   3.5. Lectins 426
   3.6. Biomimetic ligands 427
   3.7. Covalent affinity chromatography 428
4. Attachment Chemistry 429
   4.1. Activation of surface 430
   4.2. Ligand attachment 430
5. Purification Method 433
   5.1. Sample preparation 433
   5.2. Binding and wash 433
   5.3. Elution 434

References 435

Abstract

Affinity chromatography is one of the most diverse and powerful chromatographic methods for purification of a specific molecule or a group of molecules from complex mixtures. It is based on highly specific biological interactions between two molecules, such as interactions between enzyme and substrate, receptor and ligand, or antibody and antigen. These interactions, which are typically reversible, are used for purification by placing one of the

Promega Corporation, Madison, Wisconsin, USA

Methods in Enzymology, Volume 463
ISSN 0076-6879, DOI: 10.1016/S0076-6879(09)63026-3

interacting molecules, referred to as affinity ligand, onto a solid matrix to create a stationary phase while the target molecule is in the mobile phase. Successful affinity purification requires a certain degree of knowledge and understanding of the nature of interactions between the target molecule and the ligand to help determine the selection of an appropriate affinity ligand and purification procedure. With the growing popularity of affinity purification, many of the commonly used ligands coupled to affinity matrices are now commercially available and are ready to use. However, in some cases new affinity chromatographic material may need to be developed by coupling the ligand onto the matrix such that the ligand retains specific binding affinity for the molecule of interest. In this chapter, we discuss factors which are important to consider when selecting the ligand, proper attachment chemistry, and the matrix. In recent years, matrices with unique features which overcome some of the limitations of more traditional materials have been developed and these are also described. Affinity purification can provide significant time savings and several hundred-fold or higher purification, but the success depends on the method used. Thus, it is important to optimize the purification protocol to achieve efficient capture and maximum recovery of the target.

1. Introduction

Affinity chromatography is a method for selective purification of a molecule or group of molecules from complex mixtures based on highly specific biological interaction between the two molecules. The interaction is typically reversible and purification is achieved through a biphasic interaction with one of the molecules (the ligand) immobilized to a surface while its partner (the target) is in a mobile phase as part of a complex mixture. The capture step is generally followed by washing and elution, resulting in recovery of highly purified protein. Highly selective interactions allow for a fast, often single step, process, with potential for purification in the order of several hundred to thousand-fold. Additional uses of affinity chromatography include the ability to concentrate substances present at low concentration and the ability to separate proteins based on their biological function where an active form can be separated from the inactive form or a form with different biological function.

Recent decades have seen tremendous advancements in the utility of affinity chromatography, with developments in support materials such as flow-through beads, magnetic beads and monolithic materials as well as new ligands with a variety of interesting biological properties. In addition, this approach is no longer used only for purification of specific biomolecules. It is also quickly becoming a method of choice to study biological interactions and can be used for preparation of samples for mass spectrometry or for specific removal of contaminants.

While the first examples of affinity chromatography exploited binding of enzyme amylase to insoluble starch (Starkenstein, 1910), it was the development of solid support materials and the chemistry used to attach ligands to these solid supports that truly enabled affinity chromatography (Campbell *et al.*, 1951; Cuatrecasas *et al.*, 1968; Lerman, 1953a,b). Since these seminal advances, there has been an explosion in the development of more support materials, affinity ligands, and improvements in the methods that make affinity purification an attractive approach to purify biomolecules from complex mixtures. While affinity chromatography can be used to purify any biological molecule with a specific interacting ligand, this chapter will focus on purification of proteins. Several excellent reviews and books have been written on this subject (Hage *et al.*, 2006; Hirabayashi, 2008; Labrou, 2003; Ostrove, 1990; Zachariou, 2007); thus, we will not review all the proteins and methods that have been developed up to date, but will rather give general guidelines and considerations important for the selection of different support materials, ligands, attachment chemistry, and optimization of the purification and discuss some of the latest developments in affinity chromatography.

2. Selection of Affinity Matrix

Successful affinity purification depends on the selection of a suitable solid support and a suitable immobilized ligand. The affinity matrix (i.e., a solid support onto which ligand is immobilized) should ideally be macroporous with high chemical and physical stability, and should selectively capture target of interest while at the same time exhibit low nonspecific adsorption and maintain good flow properties throughout processing. They are preferably inexpensive, readily available, and simple to use. The affinity matrix may be commercially available or can be made by attaching a suitable ligand to a solid support via the appropriate chemistry (see Section 4). There are many detailed reviews on how to select a suitable affinity matrix (GE Healthcare, 2007; Gustavsson and Larsson, 2006; Zachariou, 2007), a summary of some of the key consideration points are given in the following sections.

2.1. General features of the support material

The matrix should be macroporous with uniform particle and pore size and with good flow properties. Pore size of a matrix is inversely correlated to its surface area, which in turn, directly affects the amount of immobilized ligand and thus the capacity. The pore size correlates to exclusion limit, which is the size (molecular weight) or the size range of proteins that cannot enter the pore. Large pores do not suffer from size exclusion effect and allow unhindered access of large molecules to the immobilized ligands, but have reduced surface area and lower ligand density that can result in lower capacity. According to the Renkin equation (Renkin, 1954), the size of the pores should be at least five times larger than the average size of a biomolecule for its easy access to immobilized ligands. The size of the pores should be at least 300 Å or greater if assuming an average protein size of ~60 Å. Most commonly available supports satisfy this requirement (Table 26.1). Some soft gel-based matrices are cross-linked to increase their mechanical stability, but this process can reduce porosity (pore volume), thus reducing the amount of attached ligand and binding capacity. In most cases, a compromise is made between pore size and surface area to fit a particular application.

Other important considerations are diameter of particles and particle size distribution. In theory, smaller particles are better because they allow for faster mass transfer between outer flow and interior of the particle, making higher flow rates possible while maintaining efficient affinity capture. However, they lead to higher flow resistance, greater potential for particle collapse, and increased sensitivity to contaminants such as particles and denatured proteins in the sample which can lead to high back pressure. The selection of the particle size will largely depend on the purpose and method of the separation where particles of 10 μm are best suited for HPLC separation and 400 μm for preparative purposes. The particle size distribution should be as uniform as possible so that smaller particles do not fill the void volume and restrict the flow.

2.2. Selectivity

One key feature of an affinity matrix is its selectivity. It should be specific for a protein of interest as determined by the specific ligand coupled to the matrix and inert to all other compounds present in the complex sample. Since most applications are performed in aqueous solutions, often with low ionic strength, the support should be hydrophilic and contain limited charge that may lead to undesirable ionic interaction. Many commercially available supports fulfill these requirements, either with their native structure or by coating with suitable materials. Common supports include beaded agarose and cellulose, available commercially from a number of vendors including Sepharose from GE Healthcare and Affigel from Bio-Rad (Table 26.1). Nonspecificity can come from the support itself, such as hydrophobicity associated with polystyrene beads and negative charge on the surface of silica. It can also be introduced when modifying a matrix to accept a particular ligand. In these cases, the attachment chemistry, the ligand, and the spacer between the ligand and the matrix should be carefully designed, screened and optimized for selectivity, capacity for target of interest, and low nonspecific binding.

Table 26.1 Examples of commercially available matrices

Name	Vendor	Matrix material	Particle size (μm)	Exclusion limit (Da)
Sepharose™ CL6B	GE Healthcare	Agarose	40–165	10,000–1,000,000
Sepharose™ CL4B	GE Healthcare	Agarose	40–165	30,000–5,000,000
Bio-gel A-5m medium	Bio-Rad	Agarose	75–300 (50–200 mesh)	10,000–5,000,000
Perloza MT100 medium	Iontosorb	Cellulose	100–250	2,000,000
Perloza MT50 medium	Iontosorb	Cellulose	100–250	100,000
AllTech Macrosphere	Grace	Silica	7	Pore size 60–300 Å
Bio-gel P-100 medium	Bio-Rad	Polyacrylamide	90–180	5000–100,000
Sephacryl™	GE Healthcare	Cross-linked ally dextrose	50	1000–100,000
Poros 50	Applied Biosystems	Cross-linked poly(styrene-divinylbenzene)	50	Pore size 50–100 nm

2.3. Stability

The affinity matrix must also be chemically and physically stable during the process, such that the support material itself as well as the attached ligand should not react to the solvents used in the process, nor should they be degraded or damaged by enzymes and microbes that might be present in the sample. The chemical compatibilities of commercially available affinity matrices are usually supplied by the manufacturer, which should be used as guidance for developing a successful purification protocol. Cross-linked agarose can usually withstand a wide pH range (e.g., pH 3–12), most aqueous solvents (including denaturants), many organic solvents or modifiers, and enzymatic treatments. Materials such as glass and silica are not stable at alkaline conditions due to hydrolysis; therefore, coating of these surfaces is often needed before attaching ligands. The matrix should also withstand physical stress, such as pressure, especially when packed into a column, and remain intact during the purification process. High pressure can compress the matrix, causing it to collapse. Agarose beads and other soft gel matrices are more susceptible to pressure, relative to stronger supports, such as silica, polystyrene and other highly cross-linked materials.

Monoliths are macroporous, nonbeaded single matrix that can be made from different materials such as agarose, silica, GMA/EDMA, and cryogel (Mallik and Hage, 2006; Plievaa *et al.*, 2009). It possesses many of the desired properties of an affinity support and has become popular due to the presence of large flow-through pores with no void volume permitting convective flow instead of diffusion and high flow rate for shortened run time. In addition, the matrix will not compress and has less pressure drop during column chromatography and can be made to withstand larger pH ranges and harsh chemicals. There are several limitations of monoliths as compared to traditional matrix, such as lower capacity, special processes required for making each type of monolith affinity supports and the current limited range of available affinity types (Mallik and Hage, 2006).

2.4. Magnetic affinity beads

Magnetic separation can significantly shorten the purification process by quick retrieval of affinity beads at each step (e.g., binding, wash, and elution), and reduce sample dilution usually associated with traditional column-based elution. The method can be used on viscous materials that will otherwise clog traditional columns and can therefore simplify the purification process by eliminating sample pretreatment, such as centrifugation or filtration to remove insoluble materials and particulates. The capability of miniaturization and parallel screening of multiple conditions, such as growth conditions for optimal protein expression and buffer conditions for purification, makes magnetic separation amenable to high-throughput analysis which can significantly shorten the purification process (Saiyed *et al.*, 2003).

Paramagnetic particles are available as unmodified, modified with common affinity ligands (e.g., streptavidin, GSH, Protein A, etc.), and conjugated particles with specific recognition groups such as monoclonal and polyclonal antibodies (Koneracka *et al.*, 2006). In addition to target protein purification, they can also be used to immobilize a target protein which then acts as a bait to pull down its interaction partner(s) from a complex biological mixture. See Chapter 16.

3. Selection of Ligands

Selection of the appropriate ligand requires a certain degree of knowledge and understanding of the nature of interactions between the ligand and the target molecule; where the ligand must specifically bind the target molecule and should be stable in different binding and elution conditions. Additionally, when developing affinity purification scheme it is important to consider whether the ligand is commercially available or *de novo* development of the ligand and affinity matrix will be required. The success of the second scenario will largely depend on existing knowledge of protein structure and the nature of interaction, requiring the use of molecular modeling and combinatorial organic synthesis coupled with immobilization chemistry and selective binding analysis. The time and effort to design a novel ligand and to develop appropriate coupling chemistry and matrix may prove to be too lengthy and costly, whereas the use of nonaffinity-based purification technique such as ion exchange and hydrophobic interaction schemes may be a better choice.

3.1. General considerations for ligands design and selection

Affinity between ligand and target molecule is one of the most important considerations when developing a new affinity purification material; low affinity can reduce the binding efficiency resulting in poor yield while high affinity can lead to inefficient elution or inactivation of the target protein by harsh elution conditions also giving low yield. Generally, an affinity constant in the range of 10^6–10^8 M^{-1} can be used for affinity-based purification. It is preferred that binding affinity of ligand and target protein be evaluated prior to the creation of an affinity matrix, but one has to be aware that affinity in solution may differ from affinity after immobilization and in some cases immobilization may even result in a change in specificity.

When attaching ligand to a matrix, covalent coupling is preferred due to the reduced risk of ligand leaching during purification. However, noncovalent attachment strategies such as nonspecific adsorption, biospecific interaction (biotin–streptavidin), entrapment, and most recently developed, molecular imprinting (Alexander *et al.*, 2006) may also be utilized in the absence of reactive groups or to reduce the risk of ligand denaturation. See Section 4 for more details.

Besides weak affinity, reduced binding between ligand and target can also be a result of steric hindrance caused either by the matrix or by other ligands. Introduction of a spacer between the matrix and the ligand can reduce the steric hindrance caused by the matrix (Hage *et al.*, 2006). When designing a spacer arm, consideration should be given to both its length and the hydrophobic character, as a more hydrophilic nature is desirable to reduce the nonspecific interactions with molecules other than the target. Optimization of spacer length is also important, since shorter spacers may not relieve the steric hindrance effect of the matrix and longer ones may promote nonspecific interaction or may fold back onto themselves thereby limiting specific interactions. The density of ligand on the surface should also be optimized. Very high density of a ligand may have an adverse effect and lead to loss of binding capacity either because of close proximity of binding sites causing steric hindrance or strong binding that may prevent effective elution (Hage *et al.*, 2006).

Other factors that may influence selection of the best ligand are ability to be sterilized, stability of the ligand, proper storage conditions and cost.

3.2. Characterization of immobilized ligand

An important factor determining the utility of the affinity resin is the ability to reproducibly make the resin with similar performance characteristics. Thus, it is important to determine optimal reaction conditions including the amount of the ligand needed for the synthesis to assure consistency in performance. Determining the optimal amount of ligand in the synthesis may also reduce the cost by preventing wasteful use of excess ligand.

For covalent attachment it is often desirable to determine how much ligand is attached on the matrix and to determine how coupling conditions affect the amount of immobilized ligand. A simple approach to determine ligand density is to analyze the amount of the ligand left in the reaction after coupling is complete and subtract it from the total starting amount. Depending on the nature of the ligand, different detection methods can be applied, such as spectrophotometric detection, BCA assay for proteins, Ellman's reagent can be used to detect sulfhydryl groups, fluorescent detection and others (Guilbault, 1988; Langone, 1982).

In some instances, such as amino acid or elemental analysis, the ligand can be measured directly on the support, but this leads to destruction of the material and should therefore be done on a small fraction. An indirect estimation of immobilized ligand can be achieved by estimating the binding capacity by determining the amount of target molecule retained by the affinity matrix under desired conditions. This may be the most relevant measurement of the ligand immobilization since it will directly correlate to the performance of the affinity matrix. Maximum binding capacity should be performed at equilibrium, usually at a slow flow rate or in batch binding mode. Note that dynamic binding capacity may differ from optimal binding capacity as it will be affected by the flow rate, mass transfer through matrix pores and affinity constant.

3.3. Affinity matrices carrying specific ligands

In recent years, we have witnessed an explosion in the development of affinity ligands and ready-to-use affinity matrices specific for different target molecules. An overview of commonly used, commercially available ligands is given in Table 26.2, which is by no means a comprehensive list of all the available ligands.

Most available affinity matrices carry what are often called "group specific ligands" that exhibit binding affinity for a group of structurally or functionally related proteins and thus can be used for purification of different target proteins with similar functionality. In the following sections, we describe some commonly used ligands for affinity matrices.

3.4. Immunoglobulin binding proteins

Purification of antibodies, one of the most effective and frequently used application of affinity purification methods is based on binding between the constant region (F_c) of many types of immunoglobulins and protein A or protein G. Protein A from *Staphylococcus aureus* and protein G made by *Streptococcus* are related bacterial proteins that bind the IgG class of antibodies, with differences in subclass specificity and source organism. Detailed lists of different binding affinities of protein A and G can be found in many sources including several commercial suppliers (Guss *et al.*, 1986; Hage *et al.*, 2006). Other immunoglobulin binding proteins that have lately become popular are protein B, a surface protein of group A *Streptococci* bacteria, which binds several subclasses of human IgA antibodies, and protein L (*Peptostreptococcus magnus*), which has ability to interact with kappa light chains without affecting the antigen-binding site of antibodies (Faulmann *et al.*, 1991; Hermanson, 1992). This makes protein L uniquely suited for purification of antibodies lacking F_c regions.

In most cases, good binding to antibodies can be achieved at or near neutral pH, but the optimal pH for sample application can differ depending on the protein used. Protein A binds antibodies strongest at pH 8.2, protein L at pH 7.5 and protein G at pH 5 but protein G can also be used at pH 7–7.5. For elution, solutions at acidic pH (2.5–3.0) are often applied

Table 26.2 Commonly used ligands and specificity

Ligand	Specificity
Cibacron Blue F3G-A	Albumin, kinases, dehydrogenases, enzymes requiring adenylyl-containing cofactors, NAD^+
Blue B	Kinases, dehydrogenases, nucleic acid binding proteins
Orange A	Lactate dehydrogenase
Green A	HAS, dehydrogenases
Polymixin	Endoproteins
Benzamidine	Serine proteases (thrombin, trypsin, kallikrein)
Biotin	Streptavidin, avidin
Gelatin	Fibronectin
Heparin	DNA binding proteins, serine protease inhibitors (antithrombin III), growth factors, lipoproteins, hormone receptors, coagulation factors, DNA, RNA
Lysine	Plasminogen, rRNA, dsDNA
Arginine	Serine proteases with affinity for arg, fibronectin, prothrombin
ADT	Enzymes with affinity for $NADP^+$
AMP	NAD-dependent dehydrogenases and ATP-dependent kinases
NAD, NADP	Dehydrogenases
Lectins	Glycoproteins, polysaccharides, glycolipids
Calmodulin	Calmodulin binding proteins, ATPase, adenylate cyclase, kinases, phosphodiesterase,
Protein A	Fc regions of many IgG subtypes, species dependent, weak interactions with IgA, IgM, IgD
Protein G	Fc region of many IgG subtypes, species dependent
Protein L	Kappa light chains of antibodies (Fab, single chain variable region scFv)

and samples are collected into buffers with neutral or slightly basic pH to avoid denaturation and loss of activity. In cases where biological stability is lost during elution at low pH, elution with a pH gradient in combination with salt can be explored.

3.5. Lectins

Lectins are a diverse group of proteins which bind carbohydrates with high degree of specificity where each lectin has its own specificity profile. They are often used in affinity purification or enrichment of carbohydrate moieties of complex glycoconjugates, such as polysaccharides, glycolipids, and glycoproteins. They also allow for a specific isolation of different glycolforms of a specific protein depending on the nature of glycosylation. Recently, lectins have been used not only for the purification of sugar-containing molecules but also for the enrichment of subgroups of glycoproteins for analysis in mass spectrometry (Hirabayashi, 2008). See Chapter 34.

Most commercially available lectins are of plant origin, and there are over 100 different lectins available either in conjugated or free form. Lectin from *Canavalia ensiformis*, known as Conacanavalin A (ConA) has the affinity for α-D-mannose, α-D-glucose, and *N*-acetylglucosamine and is probably the most frequently used lectin (Hermanson, 1992). Two other popular lectins are wheat germ agglutinin (WGA) and Jacalin; WGA binds to sialic acid and molecules containing *N*-acetyl-D-glucosamine residue and Jacalin binds to α-D-galactosyl groups.

Coupling of lectin onto resin is often performed at neutral pH and in the presence of sugar to preserve sugar binding site. Binding to the target molecule is also usually carried out in neutral pH; note that some lectins require the presence of divalent metal ions, Ca^{2+} and Mn^{2+} in the case of ConA, for optimal binding. Elution is accomplished by adding an access of the specific sugar molecule to the elution buffer and this can be done as stepwise or gradient elution. After elution, the free sugar should be removed by dialysis or size exclusion chromatography.

3.6. Biomimetic ligands

Most of affinity chromatography ligands are naturally occurring and the appeal of these ligands is high selectivity and capacity of binding, but they often suffer from several drawbacks such as low stability, need for purification in their own right, variability in performance from lot to lot, contamination with other biomolecules, sensitivity to sterilization methods and high cost. To overcome some of these limitations and accelerate the use of affinity chromatography in purification of therapeutic proteins, there is an increased focus on the development of synthetic or altered affinity ligands, biomimetics, which mimic the structure of and binding of natural biological ligands.

One of the most popular group of biomimetic ligands are reactive textile dyes which gained popularity because of their flexibility and ability to assume polarity and geometry of the surface of a variety of competitive biomolecules and can function in solution as a competitive inhibitor, coenzyme or effector of many proteins (Madoery and Minchiotti, 2006; Stellwagen, 1990). The majority of these ligands, including the most popular dye, Cibacron blue F3G-A, contain a triazine scaffold which can be modified for improved specificity, forming the basis for the biomimetic dye–ligands concept. Over the past decades, a plethora of dyes were developed and used in the purification of a broad spectrum of proteins including blood proteins such as albumin, oxidoreductase, decarboxylases, glycolitic enzymes, nucleases, hydrolases, lyases, synthetases, and transferases, and many dye–ligand-based affinity chromatographic materials are now commercially available materials (see Table 26.2) (Labrou and Clonis, 1994, 2002). In addition to the expansion of commercially available materials, there is a growing popularity in creating tailor-made biomimetics for better performance. These ligands are designed to target specific proteins by mimicking peptide templates, natural biological recognition motifs, or complementary surface-exposed residues and have been generated by a combination of rational design, combinatorial library synthesis, and subsequent screening (Cecília *et al.*, 2005; Labrou, 2003; Lowe *et al.*, 2001). The immobilization of triazine containing dye onto affinity matrices such as agarose, dextran, and cellulose can be achieved under alkaline conditions by using nucleophilic displacement of the dye's chlorine by hydroxyl groups on the support surface (Labrou, 2000; Labrou and Clonis, 1994; Labrou *et al.*, 1995).

3.7. Covalent affinity chromatography

Affinity chromatography typically relies on reversible interaction between the target molecule and the ligand; however, covalent interactions have also been utilized for isolation of specific target molecules (Blumberg and Strominger, 1972; GE Healthcare, 2007; Hage, 2006).

One such approach is based on the covalent binding between a thiol-containing target molecule and activated thiol immobilized onto purification matrix. Bound protein can be eluted by reducing the cysteine disulfide with 2-mercaptoethanol, TCEP, or dithiothreitol.

Recently, another covalent binding-based purification has been developed on the basis of specific and covalent interaction between a chloroalkane ligand and a protein called HaloTag, Figure 26.1 (Los *et al.*, 2008; Ohana *et al.*, 2009). HaloTag is a 34-kDa monomeric protein fusion tag which can be genetically fused to any protein of interest either at its C- or N-terminus. HaloTag protein was created by first modifying the active site of bacterial haloalkane dehydrolase so that a permanent covalent bond can be formed with the specific chloroalkane ligand. This modification was followed by additional mutagenesis to increase stability of the protein and the binding rate, which is similar to that of the biotin–streptavidin. To make use of HaloTag for purification, a chromatographic matrix, the HaloLink resin carrying the chloroalkane ligand, was created. The covalent nature of the HaloTag-fusion protein capture, combined with rapid binding kinetics, overcome the equilibrium-based limitations associated with traditional affinity purification and enables efficient capture of target proteins even at very low abundance. Furthermore, it allows for extensive and stringent washing without losing the bound protein. While the covalent association

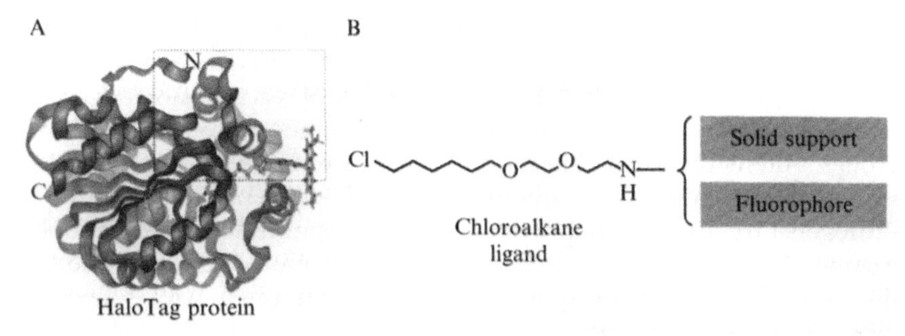

Figure 26.1 HaloTag technology comprises two components: (A) The HaloTag protein shown on the left with covalently bound HaloTag-TMR ligand. N- and C-termini are indicated. (B) The chloroalkane ligand. Different functional groups including, but not limited to, fluorescent dyes or solid support surfaces can be attached to the chloroalkane ligand, which covalently binds to HaloTag protein and imparts different functionalities including protein immobilization and fluorescent labeling.

clearly has its advantages, it also creates a challenge in eluting the protein of interest. Because the covalent bond between HaloTag and chloroalkane cannot be reversed, traditional approaches to elute proteins from the resin cannot be utilized. Instead the protein of interest can be released from HaloTag by specific protease (TEV) as the TEV recognition site is present between the two moieties (the HaloTag and fusion target protein). Upon cleavage, HaloTag stays bound to the resin while the fusion partner is released yielding highly pure protein free of tag (Urh *et al.*, 2008).

Besides using HaloTag for purification of fusion proteins described earlier, the immobilized HaloTag fusions can also be considered as affinity ligands in their own right, and can, similarly to protein G, be used to capture antibodies or other proteins which specifically bind to the proteins fused to HaloTag. The advantage of this system is that unlike other covalent immobilization techniques, where binding of protein is random and may lead to multiple attachment sites and improper orientation, immobilization using HaloTag is a single point attachment through active site and therefore oriented. Single point, oriented attachment increases capacity, effectiveness and reproducibility of the system, and HaloTag fusions covalently bound to the matrix can therefore improve purification of specific antibodies or isolation of binding partners.

4. Attachment Chemistry

This section will briefly review some of the more common chemistry for the covalent attachment of affinity ligands to conventional surfaces such as agarose, cellulose, silica, glass, and synthetic polymeric supports. Strategically, the process can be divided into three components: (1) the surface or resin (matrix); (2) the linkage or resin-activation component (spacer); and (3) the affinity ligand. The constraint that most often dictates how the ligand is attached to a surface is the type and availability of reactive moieties on the ligand. Common moieties used are primary amine, thiol, and carboxyl groups. Alternatively, modification of a ligand to create a reactive group, such as the oxidation of a carbohydrate or the covalent attachment of an orthogonal reactive group, provides additional options for ligand immobilization. Attachment strategies for small synthetic molecules tend to follow the same chemical pathway as biomolecules but often with somewhat different design considerations.

4.1. Activation of surface

The process of preparing a covalent affinity-based surface generally begins with activation of the surface to either directly accept the ligand or to accept an activated linker to physically "space" the ligand away from the surface. Representative attachment chemistries are summarized in Table 26.3, listed with established surfaces and shown in contrast with the ligand entity used to attach to the activated surface. A variety of preactivated surfaces are commercially available for most of the chemistries listed in Table 26.3. If the appropriate chemical entity is readily displayed on the ligand, the remaining chemistry is reduced to defining the optimal conditions for the coupling reaction to occur weighed against the stability constraints of the ligand. Most ligand coupling reactions are done in aqueous pH-controlled buffers for periods between 1 and 20 h at temperatures between 4 °C and room temperature. Since the reaction is heterogenous, both concentration and mixing are important to ensure an even distribution of reactants. Whenever possible keep the solution phase of the reaction to no more than twice the resin volume. Ligand concentration will vary somewhat by objective but for proteins a 2.5–10 mg/ml is a reasonable starting point. Depending on the activation chemistry and ligand reactivity, protocols exist for a fairly broad range of pH and buffer conditions (Hermanson, 1992). Efficiencies will vary from system to system and as a consequence it is essential to cap left over surface reactive groups to neutralize or reduce nonspecific interactions.

4.2. Ligand attachment

For a peptide, protein, or nucleic acid, the covalent attachment is most commonly done through an amine group on the ligand to an amine reactive functionality on the resin surface. The most common amine-reactive attachment is an imidocarbonate that results from the reaction of cyanogen bromide activated surface with a primary amine, most typically a lysine side chain (Hermanson, 1992; Porath *et al.*, 1967) (see Chapter 28). Advantages

Table 26.3 Commonly used attachment chemistries

Bead or surface	Activation and/or linkage	Affinity ligand reactive group
Soft gels: agarose, cellulose	Cyanogen bromide	Amine
Synthetic supports: Polyacrylamide beads, Trisacryl, Sephacryl, Ultragel, Azlactone beads, Methacrylate (TSK gel), Eupergit, Polystyrene (Poros supports)	Aldehyde (reductive amination)	Amine
	Activated carboxyl ester (succinimidyl ester)	Amine
	Carbonyldiimidazole	Amine
	Carboxyl (activation concurrent with coupling)	Amine
	FMP activation	Amine, thiol
	Divinyl sulfone	Amine, thiol
	Azlactone	Amine, thiol
	Epoxy (bisoxirane, epichlorohydrin)	Amine, thiol
	Tresyl chloride	Amine, thiol
	Haloacetyl (iodo or bromo)	Thiol
	Maleimide	Thiol
	Pyridyl disulfide	Thiol
	Amine	Carboxyl (activation concurrent with coupling)
	Hydrazide	Carbohydrate (periodate reduced)
Inorganics: controlled pore glass, silica, alumina, zeolites, etc.	3-(Glycidyloxypropyl) trimethoxy-silane	Amine
	3-(Aminopropyl) trimethoxysilane	Carboxyl (after activation), aldehyde

of this activation include commercially available preactivated resin and very efficient protein coupling, near 100%. While this is historically a popular method, it suffers a number of drawbacks including: (1) linkage leading to leaching of the ligand off of the resin, (2) attachment directly to the surface with no spacer, and (3) required additional safety precautions due to the toxicity of cyanogen bromide. In addition, cross-linking may be needed to help limit the leakiness of this immobilization (Korpela and Hinkkanen, 1976; Kowal and Parsons, 1980).

Covalent attachment through the formation of an amide bond is an alternative to the cyanogen bromide chemistry but requires either an activated carboxyl surface (such as a *N*-hydroxysuccinimidyl ester) or *in situ* activation of the carboxyl group with a coupling agent such as *N*-ethyl-*N'*-(3-dimethylaminopropyl)carbodiimide (EDC) (Wilchek *et al.*, 1984, 1994). Another stable alternative is the formation of a secondary amine linkage resulting from a reductive alkylation of a Schiff-base intermediate formed between a primary amine (lysine or N-terminus) to an aldehyde activated surface. This reductive alkylation attachment mechanism is a popular method to immobilize enzymes to carbohydrate surfaces (agarose or cellulose) as the conditions for coupling are mild and the immobilized enzyme has been reported to retain more activity than observed with other methods of attachment (Hermanson, 1992).

Amine reactive linkages are also a common approach to attachment of small molecule ligands to surfaces. The primary requirement being that the ligand has the predesigned amine group for attachment. A number of the chemistries that are applied to amine immobilization also apply to sulfhydryl (or thiol)-reactive immobilization which is in concept the same as amine-reactive immobilization but relies on the reaction of a cysteine with the reactive surface. For thiol specific immobilization, the two most prominent strategies are through haloacetyl (iodo or bromo) and maleimide activated surfaces (Mallik *et al.*, 2007). Sulfhydryl linkage may be advantageous in that it can also be made reversible through the use of a disulfide linkage that can be removed from the surface by treatment with a reducing agent such as DTT or TCEP (Brena *et al.*, 1993).

Immobilization of antibodies or glycosylated protein to a surface can be done by the methods described earlier but an additional option that offers a potentially more oriented attachment is also available through modification of the carbohydrate (Oates *et al.*, 1998; Vijayendran and Leckband, 2001). Immobilization through carbohydrate requires a mild oxidation (i.e., periodate) of the carbohydrate to form reactive aldehydes with the sugar residues. The aldehydes produced by oxidation are then used to immobilize the protein or antibody to a hydrazide reactive surface. This approach is commonly used for antibodies, glycoproteins, glycopolymers, and ribonucleic acids (O'Shannessy and Wilchek, 1990).

In most cases, it is sufficient to rely on the attachment of a ligand through one of the reactive groups discussed earlier, but in some cases, it may be desirable to space the linker further from the surface. One option for doing so is to orthogonally modify the ligand of interest with a spacer and reactive group selective for a second group that will preferentially recognize the orthogonal label attached to a surface. This requires modification of both the surface and the ligand with the appropriate reactive groups. A current popular embodiment of this approach is the copper(I)-catalyzed 1,3-dipolar cycloaddition of azides to terminal alkynes to form 1,2,3-triazoles known commonly now as "click chemistry" (Chandran *et al.*, 2009; Gauchet *et al.*, 2006). The advantage of this approach is that the chemistry is mild, the selectivity is unique and reactive groups can be easily switched between the ligand and the surface. Because the reaction itself requires catalysis, the reactive groups on their own are much more stable than reactive groups commonly used with thiol and amine labeling.

5. Purification Method

Purification by affinity starts with proper handling of the sample and the matrix, followed by selective binding (capture) of the target, washing to remove nonspecific background, and, finally, elution of the bound target. Successful affinity purification depends on a number of notable factors including, the amount and accessibility of the ligand on the resin, the strength of the interaction, and the integrity of protein to be immobilized. Usually, conditions are optimized to maximize the interaction between a target and immobilized ligand during the binding and wash process, then, switched to substantially weaken the interaction thus allowing for release of the target. It is recommended to perform small scale trials to select for the best purification conditions. A short summary of some common practical issues and considerations affecting affinity purification is given in the following sections.

5.1. Sample preparation

When preparing sample for purification, conditions should be selected to retain the proper fold and functionality of the target of interest. It is also highly recommended to remove insoluble materials and to reduce viscosity because both of these factors could clog the column, reduce the flow rate, and increase back pressure.

Some proteins tend to aggregate at high concentrations, which results in increased apparent molecular weight, decreased diffusion rate, and reduced capture by the affinity matrix. Dilution of the sample or cell lysate in a larger volume may be needed to reduce aggregation and increase protein capture and recovery under this circumstance. Conversely, if the sample is too dilute, binding rate and capture efficiency can be reduced especially for low-affinity binders.

5.2. Binding and wash

Efficiency of binding is related to the strength and the kinetics of protein–ligand interaction which can be affected by the nature of the interaction, the concentration of applied target, the amount of immobilized ligand, and the flow rate used for binding. The binding process can be simplified as Eq. (26.1), assuming a 1:1 molar ratio where K_a is the association equilibrium constant, $[L]$ is ligand concentration, $[T]$ is the concentration of target protein, and $[LT]$ is the concentration of the complex. K_a equals $[LT]/([L] \times [T])$, which can also be expressed as k_a/k_d, where k_a is second order association rate constant that depends on the concentrations of both L and T, and k_d is the first order dissociation constant that does not depend on ligand concentration. Higher K_a usually leads to a higher adsorption ratio, defined as the ratio of bound to total applied target, thus, better binding. Normally, the ligand concentration on the matrix is around 10^{-2}–10^{-4} M and to achieve efficient binding, the K_a value should be in the range of 10^4–10^6 M^{-1}.

$$L + T \underset{k_d}{\overset{k_a}{\rightleftarrows}} LT \qquad (26.1)$$

Affinity immobilization onto solid support can be achieved by passing the sample through a column packed with the affinity matrix, usually under ambient pressure and a slow flow rate. Generally, a higher flow rate will reduce the binding efficiency, especially, when the interaction between the ligand and protein is weak or the mass-transfer rate in the column is slow. The binding process can also be performed in batch, where the resin and sample are constantly mixed. Batch binding promotes effective contact between target and immobilized ligand and often saves time, especially when dealing with large sample volumes; however, nonspecific binding can also increase. It is a good practice to optimize the amount of resin used during purification, where saturation of the resin with target during binding is preferred since excess resin can result in an increase in nonspecific binding as well as reduced target recovery due to readsorption, unless the latter is required under special circumstances.

Following binding, protein bound by nonspecific interactions can be removed by washing. For example, ionic interactions can be reduced by increasing salt (0.1–0.5 *M*) or changing pH values, and hydrophobic interactions can be removed by decreasing salt, altering pH, or adding surfactants (such as Triton X-100). Low amounts of competitive reagents can also be used to remove contaminants with weak affinity to the ligands. The flow rate and the volume (e.g., 5–10 column volumes) of the wash buffer should also be carefully determined for maximum removal of contaminants with minimum loss of target.

5.3. Elution

Elution of bound target from the resin is essentially the reverse process of binding, where conditions are optimized to reduce the K_a, that is, weakening the interaction between target and ligand. The elution condition should not denature the target protein, unless such conditions are compatible with downstream applications.

There are two different types of elution methods, namely, specific and nonspecific elution. In specific elution, the target protein–ligand complex is challenged by agents that will compete for either the ligand or the target thereby releasing the target protein into solution. The concentration and amount (volume) of competitive reagent used for elution will depend on their affinity relative to that of the immobilized complex, where weaker competitive reagents require higher concentration and more volume as compared to higher affinity additives. A good starting point for weak competitors is to use concentration 10-fold higher than that of the ligand. The specific elution is usually milder and proteins are more likely to retain their activity, but the slow elution, broad elution peaks, and the need to remove competing agent from the recovered protein are some of the drawbacks of this approach. For nonspecific elution, solvent conditions are manipulated to reduce the association rate constant (Eq. (26.1)), which ideally should approach zero, and to increase the dissociation rate constant, thus, weakening the overall affinity (K_a) resulting in dissociation of the complex. Elution conditions can be optimized according to the mechanism of interaction between the ligand and protein, such as increasing salt concentration to reduce ionic interactions or by altering pH to change the protonation/ionization state, thus modulating the strength of hydrogen bonds, hydrophobic interactions as well electrostatic interactions. An example of elution by changing pH is the elution of antibodies from immobilized protein A or protein G, yet because the affinity of proteinA/G to antibodies is very strong, with K_a in the $10^8\,M^{-1}$ range, a combination of different elution conditions may be required for maximum antibody release. In many cases, affinity requires proper three-dimensional folding of a protein so that chaotropic reagents or reagents that will affect protein folding can be used to elute target of interest; however, care must be taken to maintain proper folding of the target after elution by quickly returning to native conditions. When the affinity is weak, binding is achieved at high concentration of the target molecule which is then eluted by dissociation of the complex through dilution. This approach is known as isocratic elution.

REFERENCES

Alexander, C., Andersson, H. S., Andersson, L. I., Ansell, R. J., Kirsch, N., Nicholls, I. A., O'Mahony, J., and Whitcombe, M. J. (2006). Molecular imprinting science and technology: A survey of the literature for the years up to and including 2003. *J. Mol. Recognit.* **19,** 106–180.

Blumberg, P. M., and Strominger, J. L. (1972). Isolation by covalent affinity chromatography of the penicillin-binding components from membranes of *Bacillus subtilis*. *Proc. Natl. Acad. Sci. USA* **69,** 3751–3755.

Brena, B., Ovsejevi, K., Luna, B., and Batista-Viera, F. (1993). Thiolation and reversible immobilization of sweet potato beta-amylase on thiolsulfonate-agarose. *J. Mol. Catal.* **84,** 381–390.

Campbell, D. H., Luescher, E., and Lerman, L. S. (1951). Immunologic adsorbents: I. Isolation of antibody by means of a cellulose-protein antigen. *Proc. Natl. Acad. Sci. USA* **37,** 575–578.

Cecília, A., Roque, A., and Lowe, C. R. (2005). Advances and applications of de novo designed affinity ligands in proteomics. *Biotechnol. Adv.* **24,** 17–26.

Chandran, S. P., Hotha, S., and Prasad, B. L. V. (2009). Tunable surface modification of silica nanoparticles through "click" chemistry. *Current Science* **95,** 1327–1333.

Cuatrecasas, P., Wilchek, M., and Anfinsen, C. B. (1968). Selective enzyme purification by affinity chromatography. *Proc. Natl. Acad. Sci. USA* **61,** 636–643.

Faulmann, E. L., Duvall, J. L., and Boyle, M. D. (1991). Protein B: A versatile bacterial Fc-binding protein selective for human IgA. *Biotechniques* **10,** 748–755.

Gauchet, C., Labadie, G. R., and Poulter, C. D. (2006). Regio- and chemoselective covalent immobilization of proteins through unnatural amino acids. *J. Am. Chem. Soc.* **128,** 9274–9275.

GE Healthcare (ed.) (2007). Affinity chromotography, principle and methods. GE Healthcare.

Guilbault, G. G. (ed.) (1988). *In* "Analytical uses of immobilized biological compounds for detection, medical and industrial uses," Dordrecht, Boston, MA.

Guss, B., Eliasson, M., Olsson, A., Uhlen, M., Frej, A. K., Jornvall, H., Flock, J. I., and Lindberg, M. (1986). Structure of the IgG-binding regions of streptococcal protein G. *EMBO J.* **5,** 1567–1575.

Gustavsson, P.-E., and Larsson, P.-O. (2006). Support materials for affinity chromatography. *In* "Handbook of Affinity Immobilization," (D. S. Hage, ed.), pp. 15–34. CRC press/ Taylor and Francis Group, Boca Raton, FL.

Hage, D. S. (ed.) (2006). *In* "Handbook of affinity chromatography," CRC press/Taylor and Francis Group, Boca Raton, FL.

Hage, D. S., Bian, M., Burks, R., Karle, E., Ohnmachi, C., and Wa, C. (2006). Bioaffinity Chromatography. *In* "Handbook of Affinity Immobilization," (D. S. Hage, ed.), pp. 101–126. CRC press/Taylor and Francis Group, Boca Raton, FL.

Hermanson, G. T. (ed.) (1992). *In* "Immobilized affinity ligand techniques," Academic Press, San Diego, CA.

Hirabayashi, J. (2008). Concept, strategy and realization of lectin-based glycan profiling. *J. Biochem.* **144,** 139–147.

Koneracka, M., Kopcansky, P., Timbo, M., Ramchand, C. N., Saiyed, Z. M., and Trevan, M. (2006). Immobilization of enzymes on magnetic particles. *In* "Immobilization of Enzymes and Cells," (J. M. Guisan, ed.), pp. 217–228. Humana Press.

Korpela, T., and Hinkkanen, A. (1976). A simple method to introduce aldehydic function to agarose. *Anal. Biochem.* **71,** 322–323.

Kowal, R., and Parsons, R. G. (1980). Stabilization of proteins immobilized on Sepharose from leakage by glutaraldehyde crosslinking. *Anal. Biochem.* **102,** 72–76.

Labrou, N., and Clonis, Y. D. (1994). The affinity technology in downstream processing. *J. Biotechnol.* **36,** 95–119.

Labrou, N. E. (2000). Dye-ligand affinity chromatography for protein separation and purification. *Methods Mol. Biol.* **147,** 129–139.

Labrou, N. E. (2003). Design and selection of ligands for affinity chromatography. *J. Chromatogr. B Analyt. Technol. Biomed. Life Sci.* **790,** 67–78.

Labrou, N. E., and Clonis, Y. D. (2002). Immobilized synthetic dyes in affinity chromatography. *In* "Theory and Practice of Biochromatograph," (M. A. Vijayalakshimi, ed.), pp. 235–251. Taylor and Francis Group, London.

Labrou, N. E., Karagouni, A., and Clonis, Y. D. (1995). Biomimetic-dye affinity adsorbents for enzyme purification: Application to the one-step purification of *Candida boidinii* formate dehydrogenase. *Biotechnol. Bioeng.* **48,** 278–288.

Langone, J. J. (1982). Applications of immobilized protein A in immunochemical techniques. *J. Immunol. Methods* **55,** 277–296.

Lerman, L. S. (1953a). A biochemically specific method for enzyme isolation. *Proc. Natl. Acad. Sci. USA* **39,** 232–236.

Lerman, L. S. (1953b). Antibody chromatography on an immunologically specific adsorbent. *Nature* **172,** 635–636.

Los, G. V., Encell, L. P., McDougall, M. G., Hartzell, D. D., Karassina, N., Zimprich, C., Wood, M. G., Learish, R., Ohana, R. F., Urh, M., Simpson, D., Mendez, J., *et al.* (2008). HaloTag: A novel protein labeling technology for cell imaging and protein analysis. *ACS Chem. Biol.* **3,** 373–382.

Lowe, C. R., Lowe, A. R., and Gupta, G. (2001). New developments in affinity chromatography with potential application in the production of biopharmaceuticals. *J. Biochem. Biophys. Methods* **49,** 561–574.

Madoery, R., and Minchiotti, M. (2006). Cibacron Blue-Eupergit, an affinity matrix for soybean (*Glycine max*) phospholipase A_2 purification. *Enzyme and Microb. Technol.* **38,** 869–872.

Mallik, R., and Hage, D. S. (2006). Affinity monolith chromatography. *J. Sep. Sci.* **29,** 1686–1704.

Mallik, R., Wa, C., and Hage, D. S. (2007). Development of sulfhydryl-reactive silica for protein immobilization in high-performance affinity chromatography. *Anal. Chem.* **79,** 1411–1424.

O'Shannessy, D. J., and Wilchek, M. (1990). Immobilization of glycoconjugates by their oligosaccharides: Use of hydrazido-derivatized matrices. *Anal. Biochem.* **191,** 1–8.

Oates, M. R., Clarke, W., Marsh, E. M., and Hage, D. S. (1998). Kinetic studies on the immobilization of antibodies to high-performance liquid chromatographic supports. *Bioconjug. Chem.* **9,** 459–465.

Ohana, R. F., Encell, L. P., Zhao, K., Simpson, D., Slater, M. R., Urh, M., and Wood, K.V (2009). Halotag7: A genetically engineered tag that enhances bacterial expression of soluble protein and improves protein purification. *Protein Expr. Purif.* **68,** 110–120.

Ostrove, S. (1990). Affinity chromatography: General methods. *Methods Enzymol.* **182,** 357–371.

Plievaa, F. M., De Seta, E., Galaevb, I. Y., and Mattiasson, B. (2009). Macroporous elastic polyacrylamide monolith columns: Processing under compression and scale-up. *Sep. Purif. Technol.* **65,** 110–116.

Porath, J., Axen, R., and Ernback, S. (1967). Chemical coupling of proteins to agarose. *Nature* **215,** 1491–1492.

Renkin, E. M. (1954). Filtration, diffusion, and molecular sieving through porous cellulose membranes. *J. Gen. Physiol.* **38,** 225–243.

Saiyed, Z., Telang, S., and Ramchand, C. (2003). Application of magnetic techniques in the field of drug discovery and biomedicine. *Biomagn. Res. Technol.* **1,** 2.

Starkenstein, E. (1910). Ferment action and the influence upon it of neutral salts. *Biochem. Z.* **24,** 210–218.

Stellwagen, E. (1990). Chromatography on immobilized reactive dyes. *Methods Enzymol.* **182,** 343–357.

Urh, M., Hartzell, D., Mendez, J., Klaubert, D. H., and Wood, K. (2008). Methods for detection of protein-protein and protein-DNA interactions using HaloTag. *Methods Mol. Biol.* **421,** 191–209.

Vijayendran, R. A., and Leckband, D. E. (2001). A quantitative assessment of heterogeneity for surface-immobilized proteins. *Anal. Chem.* **73,** 471–480.

Wilchek, M., Knudsen, K. L., and Miron, T. (1994). Improved method for preparing N-hydroxysuccinimide ester-containing polymers for affinity chromatography. *Bioconjug. Chem.* **5,** 491–492.
Wilchek, M., Miron, T., and Kohn, J. (1984). Affinity chromatography. *Methods Enzymol.* **104,** 3–55.
Zachariou, M. (ed.) (2007). *In* "Affinity Chromatography, Methods and Protocols," Humana Press, Totowa, NJ.

CHAPTER TWENTY-SEVEN

Immobilized-Metal Affinity Chromatography (IMAC): A Review

Helena Block,* Barbara Maertens,* Anne Spriestersbach,* Nicole Brinker,* Jan Kubicek,* Roland Fabis,* Jörg Labahn,† *and* Frank Schäfer*

Contents

1. Overview on IMAC Ligands and Immobilized Ions 440
2. IMAC Applications 444
   2.1. Detection and immobilization 445
   2.2. Purification of protein fractions 446
   2.3. Purification of His-tagged proteins 448
   2.4. General considerations of protein purification by IMAC 451
   2.5. Copurifying proteins on IMAC and what to do about it 454
   2.6. IMAC for industrial-scale protein production 458
   2.7. High-throughput automation of IMAC 460
   2.8. Special applications: Purification of membrane proteins 461
   2.9. Special applications: Purification of zinc-finger proteins 463
   2.10. Protein purification protocols 464
   2.11. Cleaning and sanitization 466
   2.12. Simplified metal-ion stripping and recharging protocol 467
3. Conclusions 467
Acknowledgments 468
References 468

Abstract

This article reviews the development of immobilized-metal affinity chromatography (IMAC) and describes its most important applications. We provide an overview on the use of IMAC in protein fractionation and proteomics, in protein immobilization and detection, and on some special applications such as purification of immunoglobulins and the Chelex method. The most relevant application—purification of histidine-tagged recombinant proteins—will be reviewed

* QIAGEN GmbH, Qiagen Strasse 1, Hilden, Germany
† Institute of Structural Biology (IBI-2), Research Center Jülich, Jülich, Germany

Methods in Enzymology, Volume 463
ISSN 0076-6879, DOI: 10.1016/S0076-6879(09)63027-5

in greater detail with focus of state-of-the-art materials, methods, and protocols, and the limitations of IMAC and recent advances to improve the technology and the methods will be described.

1. Overview on IMAC Ligands and Immobilized Ions

The concept of immobilized-metal affinity chromatography (IMAC) has first been formulated and its feasibility shown by Porath *et al.* (1975). It was based on the known affinity of transition metal ions such as Zn^{2+}, Cu^{2+}, Ni^{2+}, and Co^{2+} to histidine and cysteine in aqueous solutions (Hearon, 1948) and extended to the idea to use metal ions "strongly fixed" to a support to fractionate protein solutions. As the chelating ligand used to fix the metal to agarose Porath used iminodiacetic acid (IDA) which is still in use today in many commercial IMAC resins. Already in 1975, Porath speculated that the affinity of immobilized metals to histidine-containing proteins might not be the only application for IMAC—and he was right after all.

In the years following Porath's publications, the new IMAC technology was successfully evaluated by purification of a variety of different proteins and peptides summarized in a first IMAC review by Sulkowski (1985). What started as a crude fractionation of serum soon became what today is the most widely used affinity chromatography technique (Biocompare, 2006; Derewenda, 2004), if not chromatography technique in general. This development was accelerated by the fast maturation of recombinant techniques and modern molecular biology in the late 1970s and by the invention of an improved chelating ligand, nitrilotriacetic acid (NTA) in the 1980s (Hochuli *et al.*, 1987). In the meantime, the purification of recombinant proteins genetically modified on the DNA template level in order to generate oligohistidine extended (His-tagged) polypeptides using NTA-based supports (Hochuli *et al.*, 1988) represents the most important application of IMAC. The principal mechanism of the interaction of a His-tagged protein to an immobilized metal ion is presented in Fig. 27.1.

A possible model is that of an interaction of a metal ion with histidine residues n and n_{+2} of a His tag. This is confirmed by the fact that IMAC ligands can bind to His tags consisting of consecutive histidine residues as well as to alternating tags. As at least consecutive His tags are usually unstructured and thus flexible, other interaction patterns such as $n{:}n_{+1}$, $n{:}n_{+3}$, $n{:}n_{+4}$, and so on could be imagined as well, even within a single molecule (Jacob Piehler, personal communication).

The NTA ligand coordinates the Ni^{2+} with four valencies (tetradentate, coordination number 4) highlighted spherically in Fig. 27.1B, and two valencies are available for interaction with imidazole rings of histidine residues.

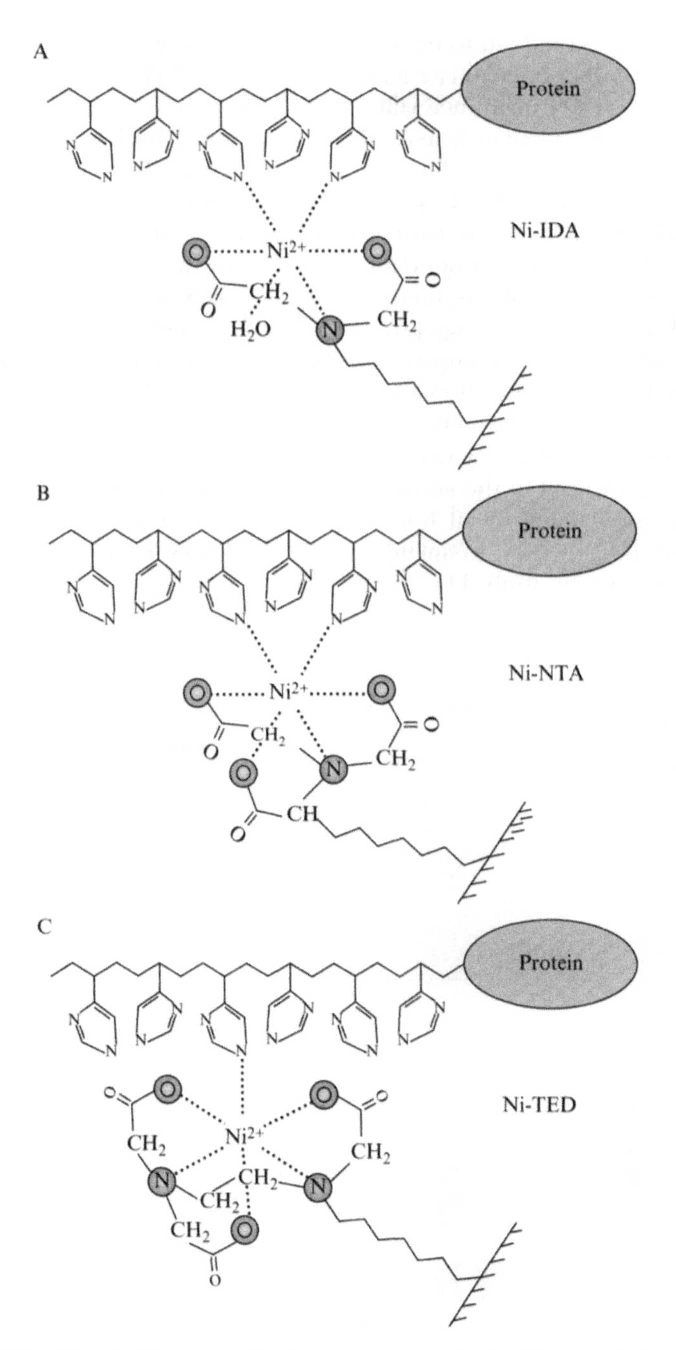

Figure 27.1 Model of the interaction between residues in the His tag and the metal ion in tri- (IDA), tetra- (NTA), and pentadentate IMAC ligands (TED).

This ratio has turned out to be most effective for purification of His-tagged proteins. Another tetradentate ligand is carboxymethyl aspartate (CM-Asp; Chaga *et al.*, 1999), commercially available as cobalt-charged Talon resin. In contrast to tetradentate ligands, IDA coordinates a divalent ion with three valencies (tridentate, coordination number 3, Fig. 27.1A) leaving three valencies free for imidazole ring interaction while it is unclear whether the third is sterically able to participate in the interaction. The coordination number seems to play an important role regarding the quality of the purified protein fraction. While protein recovery is usually similar between IDA- and NTA-based chromatography (Fig. 27.2D), a higher leaching of metal ions from IDA ligands compared to NTA is observed in general (Hochuli, 1989) and even increased under reducing conditions (Fig. 27.2C). Although the metal content in the elution fractions (E in Fig. 27.2C) is higher but still within the same order of magnitude, significantly more Ni^{2+} ions leach from the IDA resin in the equilibration and wash steps (W in Fig. 27.2C). Besides considerable metal leaching, purification of His-tagged proteins using an IDA matrix frequently results in lower purity compared to NTA-based purification (Fig. 27.2A and D).

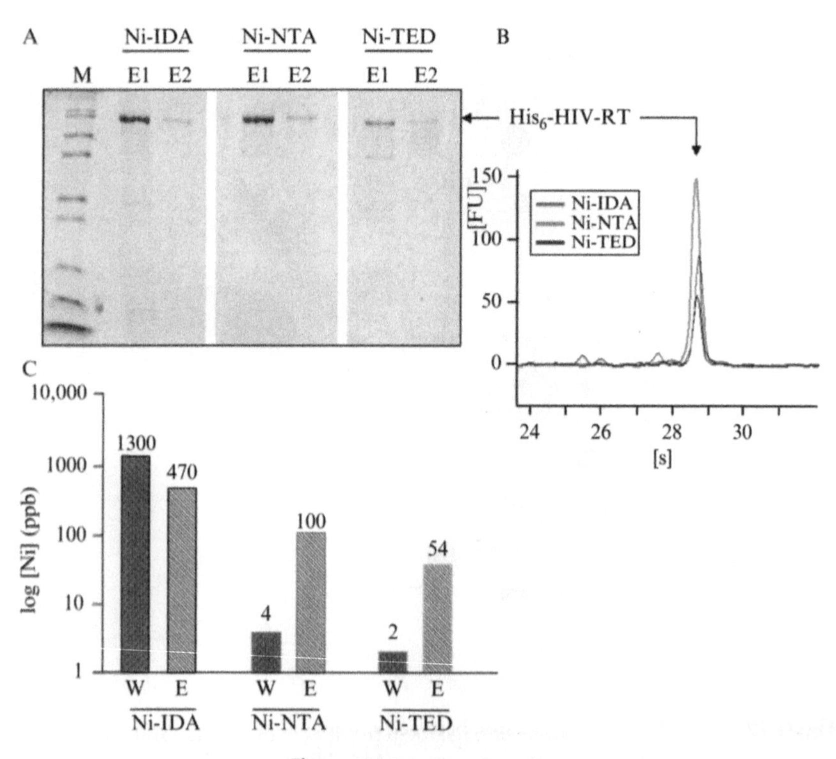

Figure 27.2 (Continued)

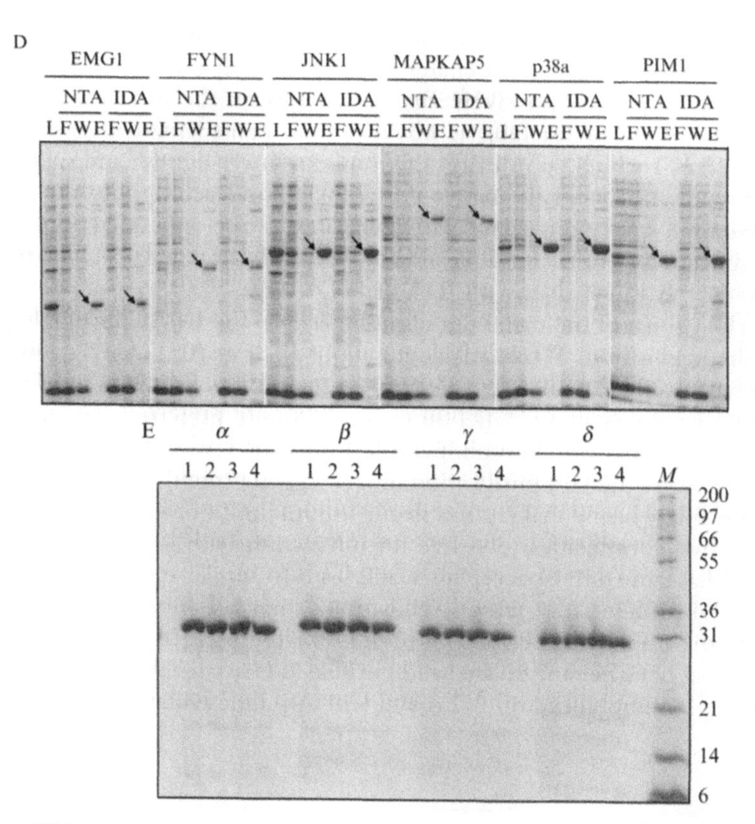

Figure 27.2 Purification of His-tagged proteins with NTA, IDA, and TED. (A) H_6-HIV-RT was expressed in *E. coli* BL21(DE3) and purified via Ni-IMAC in the presence of 1 m*M* DTT under standard conditions (IDA, NTA; see Section 2.10 for standard conditions) or according to the manufacturer's recommendations (TED). Corresponding aliquots of the IMAC elution fractions (E1, E2) were analyzed by SDS–PAGE and Coomassie staining. (B) Bioanalyzer 2100 lab-on-a-chip analysis of pooled elution fractions. The peaks from the electropherograms corresponding to H_6-HIV-RT were overlayed. Peak areas directly correlate to the protein amount in the respective pool fraction. (C) Determination of the nickel content in wash (W) and pooled elution fractions (E) of the H_6-HIV-RT purifications described in (A) and (B). Nickel was measured by ICP-MS (intercoupled plasma mass spectrometry) at Wessling Laboratories, Bochum, Germany, and values are provided in μg/l (parts per billion, ppb). (D) QIAgene constructs carrying optimized human genes were used for expression of the indicated proteins in *E. coli* BL21(DE3) LB cultures. Cleared lysates were divided for purification of His-tagged proteins via Ni-NTA (NTA) and Ni-IDA (IDA), respectively. Fractions were analyzed by SDS–PAGE and Coomassie staining as follows: L, cleared lysate; F, IMAC flow-through fraction; W, wash fraction; E, peak elution fraction. (E) Fragments of H_6-tagged proteins named α, β, γ, and δ expressed in *E. coli* were purified under standard conditions using NTA and Cm-Asp tetradentate ligands loaded with Ni^{2+} or Co^{2+} as follows: 1, Ni-NTA; 2, Co-NTA; 3, Co-CmAsp; 4, Ni-CmAsp. Aliquots from the peak elution fractions (2 μl each) were analyzed by SDS–PAGE and Coomassie staining. (See Color Insert.)

The reason for the lower purity may be that leaching of metal from the tridentate ligand generates charged groups which could act as a cation exchanger and bind positively charged groups on the surface of proteins. The lowest metal leaching is obtained if a pentadentate ligand is used (Fig. 27.2C) which coordinates the ions extremely tightly, and such resins may represent a valid alternative if low metal ion leaching into the protein preparation is very important. However, in this case only one coordination site remains for His tag binding and recovery of His-tagged protein is usually considerably lower than with IDA or NTA (Fig. 27.2B).

The choice of the metal ion immobilized on the IMAC ligand depends on the application. While trivalent cations such as Al^{3+}, Ga^{3+}, and Fe^{3+} (Andersson and Porath, 1986; Muszynska *et al.*, 1986; Posewitz and Tempst, 1999) or tetravalent Zr^{4+} (Zhou *et al.*, 2006) are preferred for capture of phosphoproteins and phosphopeptides, divalent Cu^{2+}, Ni^{2+}, Zn^{2+}, and Co^{2+} ions are used for purification of His-tagged proteins. Combinations of a tetradentate ligand that ensures strong immobilization, and a metal ion that leaves two coordination sites free for interaction with biopolymers (Ni^{2+}, Co^{2+}) has gained most acceptance and leads to similar recovery and purity of eluted protein. As a typical result using such combinations, Fig. 27.2E shows the purification of several protein fragments derived from different genes that have been expressed and purified as His-tagged proteins by Ni^{2+} and Co^{2+} immobilized on NTA and Cm-Asp tetradentate ligands.

2. IMAC Applications

Initially developed for purification of native proteins with an intrinsic affinity to metal ions (Porath *et al.*, 1975), IMAC has turned out to be a technology with a very broad portfolio of applications. On the chromatographic purification side, the range of proteins was expanded from the primary metalloproteins to antibodies, phosphorylated proteins, and recombinant His-tagged proteins. IMAC is being used in proteomics approaches where fractions of the cellular protein pool are enriched and analyzed differentially (phosphoproteome, metalloproteome) by mass spectrometrical techniques; here, IMAC formats can be traditionally bead based or the ligand can be used on functionalized surfaces such as SELDI (surface-enhanced laser desorption/ionization) chips. Other chip-based applications include surface plasmon resonance (SPR) and allow the immobilization of His-tagged proteins for quantitative functional and kinetic investigations. In addition, the IMAC principle has been used as an inhibitor depletion step prior to PCR amplification of nucleic acids from complex samples such as blood in a technology called Chelex (Walsh *et al.*, 1991). The most relevant of the numerous IMAC applications will be discussed briefly in the following sections. The main application of IMAC—the purification of His-tagged proteins—will then be discussed in detail. This section will include problems and limitations of IMAC, solutions and recent advances in this field.

2.1. Detection and immobilization

In efforts to make use of the specificity and high affinity of His-tagged proteins to immobilized metal ions, IMAC ligands have been employed for applications such as protein:protein interactions where proteins need to be stably immobilized on surfaces. Two applications—ELISA as a diagnostic tool and chip-based technologies for functional investigations—shall be briefly described.

Ni-NTA ligands attached to surfaces of microtiter plates are used to immobilize His-tagged antigens in its soluble and structurally intact form for serological studies. The directed immobilization via the His tag can be an advantage to standard ELISA where proteins are randomly adsorbed to plastic surfaces which destroys the protein structure and hides part of the protein surface and possible antibody-binding sites. In contrast, IMAC-based ELISA allows the screening of conformation-dependent monoclonal antibodies (Padan *et al.*, 1998) and immunosorbent assays with increased sensitivity (Jin *et al.*, 2004).

Immobilization of His-tagged proteins on chip surfaces for interaction studies with other molecules, for example by SPR, is a widely used protein characterization method. The factor "stability" is important to reduce "bleeding" of the immobilized molecule, and therefore, due to its favorable binding features described above the NTA ligand is frequently used for immobilization applications (Knecht *et al.*, 2009; Nieba *et al.*, 1997). However, interaction of biotinylated proteins to supports coated with streptavidin is still considerably stronger. A significant improvement in the stability of functional immobilization of His-tagged proteins on glass-type surfaces even at low concentration was achieved by the concept of multivalent chelator heads where a single ligand molecule carries three NTA moieties (tris-NTA; Lata and Piehler, 2005; Zhaohua *et al.*, 2006). This development represents a valid alternative to streptavidin/biotin-based protein immobilization and allows the use of His-tagged proteins without the need for biotin-labeling of proteins following purification. We have synthesized the tris-NTA ligand and coupled it to magnetic agarose beads in order to test whether this approach may be transferred to purification of His-tagged proteins and allow even more specific single-step recovery from complex samples. Initial data (Fig. 27.3) indicate that this may indeed be the case; AKT1 kinase separated from *Spodoptera frugiperda*-derived cell-free lysate reactions could be purified with both bead types and purity was slightly

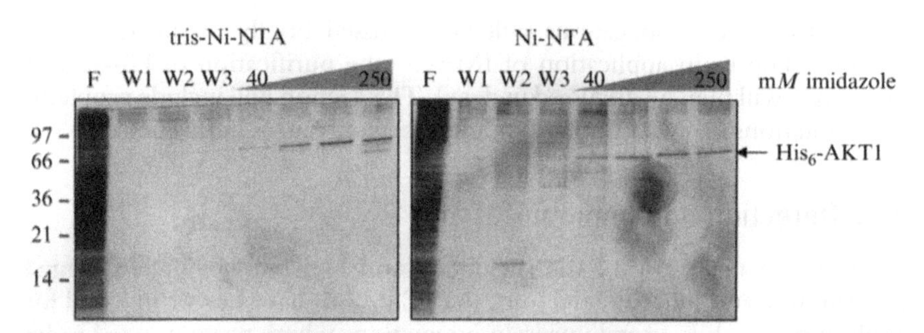

Figure 27.3 Purification of His_6-tagged AKT1 kinase using tris-Ni-NTA and Ni-NTA magnetic beads. Human C-terminally His_6-tagged AKT1 was expressed cell-free in an insect cell-based lysate (EasyXpress Insect II system) in a 100 μl reaction volume using a pIX4.0 vector construct; purification was performed using magnetic agarose beads functionalized with tris-Ni-NTA (left panel) or Ni-NTA (right panel) under standard conditions (see Section 2.10) in the presence of 0.05% (v/v) Tween-20 using a magnet-equipped tube holder. Aliquots of purification fractions as follows were analyzed by SDS–PAGE and silver staining: F, unbound protein; W1, 2, wash fractions 1 and 2 (10 m*M* imidazole); W3, wash fraction 3 (20 m*M* imidazole); 40 → 250, elution-fractions (40, 60, 100, and 250 m*M* imidazole, respectively). Identity of the purified His_6-tagged AKT1 protein was verified by anti-His Western blot analysis (data not shown).

higher using tris-Ni-NTA beads. Whether these findings are of general relevance and whether tris-NTA-based chromatography can be economic in larger scales remains to be evaluated.

IMAC ligands have also been used successfully as reporter in immunoblot type of applications replacing an antibody. His-tagged proteins transferred to nitrocellulose membrane (Western or dot blot) can be detected by probing with Ni-NTA conjugated to alkaline phosphatase or horseraddish peroxidase reporter enzymes (Lv *et al.*, 2003) in a chromogenic or chemiluminescence reaction or to quantum dots for fluorescent detection (Kim *et al.*, 2008). This represents an attractive fast and economic alternative to antibody-based detection reactions in cases where the high specificity of an antibody is not required. The specificity of NTA conjugate-based detection has been increased by generation of tris-NTA conjugates (Lata *et al.*, 2006; Reichel *et al.*, 2007).

2.2. Purification of protein fractions

IMAC had originally been developed as a group separation method for metallo- and histidine-containing proteins (Porath *et al.*, 1975). Today, these features are made use of in proteome-wide studies where the reduction of the complexity of the system (the proteome) is indispensable for sensitive analyses of low-abundance proteins. Consequently, preseparation methods such as liquid, reverse-phase, ion-exchange, and affinity chromatography—such as IMAC—have gradually been used in proteomics to enrich proteins that may otherwise be lost in detection (Loo, 2003; Stasyk and Huber, 2004). The application of IMAC in proteomics has recently been reviewed (Sun *et al.*, 2005) and is focused on the enrichment of phosphoproteins and phosphopeptides and on metal-binding proteins. In the enrichment step, a complex sample such as a cell lysate or blood is passed over the IMAC matrix, washed and the fraction of interest eluted by variation of pH or with high concentrations of imidazole. This fraction is then analyzed by mass spectrometry (MS) or fractionated further by two-dimensional gel electrophoresis followed by MS or by additional liquid chromatography coupled to MS (LC–MS).

Whereas Fe^{3+}, Al^{3+}, and Ga^{3+} are the preferred ions for phosphoprotein research and are usually immobilized to IDA, the ions useful for IMAC-based analysis of the metalloproteome are the elements copper, nickel, zinc, and iron which are essential for life. The metalloproteome is defined as a set of proteins that have metal-binding ability and several aspects of this proteomics discipline have been reviewed recently (Shi and Chance, 2008; Sun *et al.*, 2005). Proteins with metal-binding affinity can be enriched by either making use of their ability to bind to certain immobilized Me^{2+} ions (e.g., to Me^{2+}-NTA) or by making use of their bound Me^{2+} ion by catching as a Me^{2+}-protein on an uncharged IMAC ligand (e.g., NTA).

Also, chip-based proteome profiling IMAC methods have been reported (Slentz *et al.*, 2003) and are in use as a tool in clinical screening applications for phospho group- and histidine-containing proteins and peptides (SELDI–IMAC).

IMAC can also be used to bind and separate at least mono- and dinucleotides based on a complex phenomenon accounted for by differential interplay of affinities of the potential binding sites (oxygen in the phosphate group, nitrogen and oxygen on the bases, hydroxyl groups on the ribose) to the immobilized metal (Hubert and Porath, 1980, 1981).

A quite different group-specific separation application of IMAC is represented by the affinity of antibodies to immobilized metal ions. As the molecular basis for this interaction an endogenous metal-binding site on the heavy chain (Hale and Beidler, 1994) and an arrangement of histidine residues on the antibody (Porath and Olin, 1983) have been discussed. Adsorption of immunoglobulins from different sources on IMAC matrices has been reported by many authors (human IgG, Porath and Olin, 1983; humanized murine IgG, Hale and Beidler, 1994; goat IgG, Boden *et al.*, 1995). Antibody purification has been successfully performed using various IMAC formats including gels (Hale and Beidler, 1994; Vancan *et al.*, 2002), methacrylate polymer (Mészárosová *et al.*, 2003), and membraneous hollow fibers (Serpa *et al.*, 2005). The mild elution of the protein with salts, costs, and the robustness of IMAC matrices have been identified as advantageous over traditional protein A or G chromatography (Serpa *et al.*, 2005).

The Chelex method shall be mentioned here as well in order to complete the list of applications which make use of IMAC ligands. Unlike classical IMAC, however, where an immobilized metal ion is used to purify a (poly)peptide by its affinity for this transition metal, Chelex represents a nucleic acid sample preparation method that depletes metal ion inhibitors of PCR as the downstream application (Walsh *et al.*, 1991). As such, Chelex resins such as Bio-Rad's Chelex 100 are uncharged ligands like IDA coupled to a usually agarose-based matrix. In brief, the procedure works as follows: a blood or tissue sample is incubated with Chelex resin in the presence or absence of proteinase K followed by separation of beads from supernatant containing the nucleic acids. The resulting nucleic acid fraction is not pure but suitable for amplification by PCR because metal ions have been removed from the sample which may otherwise catalyze rupture of the DNA at high temperature during PCR and thus cause PCR inhibition. Chelex is mainly applied as a fast and inexpensive method to prepare small samples obtained from biopsies and puncture aspirations for amplification of small DNA fragments by PCR (García González *et al.*, 2004; Gill *et al.*, 1992).

2.3. Purification of His-tagged proteins

2.3.1. His tags and their effects on protein expression

The most important application of IMAC is purification of recombinant proteins expressed in fusion with an epitope containing six or more histidine residues, the His tag (Fig. 27.1). Due to the relatively high affinity and specificity of the His tag a single IMAC purification step in most cases leads to a degree of purity of the target protein preparation that is sufficient for many applications. The structure of the tag, that is its position, sequence, and length, can influence production of a protein on several levels: expression rate, accessibility for binding to the IMAC ligand, protein three-dimensional structure, protein crystal formation, and—although to a minor extent—solubility and activity. The most common form of a His tag consists of six consecutive histidine residues (H_6) which provides a number of six metal-binding sites high enough to shift the association/dissociation equilibrium more to the association side leading to stable binding in most cases (Table 27.1; Knecht *et al.*, 2009). In Biacore experiments, the dissociation rate of a hexahistidine-tagged protein to Ni-NTA has been determined to approximately 1×10^{-6} to 1.4×10^{-8} M at pH 7.0 to 7.4 (Knecht *et al.*, 2009; Nieba *et al.*, 1997). However, the situation on a planar chip surface is significantly different from a porous agarose bead with respect to flow characteristics, ligand density, and protein concentration. Also, the stability of the interaction of the His-tagged protein to the IMAC ligand is influenced by the grade of accessibility of the tag and by the overall number of chelating residues (histidine, cysteine, aspartate, and glutamate)

Table 27.1 Reported His tag sequences (single letter amino acid sequence code)

His tag		Reference/vector
H H H H H H	H_6	Hochuli *et al.* (1988)
H H H H H H H H	H_8	pQE-TriSystem, pTriEx
H H H H H H H H H H	H_{10}	pQE-TriSystem-5, -6
H H H H H H H H H H H H H H	H_{14}	
H Q H Q H Q H Q H Q H Q	$(HQ)_6$	Pedersen *et al.* (1999)
H N H N H N H N H N H N	$(HN)_6$	US patent 7176298
H G H G H G H G H G H Q	$(H\ G/Q)_6$	Pedersen *et al.* (1999)
H H Q H H A H H G	$(HHX)_3$	Pedersen *et al.* (1999)
K D H L I H N V H K E H A H A H N K	HAT	Imai *et al.*, 2001 andUS patent 7176298
$(H\ X^{*})_{3-6}$	$(HX^{a})_n$	US patent 5594115
H X_1 H R H X_2 H[b]	$(HXH)_2R$	US patent application 2004/0029781

[a] X can be D, E, P, A, G, V, S, L, I, and T.
[b] X_1 can be A, R, N, D, Q, E, I, L, F, P, S, T, W, V; X_2 can be A, R, N, D, C, Q, E, G, I, L, K, M, P, S, T, Y, V.

on the surface of a protein (Bolanos-Garcia and Davies, 2006; Jensen *et al.*, 2004) and is thus individual to a significant extent. In most cases, that is, if the His tag is accessible, its affinity—or better termed avidity here—to Ni-NTA is high enough for column chromatography even under stringent conditions. Table 27.1 lists the tag sequences reported in the literature and some unpublished ones tested recently in our laboratories.

A different situation compared to a "standard" purification of a soluble protein (see Section 2.10) can be encountered when a membrane protein is to be recovered in the presence of detergents because the deterent micelle may cover part of or the complete His tag. In such cases, the use of longer tag sequences or the use of a linker can be helpful to allow binding of the protein to the IMAC resin (Mohanty and Wiener, 2004). A range of variations of the His tag including alternating sequences has been proposed (Table 27.1) and improved binding to IMAC resins postulated but in our and in the hands of others (Knecht *et al.*, 2009) they have no practical advantage over the classical H_n tags. What is found to be of greater importance than the sequence of the His tag itself is its position (N- or C-terminal) and the amino acids at the N-terminus. The nature of the amino acid following the N-terminal methionine has been reported to prevent N-terminal methionine processing and to have a positive effect on the general protein expression rate in *Escherichia coli* (Dalbøge *et al.*, 1990; Hirel *et al.*, 1989). We and others have evaluated these reports and confirmed that namely, lysine and arginine in position 2 of a protein N-terminus show this effect (Pedersen *et al.*, 1999; Schäfer *et al.*, 2002a; Svensson *et al.*, 2006). The high success rate of using N-terminal His (and Strep II) tags is, however, not only based on the stimulatory effect of the second amino acid on expression but also seems to have stabilizing impact on the mRNA structure in the translation initiation region. We compared bacterial and eukaryotic expression of N- versus C-terminal positioning of the His (and Strep II) tag on several proteins and analyzed expression level and solubility (Fig. 27.4). N-terminal tags improved protein expression in most cases.

Systematic investigation of the 5′ region of mRNAs showed that hairpin loops forming in the translation initiation area frequently are the reason for low expression as they prevent the ribosome:mRNA binding (Cèbe and Geiser, 2006). Sequence optimization can be performed to destabilize hairpin formation and improve expression, and a similar result was obtained when the proteins were expressed with an N-terminal H_6 tag. The effects described in Fig. 27.4 can be explained with this initiator effect of preventing secondary structures on the mRNA in the translation initiation region. Similar observations were also made by others (Busso *et al.*, 2003; Svensson *et al.*, 2006) and this may be the reason for the attractiveness of N-terminal

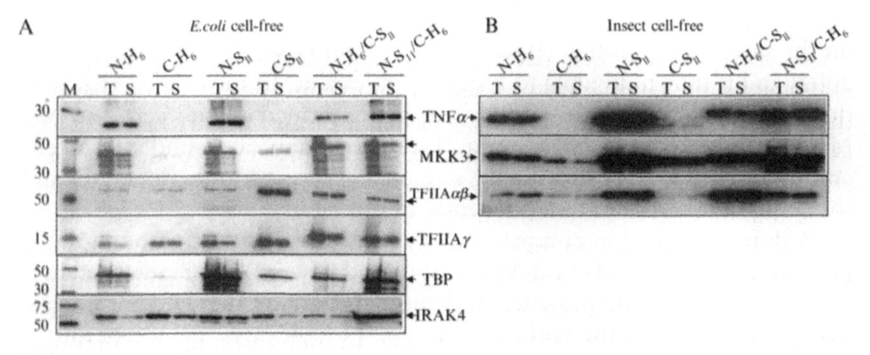

Figure 27.4 Effect of His tag position on protein expression. Proteins were expressed from PCR product templates generated by two-step PCR using the EasyXpress Linear Template Kit in *E. coli*- (A) and insect cell-derived (B) lysates (EasyXpress Protein Synthesis and EasyXpress Insect II Kits, respectively). Initiator (adapter) primers in the Linear Template Kit were designed in order to prevent formation of secondary structure in the translation initiation region on the mRNA and introduce the following tag sequence(s): N-H_6, N-terminal His_6 tag; C-H_6, C-terminal His_6 tag; N-S_{II}, N-terminal Strep II tag; C-S_{II}, C-terminal Strep II tag; N-H_6/C-S_{II}, N-terminal His_6 and C-terminal Strep II tags; N-S_{II}/C-H_6, N-terminal Strep II and C-terminal His_6 tags. Corresponding aliquots of total (T) and soluble protein (S, supernatant after centrifugation at $15{,}000 \times g$ for 10 min) were loaded on a SDS gel. Protein bands were visualized by Western blot analysis using a mixture of Penta anti-His and anti-Strep tag antibodies. Protein sizes are (kDa) TNFα, 21; TBP, 38; TFIIAαβ, 55; TFIIAγ, 12.5; MKK3, 39; IRAK4, 55. M, His-tagged protein size markers (kDa).

His tags. However, in some cases such as the one of IRAK4 the C-terminal His tag has a more pronounced effect on both expression rate and solubility (Fig. 27.4). Recently, expression of an insect toxin in *E. coli* was reported where the similar observation of higher solubility and thermostability of the C-terminally His-tagged form was made (Xu *et al.*, 2008). The authors discussed that the C-terminal tag stabilized the overall protein structure. Other groups found the His tag to contribute slightly negative to solubility when compared to the untagged protein but to improve yield when fused to the C-terminus (Woestenenk *et al.*, 2004). All in all, these data suggest that an evaluation of at least N- and C-terminally tagged variants of a protein will increase the chance to obtain reasonable expression and quality of a recombinant protein. For expression of proteins to be secreted, tags should be placed at the C-terminus to prevent interference with membrane trafficking.

2.4. General considerations of protein purification by IMAC

There are several advantages of IMAC for purification of His-tagged proteins compared to other affinity chromatography principles as the reason for being the most widely used chromatographic technique (Biocompare, 2006; Derewenda, 2004). Besides low costs and the simplicity of use the robustness of IMAC is certainly its most striking feature: (i) the His-tag: ligand interaction works under both native and denaturing conditions such as 8 *M* urea or 6 *M* guanidinium hydrochloride (Hochuli *et al.*, 1988) enabling subsequent on-column refolding (Jungbauer *et al.*, 2004) as well as (ii) both oxidizing and reducing conditions, (iii) protein binding withstands a broad spectrum of various chemicals of different types (Table 27.1 summarizes chemical compatibilities for Ni-NTA IMAC and some limitations), (iv) its relatively high affinity and specificity allows high capturing efficiency even in the presence of high protein titers, and (v) the scalability of the purification procedure.

Despite the broad compatibility, IMAC has its limitations. Obviously, the use of chelating agents has to be avoided which can be a disadvantage as EDTA, a potent inhibitor of metalloproteases, can only be applied in low concentrations. Care should also be taken with the use of other potentially chelating groups such as Tris, ammonium salts, and certain amino acids (Table 27.2).

Until recently, the use of strong reducing agents such as DTT in IMAC processing has been regarded as problematic because of reduction of nickel and, as a consequence, suspected increase of nickel concentrations in protein preparations. However, we found that moderate concentrations of DTT are fully compatible with NTA-based purification. This is shown, for example, in Fig. 27.5A for HIV-1 reverse transcriptase (RT) purified with unaffected efficiency in the presence of up to 10 m*M* DTT. Also, RT activity is not influenced by these conditions and both end-point

Table 27.2 Chemical compatibility of purification of His-tagged protein using agarose-based IMAC (Ni-NTA) resins and its limitations

IMAC chemical compatibility			
Component	**Limitation (up to)**	**Component**	**Limitation (up to)**
Buffers		**Salts**	
Na-phosphate	Recommended, limit not known	NaCl	4 *M*
Phosphate citrate	Limit not known	$MgCl_2$	4 *M*
Tris–HCl, HEPES, MOPS	100 m*M*	$CaCl_2$	5 m*M*[c]
Citrate	60 m*M*	$NaHCO_3$	Not recommended
Detergents (in 300 m*M* NaCl)		Ammonium salts	Not recommended
n-Hexadecyl-*β*-D-maltoside	0.0003% (w/v)	**Protease inhibitors**	
n-Tetradecyl-*β*-D-maltopyranoside	0.005% (w/v)	EDTA	1 m*M*[a]
n-Tridecyl-*β*-D-maltopyranoside	0.016% (w/v)	Commonly used protease inhibitors[d]	Compatible in effective concentrations
Brij 35	0.1% (v/v)	Complete cocktail (EDTA-free)	1× concentrated
Digitonin	0.6% (w/v)	**Denaturants**	
Cymal 6	1% (w/v)	Urea	8 *M*
n-Nonyl-*β*-D-glucopyranoside (NG)	1% (w/v)	Gu–HCl	6 *M*
n-Decyl-*β*-D-maltopyranoside (DM)	2% (w/v)	**Amino acids**	
n-Dodecyl-*β*-D-maltoside (DDM)	2% (w/v)	Histidine	1–2 m*M*[b]
C12–E9	1% (w/v)	Glycine	Not recommended
n-Octyl-*β*-D-glucopyranoside (OG)	1.5% (w/v)	Cysteine	Not recommended
Triton X-100, Tween, NP-40	2% (v/v)	Glutamate	Not recommended
Triton X-114	2% (v/v)	Aspartate	Not recommended
Fos-Cholines	0.05% (w/v)	Arginine	500 m*M*

Dodecyldimethyl-phosphineoxide	0.15% (w/v)	**Organic solvents**	
N,N-Dimethyldodecylamine-*N*-oxide (LDAO)	0.7% (w/v)	Isopropanol	60% (v/v)[e]
CHAPS	1% (w/v)	Ethanol	20% (v/v)
Laurosyl-sarcosine	1% (w/v)	**Reducing reagents**	
SDS	0.3% (w/v)[a]	β-ME	20 m*M*
Other		TCEP	20 m*M*
EGTA	1 m*M*[a]	DTT	10 m*M*
Imidazole	10–20 m*M*[b]	DTE	10 m*M*
Hemoglobin	Not recommended		
Glycerol	50% (v/v)		

[a] Has been used successfully in the indicated concentration but should be avoided whenever possible.
[b] Dissociates His-tagged proteins at high concentrations.
[c] Should be avoided in combination with Na-phosphate.
[d] Include, for example, Aprotinin, Leupeptin, PMSF, and related serine protease inhibitors: Pepstatin, Antipain, Bestatin, E64, Benzamidine.
[e] Compatible with Ni-NTA purification and endotoxin removal according to Franken *et al.* (2000), but in this concentration is incompatible with reuse of the chromatographic media (data not shown).

This table provides some of the most relevant tested substances and concentrations and may not be complete or represent the maximal concentrations compatible with purification of His-tagged proteins.Abbreviations: β-ME, β-mercaptoethanol; TCEP, tris(2-carboxyethyl)phosphine hydrochloride; DTT/DTE, dithio-threitol/-erythritol; Gu–HCl, guanidinium hydrochloride.

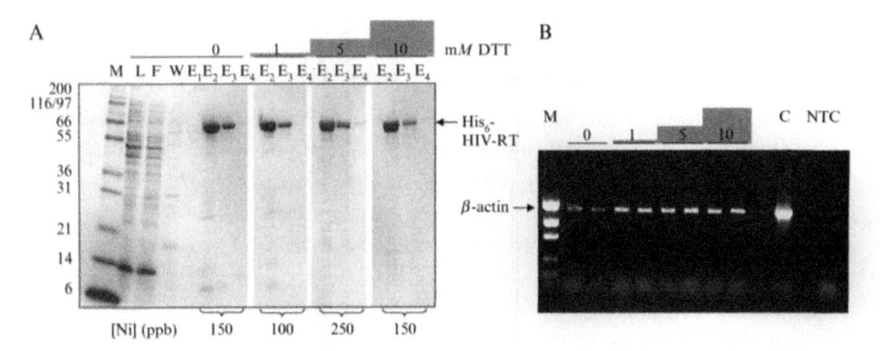

Figure 27.5 IMAC under reducing conditions. (A) His_6-tagged HIV-1 reverse transcriptase (RT) was purified under standard conditions in the presence of the indicated DTT concentrations. Aliquots of each chromatographic fraction of the purification without DTT (M, markers; L, lysate; F, flow-through; W, wash; E, elution fraction) and of corresponding volumes of elution fractions 2–4 containing DTT were analyzed by SDS–PAGE and Coomassie staining. Elution fractions were pooled and analyzed for nickel content by ICP-MS at Wessling Laboratories, Bochum, Germany; nickel concentrations are given in μg/l (parts per billion, ppb). (B) Equal amounts of HIV-1 RT purified in the presence of the indicated concentrations of DTT were used in duplicates to reverse transcribe a 1.5 kb β-actin cDNA which was subsequently amplified by PCR and the PCR products were analyzed by agarose gel electrophoresis and ethidium bromide staining. Reducing conditions of at least 1 m*M* DTT were found to be required for full RT activity (compare weaker amplification with RT purified in the absence of DTT). C, Omniscript positive control; an equal amount of Omniscript RT protein was used; NTC, no template negative control.

(Fig. 27.5B) and quantitative real-time RT-PCR (data not shown) show no inhibitory effect which could origin from high heavy-metal-ion concentrations. Even though nickel ions may become reduced by DTT leading to a color change of the resin bed they do not increasingly leach from the ligand (Fig. 27.5A), and resins processed under reducing conditions can be repeatedly reused and regenerated (data not shown). These findings suggest that despite a color change as a consequence of nickel reduction by DTT the resin is still functional. TCEP, a different, nonthio-based reducing reagent is more and more replacing DTT and β-ME in protein purification by IMAC as it seems to be more selective to reduction of disulfide bonds, is odorless and more stable in aqueous solution. We recommend the use of TCEP with Ni-NTA chromatography in a concentration of 1–5 m*M*.

2.5. Copurifying proteins on IMAC and what to do about it

IMAC leads to high protein purity upon single-step chromatographic purification in many cases (Figs. 27.2A, D, E and 27.5A; Bornhorst and Falke, 2000; Schmitt *et al.*, 1993). The higher the purity is the closer the amount of IMAC resin applied in the chromatographic step has been correlated to the amount of recombinant His-tagged protein present in the sample to be processed. The reason is that proteins naturally displaying surface motifs suitable to interact with an immobilized metal ion may bind to the resin although usually with slightly lower affinity than a flexible His tag. Most His-tagged proteins will therefore displace proteins with natural or accidental surface motifs. However, there are some proteins where the local density of chelating amino acids such as histidine is so high that they will bind to immobilized metal ions almost inevitably. In general, mammalian systems have a higher natural abundance than bacterial systems of proteins containing consecutive histidines (Crowe *et al.*, 1994), and a prominent example in human cells is the alpha subunit of transcription factor TFIIA which has seven consecutive and surface-exposed histidine residues and can be purified via IMAC from natural sources under native conditions (DeJong and Roeder, 1993; Ma *et al.*, 1993) and which is observed frequently as a signal band of 55 (the $\alpha\beta$ precursor) or 35 kDa (the α subunit) of TFIIA in Western blots using an anti-His tag antibody. Another example is the human transcription factor YY1 with 11 consecutive histidines (Shi *et al.*, 1991). In *E. coli*, the proteins observed to copurify with His-tagged target proteins can be divided into four groups: (i) proteins with natural metal-binding motifs, (ii) proteins with histidine clusters on their surfaces, (iii) proteins that bind to heterologously expressed His-tagged proteins, for example by a chaperone mechanism, and (iv) proteins with affinity to agarose-based supports (Bolanos-Garcia and Davies, 2006). Whether or not one of the *E. coli* proteins is copurified is not easily predictable. For example, a protein from group (ii) sometimes reported to copurify with Ni-NTA is the 21 kDa SlyD, however, we have never observed this one in our lab upon purification from *E. coli* BL21(DE3), DH5α, *M*15(pREP4), and other strains. This may be explained by the fact that many of these impurities are stress-responsive proteins, suggesting that the cultivation conditions and the bacterial strain have an influence on their abundance and, a consequence, their appearance as a contaminating species in the target protein preparation; it is therefore recommended to induce as little stress as possible during cultivation of *E. coli* cells (e.g., by using shake flasks without baffles). Furthermore, some copurifying proteins seem to have a binding preference for Co over Ni (or other ions) and others vice versa.

Several options to get rid of these copurified proteins or prevent their adsorption early on have been evaluated and some of them shall be discussed in the following section. These options include (i) performing additional purification steps, (ii) adjusting the His-tagged protein to resin ratio, (iii) to use an engineered host strain that does not express certain proteins, (iv) using an alternative support, and (v) tag cleavage followed by reverse chromatography.

Suitable additional purification steps include classical chromatographic techniques such as ion-exchange (IEX) and size exclusion chromatography (SEC) whereas IEX has the higher separation power. However, as SEC not only serves to separate molecules by size and helps to remove aggregates of high molecular weight but also can be used to desalt the preparation and provide conditions suitable for certain downstream applications, it is frequently used as a standardized procedure by, for example, high-throughput labs such as structural biology consortia (Acton *et al.*, 2005; Gräslund *et al.*, 2008) who perform protein crystallization or NMR spectroscopy. Although IMAC–SEC (as opposed to IMAC–IEX) can be performed as a standardized procedure without having to take into account the protein biochemistry such as pI the separation range of a given SEC column will not be suitable for all separation tasks and a range of SEC columns may have to be held in place. Another issue with the application of IEX and SEC is that in order to make use of the full power provided by these technologies, costly equipment such as an automated chromatography system is required which frequently precludes multiparallel processing leading to low throughput. Affinity purification such as IMAC can usually be done as a bind–wash–elute procedure in a bench top/gravity flow mode. By introduction of a second affinity tag (e.g., StrepII, GST, Flag) into the expression construct a two-step purification leading to high-purity protein preparations is enabled as a bench-top two-step affinity chromatography procedure (Cass *et al.*, 2005; Prinz *et al.*, 2004).

As mentioned earlier, adjusting the amount of His-tagged protein to be recovered to the binding capacity of the used IMAC resin can help to improve protein purity by preventing copurification of proteins with certain affinity to IMAC resins. However, the amount of His-tagged protein is usually unknown unless a pilot experiment is performed to estimate the content of target protein. While this is an option, the better way would be to exclude the presence of such copurifying proteins by expressing the target protein in an engineered strain where the respective genes have been knocked out. However, to our knowledge results from work with such strains have not yet been reported and the experience with knockout strains in protein production is still low. Also, it does not seem realistic that an *E. coli* strain can be generated which lacks 17 proteins reported to bind to IMAC resins including as important functions as superoxide dismutase and iron-uptake regulation (Bolanos-Garcia and Davies, 2006) and which is still well viable under stress situations such as protein overproduction.

A different approach to improve the purity of proteins recovered from IMAC has been reported that made use of dextran-coating of an agarose matrix, the constituting material of the most widely used chromatographic supports (Sepharoses, Superflow, Agaroses), and which prevented copurification of proteins with affinity to these matrices (Mateo *et al.*, 2001). Dextran-coated beads, however, are not readily commercially available as IMAC resins and this measure only helps to preclude proteins with affinity to agarose and not to the immobilized metal or to the target protein. Silica-based IMAC supports also prevent adsorption of proteins with affinity for agarose and, in addition, have good pressure stability which makes them suitable for high-resolution HPLC applications but the silica resins frequently suffer from low binding capacity and limited resistance to high pH sanitization procedures. A very recent method that avoids the need to use solid chromatographic supports completely is called affinity precipitation (Hilbrig and Freitag, 2003) and has been applied to IMAC (Matiasson *et al.*, 2007). Here, the IMAC ligand is chemically coupled to a responsive polymer which, after binding to the His-tagged protein, can be aggregated upon change of environmental conditions such as pH or temperature and can thus be precipitated by centrifugation. Protocols for its use are still relatively complicated but as soon as robust and easy to use commercial materials are available this method may have the potential to play an important role in IMAC applications. Using a ligand in solution could overcome steric hindrance of the binding of some His-tagged proteins to an immobilized ligand as well as mass transport limitations of porous chromatographic media. Moreover, it is in line with a trend in industrial-scale chromatography toward single-use disposable materials.

Yet another approach has recently been reported that can be applied for protein separation from lysates after cell-free expression (Kim *et al.*, 2006). An *E. coli*-derived lysate was preincubated with Ni-NTA magnetic agarose beads to remove proteins with affinity to Ni-NTA prior to template addition and protein expression; the expression capacity of the S30 extract was found to remain unaltered and the Ni-NTA purified His-tagged protein fractions to be of higher purity than without pretreatment.

While the aforementioned strategies to improve the purity of MAC protein preps have proven useful in many cases they are not generally applicable and successful. There is a method, however, that almost meets the criterium of universal applicability regarding improvement of purity, and it has the additional benefit of resulting in a protein native or near-native structure: proteolytic His tag cleavage using a His-tagged protease followed by reverse IMAC (Block *et al.*, 2008). This strategy overcomes the copurification issue by passing the proteolytically processed protein under similar or identical conditions over the same column, and the proteins that bound to the IMAC resin as impurities in the initial purification step will bind to the same resin again while the cleaved, that is untagged, target protein is collected in the flow-through fraction (reverse or subtractive IMAC mode). It can be performed with both exo- and endoproteases that themselves carry an (uncleavable) His-tag (Nilsson *et al.*, 1997; Polayes *et al.*, 2008) but the exoproteolytical approach has the advantage that it is faster and results in a protein with native structure with no vector-derived amino acids (Arnau *et al.*, 2006; Block *et al.*, 2008; Pedersen *et al.*, 1999). This approach is especially suitable for demanding downstream applications such as protein crystallization or biopharmaceutical production. Notably, the method requires only a single chromatography column to achieve an extremely high degree of purity. An application is shown in Fig. 27.6 where TNFα was crystallized in its IMAC one-step purified form (A, B, C, D) or in its subtractive-IMAC processed form as described above (A, E, F, G). Although already His-tagged TNFα purified on Ni-NTA appeared highly pure upon SDS–PAGE and Coomassie staining (Fig. 27.6A, lane IMAC) analysis on a silver-stained 2D gel showed impurities (Fig. 27.6B). These impurities are removed by reverse mode IMAC resulting in a protein preparation of extremely high purity (Fig. 27.6E). Figure 27.6 also demonstrates the influence the His tag can have on protein crystallization. While both the tagged and the mature native form of TNFα eluted from a SEC column as a trimer (Fig. 27.6C and F), they crystallized under significantly different conditions and resulted in different crystal forms (His_6-TNFα: tetragonal, Fig. 27.6D; TNFα: rhombohedral, Fig. 27.6G). Nevertheless, calculated structures both were in accordance with the one deposited in the pdb (data not shown). However, when we attempted to perform the same experimental workflow with His-tagged and native mature IL-1β, we were unable to crystallize the His-tagged cytokine (Block *et al.*, 2008; and data not shown). Our data confirm the observation also made by many others that it is frequently possible to crystallize tagged proteins but suggest that it makes sense to provide a tag cleavage option when designing an expression construct.

2.6. IMAC for industrial-scale protein production

IMAC for production of proteins in industrial scale, for example for use as biopharmaceutical, has not been used until quite recently due to worries regarding the potential immunogenicity of a His tag sequence and because of allergenic effects of nickel leaching from an IMAC matrix. However, nickel concentrations typically observed in protein preparations obtained from tetradentate IMAC resins are low and content in expected daily doses of a biopharmaceutical will be far below the typical daily intake of nickel and the permanent nickel body burden (Block *et al.*, 2008). Removal of artificial sequences from recombinant proteins by the use of protease has been discussed above as the way to go for use in humans but proteins carrying a His tag have been successfully used for vaccination (Kaslow and Shiloach, 1994; Stowers *et al.*, 2001) or are presently commercialized as drugs (unpublished). IMAC is a chromatography method that can simply be scaled linearly from milliliter to liter bed volumes (Block *et al.*, 2008; Hochuli *et al.*, 1988; Kaslow and Shiloach, 1994; Schäfer *et al.*, 2000) and Ni-NTA Superflow columns in dimensions up to 50 l are in use for biopharmaceutical production processes (F. Schäfer, personal communication). Compatibility of IMAC matrices with a wide range of chemicals such as chaotropics, salts, organic solvents, and detergents (Table 27.2) facilitates adaption to the specific needs of the production of the individual protein.

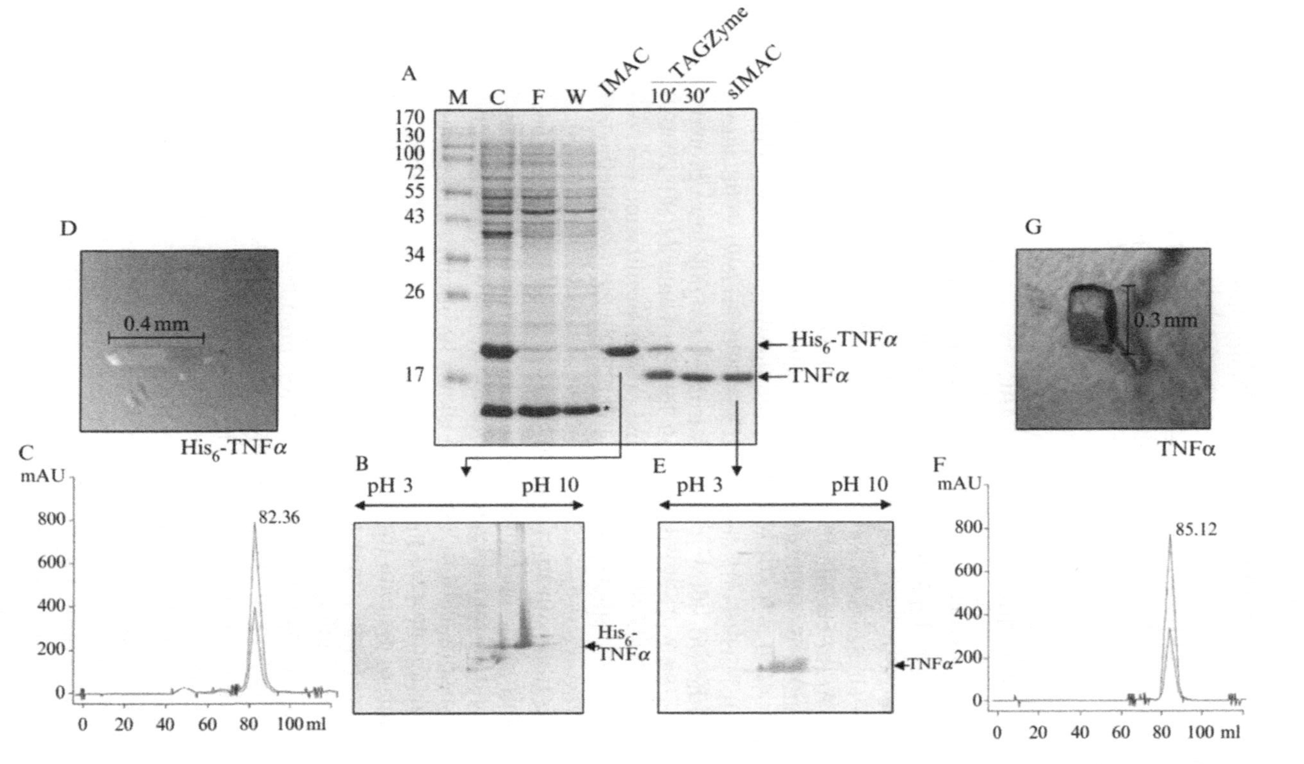

Figure 27.6 Removal of copurifying proteins by His tag cleavage and reverse IMAC. (A) His_6-TNFα expressed in *E. coli* was purified via Ni-NTA Superflow and processed using the TAGZyme exoproteolytic system as described (Schäfer *et al.*, 2002a). (B, E) 2D gel electrophoresis and silver staining of His_6-TNFα and TNFα, respectively, was performed as described (Block *et al.*, 2008). The subband pattern in the first dimension between pI 6.7 and 5.8 for both His_6-TNFα and TNFα is in accordance with the report for TNFα produced in yeast (Eck *et al.*, 1988). (C, F) Analytical size exclusion chromatography (SEC) on HR 10/30 Superdex 200 was run with 1× TAGZyme buffer (Schäfer *et al.*, 2002a). (D, G) His_6-TNFα tetragonal crystal (D) formed in 2.7 *M* $MgSO_4$, MES, pH 5.5 and diffracted to 2.5 Å (homelab X-ray source, FR591 Nonius Bruker); TNFα rhombohedral crystal (G) formed in 1.8 *M* NH_4SO_3, 200 m*M* Tris–HCl, pH 7.8 and diffracted to 2.0 Å (ESRF synchrotron). The size of typical crystals (mm) is indicated by the bar in (D) and (G).

For example, bacterial endotoxins (lipopolysaccharides) can be eliminated from the protein product during chromatography by including a wash step using a detergent (Triton X-114, Block *et al.*, 2008) or an organic solvent (60% isopropanol; Franken *et al.*, 2000). The suitability of IMAC for industrial production purposes has been demonstrated and it can be expected that IMAC-based processes become increasingly used in the future due to its robustness and relatively low requirements for individual optimization.

2.7. High-throughput automation of IMAC

Due to its robustness, universal applicability, and widespread use, IMAC is also an ideal tool for multiparallel screening for protein expression and solubility. This is mostly done in the convenient 96-well format using agarose- or magnetic bead-based IMAC resins in SBS footprint filter or microplates on 96-well magnets or plate centrifuges (Braun *et al.*, 2002; Büssow *et al.*, 2000). As the complete expression and the simple bind–wash–elute IMAC purification workflow is easily miniaturizable in microplate formats it was also suitable for hands-on free automation using liquid handling laboratory robots (see Lesley, 2001, for an overview). The automated steps covered the workflow to various extents, ranging from protein purification from manually generated *E. coli* lysates (Lanio *et al.*, 2000), *E. coli* or insect cell lysis, lysate clarification, and protein purification (Garzia *et al.*, 2003; Schäfer *et al.*, 2002b; Scheich *et al.*, 2003), to automation of complete workflows from construct cloning to protein analysis (Acton *et al.*, 2005; Hunt, 2005; Koehn and Hunt, 2009). Recently, we added another series of protocols and consumables for purification of His-tagged proteins from *E. coli* or eukaryotic cells or cell-free lysates to the list of options: a conceptually new lab automation instrument allows isolation of microgram to milligram amounts of proteins from a variable number of samples (with a random-access sample feeding option) using ready-to-go prefilled cartridges that provide enzymes, buffers, and Ni-NTA magnetic beads for lysis and purification. Figure 27.7 describes an expression and purification screening using a set of 24 optimized-gene constructs for production of human proteins. Between 1.4 and 35 μg of highly pure protein was obtained under native conditions. Proteins that could not be purified under native conditions were obtained upon purification under denaturing conditions, and Western blot analyses using an anti-His antibody showed the absence of cross-contaminations between wells (data not shown). Protein resulting from such high-throughput purification experiments can be used for functional assays (e.g., interaction studies), characterization of protein random mutagenesis, solubility analyses, and clone screening.

The next step following an expression screening is frequently a scale-up with a limited number of proteins or clones for production of milligram amounts

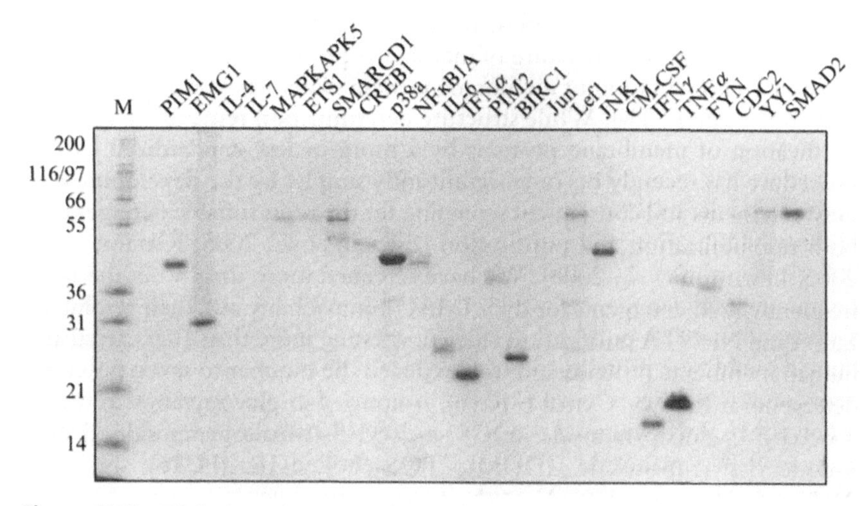

Figure 27.7 High-throughput protein purification screening on a new automation platform. *E. coli* BL21 (DE3) cells were transformed with QIAgene constructs carrying optimized human genes for expression of the indicated proteins in 1 ml LB cultures in a 96-deep well block. Cells were harvested by centrifugation and the pellet block placed onto the sample input drawer of the QIAsymphony SP instrument. Cells were resuspended and lysed and His_6-tagged proteins purified from the crude lysates using Ni-NTA magnetic agarose beads and buffer solutions provided in the QIAsymphony cartridge setup. A 5 μl of each elution fraction was analyzed by SDS–PAGE and Coomassie staining. Expected protein sizes (kDa) for the individual proteins were as follows: PIM1, 40; EMG1,30; IL-4, 15.5; IL-7,18; MAPKAPK5,55; ETS1,55; SMARCD1,60; CREB1,50; p38a, 40; NFκB1A, 40; IL-6,21; IFNα, 20; PIM2, 40; BIRC5, 30; Jun, 50; Lef1, 55; JNK1, 45; CM-CSF, 15; IFNγ, 17; TNFα, 17; FYN, 40; CDC2, 35; YY1, 66; SMAD2, 55. *M*, markers (kDa).

of protein for animal immunization, structural, or pharmacokinetic studies. Single proteins can be purified using standard ÄkTA or FPLC systems, and an ÄkTA system for slightly increased throughput has been developed (ÄkTAxpress). However, systems with a significantly higher throughput, lower complexity, and more dedicated to one-step (mainly IMAC) affinity purification have been reported (Steen *et al.*, 2006; Strömberg *et al.*, 2005) and are in use in high-throughput projects such as the human protein atlas project (Hober and Uhlén, 2008).

2.8. Special applications: Purification of membrane proteins

Membrane proteins have received the highest attention of all protein classes in the past few years due to their enormous importance as drug targets. In fact, more than 50% of all currently commercialized drugs as well as those under development target membrane proteins (Drews, 2000). Furthermore, membrane proteins account for approximately 30% of the human proteome. However, in contrast to soluble proteins, very little is known about the biology and structure of membrane proteins which is reflected by the underrepresentation of membrane protein structures in public databases such as the pdb (<1%). While structure determination remains a challenge, purification of membrane proteins by a more or less standardized IMAC procedure has recently become significantly simpler by the development of new detergents and consequent screening for the most suitable detergent for both resolubilization and purification (Eshaghi *et al.*, 2005; Klammt *et al.*, 2005; Lewinson *et al.*, 2008). We have screened more than 50 of the most frequently used detergents for their IMAC compatibility and their resolubilization and Ni-NTA purification efficiency testing more than 10 bacterial and human membrane proteins and have reduced the number to seven powerful detergents as follows: Cymal 6 (Cy6), *n*-nonyl-*β*-D-glucopyranoside (NG), *n*-octyl-*β*-D-glucopyranoside (OG), *n*-decyl-*β*-D-maltopyranoside (DM), *n*-dodecyl-*β*-D-maltoside (DDM), FOS-choline-16 (FC16), and *N*,*N*-dimethyldodecylamine-*N*-oxide (LDAO). In screenings, we achieved a positive result for efficient resolubilization from *E. coli* or insect cell membrane fractions and IMAC purification for at least one detergent in each case. In Fig. 27.8, the detergent screening and Ni-NTA purification of His-tagged human Caveolin 1 is shown as an example.

IMAC compatible detergents are listed in Table 27.2 whereas this list may not be complete. Nevertheless, binding of a His-tagged membrane protein to IMAC resins seems not to work in combination with certain

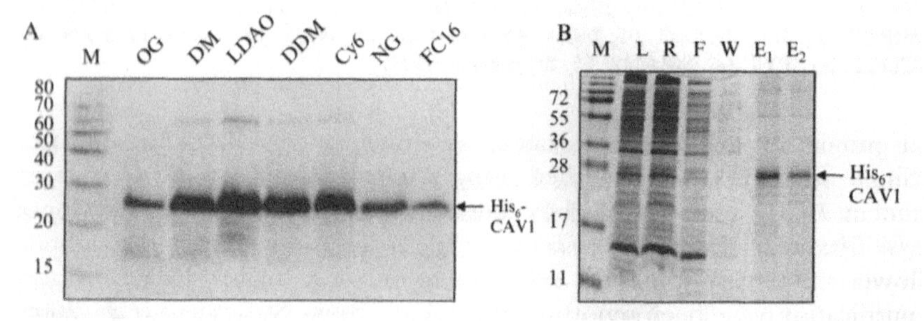

Figure 27.8 Detergent screening and IMAC purification of His_6-tagged human Caveolin 1 expressed in *E. coli* C41(DE3) in the presence of LDAO. The gene optimized for expression in *E. coli* was synthesized and cloned into pQE-T7 with an N-terminal His_6 tag. (A) Detergent screening; the membrane fraction from 200 ml *E. coli* culture divided into seven portions was mixed with detergent as indicated, centrifuged at 20,000 × *g* and aliquots of detergent-soluble protein analyzed by Western blot (Penta anti-His antibody). Detergents are given in the text. (B) Caveolin 1 purification upscale; His_6-tagged Caveolin 1 (CAV1) was purified from 200 ml *E. coli* culture volume via Ni-NTA Superflow in the presence of 30 mM LDAO. Fractions were analyzed by SDS–PAGE and Coomassie staining: M, markers (kDa); L, *E. coli* total lysate; R, supernatant from resolubilization (loaded onto IMAC column); F, IMAC flow-through fraction; W, wash fraction; E, elution fractions.

detergents, including some of the listed ones even though resolubilization is fine. These phenomena depend on the protein–detergent combination and seem to arise from the detergent micelle around the protein partially or completely hiding the His tag. The use of longer tag sequences such as a H_{10} tag has become popular and seems to overcome such limitations of membrane protein recovery and improve affinity to IMAC resins (Byrne and Jormakka, 2006; Grisshammer and Tucker, 1997; Mohanty and Wiener, 2004; Rumbley *et al.*, 1997).

2.9. Special applications: Purification of zinc-finger proteins

Another big protein superfamily which—due to technical issues—shall be mentioned in the context of IMAC is the group of proteins containing zinc-finger motifs. The C2H2 zinc-finger transcription factors alone with over 600 members represent more than 2% of the human proteome (Knight and Shimeld, 2001). In these proteins, zinc ions are coordinated by a defined spatial organization of two cysteine and two histidine residues each per finger, and a single polypeptide usually contains four or five of these motifs. Purification of such a metalloprotein via a metal ion chelated in a similar, that is tetradentate, manner deserves some reflection regarding the best way of IMAC purification.

Despite the strong interaction of metal ions with tetradentate IMAC ligands, it may not be excluded that a metal from the resin could exchange with the zinc in the zinc finger. Nickel, the most frequently used metal in IMAC protein purification has quite similar physicochemical properties to zinc and might therefore well replace it in such a metal-binding motif, and the nickel concentration in an IMAC resin (approximately 15 m*M*) is usually considerably higher than the concentration of a His-tagged target protein (μM range). In order to analyze a potential metal ion exchange between matrix and metalloprotein, we expressed the His_6-tagged C2H2 zinc-finger containing transcription factor YY1 in *E. coli* and purified the protein in parallel using Ni-NTA and Zn-NTA. YY1 could be purified to high purity by both nickel- and zinc-based IMAC (Fig. 27.9A, lane E).

Both YY1 preparations were subjected to determination of nickel and zinc content by ICP-MS (intercoupled plasma mass spectrometry), a quantitative method frequently applied to detect trace metals in biological systems (Shi and Chance, 2008). The zinc-IMAC purified protein preparation contained 35.6 μM zinc corresponding to approximately six Zn^{2+} ions per YY1 polypeptide and some Ni^{2+} which can be attributed to residual traces from the buffers (Fig. 27.9B, right). The protein recovered by nickel-IMAC, however, contained more than 25 μM Ni^{2+} and 14.2 μM Zn^{2+} (Fig. 27.9B, left), corresponding to again approximately six Me^{2+} ions per YY1 polypeptide. The Me^{2+}:polypeptide molar ratio of 6 is higher than the four zinc-finger motifs in YY1 reported in databases (http://www.uniprot.org/uniprot/P25490) which may in part be explained with additional metal

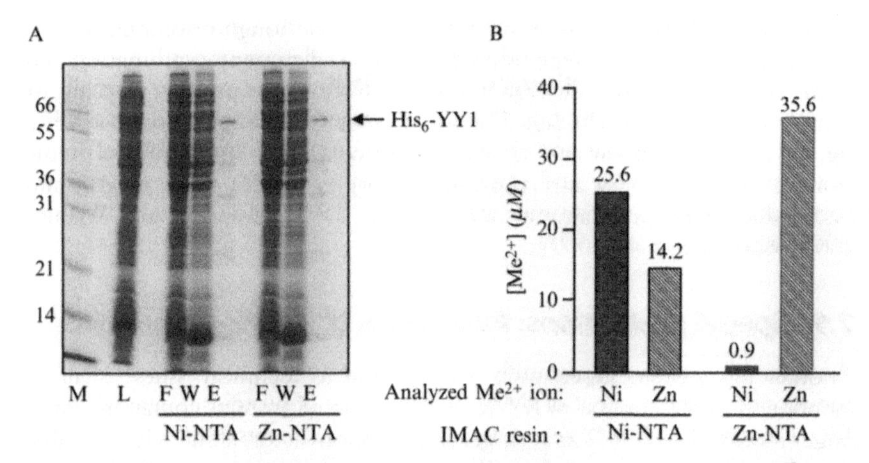

Figure 27.9 Purification of His_6-tagged zinc-finger protein YY1 by Ni- and Zn-IMAC. (A) IMAC purification. His_6-YY1 was expressed in *E. coli* Bl21 (DE3) and a cleared lysate was generated as described below. NTA resin was treated with $ZnCl_2$ as described below to generate Zn-charged NTA (Zn-NTA). Protein purifications were performed using the same cleared lysate preparation under standard conditions (see Section 2.10) and aliquots of the chromatographic fractions (L, cleared lysate; F, flow-through fraction; W, wash fraction; E, elution fraction) analyzed by SDS–PAGE and Coomassie staining. Protein concentration of elution fractions were determined by the Bradford method to 0.3 mg/ml (6.7 μM, Ni-NTA) and 0.25 mg/ml (5.6 μM, Zn-NTA). (B) Quantitative detection of metals. Both elution fractions were then dialyzed for 16 h at 4 °C versus three changes of buffer NPI-10 to remove metal ions from the solution. Dialyzed proteins and fresh dialysis buffer were analyzed for nickel and zinc content by ICPMS at Wessling Laboratories (Bochum, Germany) and the values determined for the buffer subtracted from the protein analyses. Values (ppm, mg/l) were converted to molar units using the mean relative masses for Ni (58.7) and Zn (65.4).

ions binding elsewhere on the protein. Nevertheless, the data from Fig. 27.9B suggest that in the case of YY1 there is a significant exchange of metal ions between the charged IMAC ligand and the zinc fingers. Metal-affinity purification of proteins with C2H2 zinc-finger motifs can be effectively performed but an IMAC ligand charged with zinc, if applicable, should be chosen in order to obtain a preparation with an intact and homogenous zinc-finger composition.

2.10. Protein purification protocols

In this section, the standard IMAC protocols will be described briefly. The procedures have been optimized for the use of tetradentate (i.e., Ni-NTA) resins but should be transferable to tridentate (IDA-based) resins as well. TED resins behave different and the manufacturers' recommendations should be followed (e.g., proteins elute at significantly lower imidazole concentrations and consequently imidazole level in wash buffers should be kept low).

Here, we will provide recommendations for agarose-based Ni-NTA resins (Superflow, Agarose) regarding purification, cleaning, and recharging. These can be regarded as an update of the more detailed description provided in the Ni-NTA handbook (QIAGEN, 2003). Buffers provided in the following may be supplemented according to the needs of the individual protein (e.g., in order to generate reducing conditions, to stabilize the protein with glycerol, or to provide the presence of cofactors).

2.10.1. Purification of His-tagged proteins under native conditions

1. Lyse cells using the basis buffer NPI-10 (50 m*M* NaH_2PO_4, 300 m*M* NaCl, 10 m*M* imidazole, pH 8.0) supplemented with a suitable lysis reagent.

   For *E. coli* lysis, we recommend to use standard hen egg white lysozyme at 1 mg/ml final concentration because lysis is extremely efficient (if cells had been frozen) and lysozyme is inexpensive. Lysozyme is reliably washed from IMAC resins and does not occur in elution fractions (Figs. 27.5A, 27.6A, and 27.8; Block *et al.*, 2008). Other suitable methods are based on detergents (e.g., 1 % (v/v) CHAPS or proprietary solutions) or physical treatment (sonication, high pressure/depressurization homogenization). For cultures derived from insect or mammalian cells, 1% Igepal CA-630 (former name NP-40) is recommended. For reduction of lysate viscosity, the addition of a nuclease is helpful, and Benzonase (3 units/ml bacterial culture) has been shown to be robust and its removal in wash steps from IMAC resins works reliably and can be checked by a commercial ELISA (Block *et al.*, 2008). Incubate the lysate on ice for 30 min.
2. Generate a cleared lysate by centrifugation for 30 min at $\geq 10{,}000 \times g$ and 2–8 °C. Collect the supernatant.
3. Load the cleared lysate onto the resin equilibrated with 5 bed volumes (bv) of NPI-10 and allow to flow through at approximately 1 ml/min (column of 1 ml bv) or let flow through (gravity flow application).

   A suitable linear flow rate during binding is 155 cm/h corresponding to 1 ml/min of a column with a bed diameter of ~7 mm (1 ml HisTrap and Ni-NTA Superflow Cartridges). If applicable to the workflow, we recommend to perform binding in batch mode as this is most efficient with agarose-based resins in general; for this, add the required volume of lysate to the equilibrated resin and incubate for 1 h rotating end-over-end at 2–8 °C.
4. Wash the Ni-NTA column with 10 bv wash buffer NPI-20 (50 m*M* NaH_2PO_4, 300 m*M* NaCl, 20 m*M* imidazole, pH 8.0).

   If binding had been performed in batch, pour the binding suspension into a suitable flow-through column.
5. Elute His-tagged protein with 5 bv of elution buffer NPI-500 (50 m*M* NaH_2PO_4, 300 m*M* NaCl, 500 m*M* imidazole, pH 8.0).

2.10.2. Purification of His-tagged proteins under denaturing conditions

1. Lyse cells using the basis buffer B (100 m*M* NaH_2PO_4, 10 m*M* Tris–HCl, 8 *M* urea, pH 8.0).

   E. coli as well as most eukaryotic cells are efficiently lysed by 7–9 *M* urea but occasionally, His-tagged proteins forming inclusion bodies do not completely dissolve; in such cases, we recommend to replace the chaotrope urea by guanidinium hydrochloride (buffer A: 100 m*M* NaH_2PO_4, 10 m*M* Tris–HCl, 6 *M* Gu–HCl, pH 8.0). Reduction of lysate viscosity by the addition of Benzonase (3 units/ml final concentration) is also possible and effective under denaturing conditions but here the maximum urea concentration is 7 *M* (use of Gu–HCl is not possible in combination with Benzonase). Incubate the lysate for 30 min at ambient temperature.
2. Generate a cleared lysate by centrifugation for 30 min at $\geq 10{,}000 \times g$ and room temperature. Collect the supernatant.
3. Load the cleared lysate onto the resin equilibrated with 5 bv of buffer B (or buffer A where applicable) and allow to flow through at approximately 1 ml/min (column of 1 ml bv) or by gravity flow.

   A suitable linear flow rate during binding is 155 cm/h corresponding to 1 ml/min of a column with a bed diameter of ~7 mm (1 ml HisTrap or Ni-NTA Superflow Cartridges). If applicable to the workflow and the resin format used, we recommend to perform binding in batch mode as this is most efficient with agarose-based resins in general; for this, add the required volume of resin slurry to the lysate and incubate for 1 h rotating end-over-end.
4. Wash the Ni-NTA column with 10 bv wash buffer C (100 m*M* NaH_2PO_4, 10 m*M* Tris–HCl, 8 *M* urea, pH 6.3).

   If binding had been performed in batch, pour the binding suspension into a suitable flow-through column prior to the wash step. If lysate was generated using buffer A, wash and elution steps may be performed by switching to urea-based buffers or by continuing with Gu–HCl-based buffers with pH values adjusted accordingly.
5. Elute His-tagged protein with 5 bv of elution buffer E (100 m*M* NaH_2PO_4, 10 m*M* Tris–HCl, 8 *M* urea, pH 4.5).

2.11. Cleaning and sanitization

A simple and effective cleaning-in-place (CIP) method for Ni-NTA IMAC resins used to purify proteins from "standard" samples such as *E. coli* or human cell lysates or supernatants is contacting the resin with 0.5 *M* NaOH for 30 min (Schäfer *et al.*, 2000). The resins have been stored in up to 1 *M* NaOH for several months and shown to withstand these conditions without compromising its performance even upon >100 cycles of purification/CIP cycles (data not shown). This CIP procedure reliably denatures and desorbs proteins originating from the loaded sample that might have bound unspecifically to agarose during purification and is generally suitable for sanitization (depyrogenation, viral, and microbial clearance; Levison *et al.*, 1995). Cleaning protocols may have to be adapted if more "unusual" samples such as lysates rich in lipids are loaded onto a column. Bases, acids, and other reagents that may be used for cleaning include ethanol (100%), isopropanol (30%, v/v), SDS (2%, w/v), acetic acid (0.2 *M*), NaOH (1 *M*), or detergents (see Table 27.2).

For repeated reuse of a Ni-NTA column, we recommend to perform the CIP description provided above followed by reequilibration. For long-term storage (several years), resin may be kept in either 30% (v/v) ethanol or, if an inflammable reagent is preferred, in 0.01–0.1 *M* NaOH. Storage in 10 m*M* NaN_3 is possible as well. It is usually not required to strip off the metal ions and recharge Ni-NTA, even after repeated reuse or long-term storage.

2.12. Simplified metal-ion stripping and recharging protocol

However, in cases where the resin has been seriously damaged or if binding capacity decreased over time, for example, by repeated loading of lipid-rich or samples containing chelating components, Ni-NTA may be easily stripped and recharged with nickel or a different metal ion. Starting with step 3, this simplified protocol is also suitable to initially charge NTA resin purchased uncharged.

1. Wash cleaned (see above) resin with 10 bv of deionized H_2O (dH_2O).
2. Strip off metal ions by passing 5 bv of 100 m*M* EDTA, pH 8.0 over the resin bed.
3. Wash resin with 10 bv of dH_2O.
4. Pass 2 bv of a 100 m*M* metal ion aqueous solution (e.g., $NiSO_4$ or $NiCl_2$) over the resin bed.

   Other metal ions that have been successfully and stably immobilized to NTA include copper ($CuCl_2$, $CuSO_4$), zinc ($ZnCl_2$, $ZnSO_4$), cobalt ($CoCl_2$, $CoSO_4$), and iron ($FeCl_3$, $Fe_2(SO_4)_3$).
5. Wash resin with 10 bv of dH_2O to remove any unbound metal ions.
6. Add storage buffer or equilibrate the column with at least 5 bv of starting buffer for immediate use.

3. Conclusions

We have presented the wide variety of applications the IMAC principle offers for research in general and for production of His-tagged proteins in particular. Its robustness and versatility are the reasons why IMAC has become one of the most broadly used chromatographic methods. Modifications in production procedures for both resin and ligand materials as well as optimized application protocols led to a significant improvement of IMAC performance in the recent past, well reflected by, for example, the increase of binding capacity of both NTA- and IDA-based matrices for His-tagged proteins from 5–10 to 50 mg per ml resin bv. We anticipate a continued methodological improvement and dissemination of the use of IMAC in the field of purification of recombinant proteins, for example, regarding industrial-scale production of biopharmaceuticals.

ACKNOWLEDGMENTS

The authors thank Jacob Piehler for valuable discussions regarding the manuscript and his support during the transfer of the tris-NTA ligand synthesis procedures. Furthermore, we thank Annette Zacharias-Koch for excellent technical assistance and her support in preparation of the manuscript. Part of the work presented here was performed under a grant of the German Ministry of Education and Research (BMBF, grant no. 0313965B).

REFERENCES

Acton, T. B., Gunsalus, K. C., Xiao, R., Ma, L. C., Aramini, J., Baran, M. C., Chiang, Y.-W., Climent, T., Cooper, B., Denissova, N. G., Douglas, S. M., Everett, J. K., *et al.* (2005). Robotic cloning and protein production platform of the northeast structural genomics consortium. *Methods Enzymol.* **394,** 210–243.

Andersson, L., and Porath, J. (1986). Isolation of phosphoproteins by immobilized metal (Fe^{3+})affinity-chromatography. *Anal. Biochem.* **154,** 250–254.

Arnau, J., Lauritzen, C., Petersen, G. E., and Pedersen, J. (2006). Current strategies for the use of affinity tags and tag removal for the purification of recombinant proteins. *Protein Expr. Purif.* **48,** 1–13.

Biocompare (2006). Protein chromatography: Tools for protein expression and purification. Biocompare Surveys and Reports, Biocompare Inc., July 10.

Block, H., Kubicek, J., Labahn, J., Roth, U., and Schäfer, F. (2008). Production and comprehensive quality control of recombinant human Interleukin-1β: A case study for a process development strategy. *Protein Expr. Purif.* **27,** 244–254.

Boden, V., Winzerling, J. J., Vijayalakshmi, M., and Porath, J. (1995). Rapid one-step purification of goat immunoglobulins by immobilized metal ion affinity chromatography. *J. Immunol. Methods* **181,** 225–232.

Bolanos-Garcia, V. M., and Davies, O. R. (2006). Structural analysis and classification of native proteins from *E. coli* commonly co-purified by immobilized metal affinity chromatography. *Biochim. Biophys. Acta* **1760,** 1304–1313.

Bornhorst, J. A., and Falke, J. J. (2000). Purification of protein using polyhistidine affinity tags. *Methods Enzymol.* **326,** 245–254.

Braun, P., Hu, Y., Shen, B., Halleck, A., Koundinya, M., Harlow, E., and LaBaer, J. (2002). Proteome-scale purification of human proteins from bacteria. *Proc. Natl. Acad. Sci. USA* **99,** 2654–2659.

Busso, D., Kim, R., and Kim, S.-H. (2003). *J. Biochem. Biophys. Methods* **55,** 233–240.

Büssow, K., Nordhoff, E., Lübbert, C., Lehrach, H., and Walter, G. (2000). A human cDNA library for high-throughput protein expression screening. *Genomics* **65,** 1–8.

Byrne, B., and Jormakka, M. (2006). Solubilization and purification of membrane proteins. *In* "Structural Genomics on membrane proteins" (K. Lundstrom, ed.), pp. 179–198. CRC Press, Taylor and Francis, Boca Raton.

Cass, B., Pham, O. L., Kamen, A., and Durocher, Y. (2005). Purification of recombinant proteins from mammalian cell culture using a generic double-affinity chromatography scheme. *Protein Expr. Purif.* **40,** 77–85.

Cèbe, R., and Geiser, M. (2006). Rapid and easy thermodynamic optimization of the 5′-end of mRNA dramatically increases the level of wild type protein expression in *Escherichia coli. Protein Expr. Purif.* **45,** 374–380.

Chaga, G., Hopp, J., and Nelson, P. (1999). Immobilized metal ion affinity chromatography on Co^{2+}-carboxymethylaspartate–agarose Superflow, as demonstrated by one-step purification of lactate dehydrogenase from chicken breast muscle. *Biotechnol. Appl. Biochem.* **29,** 19–24.

Crowe, J., Döbeli, H., Gentz, R., Hochuli, E., Stüber, D., and Henco, K. (1994). 6xHis-Ni-NTA chromatography as a superior technique in recombinant protein expression/purification. *Methods Mol. Biol.* **31,** 371–381.

Dalbøge, H., Bayne, S., and Pedersen, J. (1990). *In vivo* processing of N-terminal methionine in *E. coli. FEBS Lett* **266,** 1–3.

DeJong, J., and Roeder, R. G. (1993). A single cDNA, hTFIIA/α, encodes both the p35 and p19 subunits of human TFIIA. *Genes Dev.* **7,** 2220–2234.

Derewenda, Z. S. (2004). The use of recombinant methods and molecular engineering in protein crystallization. *Methods* **34,** 354–363.

Drews, J. (2000). Drug discovery: A historical perspective. *Science* **287,** 1960–1964.

Eck, M. J., Beutler, B., Kuo, G., Merryweather, J. P., and Sprang, S. R. (1988). Crystallization of trimeric recombinant human tumor necrosis factor (catechin). *J. Biol. Chem.* **263,** 12816–12819.

Eshaghi, S., Hedrén, M., Ignatushchenko Abdel Nasser, M., Hammarberg, T., Thomell, A., and Nordlund, P. (2005). An efficient strategy for high-throughput expression of recombinant integral membrane proteins. *Protein Sci.* **14,** 676–683.

Franken, K. L. M. C., Hiemstra, H. S., van Meijgaarden, K. E., Subronto, Y., den Hartigh, J., Ottenhoff, T. H. M., and Drijfthout, J. W. (2000). Purification of His-tagged proteins by immobilized chelate affinity chromatography: The benefits from the use of organic solvents. *Protein Expr. Purif.* **18,** 95–99.

García González, L. A., Rodrigo Tapia, J. P., Sánchez Lazo, P., Ramos, S., and Suárez Nieto, C. (2004). DNA extraction Using Chelex resin for the oncogenic amplification analysis in head and neck tumors. *Acta Otorrinolaringol. Esp.* **55,** 139–144.

Garzia, L., André, A., Amoresano, A., D'Angelo, A., Martusciello, R., Cirulli, C., Tsurumi, T., Marino, G., and Zollo, M. (2003). Method to express and purify nm23–H2 protein from baculovirus-infected cells. *BioTechniques* **35,** 384–391.

Gill, P., Kimpton, C. P., and Sullivan, K. (1992). A rapid polymerase chain reaction method for identifying fixed specimens. *Electrophoresis* **13,** 173–175.

Gräslund, S., Nordlund, P., Weigelt, J., Bray, J., Gileadi, O., Knapp, S., Oppermann, U., Arrowsmith, C., Hui, R., Ming, J., dhe-Paganon, S., Park, H.-W., *et al.* (2008). Protein production and purification. *Nat. Methods* **5,** 135–146.

Grisshammer, R., and Tucker, J. (1997). Quantitative evaluation of neurotensin receptor purification by immobilized metal affinity chromatography. *Protein Expr. Purif.* **11,** 53–60.

Hale, J. E., and Beidler, D. E. (1994). Purification of humanized murine and murine monoclonal antibodies using immobilized metal-affinity chromatography. *Anal. Biochem.* **222,** 29–33.

Hearon, J. (1948). The configuration of cobaltihistidine and oxy-bis (cobalthistidine). *J. Natl. Cancer Inst.* **9,** 1–11.
Hilbrig, F., and Freitag, R. (2003). Protein purification by affinity precipitation. *J. Chromatogr. B* **790,** 79–90.
Hirel, P.-H., Schmitter, J.-M., Dessen, P., Fayat, G., and Blanquet, S. (1989). Extent of N-terminal methionine excision from *Escherichia coli* proteins is governed by side-chain length of the penultimate amino acid. *Proc. Natl. Acad. Sci. USA* **86,** 8247–8251.
Hober, S., and Uhlén, M. (2008). Human protein atlas and the use of microarray technologies. *Curr. Opin. Biotechnol.* **19,** 30–35.
Hochuli, E. (1989). Genetically designed affinity chromatography using a novel metal chelate adsorbent. *Biologically Active Mol.* **411,** 217–239.
Hochuli, E., Döbeli, H., and Schacher, A. (1987). New metal chelate adsorbent for proteins and peptides containing neighbouring histidine residues. *J. Chromatogr.* **411,** 177–184.
Hochuli, E., Bannwarth, W., Döbeli, H., Gentz, R., and Stüber, D. (1988). Genetic approach to facilitate purification of recombinant proteins with a novel metal chelate adsorbent. *Biotechnology* **6,** 1321–1325.
Hubert, P., and Porath, J. (1980). Metal chelate affinity chromatography. I. Influence of various parameters on the retention of nucleotides and related compounds. *J. Chromatogr.* **198,** 247–255.
Hubert, P., and Porath, J. (1981). Metal chelate affinity chromatography. II. Group separation of mono- and dinucleotides. *J. Chromatogr.* **206,** 164–168.
Hunt, I. (2005). From gene to protein: A review of new and enabling technologies for multi-parallel protein production. *Protein Expr. Purif.* **40,** 1–22.
Imai, T., Tokunaga, A., Yoshida, T., Hashimoto, M., Mikoshiba, K., Weinmaster, G., Nakafuku, M., and Okano, H. (2001). The neural RNA-binding protein Musashi1 translationally regulates mammalian *numb* gene expression by interacting with its mRNA. *Mol. Cell. Biol.* **21,** 3888–3900.
Jensen, M. R., Lauritzen, C., Dahl, S. W., Pedersen, J., and Led, J. J. (2004). Binding ability of a HHP-tagged protein towards Ni^{2+} studied by paramagnetic NMR relaxation: The possibility of obtaining long-range structure information. *J. Biomol. NMR* **29,** 175–185.
Jin, S., Issel, C. J., and Montelaro, R. C. (2004). Serological method using recombinant S2 protein to differentiate equine infectious anemia virus (EIAV)-infected and EIAV-vaccinated horses. *Clin. Diagn. Lab. Immunol.* **11,** 1120–1129.
Jungbauer, A., Kaar, W., and Schlegl, R. (2004). Folding and refolding of proteins in chromatographic beds. *Curr. Opin. Biotechnol.* **15,** 487–494.
Kaslow, D. C., and Shiloach, J. (1994). Production, purification and immunogenicity of a Malaria transmission-blocking vaccine candidate: TBV25H expressed in yeast and purified using Ni-NTA Agarose. *Biotechnology* **12,** 494–499.
Kim, T.-W., Oh, I.-S., Ahn, J.-H., Choi, C.-Y., and Kim, D.-M. (2006). Cell-free synthesis and in situ isolation of recombinant proteins. *Protein Expr. Purif.* **45,** 249–254.
Kim, M. J., Park, H.-Y., Kim, J., Ryu, J., Hong, S., Han, S.-J., and Song, R. (2008). Western blot analysis using metal–nitrilotriacetate conjugated CdSe/ZnS quantum dots. *Anal. Biochem.* **379,** 124–126.
Klammt, C., Schwarz, D., Fendler, K., Haase, W., Dötsch, V., and Bernhard, F. (2005). Evaluation of detergents for the soluble expression of α-helical and β-barrel-type integral membrane proteins by a preparative scale individual cell-free expression system. *FEBS J.* **272,** 6024–6038.
Knecht, S., Ricklin, D., Eberle, A. N., and Ernst, B. (2009). Oligohis-tags: Mechanisms of binding to Ni^{2+}-NTA surfaces. *J. Mol. Recognit.* **22,** 270–279.
Knight, R. D., and Shimeld, S. M. (2001). Identification of conserved C2H2 zinc-finger gene families in the bilateria. *Genome Biol.* **2,** 1–8.
Koehn, J., and Hunt, I. (2009). High-throughput protein production (HTPP): A review of enabling technologies to expedite protein production. *Methods Mol. Biol.* **498,** 1–18.

In "High Throughput Protein Expression and Purification" (S. A. Doyle, ed.). Humana Press, Springer.

Lanio, T., Jeltsch, A., and Pingoud, A. (2000). Automated purification of His_6-tagged proteins allows exhaustive screening of libraries generated by random mutagenesis. *BioTechniques* **29,** 338–342.

Lata, S., and Piehler, J. (2005). Stable and functional immobilization of histidine-tagged proteins via multivalent chelator headgroups on a molecular poly(ethylene glycol) brush. *Anal. Chem.* **77,** 1096–1105.

Lata, S., Gavutis, M., Tampé, R., and Piehler, J. (2006). Specific and stable fluorescence labeling of histidine-tagged proteins for dissecting multi-protein complex formation. *J. Am. Chem. Soc.* **128,** 2365–2372.

Lesley, S. A. (2001). High-throughput proteomics: Protein expression and purification in the postgenomic world. *Protein Expr. Purif.* **22,** 159–164.

Levison, P. R., Badger, S. E., Jones, R. M. H., Toome, D. W., Streater, M., Pathirana, N. D., and Wheeler, S. (1995). Validation studies in the regeneration of ion-exchange celluloses. *J. Chromatogr. A* **702,** 59–68.

Lewinson, O., Lee, A. T., and Douglas, C. R. (2008). The funnel approach to the precrystallization production of membrane proteins. *J. Mol. Biol.* **377,** 62–73.

Loo, J. A. (2003). The tools of proteomics. *Adv. Protein Chem.* **65,** 353–369.

Lv, G. S., Hua, G.C, and Fu, X. Y. (2003). Expression of milk-derived antihypertensive peptide in *Escherichia coli*. *J. Dairy Sci.* **86,** 1927–1931.

Ma, D., Watanabe, H., Mermelstein, F., Admon, A., Oguri, K., Sun, X., Wada, T., Imai, T., Shiroya, T., Reinberg, D., and Handa, H. (1993). Isolation of a cDNA encoding the largest subunit of TFIIA reveals functions important for activated transcription. *Genes Dev.* **7,** 2246–2257.

Mateo, C., Fernandez-Lorente, G., Pessela, B. C. C., Vian, A., Carrascosa, A. V., Garcia, J. L., Fernandez-Lafuente, R., and Guisan, J. M. (2001). Affinity chromatography of polyhistidine tagged enzymes: New dextran-coated immobilized metal ion affinity chromatography matrices for prevention of undesired multipoint adsorptions. *J. Chromatogr. A* **915,** 97–106.

Matiasson, B., Kumar, A., Ivanov, A. E., and Galaev, I. Y. (2007). Metal-chelate affinity precipitation of proteins using responsive polymers. *Nat. Protocols* **2,** 213–220.

Mészárosová, K., Tishchenko, G., Bouchal, K., and Bleha, M. (2003). Immobilized-metal affinity sorbents based on hydrophilic methacrylate polymers and their interaction with immunoglobulins. *React. Funct. Polym.* **56,** 27–35.

Mohanty, A. K., and Wiener, M. C. (2004). Membrane protein expression and production: Effects of polyhistidine tag length and position. *Protein Expr. Purif.* **33,** 311–325.

Muszynska, G., Andersson, L., and Porath, J. (1986). Selective adsorption of phosphoproteins on gel-immobilized ferric chelate. *Biochemistry* **25,** 6850–6853.

Nieba, L., Nieba-Axamann, S. E., Persson, A., Hämäläinen, M., Edebratt, F., Hansson, A., Lidholm, J., Magnusson, K., Karlsson, A. F., and Plückthun, A. (1997). Biacore analysis of histidine-tagged proteins using a chelating NTA sensor chip. *Anal. Biochem.* **252,** 217–228.

Nilsson, J., Ståhl, S., Lundeberg, J., Uhlén, M., and Nygren, P.-A. (1997). Affinity fusion strategies for detection, purification, and immobilization of recombinant proteins. *Protein Expr. Purif.* **11,** 1–16.

Padan, E., Venturi, M., Michel, H., and Hunte, C. (1998). Production and characterization of monoclonal antibodies directed against native epitopes of NhaH, the Na^+/H^+ antiporter of *E. coli*. *FEBS Lett.* **441,** 53–58.

Pedersen, J., Lauritzen, C., Madsen, M. T., and Dahl, S. W. (1999). Removal of N-terminal polyhistidine tags from recombinant proteins using engineered aminopeptidases. *Protein Expr. Purif.* **15,** 389–400.

Polayes, D.A., Parks, T.D., Johnston, S.A., and Dougherty, W.G. (2008). Application of TEV protease in protein production. *Methods Mol. Med.* **13,** 169–183. *In* "Molecular Diagnosis of Infectious Diseases" (U. Reischl. ed.). Humana Press, Springer.

Porath, J., and Olin, B. (1983). Immobilized metal ion affinity adsorption and immobilized metal ion affinity chromatography of biomaterials. Serum protein affinities for gel-immobilized iron and nickel ions. *Biochemistry* **29,** 1621–1630.

Porath, J., Carlsson, J., Olsson, I., and Belfrage, G. (1975). Metal chelate affinity chromatography, a new approach to protein fractionation. *Nature* **258,** 598–599.

Posewitz, M. C., and Tempst, P. (1999). Immobilized gallium (III) affinity chromatography of phosphopeptides. *Anal. Chem.* **71,** 2883–2892.

Prinz, B., Schultchen, J., Rydzewski, R., Holz, C., Boettner, M., Stahl, U., and Lang, C. (2004). Establishing a versatile fermentation procedure for human proteins expressed in the yeasts *Saccharomyces cerevisiae* and *Pichia pastoris* for structural genomics. *J. Struct. Funct. Genomics* **5,** 29–44.

QIAGEN (2003). The QIAexpressionist. Handbook for High-Level Expression and Purification of 6xHis-Tagged Proteins. 5th edn. Hilden, Germany: QIAGEN, Hilden, Germany.

Reichel, A., Schaible, D., Al Furoukh, N., Cohen, M., Schreiber, G., and Piehler, J. (2007). Noncovalent, site-specific biotinylation of histidine-tagged proteins. *Anal. Chem.* **79,** 8590–8600.

Rumbley, J. N., Furlong Nickels, E., and Gennis, R. B. (1997). One-step purification of cytochrome bo_3 from *Escherichia coli* and demonstration that associated quinone is not required for the structural integrity of the oxidase. *Biochim. Biophys. Acta* **1340,** 131–142.

Schäfer, F., Blümer, J., Römer, U., and Steinert, K. (2000). Ni-NTA for large-scale processes—systematic investigation of separation characteristics, storage and CIP conditions, and leaching. QIAGEN. *QIAGEN News* **4,** 11–15. Available from http://www1.qiagen.com/literature/qiagennews/0400/Ni-NTA%20for%20large-scale.pdf

Schäfer, F., Schäfer, A., and Steinert, K. (2002a). A highly specific system for efficient enzymatic removal of tags from recombinant proteins. *J. Biomol. Tech.* **13,** 158–171.

Schäfer, F., Römer, U., Emmerlich, M., Blümer, J., Lubenow, H., and Steinert, K. (2002b). Automated high-throughput purification of 6xHis-tagged proteins. *J. Biomol. Tech.* **13,** 131–142.

Scheich, C., Sievert, V., and Büssow, K. (2003). An automated method for high-throughput protein purification applied to a comparison of His-tag and GST-tag affinity chromatography. *BMC Biotechnol.* **3,** 12. Available from www.biomedcentral.com/1472-6750/3/12

Schmitt, J., Hess, H., and Stunnenberg, H. G. (1993). Affinity purification of histidine-tagged proteins. *Mol. Biol. Rep.* **18,** 223–230.

Serpa, G., Augusto, E. F. P., Tamashiro, W. M. S. C., Ribeiro, M. B., Miranda, E. A., and Bueno, S. M. A. (2005). Evaluation of immobilized metal membrane affinity chromatography for purification of an immunoglobulin G_1 monoclonal antibody. *J. Chromatogr. B* **816,** 259–268.

Shi, W., and Chance, M. R. (2008). Metallomics and metalloproteomics. *Cell. Mol. Life Sci.* **65,** 3040–3048.

Shi, Y., Seto, E., Chang, L.-S., and Shenk, T. (1991). Transcriptional repression by YY1, a human GLI-Krüppel-related protein, and relief of repression by adenovirus E1A protein. *Cell* **67,** 377–388.

Slentz, B. E., Penner, N. A., and Regnier, F. E. (2003). Protein proteolysis and the multi-dimensional electrochromatographic separation of histidine-containing peptide fragments. *J. Chromatogr. A* **984,** 97–103.

Stasyk, T., and Huber, L. A. (2004). Zooming in: Fractionation strategies in proteomics. *Proteomics* **4,** 3704–3716.

Steen, J., Uhlén, M., Hober, S., and Ottosson, J. (2006). High-throughput protein purification using an automated set-up for high-yield affinity chromatography. *Protein Expr. Purif.* **46,** 173–178.

Stowers, A. W., Zhang, Y., Shimp, R. L., and Kaslow, D. C. (2001). Structural conformers produced during malaria vaccine production in yeast. *Yeast* **18,** 137–150.

Strömberg, P., Rotticci-Mulder, J., Björnestedt, R., and Schmidt, S. R. (2005). Preparative parallel protein purification (P4). *J. Chromatogr. B* **818,** 11–18.

Sulkowski, E. (1985). Purification of proteins by IMAC. *Trends Biotechnol.* **3,** 1–7.

Sun, X., Chiu, J.-F., and He, Q.-Y. (2005). Application of immobilized metal affinity chromatography in proteomics. *Expert Rev. Proteomics* **2,** 649–657.

Svensson, J., Andersson, C., Reseland, J. E., Lyngstadaas, P., and Bülow, L. (2006). Histidine tag fusions increases expression levels of active recombinant amelogenin in *Escherichia coli*. *Protein Expr. Purif.* **48,** 134–141.

Vancan, S., Miranda, E. A., and Bueno, S. M. A. (2002). IMAC of human IgG: Studies with IDA-immobilized copper, nickel, zinc, and cobalt ions and different buffer systems. *Process Biochem.* **37,** 573–579.

Walsh, P. S., Metzger, D. A., and Higuchi, R. (1991). Chelex 100 as a medium for simple extraction of DNA for PCR-based typing from forensic material. *BioTechniques* **10,** 506–513.

Woestenenk, E. A., Hammarström, M., van den Berg, S., Härd, T., and Berglund, H. (2004). His tag effect on solubility of human proteins produced in *Escherichia coli*: Comparison between four expression vectors. *J. Struct. Funct. Genomics* **5,** 217–229.

Xu, C.-G., Fan, X.-J., Fu, Y.-J., and Liang, A.-H. (2008). Effect of the His-tag on the production of soluble and functional *Buthus martensii* Karsch insect toxin. *Protein Expr. Purif.* **59,** 103–109.

Zhaohua, H., Park, J. I., Watson, D. S., Hwang, P., and Szoka, F. C. (2006). Facile synthesis of multivalent nitrilotriacetic acid (NTA) and NTA conjugates for analytical and drug delivery applications. *Bioconjug. Chem.* **17,** 1592–1600.

Zhou, H., Xu, S., Ye, M., Feng, S., Pan, C., Jiang, X., Li, X., Han, G., Fu, Y., and Zou, H. (2006). Zirconium phosphonate-modified porous silicon for highly specific capture of phosphopeptide and MALDI-TOF MS analysis. *J. Proteome Res.* **5,** 2431–2437.

CHAPTER TWENTY-EIGHT

Identification, Production, and Use of Polyol-Responsive Monoclonal Antibodies for Immunoaffinity Chromatography

Nancy E. Thompson, Katherine M. Foley, Elizabeth S. Stalder, *and* Richard R. Burgess

Contents

1. Introduction 476
2. Polyol-Responsive Monoclonal Antibodies 477
   2.1. Properties of a PR-mAb 478
   2.2. Source of mAbs 479
   2.3. Identification of PR-mAbs by ELISA-elution assay 480
   2.4. Producing mAbs in continuous culture 482
   2.5. Purification of the antibody 485
   2.6. Immobilization of PR-mAbs on a chromatography support 487
   2.7. Purification of proteins with PR-mAbs 488
   2.8. Purification of proteins using cross-reacting PR-mAbs 490
   2.9. Use of epitopes of PR-mAbs as purification tags 491
3. Conclusions 492
Disclosure 493
References 493

Abstract

Immunoaffinity chromatography is a powerful tool for purification of proteins and protein complexes. The availability of monoclonal antibodies (mAbs) has revolutionized the field of immunoaffinity chromatography by providing a continuous supply of highly uniform antibody. Before the availability of mAbs, the recovery of the target protein from immobilized polyclonal antibodies usually required very harsh, often denaturing conditions. Although harsh conditions are often still used to disrupt the antigen–antibody interaction when using a mAb, various methods have been developed to exploit the uniformity of the antigen–antibody reaction in order to identify agents or conditions that gently disrupt

McArdle Laboratory for Cancer Research, University of Wisconsin–Madison, Madison, Wisconsin, USA

Methods in Enzymology, Volume 463

ISSN 0076-6879, DOI: 10.1016/S0076-6879(09)63028-7

this interaction and thus result in higher recovery of active protein from immunoaffinity chromatography. We discuss here the use of a specific type of monoclonal antibody that we have designated "polyol-responsive monoclonal antibodies" (PR-mAbs). These are naturally occurring mAbs that have high affinity for the antigen under binding conditions, but have low affinity in the presence of a combination of low molecular weight hydroxylated compounds (polyols) and nonchaotropic salts. Therefore, these PR-mAbs can be used for gentle immunoaffinity chromatography. PR-mAbs can be easily identified and adapted to a powerful protein purification method for a target protein.

1. Introduction

All forms of affinity chromatography are defined by a specific interaction between two components that allows the purification of one of the components. Immunoaffinity chromatography is a subset of the affinity chromatography principle where the specific interaction of an antigen with an antibody is employed (for review, see Subramanian, 2002). Immunoaffinity chromatography is really a scaled up extension of an immunoprecipitation procedure, except that one of the components (generally the antigen) is recovered after the chromatography as an active protein.

The ability to produce monoclonal antibodies (mAbs) has revolutionized the field of immunochemistry (for review, see Nelson *et al.*, 2000). When considering immunoaffinity chromatography, two features give mAbs an advantage over polyclonal antibodies derived from immune serum. First, the mAb is a reproducible reagent that can be prepared in large quantities. Second, a mAb is a homogeneous population that responds uniformly to an eluting reagent. Generally, the purified antibody is conjugated to some type of bead and the antigen-containing solution is applied to the bead (in batch or in a column). After washing away unbound or loosely bound material, the antigen is eluted from the bead.

The elution step is usually the most difficult obstacle to overcome in developing an immunoaffinity chromatography procedure. Antigen–antibody interactions are generally a result of a combination of ionic, hydrophobic, and hydrogen bonds formed between amino acids in the specific antigenic determinant of the antigen (epitope) and the protein loops of the complementarity determining regions (CDRs), which are located in the variable regions of the heavy and light chain of the antibody molecule. The ideal way to gently elute the antigen is by competition with a peptide containing the epitope for the antibody. However, the epitope is not always known, and a peptide is not always available or is too expensive to synthesize in the quantities needed. In addition, sometimes the antibody reacts with a discontinuous epitope, and the epitope cannot be mimicked by a synthetic peptide. In these cases, the antigen is usually eluted with very harsh conditions (high or low pH values, denaturants such as urea, or an ionic detergent.) that can inactivate the protein.

2. Polyol-Responsive Monoclonal Antibodies

We have pioneered the use of a specific type of monoclonal antibody for use in immunoaffinity chromatography. These mAbs can be used for "gentle" immunoaffinity chromatography because the elution conditions require only a combination of a nonchaotropic salt and a low molecular weight polyhydroxylated compound (polyol), conditions that are regarded as nondenaturing to proteins. We have referred to this type of antibody as a "polyol-responsive" mAb (PR-mAb).

Our laboratory studies proteins involved in transcription. Therefore, most of the PR-mAbs that we have isolated have been mAbs that react with proteins involved in transcription in either prokaryotic or eukaryotic systems. Eukaryotic transcription systems pose a significant challenge for the separation scientist because many of the factors are actually multisubunit proteins. For example, eukaryotic RNA polymerase II (RNAP II) contains 12 subunits (12 different gene products); however, for initiation from a promoter, RNAP II also requires transcription factors TFIIA, TFIIB, TFIID, TFIIE, TFIIF, and TFIIH (for review, see Woychik and Hampsey, 2002). With the exception of TFIIB, all of these transcription factors are comprised of two or more subunits. Large protein complexes are ideal subjects for the use of PR-mAb immunoaffinity chromatography. We have developed PR-mAbs for *E. coli* RNAP (Thompson *et al.*, 1992) and eukaryotic RNAP II (Thompson *et al.*, 1990), as well several eukaryotic transcription factors (Table 28.1).

Perhaps the most notable use of PR-mAb technology was the use of our PR-mAb 8WG16 for use to purify large amounts of yeast RNAP II (Edwards *et al.*, 1990) for use in crystallization of this protein complex (Cramer *et al.*, 2000). Other laboratories have successfully isolated PR-mAbs for purifying their protein of interest (Jiang *et al.*, 1995; Lynch *et al.*, 1996; Nagy *et al.*, 2002). It is also possible to screen already established collections of mAbs for the polyol-responsive property.

Several reviews have been written concerning PR-mAbs (Burgess and Thompson, 2002; Thompson and Burgess, 1996, 2001; Thompson *et al.*, 2006). In this report, we will update some of these methods. We will also describe our method for producing large quantities of PR-mAbs by a relatively inexpensive cell culture system, thus minimizing the need to prepare the mAbs in mouse ascites fluid. Finally, because polyol-responsiveness is a property of the antibody, an unrelated protein of interest can be tagged with the epitope for

Table 28.1 Polyol-responsive monoclonal antibodies that have been used to purify proteins involved with transcription

PR-mAb	Antigen	Epitope	Reference
8WG16	Heptapeptide repeat on the C-terminus of the largest subunit of RNAP II	YSPTSPSYSPTSPS	Edwards *et al.* (1990) and Thompson *et al.* (1990)
NT73	Far C-terminus of the β' subunit of *E. coli* RNAP (1392–1404)	SLAELLNAGLGGS	Thompson *et al.* (1992, 2003)
8RB13	β-flap region of the β subunit of *E. coli* RNAP	PEEKLLRAIFGEKAS	Bergendahl *et al.* (2003), Probasco *et al.* (2007), and E. S. Stalder *et al.* (unpublished)
4RA2	C-terminal half of the α subunit of *E. coli* RNAP	Unknown	Anthony *et al.* (2003)
1TBP22	N-terminus of human TBP (1–199)	Unknown	Thompson *et al.* (2004)
IIB8	N-terminus of human TFIIB (61–68)	TKDPSRVG	Duellman *et al.* (2004) and Thompson and Burgess (1994)
1RAP1	N-terminus of RAP30 (1–118)	Unknown	Thompson and Burgess (1999)

a PR-mAb by recombinant DNA methodologies, and the protein purified by this gentle polyol-elution method.

2.1. Properties of a PR-mAb

1. By screening a large number of mAbs (218 antigen-specific wells), we have estimated that PR-mAbs represent 5–10% of mAbs (Thompson *et al.*, 1992).

2. Screening can be performed at the master-well stage in order to immediately identify PR-mAbs.
3. PR-mAbs are not limited to any subisotype of mouse IgG. PR-mAbs have also been identified in a collection of rat mAbs (R. R. Burgess, unpublished data).
4. Most PR-mAbs respond to a variety of different combinations of salt and polyol. Most commonly we use 0.75 *M* ammonium sulfate or 0.75 *M* NaCl combined with 30–40% propylene glycol.
5. PR-mAbs can be high-affinity antibodies. In fact, high affinity is preferred in order for the mAb to be effective in binding the antigen in a dilute solution. These mAbs become low-affinity antibodies under the elution conditions.
6. Most (but not all) PR-mAbs respond to a variety of combinations of salt and polyol. Salts that we have tested are ammonium sulfate, sodium chloride, sodium acetate, and potassium glutamate. Polyol that we have tested are propylene glycol, ethylene glycol, 2,3-butanediol, and in some cases, glycerol.
7. Because these salts and polyols are often used as protein stabilizers, the gentle elution results in antigens that retain biological properties and structural integrity, even for multisubunit complexes (Cramer *et al.*, 2000; Thompson *et al.*, 1990).

2.2. Source of mAbs

We produce our mAbs by a typical hybridoma procedure (Harlow and Lane, 1988); that is by the fusion of antigen-stimulated plasma cells from a mouse spleen and a myeloma cell line. We find that a hyperimmunized mouse is preferred; only about 5–10% of mAbs are PR-mAbs, a large number of original hybridomas will make it more likely that a PR-mAb will be isolated. However, it is possible to use mAbs that are produced by other methods, such as by retroviral infection of plasma cells (Largaespada *et al.*, 1996) or construction of antibody libraries by recombinant methods.

Regardless of the source of antibodies, antigen-specific antibodies must be identified. We find that a standard enzyme-linked immunosorbent assay (ELISA) works well for this. We then screen the antibodies by a modified ELISA, which we have termed "ELISA-elution assay." This procedure is based on a standard ELISA except that the specific antigen–antibody complex is treated with a combination of salt and polyol before the enzyme-conjugated secondary antibody is added. The subsequent reaction with the substrate gives an estimation of the amount of antigen–antibody complex that is dissociated by the salt/polyol treatment. A general procedure for the ELISA-elution assay is as follows.

2.3. Identification of PR-mAbs by ELISA-elution assay

1. Coat a polystyrene microtiter plate with antigen. We typically use 30–100 ng of antigen contained in 50 μl of phosphate buffered saline (PBS, pH 7.4) per well. The antigen is allowed to incubate at room temperature for 1 h to allow sufficient time for the antigen to bind to the polystyrene.
2. Each well is then blocked with 200 μl of 1% nonfat dry milk in PBS (1% BLOTTO). We typically block overnight at 4 °C, but 2 h at room temperature is sufficient.
3. The test antibody (50 μl) is then added to two adjacent wells, and incubated at room temperature for 1 h. If the antibody is contained in cell culture fluid, it usually can be used directly. However, very high affinity antibodies, or very high titer antibody preparations should be diluted so that nonsaturating levels are used.
4. The wells are washed five times with PBS containing 0.1% Tween 20 (PBST) to remove the unbound antibody.
5. To the control well, 100 μl of TE buffer (50 m*M* Tris–HCl, pH 7.9 and 0.1 m*M* EDTA) is added, and to the test well, 100 μl of TE buffer containing 0.75 *M* ammonium sulfate and 40% propylene glycol is added. The plate is incubated at room temperature for 20 min, with occasional (about every 5 min) tapping of the side of the plate to mix the contents.
6. The wells are washed five times with PBST.
7. A commercially available horseradish peroxidase-conjugated secondary antibody is diluted into 1% BLOTTO (generally 1:2000) and 50 μl is added to each well. The plate is incubated at room temperature for 1 h.
8. The wells are washed 10 times with PBST.
9. The appropriate substrate is added to the wells. We use 100 μl of 0.03% H_2O_2 and 0.4 mg/ml *o*-phenylenediamine (OPD) contained in 0.05 *M* citrate buffer (pH 5.0).
10. The plate is incubated 5–15 min at room temperature. The reactions are stopped in pairs (TE and TE + salt/polyol for each mAb) with 50 μl of 1 *M* H_2SO_4 per well.
11. The absorbance is read on a microtiter plate reader; for OPD we use 490 nm. Treatment with the polyol and salt will reduce the absorbance of a PR-mAb in the test well by about 50% compared to the control well that received just TE buffer (Fig. 28.1A) If a plate reader is not available, the approximately 50% reduction is usually evident by visual inspection.

2.3.1. Comments

1. Screening for PR-mAbs can be performed at the master-well stage, immediately after the hybridomas are screened for specific antibody

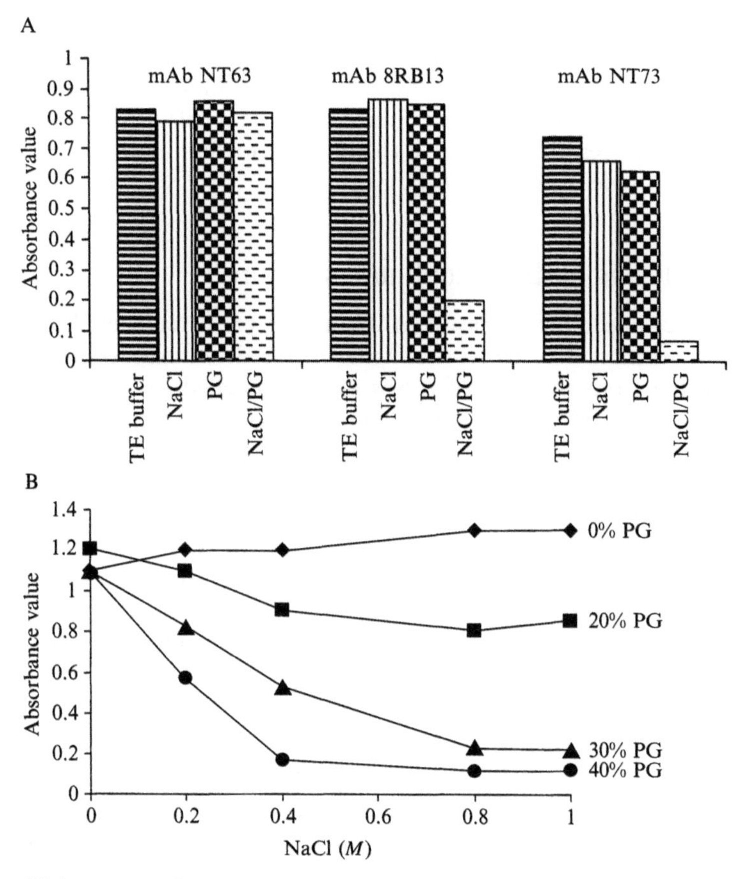

Figure 28.1 ELISA-elution assay to identify and characterize PR-mAbs. Each well of the microtiter plate was coated with 100 ng core RNAP. (A) ELISA-elution assay of previously identified PR-mAb. mAb NT63 is a control mAb that reacts with *E. coli* RNAP (Thompson *et al.*, 1992). mAbs 8RB13 and NT73 are PR-mAbs (Bergendahl *et al.*, 2003; Thompson *et al.*, 1992). mAb NT63 did not elute from the antigen with 0.75 *M* NaCl and 40% propylene glycol (NaCl/PG), but mAbs 8RB13 and NT73 did elute with the salt/polyol combination. None of the mAbs eluted well with 0.75 *M* NaCl or 40% propylene glycol (PG) alone. (B) ELISA-elution assay using mAb 8RB13 and varying concentrations of both NaCl (0–1.0 *M*) and propylene glycol (0–40%).

production. The preliminary screening can be performed with 100 μl of cell culture fluid (50 μl for the buffer control and 50 μl for the buffer containing polyol and salt). In one study (Thompson *et al.*, 1992), we screened over 200 hybridomas for PR-mAb from a single fusion at the master-well stage.

2. Binding of the antigen to the microtiter plate can result in distortion of the antigen structure. This can expose buried epitopes that are not

accessible in the protein when it is in solution. This can result in a "false-positive" because the mAb does not react with the target when it is in solution and is not a useful mAb for immunoaffinity chromatography.

3. When a few milliliters of cell culture fluid are available, it is possible to examine the response of a mAb to different concentrations of polyol and salt on a single plate (Fig. 28.1B).

2.4. Producing mAbs in continuous culture

The practice of producing mouse mAbs in ascites fluid has been discouraged in many settings. Therefore, we have explored alternatives methods to produce the large amounts of mAbs that are needed for immunoaffinity chromatography. We describe here one method that we have found to be particularly useful for this scale of antibody production. This protocol uses a commercially available cell culture chamber called a CELLine Flask 350 (CL 350) manufactured by Integra Biosciences AG (Switzerland). In the United States, this product is distributed by Argos Technologies, Inc (Elgin, IL) and Bioraco International (Framingham, MA). We found this product to be easy to use, capable of producing 10–50 mg of antibody, and reusable several times for the same hybridoma. Standard cell culture aseptic techniques are required, along with a cell culture hood and a 37 °C humidified CO_2 incubator (5%). A schematic of this culture flask is shown in Fig. 28.2A. The general protocol is as follows:

1. *Preparing the cell culture media*: Two slightly different media are prepared for the two chambers of the CL 350 flask. Dulbecco's Modified Eagle Medium (DMEM) containing glutamine and high glucose (Gibco/Invitrogen #11965), supplemented with 1 m*M* sodium pyruvate (Sigma), 100 units of penicillin/ml, and 100 μg/ml of streptomycin (Gibco/Invitrogen) is used as the base for the two media. For the cell growth chamber, DMEM with the supplements indicated above is also supplemented with 15% heat-inactivated fetal bovine serum (Hyclone). The nutrient chamber uses DMEM with the supplements indicated above and 5% fetal bovine serum. We refer to the medium for the growth chamber as "complete medium" and the medium for the nutrient chamber as "nutrient medium."
2. *Preparing the inoculum for seeding the flask*: Remove a vial containing the hybridoma from liquid N_2 and thaw rapidly at 37 °C. Plate the hybridoma in complete medium at about 2×10^4 cells/ml. We use 10-cm cell culture plates containing 20 ml of medium.
3. *Seeding the flask*: Prepare 8×10^6 to 20×10^6 viable cells in log growth phase in 5 ml of fresh complete medium. Place 25 ml of nutrient medium in the nutrient compartment (green cap) to wet the membrane between the nutrient chamber and the cell chamber before the cells are

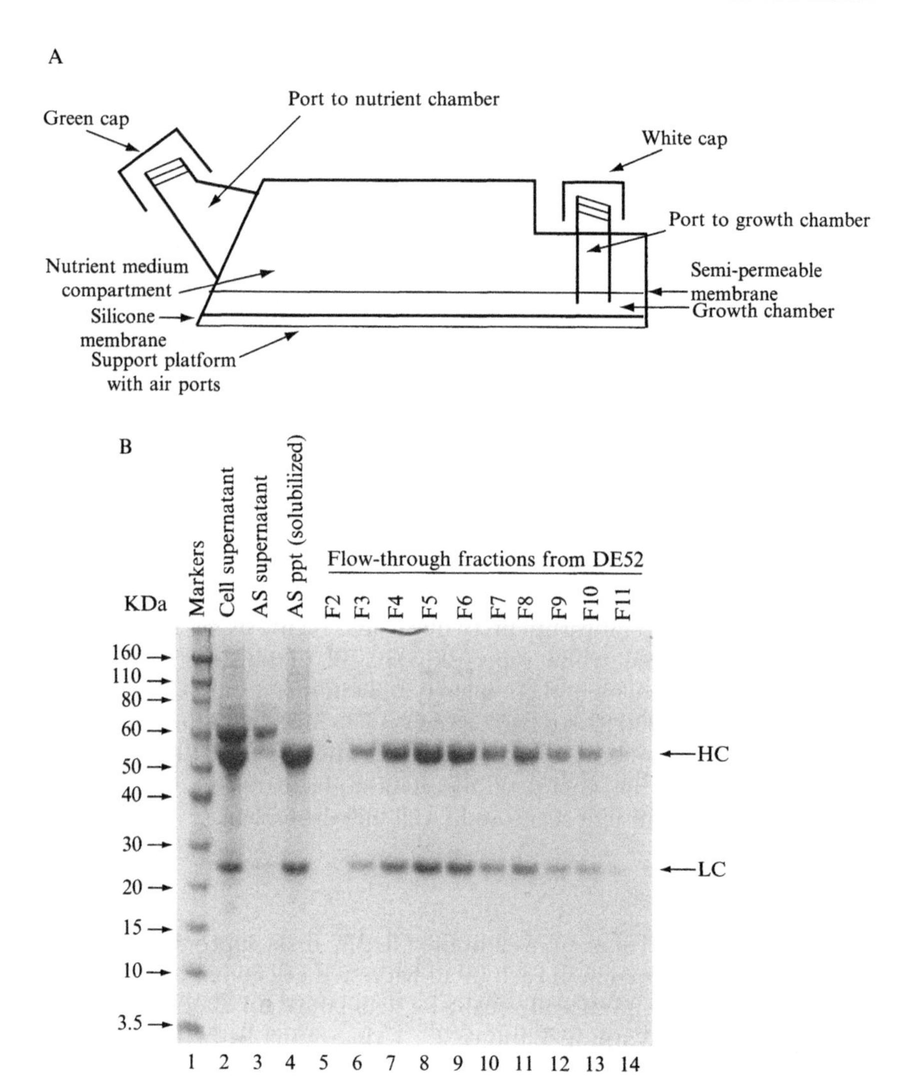

Figure 28.2 mAb production in CELLine flasks. (A) Diagrammatic representation of the CL 350 flask (adapted from the manufacturer's literature). (B) The hybridoma that produces the 8RB13 mAb was cultured in the flask for 3 weeks. The cell supernatants that were collected every 3 days were pooled (Lane 2) and purified using ammonium sulfate precipitation (Lanes 3 and 4) and chromatography on DE52 (Lanes 5–14) as described in the text. The fractions were run on a 4–12% NuPAGE gel (Invitrogen), using MES buffer; the gel was stained with GelCode (Pierce). The immunoglobulin heavy chain (HC) and light chain (LC) are indicated.

placed in the cell chamber. Suspend the cell preparation and aspirate into a 10-ml serological pipette. With the green cap loose, inoculate 5 ml suspension into cell compartment (white cap) by inserting the pipette firmly into the cell compartment port. Remove trapped air bubbles by pipetting the fluid up and down slowly to allow the bubbles to rise prior to returning only the fluid to the cell chamber. Replace and tighten the white cap. Add 350 ml of nutrient medium to the nutrient compartment and completely tighten the green cap.

4. *Cell compartment harvest and culture maintenance*: Every 3–7 days remove nutrient medium from nutrient medium compartment. With green cap loose, insert 10 ml pipette into the cell compartment and pipette fluid up and down to thoroughly mix the cells. Then remove entire cell compartment liquid volume to a centrifuge tube. Volume may be greater than 5 ml due to osmotic flux. Take a sample for a cell count and cell viability using a hemocytometer and trypan blue. Centrifuge the material removed from the cell compartment to pellet the cells. Remove and save the medium, which contains the mAbs; this can be frozen for purification at a later date. Suspend the cells with fresh complete medium. Depending upon the initial inoculation density, growth rate, and maintenance frequency the cells may need to be split back (usually 1:2–1:4) at this point. With the green cap loosened, return 5 ml of cells back to the cell compartment (white cap). Remove air bubbles and completely tighten white cap. Add 350 ml of nutrient medium to nutrient compartment and completely tighten green cap.
5. The Integra CL 350 flask is harvested every 3–7 days after the culture has been established. Harvest intervals depend on the growth rate of the hybridoma and the ability of the hybridoma to adapt to the flask environment. This trait seems to be cell line dependent.

2.4.1. Comments

1. For hybridomas that grow well in the CL 350 flask, approximately 0.5–1 mg of mAb is present in each ml of harvested cell culture supernatant. The continuous culture can usually be maintained for about 1 month.
2. *Liquid handling*: Warm medium to 37 °C in a water bath. This helps to prevent condensation on the flask and temperature shocking the cells. When adding or removing liquid from the cell compartment (white cap), always loosen the green cap of the nutrient medium compartment first to prevent air lock. Always tighten both white and green caps before placing the flask into the incubator. The use of 10 ml serological pipettes is recommended for all cell compartment manipulations. The medium in the nutrient medium chamber can be exchanged by aspirating the medium out and pouring fresh medium in.
3. The minimum cell concentration is 1.5×10^6 cells/ml for the inoculum (Step 2). We have not been successful in reducing the need for fetal bovine serum in the nutrient medium. Some of the new serum-free media on the market may be used as the nutrient medium.

4. It is important to track cell numbers and health (Step 3). This helps to determine whether to split cells to reduce numbers once the total viable cell count is greater than 100×10^6 cells. If the cell viability is greatly reduced, the maintenance frequency needs to be increased.
5. Because the cells are split back every time the culture is harvested, the percentage of viable cells (Step 4) will decrease during the continuous culture due to death of the cells. It is not unusual to end up with only 30–40% viability at the end of the culture period.
6. Some hybridomas are unstable in this long-term, continuous culture and lose the ability to produce the mAb. Therefore, the antibody production should be monitored during the culturing. We do this by a standard ELISA.
7. Information on CELLine flasks can be obtained at: www.integra-biosciences.com/celline.

2.5. Purification of the antibody

It is usually advantageous to at least partially purify the antibody in order to maximize the binding of the antibody to the available reactive sites on the support during immobilization. Antibodies can be purified by many different methods. Some of the most common, low-cost methods are by size-exclusion or ion-exchange chromatography. A common, but more costly method of purification is by affinity chromatography on a column containing either protein A or protein G, or a mixture of the two. Mouse mAbs that belong to the IgG class of immunoglobulins belong to one of four subclasses: IgG_1, IgG_{2a}, IgG_{2b}, and IgG_3. Mouse IgG_{2a}, IgG_{2b}, and IgG_3 can be purified on a column containing protein A. Mouse mAbs of the IgG_1 subclass do not bind to protein A very well. These mAbs bind better to protein G, but not as well as mAbs of the mouse IgG_{2a} and IgG_{2b} bind to either protein A or protein G.

In our experience, the majority of the mAbs identified in a typical fusion are of the IgG_1 subclass. We describe here a simple method for purifying mouse IgG mAbs on an inexpensive DEAE column (Whatman DE52). This method is particularly useful for IgG_1 mAbs. In fact, every IgG_1 mAb that we have tested can be purified to about 90% purity. In addition, many mouse IgG_{2a} and IgG_{2b} antibodies can be purified by this method.

1. Antibodies contained in either ascites fluid or concentrated cell supernatant from the CELLine Flask are precipitated by ammonium sulfate. A saturated solution of ammonium sulfate is added to the mAb until 45% saturation is reached. The slurry is allowed to stir for 20 min on ice. This precipitates the mAb but leaves most of the serum albumin in solution.
2. The precipitate is removed by centrifugation (about $6000 \times g$, 10 min). Antibody buffer (50 m*M* Tris–HCl, pH 6.9, 25 m*M* NaCl) is added to

the pellet to obtain one-fourth (for the CELLine flask-prepared antibody) to one-half (for ascites fluid) of the volume of the original antibody preparation.

3. The precipitate is allowed to solubilize for about 15 min, and clarified by centrifugation (6000×*g*, 10 min).
4. The supernatant from Step 3 is then dialyzed against 1 l of antibody buffer overnight at 4 °C.
5. The clarified supernatant is applied to a column of DE52 that has been prepared as discussed in "comments" below. The column is equilibrated in antibody buffer (pH 6.9), and at this pH most of the mAb flows through the column while most impurities bind. For 10 ml of the starting material, we use a 5-ml column of DE52.
6. Fractions (about 1 ml) of material that does not bind to the column are collected, and samples of each fraction are subjected to SDS–PAGE. The SDS–PAGE in Fig. 28.2B shows the fractions from the purification of mAb 8RB13 (an IgG_1 mAb) that was produced in a CL 350 flask. Fractions are pooled according to purity of the antibody product.
7. The column is then eluted with 5 ml of antibody buffer containing 0.5 *M* NaCl. Save this eluate until the fractions are analyzed by PAGE.

2.5.1. Comments

1. The preparation of the DE52 is particularly important. Although the manufacturer states that precycling is not necessary, we have found that precycling greatly improves the performance of the chromatography. Ten grams of resin should be resuspended in 100 ml of water and washed several times with 100 ml of water, removing the fines (the small particles that do not settle easily) with each washing. The resin is then treated with 100 ml of 0.1 *M* HCl for 30 min. The acid is decanted and the resin washed at least three times with 100 ml of water for each wash. After the last wash is decanted, the resin is treated with 0.1 *M* NaOH for 30 min. The base is decanted, and the resin is washed at least three times with 100 ml of water for each wash. The resin is then washed three times with 100 ml of the antibody buffer and resuspended in the same buffer. The pH is checked with pH paper, and NaN_3 is added to 0.02%. Resin is distributed to disposable tubes and stored in the refrigerator.
2. If the mAb is made in ascites fluid, the product is not as clean (approximately 80–90% pure), but the impurities do not seem to interfere with the performance of the immunoaffinity resin.
3. Under the conditions described above, most mouse mAbs will flow through the DE52. However, a few mAbs will bind under these conditions. Therefore, the high salt elution (Step 7) will elute the mAb. It will

be necessary to use a salt gradient to purify a mAb that binds to the DE52 column.

4. If the mAb is made in the CELLine flask, the DE52 step might not be necessary.

2.6. Immobilization of PR-mAbs on a chromatography support

Numerous resins and coupling chemistries are available through commercial suppliers. We have tested a number of these resins, but have not found that any of them work better than cross-linked agarose that has been derivatized with cyanogen bromide (CNBr).

1. The purified mAb is dialyzed against coupling buffer (100 m*M* sodium bicarbonate, 500 m*M* NaCl, pH 8.3). The antibody solution is removed from the dialysis tube and the volume adjusted to about 10 ml with coupling buffer for each gram of dried CNBr-activated Sepharose used. A sample (about 100 μl) is reserved for protein concentration analysis.
2. Dried CNBr-activated agarose is swelled in 0.1 m*M* HCl for about 20 min. Each gram of dried CNBr-activated resin makes 3.5 ml of swollen gel. The resin is then washed on a fritted glass filter, using about 100 ml of 0.1 m*M* HCl for a gram of resin.
3. The resin is quickly washed with about 20 ml of coupling buffer, and the resin is transferred to the antibody solution. This slurry is mixed end-over-end on a laboratory rotator for 2 h at 23 °C.
4. The resin is collected on the fritted glass filter, and the filtrate is saved for determining the amount of protein that did not couple.
5. The resin is transferred to about 10 ml of 1 *M* ethanolamine, pH 8.3 and mixed end-over-end to 2 h at 23 °C, to allow the ethanolamine to react with any remaining reactive CNBr groups.
6. The resin is again collected on the fritted glass filter, and washed with the coupling buffer (about 50 ml) and then washed with 100 m*M* sodium acetate buffer (pH 4.0).
7. Repeat the two washings in Step 6 at least two to three times.
8. Store the conjugated Sepharose in 10 ml of coupling buffer containing 0.02% NaN_3 at 4 °C.
9. Perform a protein concentration analysis on the antibody solutions saved before and after conjugation (samples from steps 1 and 4). A coupling efficiency can be determined from these samples.

2.6.1. Comments

1. One gram of this resin yields 3.5 ml of swollen resin. For most purposes, we find making 0.5–2.0 g at a time to be convenient. We have found

that coupling 2.5 mg of mAb to 1 ml of swelled resin works well. Therefore, 8.75 mg of mAb is needed to conjugate 3.5 ml (1 g) of resin.

2. The CNBr is activated by pH values over 8. Therefore, the transfer of the resin to the antibody solution described in Step 3 above should be performed as quickly as possible.
3. The blocking agent (Step 5) can vary with the use. For example, yeast produces an ethanolamine-binding protein that will be copurified with the target protein if ethanolamine is used for blocking. In this case 0.1 *M* glycine is a more appropriate blocking agent.
4. The resin is stored at 4 °C in 0.02% NaN_3. The resin is stable for about 6 months under these conditions. We have noticed that some antibody leaches off after about 6 months of storage. Storage of the resin in the above buffer but at 50 % glycerol allows storage at − 20 °C (where it does not freeze) and seems to allow for a longer half life.

2.7. Purification of proteins with PR-mAbs

As an example, we describe here the method of purifying RNA polymerase from *E. coli*, using either mAb NT73 or 8RB13. We have presented a one-step immunoaffinity chromatography procedure that will yield RNAP that is about 90% pure. Some of the extra proteins that coelute are RNAP-binding proteins. This protocol assumes that the starting material is a bacterial pellet containing *E. coli* cells (2–3 g wet weight) from 1 l of late-log phase culture. The SDS–PAGE shown in Fig. 28.3 is a composite gel of this purification.

1. The pellet is allowed to partially thaw on ice and resuspended in 20 ml of TEN buffer (50 m*M* Tris–HCl, pH 7.9, 0.1 m*M* EDTA, and 100 m*M* NaCl).
2. Lysozyme is added to a concentration of 250 μg/ml and the cells are incubated on ice for 20 min. Alternatively, 1500 kU of rLysozyme (EMD/Novagen #71110) can be used.
3. The cells are sonicated on ice four times for 15 s each, with 15 s rests between each sonication burst.
4. The lysate is centrifuged at 15,000 rpm (27,000×*g*) for 15 min.
5. The supernatant is applied to the immunoaffinity column (1–2 ml) at 23 °C. A flow-through fraction is saved.
6. The column is washed with TE containing 100 m*M* NaCl (about 20 ml) and then with TE containing 500 ml NaCl (about 5 ml). The column is the reequilibrated in TE containing 100 m*M* NaCl (about 10 ml).
7. The column is eluted with TE containing 0.75 *M* NaCl and 40% propylene glycol (at room temperature). The fractions are placed on ice as they are collected.
8. The peak fractions are analyzed on a SDS–PAGE gel. The SDS–PAGE in Fig. 28.3 shows the fractions obtained from the one-step

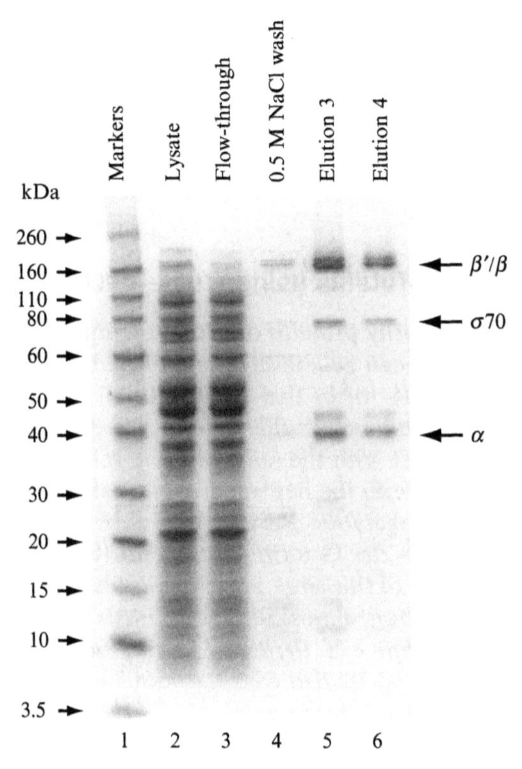

Figure 28.3 SDS–PAGE (4–12%) of RNAP purified from *E. coli* by a one-step immunoaffinity chromatography procedure, using mAbNT73-Sepharose. The cell lysate was prepared as described in the text, and the soluble material (Lane 2) was applied to a NT73-Sepharose column (2 ml). The flow-through material (Lane 3) was collected. After washing with TE buffer containing 100 m*M* NaCl, the column was washed briefly with TE containing 0.5 *M* NaCl (Lane 4). The column was re-equilibrated in TE containing 100 m*M* NaCl, and then the RNAP was eluted from the column with TE buffer containing 0.75 *M* NaCl and 40% propylene glycol, and 1-ml fractions collected (Lanes 5 and 6). RNAP subunits are indicated on the right side. Most of the extra bands are RNAP-binding proteins.

chromatography. The appropriate fractions (Lanes 5–6 in Fig. 28.3) are pooled and dialyzed against a suitable storage buffer. (For transcription proteins, we use 50 m*M* Tris–HCl, pH 7.9; 0.1 m*M* EDTA, 0.1 m*M* DTT, 50 m*M* NaCl, and 20–50% glycerol).

2.7.1. Comments

1. The lysate (Step 3) can be treated with Benzonase (EMD/Novagen #70746, Madison, WI) to help digest nucleic acids and decrease viscosity (see Chapter 18 in this volume).

2. The 500 m*M* NaCl wash (Step 6) helps to remove nucleic acids and NusA, a RNA polymerase-binding protein.
3. For reasons that are unknown to us, the elution of the target protein (Step 8) is significantly more effective if performed at room temperature than at 4 °C.

2.8. Purification of proteins using cross-reacting PR-mAbs

It is often desirable to purify proteins or protein complexes from biological materials that have not been genetically manipulated. In this case, we have found that the use of PR-mAbs that react with an epitope that is highly conserved makes the immunoadsorbent more versatile. Two of our most successful PR-mAbs react with the same enzyme across many species.

mAb 8WG16 reacts with the heptapeptide repeat on the C-terminus of the largest subunit of eukaryotic RNAP II (Table 28.1). This sequence, commonly referred to as the C-terminal domain (CTD) of RNAP II, is accessible on the surface of this large protein complex. The CTD is highly conserved on RNAP II from almost all species. mAb 8WG16 has been used to purify RNAP II from calf thymus (Thompson *et al.*, 1990), yeast (Edwards *et al.*, 1990), and human cells (Maldonado *et al.*, 1996). In fact, this PR-mAb was used to purify the yeast RNAP II that was used in the first crystallization studies of RNAP II (Cramer *et al.*, 2000). We have described a detailed procedure for using this PR-mAb to purify RNAP II (Thompson and Burgess, 1996). In most cases, the purification of RNAP II from crude material requires some bulk purification steps before the material is applied to the immunoaffinity resin.

mAb 8RB13 is a highly cross-reactive PR-mAb that has proved to be very useful for purification of bacterial core RNAP (Bergendahl *et al.*, 2003; Probasco *et al.*, 2007). This PR-mAb reacts with the highly conserved "β-flap" region of the bacterial RNAP from most bacteria (E. S. Stalder, unpublished data). Because the β-flap is one of the major binding sites for the sigma subunit (Kuznedelov *et al.*, 2002), core RNA polymerase (that lacks a sigma subunit) is purified on the 8RB13-agarose column. Lane 3 in Fig. 28.4 shows core RNAP purified from *E. coli*. Because of the cross-reactivity of this mAb, core RNAP can also be purified from *Bacillus subtilis* (Lane 4). Also shown in Fig. 28.4 is RNAP purified from *E. coli* using mAb NT73 (Lane 2); this preparation is a mixture of the holoenzyme and the core enzyme, and these two forms of the enzyme can be separated from this fraction by ion-exchange chromatography (Thompson *et al.*, 1992). The peptide band corresponding to the major sigma factor is missing from core RNAP purified from *E. coli* and *B. subtilis*.

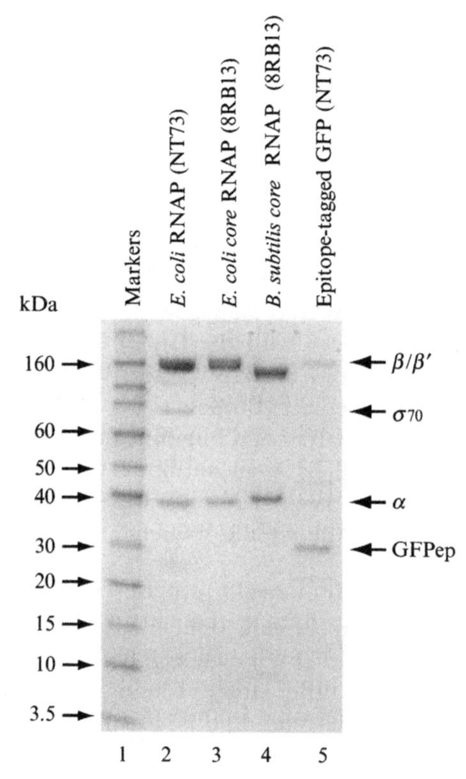

Figure 28.4 SDS–PAGE (4–12%) of RNAP and epitope-tagged GFP purified on PR-mAbs. Lane 1 contains Novex Sharp Standard markers; Lane 2 contains *E. coli* RNAP purified on mAb NT73; Lane 3 contains *E. coli* core RNAP purified on mAb 8RB13; Lane 4 contains *B. subtilis* core RNAP purified on mAb 8RB13; Lane 5 contains epitope-tagged GFP purified on mAb NT73.

2.9. Use of epitopes of PR-mAbs as purification tags

Although we believe that about 5–10% of the mAbs that are isolated from standard hybridoma techniques are potentially useful PR-mAbs, another approach is to epitope-tag a protein with the epitope for an established PR-mAb. The concept of purification tags is described in Chapter 16 in this volume. Therefore, we will only briefly describe this topic, and only as it relates to PR-mAbs and their epitopes.

The ability to tag a protein with an epitope and use the PR-mAb to purify the protein established that polyol-responsiveness is indeed a property of the antibody, and not a property of the environment in which the epitope is presented. The three PR-mAbs for which we have developed epitope-tagged systems are NT73, IIB8, and 8RB13 (Duellman *et al.*, 2004;

Thompson *et al.*, 2003; E. S. Stalder, unpublished data). We have designated these epitope tags as "Softags" (Burgess and Thompson, 2002). All of these epitopes are listed in Table 28.1, and a patent has been issued on the use of these tags (Burgess *et al.*, 2007).

All of our PR-mAbs have been isolated using the full-length protein as the immunogen; therefore, for some of these mAbs, the epitope mapping has been a laborious process. We have not attempted to isolate a PR-mAb using a synthetic peptide as an immunogen, although this should be possible. IIB8, the PR-mAb that reacts with human TFIIB, was mapped using phage display followed by site-directed mutagenesis (Duellman *et al.*, 2004). The two PR-mAbs that react with the largest subunits of *E. coli* RNA polymerase (mAbs NT73 and 8RB13) were roughly mapped using the ordered fragment ladder method (Burgess *et al.*, 2000; Rao *et al.*, 1996), followed by fine deletion analysis and oligonucleotide tagging of an unrelated protein for which we have an antibody (Thompson *et al.*, 2003; E. S. Stalder, unpublished data). The tags are constructed by fusing an oligonucleotide containing the coding sequence for the epitope in-frame with the target protein.

We have used the green fluorescent protein (GFP) as a target protein as proof-of-principle for the technique (Duellman *et al.*, 2004; Thompson *et al.*, 2003). The mAb NT73 epitope tag can be fused to either the N- or C-terminus of GFP, but for other target proteins, this might be dependent upon the accessibility of the termini. In the case of GFP, both of the termini are accessible in the crystal structure (Tsien, 1998). The epitope for mAb 8RB13 has been used as a tag in both the *E. coli* and mammalian cell culture systems (E. S. Stalder, unpublished).

While we have used the epitope tags derived from *E. coli* RNAP for purification of epitope-tagged GFP produced in the *E. coli* expression system, the endogenous RNAP is still in the lysate and some RNAP is also purified. Lane 5 in Fig. 28.4 shows GFP tagged with the epitope for NT73 expressed in *E. coli* and purified with NT73-Sepharose. The RNAP can be removed from this material by bringing the lysate to 300 m*M* NaCl and adding 0.3% polyethyleneimine (PEI) (see Chapter 20 in this volume). The resulting heavy precipitate is removed by centrifugation, and the epitope-tagged GFP remains in solution. The supernatant is then applied to the immunoaffinity column.

3. Conclusions

We have described the isolation, identification, and use of PR-mAbs for use in gentle immunoaffinity chromatography. Although the examples that we have presented here apply to our PR-mAbs and their use to purify transcription factors, the methods outlined in this chapter can be applied to any PR-mAb or to a protein that has been genetically tagged with an epitope for a PR-mAb. The most expensive component of this procedure is the isolation of the PR-mAb. Because there are now many panels of mAbs for a given protein, and these existing mAbs can be screened for polyol-responsiveness by the ELISA-elution assay, it is possible that a PR-mAb that reacts with a protein that you are interested in already exists.

DISCLOSURE

N. Thompson and R. Burgess are required by the University of Wisconsin–Madison Conflict of Interest Committee to disclose that they have financial interests in the company NeoClone which markets many of the mAbs mentioned in this chapter.

REFERENCES

Anthony, J. R., Green, H. A., and Donohue, T. J. (2003). Purification of *Rhodobacter sphaeroides* RNA polymerase and its sigma factors. *Meth. Enzymol.* **370,** 54–65.

Bergendahl, V., Thompson, N. E., Foley, K. M., Olson, B. M., and Burgess, R. R. (2003). A cross-reactive polyol-responsive monoclonal antibody useful for isolation of core RNA polymerase from many bacterial species. *Protein Expr. Purif.* **31,** 155–160.

Burgess, R. R., and Thompson, N. E. (2002). Advances in gentle immunoaffinity chromatography. *Curr. Opin. Biotechnol.* **13,** 304–308.

Burgess, R. R., Arthur, T. A., and Pietz, B. C. (2000). Mapping protein-protein interaction domains using ordered fragment ladder far-Western analysis of hexahistidine-tagged fusion proteins. *Meth. Enzymol.* **328,** 141–157.

Burgess, R. R., Thompson, N. E., and Duellman, S. J. (2007). Immunoaffinity chromatography using epitope tags to polyol-responsive monoclonal antibodies. US Patent No. 7,241,580.

Cramer, P., Bushnell, D. A., Fu, J., Gnatt, A. L., Maier-Davis, B., Thompson, N. E., Burgess, R. R., Edwards, A. M., David, P. R., and Kornberg, R. D. (2000). Architecture of RNA polymerase II and implications for the transcription mechanism. *Science* **288,** 640–649.

Duellman, S. J., Thompson, N. E., and Burgess, R. R. (2004). An epitope tag derived from human transcription factor IIB that reacts with a polyol-responsive monoclonal antibody. *Protein Expr. Purif.* **35,** 147–155.

Edwards, A. M., Darst, S. A., Feaver, W. J., Thompson, N. E., Burgess, R. R., and Kornberg, R. D. (1990). Purification and lipid layer crystallization of yeast RNA polymerase II active in transcription initiation. *Proc. Natl. Acad. Sci. USA* **87,** 2122–2126.

Harlow, E., and Lane, D. (1988). Antibodies: A Laboratory Manual. Cold Spring Harbor Press, Cold Spring Harbor, NY11724.

Jiang, Y., Zhang, S. J., Wu, S. M., and Lee, M. Y. (1995). Immunoaffinity purification of DNA polymerase delta. *Arch. Biochem. Biophys.* **230,** 297–304.

Kuznedelov, K., Minakhin, L., Niedziela-Majka, A., Dove, S. L., Rogulja, D., Nickels, B. E., Hochschild, A., Heyduk, T., and Severinov, K. (2002). A role for the interaction of the RNA polymerase flap domain with the sigma subunit in promoter recognition. *Science* **295,** 855–857.

Largaespada, D. A., Jackson, M. W., Thompson, N. E., Kaehler, D. A., Byrd, L. G., and Mushinski, J. F. (1996). The ABL-MYC retrovirus generates antigen-specific plasmacytomas by *in vitro* infection of activated B lymphocytes from spleen and other murine lymphoid organs. *J. Immunol. Methods* **197,** 85–95.

Lynch, N. A., Jiang, H., and Gibson, D. T. (1996). Rapid purification of the oxygenase component of toluene dioxygenase from a polyol-responsive monoclonal antibody. *Appl. Environ. Microbiol.* **62,** 2133–2137.

Maldonado, E., Drapkin, R., and Reinberg, D. (1996). Purification of human RNA polymerase II and general transcription factors. *Meth. Enzymol.* **274B,** 72–100.

Nagy, P. L., Griesenbeck, J., Kornberg, R. D., and Cleary, A. (2002). A trithorax-group complex purified from *Saccharomyces cerevisiae* is required for methylation of histone H3. *Proc. Natl. Acad. Sci. USA* **99,** 90–94.

Nelson, P. N., Reynolds, G. M., Waldron, E. E., Ward, E., Giannopoulos, K., and Murray, P. G. (2000). Demystified: Monoclonal antibodies. *J. Clin. Pathol. Mol. Pathol.* **53,** 111–117.

Probasco, M. D., Thompson, N. E., and Burgess, R. R. (2007). Immunoaffinity purification and characterization of RNA polymerase from *Shewanella oneidensis. Protein Expr. Purif.* **55,** 23–30.

Rao, L., Jones, D. P., Nguyen, L. H., McMahan, S. A., and Burgess, R. R. (1996). Epitope mapping using histidine-tagged protein fragments: Application to *Escherichia coli* RNA polymerase sigma70. *Anal. Biochem.* **241,** 141–157.

Subramanian, A. (2002). Immunoaffinity chromatography. *Mol. Biotechnol.* **20,** 41–47.

Thompson, N. E., and Burgess, R. R. (1994). Purification of recombinant human transcription factor IIB (TFIIB) by immunoaffinity chromatography. *Protein Expr. Purif.* **5,** 469–475.

Thompson, N. E., and Burgess, R. R. (1996). Immunoaffinity of RNA polymerase and transcription factors using polyol-responsive monoclonal antibodies. *Meth. Enzymol.* **274B,** 513–526.

Thompson, N. E., and Burgess, R. R. (1999). Immunoaffinity purification of the RAP30 subunit of the human transcription factor IIF (TFIIF). *Protein Expr. Purif.* **17,** 260–266.

Thompson, N. E., and Burgess, R. R. (2001). Preparation and use of specialized antibodies. Identification of polyol-responsive monoclonal antibodies for use in immunoaffinity chromatography. *Curr. Protoc. Mol. Biol. Sect. VI* (Suppl. 54), 11.18.1–11.18.9.

Thompson, N. E., Aronson, D. B., and Burgess, R. R. (1990). Purification of eukaryotic RNA polymerase II by immunoaffinity chromatography: Elution of active enzyme with protein stabilizing agents from a polyol-responsive monoclonal antibody. *J. Biol. Chem.* **265,** 7069–7077.

Thompson, N. E., Hager, D. A., and Burgess, R. R. (1992). Isolation and characterization of a polyol-responsive monoclonal antibody useful for gentle purification of *E. coli* RNA polymerase. *Biochemistry* **31,** 7003–7008.

Thompson, N. E., Arthur, T. M., and Burgess, R. R. (2003). Development of an epitope tag for the gentle purification of proteins by immunoaffinity chromatography: Application to epitope-tagged green fluorescent protein. *Anal. Biochem.* **323,** 171–179.

Thompson, N. E., Foley, K. M., and Burgess, R. R. (2004). Antigen-binding properties of monoclonal antibodies reactive with human TATA-binding protein (TBP) and use in immunoaffinity chromatography. *Protein Expr. Purif.* **36,** 186–197.

Thompson, N. E., Jensen, D. B., Lamberski, J. A., and Burgess, R. R. (2006). Purification of protein complexes by immunoaffinity chromatography: Application to transcription machinery. *Genet. Eng.* **27,** 81–100.

Tsien, R. Y. (1998). The green fluorescent protein. *Annu. Rev. Biochem.* **67,** 509–544.

Woychik, N. A., and Hampsey, M. (2002). The RNA polymerase II machinery: Structure illuminates function. *Cell* **108,** 453–463.

SECTION EIGHT

PURIFICATION PROCEDURES: ELECTROPHORETIC METHODS

CHAPTER TWENTY-NINE

One-Dimensional Gel Electrophoresis[1]

David E. Garfin

Contents

1. Background 498
2. Polyacrylamide Gels 500
3. Principle of Method 501
4. Procedure 502
   4.1. Stock solutions 502
   4.2. Catalyst 503
   4.3. Electrode buffer 503
   4.4. Casting gels 503
   4.5. Sample preparation 505
   4.6. Electrophoresis 505
   4.7. Comments on method 506
   4.8. Variations of method 507
5. Detection of Proteins in Gels 508
   5.1. Dye staining with Coomassie Brilliant Blue R-250 509
   5.2. Silver staining 509
   5.3. Copper staining 510
6. Marker Proteins 510
7. Molecular Weight Determination 511
8. Preparative Electrophoresis 511
References 513

Sodium dodecyl sulfate-polyacrylamide gel electrophoresis (SDS–PAGE) is an excellent method with which to identify and monitor proteins during purification and to assess the homogeneity of purified fractions. SDS–PAGE is routinely used for the estimation of protein subunit molecular weights and for determining the subunit compositions of purified proteins. SDS–PAGE can also be scaled up, for use in a preparative mode, to yield sufficient protein for further studies. In addition, two-dimensional analysis, combining isoelectric focusing with SDS–PAGE (Dunbar, 1987, this volume), is a very high resolution method for protein fractionation, enabling thousands of polypeptides to

Chemical Division, Research Products Group, Bio-Rad Laboratories, Incorporated, Richmond, California
[1] Reprinted from *Methods in Enzymology*, Volume 182 (Academic Press, 1990)

Methods in Enzymology, Volume 463
ISSN 0076-6879, DOI: 10.1016/S0076-6879(09)63029-9

be resolved in a single gel. When used in conjunction with blotting methods (Timmons and Dunbar, this volume), SDS–PAGE provides one of the most powerful means available for protein analysis.

A great many electrophoretic systems have been developed and no attempt is made to summarize them here. In particular, the distinctions between the various "continuous" and "discontinuous" buffer systems are not discussed, nor are alternative support matrices considered. Gradient gels (gels whose pore sizes vary) are also omitted from discussion, since these can be prepared by relatively straightforward adaptation of any of a number of well-known methods for forming gradients. Rather, only the most common (and most reliable) analytical SDS–PAGE procedure (Laemmli, 1970) is described. Those wishing further information on the practical or theoretical aspects of electrophoretic processes can use (Allen *et al.*, 1984; Andrews, 1986; Chrambach, 1985; Hames, 1981) to gain access to the large volume of literature in the field. Some problems may require adoption of alternative procedures (Allen *et al.*, 1984; Andrews, 1986; Blackshear, this series; Bury, 1981; Chrambach, 1985; Hames, 1981; Neville, 1971; Neville and Glossmann, this series), but for most applications the SDS–PAGE method presented here will perform satisfactorily.

1. Background

Although the detailed theory of gel electrophoresis is complicated and at present incomplete (Bier *et al.*, 1983; Chrambach and Jovin, 1983; Jovin, 1973), the fundamental concepts are easily understood. Briefly, in an electrophoretic separation, charged particles are caused to migrate toward the electrode of opposite sign under the influence of an externally applied electric field. The movements of the particles are retarded by interactions with the surrounding gel matrix, which acts as a molecular sieve. The opposing interactions of the electrical force and molecular sieving result in differential migration rates for the constituent proteins of a sample.

In general, fractionation by gel electrophoresis is based on the sizes, shapes, and net charges of the macromolecules. Systems designed to fractionate proteins in their native configurations cannot distinguish between the effects of size, shape, and charge on electrophoretic mobility. As a consequence, proteins with differing molecular weights can have the same mobility in these systems. Thus, while PAGE methods for native proteins are valuable for separating and categorizing protein mixtures, they should not be used to assess the purity of a preparation or the molecular weight of an unknown.

SDS–PAGE overcomes the limitations of native PAGE by imposing uniform hydrodynamic and charge characteristics on all the proteins in a sample mixture. During sample preparation, proteins are treated with hot SDS. The anionic detergent binds tightly to most proteins at about 1.4 mg of SDS/mg of protein, imparting a negative charge to the resultant complexes (Nielsen and Reynolds, this series). Interaction with SDS disrupts all noncovalent protein bonds, causing the macromolecules to unfold. Concomitant treatment with a disulfide-reducing agent, such as 2-mercaptoethanol or dithiothreitol, further denatures proteins, breaking them down to their constituent subunits. The electrophoretic mobilities of the resultant detergent–polypeptide complexes all assume the same functional relationship to their molecular weights. Migration of SDS derivatives is toward the anode at rates inversely proportional to the logarithms of their molecular weights (Neville, 1971; Neville and Glossmann, this series; Shapiro *et al.*, 1967; Weber and Osborn, 1969). SDS polypeptides, thus, move through gels in a predictable manner, with low-molecular-weight complexes migrating faster than larger ones. This means that the molecular weight of a protein can be estimated from its relative mobility in a calibrated SDS–PAGE gel and that a single band in such a gel is a criterion of purity.

Most electrophoresis is done in vertical chambers in gel slabs formed between two glass plates (Andrews, 1986; Hames, 1981). The slab format provides uniformity, so that different samples can be directly compared in the same gel. Gel thicknesses are established by spacers placed between the glass plates and sample wells are formed in the gels during polymerization with plastic, comb-shaped inserts. Electrophoresis cells provide means for sealing the assemblies during gel formation and for maintaining contact with the electrode buffers during runs. The better cells provide means for heat dissipation, because uneven heat distribution in the gel slab can cause band distortions.

Conventional gels are of the order of 16–20 cm long, 16 cm wide, and 0.5–3.0 mm in thickness and can accommodate about 25 samples. Thick gels have greater total protein capacity than thin ones, but are correspondingly less efficient at dissipating electrically generated heat and more difficult to stain and destain. Gel thicknesses of 0.75 or 1 mm are good compromise sizes, combining adequate protein loads and good staining speeds with minimal heat-related distortions. Typical runs take 4–5 h.

Small-format cells (minicells) allow rapid analyses and are adequate for relatively uncomplicated samples. The design of these cells allows analyses to be completed two to three times faster than is possible with conventional cells. The gels are about 7 cm long × 8 cm wide and are very easily manipulated. Each gel can hold up to about 15 samples and a typical run can be completed in less than 1 h (not counting setup and polymerization time). The resolution of complex samples may be better in conventional gels than with minicells, since the separation of protein bands is improved by increasing the lengths of SDS–PAGE gels.

2. Polyacrylamide Gels

Polyacrylamide gels are formed by copolymerization of acrylamide monomer, $CH_2{=}CH{-}CO{-}NH_2$, and a cross-linking comonomer, *N,N′*-methylenebisacrylamide, $CH_2{=}CH{-}CO{-}NH{-}CH_2{-}NH{-}CO{-}CH{=}CH_2$, (bisacrylamide) (Allen *et al.*, 1984; Andrews, 1986; Chrambach, 1985; Chrambach and Rodbard, 1971; Hames, 1981). The mechanism of gel formation is vinyl addition polymerization and is catalyzed by a free radical-generating system composed of ammonium persulfate (the initiator) and an accelerator, tetramethylethylenediamine (TEMED). TEMED causes the formation of free radicals from persulfate and these in turn catalyze polymerization. Oxygen, a radical scavenger, interferes with polymerization, so that proper degassing to remove dissolved oxygen from acrylamide solutions is crucial for reproducible gel formation.

The sieving properties of a gel are established by the three-dimensional network of fibers and pores which is formed as the bifunctional bisacrylamide cross-links adjacent polyacrylamide chains (Rodbard and Chrambach, 1970). Within limits, as the acrylamide concentration of a gel increases, its effective pore size decreases. The effective pore size of a gel is operationally defined by its sieving properties; that is, by the resistance it imparts to the migration of protein molecules. By convention, a given gel is physically characterized by the pair of figures ($\%T$, $\%C$), where $\%T$ is the weight percentage of total monomer (acrylamide + cross-linker, in g/100 ml), and $\%C$ is the proportion of cross-linker (as a percentage of total monomer) in the gel. The practical limits for $\%T$ lie between 3% and 30%. The factors governing pore size are complicated, but, in general, the pore size of a gel decreases as $\%T$ increases. For any given fixed $\%T$, pore size is at a minimum at about 5% C, increasing at both higher and lower cross-linker concentrations (Allen *et al.*, 1984; Andrews, 1986; Chrambach, 1985; Chrambach and Rodbard, 1971; Hames, 1981).

The use of high-quality reagents is a prerequisite for reproducible, high-resolution gels. This is particularly true of acrylamide, which constitutes the most abundant component in the gel–monomer mixture. Residual acrylic acid, linear polyacrylamide, and ionic impurities are the major contaminants of acrylamide preparations. Moreover, buffer components should be of reagent grade and only distilled or deionized water should be used for all phases of gel electrophoresis.

In SDS–PAGE, the quality of the SDS is of prime importance. Differential protein-binding properties of impurities such as C_{10}, C_{14}, and C_{16} alkyl sulfates can cause single proteins to form multiple bands in gels (Margulies and Tiffany, 1984). Even with pure SDS, very basic proteins, very acidic proteins, various glycoproteins, and lipoproteins, because of their unusual compositions, migrate "anomalously" during electrophoresis (Allen *et al.*, 1984; Andrews, 1986; Hames, 1981).

3. Principle of Method

The most popular electrophoretic method is the SDS–PAGE system developed by Laemmli (Allen *et al.*, 1984; Andrews, 1986; Blackshear, this series; Hames, 1981; Laemmli, 1970). This is a discontinuous system consisting of two contiguous, but distinct gels: a resolving or separating (lower) gel and a stacking (upper) gel. The two gels are cast with different porosities, pH, and ionic strength. In addition, different mobile ions are used in the gel and electrode buffers. The buffer discontinuity acts to concentrate large volume samples in the stacking gel, resulting in better resolution than is possible using the same sample volumes in gels without stackers. Proteins, once concentrated in the stacking gel, are separated in the resolving gel.

The Laemmli SDS–PAGE system is made up of four components. From the top of the cell downward, these are the electrode buffer, the sample, the stacking gel, and the resolving gel. Samples prepared in low-conductivity buffer (0.06 *M* Tris–Cl, pH 6.8) are loaded between the higher conductivity electrode (0.025 *M* Tris, 0.192 *M* glycine, pH 8.3) and stacking gel (0.125 *M* Tris–Cl, pH 6.8) buffers. When power is applied, a voltage drop develops across the sample solution which drives the proteins into the stacking gel. Glycinate ions from the electrode buffer follow the proteins into the stacking gel. A moving boundary region is rapidly formed with the highly mobile chloride ions in the front and the relatively slow glycinate ions in the rear (Allen *et al.*, 1984; Andrews, 1986; Blackshear, this series; Bury, 1981; Hames, 1981; Wyckoff *et al.*, 1977). A localized high-voltage gradient forms between the leading and trailing ion fronts, causing the SDS–protein complexes to form into a thin zone (stack) and migrate between the chloride and glycinate phases. Within broad limits, regardless of the height of the applied sample, all SDS–proteins condense into a very narrow region and enter the resolving gel as a well-defined, thin zone of high protein density. (The stacking phenomenon is strikingly demonstrated with prestained protein standards, which are mixtures of proteins derivatized with reactive dyes.) The large-pore stacking gel (4% *T*) does not retard the migration of most proteins and serves mainly as an anticonvective medium. At the interface of the stacking and resolving gels, the proteins experience a sharp increase in retardation due to the restrictive pore size of the resolving gel. (Proteins too large to enter the resolving gel will stop at the interface.) Once in the resolving gel, proteins continue to be slowed by the sieving of the matrix. The glycinate ions overtake the proteins, which then move in a space of uniform pH (pH 9.5) formed by the Tris and glycine. Molecular sieving causes the SDS–polypeptide complexes to separate on the basis of their molecular weights.

4. Procedure

Equipment and reagents for SDS–PAGE can be obtained from a variety of suppliers. Electrophoresis cells vary in design, but their operation generally follows the steps outlined below. Since the many available cells differ in size, formulations are presented in conveniently sized units for simplicity. Required volumes can be prepared using multiples of these unit sizes. Except where noted, reagents for SDS–PAGE can be prepared as concentrated stock solutions.

4.1. Stock solutions

1. *Acrylamide concentrate (30% T, 2.7% C)*: Dissolve 29.2 g of acrylamide and 0.8 g of bisacrylamide in 70 ml of deionized water. When the acrylamide is completely dissolved, add water to a final volume of 100 ml. Filter the solution under vacuum through a 0.45-μm membrane. Store stock acrylamide at 4 °C in a dark bottle for no more than 1 month. *Caution*: Acrylamide monomer is a neurotoxin. Avoid breathing acrylamide dust, do not pipette acrylamide solutions by mouth, and wear gloves when handling acrylamide powder or solutions containing it. For disposal of unused acrylamide, add bisacrylamide (if none is present), induce polymerization, and discard the solidified gel.
2. *1.5 M Tris–Cl, pH 8.8, concentrated resolving gel buffer*: Dissolve 18.2 g Tris base in ≈80 ml of water, adjust to pH 8.8 with HCl, and add water to a final volume of 100 ml. Store at 4 °C.
3. *0.5 M Tris–Cl, pH 6.8, concentrated stacking gel buffer*: Dissolve 6.1 g Tris base in ≈80 ml of water, adjust to pH 6.8 with HCl, and add water to a final volume of 100 ml. Store at 4 °C.
4. *10% (w/v) sodium dodecyl sulfate (SDS)*: Dissolve 10 g SDS in ≈60 ml of water and add water to a final volume of 100 ml.
5. *Stock sample buffer (0.06 M Tris–Cl, pH 6.8, 2% SDS, 10% glycerol, 0.025% Bromphenol Blue)*:

Water	4.8 ml
0.5 *M* Tris–Cl, pH 6.8	1.2 ml
10% SDS	2.0 ml
Glycerol	1.0 ml
0.5% Bromphenol Blue (w/v water)	0.5 ml

Store at room temperature. SDS-reducing buffer is prepared by adding 50 μl of 2-mercaptoethanol to each 0.95 ml of stock sample buffer before use.

4.2. Catalyst

1. *10% ammonium persulfate (APS)*: Dissolve 100 mg APS in 1 ml of water. Make the APS solution fresh daily.
2. *TEMED (N,N,N′,N′-tetramethylethylenediamine)*: Use TEMED undiluted from the bottle. Store cool, dry, and protected from light.

4.3. Electrode buffer

Electrode buffer: 0.025 *M* Tris, 0.192 *M* glycine, 0.1% (w/v) SDS, pH 8.3 (0.3 g Tris base, 1.4 g glycine, 1 ml 10% SDS/100 ml electrode buffer). Do not adjust the pH of the electrode buffer; just mix the reagents together and confirm that the pH is near 8.3 (± 0.2). Electrode buffer can be made as a 5× concentrate consisting of 15 g Tris base, 72 g glycine, and 5 g SDS/l. 5× electrode buffer concentrate must be stored in glass containers. To use 5× concentrate, dilute it with four parts water.

4.4. Casting gels

Thoroughly clean the glass plates, spacers, combs, and upper buffer reservoir of the gel apparatus with detergent and rinse them well. Wear gloves while assembling the equipment. The resolving gel is cast first, then overlaid with the stacking gel.

1. Assemble the casting apparatus and determine the gel volume from the manufacturer's instructions or by calculation. A 1- to 2-cm stacking gel is used above the resolving gel. Determine the height to which the resolving gel is to be poured by inserting a well-forming comb between the glass plates and marking the outer plate 1–2 cm below the teeth of the comb.
2. Prepare the monomer solution for the appropriate resolving gel by combining all of the reagents in Table 29.1 except the ammonium persulfate (APS) and TEMED; a disposable, plastic beaker is a convenient mixing vessel. The two gel recipes given in Table 29.1 cover the molecular weight ranges usually encountered. Gels of any other acrylamide concentration desired (Andrews, 1986; Blackshear, this series; Hames, 1981), can be prepared by adjusting (only) the amounts of 30% monomer stock and water used in the recipes. Deaerate the solution under vacuum (e.g., in a bell jar or desiccator) for at least 15 min.
3. Gently mix the APS and TEMED (Table 29.1) into the deaerated monomer solution. Using a pipet and bulb, add the monomer solution between the gel plates up to the mark delimiting the resolving gel. Immediately overlay the monomer solution with water-saturated 2-butanol or *tert*-amyl alcohol to exclude air, which might inhibit

Table 29.1 Formulations of SDS–PAGE resolving gels[a]

Component	7.5% *T*[b]	12% *T*[c]
Water	4.85 ml	3.35 ml
1.5 *M* Tris–Cl, pH 8.8	2.5 ml	2.5 ml
10% SDS	0.1 ml	0.1 ml
Acrylamide/bis (30% *T*, 2.7% *C*)	2.5 ml	4.0 ml
10% ammonium persulfate[d]	50 μl	50 μl (0.05%)
TEMED	5 μl	5 μl (0.05%)

[a] Any desired volume of monomer solution can be prepared by using multiples of the 10-ml recipes. Combine the first four items and deaerate the solution under vacuum for 15 min. Start polymerization by adding ammonium persulfate and TEMED.

[b] For SDS-treated proteins in the approximate molecular weight range between 40 and 250 K.

[c] For SDS-treated proteins in the approximate molecular weight range between 10 and 100 K.

[d] To make 10% ammonium persulfate (APS), dissolve 100 mg APS in 1 ml of water. Make the APS solution fresh daily.

polymerization, from the surface of the monomer mixture. Allow the gel to polymerize for 45 min to 1 h. Polymerization is evidenced by the appearance of a sharp interface beneath the overlay, which will start to become visible in about 15 min. Polymerization is essentially complete in about 90 min, but the stacking gel can be poured after about an hour (Bio-Rad Lab., Bull. No. 1156). Allow unused monomer to polymerize in the beaker and discard the gel.

4. Prepare 10 ml of stacking gel monomer solution (4% *T*, 2.7% *C*), by combining

Water	6.1 ml
0.5 *M* Tris–Cl, pH 6.8	2.5 ml
Acrylamide stock solution (30% *T*)	1.3 ml
10% SDS	0.1 ml

Deaerate the monomer solution under vacuum for at least 15 min.

5. Thoroughly rinse the top of the resolving gel with water and dry the area above it with filter paper. Place a well-forming comb between the gel plates and tilt it at a slight angle to provide a way for bubbles to escape.
6. Add 50 μl of 10% APS and 10 μl of TEMED to each 10 ml of degassed monomer solution and pour the stacking gel solution on top of the resolving gel. Align the comb in its proper position, being careful not to trap bubbles under the teeth. Visible polymerization of the stacking gel should occur in about 10 min. No overlay is required, because the comb excludes oxygen from the surfaces of the wells. Allow the gel to polymerize for 30–45 min. Allow unused monomer to polymerize in the beaker before disposing of it.

In some situations, it may be necessary or convenient to let the gel stand overnight before it is used. When this is the case, it is best to pour the stacking gel on the day of the run to maintain the ion discontinuities at the interface between the two gels. For storage, the top of the resolving gel should be rinsed thoroughly and covered with resolving gel buffer (0.375 *M* Tris–Cl, 0.1% SDS, pH 8.8) to avoid dehydration and ion depletion. Also, the tops of the gel sandwiches should be covered with plastic wrap during storage.

4.5. Sample preparation

The common biochemical buffers are usually tolerated in SDS–PAGE, so that pretreatment of samples is not generally required. Distorted band patterns, such as pinching or flaring of lanes, can be caused by excessive amounts of salt in the samples. These distortions can often be remedied by desalting the samples.

1. Prepare the volume of SDS-reducing buffer required for the number of samples to be run by adding 50 μl of 2-mercaptoethanol to each 0.95 ml of stock sample buffer (to a final concentration of 5% 2-mercaptoethanol). This step may be omitted, if reduction of disulfide bonds is not desired.
2. Dilute samples with at least 4 vol of complete SDS-reducing buffer (although as little as twofold dilution may be adequate for some samples). Sample volumes are of the order of 20–50 μl for conventional gels and 5–30 μl for minicells, depending on the widths of the wells and the thicknesses of the gels. Detection in gels requires on the order of 1 μg of protein per band for easy visibility when staining with Coomassie Blue R-250 or 0.1 μg of protein per band with silver staining (see below).
3. Heat the diluted samples at 95 °C for 4 min by suspending the sample tubes in hot water. Do not store prepared samples.

4.6. Electrophoresis

Assemble the electrophoresis cell, fill the upper and lower reservoirs with electrode buffer, and remove the comb from the stacking gel. Load the prepared samples into the wells in the stacking gel by layering them under electrode buffer using a microliter syringe or micropipet. The glycerol in the samples provides the necessary density for them to sink to the bottoms of the wells and the Bromphenol Blue tracking dye enables the samples to be seen during loading. Finally, attach leads to the unit and connect them to a power supply. The lower electrode is the anode and the upper one is the cathode, in SDS–PAGE.

During an electrophoresis run, electrical energy is converted to heat which can cause band distortion and diffusion. In general, electrophoresis should be carried out at power settings at which the run proceeds as rapidly as the chamber's ability to draw off heat will allow. In other words, the run should be as fast as possible without exceeding desired resolution and distortion limits.

Many of the power supplies which are available allow control of any electrical quantity and the choice is almost a matter of preference. Constant current conditions, as a rule, result in shorter but hotter runs than does constant voltage (Allen *et al.*, 1984). In the early stages of a run, the resistance of the gel increases as the chloride ions migrate out of it. Accordingly, voltage will rise or current will fall, depending on whether constant current or constant voltage operation is in use.

Small-format minicells, with their thin glass plates, are better able to efficiently dissipate the heat generated by the initially high currents at the beginnings of runs than are standard-sized cells. Thus, the recommendation is that gels should be run under constant current conditions (16–24 mA/mm of gel thickness) in conventional apparatus and at constant voltage (20–30 V/cm of gel length) in minicells. The use of recirculated coolant, where possible, allows higher voltages and currents to be used for shortened run times. Electrophoresis should be started immediately after the samples are loaded and is generally continued until the Bromphenol Blue tracking dye has reached the bottom of the gel.

4.7. Comments on method

The Laemmli SDS–PAGE system (Allen *et al.*, 1984; Andrews, 1986; Blackshear, this series; Hames, 1981; Laemmli, 1970) is an adaptation of an earlier method devised by Ornstein (1964) and Davis (1964) for fractionation of native serum proteins. The different (discontinuous) buffers used in the stacking and resolving gels are required for the proper functioning of the Ornstein–Davis system (Ornstein, 1964; Jovin, 1973). However, inclusion of SDS modifies the rationale of the Ornstein–Davis technique in important ways, since the properties of the detergent dominate the system (Allen *et al.*, 1984; Chrambach, 1985; Wyckoff *et al.*, 1977).

The necessary components of the Laemmli SDS–PAGE system are a Tris–Cl gel buffer, the Tris–glycine–SDS electrode buffer, and the SDS-reducing sample buffer. As a consequence of SDS in the system, it is actually not necessary to cast the stacking gels at different pH or ionic strength than the resolving gels. Similar resolution is obtained whether the stacking gel is cast as above or in resolving gel buffer (0.375 *M* Tris–Cl, pH 8.8). This is because the mobilities of SDS–polypeptide complexes are insensitive to pH in this range (Allen *et al.*, 1984). When many gels are being cast at one time for storage and later use, it is convenient to cast the stacking and resolving gels in the same buffer.

Total SDS load, on the other hand, has considerable influence on resolution (Wyckoff *et al.*, 1977). Inclusion of more than 200 μg of SDS in 30–50 μl samples in the minigel configuration can lead to broadening and spreading of protein bands. With dilute, large volume samples, it may prove advantageous to limit the total SDS in the system by dropping the final SDS concentration of the treated sample to about 0.5% and casting the gels without SDS. Because the mobility of SDS is greater than those of proteins, SDS from the electrode buffer quickly overtakes the proteins during electrophoresis. The gel is thus supplied and continuously replenished with SDS from the electrode buffer at a level sufficient to maintain the saturation of the proteins (Chrambach, 1985).

4.8. Variations of method

The complete denaturation and dissociation of proteins with the Laemmli SDS–PAGE system (Allen *et al.*, 1984; Andrews, 1986; Blackshear, this series; Hames, 1981; Laemmli, 1970) are not always desirable. For some analyses, it might be of interest to estimate the molecular weights of particular proteins in their intact, oligomeric forms. In other experiments, interest might center on the biological activities of proteins in their native, nondenatured states. Through selective use of the two denaturants, 2-mercaptoethanol and SDS, conditions can be adjusted as needed to separate proteins in the completely denatured, partially denatured, or native states.

Covalent associations between protein units can be maintained by omitting 2-mercaptoethanol from the sample buffer. In the absence of the reducing agent, the intra- and interchain disulfide bonds of sample proteins remain intact. The electrophoretic mobilities of the resultant SDS–protein complexes are correspondingly altered relative to those obtained under dissociating conditions. During electrophoresis, the mobilities of oligomeric SDS–proteins are lower than those of their fully denatured SSDS-polypeptide components. Further, the electrophoretic behaviors of single-chain polypeptides can also be affected by reduction. The intrachain disulfide bridges of single-chain proteins can hold them in compact configurations that are more or less retained in the presence of SDS. Thus, some SDS-proteins migrate faster electrophoretically in the absence of 2-mercaptoethanol than when in the extended structures brought on by reduction. Proteins often show characteristic, individual responses to reduction, so that comparisons of SDS–PAGE gels run with and without 2-mercaptoethanol can be very informative (Marshall, 1984).

To separate proteins without reduction, carry out the SDS–PAGE procedure described above, omitting 2-mercaptoethanol from the sample buffer. Note that oligomeric SDS–protein complexes migrate more slowly than their SDS-polypeptide subunits. It may, therefore, be necessary to use lower concentration (%*T*) gels than with the fully denaturing method to get oligomers to move adequate distances into the matrices. In addition, nonreduced proteins may not be completed saturated with SDS and, hence, may not bind the detergent in a constant weight ratio. This makes molecular weight determinations of these molecules by SDS–PAGE less straightforward than analyses of fully denatured polypeptides, since it is necessary that both standards and unknown proteins be in similar configurations for valid comparisons.

When both SDS and 2-mercaptoethanol are left out of the Laemmli procedure, what remains is the classical Ornstein–Davis PAGE system (Davis, 1964; Ornstein, 1964) for native proteins. This is a high-resolution native PAGE method designed for separation of the full spectrum of serum proteins. Because the system was meant to separate a wide variety of proteins, resolution may not be optimal for some restricted ranges of protein mobilities. Although there are a number of high-resolution native PAGE systems available to meet differing requirements (Allen *et al.*, 1984; Andrews, 1986; Blackshear, this series; Chrambach, 1985; Hames, 1981), the Ornstein–Davis method should perform adequately for the fractionation of the majority of commonly encountered protein mixtures. Molecular weights are more difficult to determine by native PAGE than by SDS–PAGE, since a single native system cannot distinguish the effects of charge and conformation on protein electrophoretic mobilities (Allen *et al.*, 1984; Andrews, 1986; Chrambach, 1985; Hames, 1981).

The procedure described here is readily modified for native PAGE. Merely omit 2-mercaptoethanol from the sample buffer and replace the 10% SDS in the recipes for the gel, sample, and electrode buffers with equivalent volumes of water. Follow the procedure as otherwise presented, except for sample treatment. Samples should be diluted in nondenaturing buffer (0.06 *M* Tris–Cl, pH 6.8, 10% glycerol, 0.025% Bromphenol Blue) following the same guidelines as for denaturing gels, but they should not be heated.

5. Detection of Proteins in Gels

Three of the simplest and most reliable methods for the detection of proteins in SDS–PAGE gels are presented. They should be adequate to cover the requirements of most situations. Coomassie Brilliant Blue R-250 is the most common protein stain and is recommended for routine work. Silver staining is the most sensitive method for staining proteins in gels and should be employed when electrophoresis is used to assess the purity of a preparation; for example, an antigen preparation. Copper staining is a recent development allowing rapid and sensitive staining. Discussions of other detection methods, including radiolabeling and means for quantitating proteins in gels, can be found by Dunbar (1987), Andrews (1986), Allen *et al.* (1984), Hames (1981), and Merril (this volume).

After electrophoresis, remove the gel assembly and separate the glass plates. The gel will probably stick to one of the two plates. Remove the spacers and cut off and discard the stacking gel. Place the glass plate holding the gel into fixative or staining solution and float the gel off of the plate. All of the steps in gel staining are done at room temperature with gentle agitation (e.g., on an orbital shaker platform) in any convenient container, such as a glass casserole or a photography tray. Always wear gloves when staining gels, since fingerprints will stain. Permanent records of stained gels can be obtained by photographing them or by drying them on filter paper using commercially available drying apparatus.

5.1. Dye staining with Coomassie Brilliant Blue R-250

This is the standard method of protein detection (Allen *et al.*, 1984; Andrews, 1986; Hames, 1981; Wilson, this series). Easy visibility requires on the order of 0.1–1 μg of protein per band.

1. Prepare the staining solution: 0.1% Coomassie Brilliant Blue R-250 (w/v) in 40% methanol (v/v), 10% acetic acid (v/v). Filter the staining solution after the dye has dissolved. The staining solution is reusable. Store it at room temperature.
2. Soak the gel in an excess of staining solution for 30 min.
3. Destain with a large excess of 40% methanol, 10% acetic acid. Change the destaining solution several times, until the background has been satisfactorily removed.

The acid–alcohol solutions used in this procedure do not completely fix proteins in the gel. This can lead to losses of some low-molecular-weight proteins during the staining and destaining of thin gels. Permanent fixation is obtainable by incubating the gel in 40% methanol (v/v), 10% trichloroacetic acid (w/v) for 1 h before it is immersed in the staining solution.

5.2. Silver staining

This method, developed by Merril and coworkers, can be as much as 100 times more sensitive than dye staining (Allen *et al.*, 1984; Merril *et al.*, 1981, this series). Bands containing 10–100 ng of protein can be easily seen. The reagents are available in kit form from Bio-Rad Laboratories.

Reaction times vary with the thicknesses of the gels.

1. Fix the proteins in the gel in about 400 ml of 40% methanol, 10% acetic acid (v/v) (or 40% methanol, 10% trichloroacetic acid) for 30 min to overnight.
2. Fix twice in 400 ml 10% ethanol, 5% acetic acid (v/v) for 15–30 min.

3. Soak the gel for 3–10 min in 200 ml of fresh oxidizer solution (0.0034 *M* potassium dichromate, 0.0032 N nitric acid).
4. Wash the gel three or four times for 5–10 min in 400 ml water, until the yellow color has been washed out.
5. Soak the gel in 200 ml fresh silver reagent (0.012 *M* silver nitrate) for 15–30 min.
6. Wash the gel with 400 ml water for 1–2 min.
7. Wash the gel for about 1 min in developer (0.28 *M* sodium carbonate, 1.85% paraformaldehyde).
8. Replace the developer with fresh solution and incubate for 5 min.
9. Replace the developer a second time and allow development to continue until satisfactory staining has been obtained.
10. Stop development with 5% acetic acid (v/v).

Vertical streaks and sample-independent bands in the 50–70-kDa region are sometimes seen in silver-stained gels. These artifacts have been attributed to reduction of contaminants inadvertently introduced into the samples (Ochs, 1983). They can be eliminated by adding excess iodoacetamide to sample solutions after treatment with SDS-reducing buffer (Görg *et al.*, 1987).

5.3. Copper staining

Rapid, single-step staining of SDS–PAGE gels is achieved by incubating gels in copper chloride (Lee *et al.*, 1987). The resultant, negatively stained image of the electrophoretogram is intermediate in sensitivity between Coomassie blue and silver staining.

1. Wash the gel briefly in water.
2. Soak the gel in 0.3 *M* $CuCl_2$ for 5 min.
3. Wash the gel for 2–3 min in water.

The method yields negatively stained gels showing clear protein bands on an opaque, blue–green background. The protein bands can be easily seen and photographed with the gel on a black surface. Proteins are not permanently fixed by this method and can be quantitatively eluted after chelating the copper (Lee *et al.*, 1987). The electrophoretic pattern is lost when copper-stained gels are dried so they must be photographed, restained with Coomassie Blue, or stored in water.

6. Marker Proteins

Mixtures of marker proteins are available for calibrating gels. PAGE standards are mixtures of proteins with precisely known molecular weights blended for uniform staining. They are obtainable in various molecular weight ranges. Concentrated stock solutions of the standards are diluted in sample buffer just prior to electrophoresis and treated in the same manner as the sample proteins. These proteins are suitable as reference markers for molecular weight determinations.

Prestained SDS–PAGE standards have recently become available. The coupling of dye molecules to the marker proteins changes their molecular weights significantly and unpredictably and they should not be used for molecular weight determinations. However, prestained standards are very useful for following the course of an electrophoretic run and are valuable for assessing the efficiencies of protein transfers when gels are blotted.

7. Molecular Weight Determination

Molecular weights of proteins are determined by comparison of their mobilities with those of several marker proteins of known molecular weight (Allen *et al.*, 1984; Andrews, 1986; Blackshear, this series; Chrambach, 1985; Hames, 1981). After the gel has been run, but before it has been stained, mark the position of the Bromphenol Blue tracking dye to identify the leading edge of the electrophoretic ion front. This can be done by cutting notches in the edges of the gel or by inserting a needle soaked in India ink into the gel at the dye front. After staining, measure the migration distances of each protein (markers and unknowns) from the top of the resolving gel. Divide the migration distance of each protein by the distance traveled by the tracking dye. The normalized migration distances so obtained are called the relative mobilities of the proteins (relative to the dye front) and conventionally denoted as R_f. Construct a (semilogarithmic) plot of the logarithms of the molecular weights of the protein standards as functions of the R_f values. Note that the graphs are slightly sigmoid. As long as the extremities of a molecular weight range are avoided, unknown molecular weights can be estimated by linear regression analysis or interpolation from the curves of $\log M_r$ versus R_f. Keep in mind that the molecular weights obtained using SDS–PAGE are those of the polypeptide subunits and not those of native, oligomeric proteins.

8. Preparative Electrophoresis

The most satisfactory way to recover proteins separated by SDS–PAGE for further study is to extract them from bands excised from the gels. Many attempts have been made to design continuous elution devices suitable for routine protein purification, in which bands emerging from the bottoms of electrophoresis gels are swept away to fraction collectors

(Andrews, 1986; Chrambach, 1985; Chrambach and Nguyen, 1979). The scarcity of preparative gel devices is evidence of the disappointing lack of success in developing generally useful instruments. Preparative gel electrophoresis would ideally be capable of yielding high-milligram to gram quantities of individual proteins recovered cleanly with the resolution anticipated from the corresponding analytical gels. In general, though, band distortion and poor elution have limited the resolution attainable with most apparatus so that they have only worked well with relatively simple protein mixtures. The difficulties in scaling gel electrophoresis up to preparative levels has tended to result in devices which are rather cumbersome and which require much technical skill for best results. As a consequence, proteins are usually obtained by extraction from analytical type gels (Harrington, this volume).

Gels to be run for the isolation of proteins (Andrews, 1986; Chrambach, 1985) can be cast using special preparative combs. These combs form wide sample wells spanning the widths of the gels and usually provide a separate, narrow reference well for marker proteins. The maximum amount of sample which can be loaded on a gel ultimately depends on how well the proteins of interest are separated from their neighbors in the sample mixture. Since bands become wider as the amount of material increases, as sample load is raised, the corresponding loss of resolution will eventually become unacceptable. Protein loads 10- to 50-fold greater per unit of cross-sectional area than are usually run in analytical gels are easily tolerated. Thus, with some large slab gels, proteins can be recovered in tens-of-milligram amounts.

Copper staining (Lee *et al.*, 1987) is advisable for the visualization of the bands in preparative SDS–PAGE, since this method does not employ fixative solvents. Desired bands are cut from the gel and destained by incubation in three changes (for 10 min each) of 0.25 *M* EDTA, 0.25 *M* Tris–Cl, pH 9. After destaining, gel slices are incubated in the appropriate elution buffer.

Proteins are often extracted from macerated gel slices by simple diffusion into appropriate buffers or by solubilization of the gel (Andrews, 1986; Harrington, this volume). In the latter method, cross-linkers other than bisacrylamide are copolymerized into the gels (Allen *et al.*, 1984; Andrews, 1986). For example, gels cross-linked with *N,N′*-bisacrylylcystamine (BAC) are dissolvable in 2-mercaptoethanol or dithiothreitol, while both *N,N′*-dihydroxyethylenebisacrylamide (DHEBA) and *N,N′*-diallyltartardiamide (DATD) result in gels which can be solubilized with periodic acid. Once gels have been dissolved, proteins must be separated from the large excess of gel material by gel filtration or ion-exchange chromatography.

Electrophoretic elution is an efficient method for recovering proteins from gel slices (Andrews, 1986; Chrambach, 1985; Dunbar, 1987). In the simplest versions of this method, proteins are electrophoresed out of gel pieces into dialysis sacks in the types of apparatus used for running cylindrical gel rods. Devices are available for the rapid recovery of proteins in small volumes with yields of greater than 70% in most cases. Elution takes about 3 h at 10 mA/tube in 0.025 *M* Tris, 0.192 *M* glycine, 0.1% SDS, pH 8.3 (standard SDS–PAGE electrode buffer). SDS can be removed from the eluted samples by dialysis or ion-exchange chromatography (Furth, 1980).

REFERENCES

Allen, R. C., Saravis, C. A., and Maurer, H. R. (1984). *Gel Electrophoresis and Isoelectric Focusing of Proteins: Selected Techniques*. de Gruyter, Berlin.

Andrews, A. T. (1986). *Electrophoresis: Theory, Techniques, and Biochemical and Clinical Applications*. 2nd edn. Oxford University Press, New York.

Bier, M., Palusinski, O. A., Mosher, R. A., and Saville, D. A. (1983). *Science* **219,** 1281.

Blackshear, P. J. (this series). Vol. 104, p. 237.

Bury, A. F. (1981). *J. Chromatogr.* **213,** 491.

Chrambach, A. (1985). *The Practice of Quantitative Gel Electrophoresis*. Weinheim, VCH.

Chrambach, A., and Jovin, T. M. (1983). *Electrophoresis* **4,** 190.

Chrambach, A., and Nguyen, N. Y. (1979). *In* "Electrokinetic Separation Methods", (P. G. Righetti, C. J. Van Oss, and J. W. Vanderhoff, eds.), p. 337. Elsevier, Amsterdam.

Chrambach, A., and Rodbard, D. (1971). *Science* **172,** 440.

Davis, B. J. (1964). *Ann. NY Acad. Sci.* **121,** 404.

Dunbar, B. S. (1987). *Two-Dimensional Electrophoresis and Immunological Techniques*. Plenum, New York.

Dunbar, B. S., Kimura, H., and Timmons, T. M. (this volume). Chapter 34.

Furth, A. J. (1980). *Anal. Biochem.* **109,** 207.

Görg, A., Postel, W., Weser, J., Günther, S., Strahler, J. R., Hanash, S. M., and Somerlot, L. (1987). *Electrophoresis* **8,** 122.

Hames, B. D. (1981). *In* "Gel Electrophoresis of Proteins: A Practical Approach", (B. D. Hames and D. Rickwood, eds.), p. 1. IRL Press, Oxford.

Harrington, M. (this volume). Chapter 37.

Jovin, T. M. (1973). *Biochemistry* **12,** 871, 879, 890.

Laemmli, U. K. (1970). *Nature (London)* **227,** 680.

Lee, C., Levin, A., and Branton, D. (1987). *Anal. Biochem.* **166,** 308.

Margulies, M. M., and Tiffany, H. L. (1984). *Anal. Biochem.* **136,** 309.

Marshall, T. (1984). *Clin. Chem.* **30,** 475.

Merril, C. R. (this volume). Chapter 36.

Merril, C. R., Goldman, D., Sedman, S. A., and Ebert, M. H. (1981). *Science* **211,** 1437.

Merril, C. R., Goldman, D., and Van Keuren, M. L. (this series). Vol. 104, p. 441.

Neville, D. M. Jr. (1971). *J. Biol. Chem.* **246,** 6328.

Neville, D. M., and Glossmann, H. (this series). Vol. 32, p. 92.

Nielsen, T. B., and Reynolds, J. A. (this series). Vol. 48, p. 3.

Ochs, D. (1983). *Anal. Biochem.* **135,** 470.

Ornstein, L. (1964). *Ann. NY Acad. Sci.* **121,** 321.

Rodbard, D., and Chrambach, A. (1970). *Proc. Natl. Acad. Sci. USA* **65,** 970.

Shapiro, A. L., Vinuela, E., and Maizel, J. V. Jr. (1967). *Biochem. Biophys. Res. Commun.* **28,** 815.

Timmons, T. M., and Dunbar, B. S. (this volume). Chapter 51.

Weber, K., and Osborn, M. (1969). *J. Biol. Chem.* **244,** 4406.

Wilson, C. M. (this series). Vol. 91, p. 236.

Wyckoff, M., Rodbard, D., and Chrambach, A. (1977). *Anal. Biochem.* **78,** 459.

CHAPTER THIRTY

Isoelectric Focusing and Two-Dimensional Gel Electrophoresis

David B. Friedman,* Sjouke Hoving,† *and* Reiner Westermeier‡

Contents

1. Introduction 516
   1.1. Isoelectric focusing—Basic principle 517
   1.2. Types of pH gradients for isoelectric focusing 518
   1.3. SDS polyacrylamide electrophoresis 521
   1.4. Enhancement of resolution for alkaline pH gradients 523
   1.5. Difference gel electrophoresis 524
2. Materials 527
   2.1. Equipment 527
   2.2. Solutions and reagents 527
3. Methods 528
   3.1. Protein sample preparation 528
   3.2. Sample cleanup and precipitation 530
   3.3. Isoelectric focusing using acidic range IPG gels 531
   3.4. Isoelectric focusing using alkaline range IPG gels 534
   3.5. Equilibration of IPG gels 535
   3.6. SDS–PAGE: The second dimension 536

References 538

Abstract

By far the highest resolution of all separation techniques for intact proteins in a single analytical run continues to be by the combination of isoelectric focusing (IEF) and SDS–PAGE, originally introduced by O'Farrell [(1975). *J. Biol. Chem.* 250, 4007–4021]. This analytical platform has seen a number of significant advances and applications over the past 25 years, including reproducibility using immobilized pH gradient (IPG) strips [Bjellqvist *et al.* (1982). *J. Biochem. Biophys. Methods* 6, 317–339.], resolution in alkaline IEF using hydroxyethyldisulfide (HED) [Olsson *et al.* (2002). *Proteomics* 2, 1630–1632], and quantification for differential expression proteomics on intact proteins on a global scale

* Proteomics Laboratory, Mass Spectrometry Research Center, Vanderbilt University, Nashville, Tennessee, USA
† Novartis Institutes of BioMedical Research, Basel, Switzerland
‡ Gelcompany GmbH, Paul-Ehrlich-Strasse, Tübingen, Germany

Methods in Enzymology, Volume 463
ISSN 0076-6879, DOI: 10.1016/S0076-6879(09)63030-5

[DIGE; Unlu *et al.* (1997). *Electrophoresis* 18, 2071–2077]. These major improvements will be highlighted in this chapter alongside the principle and theory of 2D gel electrophoresis, as well as detailed methods for general 2D gel electrophoresis best use protocols.

1. Introduction

For complex mixtures such as whole cell lysates or enriched subcellular fractions, two-dimensional gel electrophoresis (2D gel) can typically resolve hundreds-to-thousands of individual protein species using two orthogonal separations[1] (Fig. 30.1). The first separation is typically based on charge using denaturing IEF, and the second separation by apparent molecular mass using denaturing sodium dodecyl sulfate–polyacrylamide gel electrophoresis (SDS–PAGE).[2] 2D gel experiments are particularly powerful for visualizing protein isoforms that result from charged posttranslational modification, such as phosphorylation and sulfation (which add charge), or acetylation (which neutralize charge), among others. They are also useful in detecting splice variants and proteolytic cleavages that result in protein species with altered molecular weight (MW) and isoelectric point (p*I*). More recently, 2D gels have been used extensively in differential expression proteomics experiments, especially since the commercialization of the difference gel electrophoresis (DIGE) technology in early 2000 (Friedman and Lilley, 2009; Lilley and Friedman, 2004; Unlu *et al.*, 1997).

In modern 2D gel-based proteomics experiments, intact protein resolution by 2D gel electrophoresis is commonly coupled with protein ("spot") excision, in-gel digestion and mass spectrometry to provide for protein identification using sophisticated database searching algorithms. As with all high-end proteomics technologies, this method also requires many

[1] Despite this resolution, proteins of extreme MW and p*I* (10 > kDa < 200; 3 > p*I* < 11), very hydrophobic proteins (e.g., integral membrane proteins with multiple trans-membrane domains), and low-abundance proteins are typically difficult to resolve or detect on 2D gels. It is in these cases where complementary data can be obtained from an orthogonal technology, such as liquid chromatography coupled with tandem mass spectrometry (LC–MS/MS), that is capable of detecting many of these proteins. The added sensitivity of the LC–MS/MS approach arises from the fact that proteins are detected based on surrogate peptides from an *en masse* proteolytic digest that are resolved by liquid chromatography. But, for this very reason, the LC–MS/MS technique is greatly challenged to identify the multiple protein isoforms that arise from charged or bulky post-translational modifications that are readily detectable and quantifiable on a 2D gel (especially in a nontargeted, discovery experiment). Furthermore, the quantification of intact protein expression using the 2D gel platform can now be done using the fluorescent multiplexing technology of DIGE, whereby the requisite number of individual biological replicates from multiple experimental conditions can be easily analyzed to provide for statistically powered discovery proteomics.

[2] Although denaturing IEF is most commonly used for the highest resolution first-dimensional separations (first described by O'Farrell, 1975), other techniques such as blue native PAGE (Schagger and von Jagow, 1991) for analysis of protein complexes, and acidic PAGE in presence of a cationic detergent (Hartinger *et al.*, 1996) for separating hydrophobic membrane proteins, are often utilized for more specific investigations. However, SDS–PAGE is almost always being used for the second-dimensional separation.

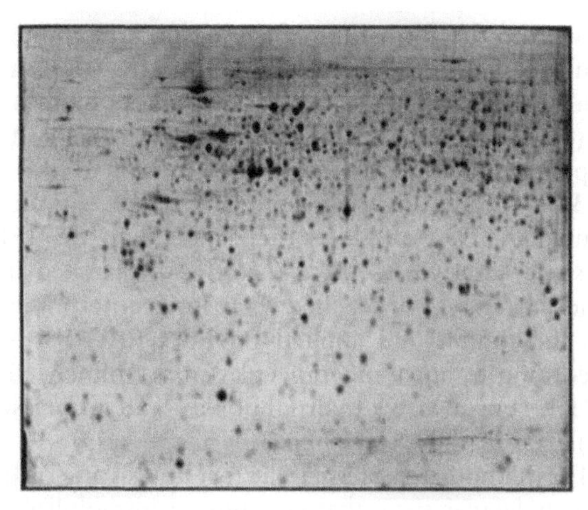

Figure 30.1 High resolution 2D gel (24 cm pH 4–7). 400 μg cellular proteins from a HeLa cell extract were separated in two orthogonal dimensions, the first based on charge by isoelectric focusing, and the second based on apparent molecular mass by sodium dodecyl sulfate polyacrylamide gel electrophoresis. The gel was stained with the fluorescent dye Sypro Ruby. Typically hundreds-to-thousands of individual protein species, including post-translationally modified isoforms and proteolytic products, can be resolved and quantified.

technically challenging manual steps in addition to specialized instrumentation. Often the success of a 2D electrophoresis experiment is dependent on the skill of the operator. By introducing stringent analysis conditions and developing the first multiple gel systems, Anderson and Anderson (1978a,b) made the technique markedly more reliable and reproducible.

1.1. Isoelectric focusing—Basic principle

Isoelectric focusing (IEF) is an electrophoretic separation method which separates amphoteric molecules such as proteins and peptides according to their charge as defined by the pK_a values of proton-accepting sites within a molecule. For proteins and peptides, these sites can be found in the free amines and carboxylic acids located at the N- and C-termini as well as on the side chains of arginine, lysine, histidine, aspartic acid, and glutamic acid residues. The isoelectric point is a specific physicochemical parameter of an amphoteric molecule defined as the pH that a molecule is in a net neutral charge state. IEF can provide very high resolving power, including the separation of protein posttranslational modifications that alter charge (e.g., phosphorylation, acetylation/deactylation). Although the method is applied for several types of amphoteric compounds, this chapter will describe the methodology for protein separations.

The sample has to be prepared and conditioned with a high chaotrope concentration, a zwitterionic detergent, a reducing thiol, and carrier ampholytes to avoid the formation of aggregates and complexes between proteins. Removal of nonprotein-based ions from the sample is also essential for high-resolution IEF, since resistance must be kept high to enable a high electric field (8000–10,000 V) while keeping the current at a minimum. Protein precipitation methods are often used to reduce such contaminants (see Section 3.2). Ideally, the only ions in a sample for IEF will be the proteins themselves. The sample is then applied to the IEF gel, which contains the same additives as the sample, and separated in the electric field. After the proteins have reached their isoelectric points, the gels are incubated in SDS buffer and applied on SDS–PAGE slab gels for the separation according to apparent molecular mass (influenced primarily by molecular weight but also by hydrophobicity and to a lesser extent by protein shape).

Modern IEF methods for 2D gel electrophoresis use a thin polyacrylamide gel as a molecular sieve that contains an immobilized pH gradient (IPG). The gels have a low concentration of acrylamide (usually 4–5% total acrylamide), because the matrix should not be restrictive to high-molecular-weight proteins. Proteins are introduced into this medium and then an electric field is applied. Proteins that are located in a position in the pH gradient that is lower (more acidic) than their p*I* will be positively charged and migrate toward the cathode in an electric field. Conversely, proteins located in a position more basic than their p*I* will migrate toward the anode. Since the p*I* is defined as the pH at which a protein is net neutral, the electric field has no effect on proteins when they are at the point in the pH gradient which is equal to the p*I*. Thus, the proteins stop migrating and are "focused" at the pH equal to their p*I*. The method has an in-built focusing effect: when a protein diffuses away from its p*I*, it will gain a charge and the electric field, therefore, forces it to migrate back to the isoelectric point.

In the end of an IEF run, the proteins become highly concentrated at their isoelectric points, resulting in a high sensitivity for detection. Even small charge differences can be differentiated, and the resolution can be increased by using longer separation distances and by employing narrow gradient intervals (Hoving *et al.*, 2000). If desired, the isoelectric points of the proteins can be estimated with a calibration curve using marker proteins.

1.2. Types of pH gradients for isoelectric focusing

1.2.1. Carrier ampholytes

The concept of separating proteins in a pH gradient built by a mixture of amphoteric buffers in free solution had been developed by Vesterberg and Svensson (1966) before it had been converted into a real method by Vesterberg a few years later (Vesterberg, 1972). Because naturally occurring amino acids and peptides have very low buffering power at their own isoelectric points, the buffers needed to be chemically synthesized.

The first reagents on the market for this purpose, Ampholines®, were mixtures of aliphatic oligo-amino/oligo-carbonic acids, of 600–700 different homologues exhibiting a spectrum of isoelectric points between pH 3 and 10. These compounds have high buffering capacities at their isoelectric points and form a pH gradient under the influence of the electric field. Optimally, they have molecular weights below 1 kDa, and do not bind to proteins, because they are highly hydrophilic. Mixtures with narrow intervals are available for higher resolution using defined isoelectric point ranges.

More recently came the development of carrier ampholytes, which are synthesized from different reagents than the Ampholines® and are available from a variety of commercial vendors. Although the function of forming a pH gradient under the influence of the electric field is the same for all of these products, the profiles of the pH gradients, the distribution of buffering power, and the number of homologues are different (Righetti *et al.*, 2007), which leads to differences in the focusing pattern.

Carrier ampholytes-based IEF is mostly performed in polyacrylamide gels, and the original 2D gel electrophoresis procedure by O'Farrell (1975) accomplishes IEF in carrier ampholyte gradients in thin gel rods ("tube gels" for the first dimension). However, these long and soft gels are not easy to handle, and the pH gradients become unstable with increased running time, resulting in cathodal drift. Although some laboratories still use the carrier ampholyte technique, and this method has mostly been supplanted by the introduction of IPG strips, which are described next.

1.2.2. Immobilized pH gradients

A major improvement in 2D gel electrophoresis methodology came from the introduction of an IPG within a polyacrylamide matrix for IEF (Bjellqvist *et al.*, 1982). This major technological development overcame several shortcomings of the carrier ampholytes system such as gradient drift (especially in the cathodal region), mechanical instability, and technical variation between runs and between laboratories. The IPG strip technology has greatly facilitated the methodology and highly increased the reproducibility within a laboratory and across different working groups (Gorg *et al.*, 2000, 2004), and is now considered to be the method of choice for IEF and is commercially available through several vendors.

For modern IPG strip technology, the pH gradient is formed by acidic and alkaline buffering groups which are copolymerized with the polyacrylamide matrix during preparing the gel. Less than 10 different acrylamide derivatives with acidic and basic pK_a values are sufficient to create any desired pH gradient. Two monomer solutions containing precalculated mixtures of acrylamide derivatives are prepared: to the solution for the acidic end of the gradient, a portion of glycerol is added to stabilize the gradient during pouring. These pH gradient gels are cast on a covalently bound film backing. After the gel has formed, polymerization catalysts and nonreacted compounds are washed out from the matrix with distilled water to produce the very low electrical conductivity necessary for IEF (Westermeier, 2005).

The IPG gels are then dried for long-term storage, and can be rehydrated as necessary shortly before use. The major benefit of IPGs is the absence of the cathodal drift: the gradient is fixed to the matrix. The major application of IPGs is IEF under denaturing conditions, serving as the first dimension of high-resolution two-dimensional electrophoresis. Carrier ampholytes are still used, and for improved conductivity and protein solubility they are added to the sample as well as to the rehydration solution of the gels with IPGs.

Precast IPG strips are now commercially available from a number of vendors with a wide variety of strip lengths and pH gradients to enable the optimal separation and display of selected protein groups within a protein lysate. In particular, pH gradients are available from a very wide range (e.g., pH 3–11), to medium ranges (e.g., pH 4–7, pH 7–11), to narrow ranges (e.g., pH 4–5, pH 5.5–6.5). Strip lengths also vary from as short as 7 cm to as long as 24 cm. Strips typically are 3 mm in width, with an average thickness (after reswelling) of approximately 0.5 mm (Fig. 30.2). Care should be taken in selection of the optimal pH range and strip length for the desired experimental goals (Hoving *et al.*, 2000). For example, 7-cm pH 3–11 IPG strips may seem best to provide the greatest range in a small gel format, but they provide overall the lowest resolution and sensitivity for proteome-wide, discovery-based experiments.

Figure 30.2 Immobilized pH gradient (IPG) strip, shown in the manual process of preparing for rehydration loading. Sample is first dispensed into a reswelling try, and then the IPG strip is carefully overlayed on the protein sample solution (face-down, with plastic backing up) using a forceps to hold one end of the IPG (the plastic backing exceeds the length of the IPG gel). The strips are then overlaid with 2–3 mL paraffin oil to prevent urea crystallization during rehydration. IPG strips are provided from commercial vendors with barcodes to aide sample-tracking.

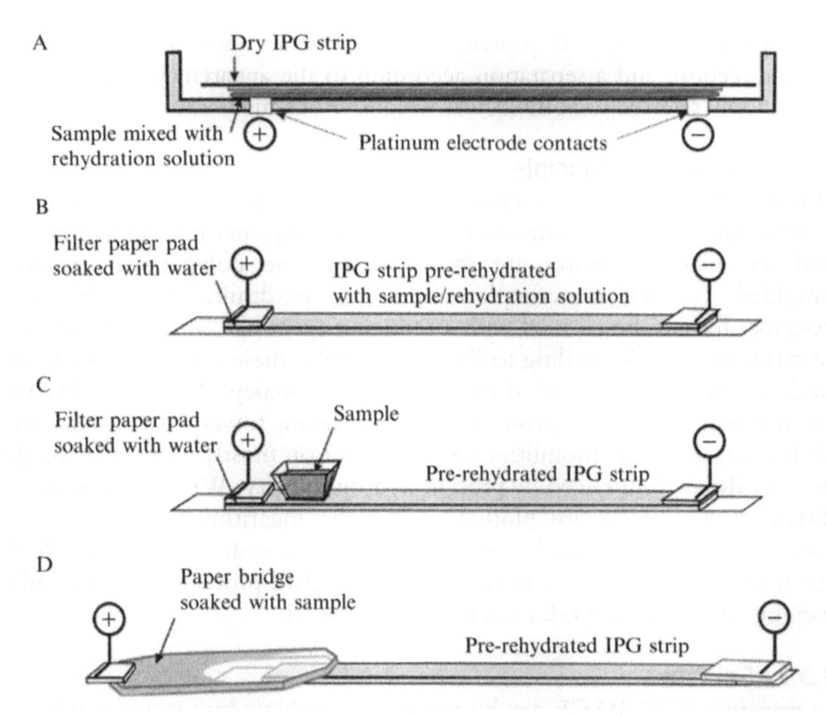

Figure 30.3 Different configurations for sample loading into immobilized pH gradient (IPG) strips for isoelectric focusing. (A) The IPG gel is placed face-down for sample loading via active rehydration loading, after which isoelectric focusing is also possible. (B) The sample is rehydrated into the IPG gel, which is then placed "sunny-side up" for isoelectric focusing. (C) The sample is introduced into a previously rehydrated IPG gel via cup loading at the anode. This configuration is especially beneficial for isoelectric focusing over alkaline pH gradients. (D) Modification of cup loading, where the sample is introduced at the anode via a paper-bridge. Reproduced with permission, Westermeier, Naven and Heopker (2008), Proteomics in Practice, 2nd edition, Wiley-VCH Verlag GmbH & Co. KGaA.

The IPG strips offer several modes of sample application. Rehydration loading and cup loading are the two major methods used, with modification for active rehydration and paper-bridge loading (Fig. 30.3). All of these methods are described in greater detail in Sections 3.3 and 3.4. The choice of the sample application method depends on the type of pH gradient and the sample. Because the IPG strips have a flat surface, they can also be applied on horizontal flatbed SDS gels if desired.

1.3. SDS polyacrylamide electrophoresis

SDS–PAGE, described by Laemmli (1970), has long been the method of choice for resolving intact proteins for a variety of biochemical analyses. Mainly because SDS is the best solubilizing detergent also for very hydrophobic proteins, all proteins, also very basic ones, migrate into the same direction, and a separation according to the apparent molecular mass (commonly referred to as molecular weight) is obtained.

1.3.1. Separation principle

When SDS is added to a protein solution in excess, the proteins form anionic micelles with a constant negative net charge per mass unit. Tertiary and secondary structures are disrupted and the polypeptides become unfolded. To achieve a complete unfolding, the disulfide bonds between cysteines have to be cleaved with a reducing agent like 2-mercaptoethanol or dithiothreitol. According to Ibel *et al.* (1990), these complexes look like necklaces with partly covered and partly open polypeptide surfaces. During electrophoresis all SDS–protein micelles migrate toward the anode, and their electrophoretic mobilities are dependent on mostly molecular weight but can also be influenced by protein hydrophobicity. When the migration distances of proteins are plotted against the logarithm of the apparent molecular mass, a sigmoidal curve results with a wide linear range. With the help of comigrated molecular weight standard proteins, the molecular weights of the polypeptides can be approximated.

1.3.2. Gel types

In principle, SDS–PAGE can be run in various basic buffer systems; however, the discontinuous buffer system according to Laemmli (1970) is the most commonly used system, employing Tris–chloride pH 8.8 in the gel and SDS–Tris–glycine in the running buffer. For one-dimensional separations, a stacking gel with pH 6.8 and lower gel concentration is applied to provide a slow sample entry without protein aggregation, and to provide a band sharpening effect at the start of the separation. However, a stacking gel typically is not included for 2D gel electrophoresis, since the proteins are preseparated and effectively stacked in the 3 mm width of the IPG strip (see Section 3.6).

The standard matrix is a homogeneous gel layer with 12.5% T (total acrylamide) and 2.6% C (bis-acrylamide for cross-linking) in the resolving gel. But some applications require optimal resolution for a certain size range, such that lower %T is required for high-molecular-weight proteins, and a higher %T provides a better resolution for low-molecular-weight proteins, so the concentration must be increased. Gradient SDS–PAGE gels can provide a greater dynamic range of resolution, but are more challenging to pour consistently and reproducibly, which is vital for quantitative proteomic studies using 2D gels. Porosity gradient gels exhibit a wider linear range of the molecular weights and improve the resolution for difficult proteins, like glycoproteins.

SDS electrophoresis can be run in vertical setups in cassettes between glass plates or on horizontal flatbed systems. The gel thicknesses vary between 1.5 mm and 0.5 mm. Thicker gels have a better mechanical stability, but are more difficult to stain and provide additional sample loss and background noise for subsequent mass spectrometry. Thinner gels can be cooled more efficiently, thus they can be run faster to obtain sharper resolution.

Of the improvements made to the 2D gel process over the past few decades, the second-dimension SDS–PAGE separations have remained largely unchanged. Vertical electrophoresis systems are more commonly used, and equipment to run individual or 12 or more gels simultaneously is available. Horizontal flatbed configurations have also been available historically and are now beginning to see a resurgence due to the low consumption of reagents (e.g., SDS running buffer), sample manipulation, as well as the potential for increased resolution afforded by short run times. Some improvements include using flexible plastic backing (including low-fluorescent media for DIGE and other fluorescence-based techniques), specialized equipment for multicasting reproducible acrylamide gradients, and different acrylamide formulations for increased shelf life and gel durability.

1.4. Enhancement of resolution for alkaline pH gradients

High-resolution IEF for alkaline pH ranges has always been more challenging than for acidic pH gradients, both due to issues of cathodic drift and also due to the loss of reductant to keep some proteins soluble (DTT is a weak acid that becomes charged at pH $>$ 8 and, therefore, leaves the alkaline region of the gradient). The first issue of cathodic drift was addressed by the introduction of nonequilibrium pH gradient electrophoresis (NEpHGE; O'Farrell *et al.*, 1977). But the problem of reductant loss at pH $>$ 8 still remained, and leads to nonspecific reactions with urea, backfolding and aggregation of polypeptides because the cysteines are no longer protected. This results in protein spot artifacts and horizontal streaking of proteins on the basic range of 2D gels.

The replacement of thiol reagents by phosphines like TBP or TCEP can partly prevent these effects, but these reagents cause additional artifacts due to their instability in the electric field. Prealkylation of proteins prior to IEF with iodoacetamide, vinylpyridine, or monomeric acrylamide also leads to additional artifacts due to incomplete alkylation of polypeptides and modifications of isoelectric points.

The introduction of the use of hydroxyethyldisulphide (HED) in excess, commercialized under the trade name DeStreak (GE Healthcare), provided a way to keep the sulfhydryls reduced (by mass action) at alkaline pH and dramatically increased the resolution of alkaline IEF using IPG strips (Olsson *et al.*, 2002). This is espeically found in conjunction with using cup loading or paper-bridge loading at the anode for sample entry, whereby the alkaline proteins all enter the IPG strip with the same acidic charge and

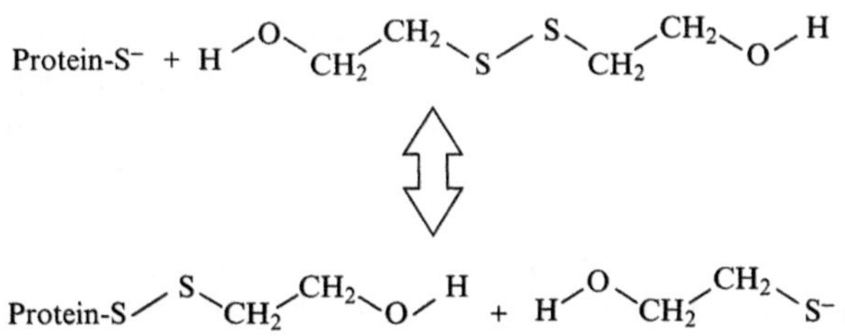

$$\text{Protein-S}^- + \text{HO-CH}_2\text{-CH}_2\text{-S-S-CH}_2\text{-CH}_2\text{-OH}$$

$$\Updownarrow$$

$$\text{Protein-S-S-CH}_2\text{-CH}_2\text{-OH} + \text{HO-CH}_2\text{-CH}_2\text{-S}^-$$

Figure 30.4 Oxidation of protein thiol groups using hydroxyethyl disulfide (HED) according to Olsson *et al.* (2002). The resulting mixed disulfides are stable and prevent the reformation of disulfide bond during isoelectric focusing over alkaline pH ranges, whereas the standard reductant DTT becomes charged and migrates away from this region of the gradient, resulting in streaking and artifacts.

then have to migrate toward the cathode to reach their isoelectric points (rather having them migrating in both directions).

When the proteins enter the IPG strip, the sulfhydryl groups on cysteine side chains become immediately oxidized to highly stable mixed disulfides. This is an equilibrium reaction with high specificity, preventing any unwanted side reactions (Fig. 30.4). Small amounts of DTT can be tolerated for the sample preparation when using DeStreak, so that proteins can stay reduced prior to focusing. However, care must be taken because excessive DTT will reduce HED to produce 2-mercaptoethanol, which results in even more horizontal streaking.[3]

1.5. Difference gel electrophoresis

Even with the increased reproducibility of IPG strips and stringent sample preparation and running conditions, gel–gel variations can occur. The success of a differential expression proteomics experiment relies on repetitive measurements across individually-procured (biological) replicates, and technical replicates (repetitive measurements on the same biological sample) are also necessary to control for analytical variation, such as sample procurement, handling, and gel–gel variation. These challenges can be addressed by applying the DIGE technique (Friedman and Lilley, 2009; Lilley and Friedman, 2004; Unlu *et al.*, 1997).

1.5.1. Staining and detection

A major advancement made recently for 2D gel electrophoresis has been in the development and application of fluorescent dyes used for the detection and quantification of the intact protein species that are resolved in the gels.

[3] When using DeStreak for IEF in alkaline pH gradients (pH 7–11, 3–11), 100 μL sample may contain up to 20 m*M* reducing agent. DeStreak can also be used for other pH ranges, with different tolerances for reductants (adapted from the DeStreak rehydration solution manual, GE Healthcare).

Fluorescent dyes typically provide detection sensitivity at least as sensitive as silver stain (*ca.* 1 ng) if not greater, and significantly increase the linear dynamic range of abundance changes to 3–4 orders of magnitude (silver and coomassie-blue stains are typically provide less than 1 order of magnitude dynamic range).

1.5.2. DIGE and quantitative analytical algorithms

The implementation of the DIGE method in 1997 (Unlu *et al.*, 1997) leverages the sensitivity and dynamic range of fluorescence labeling with multiplexing samples into the same gel to remove analytical (gel–gel) variation for the coresolved samples. It also utilizes a mixed-sample internal standard present throughout a series of DIGE gels to enable registration of migration patterns and, therefore, normalization of abundance ratios to provide multivariable experiments with exceptional statistical power (Karp and Lilley, 2005).

Briefly, the approach involves prelabeling several samples with spectrally distinct fluorescent dyes (Cy2, Cy3, and Cy5), followed by sample multiplexing and coresolution on the same 2D gel. In this way gel–gel variations are eliminated, and the mutually exclusive fluorescent images can be individually recorded and used for direct quantification of protein abundance changes for each resolved feature (Fig. 30.5). Two different labeling concepts are used: "minimal" labeling the ε-amino group of the lysine of about 5% of the total proteins and "saturation" labeling of all available cysteines of a protein mixture.

The DIGE approach is most beneficial and statistically powered when a Cy2-labeled internal standard is coresolved on a series of gels that each contains individual samples labeled with Cy3 or Cy5. Importantly, this Cy2-labeled internal standard is comprised of an equal mixture of all samples in the experiment, and is present on every gel of a multigel experiment. Because the individual samples (labeled with Cy3 or Cy5) are multiplexed with an equal aliquot of the same Cy2-standard mixture, each resolved feature can be directly related to the cognate feature in the Cy2-standard mixture within that gel. Intragel Cy3:Cy2 and Cy5:Cy2 ratios are then calculated without interference from gel–gel variation, and these ratios can then be normalized to all other measurements for that feature from the other samples across the experiment with extremely low technical (analytical) noise and high statistical power (Karp and Lilley, 2005).

The DIGE approach is also directly amenable to multivariate statistical analyses such as principle component analysis and hierarchical clustering. These additional statistical tools can be extremely beneficial in visualizing the variation within a set of experimental samples. Importantly, they can help determine if the major source of variation is describing the biology or indicating unanticipated variation between samples (or introduced during sample preparation), as well as pinpoint subsets of proteins that respond

Figure 30.5 Fluorescence image of a multiplexed DIGE gel, 24 cm pH 4–7 (12% second dimension) in false-color representation. Samples had been labeled prior to separation with Cy2 (blue; labeling 100 μg human AGS cells), Cy3 (green; labeling 100 μg human AGS cells infected with *H. pylori* strain 26695), and Cy5 (red; labeling 100 μg human AGS cells infected with *H. pylori* strain 26695 deleted for the cag pathogenicity island). Individual Cydye signals were acquired using mutually exclusive excitation and emission spectra at 100 μm resolution, and recorded at 16-bit depth using a grayscale intensity from 1 to 100,000. Using Cy2 to label a mixed-sample internal standard enables a statistically powered differential expression experiment by registering independent-replicate samples (separately labeled with Cy3 and Cy5) from multiple experimental conditions across several matched DIGE gels. (See Color Insert.)

collectively to an experimental stimulus or classification (e.g., see Franco *et al.*, 2009; Friedman *et al.*, 2006, 2007; Hatakeyama *et al.*, 2006; Suehara *et al.*, 2006; and reviewed in Friedman and Lilley, 2009; Lilley and Friedman, 2004).

1.5.3. Software analytical tools

To facilitate the quantitative analysis of 2D gel experiments (whether they be DIGE or otherwise), several software programs are now available. These programs differ mostly in the algorithms used to detect protein features (boundaries) and gel–gel alignment/registration (e.g., vector-based image warping). In general, they all provide powerful analytical tools coupled with univariate (Student's *t*-test, ANOVA) and multivariate statistical analyses (e.g., principle components analysis and hierarchical clustering) that can be extremely beneficial in evaluating abundance changes of individual protein features as well as global expression patterns that can help discern changes that describe the biological phenotype from those that arise from unanticipated variation in the experiment.

2. Materials

2.1. Equipment

1. Electrophoresis system for IEF. Several formats are available from a number of commercial vendors. Most can focus up to 12 IPG strips simultaneously.
2. Programmable Power Supply capable of producing at least 3500 V but preferably 8,000–10,000 V.
3. Multigel system for second-dimension SDS–PAGE. Several formats are available for running 2, 4, 6, 12 or more gels in a single run.
4. Thermostatic circulator (necessary for most second-dimension systems; check specific vendors).
5. Immobiline DryStrip Reswelling Tray (GE Healthcare).
6. Sample-loading cups (required for IEF in gradients that include pH > 8).
7. Rotary shaker.
8. Image capture devices (e.g. densitometry, flatbed scanners, fluorescent imager (must be compatible with Cy dyes for best results with DIGE)).

2.2. Solutions and reagents

Most reagents and ready-made solutions are available from a variety of suppliers. All solutions are prepared fresh before use, except where indicated. All solutions should be prepared using water that has a resistivity of 18.2 mΩ-cm; this is referred to as "water" throughout the text.

1. IPG strips and accompanying ampholines and/or IPG buffers are commercially available from a number of vendors in a large variety of pH ranges. This includes wide range (pH 3–11, both as a linear and a nonlinear gradient), medium range (e.g., pH 4–7, pH 7–11), and narrow range (e.g., pH 5–6, pH 5.5–6.7). Strips are also available in a variety of lengths, ranging from low-resolution 7 cm to high-resolution 24 cm.
2. Sample buffer according to Rabilloud (1998): 7 *M* urea, 2 *M* thiourea, 4% (w/v) CHAPS, 1% (w/v) DTT, 2% (v/v) Pharmalytes 3–10, and cocktail of protease inhibitors (Complete, Roche, 1 tablet/50 mL solution). For samples containing more hydrophobic proteins, the following ASB14 lysis buffer may be beneficial: 7 *M* urea, 2 *M* thiourea, 2% amidosulfobetaine-14, and 50 m*M* Tris–HCl pH 8.0. Sample buffer is stored at −80 °C in small aliquots and should be thawed only once. For DIGE experiments, use the Rabilloud buffer without DTT and

without Pharmalytes or IPG buffers; other specifics to this technique can be found in other detailed methods chapters (Friedman and Lilley, 2009).
3. Hydroxyethyldisulphide (DeStreak, GE Healthcare) is supplied either as a liquid reagent or as a complete sample buffer formulation, for IEF in alkaline pH ranges.
4. Cell lysis buffer. The above sample buffers with 7 *M* urea, 2 *M* thiourea and 4% CHAPS may be used for protein solubilization. Alternatively, TNE buffer (50 m*M* Tris–HCl pH 7.6, 150 m*M* NaCl, 2 m*M* EDTA pH 8.0, 2 m*M* DTT, 1% (v/v) NP-40), and RIPA buffer (50 m*M* Tris–HCl pH 8.0, 150 m*M* NaCl, 1% NP-40, 0.5% deoxycholic acid, 0.1% SDS), can also be used if followed by a precipitation/cleanup step (see Section 3.2).
5. Ready-made solutions for SDS–PAGE: 30% acrylamide:bis-acrylamide (37.5:1), N,N,N,N′-tetramethylethylenediamine (TEMED), and ammonium persulfate, available from several suppliers (e.g., National Diagnostics, Bio-Rad, GE Healthcare).
6. 4× separating gel buffer: 1.5 *M* Tris–HCl pH 8.8.
7. Water-saturated butanol.[4]
8. 10× SDS–PAGE running buffer (1 L): 30.25 g Tris–base, 144.13 g glycine, and 10 g SDS (0.1%).
9. Equilibration solution: 50 m*M* Tris–HCl pH 8.8, 6 *M* urea, 30% (v/v) glycerol, 2% (w/v) SDS, and 0.01% (w/v) bromophenol blue.
10. Dithiothreitol (store desiccated).
11. Iodoacetamide (store desiccated, keep in the dark).

3. Methods

3.1. Protein sample preparation

Robust sample preparation is vital for any successful analytical measurement. Buffers and materials used should be of the highest quality, and care should be taken during sample procurement to increase reproducibility and to minimize unanticipated variation that might be measured in the analysis. Small molecule protease and phosphatase inhibitors such as aprotinin, leupeptin, pepstatinA, antipain, AEBSF, sodium orthovanadate, okadaic acid, and microcystin, among others (or commercial kits that utilize these

[4] Mix equal parts butanol and water and shake vigorously. Let the two phases separate overnight, and use the butanol phase for overlay. Butanol that is not completely water saturated can extract water from the top of the gel. A more recent modification is to use a 0.1% SDS solution in a conventional spray bottle, used to carefully spray a fine mist over the top of the gels to thoroughly cover the top of the gel (the gel/overlay interface will not be as obvious).

small molecule inhibitors), should be used. In particular, do not use inhibitors (e.g., soybean trypsin inhibitor) that will resolve on the 2D gel.[5]

1. Essentially any protein extraction buffer can be used, provided that samples are subsequently precipitated to remove nonprotein-based ionic components (that can severely interfere with IEF, see Section 3.2). Samples are then resuspended with a 2D gel-compatible buffer, and in many cases protein extracts can be made directly in this buffer and analyzed without precipitation. The following buffers contain sufficient chaotropic activity without providing additional ionic composition to the sample:

   A. Standard 2D gel electrophoresis lysis buffer (Rabilloud, 1998): 7 *M* urea, 2 *M* thiourea, 4% CHAPS, 2 mg/mL DTT, and 50 m*M* Tris–HCl pH 8.0.
   B. Modification for membrane-associated proteins: 7 *M* urea, 2 *M* thiourea, 2% amidosulfobetaine-14, and 50 m*M* Tris–HCl pH 8.0.

   Buffer systems containing other salts and detergents, especially sodium dodecylsulfate, may be more efficient at protein extraction, but must be precipitated prior to IEF. For example:

   C. TNE: 50 m*M* Tris–HCl pH 7.6, 150 m*M* NaCl, 2 m*M* EDTA pH 8.0, 2 m*M* DTT, and 1% (v/v) NP-40.
   D. RIPA buffer: 50 m*M* Tris–HCl pH 8.0, 150 m*M* NaCl, 1% NP-40, 0.5% deoxycholic acid, and 0.1% SDS.

2. It is critical to resuspend the cells rapidly to prevent proteolytic activity. Extracts should be centrifuged for 10 min at 15,000 or 100,000×*g* to remove insoluble proteins and phospholipids that can interfere with IEF.
3. Sonication can, in some cases, improve sample quality by disrupting nucleic acids which are subsequently removed by sample cleanup along with phospholipids. Both of these nonprotein-based ionic components can interfere with IEF (see Section 3.2). Short bursts with a tip sonicator, on ice, are suggested.
4. At all steps, it is important to keep the system chilled, especially in the presence of urea-containing samples that should never be heated. Excessive heating of urea (above 37 °C) can accelerate the formation of isocyanate (a natural breakdown product of urea), which will in turn carbamylate free amines. When this occurs on proteins (either at the amino-terminal residue or on the epsilon-amine group of lysine residues), it prevents these sites from becoming protonated, causing an acidic shift in the isoelectric point. The end result of heavy sample carbamylation in 2D gel electrophoresis is

[5] An excellent practical 2D electrophoresis handbook with many tips and tricks with respect to basic methodology can be downloaded for free at the following site: http://www5.gelifesciences.com/aptrix/upp01077.nsf/Content/2d_electrophoresis~2delectrophoresis_handbook

beautiful charge trains of proteins that appear to be post-translationally modified, but are completely artifactual.

5. Measure protein concentration using a variety of standard methods. Take care to use a method that is compatible with the buffer that the proteins are extracted in. For example, CHAPS and thiourea (although perfectly suitable for extraction) will interfere with either the Bradford or BCA assays, making the data inaccurate and unreliable. In these cases, aliquots should be precipitated prior to quantification in a suitable buffer, or the use of a detergent compatible assay should be utilized.
6. For cell culture experiments, protein concentration can be estimated from the cell number. For example, for protein samples prepared from the 697 pre-B lymphoma cell line (available from the German Collection of Microorganisms and Cell Cultures, Deutsche Sammlung von Mikroorganismen und Zellkulturen GmbH, DSMZ No: ACC 42), 10^7 cells correspond to approximately 1 mg of extracted proteins. It is essential to wash the cells thoroughly (at least twice in PBS) in order to remove components of the growth medium (especially serum proteins) before collecting the cells. For the final cell pellet, all PBS is carefully removed with a fine pipette tip. Cell pellets (usually 10^8 cells per tube) are stored at −80 °C.[6]
7. The extract can now be applied to the dehydrated IPG strips, either by in-gel rehydration or by cup loading. Appropriate ampholytes or IPG buffers should be added prior to sample application, at 0.5% (v/v) for most applications (but up to 2% can be tolerated if necessary). If bromophenol blue was not present in the sample buffer, add a small amount (either as a few grains of dry solid or a few microliters of a solution in water) to use as a tracking dye during IEF as well as a visual aid during cup loading. Alternatively, samples can be stored at −80 °C indefinitely.

3.2. Sample cleanup and precipitation

As already mentioned, the presence of nonprotein-based ions (e.g., salts, phospholipids) can sometimes interfere with IEF by reducing the resistance in the IPG strip such that it becomes difficult to attain the high voltages necessary for high-resolution focusing without overheating the strips. Since most commercial IEF units focus multiple strips together in a parallel circuit, one strip of significantly different resistance can adversely affect the resolution of the other strips.

Often a cleanup or precipitation step can remove these interfering ions, or at least normalize all samples to be of similar resistance. Precipitation not

[6] For protein extraction from cell pellets using 2D gel sample buffer, it is useful for the solution to contain 2% (v/v) Pharmalytes 3–10 to prevent aggregation of DNA. Cells (e.g., 10^8 cells) are best disrupted by rapid addition of 1 mL sample buffer (Rabilloud, 1998) to the cell pellet. Nucleases (DNase, RNase, and Benzonase) are commonly added as well, but their efficacy is questionable in this highly denaturing buffer.

only removes several contaminants at the same time, but also disrupts complexes efficiently and prevents protease activities irreversibly. The most effective methods for precipitation are using methanol and chloroform according to Wessel and Flugge (1984) or TCA, deoxycholate, and acetone according to Arnold and Ulbrich-Hofmann (1999). Before IEF, the proteins must be redissolved in solubilization solution (see Section 3.1). In addition, several protein precipitation kits for 2D gel analysis are available from a number of commercial vendors. An adaption of the method of Wessel and Flugge (1984) is described as follows:

1. Bring up predetermined amount of protein extract to 100 μL with water.
2. Add 300 μL (3-volumes) water.
3. Add 400 μL (4-volumes) methanol.
4. Add 100 μL (1-volume) chloroform.
5. Vortex vigorously and centrifuge; the protein precipitate should appear at the interface.
6. Remove the water/MeOH mix on top of the interface, being careful not to disturb the interface. Often the precipitated proteins do not make a visibly white interface, and care should be taken not to disturb the interface.
7. Add another 400 μL methanol to wash the precipitate.
8. Vortex vigorously and centrifuge; the protein precipitate should now pellet to the bottom of the tube.
9. Remove the supernatant and briefly dry the pellets in a vacuum centrifuge.
10. Resuspend the pellets in a suitable amount of 2D gel-compatible buffer (see Section 3.1).

When using precipitation as a cleanup step, it is recommended to have a starting protein concentration in the range of 1–10 mg/mL. When protein samples are too diluted, it will be difficult to quantitatively recover proteins following precipitation cleanup. Freeze/thawing should also be kept to a minimum; freezing samples in 1 mL aliquots or less will usually suffice. For DIGE experiments, this precipitation step greatly facilitates the requirement for sample labeling to be performed in 2D electrophoresis sample buffer that is devoid of free amines and pH balanced.

3.3. Isoelectric focusing using acidic range IPG gels

Most commercial IEF units can accommodate up to 12 individual IPG strips per run. It is important to keep any IPG strips that are resolved in the same run to be as equally matched as possible, because in most configurations the individual IPG strips form a parallel circuit between the anode and cathode electrodes. Any IPG strip that is significantly different from the others in composition (especially with respect to ionic content) can adversely affect the resolution of all of the strips, sometimes dramatically. It is, therefore, strongly recommended to also cofocus only IPG strips of the same pH gradient and strip length, using samples derived from the same experiment (e.g., sample types and composition should be a similar as possible).[7]

The following method is presented for 24 cm IPG strips pH 4–7:

1. IPG strips are rehydrated in the 2D gel sample buffer (see Section 3.1) with total protein amounts as high as 0.5–2 mg protein, diluted or resuspended in a final volume of 450 μL solution per 24 cm strip in a reswelling tray. The strips are overlaid with 2–3 mL paraffin oil to prevent urea crystallization.[8]
2. Rehydration of the strips in the presence of sample should take at least 12 h, but overnight rehydration is recommended, at room temperature. Some proteins, especially high-molecular-weight proteins, need more time to enter into the strip during the rehydration step.
3. After complete rehydration, the strips are moved to a flatbed electrophoresis system of which the temperature of the cooling block is held constant at 20 °C.
4. Humid paper wicks are applied on the ends of the rehydrated IPG strips, and electrodes are positioned in place and the strips are covered with paraffin oil.[9]
5. See Table 30.1 for typical focusing program.[10]

Rehydration loading is the easiest method for sample introduction for 2D gel electrophoresis, where protein samples in sample buffer are introduced passively while the IPG strip takes up the sample solution, and the

[7] When using the DIGE technique where different combination of samples are multiplexed and coresolved using a series of gels, the potential of the different strips containing different sample amounts is minimized by properly randomizing the loading of the samples such that each IPG strip contains the same variety of samples. For example, in a 3-dye experiment of control versus treated samples using four biological replicates (using Cy3 and Cy5 to randomly label each of the eight individual samples and Cy2 for the mixed-sample internal standard), each of the four gels would contain one control and one treated sample along with an aliquot of the Cy2-labeled mixed-sample internal standard.

[8] It is very important to make sure that the rehydration solution is evenly distributed across the entire strip length. The sample is first distributed along the length of a rehydration well and then the dehydrated IPG strip is carefully overlayed (with the plastic backing facing up) to ensure that the sample is evenly distributed beneath the overlayed strip. The IPG strips are covered with paraffin oil to avoid crystallization of the highly concentrated urea (*ca.* 10 *M* urea is saturation at room temperature).

[9] Placing humid paper wicks (filter paper wetted with water, but not oversaturated) between the electrodes and both ends of the IPG strips will provide a sink for excess ions as well as proteins with isoelectric points outside the pH range of the IPG strip. In extreme cases, it is possible to exchange the paper wicks during the IEF run. The paraffin oil prevents drying out of the strip and crystallization of the urea/thiourea present, and also serves to prevent carbon dioxide to enter the system. Carbon dioxide from the air can dissolve into the IPG gel, acting as a buffer with p*K* 6.3 and thus changing the pH gradient.

[10] IEF is mostly performed in a horizontal flatbed type of equipment. The focusing effect requires high field strength; therefore, a power supply with high voltage is required. IEF must be performed at an exactly controlled temperature, because the p*K* values of the proteins as well as the buffers are highly temperature dependent. Thus a reliable cooling system is necessary for this method to be reproducible. It is important to start with a low voltage to avoid aggregation of proteins. It is very useful to apply a voltage/time gradient on the strips in order to minimize differences in conductivities coming from varying salt loads between the IPG strips. The best results are obtained with high-end voltages of 8,000–10,000 V, with the overall amount of total volt-hours being ultimately important for high-resolution focusing. Optimal gradients and total Vh vary with strip length, pH range, and commercial formulations.

Table 30.1 Recommended isoelectric focusing programs using IPG gels[a]

Voltage program for 24 cm IPG 4–7			
Mode	Voltage (V)	Time (h)	Vh
Step and hold	500	1	
Gradient	1000	1	
Gradient	10,000	3	
Step and hold	10,000	5	
Total		10	65.0 kV h
Voltage program for 24 cm IPG 3–11NL or IPG 3–10NL			
Step and hold	150	2	
Step and hold	300	2	
Step and hold	500	2	
Gradient	1000	4.5	
Gradient	10,000	4.5	
Step and hold	10,000	1.5	
Total		16.5	45.0 kV h
Voltage program for 24 cm IPG 7–11NL			
Step and hold	150	2	
Step and hold	300	2	
Step and hold	500	2	
Gradient	1000	3.5	
Gradient	10,000	3.5	
Step and hold	10,000	5	
Total		18.0	73.8 kV h

[a] For best results, IEF should start at low voltage, and voltages should then slowly ramp up to their maximum as indicated above. In some circumstances, it may be beneficial to add additional low-voltage steps to the 4–7 program. Ultimately, it is the final kV h that becomes the most important factor for high-resolution IEF.

proteins are evenly distributed across the entire pH gradient. In some commercial IEF units, the IPG strip rehydration and focusing can be performed in the same apparatus, eliminating extra manual steps. This configuration also allows for the process of a so-called active rehydration, whereby applying a low voltage (30–50 V) across the strip during rehydration loading can, in some cases, improve the sample application, because this drives salt ions toward the electrodes, removes proteases from the proteins, and helps large molecules from entering the gel. However, this configuration of the IPG strip rehydration ("facedown") during IEF can also produce lower resolution in some cases due to the extra mechanical stress of the IEF gel being sandwiched between the focusing tray and the plastic gel backing, especially in the case of very abundant proteins. In these cases, using an apparatus that performs the focusing with the plastic backing directly in contact with the focusing tray (the gel is "sunny-side up") can be advantageous for more difficult samples (Fig. 30.3).

6. Optional: active rehydration (available with some instrument configurations). Sample rehydration occurs with the IPG strip face down and in contact with the electrodes, which supply a low (typically 30–50 V) electric field across the strip during rehydration. In some cases, active rehydration can be found to facilitate the entry of high-molecular-weight proteins into the IPG strip. IEF units with this configuration typically commence with the focusing with the IPG strips in the face-down orientation and do not require user intervention at this point.

3.4. Isoelectric focusing using alkaline range IPG gels

The great advantage of rehydration loading is the reduction of handling steps. However, it can also have some disadvantages especially in alkaline pH ranges where proteins can aggregate on the strip surface and cause horizontal streaks in the second-dimension gel.

The conditions for running alkaline IPG strips have to be changed compared to the acidic range. During IEF at alkaline pH, water transport and migration of the reducing agent DTT take place. To minimize these effects, protein samples are always applied by cup loading at the anodic side and not by in-gel rehydration (Fig. 30.3). The maximum protein load on alkaline gels is lower than on the acidic gels. It is strongly suggested not to load more than 0.5 mg protein per cup to prevent streaking.

The following method is presented for 24 cm IPG strips pH 3–11 and 7–11:

1. Ensure that the final sample volume will not exceed the volume of the cup (typcially between 100 and 120 μL, check manufacturer).[11]
2. The IPG strip is rehydrated (without protein sample) with the rehydration buffer containing HED (DeStreak; Olsson *et al.*, 2002) in the reswelling cassette. IPG buffers can be added at this stage to 0.5% final concentration for flatbed IEF and vertical SDS–PAGE. Since no proteins are present during rehydration, the time can be as short as 6 h (to enable rehydration and focusing to be done on the same day), but rehydration for >12 h is generally recommended.
3. After complete rehydration, the IPG strips are placed on a flatbed electrophoresis system of which the temperature of the cooling block is held constant at 20 °C.
4. Humid paper wicks are applied on the ends of the rehydrated IPG strips, and electrodes are positioned in place. Application cups are then

[11] The protein concentration in the sample cup should not be too high (<10 mg/mL) to avoid protein precipitation at the entry point. If possible, the sample should be diluted with the sample buffer.

carefully positioned just below the anode (on the acidic side of the gradient) such that the bottom of the cup is in contact with the top of the IPG gel and makes a seal, but not crushing the gel. The strips are then covered with paraffin oil.[9]

5. Ensure that oil does not leak into the cups before proceeding. Then add approximately 10–20 μL paraffin oil to the cup, and carefully pipet the sample under the oil (check manufacturer for maximum sample volume per cup).[12]
6. See Table 30.1 for typical focusing program.[10]

3.5. Equilibration of IPG gels

After the IEF step is completed, it is necessary to equilibrate the IPG strips in two steps prior to the second dimension. The equilibration buffer contains 6 *M* urea and 30% glycerol to diminish electroendosmotic effects (Sanchez *et al.*, 1997), which are responsible for poor transfer from first to second dimension. Further, the equilibration buffer contains 2% (w/v) SDS to make the proteins are negatively charged for the SDS–PAGE. Loss of proteins during the equilibration step and subsequent transfer from the first to the second dimension are mainly due to proteins which are strongly absorbed to the IPG gel matrix and due to wash-off effects. The majority of these proteins appear to be lost during the first minutes of equilibration, whereas protein losses during the second equilibration step are only minimal (Sanchez *et al.*, 1997).

1. After focusing is complete, the IPG strips are removed from the paraffin oil and incubated in the equilibration buffer. This can be done in a variety of vessels, sealed tubes, and even in the flatbed focusing IPG strip holder to minimize strip handling.
2. The strips are first equilibrated in 25–100 mL equilibration buffer supplemented with 1–2% (w/v) DTT (15–20 min) to reduce disulfide bonds.
3. The solution is discarded, and replaced with an equal volume of equilibration buffer supplemented with iodoacetamide (2.5 × w/w more than what was used for DTT) for an additional 15–20 min to alkylate the formed sulfhydryl groups with a carbamidomethyl group (net mass increase of 57 Da). This is to prevent reoxidation of the sulfhydryl groups during the second-dimension SDS–PAGE in order to prevent streaking of proteins.[13]

[12] Paper-bridge loading is a modification of the cup-loading technique that uses a piece of clean filter cardboard containing the protein sample solution and applied at one of the ends of the IPG strip. Both this as well as the conventional cup-loading technique needs more handling steps and can lead to losses of proteins with their isoelectric points close to the pH value of the application point. However, with these techniques all proteins have the same charge sign and will not form complexes during electrophoretic migration into the IPG strip, and so in some cases it can deliver better results especially for IEF at pH ranges greater than pH 8.

[13] Although suggested in several scientific papers, it is generally not recommended to prealkylate proteins prior to IEF, because it is often not easy to guarantee 100% alkylation.

4. Equilibrated IPG strips can now be run on the second-dimension SDS–PAGE gels, or can be double wrapped in plastic wrap and frozen for storage at − 20 or − 80 °C for up to several months. For samples labeled with DIGE, storage in this way for at least a week does not seem to affect the performance of the Cydyes. If necessary, IPG strips can be frozen in this way directly after focusing, leaving the equilibration and reduction/alkylation steps until the second-dimensional separation is ready to be performed.

3.6. SDS–PAGE: The second dimension

When equilibration is completed, the strips are ready for the second dimension (SDS–PAGE). When more than one gel is to be run (which is always the case in a comparative proteomics study), the most reproducible results can be obtained when gels are run together. Typically, electrophoresis chambers holding 10–12 large-format (24 × 20 cm) SDS–PAGE gels are used, and are available from a number of commercial vendors. For enhanced reproducibility, and to reduce depletion of SDS during the run, it is recommended to utilize a cooling system to keep the buffer at 20 °C.

Homemade gels are commonly used and are very economical but also very labor-intensive to prepare reproducibly. Readymade precast gels are now available on the market from a number of vendors, including using a low-fluorescence backing medium for DIGE. A variety of acrylamide concentrations with varying cross-linker ratios (acrylamide:bis-acrylamide) can be used. For general purposes, 12% homogeneous pore size gels (12% T; 2.6% C) afford the best compromise between ease of preparation and resolution. Gradient gels can provide better separation and resolution over a broader mass range, but are slightly more difficult to prepare reproducibly on a large scale. Dedicated gel casters for reproducible gradient gels are commercially available.

Hand-cast gels can either be made individually or in bulk using commercial multigel casters. The total volume of acrylamide solution necessary per multigel casting should be determined for each individual setup and equipment ahead of time. The following is an example for multiple gel casting where the final volume is 2.3 L. For smaller volumes, adjust the amounts proportionally.

1. Typically, glass cassettes with fixed spacers of 1.0 or 1.5 mm thickness are separated by plastic sheets to prevent the plates from sticking to each other after polymerization. If desired, each gel can be given an unique serial identity number by placing a piece of printed Whatman paper between the glass plates prior to casting.
2. In a vacuum flask is prepared 920 mL 30% (w/v) acrylamide:bisacrylamide solution (37.5:1), 575 mL 1.5 *M* Tris–HCl, pH 8.8 and 770 mL water (see Step 4).

3. The mixture is degassed for 10 min to remove microair bubbles that can interfere with polymerization and cause point streaking after protein staining in some cases.
4. Optional: after degassing, SDS can be added to the acrylamide solution to a final concentration of 0.01% (add 2.3 g in 35 mL water). However, many have shown that excellent resolution can be obtained without SDS at this stage, most likely because the SDS that carries proteins through the gel is already associated with the protein during the equilibrium stage and/or carried through by the running buffer (and any SDS present in the gel at the start of the run would leave the gel when the electric field is applied, before proteins entered the gel). Many commercial vendors now offer precast gels made both with and without SDS. If omitting SDS at this stage, then add 805 mL water instead of 770 mL in Step 2 (see Step 2).
5. Just prior to pouring, using gentle mixing or a stir bar, add 700 mg ammonium persulfate (APS, made fresh) and 300 μL TEMED to the acrylamide mixture. (Use proportionally less APS and TEMED when using smaller volumes of acrylamide mixtures for different pouring configurations).
6. Multigel casting chambers are typically filled from the bottom to a height of about 2 cm below the top of the glass plates. The gels are carefully overlaid with 1.0–1.5 mL buffer-saturated iso-butanol to allow for polymerization.[14]
7. After full polymerization, clean the cassettes of residual polyacrylamide residues, overlay with water or SDS running buffer, and double seal with plastic wrap for storage up to several weeks at 4 °C.
8. If first-dimensional IPG strips (equilibrated and reduced/alkylated) were previously frozen, allow a few minutes for the strips to thaw completely prior to unsealing and handling them. Rinse and remove the water or SDS running buffer out of the top well space of the second-dimension cassette.
9. The IPG strip should be dipped in SDS running buffer and then placed inside the top well space of the SDS–PAGE gel assembly. The plastic backing of the IPG strip should adhere to the inside surface of one of the glass plates (typically the longer plate if possible), and can then be easily tapped into place using a thin card or ruler (making sure that

[14] Casting a large number of gels requires some practice. The amount of catalyst that is used is minimal to prevent the gels from polymerizing too rapidly, which leads to excessive heating of the casting chamber. Ideally, initial polymerization, which can be seen by the development of a distinct gel surface below the butanol layer, should take 30–60 min. Subsequently, gels can be washed, covered with gel buffer, and stored at room temperature. The amounts of catalyst can be increased by 10% if necessary. "Fast" polymeration should occur within an hour or two, after which the isopropanol can be replaced with water or SDS running buffer. However, gels should ideally be left to polymerize overnight to enable complete polymerization, and the top of the gels should be lightly sealed with plastic wrap to prevent evaporation of the overlay (especially if the butanal is left on overnight).

pressure is applied to the plastic backing, not the IPG gel). Once the SDS running buffer is overlayed within the running chamber, good contact and removal of air bubbles can be easily achieved using the same tapping procedure. Once placed, the IPG strip should not move during the second-dimension run.

10. Optional: to ensure good contact between the strip and the gel, an agarose solution can be used to keep the IPG strip in place. The agarose solution is kept at 65°C and added first on top of the gel. Immediately after, the equilibrated strip is placed on the gel. This optional procedure is favored by some, but is cumbersome and can be difficult if the solution solidifies before the strip is in place and any air bubbles removed.[15]
11. Optional: a molecular weight marker can be run at the side (loaded through a small paper wick).
12. The current is set to 5–10 mA/gel initially to let the proteins enter the gel and subsequently set to 20 mA/gel over night at 15 °C until the blue front reach the end of the gel. If available, using constant wattage, setting less than 1 watt per gel for at least the first hour followed by no more than 15 watt per gel until completion is preferred because the voltage and amperage will vary during the run as salts and ions leave the gel.

After completion, gels can be fixed and stained using a variety of visual and fluorescent staining techniques. DIGE gels can be imaged on a fluorescent imager immediately after the second dimension is completed. Gels can then be stored in the refrigerator or in a cold room at 4 °C for months[16]. It is no problem to identify proteins using mass spectrometry from 2D gels that were stored over a long period of time.

REFERENCES

Anderson, N. G., and Anderson, N. L. (1978a). Analytical techniques for cell fractions. XXI. Two-dimensional analysis of serum and tissue proteins: Multiple isoelectric focusing. *Anal. Biochem.* **85,** 331–340.

Anderson, N. L., and Anderson, N. G. (1978b). Analytical techniques for cell fractions. XXII. Two-dimensional analysis of serum and tissue proteins: Multiple gradient-slab gel electrophoresis. *Anal. Biochem.* **85,** 341–354.

[15] The agarose overlay is prepared in such a way that it remains fluid at relative low temperatures. To realize this a mixture of low-melting agaroses is used (e.g., 0.4% (w/v) standard low M_r from Bio-Rad Laboratories, and 0.1% (w/v) type VII-A-low gelling temperature from Sigma), dissolved in 1× SDS running buffer. A trace of bromophenol blue can be added as tracking dye.

[16] For long-term storage, 0.02% (w/v) sodium azide can be added (optional) to prevent bacterial and fungal growth. For DIGE gels and other applications where the gel is affixed to one of the glass plates, the gels can be stored with the glass cassette intact to minimize dehydration during storage.

Arnold, U., and Ulbrich-Hofmann, R. (1999). Quantitative protein precipitation from guanidine hydrochloride-containing solutions by sodium deoxycholate/trichloroacetic acid. *Anal. Biochem.* **271,** 197–199.

Bjellqvist, B., Ek, K., Righetti, P. G., Gianazza, E., Gorg, A., Westermeier, R., and Postel, W. (1982). Isoelectric focusing in immobilized pH gradients: Principle, methodology and some applications. *J. Biochem. Biophys. Methods* **6,** 317–339.

Franco, A. T., Friedman, D. B., Nagy, T. A., Romero-Gallo, J., Krishna, U., Kendall, A., Israel, D. A., Tegtmeyer, N., Washington, M. K., and Peek, R. M. (2009). Delineation of a carcinogenic *Helicobacter pylori* proteome. *Mol. Cell. Proteomics* **25,** 25.

Friedman, D. B., and Lilley, K. S. (2009). Difference gel electrophoresis (DIGE). *In* "The Protein Protocols Handbook" (Walker John, ed.), 3rd edn. pp. 379–408. Humana Press, Totowa, NJ.

Friedman, D. B., Stauff, D. L., Pishchany, G., Whitwell, C. W., Torres, V. J., and Skaar, E. P. (2006). *Staphylococcus aureus* redirects central metabolism to increase iron availability. *PLoS Pathog.* **2,** e87.

Friedman, D. B., Wang, S. E., Whitwell, C. W., Caprioli, R. M., and Arteaga, C. L. (2007). Multi-variable difference gel electrophoresis and mass spectrometry: A case study on TGF-beta and ErbB2 signaling. *Mol. Cell. Proteomics* **6,** 150–169.

Gorg, A., Obermaier, C., Boguth, G., Harder, A., Scheibe, B., Wildgruber, R., and Weiss, W. (2000). The current state of two-dimensional electrophoresis with immobilized pH gradients. *Electrophoresis* **21,** 1037–1053.

Gorg, A., Weiss, W., and Dunn, M. J. (2004). Current two-dimensional electrophoresis technology for proteomics. *Proteomics* **4,** 3665–3685.

Hartinger, J., Stenius, K., Hogemann, D., and Jahn, R. (1996). 16-BAC/SDS-PAGE: A two-dimensional gel electrophoresis system suitable for the separation of integral membrane proteins. *Anal. Biochem.* **240,** 126–133.

Hatakeyama, H., Kondo, T., Fujii, K., Nakanishi, Y., Kato, H., Fukuda, S., and Hirohashi, S. (2006). Protein clusters associated with carcinogenesis, histological differentiation and nodal metastasis in esophageal cancer. *Proteomics* **6,** 6300–6316.

Hoving, S., Voshol, H., and van Oostrum, J. (2000). Towards high performance two-dimensional gel electrophoresis using ultrazoom gels. *Electrophoresis* **21,** 2617–2621.

Ibel, K., May, R. P., Kirschner, K., Szadkowski, H., Mascher, E., and Lundahl, P. (1990). Protein-decorated micelle structure of sodium-dodecyl-sulfate–protein complexes as determined by neutron scattering. *Eur. J. Biochem.* **190,** 311–318.

Karp, N. A., and Lilley, K. S. (2005). Maximising sensitivity for detecting changes in protein expression: Experimental design using minimal CyDyes. *Proteomics* **5,** 3105–3115.

Laemmli, U. K. (1970). Cleavage of structural proteins during the assembly of the head of bacteriophage T4. *Nature* **227,** 680–685.

Lilley, K. S., and Friedman, D. B. (2004). All about DIGE: Quantification technology for differential-display 2D-gel proteomics. *Expert Rev. Proteomics* **1,** 401–409.

O'Farrell, P. H. (1975). High resolution two-dimensional electrophoresis of proteins. *J. Biol. Chem.* **250,** 4007–4021.

O'Farrell, P. Z., Goodman, H. M., and O'Farrell, P. H. (1977). High resolution two-dimensional electrophoresis of basic as well as acidic proteins. *Cell* **12,** 1133–1141.

Olsson, I., Larsson, K., Palmgren, R., and Bjellqvist, B. (2002). Organic disulfides as a means to generate streak-free two-dimensional maps with narrow range basic immobilized pH gradient strips as first dimension. *Proteomics* **2,** 1630–1632.

Rabilloud, T. (1998). Use of thiourea to increase the solubility of membrane proteins in two-dimensional electrophoresis. *Electrophoresis* **19,** 758–760.

Righetti, P. G., Simo, C., Sebastiano, R., and Citterio, A. (2007). Carrier ampholytes for IEF, on their fortieth anniversary (1967–2007), brought to trial in court: The verdict. *Electrophoresis* **28,** 3799–3810.

Sanchez, J. C., Rouge, V., Pisteur, M., Ravier, F., Tonella, L., Moosmayer, M., Wilkins, M. R., and Hochstrasser, D. F. (1997). Improved and simplified in-gel sample application using reswelling of dry immobilized pH gradients. *Electrophoresis* **18,** 324–327.

Schagger, H., and von Jagow, G. (1991). Blue native electrophoresis for isolation of membrane protein complexes in enzymatically active form. *Anal. Biochem.* **199,** 223–231.

Suehara, Y., Kondo, T., Fujii, K., Hasegawa, T., Kawai, A., Seki, K., Beppu, Y., Nishimura, T., Kurosawa, H., and Hirohashi, S. (2006). Proteomic signatures corresponding to histological classification and grading of soft-tissue sarcomas. *Proteomics* **6,** 4402–4409.

Unlu, M., Morgan, M. E., and Minden, J. S. (1997). Difference gel electrophoresis: A single gel method for detecting changes in protein extracts. *Electrophoresis* **18,** 2071–2077.

Vesterberg, O. (1972). Isoelectric focusing of proteins in polyacrylamide gels. *Biochim. Biophys. Acta* **257,** 11–19.

Vesterberg, O., and Svensson, H. (1966). Isoelectric fractionation, analysis, and characterization of ampholytes in natural pH gradients. IV. Further studies on the resolving power in connection with separation of myoglobins. *Acta. Chem. Scand.* **20,** 820–834.

Wessel, D., and Flugge, U. I. (1984). A method for the quantitative recovery of protein in dilute solution in the presence of detergents and lipids. *Anal. Biochem.* **138,** 141–143.

Westermeier, R. (2005). Electrophoresis in Practice (4th edn). Method 10 - IEF in immobilized pH gradients, pp. 257–273. Wiley-VCH.

CHAPTER THIRTY-ONE

Protein Gel Staining Methods: An Introduction and Overview

Thomas H. Steinberg

Contents

1. Introduction	542
2. General Considerations	543
3. Instrumentation: Detection and Documentation	545
4. Total Protein Detection	545
4.1. Colorimetric total protein stains	545
4.2. Fluorescent total protein stains	549
4.3. Preelectrophoresis sample labeling	556
5. Phosphoprotein Detection	556
5.1. Pro-Q Diamond phosphoprotein gel stain	557
5.2. Phos-tag phosphoprotein stains	557
6. Glycoprotein Detection	557
6.1. General glycoprotein detection	557
6.2. *O*-GlcNAc detection	559
References	559

Abstract

Laboratory scientists who encounter protein biochemistry in many of its myriad forms must often ask: is my protein pure? The most frequent response: run a denaturing SDS polyacrylamide gel. Running this gel raises another series of considerations regarding detection, quantitation, and characterization and so the next questions invariably center on suitable protein gel staining and detection methods. A total protein profile can be determined with the colorimetric methods embodied in Coomassie Blue and silver staining methods, or increasingly, with fluorescent stains. Protein quantitation can be done following staining, with fluorescence- and instrumentation-based methods offering the greatest sensitivity and linear dynamic range. Protein posttranslational modifications such as phosphorylation and glycosylation can be reliably determined with several fluorescence-based protocols. Staining and detection with two or more different stains can be done in series to establish relative profiles of modified versus total protein or to assess purity at two levels of quantitative

McArdle Laboratory for Cancer Research, University of Wisconsin–Madison, Madison, Wisconsin, USA

Methods in Enzymology, Volume 463
ISSN 0076-6879, DOI: 10.1016/S0076-6879(09)63031-7

sensitivity. The choice of staining method and protocol depends on the required rigor of detection and quantitation combined with available instrumentation and documentation capabilities. Other considerations for staining methods include intended downstream analytical procedures such as mass spectrometry or peptide sequencing, which preclude some methods. Nonfixative staining methods allow western blotting after gel staining. Laboratory custom and budget or intellectual curiosity may be the ultimate determinate of the chosen gel staining protocol.

1. Introduction

This chapter focuses on electrophoresis-based protein detection with emphasis on current methods for detection of total protein and detection of the most frequently encountered protein posttranslational modifications (PTMs), phosphorylation, and glycosylation. In the 20 years since the publication of first edition of this volume, denaturing sodium dodecyl sulfate polyacrylamide gel electrophoresis (SDS–PAGE) followed by protein staining has remained a fundamental laboratory procedure to determine the composition and characterization of protein-containing mixtures at all stages of a purification scheme. It is standard practice to display protein profiles from a sampling of all purification steps—cell lysate to the final product—to evaluate polypeptide composition, purity, and quantitation. Thus a sample from each step should reveal a progressively simpler polypeptide profile showing a greater proportion of the target protein.

The previously summarized methods and principles of protein electrophoresis and detection remain relevant (Dunbar *et al.*, 1990; Garfin, 1990a,b; Merril, 1990).

The widespread use of SDS–PAGE protein analytical methods has driven the commercial development and common use of standardized electrophoresis apparatus and reagents, precast multiwell 1D and 2D slab gels (minigels; *ca.* 8×8 cm, $\times$ 1.0 mm thick), and ready-to-use staining formulations or kits for simplified, rapid staining. Additionally, proteomics-oriented electrophoresis techniques have emerged, directed toward analysis of complex protein mixtures by large format 2-D gel electrophoresis followed by mass spectrometry (MS) of excised protein spots. The utility of gel-based methodology in proteome analysis has spurred further commercial development of extant colorimetric stains for total protein and development of new fluorescent stains and modalities for total protein and for protein PTMs, driving the coevolution of instrumentation for sensitive detection, easy documentation, and accurate quantitation of both fluorescent and colorimetric signals. A recent compendium of proteomics protocols describes methods and applications for protein detection in 2-D and

1-D gels and subsequent analytical methods (Walker, 2005). Recent reviews of protein detection in gels for proteomics summarize the history and scope of gel staining and detection methods (Miller *et al.*, 2006; Patton, 2000, 2002; Smejkal, 2004). Clearly, there is convergence in the considerations and methods for protein identification in gels for proteomic analysis and protein purification.

Zymography—in-gel enzymatic activity analysis—is a complementary, gel-based protein characterization methodology beyond the scope of this chapter. A comprehensive handbook of zymographic protocols has recently been published (Manchenko, 2003). A related, emerging field—activity-based proteomics—utilizes small molecule probes to covalently link reporter moieties to catalytic sites of specific classes of active enzymes in complex mixtures such as lysates or whole cells. This increasingly popular approach can be used to in-gel assays to identify enzymes and to track their purification (Jessani and Cravatt, 2004; Paulick and Bogyo, 2008).

A detection method that accurately evaluates the total protein profile is usually sufficient for a general purification protocol. Protein detection in SDS–PAGE can be done by labeling samples—by covalent modification with a fluorescent dye—prior to running gels or, more commonly, by non-covalent postelectrophoresis staining with organic dyes or by metal deposition techniques. Posttranslational modifications can be measured by specialized staining formulations and protocols that target the modified moieties in a background of total protein. PTM detection methods always require a control lane displaying proteins known to contain and others known to be devoid of the targeted PTM. If selective PTM detection utilizes fluorescence, bear in mind that intrinsic protein fluorescence may be a confounding factor, especially if ultraviolet (UV) light-excited, blue or green light-emitting fluorophores are used. Therefore, an additional control—chemical or enzymatic treatment of a sample—to remove the postulated PTM may be necessary.

2. General Considerations

Common features of all gel staining protocols are grounded in good laboratory practices: cleanliness, careful manipulations, and attention to detail. Postelectrophoresis gel staining is generally accomplished in covered polycarbonate or polypropylene dishes. Accumulation of residual stain may compromise results, particularly with fluorescent stains. Dishes should be cleaned with 70–100% ethanol or methanol followed by a water rinse. The gels are incubated on a rocking platform or an orbital shaker or at a moderate speed, for example, 60 rpm. The gel should float freely in solution, neither stuck to the bottom of the dish nor float on top of the fixing, washing, and staining solutions. If a stacking gel is used, it should be removed; if a commercial, "prepoured" gel is used, the wells and the foot of the gel should be trimmed away to reduce staining deposition artifacts, particularly if the SDS bolus from the sample buffer remains in the gel—this typically travels with the Bromophenol Blue tracking dye. For all methods and protocols presented below, pure—dionized or double-distilled—water is used.

Historically, standard gel staining protocols required several sequential procedures, some requiring more than one step and some requiring intervening washes with water or buffer; common to all were:

1. The gel is fixed to immobilize protein in the gel and to remove SDS, buffers, or other interfering components.
2. The gel is stained with organic dyes or silver deposition formulations.
3. Quenching a staining reaction or destaining to reduce background was required.

Fixation generally has been accomplished with variations of acidic alcohol mixtures (e.g., 7–10% acetic acid, 30–50% methanol); organic dyes and destaining solutions were often formulated with the same solvents as for fixation. Given that literally hundreds of variations of protein stains and staining protocols have been introduced in the literature over the years—often with lengthy, painstaking protocols and competing claims of detection sensitivity (or other features of utility)—the commercialization of stains, kits, and protocols have been something of a boon to the yeoman laboratory scientist, although the proprietary considerations that accompany commercial development may hinder clear description and understanding of some of these reagents. Trade names that are useful for broad product recognition can be a bit confusing. For example, "Gelcode" refers to a product line that encompasses a range of total and PTM-specific stains, while "SYPRO" refers to a product line of several total protein gel stains that are chemically diverse. For the curious, material safety data sheets may give clues to buffer compositions and the patent literature may allow gleaning of critical physical or chemical information needed to fully understand a staining method or material composition. Many kits have their basis on the open literature, and vendors that reveal more rather than less about the intellectual source and composition of the product are preferred.

Gel staining protocols can vary even within the same staining formulations, depending on the time considerations, and desired quantitative rigor of detection. Not surprisingly, there is a strong market drive for: (1) instant gratification and reasonable quantitation and (2) staining formulations with protocols that avoid strong acids, organic solvents, or heavy metals, due to environmental and disposal concerns. Thus vendors generally offer "rapid protocols" for total protein detection formulations. These may involve truncation of fixation or staining steps and, frequently, acceleration of staining or destaining by judicious heating in a microwave oven. In considering the use of commercialized stains, it is important to consult product manuals and company web sites, since products are frequently updated and supported by several alternate, one hopes, detailed, accurate protocols.

3. Instrumentation: Detection and Documentation

Obviously, detection by visual inspection has been the basis of protein gel staining; documentation has been accomplished by photography of stained gels or by drying stained gels and pasting them into a laboratory notebook. Unless quantitative data are required, most purification detection and documentation requirements now can be met with relatively low-cost flatbed scanners or with camera mounted visible or UV light box stations. In the past two decades, instrumentation for detection and documentation (often with quantitation software), driven by fluorescence detection requirements, has become widely available for both fluorescent and colorimetric protein detection modalities. Fluorescence detection can be as simple as visual inspection with a handheld light-emitting diode (LED) "keychain" flashlight or as complex as fully automated multiple mode imaging stations with linear detection capabilities spanning 4 or 5 orders of magnitude and correspondingly powerful image analysis and quantitation software for advanced proteomics applications.

The manufactures of advanced instrumentation generally provide good information regarding the best acquisition modes and data analysis for protein colorimetric and fluorescent stains, often providing validated protocols and examples of applications of commercial or open-source staining formulations. For fluorescent applications, excitation and emission setting are frequently identified not only by the wavelength settings, but also by the names of frequently used fluorescent dyes or products for DNA or protein stains (e.g., Cy dyes, ethidium bromide, SYBR dyes, SYPRO dyes). In recent years, new varieties of simple transilluminators with LED or other inexpensive filtered monochromatic light sources, paired with an orange-filtering emission cover, offer inexpensive, safe modes of detection of many of the UV-excitable dyes.

4. Total Protein Detection

4.1. Colorimetric total protein stains

Simplicity of detection by visual inspection, relative ease of use, and widespread familiarity among the user base continue to ensure that staining with Coomassie Brilliant Blue (CBB) is the most commonly used total protein gel stain, and silver staining continues to be the alternative colorimetric method, for increased detection sensitivity over Coomassie staining. If mass spectrometry of excised protein is desired, staining methods that are do not introduce covalent protein modifications are preferred.

4.1.1. Coomassie Brilliant Blue staining

CBB R-250 and the dimethyl derivative CBB G-250 are disulfonated triphenylmethane dyes that stain protein bands bright blue. The dyes bind via electrostatic interaction with protonated basic amino acids (lysine, arginine, and histidine) and by hydrophobic associations with aromatic residues. The dye does not bind to the polyacrylamide with high affinity but does penetrate the gel matrix and bind with low affinity, necessitating a destaining step, unless the stain is a colloidal formulation. Coomassie stains are generally treated as endpoint stains, allowing significant time flexibility for fixation and staining steps. Coomassie stains are noncovalent and are reversible and do not interfere with subsequent mass spectrometry of excised protein bands.

Staining with CBB R-250 has been done with 0.025–0.10% (w/v dye) in an acidic alcohol formulation: 30–50% (v/v) methanol (or, less frequently, ethanol) with 7–10% (v/v) acetic acid solution. The dye is dissolved, and the solution is then filtered through Whatman #1 paper to produce a stable formulation. Fixation and staining are typically done in ~10 gel volumes of solution, for example, 50 ml for a standard minigel. For the most sensitive and consistent results, the gel is fixed and stained in the same acidic alcohol solution; under these conditions, the gel shrinks. Destaining is in 7% acetic acid, minus the alcohol to return the gel to full size. Destaining generally requires several solution changes and can be accelerated by addition of dye-adsorbent paper or foam rubber/plastic.

Thorough analyses of Coomassie staining methods have resulted in colloidal formulations such that the stain does not effectively enter the gel matrix while binding specifically to protein bands, allowing low-background staining with reliable quantitation and increased sensitivity relative to the standard CBB-R 250 staining method, above. For colloidal staining, CBB G-250 is preferred to CBB R-250; initial formulations contained 0.1% CBB G-250 in 2% (w/v) phosphoric acid, 6% (w/v) ammonium sulfate (Neuhoff *et al.*, 1985). An improved, widely used formulation is prepared by dissolving 10% (w/v) ammonium sulfate in 2% (w/v) phosphoric acid followed by addition of CBB G-250 to 0.1% (w/v) from a 5% (w/v) aqueous stock solution; this is combined with 20% (v/v) methanol prior to staining (Neuhoff *et al.*, 1988). The addition of methanol may increase the proportion of monodisperse dye, resulting in more rapid staining and more intense bands, but with some background; increasing ammonium sulfate concentrations drives the dye toward the colloidal state. Another, more recent formulation recommends greater dye and phosphoric acid concentrations in a final formulation of 0.12% (w/v) dye, 10% (w/v) ammonium sulfate, 10% (w/v) phosphoric acid, and 20% (v/v) methanol in a single stable staining solution, prepared by sequentially combining desired amounts of phosphoric acid, ammonium sulfate, and powdered dye in an aqueous solution to 80% of the desired volume, then adding methanol to the final volume (Candiano *et al.*, 2004). Colloidal staining solutions should not be filtered.

Most of the commercially available Coomassie staining formulations are proprietary derivatives of the various formulations and protocols referenced above.

Recommendations regarding fixation vary, but good results can be obtained with a simple procedure in which following electrophoresis the gel is washed in water to remove most of the SDS placed in the staining solution for several hours to overnight. Protein bands are visible within the first few minutes, and background is initially clear, and eventually weakly blue. Following staining the gel is washed with water to reduce background. A useful, recent succinct summary of Coomassie staining protocols includes a "hot" Coomassie staining protocol utilizing an acetic acid-based CBB R-250 formulation at 90 °C or a "fast" colloidal TCA-based CBB G-250 formulation at 50 ° C (Westermeier, 2006). Several of the commercially available formulations include simple microwave-based protocols to accelerate staining (and destaining) with CBB G-250 colloidal formulations.

4.1.2. Silver staining

Silver staining, widely regarded as the standard by which all other "ultrasensitive" staining methods are judged, is the most complex and variable protein gel staining methodology, with many dozens of published protocols, all of them requiring several stepwise procedures; the basis of protein detection is reduction of protein-bound silver ions to metallic silver and, to a lesser extent, in some protocols, localized deposition of silver sulfide. Silver staining may set the standard of rigor for ultimate detection sensitivity but quantitation is not a simple matter, due to the complex nature of the color development step and difference between proteins for silver signal sensitivity. There are commercially available silver staining kits that variously employ validated protocols and reagent mixtures from the literature that allow reproducible results for first-time users and thus can be an important teaching and benchmarking tool for those who will develop and optimize "home brew" silver staining reagents and protocols for routine in-house use. Silver staining principles, methods, and protocols are reviewed and summarized by Merril (1990), Rabilloud (1990), and Poland *et al.* (2005).

The key steps are, generally:

1. The gel is fixed to immobilize protein bands and to remove from the gel interfering substances such as SDS, buffers, and salts that bind silver and would contribute to background staining.

2. The gel is incubated with substances that bind proteins and enhance silver binding or that interfere with background staining by residual unbound silver. Taken together, these various strategies are referred to as sensitization.
3. The gel is impregnated with silver ions.
4. The colorimetric silver stain signal is developed by reduction of bound silver ions to insoluble and visible metallic silver.
5. Development is stopped, to prevent background staining in the gel matrix that would obscure protein bands. The time dependency of the latter two steps in particular contribute to the complexity of the technique, as does the necessity for several time-dependent washes among steps 2–4.

Silver staining protocols are broadly distinguished on the basis of the silvering agents and corresponding development conditions. Alkaline methods use a silver diamine complex in an alkaline environment as a silvering agent with development in an acidic formaldehyde solution, whereas acidic methods use weakly acidic silver nitrate as a silvering agent with development in an alkaline formaldehyde solution. Acidic staining methods have recently become more popular on the basis of reduced cost and the perception that background staining is more easily controlled, relative to alkaline methods.

There are a wide variety of sensitizing agents. Aromatic sulfonates such as naphthalene disulfonate or sulfosalicylic acid bind to proteins and also bind silver; residual protein-bound SDS may also contribute to increased silver binding. In a similar vein, CBB-stained gels that are poststained with silver stain also show enhanced silver staining intensity. Glutaraldehyde is widely used as a sensitizing agent; the basis of utility is binding to free amino groups in proteins and thus the introduction of protein-bound reducing aldehyde groups. Sulfiding reagents such as tetrathionates or thiosulfates can introduce in close proximity to protein a free S^{2-} ion that immediately reacts with silver ion to form insoluble silver sulfide.

Mass spectrometry of proteins following silver staining may be problematic due to protein modification resulting from use of glutaraldehyde in fixation and sensitization steps. Protocols that are compatible with mass spectrometry omit glutaraldehyde, which may reduce relative sensitivity. The sensitizing steps rely on tetrathionate and thiosulfate, and the development step is also correspondingly extended. In that regard, Colloidal Coomassie Blue staining followed by a rapid glutaraldehyde-free silver staining protocol may be a convenient two-step colorimetric approach to evaluate the purity of a protein sample when subsequent mass spectrometry is desired.

4.1.3. Zn^{2+} reverse staining

There are several staining methods that use metal cations to rapidly stain proteins after SDS–PAGE without recourse to fixatives or organic dyes. Zinc ion "reverse" staining utilizes the ability of proteins and protein–SDS complexes to bind and sequester Zn^{2+} in a milieu where the precipitation reaction between imidazole and zinc ion produces zinc imidazolate ($ZnIm_2$) to create an opaque background, contrasting with transparent protein–SDS–Zn^{2+} zones. The technique is useful as a nonfixative procedure where subsequent microanalyses such as biological or enzymatic assays of eluted bands are anticipated and is compatible with mass spectrometry or microsequencing methods. The method is rapid, with detection sensitivity reported to rival Coomassie Blue staining.

A rapid protocol suitable for 1 mm thick gels is as follows:

1. Following electrophoresis, the gel is incubated in 0.2 *M* imidazole, 0.1% SDS for 15 min.
2. The imidazole solution is discarded and the gel is developed by incubation in 0.3 *M* zinc sulfate for 30–45 s.
3. The developer is discarded and the gel is washed several times, with water, 1 min per wash.
4. The gel is stored in 0.5% (w/v) sodium carbonate.

Overdevelopment can be problematic, but can be corrected by incubation in 100 m*M* glycine to dissolve excess $ZnIm_2$. Though the transparent protein bands against the opaque background do not provide the striking colorimetric contrast of Coomassie or silver staining, imaging can be done with the gel illuminated against a black background (Fernandez-Patron, 2005; Fernandez-Patron *et al.*, 1998).

4.2. Fluorescent total protein stains

Fluorescent staining methods can combine detection sensitivity that rivals silver staining with workflow advantages similar to Coomassie or zinc ion staining, and offer linear quantitation ranges 10–100-fold greater than the colorimetric methods. Detection is instrumentation dependent, requiring a monochromatic excitation light source, selective optical filtration to separate the longer wavelength emitted light from the shorter wavelength (and much brighter) excitation light, and a detection mode. For many fluorescent stains, detection can also be by visual inspection, but with reduced sensitivity in comparison to photographic or instrumentation methods. With any fluorescent compound, some degree of photobleaching is a consequence of light exposure. Many of the commercially available fluorescent stains have been developed to be relatively photostable. Nevertheless, it is prudent to avoid prolonged exposure to intense ambient light prior to visual inspection and image acquisition. Excitation and emission maxima of the fluorescent stains and dyes discussed in this chapter are presented in Table 31.1. Customarily, and for the purpose of discussion below, light excitation and emission colors are frequently distinguished in broad categories: ultraviolet (UV), 250—400 nm; blue,

Table 31.1 Fluorescent protein gel stains

Fluorescent protein gel stain	Target	Excitation peaks (nm)	Emission peak (nm)	Vendor
Nile red	Total protein	270, 550	600	Sigma-Aldrich
SYPRO Orange	Total protein	280, 470	569	Life Technologies
SYPRO Red	Total protein	280, 547	631	Life Technologies
SYPRO Tangerine	Total protein	280, 490	640	Life Technologies
SYPRO Ruby	Total protein	280, 450	610	Life Technologies
Deep Purple	Total protein	400, 500	610	GE Healthcare
Krypton	Total protein	520	580	Thermo Fisher
Krypton Infrared	Total protein	690	718	Thermo Fisher
Flamingo	Total protein	270, 512	535	Bio-Rad
LUCY 506	Total protein	505	515	Sigma-Aldrich
LUCY 569	Total protein	569	580	Sigma-Aldrich
C16-FL	Total protein	470	530	Life Technologies
Pro-Q Diamond	Phosphoprotein	555	580	Life Technologies
Phos-tag 300/460	Phosphoprotein	300, 460	630	Perkin Elmer
Phos-tag 540	Phosphoprotein	540	570	Perkin Elmer
Pro-Q Emerald 300	Glycoprotein	288	533	Life Technologies
Pro-Q Emerald 488	Glycoprotein	512	525	Life Technologies
Krypton glycoprotein	Glycoprotein	654	673	Thermo Fisher
Glycoprofile III	Glycoprotein	430	480	Sigma-Aldrich
TAMRA alkyne	*O*-GlcNAc	545	580	Life Technologies
Dapoxyl alkyne	*O*-GlcNAc	370	580	Life Technologies

Fluorescent protein gel stains discussed in the text are summarized regarding stain specificity, excitation and emission maxima, and vendor. The web sites are as follows, with parenthetical indications regarding the division specializing in protein stains, if applicable: Bio-Rad Laboratories, http://www.bio-rad.com; GE Healthcare (Amersham), http://www.gehealthcare.com; Life Technologies (Invitrogen, Molecular Probes), http://www.lifetechnologies.com; Perkin Elmer, http://www.perkinelmer.com; Sigma-Aldrich, http://www.sigma-aldrich.com; Thermo Fisher Scientific (Pierce), http://www.thermofisher.com.

400–500 nm; green, 500–550 nm; yellow/orange, 550–580 nm; red, 580–650 nm; near infrared, 650–850 nm.

Fluorescent stains fall into two general categories: (a) fluorogenic stains that show significant fluorescent enhancement corresponding to localization with protein bands and (b) intrinsically fluorescent stains that bind selectively to protein bands and do not bind to the gel matrix. The first commercialized fluorescent total protein stains, SYPRO Orange and SYPRO Red gel stains, were developed in the 1990s in the context of routine fluorescence detection of stained DNA in gels by 300 nm UV transillumination followed by Polaroid photography, with the intent of introducing a simple, one-step total protein-specific staining and documentation work flow similar to ethidium bromide or SYBR Green DNA staining methods, with detection sensitivity exceeding Coomassie Blue staining methods. Further development of total protein stains has resulted in several commercial products or formulations with detection sensitivity limits matching or exceeding silver staining detection limits. Fluorescent gel stains that do not require preelectrophoresis covalent sample labeling are compatible with subsequent mass spectrometry of eluted protein bands.

4.2.1. Nile red protein gel stain

Nile red is a phenoxazone dye that shows strong fluorescence enhancement upon transition from aqueous to hydrophobic environments such as SDS micelles or protein–SDS complexes. Nile red does not interact significantly with SDS monomers. This property has been exploited to develop a rapid, nonfixative total protein staining method for SDS gels. The protocol relies upon electrophoresis in nonstandard conditions with running buffer SDS at 0.05% (w/v), below the detergent's critical micelle concentration, rather than 0.1% (w/v) SDS typically used in 1D and 2D SDS–PAGE. Protein sample preparation is standard (as in Garfin, 1990a,b) such that, it is argued, SDS–protein complexes remain stable during electrophoresis and protein band migration is considered to be normal. The protocol is simple and rapid: Nile red is diluted from a stock solution (0.4 mg/ml in DMSO) 200-fold into water to 2 μg/ml, and a 10-fold volume excess (e.g., 50 ml staining solution for a 5 ml minigel) is added to a gel with immediate and thorough agitation. The dye precipitates quickly in water such that staining is optimal in 2–5 min and does not improve thereafter. Following staining, the gels are briefly washed with water. Excitation can be with UV light or with green light sources; protein bands appear pale red. Detection sensitivity is similar to that obtained with Coomassie stains. Because there is no fixation, Nile red-stained gels can be subsequently electroblotted with good transfer efficiency (Daban, 2001; Daban *et al.*, 1991). High fluorescent background in the gel matrix can be problematic due to dye precipitation, and the affinity of this stain for SDS micelles combined with dye insolubility precludes use of this after dye SDS–PAGE done with running buffer containing 0.1% SDS. Photostability can also be problematic.

4.2.2. SYPRO Orange, SYPRO Red, and SYPRO Tangerine protein gel stains

SYPRO Red and SYPRO Orange protein gel stains are described by Steinberg *et al.* (1996a,b) and Haugland *et al.* (1997); SYPRO Tangerine protein gel stain is described by Steinberg *et al.* (2000b) and Yue *et al.* (2003); all three stains are also discussed by Steinberg *et al.* (2005). These dyes contain a hydrophilic functional group, an aromatic fluorophore, and an aliphatic tail, facilitating both good aqueous solubility and intercalation into protein–SDS complexes, SDS micelles, or membranes, with strong fluorescent enhancement in nonpolar environments. These features, combined with good chemical and photostability allow simple, versatile staining protocols following SDS–PAGE under standard buffer conditions (0.1% SDS) with detection sensitivity surpassing Colloidal Coomassie Blue G-250. Staining sensitivity can be increased a further fourfold with nonstandard buffer conditions (0.05% SDS; cf. Nile red) but that is not a required feature for utility with these dyes.

SYPRO Orange and SYPRO Red, proprietary sulfopropylaminostyryl dyes, are commercially available as 10 m*M* stock solutions in DMSO. The staining protocol is simple:

1. Following SDS–PAGE, the gel is placed in staining solution.
2. Prior to image acquisition, the gel is washed briefly with water.

To prepare a staining solution, the dye is diluted 5000-fold to 2 μM in dilute acetic acid (7%, v/v is standard; 2–10% is equally effective). The staining solution typically is prepared fresh, but remains stable for many months. Following electrophoresis the gel is placed in 10–20 gel volumes (50–100 ml for a 5-ml volume minigel) of staining solution, typically in a polypropylene or polycarbonate dish with continuous gentle agitation. Staining can be monitored periodically by placing the dish—with the gel remaining in the staining solution—on a UV light box or on a blue-light transilluminator, or staining can be monitored with a handheld UV light or a blue LED. For a typical 10 μg cell lysate, fluorescent bands of abundant proteins can be detected with 10 or 15 min against a fluorescent background. As staining progresses, protein band signal increases while background decreases, and the less-abundant proteins become more visible. For a 1-mm-thick gel, staining is complete in 1 h and is stable for many days with the gel remaining in the staining solution. It appears that in dilute acetic acid the protein–SDS–dye complexes are stabilized, while the remaining SDS diffuses out of the gel, resulting in low background. Destaining thus is minimal. Prior to imaging acquisition, the gel is washed with two changes of water for 2–3 min per wash to remove from the gel surface residual dye, SDS, and acetic acid.

SYPRO Tangerine protein gel stain, a proprietary carbazolylvinyl stain, is distinguished from SYPRO Orange and SYPRO Red protein gel stains on the basis of recommended staining diluent and intended utility. This protein stain can be used in a neutral pH buffered salts diluent—50 m*M* phosphate, 150 m*M* NaCl, pH 7.0—such that stained protein bands are not fixed and can be subsequently subjected to zymography, elution and renatured for *in vitro* activity assays, or electroblotted. The dye is commercially available as a 10 m*M* stock solution; staining solutions are prepared fresh with 5000-fold dilution to 2 μM. Staining proceeds as with SYPRO Orange, and is complete within 1 h; the gel is washed briefly with water before image acquisition. Excitation is with UV light or a blue light source.

4.2.3. SYPRO Ruby protein gel stain and other Ruthenium-based formulations

Organometallic ruthenium ion-based luminescent stains for protein detection in gels and on blots (Bhalget *et al.*, 2001), introduced as SYPRO Ruby protein gel stains, were developed to provide fluorescent detection sensitivity rivaling or exceeding silver staining methods with simple end-point staining protocols, and were introduced in ready-to-use formulations for staining after SDS–PAGE (Berggren *et al.*, 2000) or after isoelectric focusing electrophoresis (Steinberg *et al.* 2000a). The development and formulation of these stains had its basis in colloidal Coomassie staining strategies whereby the organic component chelates luminescent ruthenium(II) and also provides the basis for noncovalent protein in much the same manner as Coomassie staining primarily by ionic interactions with basic amino acids and secondary by hydrophobic interactions. An explicit description of a ruthenium II tris (bathophenanthroline disulfonate) preparation and staining protocol and comparison to SYPRO Ruby protein gel stain (Rabilloud *et al.*, 2001) has led to the suggestion that the staining principle in SYPRO Ruby is a very similar compound; mass spectrometry has revealed some minor differences, a consequence of proprietary chemistry. A reformulated SYPRO Ruby gel stain, also a ready-to-use solution, with improved properties over the original formulation, suitable for both SDS–PAGE and isoelectric focusing gels, was then developed (Berggren *et al.*, 2002). Thus SYPRO Ruby is distinguished on the basis of the fluorescent chemical moiety and, importantly, also on the basis of the staining formulation that allows a stable ready-to-use solution and a relatively simple protocol:

1. The gel is fixed in 50% (v/v) methanol, 10% (v/v) acetic acid; fixation can be 30 min to overnight.
2. The gel is stained with SYPRO ruby stain. This is an endpoint stain; staining can be 3 h to overnight, and is stable.
3. The gel is briefly destained with 10% (v/v) methanol, 7% (v/v) acetic acid.

Staining can be accelerated with a microwave-based protocol. SYPRO Ruby has a relatively high extinction coefficient and quantum yield, so it is intrinsically "bright" to the eye and is chemically stable and photostable. Excitation can be with UV light or with blue light sources; the protein bands appear orange-red to the eye. The basis for fluorescent protein staining is differential binding to protein bands relative to the gel matrix, which does not bind the dye. Thus the protein staining is not fluorogenic *per se*. The brightness, stability, detection sensitivity, and ease of use of this product, combined with extensive literature coverage for proteomics applications and an aggressive marketing campaign has led to SYPRO Ruby becoming the comparison standard against subsequently developed protein gel stains, just as silver stain also remains a detection sensitivity benchmark.

4.2.4. Epicocconone: Deep Purple, Lighting Fast protein gel stains

Epicocconone, a fluorophore from the fungus *Epicoccum nigrum*, has been utilized in total protein staining kits under several trade names, including the Deep Purple and Lightning Fast total protein gel stains, with several protocol variations. In one protocol:

1. An SDS–PAGE gel is fixed for 1 h in 7.5% (v/v) acetic acid.
2. The gel is washed with water (2 × 30 min).
3. The gel is stained for 1 h in an aqueous solution into which an aliquot of the fluorophore stock solution is diluted.
4. The gel is incubated 0.05% in (v/v) ammonia (3 × 10 min).
5. Image acquisition is immediate; excitation can be with UV or blue or green visible light.

The stain sensitivity has been described to rival or exceed SYPRO Ruby (Bell and Karuso, 2003; Mackintosh *et al.*, 2003). The staining is reversible and is compatible with mass spectrometry. The dull red signal is not strikingly bright, and the basis of sensitivity is the fluorogenic nature of staining, such that background is very low. The staining seems to require trace amounts of SDS that remain complexed with protein after the acetic acid fixation. The stain is only weakly fluorescent in water (green) but in the presence of SDS-treated protein fluorescence increases and is redshifted. Epicocconone has been shown to react with lysine amines, to form a fluorescent adduct that can then be hydrolyzed by base, accounting for the fluorogenic, reversible nature of this product (Coghlan *et al.*, 2005). The concentrated, staining stock solution must be stored frozen and thawed before use. Image acquisition should follow immediately upon completion of staining, as stability may be an issue.

4.2.5. Fluorescein derivatives

Fluorescein derivatives containing hydrocarbon tails have been shown to be effective protein gel stains. A comparison of 5-dodecanoyl amino-(C12-FL), 5-hexadecanoyl amino-(C16-FL), and 5-octadecanoyl amino-fluorescein (C18-FL), in several simple staining protocols analogous to classic CBB R-250 protocols, demonstrated staining comparable to silver staining sensitivity. The most effective fluorophore was C16-FL. In one example, gels were stained in two changes of 30% (v/v) ethanol, 7.5% (v/v) acetic acid, 1 μM dye, washed with two changes of water, and destained with 7.5% acetic acid. The stain is fluorogenic, apparently on the basis of binding to residual SDS in a complex with protein; mass spectrometry compatibility has been demonstrated (Kang *et al.*, 2003). Excitation and emission are in the blue and green, respectively, as for fluorescein.

4.2.6. Krypton protein gel stains

Krypton protein gel stain is a proprietary hydroxyquinolone formulation with green excitation and orange fluorescence (Wolf *et al.*, 2007). Krypton Infrared protein gel stain is a proprietary formulation containing coumarine/chinolone and/or Martina-type dyes with red/near infrared excitation and near infrared emission (Czerney *et al.*, 2008). These products are available in 10× formulations that are diluted into a working solution and have been demonstrated to be competitive with SYPRO Ruby and Deep Purple protein gel stains for detection sensitivity and dynamic linear range. The staining protocol follows a standard fix–stain–rapid destain work flow and have been developed for rapid staining such that staining can be done in 1–4 h overall; for maximum sensitivity and signal linearity the longer protocol is preferred. These stains are compatible with mass spectrometry.

4.2.7. Flamingo protein gel stain

Flamingo protein gel stain is a proprietary coumarin-based cyanine dye formulation (Berkelman, 2006). The staining protocol requires fixation prior to staining; the dye is available in a 10× stock solution that is diluted into water just before use. The stain shows fluorescence enhancement upon binding to protein and is reported to have a low background that can be further reduced with a short destaining step. Excitation is in the green range and the signal is orange/red. The stain has been reported to be competitive with silver stain and with SYPRO Ruby and is mass spectrometry compatible.

4.2.8. LUCY protein gel stains

LUCY protein gel stains are proprietary trimethincyanine dyes that show fluorescent enhancement in SDS/protein complexes. Staining is most frequently done in dilute acetic acid, in same manner as with SYPRO Orange. The stains are reported to be mass spectrometry compatible (Kovalska *et al.*, 2006).

4.3. Preelectrophoresis sample labeling

The development and use of succinimidyl esters of charge-balanced cyanine dyes (Cy dyes) for covalent amine labeling of protein samples prior to electrophoresis has been the basis of fluorescent two-dimensional differential gel electrophoresis (2-D DIGE), a very important proteomics technology (Minden *et al.*, 2002; Tonge *et al.* 2001; Waggoner *et al.*, 1993). DIGE is discussed in more detail in Chapter 30. There are many succinimidyl esters of florescent compounds, and other reactive florescent covalent protein labels are also widely used (Haugland, 2005). With the broad instrumentation base now available, gel-based analysis of the composition of any fluorescently labeled protein preparation should become a routine procedure.

5. Phosphoprotein Detection

The importance of reversible protein phosphorylation at selected amino acid residues as a fundamental cell signaling mechanism is beyond dispute. Current phosphoprotein stains are proprietary formulations with phosphate-binding moieties covalently attached to fluorophores. The mode of detection is selective binding to phosphorylated amino acids, but with no fluorescent enhancement. Many soluble fluorescent compounds can be, to some degree, total protein stains. A selective phosphoprotein staining solution combines a phosphate-binding moiety, a fluorophore that shows low intrinsic total protein staining, a formulation that suppresses nonspecific, total protein staining, and a destaining protocol that favors dissociation of residual nonspecific stain. There are only a few phosphates per polypeptide relative to the number of amino acids targeted by total protein stains, so phosphoprotein detection sensitivity is generally at the same protein mass sensitivity level as detection by total proteins as colloidal Coomassie staining or SYPRO Orange staining, and not as sensitive as with silver or SYPRO Ruby stain. Protein phosphorylation determination in gels must always be done with appropriate controls, including gel lanes displaying proteins of known positive or negative phosphorylation status and, if possible, control samples treated with an effective phosphatase. After the phosphoprotein signal is detected, the gel should be poststained with a total protein stain. The commercially available phosphoprotein stains are compatible with subsequent staining with Coomassie, Silver, or total fluorescent stains, such as SYPRO Ruby, but not with SDS-dependent fluorogenic stains such as SYPRO Orange. The phosphoprotein stains are compatible with subsequent analytical procedures such as mass spectrometry or peptide sequencing.

5.1. Pro-Q Diamond phosphoprotein gel stain

Pro-Q Diamond phosphoprotein gel stain (Life Technologies), a ready-to-use formulation, was developed from principles of immobilized metal affinity chromatography for phosphopeptides, utilizing a fluorophore-conjugated metal chelating moiety in a solution containing a metal ion, salts, and a water-miscible organic solvent, buffered to pH 4 (Agnew *et al.*, 2006). The staining protocol requires fixation in methanol/acetic acid, washing with water to remove the fixative, staining with the Pro-Q Diamond formulation and subsequent destaining in a mixture containing 1,2-propanediol or, more effectively, acetonitrile (e.g., 50 m*M* sodium acetate, pH 4, 15–20% 1,2-propanediol, or 15–20% acetonitrile). Detection of phosphoproteins in the minigel format is of as little as 1–2 ng for β-casein, a pentaphosphorylated protein, or 8 ng for pepsin, a monophosphorylated protein, with 1000-fold linear dynamic detection range. *In vitro* protein phosphorylation or dephosphorylation can be demonstrated with Pro-Q Diamond staining followed by total protein staining (Steinberg *et al.*, 2003).

5.2. Phos-tag phosphoprotein stains

The Phos-tag phosphoprotein stains (Perkin Elmer), available in kit form, employ the same general staining strategy, albeit with different buffer compositions, as Pro-Q Diamond stain. These stains contain the well-described Phos-tag-binding moiety, an alkoxide-bridged dinuclear zinc(II) complex; the basis for the molecular design was from the enzymology of reversible phosphate–Zn(II) coordination. A Phos-tag-phenyl-phosphate complex with a dissociation constant of 25 n*M* has been described, in contrast to micomolar dissociation constant for other phosphate-binding moieties (Kinoshita *et al.*, 2004; Koike *et al.*, 2007). Phos-tag technology has been shown to be useful for gel-based phosphoprotein analysis in other contexts than direct in-gel staining of phosphoproteins (Kinoshita *et al.*, 2006; Yamada *et al.*, 2007). Phos-tag gel stain kits are available as two fluorophore choices, identified by the excitation maxima: Phos-tag phosphoprotein gel stains 300/460 (dual excitation maxima) or 540. Use of Phos-tag 300/460 stain for detection of a bacterial response regulator aspartate phosphorylation has recently been reported (Barbieri and Stock, 2008).

6. Glycoprotein Detection

6.1. General glycoprotein detection

Glycosylation is the most frequent protein posttranslational modification in eukaryotes. Oligosaccharides are usually linked to asparagine side chains (*N*-linked glycosylation) or to serine and threonine hydroxyl side chains

(*O*-linked glycosylation). Postelectrophoretic glycoprotein staining in gels generally utilizes periodate oxidation of vicinal glycol residues followed by hydrazide conjugation by a Schiff's base mechanism.

6.1.1. Chromogenic: Acid fuchsin dye

A chromogenic method that uses acid fuchsin dye (Zacharius *et al.*, 1969) has been the basis of several commercially available kits, including the Gelcode (Thermo Scientific) and Glyco-Pro (Sigma-Aldrich) glycoprotein stain kits. The gel is fixed in 50% methanol, washed, incubated in 1% periodate to oxidize vicinal glycols, washed, incubated in fuchsin sulfite reagent, incubated in a reducing reagent (sodium metabisulfite or sodium borohydride), washed, and stored in dilute acetic acid. Glycoproteins are indicated by magenta bands that begin to appear during the staining reaction and slowly intensify thereafter.

6.1.2. Fluorescent general glycoprotein stains

Pro-Q Emerald 300 and Pro-Q Emerald 488 glycoprotein gel stain kits (Life Technologies; the numerals refer to commonly used excitation sources, respectively), Krypton glycoprotein staining kit (Thermo Scientific Pierce), and Glycoprofile III fluorescent glycoprotein staining kit (Sigma-Aldrich) all utilize periodate oxidation followed by conjugation with a fluorescent hydrazide. With these kit reagents, subsequent reduction, for example, with sodium metabisulfite, is not needed.

Most available fluorescent hydrazides are intrinsically fluorescent. The brightly green fluorescent Pro-Q Emerald 300 hydrazide reagent is nonfluorescent in solution and is weakly fluorogenic upon conjugation to small molecule aldehydes. The stained glycoprotein bands are bright green; the conjugated dye may act as nucleation sites for the deposition of free reagent, which is in solution at near saturating concentration (Haugland *et al.*, 2005). The utility of Pro-Q Emerald 300 and Pro-Q Emerald 488 glycoprotein staining kits has been described in some detail (Hart *et al.*, 2003; Steinberg *et al.*, 2001). In these studies, Pro-Q Emerald 300 glycoprotein staining was found to be the most sensitive selective glycoprotein stain, but is suited for use only with a UV light source. Data and observations in both studies emphasize that many fluorescent hydrazides exhibit nonspecific staining of total proteins. Moreover, selectivity is also strongly influenced by the composition of the staining diluent. It is helpful to include a control lane containing proteins of known glycosylation state, by choosing well-characterized proteins and/or by glycosidase digestion. A parallel gel that is not subject to periodate oxidation may also reveal either intrinsic nonspecific protein staining or staining of endogenous aldehydes or ketones due to other oxidative events. Finally, it may be noted that acidic fuchsin sulfite-stained protein bands could be detected by fluorescence scanning with a

532-nm laser and a 675-nm long-pass filter, with a two- to fourfold increase in sensitivity over the observable colorimetric limit.

6.2. *O*-GlcNAc detection

6.2.1. Azide–alkyne "click chemistry" reagents

Many proteins are dynamically modified by attachment of *O*-linked *N*-acetylglucosamine (*O*-GlcNAc) in a β-linkage to serine and threonine residues. This PTM has shown site-specific reciprocity to phosphorylation in a regulated manner. *O*-GlcNAcylation thus may have a dynamic "Yin–Yang" role mediating phosphorylation-modulated signaling as well as other specific protein signaling functions (Golks and Guerini, 2008; Hart, 1999). Historically, gel-based *O*-GlcNAc detection has been problematic, but sensitive *O*-GlcNAc-specific fluorescent labeling for detection in gels now can be accomplished.

The method presented has its basis in the copper(I)-catalyzed Huisgen [3 + 2] cycloaddition between azides and alkynes, a "click chemistry" reaction (Agnew *et al.*, 2008; Clark *et al.*, 2008). Proteins (at any stage of purity) are labeled enzymatically (overnight, 4° C) with the Click-iT *O*-GlcNAc Enzymatic Labeling System (Life Technologies) utilizing a permissive mutant β-1,4-galactosyltransferase (Gal-T1 Y289L) that transfers azido-modified galactose (GalNAz) from UDP-GalNAz to *O*-GlcNac residues on the target proteins (Ramakrishnan and Qasba, 2002). The azide-modified *O*-GlcNAcylated proteins are then fluorescently labeled by the Cu(I)-catalyzed click reaction with tetramethylrhodamine–alkyne (TAMRA–alkyne) or with Dapoxyl alkyne dyes, available with the catalytic reagents in the Click-iT TAMRA or Click-iT Dapoxyl protein analysis detection kits, respectively (Life Technologies). The labeled proteins are separated by SDS–PAGE and image acquisition is done with appropriate instrumentation. The enzymatic labeling kit contains a positive control protein, α-crystalline. The technique has been demonstrated to detect 10–40 fmol of *O*-GlcNAc in a minigel format with the TAMRA alkyne detection reagent, and is compatible with downstream mass spectrometry analyses. Labeled proteins can also be poststained with a total protein stain or with other PTM-specific stains such as Pro-Q Diamond phosphoprotein stain or Pro-Q Emerald 300 glycoprotein stain.

REFERENCES

Agnew, B., Beechem, J., Gee, K., Haugland, R., Liu, J., Martin, V., Patton, W., and Steinberg, T. (2006). *Compositions and methods for detection and isolation of phosphorylated molecules.* United States Patent 7102005.

Agnew, B., Buck, S., Nyberg, T., Bradford, J., Clarke, S., and Gee, K. (2008). Click chemistry for labeling and detection of biomolecules. *Proc. SPIE* **6867,** 686708.

Barbieri, C. M., and Stock, A. M. (2008). Universally applicable methods for monitoring response regulator aspartate phosphorylation both *in vitro* and *in vivo* using Phos-tag base reagents. *Anal. Biochem.* **376,** 73–82.

Bell, P. J. L., and Karuso, P. (2003). Fluorescent compounds. United States Patent Application 20030157518.

Berggren, K., Chernokalskaya, E., Steinberg, T. H., Kemper, C., Lopez, M. F., Diwu, Z., Haugland, R. P., and Patton, W. F. (2000). Background-free, high sensitivity staining of proteins in one- and two-dimensional sodium dodecyl sulfate-polyacrylamide gels using a luminescent ruthenium complex. *Electrophoresis* **21,** 2509–2521.

Berggren, K. N., Schulenberg, B., Lopez, M. F., Steinberg, T. H., Bogdanova, A., Smejkal, G., Wang, A., and Patton, W. F. (2002). An improved formulation of SYPRO Ruby protein gel stain: Comparison with the original formulation and with a ruthenium II tris (bathophenanthroline disulfonate) formulation. *Proteomics* **2,** 486–498.

Berkelman, T. R. (2006). *Coumarin-based cyanine dyes for non-specific protein binding.* United States Patent Application 20060166368.

Bhalget, M. K., Diwu, Z., Haugland, R. P., and Patton, W. F. (2001). *Luminescent protein stains and their method of use.* United States Patent 6316267.

Candiano, G., Bruschi, M., Musante, L., Santucci, L., Ghiggeri, G. M., Carnemolla, B., Orecchia, P., Zardi, L., and Righetti, P. G. (2004). Blue silver: A very sensitive colloidal Coomassie G-250 staining for proteome analysis. *Electrophoresis* **25,** 1327–1333.

Clark, P. M., Dweck, J. F., Mason, D. E., Hart, C. R., Buck, S. B., Peters, E. C., Agnew, B. J., and Hsieh-Wilson, L. C. (2008). Direct in-gel fluorescence detection and cellular imaging of O-GlcNAc-modified proteins. *J. Am. Chem. Soc.* **130,** 11576–11577.

Coghlan, D. R., Mackintosh, J. A., and Karuso, P. (2005). Mechanism of reversible staining of protein with epicocconone. *Org. Lett.* **7,** 2401–2404.

Czerney, P. T., Desai, S., Lehmann, F. G., Murtaza, Z. S., Schweder, B. G., Wenzel, M. S., and Wolf, B. D. (2008). *Protein probe compounds, compositions, and methods.* United States Patent Application 20080026478.

Daban, J.-R. (2001). Fluorescent labeling of proteins with Nile red and 2-methoxy-2, 4-diphenyl-3(2H)-furanone: Physicochemical basis and application to the rapid staining of sodium dodecyl sulfate polyacrylamide gels and Western blots. *Electrophoresis* **22,** 874–880.

Daban, J.-R., Bartolome, S., and Samso, M. (1991). Use of the hydrophobic probe Nile red for the fluorescent staining of protein bands in sodium dodecyl sulfate-polyacrylamide gels. *Anal. Biochem.* **199,** 169–174.

Dunbar, B. S., Kimura, H., and Timmons, T. M. (1990). Protein analysis using high-resolution two-dimensional polyacrylamide gel electrophoresis. *In* "Meth. Enzmyol." (M. P. Deutscher, ed.) **182,** pp. 441–459. Academic Press, San Diego, CA.

Fernandez-Patron, C. (2005). Zn^{2+} reverse staining technique. *In* "The Proteomics Protocol Handbook" (J. M. Walker, ed.), pp. 215–222. Humana Press, Totowa, NJ.

Fernandez-Patron, C., Castellanos-Serra, L., Hardy, E., Guerra, M., Estevez, E., Mehl, E., and Frank, R. W. (1998). Understanding the mechanism of the zinc-ion stains of biomacromolecules in electrophoresis gels: generalization of the reverse-staining technique. *Electrophoresis* **19,** 2398–2406.

Garfin, D. E. (1990a). One-dimensional gel electrophoresis. *In* "Meth. Enzmyol." (M. P. Deutscher, ed.) **182,** pp. 425–441. Academic Press, San Diego, CA.

Garfin, D. E. (1990b). Isoelectric focusing. *In* "Meth. Enzmyol." (M. P. Deutscher, ed.) **182,** pp. 459–478. Academic Press, San Diego, CA.

Golks, A., and Guerini, D. (2008). The O-linked *N*-acetylglucosamine modification in cellular signaling and the immune system. *EMBO Rep.* **9,** 748–753.

Hart, G. W. (1999). The O-GlcNAc modification. *In* "Essentials of Glycobiology" (A. Varki, J. Esko, H. Freeze, G. Hart, and J. Marth, eds.), 2nd edn. p. 653. Cold Spring Harbor Laboratory Press, Cold Spring Harbor, NY.

Hart, C., Schulenberg, B., Steinberg, T. H., Leung, W.-Y., and Patton, W. F. (2003). Detection of glycoproteins in polyacrylamide gels and on electroblots using Pro-Q Emerald 488 dye, a fluorescent periodate Schiff-base stain. *Electrophoresis* **24,** 588–598.

Haugland, R. P. (2005). *The Handbook: A Guide to Fluorescent Probes and Labeling Technologies.* 10th edn. Invitrogen, Carlsbad, CA, 1126pp.

Haugland, R. P., Singer, V. L., Jones, L. J., and Steinberg, T. H. (1997). *Non-specific protein staining using merocyanine dyes.* United States Patent 5616502.

Haugland, R. P., Steinberg, T. H., Patton, W. F., and Diwu, Z. (2005). *Reagents for labeling biomolecules having aldehyde or ketone moieties.* United States Patent 6967251.

Jessani, N., and Cravatt, B. F. (2004). The development and application of methods for activity-based protein profiling. *Curr. Opin. Chem. Biol.* **8,** 54–59.

Kang, C., Kim, H. J., Kang, D., Jung, D. Y., and Suh, M. (2003). Highly sensitive and simple fluorescence staining of proteins in sodium dodecyl sulfate-polyacrylamide-based gels by using hydrophobic tail-mediated enhancement of fluorescein luminescence. *Electrophoresis* **24,** 3297–3304.

Kinoshita, E., Takahashi, M., Takeda, H., Shiro, M., and Koike, T. (2004). Recognition of the phosphate monoester dianion by an alkoxide-bridged dinuclear zinc(II) complex. *Dalton Trans.* 1189–1193.

Kinoshita, E., Kinoshita-Kikuta, E., Takiyama, K., and Koike, T. (2006). Phosphate-binding tag, a new tool to visualize phosphorylated proteins. *Mol. Cell. Proteomics* **5,** 749–757.

Koike, T., Kawasaki, A., Kobashi, T., and Takahagi, M. (2007). *Method for labeling phosphorylated peptides, method for selectively adsorbing phosphorylated peptides, complex compounds used in the methods, process for producing the complex compounds, and raw material compounds for the complex compounds.* United States Patent 7202093.

Kovalska, V., Kryvorotenko, D., Losytskyy, M., Nording, P., Rueck, A., Schoenenberger, B., Yarmoluk, S., and Wahl, F. (2006). *Detection of polyamino acids using trimethincyanine dyes.* United States Patent Application 20060207881.

Mackintosh, J. A., Choi, H.-Y., Bae, S.-H., Veal, D. A., Bell, P. J., Ferrari, B. C., Van Dyk, D. D., Verrills, N. M., Paik, Y.-K., and Karuso, P. (2003). A fluorescent natural product for ultrasensitive detection of proteins in one-dimensional and two-dimensional gel electrophoresis. *Proteomics* **3,** 2273–2288.

Manchenko, G. P. (2003). *Handbook of Detection of Enzymes on Electrophoretic Gels.* 2nd edn. CRC Press, LLC, Boca Raton, xiv+554pp.

Merril, C. R. (1990). Gel-staining techniques. *In* "Meth. Enzmyol." (M. P. Deutscher, ed.) **182,** pp. 477–488. Academic Press, San Diego, CA, xxix+894pp.

Miller, I., Crawford, J., and Gianazza, E. (2006). Protein stains for proteomic applications: Which, when, why? *Proteomics* **6,** 5385–5408.

Minden, J., Waggoner, A., and Fowler, S. J. (2002). *Difference detection methods using matched multiple dyes.* United States Patent Application 20020177122.

Neuhoff, V., Stamm, R., and Hansjorg, E. (1985). Clear background and highly sensitive protein staining with Coomassie Blue dyes in polyacrylamide gels: A systematic analysis. *Electrophoresis* **6,** 427–488.

Neuhoff, V., Arold, N., Taube, D., and Ehrhardt, W. (1988). Improved staining in polyacrylamide gels including isoelectric focusing gels with clear background at nanogram sensitivity using Coomassie Brilliant Blue G-250 and R-250. *Electrophoresis* **9,** 255–262.

Patton, W. F. (2000). A thousand points of light: The application of fluorescence detection technologies to two-dimensional gel electrophoresis and proteomics. *Electrophoresis* **21,** 1123–1144.

Patton, W. F. (2002). Detection technologies in proteome analysis. *J. Chromatogr. B* **771,** 3–31.

Paulick, M. G., and Bogyo, M. (2008). Application of activity-based probes to the study of enzymes involved in cancer progression. *Curr. Opin. Genet. Dev.* **18,** 97–106.

Poland, J., Rabilloud, T., and Sinha, P. (2005). Silver staining of 2-D gels. *In* "The Proteomics Protocol Handbook" (J. M. Walker, ed.), pp. 215–222. Humana Press, Totowa, NJ.

Rabilloud, T. (1990). Mechanisms of protein silver staining in polyacrylamide gels: A 10-year synthesis. *Electrophoresis* **11,** 785–794.

Rabilloud, T., Strub, J. M., Luche, S., van Dorsselaer, A., and Lunardi, J. (2001). A comparison between Sypro Ruby and ruthenium II tris (bathophenanthroline disulfonate) as fluorescent stains for protein detection in gels. *Proteomics* **1,** 699–704.

Ramakrishnan, B., and Qasba, P. K. (2002). Structure-based design of beta-1, 4-galactosyltransferase-I (beta 4Gal-T1) with equally efficient *N*-acetylgalactosaminyltransferase activity: Point mutation broadens beta 4Gal-T1 donor specificity. *J. Biol. Chem.* **277,** 20833–20839.

Smejkal, G. B. (2004). The Coomassie Chronicles: Past, present and future perspectives in polyacrylamide gel staining. *Expert Rev. Proteomics* **1,** 381–387.

Steinberg, T. H., Jones, L. J., Haugland, R. P., and Singer, V. L. (1996a). SYPRO orange and SYPRO red protein gel stains: One-step fluorescent staining of denaturing gels for detection of nanogram levels of protein. *Anal. Biochem.* **239,** 223–237.

Steinberg, T. H., Haugland, R. P., and Singer, V. L. (1996b). Applications of SYPRO orange and SYPRO red protein gel stains. *Anal. Biochem.* **239,** 238–245.

Steinberg, T. H., Chernokalskaya, E., Berggren, K., Lopez, M. F., Diwu, Z., Haugland, R. P., and Patton, W. F. (2000a). Ultrasensitive fluorescence protein detection in isoelectric focusing gels using a ruthenium metal chelate stain. *Electrophoresis* **21,** 486–496.

Steinberg, T. H., Lauber, W. M., Berggren, K., Kemper, C., Yue, S., and Patton, W. F. (2000b). Fluorescence detection of proteins in sodium dodecyl sulfate-polyacrylamide gels using environmentally benign, nonfixative, saline solution. *Electrophoresis* **21,** 497–508.

Steinberg, T. H., Pretty On Top, K., Berggren, K., Kemper, C., Jones, L., Diwu, Z., Haugland, R. P., and Patton, W. F. (2001). Rapid and simple single nanogram detection of glycoproteins in polyacrylamide gels and on electroblots. *Proteomics* **1,** 841–855.

Steinberg, T. H., Agnew, B. J., Gee, K. R., Leung, W.-Y., Goodman, T., Schulenberg, B., Hendrickson, J., Beechem, J. M., Haugland, R. P., and Patton, W. F. (2003). Global quantitative phosphoprotein analysis using Multiplexed Proteomics technology. *Proteomics* **3,** 1128–1144.

Steinberg, T. H., Hart, C. R., and Patton, W. F. (2005). Rapid, sensitive detection of proteins in minigels with fluorescent dyes: Coomassie Fluor Orange, SYPRO Orange, SYPRO Red, and SYPRO Tangerine stains. *In* "The Proteomics Protocol Handbook" (J. M. Walker, ed.), pp. 215–222. Humana Press, Totowa, NJ.

Tonge, R., Shaw, J., Middletown, B., Rowlinson, R., Rayner, S., Young, J., Pognan, F., Hawkins, E., Currie, I., and Davison, M. (2001). Validation and development of fluorescence two-dimensional differential gel electrophoresis proteomics technology. *Proteomics* **1,** 377–396.

Waggoner, A. S., Ernst, L. A., and Mujumdar, R. B. (1993). *Method for labeling and detecting materials employing arylsulfonate cyanine dyes.* United States Patent 5268486.

Walker, J. M. (ed.) (2005). *In* "The Proteomics Protocols Handbook", Humana Press, Totowa, NJ, xvii+988pp.

Westermeier, R. (2006). Sensitive, quantitative, and fast modifications for Coomassie blue staining of polyacrylamide gels. *Pract. Proteomics* **6**(Suppl. 2), 61–64 (online journal).

Wolf, B.D, Desai, S., Czerney, P. T., Lehmann, F. G., Schweder, B. G., and Wenzel, M. S. (2007). *Protein detection and quantitation using hydroxyquinolone dyes.* United States Patent Application 20070281360.

Yamada, S., Nakamura, H., Kinoshita, E., Kinoshita-Kikuta, E., Koike, T., and Shiro, Y. (2007). Separation of a phosphorylated histidine protein using phosphate affinity polyacrylamide gel electrophoresis. *Anal. Biochem.* **360,** 160–162.

Yue, S. T., Steinberg, T. H., Patton, W. F., Cheung, C.-Y., and Haugland, R. P. (2003). *Carbazolylvinyl dye protein stains.* United States Patent 6579718.

Zacharius, R. M., Zeli, T. E., Morrison, J. H., and Woodlock, J. J. (1969). Glycoprotein staining following electrophoresis on acrylamide gels. *Anal. Biochem.* **30,** 148–152.

CHAPTER THIRTY-TWO

Elution of Proteins from Gels

Richard R. Burgess

Contents

1. Introduction 565
2. Elution of Proteins from Gels by Diffusion 566
   2.1. Preparing the gel and protecting the protein 567
   2.2. Locating protein in the gel 567
   2.3. Elution by diffusion 568
   2.4. SDS removal and concentration 568
   2.5. Renaturation of enzyme activity 569
   2.6. Preliminary tests on your enzyme 569
   2.7. Limitations of the method 569
   2.8. Many applications 569
3. Replacing the SDS Gel with a Reverse Phase HPLC 570
4. Electrophoretic Elution 570
5. Conclusion 571
References 571

Abstract

New protein biochemical technologies have been developed that allow one to learn more and more about a protein with less and less material (on the order of μg). Polyacrylamide gel electrophoresis, which was originally strictly an analytical method, has in some cases become a high-resolution preparative method. This chapter will focus on the elution of proteins from SDS gels, with an emphasis on recovering enzyme or biological activity of the resulting protein.

1. Introduction

SDS polyacrylamide gel electrophoresis has proved to be an incredibly useful analytical method for determining the number and sizes of polypeptides in a sample. When performed skillfully it has the ability to resolve many individual sized proteins. Many of us in the early days of gel

McArdle Laboratory for Cancer Research, University of Wisconsin-Madison, Madison, Wisconsin, USA

Methods in Enzymology, Volume 463
ISSN 0076-6879, DOI: 10.1016/S0076-6879(09)63032-9

electrophoresis wished we could take advantage of the high resolution of the gels to obtain small quantities of pure proteins. Many methods have been developed that allow us now to do this. They include:

1. *Transfer of proteins out of the gel and onto a nitrocellulose or PVDF membrane.* Subsequent detection and quantification of the protein can be achieved by Western blot analysis with an appropriate antibody (see Chapter 33). Sometimes one can renature at least some of the protein to regain enzymatic function or the ability to bind another protein by far-Western analysis (e.g., see Burgess *et al.*, 2000). Edman degradation of the protein on the membrane can give limited N-terminal amino acid sequence information.
2. *Enzyme assays of proteins while still in the gel matrix.* This has been covered extensively (Manchenko, 2003).
3. *Preparative gel electrophoresis.* In this procedure the proteins are electrophoresed off the end of a gel and into a chamber where fractions can be collected. Several commercial products for preparative electrophoresis are now available.
4. *Elution of the protein from the gel and recovery of the protein.* This approach has been the subject of several extensive reviews (Harrington, 1990; Seelert and Krause, 2008). Proteins eluted from gels have been used for a wide variety of purposes including: protein chemistry, proteolytic cleavage; amino acid composition, and sequence determination; identifying a polypeptide by trypsin digestion and MALDI-TOF mass spectroscopy; antigen for antibody production; and identifying a polypeptide corresponding to an enzyme activity.

This chapter will focus on this fourth approach, that of eluting protein from SDS gels, particularly on recovering enzymatic or biological activity of the eluted protein.

2. Elution of Proteins from Gels by Diffusion

The early work on removal of SDS from protein was published by Weber and Kutter (1971), but when applied to microgram amounts of protein eluted from gels it proved cumbersome and usually resulted in loss of most of the protein.

Much of the early work in this area has been published (Hager and Burgess, 1980), but bears revisiting. The basic procedure involves the gel electrophoresis itself, locating the protein of interest on the gel, eluting the protein from the gel, removing SDS, and renaturing the protein for subsequent study or use.

2.1. Preparing the gel and protecting the protein

To the first approximation, any of a large variety of gel recipes can be used. There are several important considerations. The polymerization step of polyacrylamide gels involves the generation of free radicals to initiate the polymerization reaction. Therefore, if you are pouring your own gels you should wait at least 12 h to allow complete polymerization. Since the initiation of the reaction involves creating an oxidizing environment, one should take precautions to protect the protein against oxidation. This can readily be accomplished by adding to the sample applied to the gel an internal carrier protein such as β-lactoglobulin that is inexpensive, small, and runs ahead of most proteins and that will take the brunt of any oxidizing chemicals remaining in the gel. You should also add an anionic thiol compound to the upper reservoir buffer (we use 0.1 m*M* sodium thioglycolate) that will run through the gel, again destroying any oxidizing potential and residual free radicals within the gel. The presence of the thiol in the gel may also help reduce the reaction of protein cysteines with free acrylamide to produce cysteinyl-*S*-propionamide (Chiari *et al.*, 1992).

You should use high-quality SDS, containing only the dodecyl form (C12), because impure SDS can contain variable amounts of C14 or C16 that bind much more tightly to proteins and can be very difficult to remove (Kunitani and Kresin, 1989).

2.2. Locating protein in the gel

Once the gel has been run, one needs to locate the region of the gel that contains the band of interest or the band of a particular MW. Of course, if you do not know the size of your enzyme band, you can cut the gel into sections, elute protein from each section, and assay all for enzyme activity. Originally, we located our band of interest by soaking the gel for 5 min in cold 0.25 *M* KCl and then destaining with cold distilled water for 1 h. Now we know that you can use almost any convenient way to stain the gel, including with zinc-imidazole (Castellanos-Serra and Hardy, 2001), or with new very sensitive fluorescent protein dyes like SYPRO Ruby (see Chapter 31). You can even use Coomassie Blue stain. In one case, we succeeded in recovering activity from a band that was stained with Coomassie, destained, dried down, and stored for 10 years! Another method is to run colored MW markers (available from many companies) in flanking lanes of the gel. That way you can be guided to cut out particular sections of the gel based on prior molecular weight information.

2.3. Elution by diffusion

People have tended to ignore this simple procedure that is inexpensive, effective, and can be done on many samples at the same time. A typical protocol below is based on and adapted from Hager and Burgess (1980).

1. Locate a band on an SDS gel by rinsing the gel with cold ddH_2O and staining for 5 min with ice-cold 0.25 *M* KCl and 1 m*M* DTT. Rinse with ddH_2O and destain for 10–60 min with cold ddH_2O and 1 m*M* DTT. Alternatively, one can divide a gel lane into many 3–5 mm sections, put each section of gel into a tube, and wash as above.
2. Crush the gel slice in 1 ml elution buffer. We originally used a 3.5-ml siliconized Pyrex test tube (10 × 75 mm) and a small Teflon pestle (Kontes, K886001 size 19), but now use 0.3 ml of elution buffer and a 1.5-ml polypropylene microfuge tube and a disposable polypropylene pestle (such as Kontes pellet pestle, K749521-1500). The elution buffer is 50 m*M* Tris, pH 7.9, 0.1 m*M* EDTA, 1 m*M* DTT, 0.15 *M* NaCl, 25–100 μg/ml BSA (can be omitted), and 0.1% SDS.
3. Once the gel section has been crushed, one incubates the small gel fragments for 1–8 h on a rotator to allow passive diffusion. In Hager and Burgess (1980), the elution kinetics indicate half lives of elution from a 8.75% polyacrylamide gel matrix to be less than 30 min for a 36-kDa polypeptide and 1–1.5 h for 150 kDa polypeptides (complete elution in 4 and 16–24 h, respectively).
4. The mixture is centrifuged at maximum speed in a microfuge for 2 min to pellet the crumbled gel. The supernatant (the protein eluate) is then transferred into a clean microfuge tube.

2.4. SDS removal and concentration

Since elution is most efficient when 0.1% SDS is present in the elution buffer, one must remove SDS after elution. This can be most effectively accomplished by acetone precipitation that not only removes the SDS but also concentrates the protein. A typical protocol below is based on Hager and Burgess (1980).

1. Four volumes of cold acetone (−20 °C) are added to the protein eluate and the sample allowed to precipitate for 30 min in a dry ice–ethanol bath. Since the samples freeze in the dry ice–ethanol bath, we thaw the samples by placing them briefly in an ice water bath just before centrifugation.
2. Centrifuge at maximum speed in a microfuge for 5 min. The supernatant is removed and discarded. In test experiments with 0.16 μg of radioactive RNA polymerase, we found that our recovery of polymerase was

71%, 90%, and 99% when carrier BSA concentration in the elution buffer was 0, 15, and 100 μg/ml, respectively (Hager and Burgess, 1980).
3. Tests showed that more than 99.9% of the SDS remained in the acetone supernatant. The residual SDS can be removed by washing the pellet briefly with 1 ml of ice-cold 80% acetone and recentrifuging.

2.5. Renaturation of enzyme activity

1. The acetone precipitate is allowed to dry for about 10 min.
2. The precipitate is solubilized and denatured in 20 μl of 6 *M* guanidine hydrochloride (GuHCl) in dilution buffer (dilution buffer is 50 m*M* Tris, pH 7.9, 20% glycerol, 0.1 m*M* EDTA, 1 m*M* DTT, 0.15 *M* NaCl, 25–100 μg/ml BSA (can be omitted), and 0.1% SDS). Solubilization takes place within 20 min at room temperature.
3. The solubilized protein is rapidly diluted 50-fold by adding 1.0 ml of dilution buffer and permitted to renature 1–12 h at room temperature.
4. The renatured protein is assayed by an appropriate assay.

2.6. Preliminary tests on your enzyme

One can test the ability of a given protein to be renatured by this method by subjecting the enzyme of interest to preliminary testing to see whether it is able to regain activity after dissolving in 6 *M* GuHCl and diluting. If recovery is good, then test for recovery of activity after denaturing in SDS elution buffer, acetone precipitating, dissolving in 6 *M* GuHCl, and diluting.

2.7. Limitations of the method

While this method is quite versatile, it cannot be used for all enzymes. It cannot be used if the enzyme activity is dependent on two or more different sized polypeptides or requires a dissociable cofactor such as heme that will be separated from the catalytic protein during the SDS gel electrophoresis. It might also be difficult if the protein contains posttranslational modifications such as glycosylation or proteolytic processing that affect refolding or if there are essential disulfides that are difficult to reform.

2.8. Many applications

Even given the limitation noted above, many proteins have been successfully renatured using this method. This includes DNA topoisomerases, DNA ligase, bacterial sigma factors (Haldenwang *et al.*, 1981; Wiggs *et al.*, 1981), eukaryotic transcription factors, H-Ras GTPase, methyl reductases, centromere binding protein, and the protein component of the protein/RNA RNases. It has been used with enzymes from bacteria, yeast, rat, human, *Drosophila*, and many other organisms. It has been used with enzyme containing multiple identical subunits and with proteins with multiple different subunits if the appropriate gel slices have been mixed together. It has been used with proteins that have essential S–S bridges.

3. Replacing the SDS Gel with a Reverse Phase HPLC

A clever variation of the method of Hager and Burgess described above was published by Prokipcak *et al.* (1994). They replaced the SDS gel and acetone precipitation with RP-HPLC and lyophilization. The protein was applied to a C4 RP-HPLC, and eluted with a gradient of 0–50% acetonitrile/0.1% TFA. The fractions were lyophilized, resuspended in a small volume of 6 *M* GuHCl, diluted to refold, and assayed for enzymatic activity.

4. Electrophoretic Elution

While the above protocols describe in detail the elution of protein from gels by diffusion, many of the operations would be perfectly applicable to electrophoretic elution. While these approaches may result in slightly higher elution efficiency in slightly shorter time than elution by diffusion, they are more expensive and harder to scale up to many gel sections.

The *Schleicher and Schuell Elutrap*™ and the *Bio-Rad Model 422 Electro-eluter*™ are the two most widely used commercial electrophoretic elution apparati. Both involve the placing of a section of a gel containing a protein band of interest into the apparatus and eluting the protein by electrophoresis of the protein out of the gel piece, through a large-pore membrane or frit and into a chamber with a small-pore membrane that will only pass small molecules or proteins smaller than about 5 kDa. The protein of interest is pipetted out of the chamber and used as desired. The detailed methods for using these commercial products can be found in excellent reviews by Harrington (1990) and Seelert and Krause (2008) and in company literature.

Another commercial approach is the *ProteoPLUS*™ *electroeluter.* A small screw capped tube is constructed with upstream and downstream protein retention membranes that permit the passage of electric current when the tubes are placed in an electrophoresis tank. After the protein has eluted from the gel piece, it can be dialyzed in the same tube for subsequent use. An example of the application of this approach is given by Lei *et al.* (2007).

The *Bio-Rad Whole Gel Eluter* allows one to electrophoretically transfer a whole slab gel into 26 fractions. Typically a protein sample is applied into one gel-wide lane and subjected to SDS gel electrophoresis. The 26 grooves of the Eluter are aligned in parallel to the protein bands and the protein transferred out of the gel, into the grooves and into the harvesting box.

Another recent advance includes the ability to elute a whole slab gel into numerous individual fractions suitable for proteomic applications in multiwall plates (Antal *et al.*, 2007).

5. Conclusion

Protein bands separated by high-resolution gel electrophoresis can be recovered from the gel and used for a multitude of applications. The protein can be eluted out of the gel section by crushing the gel and allowing the protein to diffuse out of the gel or it can eluted by applying an electric field to the gel section and trapping the eluted protein in an appropriate membrane bounded apparatus. Submicrogram to 100 μg amounts of protein can be obtained and often renatured to regain enzymatic activity.

REFERENCES

Antal, J., Banyasz, B., and Buzas, Z. (2007). Shotgun electrophoresis: A proteomic tool for simultaneous sample elution from whole SDS-polyacrylamide gels slabs. *Electrophoresis* **28,** 508–511.

Burgess, R. R., Arthur, T. M., and Pietz, B. C. (2000). Mapping protein–protein interaction domains by order fragment ladder far-Western analysis using His-tagged proteins. *Meth. Enzymol.* **328,** 141–157.

Castellanos-Serra, L., and Hardy, E. (2001). Review. Detection of biomolecules in electrophoresis gels with salts of imidazole and zinc II: A decade of research. *Electrophoresis* **22,** 864–873.

Chiari, M., Righetti, P. G., Negri, A., Ceciliani, F., and Ronchi, S. (1992). Preincubation with cysteine prevents modification of sulfhydryl groups in proteins by unreacted acrylamide in a gel. *Electrophoresis* **13,** 882–884.

Hager, D. A., and Burgess, R. R. (1980). Elution of proteins from sodium dodecyl sulfate-polyacrylamide gels, removal of SDS, and renaturation of enzymatic activity: Results with sigma subunit of *E. coli* RNA polymerase, wheat germ DNA topoisomerase, and other enzymes. *Anal. Biochem.* **109,** 76–86.

Haldenwang, W. G., Lang, N., and Losick, R. (1981). A sporulation-induced sigma-like regulatory protein from *B. subtilis*. *Cell* **23,** 615–624.

Harrington, M. G. (1990). Elution of protein from gels. *Meth. Enzymol.* **182,** 488–495.

Kunitani, M. G., and Kresin, L. M. (1989). Analysis of alkyl sulfates in protein solutions by isocratic and gradient ion chromatography. *Anal. Biochem.* **182,** 103–108.

Lei, Z., Anand, A., Mysore, K. S., and Sumner, L. W. (2007). Electroelution of intact proteins from SDS-PAGE gels and their subsequent MALDI-TOF MS analysis. *Methods Mol. Biol.* **355,** 353–363.

Manchenko, G. P. (2003). Handbook of detection of enzymes on electrophoresis gels. 2nd edn. CRC Press LLC, Boca Raton.

Prokipcak, R. D., Harris, D. J., and Ross, J. (1994). Purification and properties of a protein that binds to the C-terminal coding region of human c-Myc mRNA. *J. Biol. Chem.* **269,** 9261–9269.

Seelert, H., and Krause, F. (2008). Review: Preparative isolation of protein complexes and other bioparticles by elution from polyacrylamide gels. *Electrophoresis* **29,** 2617–2636.

Weber, K., and Kutter, K. (1971). Reversible denaturation of enzymes by sodium dodecyl sulfate. *J. Biol. Chem.* **246,** 4504–4509.

Wiggs, J., Gilman, M., and Chamberlin, M. (1981). Heterogeneity of RNA polymerase in *Bacillus subtilis*: Evidence for an additional sigma factor in vegetative cells. *Proc. Natl. Acad. Sci. USA* **78,** 2762–2766.

CHAPTER THIRTY-THREE

Performing and Optimizing Western Blots with an Emphasis on Chemiluminescent Detection

Alice Alegria-Schaffer, Andrew Lodge, *and* Krishna Vattem

Contents

1. Western Blotting 574
2. Types of Western Blots 575
   2.1. Direct and indirect 575
   2.2. Far-Western 577
   2.3. Semiquantitative Western blotting 577
3. Detection Methods 579
   3.1. Enzyme conjugate 579
   3.2. Colorimetric detection 581
   3.3. Fluorescence detection 581
   3.4. Chemiluminescence detection 582
4. The Chemiluminescence Signal 583
   4.1. Signal capture 583
   4.2. Signal intensity and duration 583
   4.3. Optimizing a Western blot 585
5. Common Problems and their Explanations 588
   5.1. No signal 588
   5.2. Signal fades quickly 588
   5.3. High background 589
   5.4. New bottle of substrate does not produce a signal 589
   5.5. Brown or yellow bands on the membrane 589
   5.6. Bands or entire blot glowing in the darkroom 590
   5.7. Ghost/hollow bands 590
6. Blotting and Optimization Protocols using Chemiluminescent Substrates 593
   6.1. Western blotting protocol using chemiluminescent substrates 593
   6.2. Western blot stripping protocol 595
   6.3. Optimizing antigen concentration 596

Thermo Fisher Scientific, Pierce Protein Research, Rockford, Illinois, USA

Methods in Enzymology, Volume 463
ISSN 0076-6879, DOI: 10.1016/S0076-6879(09)63033-0

6.4. Optimizing membrane blocking 596
6.5. Optimizing the primary antibody concentration 597
6.6. Optimizing membrane washing 597
6.7. Optimizing enzyme conjugate concentration 597
6.8. Optimizing the detection method 598
References 598

Abstract

Immunodetection refers to any detection method that exploits the interaction of an antibody and antigen. The choice of detection method, such as enzyme-linked immunosorbent assays (ELISAs) or Western blotting, depends on the researcher's preferences and requirements. If a researcher wants to quantify a low-abundance target protein then a chemiluminescent ELISA is used. If a researcher wants to identify a protein that is in high abundance, a colorimetric Western blot will suffice. If there are multiple targets within an assay, then multiplex fluorescence is typically used.

This article focuses on Western blotting. Although colorimetric and fluorescent detection methods are discussed, chemiluminescent detection is used most often and is, therefore, discussed in great detail. Included is specific information about the chemiluminescent signal and factors that affect its intensity and longevity. We also describe types of blotting and present data and suggestions for obtaining semiquantitative data. Although classical Western blotting is typically used for qualitative purposes, we present information about effective quantitative analysis using specific controls. Common occurrences within the methodology and their possible explanations are also detailed. One frequent result is the appearance of ghost bands, which, based on our research, can be caused by high amounts of target or antibody cross-reactivity. Also included are the basic Western blot protocol and protocols for troubleshooting common problems and optimizing many of the specific factors that influence results.

1. Western Blotting

Western blotting is a powerful and commonly used tool to identify and quantify a specific protein in a complex mixture (Towbin *et al.*, 1979). The technique enables indirect detection of protein samples immobilized on a nitrocellulose or polyvinylidene fluoride (PVDF) membrane. In a conventional Western blot, protein samples are first resolved by sodium dodecyl sulfate polyacrylamide gel electrophoresis (SDS–PAGE) and then electrophoretically transferred to the membrane. Following a blocking step using a nonrelevant protein, the membrane is probed with a primary antibody (poly- or monoclonal) that was raised against the target antigen. The membrane is then washed and incubated with an enzyme-conjugated secondary antibody that is reactive toward the primary antibody. The membrane is washed again and incubated with an appropriate enzyme substrate. The signal is either visually evaluated, if a colorimetric substrate was used, or is detected with X-ray film or imaging instrumentation for chemiluminescence and fluorescence.

The most significant advances in Western blotting methodology are highly sensitive-enhanced chemiluminescent substrates, imaging systems, and, most recently, the availability of a wide variety of photostable fluorophores. The widespread use of extremely sensitive chemiluminescent substrates (Mattson and Bellehumeur, 1996; Walker *et al.*, 1995) has resulted in nearly eliminating the use of radioisotope-labeled probes. Protein A or G labeled with ^{125}I was once commonly used as a secondary detection reagent; however, the enhanced chemiluminescent substrates can detect proteins down to the low-femtogram level with high signal-to-noise ratios.

Charged-coupled device (CCD) cameras are commonplace in most labs. Imagers have a large dynamic range and a high degree of exposure control, which allows background signal adjustments. Furthermore, their accompanying analytical software enables densitometry analysis. Although X-ray film is still widely used and more sensitive, CCD imaging systems eliminate the hassles of film handling and developing and the associated chemical wastes.

Recently, there is a rapid trend toward fluorescent dyes. This trend is attributed to the new generation of fluorophores that offer tremendous improvements in brightness, photostability, and pH sensitivity compared to the traditional fluorophores. Western blotting detection via fluorescence is typically performed when there are two different targets on a single blot. A fluorophore pair is chosen based on their separate and distinct excitation–emission spectra. The production of an easily identifiable color difference enables multiplexing experiments. Most notably, these new fluorescent dyes coupled with software have allowed detection and quantitation of cell signaling pathways (Choudhary *et al.*, 2007; Tipsmark *et al.*, 2008).

2. Types of Western Blots

2.1. Direct and indirect

A direct Western blot refers to a detection method that uses a reporter-labeled primary antibody that directly binds to the target protein. Indirect detection uses a labeled secondary antibody (Ramlau, 1987) that binds to a nonlabeled primary antibody (Fig. 33.1). Because incubation with a secondary antibody is eliminated in direct detection, this strategy is performed in less time than an indirect Western blot. Additionally, background signal from secondary antibody cross-reactivity is eliminated (Bergendahl *et al.*, 2003). Direct

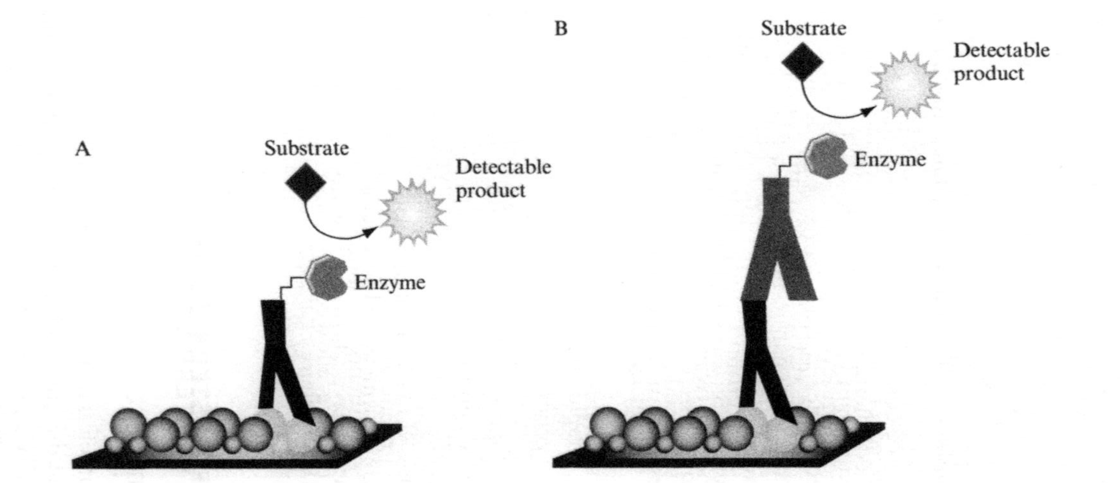

Figure 33.1 Schematic of direct and indirect Western blotting methods. In the direct detection method, labeled primary antibody binds to antigen on the membrane and reacts with the substrate, creating a detectable signal. In the indirect detection method, unlabeled primary antibody binds to the antigen. Then a labeled secondary antibody binds to the primary antibody and reacts with the substrate. (See Color Insert.)

detection also enables probing for multiple targets simultaneously. Labeling a primary antibody, however, sometimes has an adverse effect on its immunoreactivity, and even in the best of circumstances, a labeled primary antibody cannot provide signal amplification. Consequently, the direct method is generally less sensitive than indirect detection and is best used only when the target is relatively abundant. An indirect method that amplifies the signal and eliminates the secondary antibody is primary antibody biotinylation. Labeling with biotinylation reagents typically results in more than one biotin moiety per antibody molecule. Each biotin moiety is capable of interacting with an enzyme-conjugated avidin, streptavidin, or Thermo Scientific NeutrAvidin[1] Protein. These multiple enzymes catalyze the conversion of appropriate substrate to amplify the signal. Essentially, the avidin conjugate replaces a secondary antibody and its appropriate molar concentration is the same as if a secondary antibody were used. Be aware, however, if the sample applied to the gel is naturally biotinylated, there is the potential for signal generation that could interfere with target protein detection, especially when using highly sensitive substrates.

2.2. Far-Western

Occasionally, an antibody to a specific antigen is unsuitable for Western blot analysis or simply unavailable. Blotting is still possible if a binding partner to the target protein is available for use as a probe. This type of application is referred to as a far-Western blot and is routinely used for the discovery or confirmation of a protein–protein interaction. Variations on this theme are myriad and involve all the previously mentioned strategies. As with primary antibodies, labeled binding partners are used, which are frequently labeled in an *in vitro* translation reaction with ^{35}S. Biotinylating the probe and detecting with an avidin or avidin-like conjugate is another possibility and has the added effect of amplifying the signal. Care must be taken to ensure the probe is not overlabeled, which can compromise its ability to interact with the target. Alternatively, expressing a recombinant probe in bacteria with a tag, such as GST, HA, c-Myc, or FLAG, enables detection via a labeled antibody to the particular tag (Burgess *et al.*, 2000).

2.3. Semiquantitative Western blotting

Although Western blotting is often considered qualitative, it can be a quantitative method provided that specific controls are included. Semiquantitative Western blotting is advantageous when no ELISA is available for a specific sample or when a component in a biological sample interferes

[1] DyLight™, MemCode®, NeutrAvidin®, Pierce®, Restore® and SuperSignal® are trademarks of Thermo Fisher Scientific.

with an ELISA. Often antibodies directed against one protein may interact with equal specificity with another closely related protein. In such situations, an ELISA may generate false positives or overestimation of the target protein abundance. Because Western blotting involves resolution of proteins by gel electrophoresis, variations in molecular weight can be exploited to distinguish and quantify the target protein alone (Sato *et al.*, 2002; Xing and Imagawa, 1999).

To evaluate the effectiveness and accuracy of quantitative Western blotting, we used a purified target protein as an internal control and to create a standard curve. The sample must contain sufficient target protein such that its amount was within the standard curve range. To convert band intensity into a quantitative measurement, the Western blot was analyzed densitometrically using a CCD camera and imaging system. Finally, an ELISA was used to confirm the trends observed by Western blot (Mathrubutham and Vattem, 2005).

2.3.1. Assay methods

A431 cells were cultured for analyzing IκBα and p53. For IκBα, cells treated with epidermal growth factor (EGF; 10 ng/ml). For p53, cells were treated with either 50 μM cisplatin (Cis) or 5 μg/ml of doxorubicin (Dox) for 1 h. Cells were collected at time intervals and lysed. Total cellular protein was estimated using the Pierce Micro BCA Protein Assay (Thermo Fisher Scientific, Rockford, IL).

Western blots were prepared using pure protein or cell lysates. A standard curve was included with each Western blot. The linear range for IκBα and p53 was determined using recombinant IκBα (Upstate Biotechnology, Lake Placid, NY) from 0.0015 to 50 ng and recombinant p53 (Active Motif, Carlsbad, CA) from 1.8750 to 60 ng. After blocking, blots were incubated overnight at 4 °C with anti-IκBα (Upstate Biotechnology) or anti-p53 (Active Motif) antibodies at 1 μg/ml prepared in blocking buffer. Blots were washed and then incubated with horseradish peroxidase (HRP)-conjugated secondary antibodies (1:1000 dilution from 10 μg/ml stock) for 1 h at room temperature and then washed again. Signal was detected using SuperSignal[2] West Dura Substrate (Thermo Fisher Scientific). Blots were imaged using the Kodak 20000MM Image Station, and protein band densities were analyzed using the spot density analysis software provided with the image station.

The quantitative Western blot data were confirmed using p53 and IκBα ELISA assays (Assay Designs, Ann Arbor, MI). The ELISAs were performed according to the manufacturer's instructions.

[2] SuperSignal® Technology is protected by U.S. Patent # 6,432,662.

2.3.2. Results and discussion

Densitometric analysis of the Western blots indicated that the linear range for IκBα was 0.097–3.12 ng and for p53 was 1.87–60 ng. This range served as an excellent internal positive control and counteracted variations in Western blotting efficiency.

In the Western blot analysis, IκBα levels remained constant for the first 5 min after EGF treatment, rapidly dropped by 30% after 30 min and finally gradually increased during the next 24 h (Fig. 33.2). Western blot densitometric analysis revealed that p53 levels changed slightly over the first 8 h after treatment with Dox and then rapidly increased during the next 20–30 h. Treatment with Cis resulted in a gradual increase in p53 for 30 h (data not shown).

The trend in IκBα (Fig. 33.2) and p53 levels was confirmed by ELISA and also correlated well with published data (Kwok *et al.*, 1994; Sun and Carpenter, 1998). With respect to p53, although the Western blot data were confirmed by ELISA, closer comparison of the data revealed that the Western blot was more sensitive to quantity changes of p53 between 20 and 30 h. At the 20–30 h time point there was a large amount of p53 present, but the ELISA did not detect the gradual increase of p53 between these times, whereas the Western blot did. In contrast, changes in p53 during the first 8 h, when levels of p53 are relatively low, were detected better by ELISA than by quantitative Western blot. Therefore, the ELISA is more sensitive to changes in p53 at low levels, but the Western blot is more sensitive to p53 changes at high levels.

Although Western blotting and ELISA are immunodetection methods, the applications are conducted differently and usually use different immunoreagents. Variation in absolute protein amounts might be caused by differences in the recombinant protein used for controls, primary antibody specificity, and variations in protein interactions within each technique.

3. Detection Methods

3.1. Enzyme conjugate

Alkaline phosphatase (AP, 140 kDa) used to be the enzyme of choice and was typically detected with precipitating chromogenic substrates. Colorimetric reactions proceed at a steady rate, allowing accurate control of relative sensitivity and reaction development. As protein research progressed, HRP (40 kDa) became more popular because of its stability and smaller size, which enables more molecules conjugated per IgG and greater sensitivity. Furthermore, chemiluminescent substrates for HRP enabled even higher sensitivity.

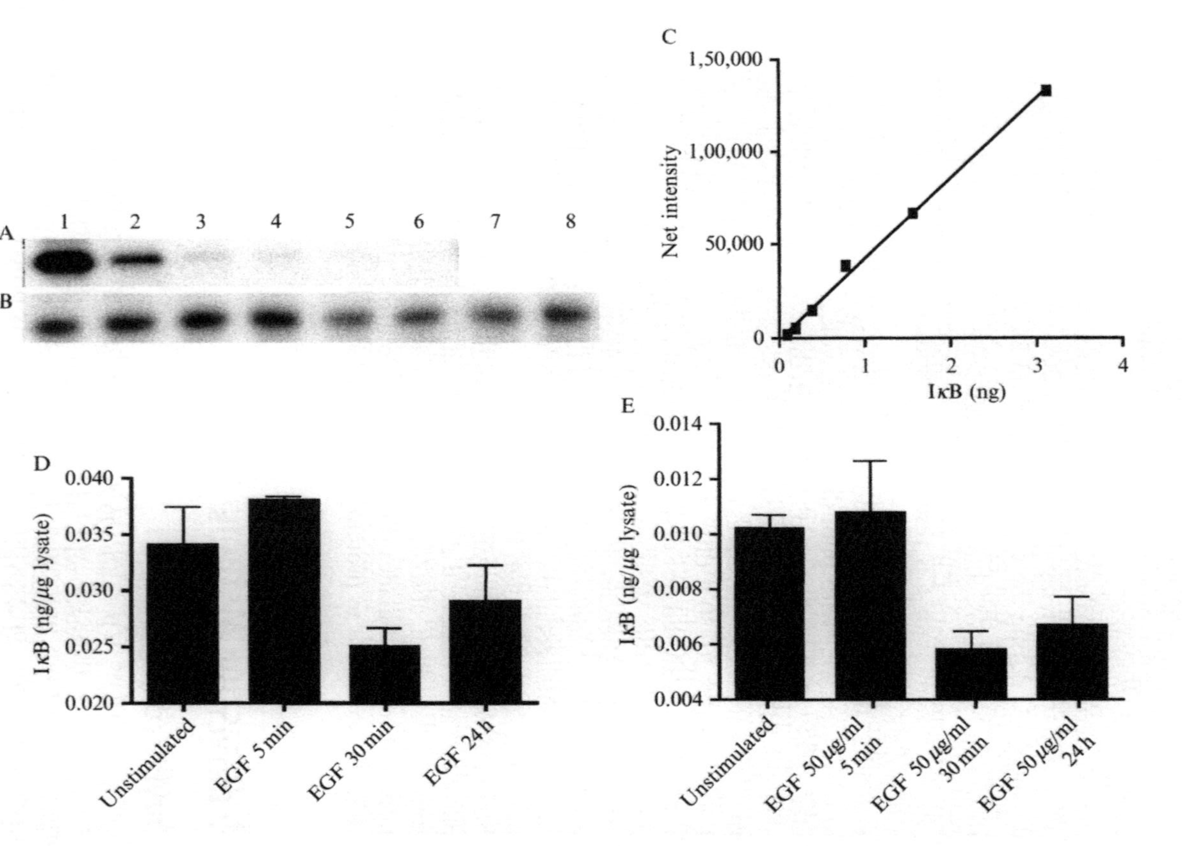

Figure 33.2 Quantitation of IκBα in EGF-treated cells. Panel A: Known amounts of recombinant IκBα (rIκBα) were Western blotted to generate a standard curve. Lanes 1–6 contain 0.15–0.012 ng of rIκBα. Panel B: Lysates of EGF-treated A431 cells were Western blotted for IκBα. Treatments were as follows: nontreated (lanes 1, 2), 50 ng for 5 min (lanes 3, 4); 50 ng for 30 min (lanes 5, 6); and 50 ng for 24 h (lanes 7, 8). Panel C: Recombinant IκBα standard curve ($r^2 = 0.99$). Panel D: Variation in the amount of IκBα with EGF exposure time as determined by quantitative Western blotting. Panel E: Quantitation by ELISA.

3.2. Colorimetric detection

Colorimetric or chromogenic substrates are perhaps the simplest and most cost-effective method of detection. When these substrates come in contact with the appropriate enzyme, they are converted to insoluble, colored products that precipitate onto the membrane and require no special equipment for processing or visualizing. Substrates such as TMB (3,3′,5,5′-tetramethylbenzidine), 4-CN (4-chloro-1-naphthol), and DAB (3,3′-diaminobenzidine tetrahydrochloride) are used with HRP. Substrates for AP include BCIP (5-bromo-4-chloro-3′-indolylphosphate-*p*-toluidine salt) and NBT (nitro-blue tetrazolium chloride), which yields an insoluble black-purple precipitate, and Fast Red (naphthol AS-MX phosphate + Fast Red TR Salt). The performance of a particular substrate may vary dramatically when obtained from different suppliers because results can be affected by the concentration and purity of the substrate and by additives and buffer components that are a part of the formulation.

3.3. Fluorescence detection

Western blotting detection via fluorescence is typically performed when there are two different targets on a single blot and high sensitivity is required. Fluorescent dyes (often called "fluorophores" or simply "fluors") are molecules whose chemical bonds excite when absorbing a photon of light at one wavelength, causing them to emit a photon of light at a longer wavelength (lower energy) as the molecule relaxes to the ground state. Small, chemically stable fluorophores that have convenient ranges of efficient (intense) excitation and emission wavelengths are useful as detectable chemical tags or labels for antibodies and other biomolecular probes.

Some fluorescence-based systems use fluorescent proteins (e.g., phycoerythrin) or bioluminescent reporter systems; however, these techniques are time-consuming, limited in their ability to detect multiple targets and do not typically provide the level of photostability and sensitivity offered by synthetic fluorescent dyes. Fluorescence techniques that use specific probes labeled with carefully selected sets of fluorescent dyes enable the detection of multiple targets and provide greater compatibility with a wide range of fluorescence instrumentation.

Although fluorescein, rhodamine, and amino-methyl-coumarin-acetate (AMCA) dyes are the traditionally used fluorophores, they have numerous limitations. Most notably, these traditional dyes have relatively low-fluorescence intensity and tend to photobleach. A new generation of fluorophores has overcome such limitations (Table 33.1) and offers tremendous improvements in brightness, photostability, and pH sensitivity. These new fluorophores cover the entire visible spectrum and much of the infrared spectrum.

Table 33.1 Spectral properties of commonly used fluorescent dyes*

Fluorescent dye	Emission	Ex/Em[a]	ε[b]
DyLight Fluor 405, Alexa Fluor[c] 405, Cascade Blue	Blue	400/420	30,000
DyLight Fluor 488, Alexa Fluor 488, fluorescein, FITC	Green	493/518	70,000
DyLight Fluor 549, Alexa Fluor 546, Alexa 555, Cy Dye 3[d], TRITC	Yellow	560/574	150,000
DyLight Fluor 594, Alexa Fluor 594, Texas Red	Red	593/618	80,000
DyLight Fluor 633, Alexa Fluor 633	Red	638/658	170,000
DyLight Fluor 649, Alexa Fluor 647, Cy Dye 5	Red	654/673	250,000
DyLight Fluor 680, Alexa Fluor 680, Cy Dye 5.5	Near IR	692/712	140,000
DyLight Fluor 750, Alexa Fluor 750	Near IR	752/778	220,000
DyLight Fluor 800, IRDye 800	Infrared	777/790	270,000

* The listed extinction coefficients are representative. The extinction coefficient for a given fluorescent dye and its analogs varies with the dye's purity, solvent and molecular structure changes, including the position and makeup of reactive groups and other moieties.

[a] Excitation and emission maxima in nanometers.

[b] Molar extinction coefficient (M^{-1} cm^{-1}).

[c] Alexa Fluor® is a trademark of Molecular Probes Inc.

[d] Cy® is a trademark of Amersham Pharmacia Biotech UK Limited Corp.

When performing a fluorescent Western blot, typically low-fluorescence (treated) membranes are used because membrane polymers autofluoresce in the visible range of the spectrum, which can interfere with detection. Using fluors with nonoverlapping excitation–emission spectra is critical for identifying targets. A typical fluor pair includes Thermo Scientific DyLight[3] Fluors 549 and 649 or the DyLight Near-infrared (IR) Fluors, 680 and 800. The near-IR fluors are especially useful because protein samples and membrane polymers are less likely to autofluoresce within these spectral ranges, resulting in lower background and higher sensitivity (Patonay and Antoine, 1991; Sowell *et al.*, 2002).

3.4. Chemiluminescence detection

The most popular Western blotting substrates are luminol-based and produce a chemiluminescent signal. Chemiluminescence is a chemical reaction that produces energy released in the form of light (Fig. 33.3). In the presence

[3] DyLight™, MemCode®, NeutrAvidin®, Pierce®, Restore® and SuperSignal® are trademarks of Thermo Fisher Scientific.

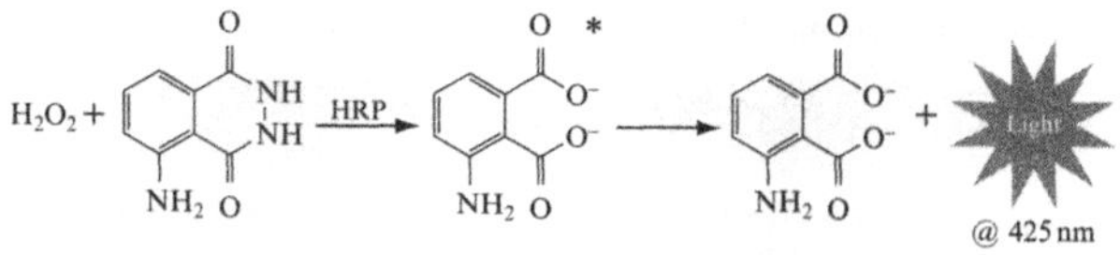

Figure 33.3 Chemiluminescence reaction scheme. Luminol is oxidized in the presence of HRP and hydrogen peroxide to form an excited state product (3-aminophthalate) that emits light at 425 as it decays to the ground state.

of HRP and a peroxide buffer, luminol oxidizes and forms an excited state product that emits light as it decays to the ground state. Light emission occurs only during the enzyme–substrate reaction and, therefore, once the substrate in proximity to the enzyme is exhausted, signal output ceases. In contrast, colorimetric substrates, such as TMB, produce precipitate that remains visible on the membrane even after the reaction has terminated.

4. The Chemiluminescence Signal

4.1. Signal capture

Although chemiluminescent Western blotting has become commonplace, attempting to capture that elusive signal can be frustrating. Because a Western blot is composed of a series of linked techniques that require skill to perform, failure to capture signal can be caused by many factors. With so many variables (Table 33.2), troubleshooting a problem blot can be equated to that proverbial needle in a haystack. The classical protocol is often ineffective in detecting a particular protein. For example, the primary antibody may not recognize the immobilized antigen in its denatured state. Although the protein can be kept in its native state by using nondenaturing conditions, this makes determination of target molecular weight more challenging. Furthermore, exceptionally large or hydrophobic proteins often preclude effective membrane transfer. On the other hand, small proteins (e.g., $\leq$10 kDa) can flow through the membrane pores, failing to bind. Towbin's original protocol is therefore often modified to ensure target detection. Such modifications may involve using different reagents for indirect detection or a labeled primary probe for direct detection. Sometimes transfer is bypassed altogether and detection is accomplished in-gel (Desai *et al.*, 2001).

4.2. Signal intensity and duration

When all Western blotting factors are optimal, a chemiluminescent signal can last for 6–24 h, depending on the specific substrate used. How much light is generated and for how long depends on the specific substrate being

Table 33.2 Factors that affect Western blotting results

Factor	Variable characteristic
Target antigen	Conformation, stability, available epitope(s), polypeptide size
Polyacrylamide gel	Manufacturer, percent polyacrylamide, age, lot
Membrane	Manufacturer, type, lot
Primary antibody	Specificity, titer, affinity, incubation time, and temperature
HRP conjugate	Enzyme activation level and activity, source animal, concentration
Blocking buffer	Protein type, concentration, cross-reactivity
Washes	Buffer, volume, duration, frequency
Substrate	Sensitivity, manufacturer lot, age
Detection method	Film age, imaging instrument manufacturer, exposure time

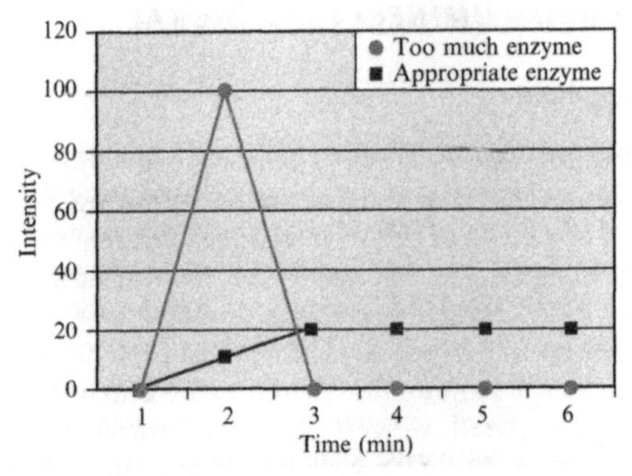

Figure 33.4 Example signal emission curves. When there is too much enzyme present in a chemiluminescent Western blot system, signal output peaks soon after substrate application and rapidly exhausts the substrate. In an optimal system, the signal emission peaks approximately 3 min after applying the substrate and plateaus for several hours.

used and the enzyme-to-substrate ratio present in the system. Although the amount of substrate on a blot is relatively constant, the amount of enzyme present depends on how much was added and other factors (Table 33.1). Too much enzyme conjugate applied to a Western blot system is the single greatest cause of signal variability, high background, short signal duration, and low sensitivity.

A signal emission curve that decays slowly (Fig. 33.4) is desirable because it demonstrates that each component of the system has been optimized and allows reproducible results. A signal that decays too quickly can cause variability, low sensitivity and lack of signal documentation. A long-lasting signal minimizes variability with transfer efficiency, different manufacturer lots of substrate and other factors.

Although HRP continues its activity for as long as substrate is available, HRP can become inactive with prolonged exposure to substrate. Free radicals produced during the oxidation reaction can bind to HRP in such a way that the enzyme can no longer interact with the substrate. An abundance of HRP in the system in turn produces an abundance of free radicals that increases the probability of HRP inactivation. Free radicals also can damage the antigen, antibodies and the membrane, prohibiting reprobing effectiveness.

4.3. Optimizing a Western blot

Each Western blot system must be optimized to obtain consistent results. Many factors influence the intensity and longevity of a signal, and each of these factors can be optimized as discussed in the following sections. The protocol section at the end of this chapter includes specific instructions and suggestions about optimization procedures.

4.3.1. Blotting membrane

HRP–luminol interactions and subsequent signal generation are likely unaffected by membrane composition. Nitrocellulose and PVDF membranes, however, do differ in their protein-binding properties. Generally, nitrocellulose binds proteins better, often produces crisper bands and sometimes results in greater sensitivity. PVDF is more hydrophobic, is difficult to wet, and sometimes results in more background signal; however, it has high tensile strength and excellent handling characteristics. For best results, empirically determine which membrane type, manufacturer and lot is optimal for each Western blotting system. Once a specific membrane has proven effective in a system, it may be beneficial to use the same lot throughout the course of the study.

4.3.2. Target protein

Transfer efficiency can vary dramatically among proteins. Proteins differ in their ability to migrate from the gel and their propensity to bind to the membrane using a particular set of conditions. Transfer efficiency depends on factors such as gel composition, the gel-membrane contact, position of the electrodes, transfer duration, protein size and composition, field strength, and the presence of detergents. Optimal transfer of most proteins is obtained in low-ionic strength buffers and with low electrical current. Transfer efficiency can be evaluated by staining the membrane with an immunoblot-compatible or reversible stain (Sasse and Gallagher, 2008).

Some common protein stains for membranes include: Ponceau S, a red stain that fades with time; coomassie dye, a sensitive blue stain that tightly binds to proteins and can interfere with Western blotting; and Thermo Scientific MemCode[4] Stain (Thermo Fisher Scientific), an easily reversible blue stain (Antharavally *et al.*, 2004).

4.3.3. Blocking buffer

Many different blocking reagents are available for Western blotting. Because no blocking reagent is appropriate for all systems, empirical testing is essential (Spinola and Cannon, 1985). An optimal blocking buffer maximizes the signal-to-noise ratio and does not react with the system's antibodies or target. For example, using 5% nonfat milk as a blocking reagent when using avidin–biotin systems results in high background because milk contains variable amounts of endogenous biotin, which binds the avidin. When switching substrates, antibodies or the target, a diminished signal or increased background can result simply because the blocking buffer was not optimal for the new system.

Some systems may benefit from adding a surfactant, such as Tween-20[5] detergent, to the blocking buffer. Surfactants can minimize background by preventing the blocking reagent from nonspecifically binding to the target or by blocking hydrophobic sites on the membrane to which antibodies can bind. Adding too much detergent, however, can prevent adequate blocking. Typically, a final concentration of 0.05% detergent is used; however, for best results, determine if detergents enhance a specific system and their optimal concentration. Always use a high-quality detergent that is low in contaminants.

4.3.4. Antibodies

Not only is the affinity of the primary antibody for the antigen important, but primary and secondary antibody concentrations also have a profound effect on signal intensity. Too much HRP on the blot can be caused by either primary and secondary antibody concentrations or both. Minimal primary antibody is advantageous, as it promotes target-specific binding and low background.

If a blot failed to generate an adequate signal, removing all detection reagents (stripping) from the blot and reprobing with either a different primary antibody or different concentrations of antibodies often conserves valuable sample and time; however, insufficient stripping can leave active HRP on the blot that will produce a signal. Applying substrate on the stripped blot and subsequent detection will reveal if active HRP remains on

[4] DyLight™, MemCode®, NeutrAvidin®, Pierce®, Restore® and SuperSignal® are trademarks of Thermo Fisher Scientific.

[5] Tween® is a trademark of ICI Americas.

the blot. Also, an abundance of inactive HRP molecules not removed by stripping can inhibit the primary antibody from binding to the target. Stripping and reprobing blots is an effective method to gain information about a specific system, but it is not a definitive way to determine the optimal system parameters.

4.3.5. Detection method

Traditionally, film has been used to detect a Western blot chemiluminescent signal. Film requires no expensive equipment and provides excellent sensitivity, but suffers from a narrow range of linear detection intensity. Adjusting film exposure time is frequently required to obtain a publication-quality blot, which can be time-consuming and generate waste. Immersing the developed film in a set of Thermo Scientific Reagents (Fig. 33.5) will remove excessive signal caused by overexposure. These reagents evenly remove silver from the film, which maintains the signal-to-noise ratio while improving the film's appearance.

CCD cameras and their accompanying analytical software can adjust background levels and perform densitometry. The imager has advantages over film-based detection because the large dynamic range and the high degree of exposure control, allowing for the best possible image documentation without issues of high background or high-signal intensity obscuring data. Additionally, optimization of exposure time avoids band signal

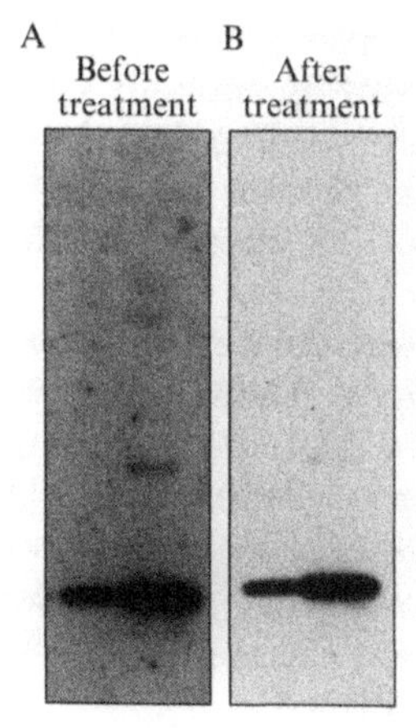

Figure 33.5 Reduction of background signal by chemical treatment of the X-ray film. Recombinant human TNFα was separated by SDS–PAGE and transferred to a nitrocellulose membrane. The blot was probed with mouse antihuman TNFα and goat antimouse-HRP antibodies and detected with SuperSignal West Dura Substrate. Panel A: The blot was exposed to film for 30 s, resulting in considerable background speckling. Panel B: The film was then treated with Pierce Background Eliminator for 2 min to remove speckling.

saturation and allows observation of minor variations in density. In contrast, film has a small dynamic range, and band signal can quickly reach saturation. When signal intensity is high, film's low dynamic range, propensity to reach saturation quickly and exposure control limitations often result in overexposed images.

5. Common Problems and their Explanations

5.1. No signal

An initial exposure that fails to capture the chemiluminescent signal indicates a Western blotting system that requires optimization. Frequently, lack of signal is caused by too much enzyme (i.e., HRP) in the system. It may seem counterintuitive to use less enzyme conjugate when a signal cannot be detected; however, for successful signal documentation, the correct balance of enzyme and substrate must be present. Substrate oxidation by the enzyme is irreversible and, therefore, once the substrate is oxidized, it can no longer interact with the enzyme to generate light. Because enzyme activity persists, the substrate is the limiting factor and once exhausted, signal output ceases. Rarely, lack of signal is caused by an insufficient amount of active enzyme present. Too much or not enough enzyme can be caused by any of the factors involved in the Western blot system.

To produce a signal that can be captured, adjust the system's parameters. For reproducible results, prepare a new gel and apply less sample or titrate the antibodies. When optimizing antibody concentrations, image the blot twice: once immediately after adding the substrate and a second time at an interval (e.g., 1 h) after incubating the substrate. The second detection provides information about the optimal enzyme concentration and helps optimize parameters.

Also, if the initial exposure did not capture a signal, a second incubation in substrate may yield a signal if some HRP remains active. Stripping all detection reagents from the blot and reprobing can save valuable sample while optimizing parameters. Perform an additional incubation in the substrate and strip the blot only to recover some information about the system. If blot-to-blot consistency or comparison is desired, the same conditions must be used and the same procedure must be followed each time the experiment is performed.

5.2. Signal fades quickly

When a particular system produces a chemiluminescent signal that fades quickly, the Western blotting system requires optimization, as described earlier. The good news is that a signal was obtained, indicating that the blot is almost optimized. Sometimes a particular system produces a signal that fades more quickly than usual although all parameters are the same. This type of result is minimized in a fully optimized system. A successful but suboptimal system is subject to the slight variations inherent to the method, such as transfer efficiency and changes in sample and antibody activity during storage and handling.

5.3. High background

High background signal is the result of either insufficient blocking, antibodies cross-reacting with the blocking proteins, or the use of too much enzyme conjugate. Researchers sometimes believe that a particular substrate causes background or can increase background. The substrate in itself typically does not produce signal without the enzyme being present. When using a substrate with greater sensitivity than what was previously used, high background often results if the parameters were not altered to compensate for the substrate's sensitivity. Using optimal concentrations of antibodies promotes target-specific binding and low background.

5.4. New bottle of substrate does not produce a signal

Occasionally, a signal cannot be captured when the only variable that has changed in a particular system is a new bottle or different lot of substrate. Typically, this result is caused by a Western blot system that has not been fully optimized. Western blotting substrates are inherently variable. Many manufacturers simply control for a minimum sensitivity, and it is possible that the new substrate is more sensitive than the previously used lot. In a fully optimized blotting system, substrate sensitivity variations, as well as other variables, are minor or unnoticeable.

5.5. Brown or yellow bands on the membrane

HRP becomes brown when it is oxidized and inactive. Within a given amount of enzyme conjugate, there always exists a portion that is oxidized. In an optimized system, the amount of oxidized HRP is miniscule and cannot be visualized on the blot. The appearance of yellow or brown bands indicates the presence of a large amount of HRP and, therefore, the oxidized and inactive portion is visible. A blotting system that results in yellow bands requires optimizing using much less enzyme conjugate. Additionally, too much HRP in a localized area produces an abundance of free radicals during enzymatic activity. Free radicals can inactivate HRP and damage antibodies, target and the membrane, prohibiting effective reprobing.

5.6. Bands or entire blot glowing in the darkroom

If a pattern of bands or the entire blot is glowing after incubation in the substrate, then there is too much HRP present in the system. This occurrence indicates that further dilution of the secondary HRP-conjugated antibody is required and possibly the primary antibody as well. Too much enzyme can be caused by many of the factors involved in the Western blotting system. If the entire blot is glowing, optimization of blocking and washing also might be necessary.

5.7. Ghost/hollow bands

Protein bands that appear as a halo with no signal in the middle of the band or an entire band that appears white in a dark background are typically referred to as ghost bands. This result occurs when the substrate is depleted within the white area.

The specific cause of ghost bands is not well-understood. Therefore, we tested several factors that could contribute to this result. Our study, briefly described in the following sections, revealed that some common causes for ghost-band effects include applying too much target protein to the gel and using antibodies that cross-react with component(s) in the blocking solution (Vattem and Mathrubutham, 2005).

5.7.1. Assay methods

A431 cells (ATCC) were cultured and treated with cisplatin. Cells were lysed, sonicated, and clarified by centrifugation. Total protein concentration was estimated using Thermo Scientific Pierce[6] BCA Protein Assay (Thermo Fisher Scientific).

Various amounts of purified NFκB protein (Upstate Biotechnology) or 30 μg of A431 lysates were transferred to nitrocellulose membranes. The membranes were incubated overnight with either rabbit anti-NFκB antibody (1 mg/ml stock; diluted 1:5000; Upstate Biotechnology) or rabbit anti-p53 (1 mg/ml stock diluted 1:1000; GENEKA). HRP-conjugated secondary antibodies (10 μg/ml stock diluted 1:1000) were applied to the blots and incubated for 1 h at room temperature. Either Thermo Scientific SuperSignal[1] West Pico or West Dura Chemiluminescent Substrate (Thermo Fisher Scientific) was added (0.5 and 5.0 ml) and incubated for various periods. Membranes were exposed to X-ray film for various periods.

[6] DyLight™, MemCode®, NeutrAvidin®, Pierce®, Restore® and SuperSignal® are trademarks of Thermo Fisher Scientific.

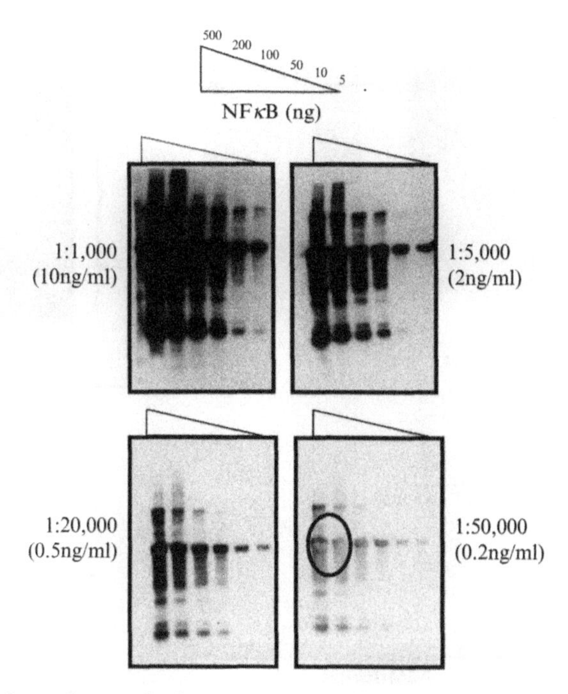

Figure 33.6 Secondary antibody concentration affects the appearance of ghost bands. Western blots were performed using various concentrations of secondary antibody. Ghost bands (oval) are present in lanes containing >200 ng of NFκB. Membranes were incubated in SuperSignal West Pico Substrate for 5 min and exposed to X-ray film for 30 s.

5.7.2. Results and discussion

Protein bands on membranes treated with 1:1000, 1:5000 or 1:20,000 dilution of a 10 μg/ml stock of secondary antibody were not significantly different; however, 1:50,000 dilution (0.2 ng/ml) of secondary specifically enhanced the appearance of ghost bands. This effect was more evident in lanes that contained >200 ng NFκB (Fig. 33.6). Also, 64 h after completing the experiment, when signal intensity had slightly reduced, ghost bands became more evident in all lanes having >200 ng NFκB, irrespective of secondary antibody dilution (data not shown).

A short substrate incubation of 30 s resulted in ghost bands in lanes containing >200 ng of protein. Generally, longer substrate incubation prolongs signal duration. Using either the West Pico or West Dura substrate enabled detection down to 10 ng of proteins, even after 64 h of completing the Western blot. West Dura is a more sensitive and longer lasting substrate than West Pico. Although West Dura produced a significantly stronger signal when detecting 5 ng of proteins, lanes with greater than 100 ng resulted in rapid signal loss, which emphasizes the importance of using an

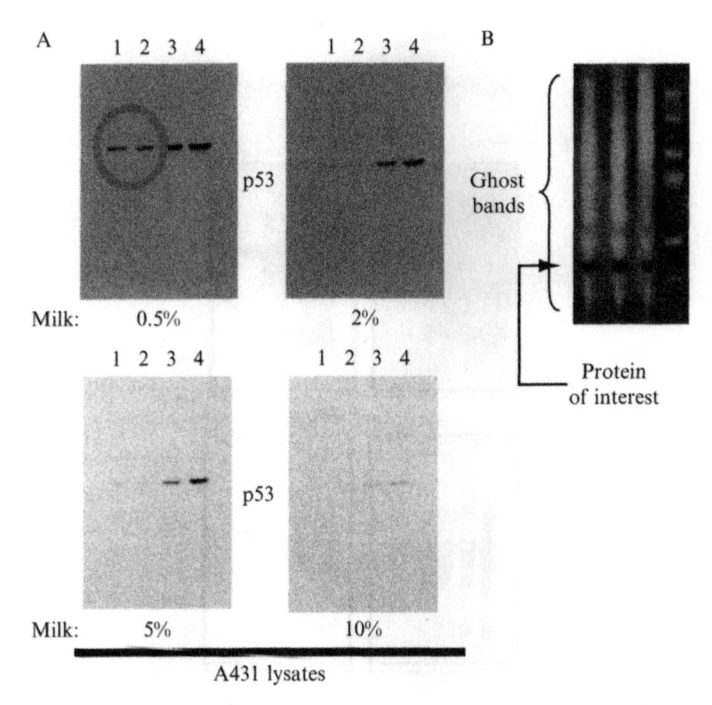

Figure 33.7 Low concentration of nonfat milk enhances the detection of low-abundance proteins, but antibodies cross-reacting with milk proteins result in ghost bands. Panel A: Lanes 1–4 are lysate samples from different cell harvest times (0, 8, 20, and 30 h) after treatment with 50 μM cisplatin. Different concentrations of nonfat milk were used for blocking and antibody incubation. The low concentration of milk (0.5%) enhanced detection of p53. Panel B: Membrane was incubated with 5% nonfat milk blocking buffer for 48 h. The primary antibody bound to the blocked regions, producing ghost bands.

appropriate amount of target protein with highly sensitive detection systems.

Incubating the blot with a nonfat milk solution is commonly used to minimize background signal. Varying the concentration of milk in the blocking solution did not result in ghost bands. Surprisingly, using 0.5% nonfat milk in the blocking solution enhanced detection of low-abundance proteins (Fig. 33.7, panel A); however, antibody cross-reacting with the blocking protein did result in ghost bands. If the primary or secondary antibody binds to milk proteins, these blocked sites chemiluminescence, producing a dark background with white lanes where sample proteins populate the membrane surface (Fig. 33.7, panel B).

Rapid signal loss may result from enhanced catalysis of substrate by a high density of enzyme-conjugated secondary antibody binding to a high amount of target protein, causing rapid substrate depletion in these regions.

A high amount of transferred protein binds high amounts of HRP-conjugated secondary antibody. Excessive product (i.e., signal) produced in these concentrated areas of immune complexes can result in a rapid substrate depletion, resulting in ghost bands. In conclusion, excess loading of gels, high concentrations of secondary antibody, suboptimal substrate incubation times, and antibodies cross-reacting with blocked areas can cause ghost bands. Optimizing Western blot parameters is especially critical to prevent ghost bands when using highly sensitive detection systems.

6. Blotting and Optimization Protocols using Chemiluminescent Substrates

6.1. Western blotting protocol using chemiluminescent substrates

1. Separate the proteins in the sample by gel electrophoresis.
2. *Prepare the transfer buffer*: Use Tris–glycine transfer buffer dissolved in 400 ml of ultrapure water plus 100 ml methanol. Use and store the transfer buffer at 4 °C. *Note*: Transfer buffer: 25 m*M* Tris, 192 m*M* glycine, pH 8.0, 20% methanol.
3. Construct a gel "sandwich" (Fig. 33.8) for wet transfer. For semidry transfer, prepare the sandwich in the same order between the anode and cathode. *Note*: Soak the pads and filters in transfer buffer before sandwich assembly.
4. Transfer proteins from the gel to a membrane. For wet transfer using a mini-transfer apparatus designed for an 8 × 10 cm gel, transfer at 40 V for 90 min keeping the buffer temperature at 4 °C. For semidry transfer use 15 V for 90 min. *Note*: Determine transfer success by staining the gel or reversibly staining the membrane.
5. Remove the membrane and block nonspecific binding sites with a blocking buffer for 20–60 min at room temperature (RT) with shaking.
6. Incubate the blot with the primary antibody solution (see Table 33.3) containing 10% blocking solution with rocking for 1 h. If desired, incubate the blot overnight at 2–8 °C.
7. Wash the membrane three times for 5 min each with Tris-buffered saline (TBS), phosphate-buffered saline (PBS) or other physiological wash buffer containing 0.05% Tween-20 detergent. If using an enzyme-conjugated primary antibody, proceed to step 10.
8. Incubate the blot with the enzyme conjugate (see Table 33.3) containing 10% blocking solution for 1 h with rocking at room temperature.
9. Wash the membrane five times for 5 min each in wash buffer to remove any nonbound conjugate. It is crucial to thoroughly wash the membrane after incubation with the enzyme conjugate.

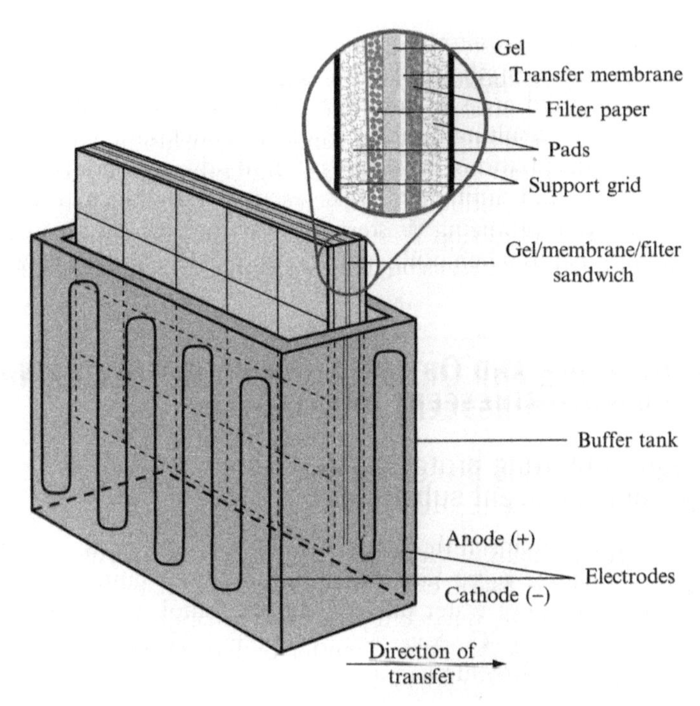

Figure 33.8 Electrophoretic transfer unit and blot setup.

Table 33.3 Primary and secondary antibody concentrations to use with Thermo Scientific Chemiluminescent Substrates

Substrate	ECL	West Pico[a]	West Dura[a]	West Femto[a]
Primary antibody concentration (μg/ml)	0.2–10	0.2–1.0	0.02–1.0	0.01–0.2
Secondary antibody concentration (ng/ml)	67–1000	10–50	4–20	2–10

[a] Part of the Thermo Scientific SuperSignal Product Family.

10. Prepare the substrate. Use a sufficient volume to ensure that the blot is completely wetted with substrate and the blot does not become dry (0.1 ml/cm^2).
11. Incubate the blot with substrate for 1 min when using ECL or 5 min when using SuperSignal Substrates.

12. Remove the blot from the substrate and place it in a plastic membrane protector. A plastic sheet protector works well, although plastic wrap also may be used. Remove all air bubbles between the blot and the surface of the membrane protector.
13. Image the blot using film or a cooled CCD camera.

6.2. Western blot stripping protocol

One of the many advantages of using a chemiluminescent substrate is that it allows the removal of all detection reagents (Kaufmann *et al.*, 1987). These stripping methods are ineffective for precipitating substrates and produce variable results with fluorescent dyes.

1. Prepare the stripping buffer. Use one of the following suggested stripping buffers:
   (a) Restore Western Blot Stripping Buffer (Thermo Fisher Scientific)
   (b) 0.1 *M* glycine–HCl (pH 2.5–3.0)
   (c) 50 m*M* Tris–HCl (pH 7), 2% sodium dodecyl sulfate (SDS), 50 m*M* dithiothreitol (DTT). *Note*: Prepare this buffer immediately before use.
2. Place the blot to be stripped in stripping buffer. Use a sufficient volume to ensure that the blot is completely wetted (approximately 20 ml for an 8 × 10 cm blot). *Note*: Some denaturation and loss of target protein may occur.
3. Incubate for 5–15 min at room temperature. Optimization of both incubation time and temperature is essential for best results. Some interactions require at least 15 min and may require incubation at 37 °C. If using buffer C, incubate for 30 min at 70 °C.
4. Remove the blot from the stripping buffer and wash using wash buffer (PBS/TBS or other physiological buffer containing 0.05% Tween-20 detergent).
5. To test for complete removal of the enzyme conjugate and primary antibody, perform the tests listed below. If signal is detected in either case, repeat steps 2–4, stripping for an additional 5–15 min or increasing the temperature to 37 °C. Optimize stripping time and temperature to ensure complete removal of antibodies while preventing damage to the antigen.
   - To test for complete removal of the enzyme conjugate, incubate the membrane with substrate and image the blot. If no signal is detected after a 5-min film exposure, the enzyme conjugate has been successfully removed.
   - To test for complete removal of the primary antibody, incubate the membrane with enzyme conjugate then wash with wash buffer. Incubate with substrate and image the blot. If no signal is detected after a 5-min film exposure, the primary antibody has been successfully removed.

After determining that the membrane is properly stripped, commence the second probing experiment. Typically, a blot can be stripped and reprobed several times but may require longer exposure times or a more sensitive substrate. Subsequent reprobing may result in decreased signal if the antigen is labile. Analysis of the individual system is required. *Note*: Reblocking a membrane is generally not necessary after stripping, but may be required in some systems.

6.3. Optimizing antigen concentration

1. Prepare different concentrations of the protein sample in SDS–PAGE sample buffer. Test a wide range of sample concentration, keeping in mind the detection limit of the substrate being used.
2. Apply an equal volume of each concentration on the gel and separate by electrophoresis. Transfer the samples to a membrane.

Note: For a rough indication of optimal concentration, perform a dot blot. Dot antigen dilutions onto dry membrane strips, using the smallest possible volume. Allow the membranes to dry 10–15 min or until no visible moisture remains and proceed with step 3.

3. Block the membrane with a standard blocking reagent and probe with primary antibody followed by enzyme conjugate. If optimized dilutions have not yet been determined, use a mid-range value according to the sensitivity of the substrate.
4. Wash membrane and add the substrate. Image the blot as desired.

6.4. Optimizing membrane blocking

1. Separate the protein sample by electrophoresis and transfer to a membrane or dot protein samples onto the membrane as described previously.
2. Cut strips from the membrane according to the number of conditions being tested. The following combinations should be tested with each blocker:
   - Blocker + primary antibody + enzyme conjugate + substrate
   - Blocker + enzyme conjugate + substrate
   - Blocker + substrate
3. Add the strips to various blocking solutions, ensuring the strip is completely immersed in the solution. Incubate each strip for 1 h at room temperature with shaking.
4. Add primary antibody and/or enzyme conjugate solutions containing 10% blocking agent to appropriate groups. If optimized dilutions have not yet been determined, use a mid-range value according to the sensitivity of the substrate.
5. Wash membrane and add the substrate. Image the blot as desired.

Note: To check for endogenous peroxidase activity in the blocker or in the sample, incubate blocked membranes with substrate. If signal is detected, use a different blocking buffer or use a peroxidase suppressor after the blocking step. Large blocking buffer volumes minimize nonspecific signal.

6.5. Optimizing the primary antibody concentration

1. Separate the protein sample by electrophoresis and transfer to the membrane. Alternatively, dot the protein sample onto the membrane as described in Section 6.3. Block the membrane using an appropriate blocking reagent. Cut strips from the membrane according to the number of primary antibody conditions being tested.
2. Prepare dilutions of primary antibody in wash buffer containing 1/10 volume of blocking agent and apply to the membrane strips. Incubate for 1 h at room temperature.
3. Wash the strips and incubate with enzyme conjugate for 1 h at room temperature. Wash again and develop signal using an appropriate substrate. Detect signal using film or a CCD camera.

Note: To check for nonspecific binding of primary antibody to blocker, incubate membrane in primary antibody working solution followed by enzyme conjugate working solution. If signal is observed, correct the problem by using a different blocking buffer or use less primary antibody.

6.6. Optimizing membrane washing

1. Use a wash buffer such as PBS or TBS or other physiological buffer containing 0.05% Tween-20.
2. Wash the membrane by agitating at least three times for 5 min each after primary antibody incubation, and at least five times for 5 min each after incubating with enzyme conjugate.
3. If nonspecific background appears upon final detection, use larger volumes of wash buffer or increase the number and time of each wash. If no improvement occurs, the problem lies with another variable.

6.7. Optimizing enzyme conjugate concentration

When determining the optimal concentration for a new Western blotting system, a simple experiment often saves much frustration with signal variability.

1. Apply the same amount of target in three (or more) wells of the gel.
2. Separate the protein sample by electrophoresis and transfer to the membrane. Block the nonspecific binding sites and probe with primary antibody.

3. After washing, cut the blot into strips containing the target.
4. Probe each strip with a different enzyme conjugate concentration. For example, for SuperSignal West Pico Substrate use 1:40,000, 1:60,000, and 1:80,000 dilutions (from a 1-mg/ml stock). Incubate strips for 1 h at room temperature with rocking.
5. Wash the strips and add the substrate. After substrate incubation, image the strips.
6. Wait 1–2 h and image the strips a second time.
7. Evaluate the results. Use the dilution that produces the strongest signal.

6.8. Optimizing the detection method

1. Separate the protein sample by electrophoresis and transfer to membrane or dot the protein sample onto the membrane as described in Section 6.3.
2. Block nonspecific binding sites and probe with primary antibody and enzyme conjugate containing 1/10 volume of blocking agent.
3. If antibody concentrations have not been optimized, choose a mid-range value. Wash the membrane after each incubation.
4. Cut strips from the membrane according to the number of substrate exposure conditions being tested. Prepare working solution of substrate to be tested.
5. Incubate the membrane strips with the substrate for a time period consistent with the manufacturer's instructions.
6. Remove the strips from the substrate using forceps, and gently tap edge onto a paper towel to remove excess substrate.
7. Place the strips in a plastic cover and image blot for varying lengths of time. Select a time with clear signal and low background. *Note*: A CCD camera might require slightly longer exposure times than film.

REFERENCES

Antharavally, B. S., Carter, B., Bell, P. A., and Mallia, K. A. (2004). A high-affinity reversible protein stain for Western blots. *Anal. Biochem.* **329**(2), 276–280.

Bergendahl, V., Glaser, B. T., and Burgess, R. R. (2003). A fast western blot procedure improved for quantitative analysis by direct fluorescence labeling of primary antibodies. *J. Immunol. Methods* **277,** 117–125.

Burgess, R. R., Arthur, T. M., and Pietz, B. C. (2000). Mapping protein–protein interaction domains using ordered fragment ladder far-Western analysis of hexahistidine-tagged fusion proteins. *Methods Enzymol.* **328,** 141–157.

Choudhary, S., Lu, M., Cui, R, and Brasier, A. R. (2007). Involvement of a novel Rac/RhoA guanosine triphosphatase-nuclear factor-kappaB inducing kinase signaling pathway mediating angiotensin II-induced RelA transactivation. *Mol. Endocrinol.* **21**(9), 2203–2217.

Desai, S., Dworecki, B., and Cichon, E. (2001). Direct immunodetection of antigens within the precast polyacrylamide gel. *Anal. Biochem.* **297,** 94–98.

Kaufmann, S. H., Ewing, C. M., and Shaper, J. H. (1987). The erasable Western blot. *Anal. Biochem.* **161,** 89–95.

Kwok, T. T., Mok, C. H., and Menton-Brennan, L. (1994). Up-regulation of a mutant form of p53 by doxorubicin in human squamous carcinoma cells. *Cancer Res.* **54,** 2834–2836.

Mathrubutham, M., and Vattem, K. (2005). Methods and considerations for quantitative Western blotting using SuperSignal® Chemiluminescent Substrates. *In* "Thermo Fisher Scientific Application Note # 12". www.thermo.com/pierce.

Mattson, D. L., and Bellehumeur, T. G. (1996). Comparison of three chemiluminescent horseradish peroxidase substrates for immunoblotting. *Anal. Biochem.* **240,** 306–308.

Patonay, G., and Antoine, M. D. (1991). Near-infrared fluorogenic labels: New approach to an old problem. *Anal. Chem.* **63,** 321A–327A.

Ramlau, J. (1987). Use of secondary antibodies for visualization of bound primary reagents in blotting procedures. *Electrophoresis* **8,** 398–402.

Sasse, J., and Gallagher, S. R. (2008). Detection of proteins on blot transfer membranes. *Curr. Protoc. Immunol.* **Chapter 8,** Unit 8.10B.

Sato, M., Nishi, N., Shoji, H., Kumagai, M., Imaizumi, T., Hata, Y., Hirashima, M., Suzuki, S., and Nakamura, T. (2002). Quantification of Galectin-7 and its localization in adult mouse tissue. *J. Biochem.* **131,** 255–260.

Sowell, J., Strekowski, L., and Patonay, G. (2002). DNA and protein applications of near-infrared dyes. *J. Biomed. Opt.* **7,** 571–575.

Spinola, S. M., and Cannon, J. G. (1985). Different blocking agents cause variation in the immunologic detection of proteins transferred to nitrocellulose membranes. *J. Immunol. Meth.* **81,** 161.

Sun, L., and Carpenter, G. (1998). Epidermal growth factor activation of NF-kappaB is mediated through IkappaBalpha degradation and intracellular free calcium. *Oncogene* **23,** 2095–2102.

Tipsmark, C. K., Strom, C. N., Bailey, S. T., and Borski, R. J. (2008). Leptin stimulates pituitary prolactin release through an extracellular signal-regulated kinase-dependent pathway. *J. Endocrinol.* **196**(2), 275–281.

Towbin, H., Staehelin, T., and Gordon, J. (1979). Electrophoretic transfer of proteins from polyacrylamide gels to nitrocellulose sheets: Procedure and some applications. *Proc. Natl. Acad. Sci. USA* **76,** 4350–4354.

Vattem, K., and Mathrubutham, M. (2005). Factors that cause the appearance of ghost bands when using chemiluminescent detection systems in a Western blot. *In* "Thermo Fisher Scientific Application Note # 11". www.thermo.com/pierce.

Walker, G. R., Feather, K. D., Davis, P. D., and Hines, K. K. (1995). SuperSignal™ CL-HRP: A new enhanced chemiluminescent substrate for the development of the horseradish peroxide label in Western blotting applications. *J. NIH Res.* **7,** 76.

Xing, C., and Imagawa, W. (1999). Altered MAP kinase (ERK1, 2) regulation in primary cultures of mammary tumor cells: Elevated basal activity and sustained response to EGF. *Carcinogenesis* **20,** 1201–1208.

SECTION NINE

PURIFICATION PROCEDURES: MEMBRANE PROTEINS AND GLYCOPROTEINS

CHAPTER THIRTY-FOUR

Detergents: An Overview

Dirk Linke

Contents

1. Introduction 604
2. Detergent Structure 604
3. Properties of Detergents in Solution 605
   3.1. Phase diagrams and critical micellization concentration 609
   3.2. Effects of temperature: Detergent solubility, krafft point, cloud point, and phase separation 611
4. Exploiting the Physicochemical Parameters of Detergents for Membrane Protein Purification 612
5. Detergent Removal and Detergent Exchange 613
6. Choosing the Right Detergent 613
   6.1. Optical spectroscopy 614
   6.2. Mass spectrometry and nuclear magnetic resonance (NMR) measurements 615
   6.3. Protein crystallization 615
7. Conclusions 615
Acknowledgments 616
References 616

Abstract

Detergents are used in molecular biology laboratories every day. They are present in cell lysis buffers (e.g., in kits for plasmid isolation), in electrophoresis and blotting buffers, and, most importantly, they are used for cleaning laboratory glassware and the hands of the laboratory staff. For these routine applications, a detailed knowledge of detergent properties is not necessary—they just work. When it comes to the isolation and purification of membrane proteins, one cannot rely on routine protocols. Many membrane proteins are only stable in a small number of different detergent buffer systems, and worst of all, different membrane proteins prefer very different detergents. Unfortunately, detergent properties are considered the domain of colloid science or physical chemistry, and thus, while the available amount of physico-chemical data on detergents is astounding, this data is rarely compiled in a way that is useful to biochemists. The aim of

Department I, Protein Evolution, Max Planck Institute for Developmental Biology, Tübingen, Germany

Methods in Enzymology, Volume 463
ISSN 0076-6879, DOI: 10.1016/S0076-6879(09)63034-2

this chapter is to provide an overview of the physical and chemical properties of detergents commonly used in membrane protein science and to explain how these properties can be exploited for protein purification.

1. Introduction

What is a detergent? The term detergent is used in many different contexts. A household detergent is a complex mixture of surfactants (to dissolve grease), abrasives (to scour), chelating agents (to counter the effect of ions that contribute to water hardness), oxidants (for bleaching), enzymes (to degrade fats, proteins, or complex carbohydrates), colors, perfumes, optical brighteners, buffer substances (to stabilize the pH), and a number of other stabilizing ingredients, for example, to modify the foaming properties or to inhibit bacterial or fungal growth. The term is sometimes also used to distinguish soap from other cleaning agents (detergents). In molecular biology laboratories, the term detergent is typically used as a synonym for surfactant. Surfactants ("surface acting agents") are wetting agents that lower the surface tension of a liquid, and the interfacial tension between two liquids. The chemical and physical basis of these properties is discussed in detail in this chapter. Obviously, surfactant is the more precise term, but in scientific literature, it is only used in the context of physical chemistry and colloid science; in biochemistry, the term detergent is widespread, and the author of this chapter is not going to try and change this—from here on, detergent is used as a synonym for surfactant.

Historically, the first detergents used were saponins and soap (or rather, the fatty acid salts therein). Soap is made from vegetable oils or animal fats which are hydrolyzed by lye to yield fatty acid salts and glycerol; this process was already used in ancient Babylon and Egypt, and was brought to perfection by chemists during the early middle ages in the Middle East. Saponins are natural detergents that can be extracted from plants, for example from Soapwort (*Saponaria officinalis*) or Soapnut (*Sapindus*). While soaps are not used in membrane protein science, the saponin Digitonin from Purple Foxglove (*Digitalis purpurea*) is. The first synthetic detergents were developed and used during World War I in Germany, when soap was scarce. Synthetic detergents have replaced soap in many household applications since then. Most detergents used in laboratory applications today were originally developed as constituents of laundry or industry detergents.

2. Detergent Structure

Detergents (or surfactants) are organic compounds of very diverse structure. Generally speaking, detergent molecules consist of two parts: an extended, hydrophobic hydrocarbon moiety, and a polar or charged headgroup. In the simplest case, the hydrocarbon part of a detergent is an unbranched, saturated alkane; hydrophobicity is then increased with increasing chain length. Alternatively, unsaturated or branched-chain alkanes, aromatic hydrocarbons or steroid moieties are found in the lipophilic group of detergents, sometimes in combination. The hydrophilic headgroup is even more variable.

The most simple classification of detergents is based on the four basic types of headgroups, namely nonionic, anionic, cationic, or zwitterionic detergents; representative members of the four classes are shown in Table 34.1. Various other classification systems have been introduced to grasp the complexity of detergent structures and properties. One of the standard classifications is based on hydrophile–lipophile balance number (HLB). It describes the balance between type and size of the hydrophilic and lipophilic parts of a detergent molecule and determines the water solubility of a detergent (Neugebauer, 1990). HLB values of surface-active substances range from 0 to 40, but detergents typically have an HLB of 12–15. Nonionic detergents with a lower HLB (<10) are not water-soluble and are used as antifoaming agents or to emulsify water in oil, while those with a higher HLB (>16) are used as stabilizers. Note that HLB is a numerically calculated number based on the detergent's molecular structure; it is not a measured parameter. The HLB is a useful tool to classify nonionic detergents, but it does not help to compare ionic with nonionic detergents. Sodium dodecylsufate (SDS), for example, has a calculated HLB of 40 and is a good laboratory detergent, while any nonionic substance with this high HLB would not be a useful detergent.

3. Properties of Detergents in Solution

Detergents in aqueous solutions can form spherical aggregates of a defined size, called micelles. In these micelles, the hydrophobic tails of the detergent molecules form the core of the sphere, while the hydrophilic headgroups occupy the interface to the aqueous solution. The number of detergent molecules in a micelle is called the aggregation number. The overall shape of a detergent molecule determines its propensity to form micelles; this is expressed as the packing parameter P, defined as

$$P = \frac{v}{al}$$

where v is the volume of the detergent chain, a is the cross-sectional area of the headgroup, and l is the length of the hydrophobic chain. If P is small ($<1/3$), spherical micelles are formed, while for bigger P values ($>1/2$),

Table 34.1 Common laboratory detergents and their properties

Detergent	Headgroup	Tail	CMC (m*M*) (at 25 °C)	Aggregation number	MW (Da)	Micellar weight (kDa)	Cloud point (°C) in low-salt buffers[a]	PubChem substance IDs[b]	Comments
Nonionic detergents									
Big CHAP	Gluconamidopropyl moieties (2×)	Cholesterol derivative	2.9–3.4	10	878.1	9	–	SID: 26758300	
Deoxy-Big CHAP	Gluconamidopropyl moieties (2×)	Cholesterol derivative	1.1–1.4	11	862.1	11	–	SID: 26758553	
Digitonin	Complex polysaccharide	Cholesterol derivative	–	60–70	1229.3	–	–	SID: 168187	Natural compound with high lot-to-lot variability
Brij-35 (C12E23)	Linear PEG (23×)	Linear hydrocarbon alcohol (C12)	0.09	40	1199.6	48	>100	SID: 24898176	
C12E8	Linear PEG (8×)	Linear hydrocarbon alcohol (C12)	0.11	123	538.8	66	74–79	SID: 24898996	
C10E4	Linear PEG (4×)	Linear hydrocarbon alcohol (C10)	0.64–0.81	54	334.5	18	20	SID: 3729667	
C10E6	Linear PEG (6×)	Linear hydrocarbon alcohol (C10)	–	74	422.6	31	–	SID: 24874132	
C8E4	Linear PEG (4×)	Linear hydrocarbon alcohol (C8)	6.5–8.5	–	306.4	26	35–40	SID: 24900138	
C8POE	Linear PEG	Linear hydrocarbon alcohol (C8)	6.6	–	330 (average)	–	58		Mixture of molecules with different headgroup lengths ($E=2$–9)
Triton X-100	Linear PEG	*p*-(2,2,4,4-Tetramethylbutyl) phenol	0.17–0.3	100–150	630 (average)	80	64–65	SID: 7889640	Mixture of molecules with different headgroup lengths (average 9.6), strong UV absorption

Triton X-114	Linear PEG	*p*-(2,2,4, 4-Tetramethylbutyl) phenol	0.2–0.35	–	540 (average)	–	20–25	SID: 24902129	Mixture of molecules with different headgroup lengths (average 7–8) classic detergent for phase separation, strong UV absorption
NP-40	Linear PEG	*p*-(2,2,4, 4-Tetramethylbutyl) phenol	0.3	100–150	600 (average)	60–90	63–67	SID: 813297	Mixture of molecules with different headgroup lengths (average 9.0), strong UV absorption
Tween-20	Polysorbate	Linear fatty acid (C12)	0.059	–	1230 (average)	–	76	SID: 47207229	Mixture of molecules with different headgroup lengths
Tween-80	Polysorbate	Linear fatty acid, unsaturated (C18:1)	0.012	58	1310 (average)	76	93	SID: 7848130	Mixture of molecules with different headgroup lengths
β-Dodecylmaltoside	β-Glycosidic maltose	Linear hydrocarbon alcohol (C12)	0.15	98	510.6	70	<0	SID: 691815	
β-Decylmaltoside	β-Glycosidic maltose	Linear hydrocarbon alcohol (C10)	1.6	–	482.6	–	<0	SID: 24894148	
β-Octylglucoside	β-Glycosidic glucose	Linear hydrocarbon alcohol (C8)	20–25	84	292.4	25	<0	SID: 205023	
β-Octylthioglucoside	β-Glycosidic glucose	Linear hydrocarbon thiol (C8)	9	–	308.4	–	7	SID: 698369	
MEGA-8	*N*-Methylglucamine	Linear fatty acid (C8)	58	–	321.5	–	–	SID: 737553	
MEGA-9	*N*-Methylglucamine	Linear fatty acid (C9)	19–25	–	335.5	–	–	SID: 737552	
MEGA-10	*N*-Methylglucamine	Linear fatty acid (C10)	6–7	–	349.5	–	–	SID: 751715	
LDAO	*N*-Oxide	Linear hydrocarbon amine	2.1–8.3	74	229.4	17	–	SID: 158752	Can also be considered as a zwitterionic detergent

(continued)

Table 34.1 *(continued)*

Detergent	Headgroup	Tail	CMC (m*M*) (at 25 °C)	Aggregation number	MW (Da)	Micellar weight (kDa)	Cloud point (°C) in low-salt buffers[a]	PubChem substance IDs[b]	Comments
Zwitterionic detergents									
CHAPS	Dimethylammonium-1-propanesulfonate	Cholesterol derivative	6.5	4–14	614.9	6		SID: 684915	
CHAPSO	Dimethylammonium-1-propanesulfonate	Cholesterol derivative	8	11	630.9	7		SID: 698662	
SB-10	Dimethylammonium-1-propanesulfonate	Linear hydrocarbon alcohol (C10)	25–40	41	307.6	12.5		SID: 738617	Zwittergent 3-10
SB-12	Dimethylammonium-1-propanesulfonate	Linear hydrocarbon alcohol (C12)	2–4	55	335.6	18.5		SID: 661588	Zwittergent 3-12
SB-14	Dimethylammonium-1-propanesulfonate	Linear hydrocarbon alcohol (C14)	0.1–0.4	83	363.6	30		SID: 661589	Zwittergent 3-14
Anioinic detergents									
Cholate, sodium salt	Carboxylic acid	Cholesterol derivative	9–15	2	430.6	0.9		SID: 152893	Typically as sodium salt
Desoxycholate, sodium salt	Carboxylic acid	Cholesterol derivative	4–8	4–10	414.6	1.7–4.2		SID: 668198	Typically as sodium salt
Lauroylsarcosine ("Sarkosyl"), sodium salt	Methylaminoacetate	Linear fatty acid (C12)		2	293.4	0.6		SID: 151872	Typically as sodium salt
Sodium dodecylsulfate (SDS)	Sulfate	Linear hydrocarbon alcohol (C12)	7–10	62	288.4	18		SID: 152210	Also available as lithium salt, with better solubility at cold temperatures
Cationic detergents									
Cetyltrimethylammonium bromide (CTAB)	Quarternary amine	Linear hydrocarbon (C16)	0.9	170	364.5	62		SID: 148809	Typically as bromide salt or chloride salt
Dodecyltrimethylammonium bromide (DTAB)	Quarternary amine	Linear hydrocarbon (C12)	15	70	280.3	20		SID: 159639	Typically as bromide salt or chloride salt

[a] Note that cloud points are influenced by the buffer composition, see text.
[b] Link to PubChem structures and annotations (http://www.ncbi.nlm.nih.gov/sites/entrez?db=pcsubstance).
Data were compiled from several reviews (Arnold and Linke, 2007, 2008; Helenius *et al.*, 1979; Hinze and Pramauro, 1993; Neugebauer, 1990), and from datasheets of commercial suppliers of detergents.MW, molecular weight; CMC, critical mizellization concentration.

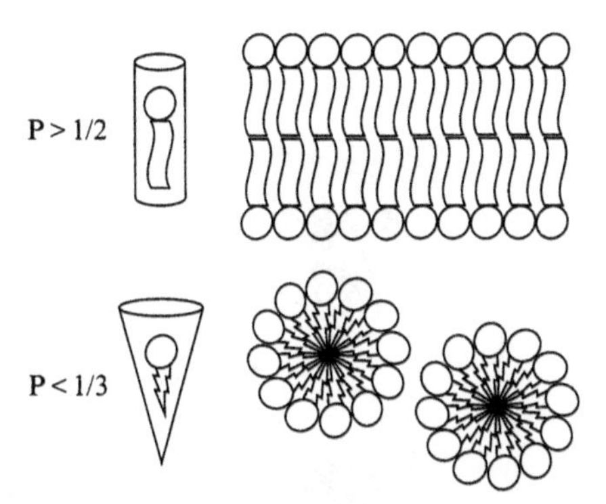

Figure 34.1 The formation of micelles depends on the molecular shape of the detergent. It is best described by the packing parameter *P* that is calculated by comparing the headgroup volume with the hydrophobic chain volume and length.

detergent molecules arrange preferentially into lamellar aggregates (Fig. 34.1) (Neugebauer, 1990). Intermediate forms, for example cylindrical micelles, exist. It has been shown that the packing parameter *P* generally correlates to the HLB parameter described above, with smaller HLB values resulting in bigger *P* values.

It is the formation of micelles that is the basis for membrane protein solubilization. The hydrophobic parts of the membrane proteins are covered by detergent molecules, such that the protein sticks out of the surrounding micelle only by its hydrophilic extensions. It is thus important to understand all physical parameters that influence the formation and destruction of micelles. In addition to the packing parameter *P* mentioned above, the concentration of the detergent, as well as temperature, pH, ionic strength, and presence of other additives in the buffer play an important role.

3.1. Phase diagrams and critical micellization concentration

Detergents only form micelles in a defined concentration and temperature range. This is best described by a graph, called the phase diagram (Fig. 34.2). At a given temperature, the minimal detergent concentration at which micelles are observed is called the critical micellization concentration (CMC). Below this concentration, only detergent monomers exist in the solution; above the CMC, detergent monomers are in equilibrium with the detergent micelles. At high detergent concentrations, other, nonmicellar phases exist, which are typically immiscible with water. These phases are usually liquid-crystalline in nature; they can be hexagonal, reverse hexagonal, or lamellar in structure (Fig. 34.2 shows a lamellar structure).

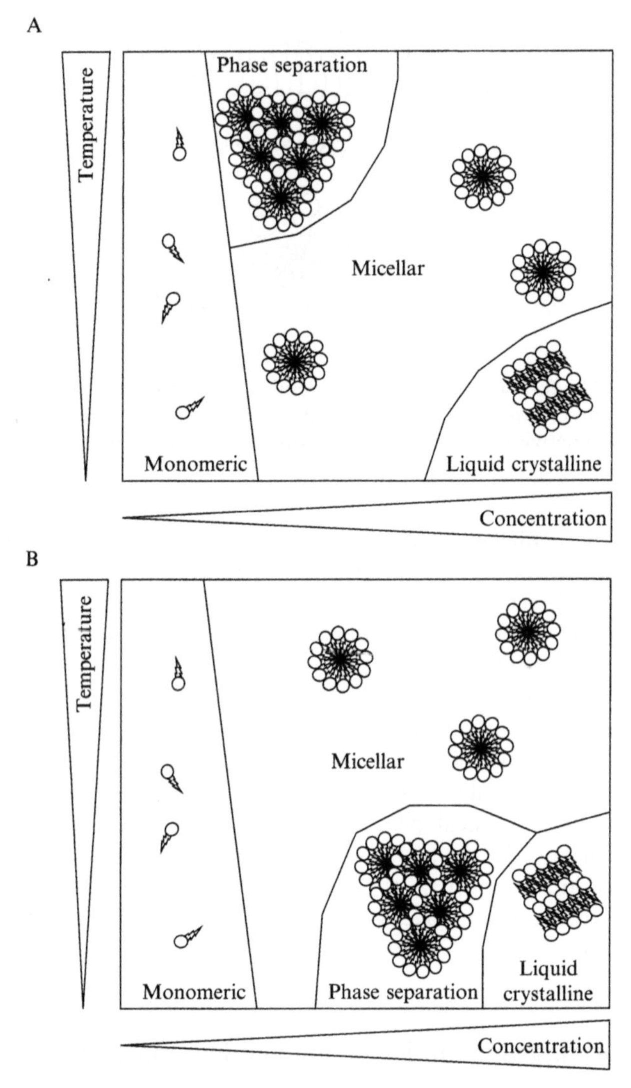

Figure 34.2 Simplified phase diagrams. Panel A shows a detergent with a lower consolute boundary. Most nonionic detergents fall into this group. Panel B shows a detergent with an upper consolute boundary. A number of glycosidic and zwitterionic detergents fall into this group. Note that the liquid-crystalline phase can also consist of hexagonally packed cylindrical micelles.

An alternative phase type which usually occurs at higher temperatures is referred to as phase separation, and describes disordered aggregates of micelles.

3.2. Effects of temperature: Detergent solubility, krafft point, cloud point, and phase separation

Temperature has drastic effects on the solubility and phase behavior of detergents, and so has the detergent concentration in aqueous solution. As membrane proteins are typically solubilized with the help of micelles, it is important to know the temperature and concentration ranges in which a given detergent will form micelles at all. The Krafft point describes the temperature above which detergent solubility increases to above the CMC (i.e., at temperatures below the Krafft point, no micelles are formed). The Krafft point is a temperature value that is specific for each detergent (Gu and Sjoblom, 1992) (Fig. 34.3). The same is true for the cloud point, which describes the temperature at which micelles start to aggregate and become immiscible with water.

The line in the phase diagram that forms the border between the micellar solution and the region of phase separation is called the consolute boundary (Arnold and Linke, 2007). While for most detergents, the cloud point is at a relatively high temperature that has to be exceeded for phase separation (crossing the so-called lower consolute boundary of the detergent), some detergents have a very low cloud point, and phase separation has to fall below that temperature value for phase separation to occur (crossing the so-called higher consolute boundary of the detergent) (Fig. 34.2). Note that the properties of detergents (phase diagram, Krafft point, cloud point, consolute boundaries) change dramatically with increasing ionic strength,

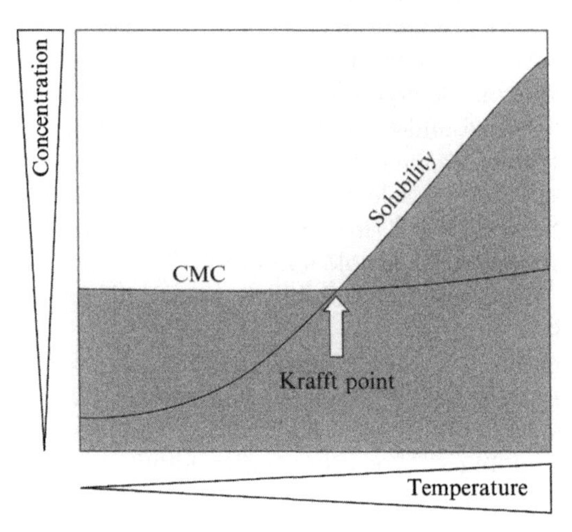

Figure 34.3 Definition of the Krafft point. The gray area describes the possible concentrations of detergent soluble at a given temperature. Above the Krafft point, detergent solubility increases dramatically as micelles can be formed.

and thus in many cases have to be determined experimentally for a given buffer system. The same is true for the presence of nonionic solutes like polyethylene glycols, sugars, or glycerol, among others.

4. Exploiting the Physicochemical Parameters of Detergents for Membrane Protein Purification

As mentioned above, the presence of micelles is essential for keeping membrane proteins in solution. Thus, a membrane protein solution needs to contain detergent at a concentration above the CMC; additionally, enough detergent needs to be available to solubilize all membrane proteins in the solution (simply speaking, there should be at least one micelle per protein molecule). This can be a problem at high protein concentrations, or when working with detergents that have a very low CMC (e.g., β-dodecylmaltoside). The molar concentration of micelles in the solution can be calculated from two specific detergent properties, the aggregation number (the number of detergent molecules in a micelle which depends on the packing parameter), and the molecular weight of the detergent monomer. Table 34.1 lists common laboratory detergents with all relevant parameters, including the molecular weight of the micelles.

The first step in membrane protein purification is typically the solubilization of a native membrane. Again, enough detergent needs to be available in the solution to accommodate all proteins in micelles; moreover, the membrane lipids form mixed micelles with the detergent used, changing the properties of the micelles. For each solubilization process, the optimal detergent:protein and detergent:lipid ratio needs to be determined, which unfortunately is a trial-and-error process. Which detergent to use strongly depends on the downstream processing planned for the membrane proteins. Note that some detergents selectively solubilize certain types of membranes. Examples are Sarkosyl and Triton X-100, which solubilize bacterial inner, but not outer membranes (Arnold and Linke, 2008), and Digitonin, which can be used to selectively solubilize eukaryotic plasma membranes (Mooney, 1988).

Phase separation effects can be exploited for membrane protein separation (Arnold and Linke, 2007). In principle, this works for all detergents; the basic idea is to shift the cloud point of the detergent used to ambient temperature by adding salts or other buffer additives to the system, or to shift the temperature to cross the consolute boundary. Care has to be taken that the high salt concentrations or high temperatures involved do not harm the membrane protein which is to be purified. After phase separation has occurred, the detergent-rich phase that contains the membrane proteins can be collected, while soluble proteins and other hydrophilic substances remain in the aqueous phase. As the volume of the detergent-rich phase is only a fraction of the total volume, this method can also be used to concentrate membrane proteins. After the procedure, excess detergent needs to be removed, for example, by dialysis.

5. Detergent Removal and Detergent Exchange

While the solubilization of membranes requires high amounts of detergent, such high concentrations are usually not wanted in downstream applications. Thus, the detergent concentration has to be lowered, which can of course be easily achieved by dilution. In cases where low concentrations of detergent are wanted, several methods to specifically reduce the detergent amount are available. Size exclusion chromatography using a low-detergent buffer is efficient for detergent removal or exchange if the micellar size is significantly different from the size of the membrane protein in question (Furth *et al.*, 1984). Dialysis can be used to remove detergent or exchange it with a different one, provided that the detergents used are dialyzable. Detergents can only be dialyzed if their micelles are sufficiently small to pass the dialysis tubing as a whole, or if the detergents have a high CMC so that a relatively high amount of detergent is present in a monomeric, not micellar, form. Other efficient methods to remove detergents are the use of BioBeads or similar hydrophobic resins that will bind detergents (Rigaud *et al.*, 1998), or the additions of cyclodextrins that can form inclusion complexes with monomeric detergent molecules (DeGrip *et al.*, 1998). Alternatively, affinity or ion-exchange chromatography can be used to bind the protein to a column, wash away excess detergent, and elute with a suitable low-detergent buffer (Arnold and Linke, 2008).

6. Choosing the Right Detergent

Unfortunately, there are no detergents that are suitable for all membrane proteins. While this needs to be tested for every membrane protein/detergent combination, some general rules apply. Typically, ionic detergents are harsher on proteins than nonionic or zwitterionic ones. Complete denaturation of proteins by detergents like SDS or CTAB may be a welcome effect in polyacrylamide electrophoresis or in DNA isolation techniques, but for the isolation and purification of intact membrane proteins, these detergents are rarely used. As a rule of thumb, detergents with bigger headgroups and longer hydrocarbon chains have milder properties, while short-chain detergents have a higher propensity to denature proteins or to disrupt membrane protein complexes (Privé, 2007).

Even if a detergent is suitable for a membrane protein in question (i.e., does solubilize it without denaturing it), it might be incompatible with downstream applications. Even simple biochemical assays are influenced by the presence of detergents. A number of known incompatibilities are listed in this section. This list is far from complete, and thus, whenever using detergent buffers, a possible detrimental effect on the experiment has to be taken into account. Note also that the purity of the detergents used can pose significant problems. Many detergents contain trace amounts of peroxides, or residual organic solvents from the synthesis procedure (Ashani and Catravas, 1980). For many sensitive applications, it is highly recommended to use only detergents of the highest available purity, and to always prepare detergent solutions freshly to prevent oxidation or hydrolysis.

6.1. Optical spectroscopy

Protein concentration measurements using UV absorption are not possible in the presence of Triton X-100 and related detergents that contain aromatic rings in their structure (NP-40, e.g., is a derivative of Triton X-100), because they strongly absorb UV light at 280 nm wavelength. Many colorimetric assays used for protein concentration measurements are not compatible with detergents. Generally speaking, the BCA assay is more resistant to detergents than either Bradford or Lowry assays, but in many cases, detergents have to be removed, for example, by precipitation. Commercial suppliers have developed numerous variations of the known colorimetric assays to overcome these problems at least partially, and it is strongly recommended to refer to the instruction booklets of these assays when samples containing detergents are to be measured. It is noteworthy that recently, a new assay technology based on a transition metal complex for colorimetric readout at 660 nm wavelength (patent pending) has been developed, which is supposed to be fully compatible with most detergents even in the presence of reducing and chelating agents (Antharavally *et al.*, 2009).

Many other spectroscopic techniques are influenced by detergents, including linear and circular dichroism spectroscopy, where detergents with chiral headgroups can produce a significant signal, or infrared spectroscopy, where the hydrophobic as well as the hydrophilic groups absorb at wavelengths relevant to protein biochemistry. Detergents strongly bind to hydrophobic dyes, influencing, for example, their fluorescence spectra, or their behavior in other colorimetric assays.

6.2. Mass spectrometry and nuclear magnetic resonance (NMR) measurements

Detergents are generally detrimental to mass spectrometry applications, as they themselves can be ionized, leading to strong signals in different detector systems. As a rule of thumb, MALDI-MS is more resistant to detergents than ESI-MS, but in general, detergents need to be removed. Some specialty detergents have been developed to overcome these problems, including cleavable detergents for MALDI-MS (Norris *et al.*, 2003) or perfluorinated detergents for ESI-MS (Ishihama *et al.*, 2000). Special detergents have also become available for NMR spectroscopy, where the detergent micelle adds to the apparent molecular weight of the protein to be analyzed, leading to slower motion in solution, and thus, to more isotropic signals. Small, lipid-like detergents like dihexanoylphosphatidylcholine have been successfully employed in NMR studies of membrane proteins (Fernandez and Wuthrich, 2003).

6.3. Protein crystallization

Detergents produce a number of problems when using them in protein crystallization assays. The most basic problem is that detergents are simply another parameter that needs to be checked in empirical screens. Thus, they simply add another level of complexity; the optimal detergent, and its optimal concentration for crystallization of a membrane protein has to be found. Moreover, detergents tend to phase-separate under the high-salt or high-PEG conditions typically used in protein crystallization screens. And last but not least, detergent purity can be an issue; as an example, it has been shown that even trace amounts of α-dodecylmaltoside in β-dodecylmaltoside buffers can effectively ruin crystal quality (Fromme and Witt, 1998).

7. Conclusions

The most important conclusion, when talking about detergents in membrane protein research, is that one always has to consider or even test whether a certain membrane protein–detergent combination is suitable for the isolation and purification procedure, but also for the downstream applications planned. To even increase the complexity of the problem, some membrane proteins are not stable without at least a little amount of native membrane lipids; this implies that solubilization and purification should in some cases not be complete. It also implies that sometimes mixtures of detergents (or lipids) might be more suitable than a pure substance. This is, in part, realized by commercially available detergent mixtures, that are indeed sometimes useful in stabilizing membrane proteins. Prominent examples are Digitonin, which is never pure as it is isolated from a plant, or Triton X-100, which actually consists of several different species with different polyethyleneglycol headgroup lengths. As biochemists have a tendency to keep their buffer recipes as simple as possible (and the author of this chapter is no exception from that rule), deliberate mixtures of, for example, ionic and nonionic detergents are rarely found in literature. One reason for this surely is that parameters like CMC, cloud point, or micellar size are impossible to predict for mixtures of detergents. Thus, finding the right detergents to use is still, at least to some extent, a trial-and-error process.

ACKNOWLEDGMENTS

The author thanks Andrei Lupas for continuing support. Funding by the Max Planck Society, the German Science Foundation (SFB766/B4), and the Bill & Melinda Gates Foundation (Grand Challenges Explorations Program) is gratefully acknowledged.

REFERENCES

Antharavally, B. S., Mallia, K. A., Rangaraj, P., Haney, P., and Bell, P. A. (2009). Quantitation of proteins using a dye-metal based colorimetric protein assay. *Anal. Biochem.* **385,** 342–345.

Arnold, T., and Linke, D. (2007). Phase separation in the isolation and purification of membrane proteins. *Biotechniques* **43,** 427–440.

Arnold, T., and Linke, D. (2008). The use of detergents to purify membrane proteins. *Curr. Protoc. Protein Sci.* **Chapter 4,** 4.8.1–4.8.30.

Ashani, Y., and Catravas, G. N. (1980). Highly reactive impurities in Triton x-100 and Brij 35: Partial characterization and removal. *Anal. Biochem.* **109,** 55–62.

DeGrip, W. J., van Oostrum, J., and Bovee-Geurts, P. H. M. (1998). Selective detergent-extraction from mixed detergent/lipid/protein micelles, using cyclodextrin inclusion compounds: A novel generic approach for the preparation of proteoliposomes. *Biochem. J.* **330,** 667–674.

Fernandez, C., and Wüthrich, K. (2003). NMR solution structure determination of membrane proteins reconstituted in detergent micelles. *FEBS Lett.* **555,** 144–150.

Fromme, P., and Witt, H. T. (1998). Improved isolation and crystallization of photosystem I for structural analysis. *Biochim. Biophys. Acta Bioenerg.* **1365,** 175–184.

Furth, A. J., Bolton, H., Potter, J., and Priddle, J. D. (1984). Separating detergent from proteins. *Methods Enzymol.* **104,** 318–328.

Gu, T. R., and Sjoblom, J. (1992). Surfactant structure and its relation to the Krafft point, cloud point and micellization—Some empirical relationships. *Colloids Surf.* **64,** 39–46.

Helenius, A., McCaslin, D. R., Fries, E., and Tanford, C. (1979). Properties of detergents. *Methods Enzymol.* **56,** 734–749.

Hinze, W. L., and Pramauro, E. (1993). A critical review of surfactant-mediated phase separations (cloud-point extractions)—Theory and applications. *Crit. Rev. Anal. Chem.* **24,** 133–177.

Ishihama, Y., Katayama, H., and Asakawa, N. (2000). Surfactants usable for electrospray ionization mass spectrometry. *Anal. Biochem.* **287,** 45–54.

Mooney, R. A. (1988). Use of digitonin-permeabilized adipocytes for cAMP studies. *Methods Enzymol.* **159,** 193–202.

Neugebauer, J. M. (1990). Detergents: An overview. *Methods Enzymol.* **182,** 239–253.

Norris, J. L., Porter, N. A., and Caprioli, R. M. (2003). Mass spectrometry of intracellular and membrane proteins using cleavable detergents. *Anal. Chem.* **75,** 6642–6647.

Privé, G. G. (2007). Detergents for the stabilization and crystallization of membrane proteins. *Methods* **41,** 388–397.

Rigaud, J. L., Levy, D., Mosser, G., and Lambert, O. (1998). Detergent removal by non-polar polystyrene beads—Applications to membrane protein reconstitution and two-dimensional crystallization. *Eur. Biophys. J. Biophys. Lett.* **27,** 305–319.

CHAPTER THIRTY-FIVE

PURIFICATION OF MEMBRANE PROTEINS

Sue-Hwa Lin* *and* Guido Guidotti[†]

Contents

1. Introduction 619
2. Preparation of Membranes 620
3. Solubilization of Native Membrane Proteins 622
4. Purification of Membrane Proteins 625
   4.1. Lectin-affinity chromatography 626
   4.2. Ligand-affinity chromatography 627
   4.3. Antibody-affinity chromatography 627
5. Detergent Removal and Detergent Exchange 628
6. Expression and Purification of Recombinant Integral Membrane Proteins 628
References 629

Abstract

Membrane proteins are pivotal players in biological processes. In order to understand how a membrane protein works, it is important to purify the protein to fully characterize it. Membrane proteins are difficult to purify because they are present in low levels and they require detergents to become soluble in an aqueous solution. The selection of detergents suitable for the solubilization and purification of a specific membrane protein is critical in the purification of membrane proteins. The aim of this chapter is to provide an overview for the isolation of plasma membranes, selection of detergents for solubilization of membrane proteins, and how the choice of detergents may affect membrane protein purification.

1. INTRODUCTION

Membrane proteins are pivotal players in biological processes. They are responsible for connecting cells to each other and to the cell matrix, for organizing the shape of the organelles and the cells, and for the transport of

* Department of Molecular Pathology, University of Texas, Houston, Texas, USA
† Department of Molecular and Cellular Biology, Harvard University, Cambridge, Massachusetts, USA

Methods in Enzymology, Volume 463
ISSN 0076-6879, DOI: 10.1016/S0076-6879(09)63035-4

ions, metabolites, and proteins across plasma membranes, and RNA transport across nuclear membrane. Given the importance of membrane proteins for a plethora of cellular functions, it is not surprising that around 50% of the current drugs for a variety of diseases target membrane proteins.

In order to understand how a membrane protein works and to generate drugs that target specific sites within the protein, it is important to purify the protein to fully characterize it. However, of all known proteins that were purified and identified biochemically, only very few were membrane proteins. This is in a large part due to the fact that membrane proteins are more difficult to purify for the following reasons. First, most membrane proteins are present at low levels; and second, they are embedded in the lipid bilayer and require detergents to become soluble in aqueous systems. The selection of detergents suitable for the solubilization and purification of the specific membrane protein is critical in the purification of membrane proteins. In general, the basic principle for the purification of membrane proteins are the same as those used for soluble proteins, but modifications in the purification scheme is required in order to accommodate the requirement of detergents that are essential in maintaining the solubility and function of these proteins.

Unlike DNA or RNA, it is impossible to present a single set of methods for the purification of all proteins, especially membrane proteins. Each membrane protein possesses a unique set of physical characteristics, preference for detergent for solubilization, and conditions for purification. An overview of the properties of detergents is described in Chapter 41. This chapter describes methods for the isolation of plasma membranes, and for solubilization and purification of membrane proteins.

2. Preparation of Membranes

Isolation of plasma membranes from cells or tissues is the first step in the purification of a plasma membrane protein. Because of the limited biochemical fractionation methods available for effective separation of detergent solubilized membrane proteins, the time invested in purifying plasma membrane fractions will improve the results at the later stages.

Due to the low levels of most membrane proteins, it is important to select a tissue or cell line that is easy to obtain a large amount of starting material for purification and that also expresses high specific activity of the protein of interest. Recently, there has been increased interest in identifying cell surface proteins as markers for different cell lineages or stem cells. As a result, obtaining a sufficient amount of plasma membranes from a limited amount of cells with relatively good enrichment of plasma membrane marker enzyme activity has become a new focus of membrane isolation.

The most commonly used method to break up tissues or cells for preparation of the membrane fraction is to homogenize them in buffered isotonic sucrose (0.25 *M*, pH 7–8) with a Dounce homogenizer. Membrane proteins are relatively stable while they are embedded in the membrane. The major potential concern for losing activity during the membrane purification step is proteolysis. During cell disruption, proteases may be released. A cocktail of protease inhibitors in a convenient tablet form is commercially available, for example, Complete Protease Inhibitor Cocktail (Roche, Indianapolis, IN). Extracellular domains of plasma membrane proteins are in direct contact with an oxidative environment and most of the sulfhydryl groups are in disulfide bonds. These disulfide bonds are formed during the synthesis of these proteins in the endoplasmic reticulum. Thus, reducing agents may not be needed in this step. In fact, reducing agents, when present in high concentration, may change the existing disulfide bonded conformation, resulting in inactivation of the enzymatic or ligand-binding activity of membrane proteins. Due to the disulfide bonded structure and glycosylation, the extracellular domain of membrane protein is relatively resistant to proteolysis. However, there are exceptions to this. For example, the Ca^{2+}-dependent cell adhesion molecules, cadherins, are susceptible to proteolysis when Ca^{2+} ions are removed by EDTA. In contrast, the cytoplasmic domains, which mediate the signal transduction of membrane proteins, are usually susceptible to proteolysis. For example, the receptor tyrosine kinase activity of insulin receptor is largely lost if protease inhibitors are not present in sufficient quantities during homogenization and the subsequent membrane fractionation steps.

The most frequently used membrane fractionation methods employ a combination of differential centrifugation and sucrose density gradient centrifugation steps. Due to the differences in lipid and protein compositions, membranes have different densities that allow them to be separated from other organelles. Differential centrifugation can remove soluble proteins, the majority of mitochondria and nuclei from the cell homogenates. Sucrose density gradients can further separate membranes with different densities. However, these multiple steps of centrifugation are lengthy and also yield only a small percentage of the plasma membranes. Often, much of the plasma membrane is lost in the early steps of centrifugation. Thus, this method is more suitable for isolating membranes from tissues that are relatively easy to obtain. Rat liver is one of the tissues that are most commonly used in the isolation of membranes for biochemical studies. Many methods have been developed for isolating various membranes from rat livers. A method developed by Neville (1968) that homogenizes livers in hypotonic solution followed by a discontinuous sucrose gradient centrifugation gave a very good recovery of liver plasma membranes and has been commonly used in the preparation of liver plasma membranes.

In the situations, when only a limited amount of tissue culture cells are used, an increase in the recovery of plasma membrane proteins without sacrificing the purity of the membrane will be needed. Affinity matrices provide a simple and quick method to use in the purification of membranes. The conventional agarose- or acrylamide-affinity matrices cannot be used to isolate membranes because they sediment with the organelles (e.g., nuclei) that have relatively high densities. Chemically derived magnetic beads can be coupled with various proteins, and they have become a new form of affinity matrix. In contrast to conventional agarose- or acrylamide-based matrices, magnetic beads can be conveniently separated from the mixture by using a magnet and thus can be used to separate organelles independent of their densities. By simply holding the tubes near the magnet, the magnetic beads can be recovered at the sides of the tubes, allowing easy recovery of the beads from the mixture. Thus, magnetic beads can be used as a substitute for centrifugation. This property is likely to have great advantages in separating plasma membranes from other organelles. We have recently used magnetic beads with immobilized lectin for the purification of plasma membrane proteins from cultured epithelial cells (Lee *et al.*, 2008). This procedure takes advantage of the fact that some of the membrane proteins are glycosylated and can bind to lectins, the carbohydrate binding proteins. In this procedure, the lectin Concanavalin A (ConA) is immobilized onto magnetic beads by binding biotinylated ConA to streptavidin magnetic beads. The ConA-magnetic beads were mixed with homogenized cell lysates, and the beads were recovered at the sides of the tubes, allowing removal of other organelles that are not associated with the magnetic beads. The membranes were found to be associated with the ConA-magnetic beads as there was an enrichment in the activity of 5′-nucleotidase, a membrane protein, from a total cell lysate of prostate PC-3 cells or cervical HeLa cells. One caveat of this lectin magnetic bead method is that we could not elute membranes from the ConA-magnetic beads by using the competitive sugar alpha-methyl mannoside, possibly because the competing sugar cannot access the binding sites between the plasma membranes and ConA-magnetic beads. As a result, we used detergents to solubilize the membrane proteins from the ConA-magnetic beads. The use of detergents for solubilization of membrane proteins is described in Section 3.

3. Solubilization of Native Membrane Proteins

Membrane proteins are embedded in the lipid bilayer. Integral membrane proteins have at least one stretch of protein sequence embedded in the membrane while peripheral proteins are associated with the membrane through electrostatic and in some cases hydrophobic interaction. Peripheral membrane proteins can be dissociated from membranes by using high salt or high pH solutions (Schindler *et al.*, 2006), for example 0.5 *M* NaCl. Because detergents are not used in the procedure, the peripheral membrane proteins can be purified by methods similar to those applied to soluble proteins.

Integral membrane proteins need to be solubilized from the lipid bilayer to become individual proteins before purification. Detergents that possess amphipathic properties are commonly used to solubilize integral membrane proteins from membranes. Detergents may be grouped into three classes, ionic, nonionic, and zwitterionic. A discussion of the different types of detergents is described in Chapter 41 of this volume. Thus, we will only discuss their use in terms of purification of membrane proteins.

Proteins are considered "solubilized" from the membrane if the proteins appear in the supernatant fraction after the detergent treated membranes are centrifuged at $105{,}000 \times g$ for 1 h at 4 °C. The process of solubilization of membrane proteins by detergents can be divided into several phases. In the first phase, the detergent binds to membranes. As the amount of detergent is increased, the detergent starts to lyse membranes. Further increase in detergent will lead to the formation of lipid/protein/detergent complexes. At this stage, the membrane proteins are "solubilized." Additional amount of detergent will be needed to "delipidate" the complexes to protein/detergent and lipid/detergent complexes. Usually, a detergent-to-protein ratio of around 1–2 is sufficient to solubilize the membrane proteins into lipid/protein/detergent complex. A detergent-to-protein ratio of around 10 or higher will lead to delipidation (Hjelmeland, 1990). The optimal detergent to protein ratio that is required to solubilize a specific membrane protein needs to be empirically determined.

The choice of detergents can be simply stated as the detergent that works for your protein of interest. Nondenaturing detergents are those detergents that solubilize the membrane proteins without significantly inactivating the activity or function of the proteins. Among detergents that are available, Triton X-100, sodium cholate, CHAPS, octylglucoside are "nondenaturing" most of the time, although a loss of certain fraction of activity during solubilization is expected.

The presence of detergents will affect several aspects of protein purification. The detergents may affect the assay for the activity, for example transport activity for membrane transporter and ligand-binding assay for a receptor. As the proteins will no longer be associated with the membranes, reconstitution of solubilized membrane proteins into phospholipid vesicles will be needed for measuring transporter activity, and methods that separate unbound ligand from the ligand–receptor complex will be needed to be developed for the specific receptor. These requirements may limit the type of detergents to be used for solubilization. For example, detergents with high critical micelle concentration (sodium cholate, CHAPS, octylglucoside), which are easier to be removed by dialysis, should be used if a subsequent reconstitution of proteins into phospholipid vesicles is planned.

Some types of detergents may interfere with certain protein assays. The polyoxyethylene derivatives, for example Triton X-100, C12E9, and Tween series, give false positive in Commassie blue-G250 dye-binding assays (also known as Bradford assay) (Bradford, 1976). Sodium cholate or sodium deoxycholate forms precipitate in the Bradford assay. Triton X-100 and NP-40 absorb at 280 nm and interfere with the use of ultraviolet absorbance method to monitor the chromatographic elution of proteins. This will require either a change of the protein assay method or choice of detergents that are compatible with the protein assay. The Bicinchoninic acid (BCA) protein assay method (Smith *et al.*, 1985) is compatible with certain amounts of detergents. A modified method based on the Lowry protein assay (Peterson, 1977) avoids the interference of detergents by first precipitating the proteins from the solution with deoxycholate and trichloroacetic acid (TCA), and can be used to measure protein concentration throughout the purification steps. However, the modified Lowry method takes a longer time compared with the Bradford method to obtain protein concentrations.

In some cases, whether the detergent is chemically pure is critical. For example, if the protein is to be crystallized for X-ray structural analysis or analyzed by mass spectrometry, the impurity in the detergents may interfere with these analyses. Polyoxyethylene derivatives, for example Triton X-100, Lubrol PX, the Tween, and Brij series, contain varied polymer lengths and are not chemically homogeneous. Octylglucoside, dodecyl maltoside, and CHAPS have well-defined chemical composition and can be obtained in relatively high purity. Noninonic detergents, including the polyoxyethylene derivatives, for example Triton X-100 and the Tween series, are less effective at dissociating protein complexes, but many proteins are more stable in nonionic detergents than in ionic detergents.

In screening of detergents for solubilization of integral membrane proteins, one should first identify the detergents that can solubilize the protein without inactivating its activity. Among those detergents that fulfill these requirements, one should then consider other factors, including the compatibility with the methods used for protein determination, characterization, and chromatography methods. The detergent interference with various chromatography methods will be discussed in Section 4.

Detergents are screened by preparing membrane fractions at a specific protein concentration, for example 1 mg/ml. The solubilization solutions with a range of detergent concentrations, for example 0.2–20 mg/ml, will then be added to the membrane preparation. Solubilization is usually achieved at detergent/protein ratios of 0.1–10 (w/w). Solutions are incubated at 0–4 °C for 30–60 min and then centrifuged at 105,000×*g* for 1 h at 4 °C. The activities of the protein of interest in both the supernatant and the pellet fractions are then measured. If the majority of the activity is found to be present in the supernatant fraction, this detergent is suitable for the membrane protein. If the activity is not detected in the supernatant fraction but remains in the pellet fraction, the detergent either cannot solubilize the protein from the membrane or the amount of the detergent used may not be sufficient. If the activity is not detected in either the supernatant or the pellet fractions, the detergent has "denatured" the protein.

4. Purification of Membrane Proteins

Once a suitable detergent has been used to solubilize membrane proteins from the membranes, fractionation methods can be used to isolate the specific protein of interest. Conventional chromatographic techniques including gel filtration, affinity, ion exchange, and chromatofocusing can be applied in the purification of membrane protein. However, the following precautions are noted when performing chromatography in the presence of detergents.

1. Include a level of detergent sufficient to keep the integral membrane proteins in a soluble form in the buffer and prevent protein aggregation.
2. Due to the hydrophobic nature of most detergents, protein separation methods that are based on hydrophobicity of proteins, for example phenyl-Sepharose and reverse-phase, may not be suitable for membrane protein purification.
3. Solubilization of membrane proteins with ionic detergents, for example cholate and deoxycholate, are not suitable for ion-exchange chromatography. Nonionic or zwitterionic detergents should be used for charge-based preparative procedures including ion-exchange chromatography and preparative electrophoresis.
4. Detergents that contain a sugar moiety may interfere with specific lectin chromatography, for example octylglucoside may interfere with ConA chromatography.
5. Because the solubilized membrane proteins are present in detergent micelles, membrane proteins have much bigger apparent molecular weight in gel filtration. Detergents that form large micellar molecular weight, for example Triton X-100, may add 60–100 kDa to the size of solubilized membrane proteins. Thus, most of the proteins will appear in the high molecular weight fractions, leading to difficulty in separating proteins based on molecular sizes.
6. Association of membrane proteins with detergent micelles, especially those with large micellar sizes or that are ionic in nature, obscures the charge of the membrane proteins. Thus, the ability of ion-exchange chromatography in separating membrane proteins may not be as good as for nonmembrane proteins.

7. In general, affinity chromatography is by far the most useful and successfully applied method for purification of integral membrane proteins and can be used at various purification steps. Because ion-exchange chromatography is sensitive to the ionic strength of the buffer, and gel filtration requires a concentrated sample of relatively small volume, affinity chromatography can be used to purify, concentrate, and perform salt exchange between different chromatography steps. Several affinity chromatographies that are commonly used in membrane protein purification are described in the following sections.

4.1. Lectin-affinity chromatography

There are three types of affinity chromatography, including the use of a general ligand (e.g., lectins), specific ligands (e.g., enzyme inhibitors, hormones), and antibodies. Immobilized lectin is a general form of affinity chromatography. Lectins are carbohydrate-binding proteins, which offer a rapid and mild method to purify plasma membrane glycoproteins. Lectin–glycoprotein interactions are reversible and can be inhibited by simple sugars. Because membrane proteins are frequently glycosylated, lectin-affinity chromatography is very useful for membrane protein purification. Numerous lectins have been identified with the most widely used lectins to be concanavalin A (binds α-D-mannose) and wheat germ agglutinin (binds sialic acid and β-D-GlcNAc). Several aspects about using lectin-affinity chromatography need to be mentioned here. First, whether the protein of interest will bind to a specific lectin needs to be empirically tested. Sometimes, due to different glycosylation enzymes present in different tissues, the same glycoprotein expressed in different tissues of the same animal may not have the same lectin-binding specificity. Second, a specific lectin-affinity chromatography purifies a group of proteins with a specific type of glycosylation, thus this method is unable to achieve the extent of purification as ligand or antibody chromatography. Third, lectins are sensitive to certain types of detergents. Although nonionic detergents, for example Triton X-100 (up to 2.5%, w/v), have negligible effects on the binding of concanavalin A or wheat germ agglutinin with their ligands, some ionic detergents, for example SDS, may inactivate lectins.

A protocol for a typical use of a lectin column for the purification of membrane proteins is described here.

1. Wash the WGA-agarose affinity matrix by transferring 2 ml of wheat germ agglutinin (WGA)-agarose to a disposable plastic column (Poly-Prep, Bio-Rad Laboratories). Fill the column with 10 ml wash buffer (50 m*M* HEPES, pH 7.4, containing 0.1% detergent). Let wash run through column to remove free uncoupled WGA and the storage buffer.

2. Add washed WGA-agarose to the solubilized membrane extract. Mix WGA-agarose and solubilized membrane extract by rotation or on a rocker for 30 min at room temperature.
3. Spin at 600 rpm for 1 min to pellet WGA-agarose. Remove supernatant with a pipette. Add 10 volumes (based on the volume of WGA-agarose) of wash buffer, invert several times.
4. Repeat step 3 two more times.
5. Add 2 ml wash buffer to the WGA-agarose and transfer slurry of WGA-agarose back to the column. Wash with 5–10 column volumes of wash buffer. Check periodically with protein assay, for example Coommassie Blue protein assay, until the protein levels drop to background, that is similar to that in the wash buffer.
6. Elute the glycoproteins from the WGA-agarose by adding half column volume of elution buffer, for example 50 m*M* HEPES, pH 7.4, 0.1% detergent, 0.25 *M* *N*-acetyl glucosamine. Collect eluted fractions.
7. Add half column volume of elution buffer. Collect eluted fractions.
8. Repeat step 7 four more times.
9. Detect protein concentration in the eluted fractions. For example, take 5 μl from each eluted fraction and determine the protein concentration with Coomassie Blue assay. Pool appropriate fractions.

4.2. Ligand-affinity chromatography

The key to a successful ligand-affinity chromatography is that the affinity between the ligand and the receptor needs to be sufficiently high as to allow for binding and washing during the purification steps. The ligand is usually immobilized onto the affinity support through cross-linking. The affinity between the ligand (or inhibitor) and receptor may be altered due to the presence of detergent. As such, choosing a detergent that does not significantly lower the receptor affinity for its ligand will be critical if ligand-affinity column is to be used for its purification.

4.3. Antibody-affinity chromatography

If the antibody is available, using immobilized antibody against the specific membrane protein is the most powerful method of purification. Antibodies are relatively stable in nonionic detergents and thus are compatible with the presence of detergents in solubilized membrane preparations. The challenge may be the elution of the protein from the immunoaffinity column (see Chapter 28).

5. Detergent Removal and Detergent Exchange

Most membrane proteins will aggregate and precipitate if detergents are completely removed. Thus, complete removal of one detergent without exchanging to another detergent is seldom done if the function of the membrane protein is to be maintained. In some case, an excess amount of a detergent present in the samples, for example during initial solubilization, may interfere with protein activity or concentration measurement, and removal of the excess amount of detergent may be desirable. In other cases, the detergent initially used for solubilization may not be appropriate for subsequent chromatographic or analytical procedure, and exchange to another detergent may be necessary. For example, the presence of ionic detergent is not compatible with ion-exchange chromatography or performing isoelectric focusing, and it will be desirable to switch to non-ionic or zwitterionic detergents. If the membrane protein needs to be reconstituted into phospholipid vesicles, it will be desirable to change to a detergent with high critical micelle concentration (cmc) value so that the detergent can be removed by dialysis to allow for the membrane proteins to reconstitute into phospholipids vesicles. Detergent properties have significant effects on the ease of detergent removal. Detergents that have a high (>1 m*M*) cmc, for example cholate and octylglucoside, can be easily removed by methods such as dialysis or through ultrafiltration membranes. Detergents with low (<1 m*M*) cmc values, such as the nonionic detergents Triton X-100, C12E9, Brij, Tween, are difficult to remove by dialysis and adsorption onto chromatography matrix, gel filtration, equilibrium method can be used. Methods for detergent removal and detergent exchange are described in Chapter 41.

6. Expression and Purification of Recombinant Integral Membrane Proteins

Expression and purification of recombinant integral membrane proteins are commonly used to obtain large amounts of membrane proteins for structure studies or generation of antibodies against the membrane proteins. Membrane proteins are usually expressed in mammalian cells or insect cells. The presence of a signal sequence is essential to allow targeting of the protein to the endoplasmic reticulum for protein synthesis and posttranslational modifications. Thus, tags are usually placed at the end of the sequence to avoid disturbing the membrane targeting process. Sometimes, a signal sequence from another protein may be used to improve membrane targeting efficiency. Small tags, for example, 6-histidine or small peptide tags, do not introduce too much change to the proteins and can increase the likelihood of expressing the membrane proteins in heterologous cell types. Metal-affinity chromatography, which has high binding capacity and where its binding is not affected by the presence of detergents, is a good choice for expressing and purification of membrane proteins. Other affinity epitope tags, for example, FLAG, myc, or HA, can also be used. However, purification of proteins using immobilized antibody has limited capacity and the cost is much higher than using metal-affinity chromatography.

Examples of expression and purification of membrane proteins can be found in several articles. One example is the purification of a cell adhesion molecule CEACAM1 expressed in insect cells (Phan *et al.*, 2001). In this study, the integral membrane protein was expressed with a 7-histidine tag at its C-terminus. The original signal sequence was used. The recombinant CEACAM1 protein was found to be localized on the plasma membrane. The cells were lysed and the membranes solubilized by Triton X-100 and the his-tagged CEACAM1 purified in one-step using metal-affinity chromatography.

REFERENCES

Bradford, M. M. (1976). A rapid and sensitive method for the quantitation of microgram quantities of protein utilizing the principle of protein-dye binding. *Anal Biochem.* **72,** 248–254.

Hjelmeland, L. M. (1990). Solubilization of native membrane proteins. *Methods Enzymol.* **182,** 253–264.

Lee, Y. C., Block, G., Chen, H., Folch-Puy, E., Foronjy, R., Jalili, R., Jendresen, C. B., Kimura, M., Kraft, E., Lindemose, S., Lu, J., McLain, T., Nutt, L., Ramon-Garcia, S., Smith, J., Spivak, A., Wang, M. L., Zanic, M., and Lin, S. H. (2008). One-step isolation of plasma membrane proteins using magnetic beads with immobilized concanavalin A. *Protein Expr. Purif.* **62,** 223–229.

Neville, D. M. J. (1968). Isolation of an organ specific protein antigen from cell surface membrane of rat liver. *Biochim. Biophys. Acta* **154,** 540–552.

Peterson, G. L. (1977). A simplication of the protein assay method of Lowry *et al.* which is more generally applicable. *Anal. Biochem.* **83,** 346–356.

Phan, D., Han, E., Birrell, G., Bonnal, S., Duggan, L., Esumi, N., Gutstein, H., Li, R., Lopato, S., Manogaran, A., Pollak, E. S., Ray, A., *et al.* (2001). Purification and characterization of human cell-cell adhesion molecule 1 (C-CAM1) expressed in insect cells. *Protein Expr. Purif.* **21,** 343–351.

Schindler, J., Jung, S., Niedner-Schatteburg, G., Friauf, E., and Nothwang, H. G. (2006). Enrichment of integral membrane proteins from small amounts of brain tissue. *J. Neural Transm.* **113,** 995–1013.

Smith, P. K., Krohn, R. I., Hermanson, G. T., Mallia, A. K., Gartner, F. H., Provenzano, M. D., Fujimoto, E. K., Goeke, N. M., Olson, B. J., and Klenk, D. C. (1985). Measurement of protein using bicinchoninic acid. *Anal. Biochem.* **150,** 76–85.

CHAPTER THIRTY-SIX

Purification of Recombinant G-Protein-Coupled Receptors

Reinhard Grisshammer

Contents

1. Introduction 632
2. Solubilization: General Considerations 633
3. Purification: General Considerations 634
   3.1. Stability of GPCRs in detergent solution 634
   3.2. General affinity purification 634
   3.3. Receptor-specific ligand affinity chromatography 636
   3.4. Analysis of detergent-solubilized GPCRs 636
4. Solubilization and Purification of a Recombinant Neurotensin Receptor NTS1 637
   4.1. Solubilization of the NTS1 fusion protein 638
   4.2. Purification of the NTS1 fusion protein by immobilized metal affinity chromatography 640
   4.3. Purification of the NTS1 fusion protein by a neurotensin column 640
5. Analysis of Purified NTS1 641
6. Conclusions 642
Acknowledgments 642
References 642

Abstract

Structural and functional analysis of most G-protein-coupled receptors (GPCRs) requires their expression and purification in functional form. The produced amount of recombinant membrane-inserted receptors depends on the optimal combination of a particular GPCR and production host; optimization of expression is still a matter of trial-and-error. Prior to purification, receptors must be extracted from the membranes by use of detergent(s). The choice of an appropriate detergent for solubilization and purification is crucial to maintain receptors in their functional state. The initial enrichment can be carried out by affinity chromatography using a general affinity tag (e.g., poly-histidine tag).

Department of Health and Human Services, National Institutes of Health, National Institute of Neurological Disorders and Stroke, Rockville, Maryland, USA

Methods in Enzymology, Volume 463
ISSN 0076-6879, DOI: 10.1016/S0076-6879(09)63036-6

If the first purification step does not yield pure receptor protein, purification to homogeneity can often be achieved by use of a subsequent receptor-specific ligand column. If suitable immobilized ligands are not available, size exclusion chromatography or other techniques need to be applied. Many GPCRs become unstable upon detergent extraction from lipid membranes, and measures for stabilization are discussed. As an example, the purification of a functional neurotensin receptor to homogeneity in milligram quantities is given below.

1. Introduction

Structure determination and functional analysis of integral membrane proteins, which are not naturally abundant, require (i) a recombinant production system and (ii) a purification strategy to allow the isolation of functional rather than nonfunctional, incorrectly folded membrane protein. Expression and purification of prokaryotic and eukaryotic membrane proteins has been covered in the literature. The reader is referred, for example, to Grisshammer and Tate (1995, 2003) and Grisshammer and Buchanan (2006). In addition, the reader may consult Chapter 35 of this volume. This chapter focuses exclusively on G-protein-coupled receptors (GPCRs) which are eukaryotic integral membrane proteins involved in cell-to-cell communication and sensory signal transduction (see Gether and Kobilka, 1998).

It is beyond the scope of this chapter to discuss in any great detail all the possible expression strategies for integral membrane proteins such as GPCRs. However, a few key points are summarized here. (i) A universal strategy for the high-level recombinant expression of functional receptors is currently unavailable. Some GPCRs accumulate in the membrane to high levels, whereas other often closely related receptors are hardly detected. Despite their assumed similarities, individual GPCRs behave quite differently in a given expression host and recombinant production is still a matter of trial-and-error. Comparative expression studies have, for example, been performed using the methylotrophic yeast *Pichia pastoris* (Andre *et al.*, 2006), the baculovirus insect cell system (Akermoun *et al.*, 2005), and the Semliki Forest virus system (Hassaine *et al.*, 2006). A survey of GPCR production in commonly used expression hosts has been summarized (Sarramegna *et al.*, 2003). (ii) The use of eukaryotic hosts seems generally better for producing functional, membrane-inserted GPCRs than prokaryotic hosts (Grisshammer, 2006), although there are exceptions. For example, the bacterium *Escherichia coli* was successfully used for the expression of the neurotensin receptor NTS1 (Grisshammer *et al.*, 1993; White *et al.*, 2004), the M1 muscarinic acetylcholine receptor (Hulme and Curtis, 1998), the adenosine A2a receptor (Weiß and Grisshammer, 2002), and the cannabinoid CB2 receptor

(Calandra *et al.*, 1997; Yeliseev et al., 2005). (iii) Numerous descriptive reports have been published on the recombinant expression of integral membrane proteins. However, few publications (see Bonander *et al.*, 2005, 2009; Griffith *et al.*, 2003; Wagner *et al.*, 2006, 2007) are currently available to understand the underlying mechanisms of how a given host cell responds to membrane protein overproduction.

The purification of receptors can be conceptually divided into two steps: extraction from membranes with a suitable detergent (solubilization), and subsequent purification by use of general affinity tags, receptor-specific ligand columns, size exclusion chromatography, and other methods. Of utmost importance, care must be taken to choose experimental conditions to maintain the membrane protein in its active state throughout the purification procedure. This latter aspect cannot be overemphasized because many GPCRs become unstable upon detergent extraction from lipid membranes.

2. Solubilization: General Considerations

The extraction of membrane-inserted receptors is accomplished with the use of detergents (see Chapter 34). The correct choice of detergent is crucial to maintain solubilized receptors in their functional state for a prolonged period of time. *N*-Dodecyl-β-D-maltoside (DDM), a mild nonionic detergent, is commonly used for the solubilization of GPCRs. Lipid-like cholesteryl hemisuccinate may be added to increase receptor stability. The use of shorter chain detergents, which are usually harsher than longer chain detergents, is appropriate as long as these detergents do not compromise the integrity of the GPCR under investigation.

Solubilized receptors must be considered as detergent–lipid–receptor complexes rather than just as receptor protein. This implies that the biochemical properties such as size, shape, or isoelectric point of solubilized receptors differ from those computed solely by the amino acid sequence. Likewise, detergent–lipid–receptor complexes cannot necessarily be regarded as homogeneous particles. Because lipids as well as receptors sequester detergent during the solubilization step, the ratio of detergent to membrane will determine how much detergent and lipid are bound around the receptor. The amount of bound lipid and detergent will change during the course of purification.

The solubilization procedure must be optimized to yield the highest extraction efficiency while maintaining the best stability of a given GPCR in solution. The effect of the systematic variation of detergent to membrane input can be monitored by performing radioligand-binding assays and determining the total protein content. This allows the calculation of an experimental B_{max} value (nmol functional receptor/mg of protein) and comparison of that value with the theoretical value for specific binding of pure functional receptor. If no test for functionality is available, then good biochemical behavior such as a symmetric size exclusion chromatography profile can be used as an indicator for integrity.

The preparation of membranes prior to solubilization constitutes an initial purification step because soluble proteins are removed before receptor extraction and the ratio of target receptor to contaminants is therefore higher. However, receptors can also be enriched from total cell lysate rather than from solubilized membranes during the first purification step.

3. Purification: General Considerations

3.1. Stability of GPCRs in detergent solution

Many GPCRs are not very stable in detergent solution (except the visual pigment rhodopsin as long as the receptor is kept in its nonsignaling dark state (De Grip, 1982)). One possible explanation for this instability may be inherent structural flexibility, that is, the receptor can adopt several/many conformations in detergent solution, some of which may aggregate. Removal of lipid during the purification process may also cause destabilization. A number of measures have been taken to increase the stability of GPCRs in detergent solution. For example, the addition of an inverse agonist/antagonist ligand (Cherezov *et al.*, 2007; Hanson *et al.*, 2008; Jaakola *et al.*, 2008; Warne *et al.*, 2008) will cause the receptor to assume its nonsignaling inactive state which is generally considered more stable than the activated state(s) (see Kobilka and Deupi, 2007). Lipids or lipid-like substances such as cholesteryl hemisuccinate (Jaakola *et al.*, 2008; Tucker and Grisshammer, 1996; Weiß and Grisshammer, 2002) and glycerol (Tucker and Grisshammer, 1996) have been included throughout the purification to improve receptor stability. Site-directed mutagenesis has also been used to generate GPCRs with increased stability (Magnani *et al.*, 2008; Roth *et al.*, 2008; Sarkar *et al.*, 2008; Serrano-Vega *et al.*, 2008; Shibata *et al.*, 2009) and hence more tolerant to a wider range of detergents.

3.2. General affinity purification

The biochemical and pharmacological properties of GPCRs in detergent solution may arbitrarily be grouped into two categories. (i) Receptors which can be purified to homogeneity in functional form under optimized buffer conditions as assessed by radioligand-binding assays and G-protein nucleotide exchange experiments (White *et al.*, 2007) are referred to as being functional. (ii) In some cases (Kobilka, 1995; White *et al.*, 2004), a detergent-soluble species has been observed which does not bind receptor-specific ligands.

I refer to this latter species as nonfunctional, incorrectly folded (at least in respect to ligand recognition) but still detergent soluble.

Recombinant cloning techniques have made it easy to introduce general affinity tags at either the N-terminus or C-terminus of a given receptor. Examples are a Flag epitope tag at the receptor N-terminus for use with an M1 antibody affinity column (Kobilka, 1995) or poly-histidine tails at the receptor C-terminus for immobilized metal affinity chromatography (IMAC) (Grisshammer and Tucker, 1997; Hanson *et al.*, 2008; Hulme and Curtis, 1998; Jaakola *et al.*, 2008; Klaassen *et al.*, 1999; Kobilka, 1995; Warne *et al.*, 2003; Weiß and Grisshammer, 2002; Yeliseev *et al.*, 2005). An antibody column (1D4 antibody; Molday and MacKenzie, 1983) recognizing the extreme C-terminus of bovine rhodopsin has been used for the single-step purification of rhodopsin (Oprian *et al.*, 1987; Reeves *et al.*, 1999) and of a β-adrenergic receptor with the 1D4 epitope tag fused to its C-terminus (Chelikani *et al.*, 2006).

The exact position of an affinity tag (i.e., well removed from the receptor transmembrane core or close to a transmembrane helix) determines whether binding of the receptor to the affinity resin must be done in batch or can be done in column mode. An affinity tag too close to the transmembrane core may be partially masked by the detergent belt around the receptor protein and hence will be less accessible to the affinity resin. Batch loading for a prolonged time will capture receptors efficiently in that case (Weiß and Grisshammer, 2002). Receptors with well exposed affinity tags can be loaded onto resin packed into a column with a short exposure time of the purification tag to the affinity resin. Batch purification procedures have to be performed manually, whereas column-loading procedures can be automated (White *et al.*, 2004). As general guideline, detergents with low critical micelle concentration (cmc) values form larger detergent belts around a membrane protein than detergents with high cmc values (see, e.g., Bamber *et al.*, 2006) although the exact amount of protein-bound detergent will depend on the properties of the respective detergent tested. The accessibility of an affinity tag to the resin may therefore be reduced in low cmc detergents compared to high cmc detergents.

Many laboratories utilize IMAC as a first enrichment step. Several resin types are commercially available such as Ni^{2+}-NTA resin (Ni^{2+}-nitrilotriacetate, Qiagen), Talon resin (Co^{2+}-carboxymethylaspartate, Clontech), and IDA resin (iminodiacetate, Zn^{2+}, Ni^{2+}, Co^{2+}, GE Healthcare). However, the properties of the various IMAC resins are slightly different (Weiß and Grisshammer, 2002). For example, Ni^{2+}-NTA resin not only binds the target membrane protein tighter but also binds more contaminants compared to Talon resin. The receptor expression level (i.e., the ratio of the target receptor to contaminants) determines the efficiency of this first purification step; the higher the receptor expression level, the more efficient the first purification step. A purity of >90% has been achieved for a mutant

β_2-adrenergic receptor using a single IMAC step (Hanson *et al.*, 2008). In some cases, the binding capacity of IMAC resin has been found to be lower for receptors in detergent solution than that for soluble proteins (Weiß and Grisshammer, 2002; White *et al.*, 2004).

The time needed for purification should be kept to a minimum, as should the number of purification steps because of incurring protein losses at all stages.

3.3. Receptor-specific ligand affinity chromatography

After the enrichment of receptors by use of a general affinity tag, a second purification step may be required to isolate pure, functional receptor protein for a number of reasons. (i) Receptors from the first purification step are not yet pure and other contaminants are still present. For example, a major contaminant was removed from the adenosine A2a receptor by use of a XAC (xanthine amine congener, antagonist) column step (Weiß and Grisshammer, 2002). Likewise, the neurotensin receptor NTS1 was purified to homogeneity by use of a NT (neurotensin, agonist) column (White *et al.*, 2004) and an alprenolol (nonselective β-adrenergic receptor antagonist) Sepharose column was used for the purification of a β-adrenergic receptor (Warne *et al.*, 2003). If receptors, eluted from the general tag affinity resin, are a mixture of functional and nonfunctional, incorrectly folded receptors, then this subsequent receptor-specific ligand affinity chromatography step will also remove the nonfunctional receptor population. (ii) The use of one (or two subsequent) general affinity tag step(s) produces almost pure receptor which are, however, a mixture of functional and nonfunctional species. The use of general affinity tags will hence not resolve correctly folded from incorrectly folded but still detergent-soluble protein. Incorrect folding of receptors may occur within the expression host, and/or may arise because of the instability of receptors in detergent solution. Application of a receptor-specific ligand affinity column will resolve functional from incorrectly folded receptor species. For example, an alprenolol column has been used for the isolation of functional β_2-adrenergic receptor (Kobilka, 1995). (iii) A single-step purification protocol led to a highly enriched preparation of a pituitary adenylate cyclase-activating polypeptide (PACAP) receptor by adding biotinylated peptide ligand (PACAP38) and avidin resin to the crude-solubilized receptor (Ohtaki *et al.*, 1998).

3.4. Analysis of detergent-solubilized GPCRs

The amount of functional, detergent-solubilized receptor can be determined by radioligand-binding experiments. Ideally, the radioligand should bind to the receptor with high affinity and should display a slow off-rate to avoid artifacts possibly arising because of nonequilibrium conditions during the process of separation of the ligand–receptor–detergent complex from free ligand. An outline into receptor-binding studies is given by Hulme (1990). Radioligand-binding analyses work best with labeled ligands which are hydrophilic and do not incorporate into empty detergent micelles. In contrast, hydrophobic ligands can insert into empty detergent micelles leading to high nonspecific background signals.

It is recommended to determine the protein content by the Amido Black assay (Schaffner and Weissmann, 1973). Many methods for protein content determination are inaccurate in the presence of detergents and lipids. The Amido Black assay includes a precipitation step to remove detergents, lipids, or other buffer components. Note that not all proteins bind dyes such as Amido Black equally well, meaning the protein concentration determinations may be biased if the dye binding is different for the receptor compared to the reference protein (BSA).

From ligand-binding data and protein content analysis, one can calculate an experimental B_{max} value (nmol functional receptor/mg of protein) and compare this value with the theoretical value for specific binding of pure functional receptor. Note that the theoretical B_{max} value is specific for each GPCR because it depends on its amino acid composition. Purification of functional receptors is achieved if the experimental value matches the theoretical B_{max} value.

4. Solubilization and Purification of a Recombinant Neurotensin Receptor NTS1

A more detailed description of the solubilization, purification, and analysis of NTS1 has been described by White and Grisshammer (2007) and is presented here in an abbreviated form. The purification to homogeneity of the NTS1 fusion protein is achieved by IMAC followed by a NTS1-specific ligand column step.

A maltose-binding protein (MBP) fusion approach is used for the expression of functional, membrane-inserted NTS1 in *E. coli* (Grisshammer *et al.*, 1993). The expression plasmid encodes MBP with its signal peptide, followed by the receptor. After removal of the signal peptide by the *E. coli* leader peptidase, the NTS1 fusion protein MBP-T43NTR-TrxA-H10 (NTS1-624), used for purification, starts with the mature *E. coli* MBP (Lys^1 to Thr^{366}), followed by the N-terminally truncated rat neurotensin type I receptor NTS1 (T43NTR, Thr^{43} to Tyr^{424}) (Tanaka *et al.*, 1990), the *E. coli* thioredoxin (TrxA, Ser^2 to Ala^{109}) (the presence of TrxA increases expression, see Tucker and Grisshammer, 1996), and a decahistidine tag (H10) (Grisshammer and Tucker, 1997). Expression is driven from a weak promoter of the low-copy number expression plasmid at low temperature (22 °C). This avoids the possible overloading of the *E. coli* translocation and membrane insertion machinery by the nascent receptor chain. Few protein molecules are produced at any given time but accumulation of correctly folded receptors is observed over 2 days. Purification of 10 mg of the NTS1 fusion protein typically requires 250 g of *E. coli* wet cells which is equivalent to 50 l of cell culture (White *et al.*, 2004). The growth of *E. coli* at this scale is most easily done by fermentation.

The expression levels of the NTS1 fusion protein in *E. coli* are moderate (i.e., the ratio of contaminants to target receptor is high). Hence, this requires an optimized protocol for IMAC to enrich the receptor fusion protein efficiently. To accomplish this, NTS1 fusion proteins with C-terminal tails of 10 histidine residues rather than six histidine residues (Grisshammer and Tucker, 1997) are used in combination with Ni^{2+}-NTA resin. The tight binding to Ni^{2+}-NTA resin of decahistidine-tagged receptors allows stringent washing steps using imidazole at a concentration of 50 m*M*. Imidazole at this concentration eliminates binding to the Ni^{2+}-NTA resin of most of the *E. coli* contaminants, but does not cause the elution of the fusion protein. This strategy not only results in efficient purification of receptors from crude membranes, but works well for the purification from total cell lysate (Grisshammer and Tucker, 1997).

The apparent affinity of NT for NTS1 is reduced in the presence of the high concentrations of sodium ions and imidazole in the NiB buffer (Grisshammer *et al.*, 1999) precluding the direct binding of functional receptors in the Ni^{2+}-NTA column eluate onto the NT column. The concentration of NaCl and imidazole must therefore be reduced from 200 to 70 m*M* by dilution with buffer to allow binding of functional NTS1 to the NT column. Because binding of NTS1 to NT is NaCl sensitive, receptors can be efficiently eluted from the NT column with NaCl at high concentration (Fig. 36.1).

4.1. Solubilization of the NTS1 fusion protein

1. All steps are performed at 4 °C or on ice unless stated otherwise.
2. At room temperature, crush 250 g of frozen cell paste between plastic sheets with a hammer to obtain small pieces. Place cells into a Waring blender.
3. Add 500 ml of cold 2× solubilization buffer (100 m*M* Tris–HCl, pH 7.4, 60% (v/v) glycerol, 400 m*M* NaCl) to the cells. Operate the Waring blender until all cells are suspended.
4. Transfer the cell suspension into a beaker with a magnetic stir bar. It is difficult to determine the volume at this stage because of the air introduced by the Waring blender into the suspension. The final volume is therefore adjusted in step 11.

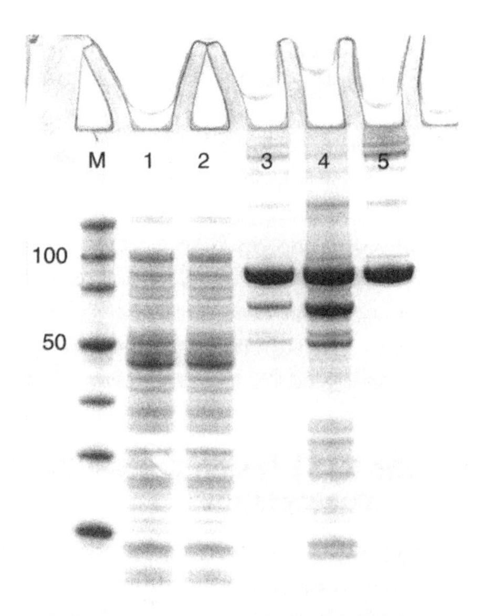

Figure 36.1 Purification of the neurotensin receptor NTS1. The fusion protein NTS1-624 was purified on a 100-ml Ni^{2+}-NTA column followed by a 20-ml NT column, starting from 250 g of *E. coli* cells. The progress of purification was monitored by SDS–PAGE (NuPAGE 4–12% Bis–Tris gel, Invitrogen, 1× MES buffer) and Coomassie R-250 staining. Lane M, Novagen Perfect Protein Marker (15–150 kDa); lane 1, 10 μg of supernatant; lane 2, 10 μg of Ni^{2+}-NTA column flow through; lane 3, 5 μg of Ni^{2+}-NTA column eluate; lane 4, 10 μg of NT column flow through; lane 5, 5 μg of NT column eluate. The incorrectly folded receptors in the NT column flow through are initially detergent soluble but aggregate over time (R. Grisshammer, unpublished results). Reprinted from White *et al.* (2004).

5. While stirring, add 1 ml of protease inhibitor stock solutions (phenyl methyl sulphonyl fluoride, PMSF, at 70 mg/ml in ethanol; leupeptin at 1 mg/ml in H_2O; pepstatin A at 1.4 mg/ml in methanol), 5 ml of 1 *M* $MgCl_2$ (final concentration of 5 m*M*), 0.6 ml of DNaseI solution (10 mg/ml, Sigma D-4527), and 50 ml of cold H_2O.
6. While stirring, add dropwise 100 ml of a CHAPS/CHS stock solution (6% (w/v) (3-[(3-cholamidopropyl)dimethylammonio]-1-propanesulfonate)/1.2% (w/v) cholesteryl hemisuccinate Tris salt in H_2O). CHS is not water soluble on its own and needs to be dissolved in CHAPS.
7. While stirring, add dropwise 100 ml of a DDM stock solution (10% (w/v) *n*-dodecyl-β-D-maltoside in H_2O).
8. Continue to stir for 15 min.
9. Sonicate for 33 min (8 s/g of cells), 1 s on, 2 s off, level 4 (Misonix sonicator 3000, 1/2 in. flat tip). Keep the sample in an ice/water bath to avoid local heating during sonication. This mild sonication step enhances the receptor solubilization efficiency.

10. Add an additional 1 ml of each protease inhibitor stock.
11. Determine volume and add cold H_2O dropwise under stirring to a final volume of 1 l.
12. Stir sample for an additional 30 min.
13. Ultracentrifuge sample for 1 h (Beckman 45Ti rotors or equivalent, at 45,000 rpm (235,000 × g at r_{max})).
14. Retrieve the solubilized receptors (supernatant).
15. Add an imidazole stock solution (2 *M*, adjusted to pH 7.4 with HCl) dropwise while stirring (final concentration of 50 m*M*). Pass sample through 0.22 μm filter. The supernatant is now ready for purification of receptors by IMAC followed by a neurotensin affinity column.

4.2. Purification of the NTS1 fusion protein by immobilized metal affinity chromatography

The purification is carried out in the cold room using an Äkta Purifier (GE Healthcare) chromatography system in a fully automated two-column mode; for details see White and Grisshammer (2007) and White *et al.* (2004). The Purifier is equipped with air sensors, a sample pump P950, a modified injection valve, a 100-ml Ni^{2+}-NTA column, a 20-ml NT column, and a fraction collector Frac950. However, all purification steps can also be performed in a simpler setting. The IMAC column can be processed first, and the Ni^{2+}-NTA column eluate is then diluted and loaded onto the NT column.

1. Equilibrate a 100-ml Ni^{2+}-NTA Superflow (Qiagen) column (XK50, GE Healthcare) with NiA buffer (50 m*M* Tris–HCl, pH 7.4, 30% (v/v) glycerol, 50 m*M* imidazole, 200 m*M* NaCl, 0.5% (w/v) CHAPS/0.1% (w/v) CHS, 0.1% (w/v) DDM).
2. Load the supernatant at a flow rate of 2 ml/min onto the Ni^{2+}-NTA column using the sample pump equipped with an air sensor to detect the end of loading. Collect the Ni^{2+}-NTA column flow through for analysis.
3. After loading is completed, wash the Ni^{2+}-NTA column with 15 column volumes of buffer NiA at 2 ml/min.
4. For elution, pass 4 column volumes of buffer NiB (buffer NiA but with 200 m*M* imidazole) at 2 ml/min over the Ni^{2+}-NTA column.
5. The total volume of fractions containing the eluted NTS1 fusion protein is ~ 120 ml.

4.3. Purification of the NTS1 fusion protein by a neurotensin column

Neurotensin has nanomolar affinity for its receptor in the presence of detergents; therefore, this property can be used for an efficient affinity purification step (Grisshammer *et al.*, 1999; Tucker and Grisshammer,

1996). The NT resin is based on N-terminally biotinylated NT (biotin-βAla-βAla-Gln-Leu-Tyr-Glu-Asn-Lys-Pro-Arg-Arg-Pro-Tyr-Ile-Leu-OH), bound to tetrameric avidin resin. Avidin has a high isoelectric point and must be succinylated to reduce the binding of *E. coli* proteins to the affinity matrix (Tucker and Grisshammer, 1996). Biotinylated NT is made by solid-phase peptide synthesis.

1. Equilibrate a 20-ml NT column (XK26) with NT70 buffer (50 m*M* Tris–HCl, pH 7.4, 30% (v/v) glycerol, 1 m*M* EDTA, 70 m*M* NaCl, 0.5% (w/v) CHAPS/0.1% (w/v) CHS, 0.1% (w/v) DDM).
2. Dilute the Ni^{2+}-NTA column eluate with buffer NT0 (50 m*M* Tris–HCl, pH 7.4, 30% (v/v) glycerol, 1 m*M* EDTA, 0.5% (w/v) CHAPS/0.1% (w/v) CHS, 0.1% (w/v) DDM) to reduce the imidazole and NaCl concentrations to 70 m*M*, respectively. The dilution step is performed by the Äkta Purifier (White *et al.*, 2004), but can also be done manually.
3. Load the diluted Ni^{2+}-NTA column eluate onto the NT column at a flow rate of 0.4 ml/min. Collect the NT column flow through for analysis.
4. Wash the NT column at 0.7 ml/min with 8 column volumes of buffer NT70.
5. Elute receptors at a flow rate of 0.5 ml/min with buffer NT1K (NT0 buffer with 1 *M* NaCl).
6. After the purification is completed, wash the NT column extensively with buffer A3 (50 m*M* Tris–HCl, pH 7.4, 1 m*M* EDTA, 1 *M* NaCl) followed by buffer B3 (50 m*M* Tris–HCl, pH 7.4, 1 m*M* EDTA, 3 m*M* NaN_3). Buffer B3 is used as storage buffer.

5. Analysis of Purified NTS1

The quality of the NTS1 preparation can be assessed by radioligand-binding assay using tritium-labeled neurotensin ([^{3}H]NT). [^{3}H]NT is very hydrophilic and not "sticky," resulting in low background binding. The detergent-solubilized NTS1 fusion protein binds the agonist with nanomolar affinity. The off-rates are slow, which allows the separation of the ligand–receptor–detergent complex from free ligand (which does not incorporate into free detergent micelles) by centrifugation-assisted gel filtration (White *et al.*, 2004). The size difference between [^{3}H]NT and the ligand–receptor–detergent complex is large enough to avoid "leakage" of free [^{3}H]NT. For routine analyses, the [^{3}H]NT concentration is set to 2 n*M*. With some assumptions, one can correct for fractional occupancy (the law of mass action predicts that the fractional receptor occupancy at equilibrium is a function of the ligand concentration: fractional occupancy $=$ [ligand]/([ligand] $+ K_D$)). A theoretical value for specific binding (B_{max})

of 10.4 nmol/mg is calculated for the NTS1-624 fusion protein (molecular mass of 96.5 kDa), assuming one ligand-binding site per receptor molecule. The protein content is determined by the Amido Black assay (Schaffner and Weissmann, 1973).

Starting from 250 g of wet *E. coli* cells, the following results can be anticipated: *supernatant* (~ 930 ml, ~ 16 mg protein/ml, ~ 0.2 nmol NTS1/ml, ~ 10–12 pmol NTS1/mg protein); *$Ni^{2\pm}$-NTA column eluate* (~ 120 ml, ~ 0.25 mg protein/ml, ~ 1.2 nmol NTS1/ml, ~ 5 nmol NTS1/mg protein); *neurotensin column eluate* (~ 14 ml, ~ 0.7 mg protein/ml, ~ 7 nmol NTS1/ml, ~ 10 nmol NTS1/mg protein).

6. Conclusions

The NTS1 receptor is expressed in functional form in *E. coli* as a fusion protein. Receptors are solubilized by a detergent mixture from total cell extract (rather than from a membrane preparation) and are enriched by immobilized metal affinity chromatography. Contaminants present in the Ni^{2+}-NTA column eluate are removed by the use of a subsequent neurotensin column. The above purification protocol is simple and robust and yields pure, functional NTS1 fusion protein.

Acknowledgments

The research of R. G. is supported by the Intramural Research Program of the NIH, National Institute of Neurological Disorders and Stroke.

References

Akermoun, M., Koglin, M., Zvalova-Iooss, D., Folschweiller, N., Dowell, S. J., and Gearing, K. L. (2005). Characterization of 16 human G protein-coupled receptors expressed in baculovirus-infected insect cells. *Protein Expr. Purif.* **44,** 65–74.

Andre, N., Cherouati, N., Prual, C., Steffan, T., Zeder-Lutz, G., Magnin, T., Pattus, F., Michel, H., Wagner, R., and Reinhart, C. (2006). Enhancing functional production of G protein-coupled receptors in *Pichia pastoris* to levels required for structural studies via a single expression screen. *Protein Sci.* **15,** 1115–1126.

Bamber, L., Harding, M., Butler, P. J., and Kunji, E. R. (2006). Yeast mitochondrial ADP/ATP carriers are monomeric in detergents. *Proc. Natl. Acad. Sci. USA* **103,** 16224–16229.

(Epub ahead of print)Bonander, N., Darby, R. A., Grgic, L., Bora, N., Wen, J., Brogna, S., Poyner, D. R., O'Neill, M. A., and Bill, R. M. (2009). Altering the ribosomal subunit ratio in yeast maximizes recombinant protein yield. *Microb. Cell Fact* **8,** 10.

Bonander, N., Hedfalk, K., Larsson, C., Mostad, P., Chang, C., Gustafsson, L., and Bill, R. M. (2005). Design of improved membrane protein production experiments: Quantitation of the host response. *Protein Sci.* **14,** 1729–1740.

Calandra, B., Tucker, J., Shire, D., and Grisshammer, R. (1997). Expression in *Escherichia coli* and characterisation of the human central CB1 and peripheral CB2 cannabinoid receptors. *Biotechnol. Lett.* **19,** 425–428.

Chelikani, P., Reeves, P. J., Rajbhandary, U. L., and Khorana, H. G. (2006). The synthesis and high-level expression of a beta2-adrenergic receptor gene in a tetracycline-inducible stable mammalian cell line. *Protein Sci.* **15,** 1433–1440.

Cherezov, V., Rosenbaum, D. M., Hanson, M. A., Rasmussen, S. G., Thian, F. S., Kobilka, T. S., Choi, H. J., Kuhn, P., Weis, W. I., Kobilka, B. K., *et al.* (2007). High-resolution crystal structure of an engineered human beta2-adrenergic G protein-coupled receptor. *Science* **318,** 1258–1265.

De Grip, W. J. (1982). Thermal stability of rhodopsin and opsin in some novel detergents. *Methods Enzymol.* **81,** 256–265.

Gether, U., and Kobilka, B. K. (1998). G Protein-coupled receptors II. Mechanism of agonist activation. *J. Biol. Chem.* **273,** 17979–17982.

Griffith, D. A., Delipala, C., Leadsham, J., Jarvis, S. M., and Oesterhelt, D. (2003). A novel yeast expression system for the overproduction of quality-controlled membrane proteins. *FEBS Lett.* **553,** 45–50.

Grisshammer, R. (2006). Understanding recombinant expression of membrane proteins. *Curr. Opin. Biotechnol.* **17,** 337–340.

Grisshammer, R., and Buchanan, S. K. (eds.) (2006). "Structural Biology of Membrane Proteins", The Royal Society of Chemistry, Cambridge, UK.

Grisshammer, R., and Tate, C. G. (1995). Overexpression of integral membrane proteins for structural studies. *Q. Rev. Biophys.* **28,** 315–422.

Grisshammer, R., and Tate, C. G. (eds.) (2003). "Special Issue on Overexpression of Integral Membrane Proteins". *Biochim. Biophys. Acta* **1610,** pp. 1–153.

Grisshammer, R., and Tucker, J. (1997). Quantitative evaluation of neurotensin receptor purification by immobilized metal affinity chromatography. *Protein Expr. Purif.* **11,** 53–60.

Grisshammer, R., Duckworth, R., and Henderson, R. (1993). Expression of a rat neurotensin receptor in *Escherichia coli*. *Biochem. J.* **295,** 571–576.

Grisshammer, R., Averbeck, P., and Sohal, A. K. (1999). Improved purification of a rat neurotensin receptor expressed in *Escherichia coli*. *Biochem. Soc. Trans.* **27,** 899–903.

Hanson, M. A., Cherezov, V., Griffith, M. T., Roth, C. B., Jaakola, V. P., Chien, E. Y., Velasquez, J., Kuhn, P., and Stevens, R. C. (2008). A specific cholesterol binding site is established by the 2.8 Å structure of the human beta2-adrenergic receptor. *Structure* **16,** 897–905.

Hassaine, G., Wagner, R., Kempf, J., Cherouati, N., Hassaine, N., Prual, C., Andre, N., Reinhart, C., Pattus, F., and Lundstrom, K. (2006). Semliki Forest virus vectors for overexpression of 101 G protein-coupled receptors in mammalian host cells. *Protein Expr. Purif.* **45,** 343–351.

Hulme, E. C. (ed.) (1990). "Receptor Biochemistry—A Practical Approach", IRL Press at Oxford University Press, Oxford.

Hulme, E. C., and Curtis, C. A. M. (1998). Purification of recombinant M1 muscarinic acetylcholine receptor. *Biochem. Soc. Trans.* **26,** S361.

Jaakola, V. P., Griffith, M. T., Hanson, M. A., Cherezov, V., Chien, E. Y., Lane, J. R., Ijzerman, A. P., and Stevens, R. C. (2008). The 2.6 angstrom crystal structure of a human A2A adenosine receptor bound to an antagonist. *Science* **322,** 1211–1217.

Klaassen, C. H., Bovee-Geurts, P. H., Decaluwe, G. L., and Degrip, W. J. (1999). Large-scale production and purification of functional recombinant bovine rhodopsin with the use of the baculovirus expression system. *Biochem. J.* **342,** 293–300.

Kobilka, B. K. (1995). Amino and carboxyl terminal modifications to facilitate the production and purification of a G protein-coupled receptor. *Anal. Biochem.* **231,** 269–271.

Kobilka, B. K., and Deupi, X. (2007). Conformational complexity of G-protein-coupled receptors. *Trends Pharmacol. Sci.* **28,** 397–406.

Magnani, F., Shibata, Y., Serrano-Vega, M. J., and Tate, C. G. (2008). Co-evolving stability and conformational homogeneity of the human adenosine A2a receptor. *Proc. Natl. Acad. Sci. USA* **105,** 10744–10749.

Molday, R. S., and Mackenzie, D. (1983). Monoclonal antibodies to rhodopsin: Characterization, cross-reactivity, and application as structural probes. *Biochemistry* **22,** 653–660.

Ohtaki, T., Ogi, K., Masuda, Y., Mitsuoka, K., Fujiyoshi, Y., Kitada, C., Sawada, H., Onda, H., and Fujino, M. (1998). Expression, purification, and reconstitution of receptor for pituitary adenylate cyclase-activating polypeptide. Large-scale purification of a functionally active G protein-coupled receptor produced in Sf9 insect cells. *J. Biol. Chem.* **273,** 15464–15473.

Oprian, D. D., Molday, R. S., Kaufman, R. J., and Khorana, H. G. (1987). Expression of a synthetic bovine rhodopsin gene in monkey kidney cells. *Proc. Natl. Acad. Sci. USA* **84,** 8874–8878.

Reeves, P. J., Klein-Seetharaman, J., Getmanova, E. V., Eilers, M., Loewen, M. C., Smith, S. O., and Khorana, H. G. (1999). Expression and purification of rhodopsin and its mutants from stable mammalian cell lines: Application to NMR studies. *Biochem. Soc. Trans.* **27,** 950–955.

Roth, C. B., Hanson, M. A., and Stevens, R. C. (2008). Stabilization of the human beta2-adrenergic receptor TM4-TM3-TM5 helix interface by mutagenesis of Glu122(3.41), a critical residue in GPCR structure. *J. Mol. Biol.* **376,** 1305–1319.

Sarkar, C. A., Dodevski, I., Kenig, M., Dudli, S., Mohr, A., Hermans, E., and Plückthun, A. (2008). Directed evolution of a G protein-coupled receptor for expression, stability, and binding selectivity. *Proc. Natl. Acad. Sci. USA* **105,** 14808–14813.

Sarramegna, V., Talmont, F., Demange, P., and Milon, A. (2003). Heterologous expression of G-protein-coupled receptors: Comparison of expression systems from the standpoint of large-scale production and purification. *Cell Mol. Life Sci.* **60,** 1529–1546.

Schaffner, W., and Weissmann, C. (1973). A rapid, sensitive, and specific method for the determination of protein in dilute solution. *Anal. Biochem.* **56,** 502–514.

Serrano-Vega, M. J., Magnani, F., Shibata, Y., and Tate, C. G. (2008). Conformational thermostabilization of the beta1-adrenergic receptor in a detergent-resistant form. *Proc. Natl. Acad. Sci. USA* **105,** 877–882.

Shibata, Y., White, J. F., Serrano-Vega, M. J., Magnani, F., Aloia, A. L., Grisshammer, R., and Tate, C. G. (2009). Thermostabilization of the neurotensin receptor NTS1. *J. Mol. Biol.* **390,** 262–277.

Tanaka, K., Masu, M., and Nakanishi, S. (1990). Structure and functional expression of the cloned rat neurotensin receptor. *Neuron* **4,** 847–854.

Tucker, J., and Grisshammer, R. (1996). Purification of a rat neurotensin receptor expressed in *Escherichia coli*. *Biochem. J.* **317,** 891–899.

Wagner, S., Bader, M. L., Drew, D., and De Gier, J. W. (2006). Rationalizing membrane protein overexpression. *Trends Biotechnol.* **24,** 364–371.

Wagner, S., Baars, L., Ytterberg, A. J., Klussmeier, A., Wagner, C. S., Nord, O., Nygren, P. A., Van Wijk, K. J., and De Gier, J. W. (2007). Consequences of membrane protein overexpression in *Escherichia coli*. *Mol. Cell Proteomics* **6,** 1527–1550.

Warne, T., Chirnside, J., and Schertler, G. F. (2003). Expression and purification of truncated, non-glycosylated turkey beta-adrenergic receptors for crystallization. *Biochim. Biophys. Acta* **1610,** 133–140.

Warne, T., Serrano-Vega, M. J., Baker, J. G., Moukhametzianov, R., Edwards, P. C., Henderson, R., Leslie, A. G., Tate, C. G., and Schertler, G. F. (2008). Structure of a beta1-adrenergic G-protein-coupled receptor. *Nature* **454,** 486–491.

Weiß, H. M., and Grisshammer, R. (2002). Purification and characterization of the human adenosine A2a receptor functionally expressed in *Escherichia coli*. *Eur. J. Biochem.* **269,** 82–92.

White, J. F., and Grisshammer, R. (2007). Automated large-scale purification of a recombinant G-protein-coupled neurotensin receptor. *Curr. Protoc. Protein Sci.* 6.8.1–6.8.31.

White, J. F., Trinh, L. B., Shiloach, J., and Grisshammer, R. (2004). Automated large-scale purification of a G-protein-coupled receptor for neurotensin. *FEBS Lett.* **564,** 289–293.

White, J. F., Grodnitzky, J., Louis, J. M., Trinh, L. B., Shiloach, J., Gutierrez, J., Northup, J. K., and Grisshammer, R. (2007). Dimerization of the class A G protein-coupled neurotensin receptor NTS1 alters G protein interaction. *Proc. Natl. Acad. Sci. USA* **104,** 12199–12204.

Yeliseev, A. A., Wong, K. K., Soubias, O., and Gawrisch, K. (2005). Expression of human peripheral cannabinoid receptor for structural studies. *Protein Sci.* **14,** 2638–2653.

CHAPTER THIRTY-SEVEN

Cell-Free Translation of Integral Membrane Proteins into Unilamelar Liposomes

Michael A. Goren, Akira Nozawa, Shin-ichi Makino, Russell L. Wrobel, *and* Brian G. Fox

Contents

1. Introduction 648
2. Overview of Cell-Free Translation 650
3. Expression Vectors 651
4. Gene Cloning 652
 4.1. Materials and reagents 654
5. PCR Product Cleanup 655
 5.1. Materials and reagents 655
6. Flexi Vector and PCR Product Digestion Reaction 656
 6.1. Reagents 656
7. Ligation Reaction 657
 7.1. Reagents 657
8. Transformation Reaction 658
 8.1. Materials and reagents 658
9. Purification of Plasmid DNA 659
 9.1. Reagents 659
10. Preparation of mRNA 660
 10.1. Reagents 660
11. Preparation of Liposomes 661
 11.1. Materials and reagents 661
12. Wheat Germ Translation Reaction 662
 12.1. Materials and reagents 663
13. Purification by Density Gradient Ultracentrifugation 665
 13.1. Materials and reagents 665
14. Characterization of Proteoliposomes 667
15. Considerations for Scale-Up 669
16. Isotopic Labeling for Structural Studies 669

Department of Biochemistry, Dr. Brian Fox Lab, University of Wisconsin-Madison, Madison, Wisconsin, USA

Methods in Enzymology, Volume 463
ISSN 0076-6879, DOI: 10.1016/S0076-6879(09)63037-8

17. Conclusions 670
Acknowledgments 670
References 670

Abstract

Wheat germ cell-free translation is shown to be an effective method to produce integral membrane proteins in the presence of unilamelar liposomes. In this chapter, we describe the expression vectors, preparation of mRNA, two types of cell-free translation reactions performed in the presence of liposomes, a simple and highly efficient purification of intact proteoliposomes using density gradient ultracentrifugation, and some of the types of characterization studies that are facilitated by this facile preparative approach. The *in vitro* transfer of newly translated, membrane proteins into liposomes compatible with direct measurements of their catalytic function is contrasted with existing approaches to extract membrane proteins from biological membranes using detergents and subsequently transfer them back to liposomes for functional studies.

1. Introduction

Membrane proteins provide the molecular mechanisms through which useful molecules gain controlled entry into a cell, and likewise provide the portals through which cellular products are exported from the cell. Membrane enzymes synthesize the molecules that make up cellular membranes (e.g., saturated and unsaturated fatty acids, phospholipids, glycerolipids, sphingolipids, sterols, polyisoprenes). Membrane enzyme complexes are responsible for electron transport and generate electrochemical gradients, and an exquisite membrane motor ATPase uses these gradients to generate ATP. They harvest light, provide allosteric receptors that transduce the information from binding of external molecules into cellular responses via signaling cascades, provide essential surface contacts as differentiating embryonic stem cells begin to assemble into more complex tissues, and help to elicit the antigenic responses observed in response to pathogen infection. These few examples do not do justice to the incredible diversity of functions and the essential roles that membrane proteins and enzymes contribute to cellular function.

Achieving control of integral membrane protein expression, transfer into the lipid bilayer, and cofactor incorporation are significant experimental challenges, and the ability to manipulate these events would be of great scientific utility. Furthermore, identification of novel ways to address increasing structural complexity, leading to the expression and facile purification of fully folded, functional membrane proteins or complexes embedded in easily handled and manipulated lipid environments or in functionally compatible detergent mixtures would also be of great utility.

Recent studies arising from structural genomics efforts suggest that cell-free translation may have unique possibilities for addressing these challenges. Although known for a long time, cell-free protein translation is undergoing a renaissance (Endo and Sawasaki, 2006; Klammt *et al.*, 2004; Schwarz *et al.*, 2007; Spirin and Swartz, 2008; Vinarov *et al.*, 2006; Yokoyama, 2003). This approach to protein synthesis, using extracts from either prokaryotic (Boyer *et al.*, 2008; Kigawa *et al.*, 2004; Klammt *et al.*, 2006; Schwarz *et al.*, 2007; Wuu and Swartz, 2008; Yokoyama, 2003) or eukaryotic (Endo and Sawasaki, 2006; Madin *et al.*, 2000) sources, offers an alternative to *E. coli* or other living cell-based platforms that are the mainstay of most recombinant protein expression efforts. Because it decouples the production of proteins and enzymes from cellular homeostasis, cell-free translation removes variability associated with the use of living expression hosts (Hall *et al.*, 2005). Cell-free systems permit the production of proteins that form complexes in the cell (Matsumoto *et al.*, 2008), inclusion bodies (Klammt *et al.*, 2004; Schwarz *et al.*, 2007), that undergo proteolysis in cells (Goren and Fox, 2008; Yokoyama, 2003), or that are toxic to cells (Klammt *et al.*, 2004; Madin *et al.*, 2000; Schwarz *et al.*, 2007). Cell-free translation can offer dramatic increases in the speed of evaluation of different conditions for expression, considerable simplification of the steps needed to purify a recombinant protein, and direct possibilities for automation of all steps after sequence-verified cloning to catalytic assay of an expressed and purified protein.

Because it is an open system, cell-free translation allows the addition of many types of reagents to the protein synthesis reaction in order to address biological complexity. Some examples of added reagents include detergents to solubilize proteins (Klammt *et al.*, 2007), liposomes to capture functional forms of membrane proteins (Goren and Fox, 2008; Nozawa *et al.*, 2007), cofactors, coenzymes, and metal ions to reconstitute catalytic function (Abe *et al.*, 2004; Boyer *et al.*, 2008; Goren and Fox, 2008), multiple mRNAs to yield cotranslation of heteromeric proteins and protein–protein complexes (Matsumoto *et al.*, 2008), or other enzymes to effect posttranslational modifications (Goerke and Swartz, 2008; Kanno *et al.*, 2007). Moreover, many different types of amino acid (AA) labeling strategies can be accomplished in cell-free translation by a simple substitution of unlabeled AAs with AAs containing isotopic substitutions such as ^{2}H, ^{13}C, or ^{15}N for NMR or selenomethionine (SeMet) for determination of X-ray diffraction phasing (Klammt *et al.*, 2004; Matsuda *et al.*, 2007; Vinarov and Markley, 2005). Residue type selective labeling can be carried out without the need for auxotrophs or specialized feeding, and unnatural AAs can be incorporated in a straightforward manner (Kiga *et al.*, 2002). Another advantage is that wheat germ cell-free translation has been successfully automated (Vinarov and Markley, 2005; Vinarov *et al.*, 2006) in 24-, 96-, and 384-well formats for screening in volumes as low as 50 μL, and 6- and 24-well formats for protein production in volumes up to 6 mL.

2. Overview of Cell-Free Translation

This chapter describes the use of wheat germ extract, a highly stable and productive eukaryotic system for cell-free translation (Kanno *et al.*, 2007; Madin *et al.*, 2000). Since wheat germ extract is prepared from a eukaryote, it provides differences in the mechanisms for translation and folding of nascent proteins as compared to bacterial translation (Endo and Sawasaki, 2006). Moreover, the eukaryotic system may provide unique chaperones and translocation factors that may aid in the expression of some proteins (Goren and Fox, 2008), possibly including the examples described here.

Figure 37.1 provides a schematic summary of the steps used for cell-free translation and shows time-course photographs of an expression of green fluorescent protein (GFP, photo images provided courtesy of our collaborator and mentor, Prof. Yaeta Endo, Cell-Free Science and Technology Research Center, Ehime University, Japan). For this work, the gene of interest is cloned into a specialized expression vector that adds 5′-translation enhancer and 3′-UTR sequences to the transcribed mRNA. Genes are transcribed using SP6 RNA polymerase, and the highly purified mRNA is used in the translation reaction. The quality and quantity of the expression

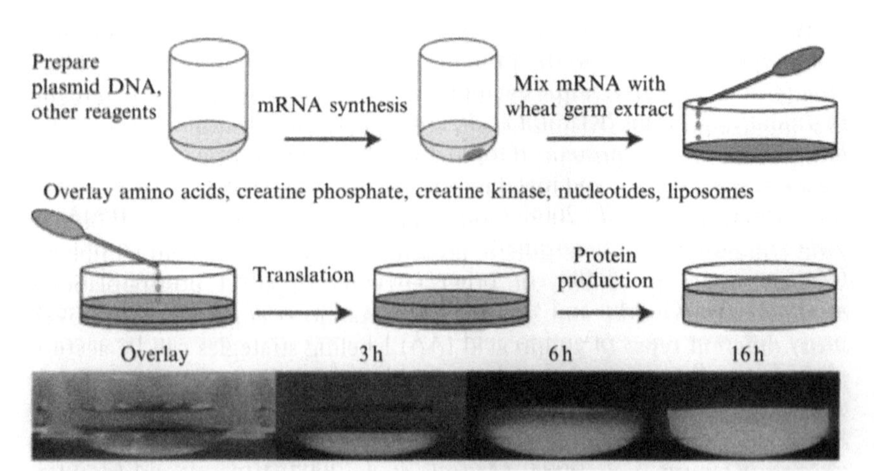

Figure 37.1 A schematic representation of the steps used for cell-free translation in a bilayer mode reaction. A customized plasmid is used to prepare reagent mRNA, which is added to the translation mixture, overlaid with amino acids, other substrates, and desired additives, and then the reaction can begin. The photos show an experimental demonstration of GFP translation over a 16 h period. (See Color Insert.)

plasmid and mRNA preparations are critical, and often underappreciated inputs to the success of high-yield cell-free translation (Tyler *et al.*, 2005; Vinarov *et al.*, 2006). Along with the mRNA template, the cell-free translation utilizes exogenously added AAs as substrates for the protein synthesis.

3. Expression Vectors

Table 37.1 summarizes specialized vectors used for this work. pEU, the original cell-free translation vector optimized for wheat germ cell-free translation (Madin *et al.*, 2000), was modified for high-throughput Flexi Vector cloning (Promega) to contain 5′-SgfI and 3′-PmeI restriction sites and a toxic selection cassette in the multicloning site (Blommel *et al.*, 2006). The vector named pEU-FV produces a protein with no purification tag, while pEU-His-FV and pEU-HSBC produces a protein with an N-terminal His6 purification tag. Vectors pEU-NGFP and pEU-GFPC produce fusions of GFP to either the N- or C-termini of the target protein. These vectors are useful for control studies and as vehicles for simplified detection of the translated fusion proteins during purification (Drew *et al.*, 2001).

Table 37.1 Flexi Vector compatible vectors for cell-free translation

Name	Product[a]	Utility
pEU	Target	Expression screening, native protein
pEU-FV	Target	Expression screening
pEU-His-FV	His6-Target	Expression screening; His6 purification
pEU-HSBC	His6-Target	Expression screening; His6 purification
pEU-NGFP	GFP-target	Expression screening; detection
pEU-GFPC	Target-GFP	Expression screening; detection
pEU-Nb5R	Target-(cyt b5 reductase N-terminal anchor peptide)[b]	Spontaneous association of N-terminal anchor peptide to liposomes
pEU-Cb5	Target-(cyt b5 C-terminal anchor peptide)[c]	Spontaneous association of C-terminal anchor peptide to liposomes

[a] The sequence of targets, domains, or tags found in the translated protein. For example, pEU-GFPC produces a target protein with GFP fused to the C-terminus, that is, target-GFP.
[b] The N-terminal anchor peptide sequence is GAQLSTLGHMVLFPVWFLYSLLM.
[c] The C-terminal anchor peptide sequence is TLITTIDSSSSWWTNWVIPAISAVAVALMYRLYMAED.

With proper design of the primer sequences, the GFP tag can be removed by treatment with tobacco etch virus (TEV) protease (Blommel and Fox, 2007; Sobrado *et al.*, 2008). The vectors pEU-Nb5R and pEU-Cb5 create fusion proteins with the membrane anchor signals from human cytochrome b_5 reductase and human cytochrome b_5, respectively. Fusion proteins containing these tags spontaneously associate with liposomes upon translation (Nomura *et al.*, 2008), and thus become amenable to purification by density gradient ultracentrifugation (Goren and Fox, 2008). Other posttranslational modifications may also be used to target proteins to liposomes (Nosjean and Roux, 2003). The vectors described herein are available from the NIH Protein Structure Initiative Material Repository (http://psimr.asu.edu).

4. Gene Cloning

This approach requires a two-step PCR procedure (Blommel *et al.*, 2006; Thao *et al.*, 2004). Figure 37.2 provides a vector map for pEU-HSCB and an example of primers designed for cloning of His6-bacteriorhodopsin (Blommel and Fox, 2007). By using this vector and primer design, the sequence MGHHHHHHAIAENLYFQ- can be liberated from the translated protein upon treatment with TEV protease to leave Ser as the first residue of the mature protein. The tobacco mosaic virus omega sequence is a translation enhancement sequence (Sawasaki *et al.*, 2000). Incorporation of the SgfI and PmeI restriction sites in the indicated positions in the respective primers allows the cloned gene to be transferred by Flexi Vector cloning (Blommel *et al.*, 2006) into any (or all) of the vectors described in Table 37.1, along with a number of other commercially available vectors (see http://www.promega.com for other examples). The sacB-CAT cassette shown in Fig. 37.2A provides toxic selection when transformants containing this construct are grown on plates containing 5% sucrose. In this manner, positive selection for cloning the desired gene can be enforced. The insert also confers resistance to chloramphenicol. The pF1K homology region enhances the efficiency of transfer of cloned genes between different Flexi Vectors (Blommel *et al.*, 2006).

In the example shown in Fig. 37.2B, the first-step PCR uses a 5′ forward primer containing gene-specific nucleotides. An invariant sequence is added to the 5′ end of the first-step PCR forward primer to encode a portion of the TEV protease site. The first-step PCR also uses a 3′ reverse complement primer as shown in Fig. 37.2C, which contains gene-specific nucleotides and the PmeI site. Other genes may be cloned by substitution of the gene-specific sequence in the designated primer sequences.

For the second-step PCR, a universal forward primer (Fig. 37.2B) is used to add the nucleotides required to complete the TEV site and the SgfI

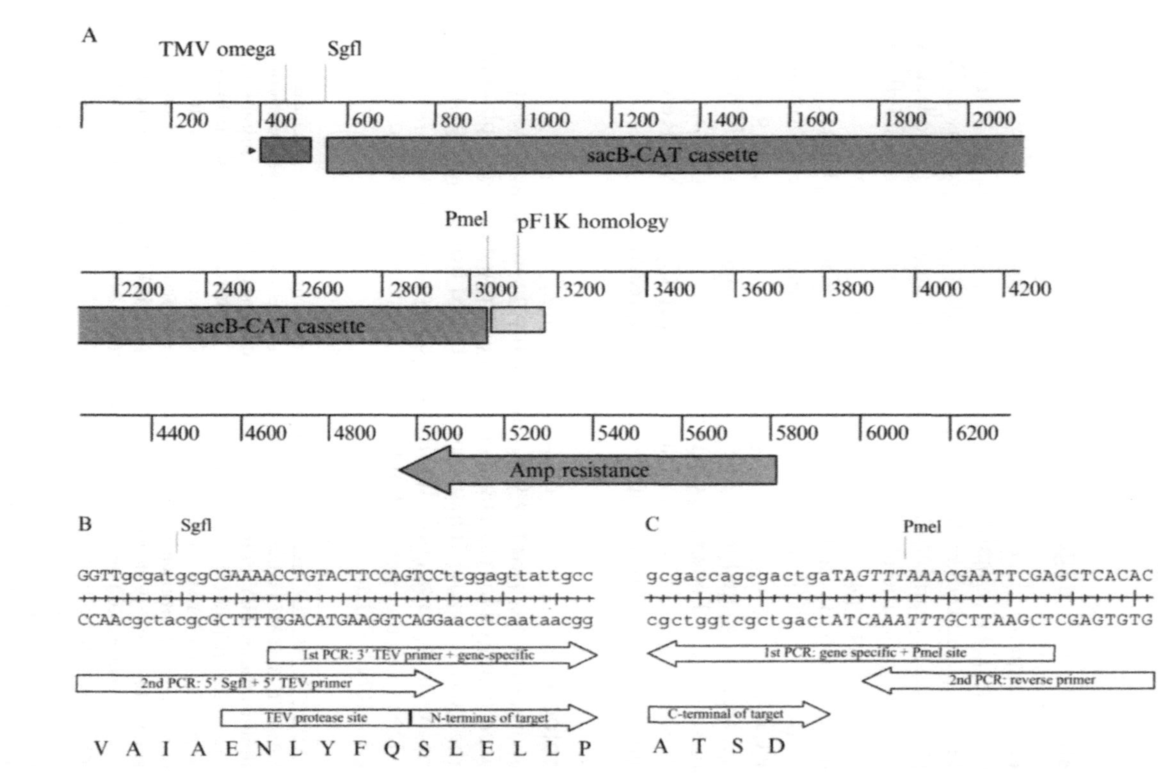

Figure 37.2 (A) Linear map of pEU-HSCB showing important features of the vector construction. The TMV omega sequence (blue box) enhances translation efficiency. The pF1K homology sequence (yellow box) prevents ligation of two plasmid backbones. (B) Example of the 5′ primer designed for the cloning of *Halobacterium salinarium* bacteriorhodopsin (GenBank M11720.1). The first-step PCR forward primer is

site. A universal reverse PCR primer is used to duplicate the PmeI site and add additional nucleotides (Fig. 37.2C). For the second-step PCR, a portion of the first-step PCR reaction is added into a new PCR reaction containing the universal forward primer and reverse primers to obtain the PCR product properly functionalized for Flexi Vector cloning.

4.1. Materials and reagents

Gene-specific primers (25 nmol synthesis with standard desalting) can be obtained from IDT (Coralville, IA).
The dNTP mix (10 m*M* of each nucleotide) is from Promega (Madison, WI).
Pfu Ultra II fusion hotstart DNA polymerase is from Stratagene (La Jolla, CA).
A 10× Flexi Enzyme Blend containing restriction endonucleases SgfI and PmeI is from Promega.
PCR plates (T-3069-B) are from ISC Bioexpress (Kaysville, UT).
Adhesive covers for PCR plates (4306311) are from Applied Biosystems (Foster City, CA).
PCR plates are centrifuged in an Allegra 6R centrifuge with a GH3.8 rotor (Beckman Coulter, Fullerton, CA).
An MJ DNA Engine, DYAD, Peltier Thermal Cycler (MJ Research, Waltham, MA) can be used for thermocycling reactions.

4.1.1. Protocol

The following steps are used to PCR amplify the desired genes and append the sequences required for cloning. A sequence-verified gene in a plasmid is used for the PCR template. These may be obtained from a variety of other sources or research projects. This PCR protocol can also be used on genomic DNA, if the gene of interest has no introns, or on reverse transcribed mRNA (Thao *et al.*, 2004).

5′-ACCTGTACTTCCAGTCCttggagttattgcc, with the upper case nucleotides corresponding to the 3′ TEV primer and the lower case nucleotides corresponding to the gene-specific sequence. The second-step universal forward primer is 5′-GGTTgcgatcgcCGAAAACCTGTACTTCCAGTCC, with the SgfI site in lower case. There is an 18 bp overlap between the first-step forward primer and universal forward primer. The TEV protease recognition sequence is ENLYFQ/S, with proteolysis between Q and S. (C) Example of the 3′ primer designed for the cloning of *Halobacterium salinarium* bacteriorhodopsin (GenBank M11720.1). The first-step PCR reverse complement primer is 5′-GCTCGAATTC*GTTTAAAC*TAtcagtcgctggtcgc, with the upper case, italic nucleotides corresponding to the PmeI site and the lower case nucleotides corresponding to the gene-specific sequence. The second-step universal reverse primer is 5′-GTGTGAGCTCGAATTCGTTTAAAC. There is an 18 bp overlap between the first-step reverse primer and universal reverse primer. (See Color Insert.)

If the plasmid being used for the template and the Flexi Vector that the PCR product will be cloned into have the same antibiotic resistance, DpnI can be added to the PCR reaction followed by incubation for 1 h at 37 °C. The DpnI will digest the template plasmid while leaving the PCR product fully intact.

1. Create a PCR Primers plate by combining forward and reverse primers for each target gene to 10 μM each in a well of an ISC PCR plate.
2. Create a PCR Master Mix consisting of 2.195 mL of water, 250 μL of 10× Pfu Ultra II Buffer, 25 μL of dNTPs (10 μM each), and 50 μL of Pfu Ultra II Hotstart DNA polymerase This Master Mix will provide up to 96 reactions, and can be scaled as appropriate. Discard the unused mix.
3. Aliquot 23.0 μL of PCR Master Mix to each well of an ISC PCR plate that will be used. This is the PCR Reaction plate.
4. Add 1 μL of the mixture from each used well of the PCR primers plate to each used well of the PCR Reaction plate.
5. Add 1 μL (~ 100 ng) of purified plasmid DNA for each gene to be cloned to a separate well of the PCR Reaction plate.
6. Centrifuge the PCR Reaction plate briefly in an Allegra 6R centrifuge and 6H3.B rotor to get liquid to the bottom of the wells and then cover the plate with sealing tape.
7. Put the PCR Reaction plate in the thermocycler and cycle using the following parameters: (1) 95 °C for 2.00 min; (2) 94 °C for 20 s; (3) 50 °C for 20 s; (4) 72 °C, 15 s/kb; (5) repeat steps 1–4 for 19 more times.
8. For the second-step PCR, 20% of the volume of the first-step reaction is added into a new PCR reaction along with 0.2 μM of each of the universal forward and reverse primers. The second-step reaction is then completed using the following cycling parameters: (1) 95 °C for 2.00 min; (2) 94 °C for 20 s; (3) 50 °C for 20 s; (4) 72 °C, 15 s/kb; (5) repeat steps 2–4 for four more times; (6) 94 °C for 20 s; (7) 55 °C for 20 s; (8) 72 °C for 15 s/kb; (9) repeat steps 6–8 for 24 more times; (10) 72 °C for 3.00 min; and (11) 4 °C and hold.
9. Analyze the completed PCR reactions for the appropriately sized products using agarose gel electrophoresis.

5. PCR Product Cleanup

The PCR products are purified prior to SgfI/PmeI digestion.

5.1. Materials and reagents

Quickstep 2 PCR purification kit (EdgeBio, Gaithersburg, MD), containing SOPE resin, Secure Seal Sterile tape, Performa Ultra 96-well Plate and V-Bottom 96-well Plate.

5.1.1. Protocol

1. Add 4 μL of well-mixed SOPE resin to 20 μL of PCR product from the FLEXI-TEV PCR protocol.
2. Cover the plate with Secure Seal Sterile tape and vortex. Let the suspension stand at room temperature while preparing the Performa Ultra 96-Well Plate.
3. Remove adhesive plate sealers from the top and bottom of the Performa Ultra 96-Well Plate. Cover with a lid.
4. Stack the Performa Ultra 96-Well Plate on top of the 96-well flat-bottom microplate.
5. Place the assembly in a cushioned centrifuge plate carrier.
6. Centrifuge for 5 min at 850×*g*.
7. Briefly spin the SOPE/PCR mix to the bottom of the wells. Transfer the SOPE/PCR reaction mixture by slowly pipetting directly into the wells of the Performa Ultra 96-Well Plate. Be sure the fluid runs into the gel matrix. Cover with a lid.
8. Stack the Performa Ultra 96-Well Plate on top of the 96-well V-bottom microplate.
9. Place the assembly in the centrifuge plate carrier and centrifuge for 5 min at 850×*g*. Retain the eluates, which contain the purified PCR products. PCR products can be stored at −20 °C until ready for use.

6. Flexi Vector and PCR Product Digestion Reaction

The following steps are used to digest the pEU vector variant and the purified PCR products with SgfI and PmeI prior to the ligation. Additional descriptions of cloning using the Flexi Vector system are presented elsewhere (Blommel and Fox, 2007; Blommel *et al.*, 2006). For success in Flexi Vector cloning, it is important to avoid overdigesting either the PCR product or the vector. SgfI has star activity so overdigestion removes the overhanging nucleotide sequence leaving bunt ends on the digested vector that efficiently ligate and yield a high background of clones with no insert after transformation.

6.1. Reagents

5× Flexi Digest Buffer (Promega)
10× SgfI/PmeI Enzyme Blend (Promega)
pEU vector variant (see Table 37.1)
Purified PCR products from PCR Product Cleanup protocol, step 9

6.1.1. Protocol

1. Create a pEU Vector Digest Master Mix consisting of 158.3 μL of sterile, deionized water, 44.0 μL of 5× Flexi Digest Buffer, 2.20 μL of 10× SgfI/PmeI Enzyme Blend, and 13.5 μL of the pEU vector variant desired (e.g., purified pEU-His-FV at a concentration of 150 ng/μL). The enzyme blend is dense and tends to settle, so must be mixed thoroughly into the remainder of the mix.
2. Place the pEU Vector Digest Master Mix in the thermocycler and cycle as follows: (1) 37 °C for 40.00 min; (2) 65 °C for 20.00 min; and (3) hold at 4 °C until needed.
3. Create the PCR Product Digest Master Mix consisting of 638 μL of sterile, deionized water, 220 μL 5× of Flexi Digest Buffer, and 22 μL of 10× SgfI/PmeI Enzyme Blend. This Master Mix will provide up to 96 reactions, and can be scaled as appropriate. Discard the unused mix.
4. Add 8.0 μL of the PCR Product Digest Master Mix to each well of an ISC PCR plate. This is the PCR Digest plate.
5. Add 2.0 μL of the purified PCR products in the PCR Reaction plate from Gene Cloning protocol, step 8 to each used well of the PCR Digest plate.
6. Place the PCR Digest plate in the thermocycler and cycle as follows: (1) 37 °C for 40.00 min; (2) 65 °C for 20.00 min; and (3) hold at 4 °C until needed. If transformation yields no colonies, or colonies containing vector without the desired insert, decrease the 37 °C incubation time to minimize star activity.

7. Ligation Reaction

Digested and purified PCR products and pEU vector variants are ligated in this step. For the best efficiency in this ligation reaction, it is important to use a high concentration of ligase.

7.1. Reagents

10× T4 DNA ligase buffer (Promega)
High concentration T4 DNA ligase (Promega)
pEU vector variant from Flexi Vector Digestion Reaction, step 2
Purified PCR product in PCR Cleanup plate from PCR Product Digestion Reaction Cleanup, step 6

7.1.1. Protocol

1. Create a Ligation Master Mix containing 225 μL of sterile, deionized water, 110 μL of 10× T4 ligase buffer, and 50 μL of high-concentration T4 DNA ligase.

2. Add 5.0 μL of purified PCR product from the PCR Cleanup plate, 2.0 μL of the pEU vector digest, and 3.5 μL of Ligation Master Mix to each well of a new PCR plate. This is the Ligation plate.
3. Incubate the Ligation plate in a thermocycler at 25 °C for 3 h.
4. Immediately proceed to the transformation step or store the Ligation plate overnight at 4 °C.

8. Transformation Reaction

The material from the ligation reaction is used to transform competent cells by the following steps. Competent cells from Invitrogen and Promega have also been used successfully by following the manufacturers protocols.

8.1. Materials and reagents

10G cells, chemically competent for transformation (Lucigen, Middleton, WI)
10G cells recovery medium (Lucigen).
Ligation Plate from Ligation Reaction, step 4.
Fisherbrand sterile disposable Petri dishes (60 × 15 mm) (Fisher Scientific, Pittsburgh, PA).
YT agar plates containing 0.5% (w/v) glucose, 50 μg/mL of kanamycin, and 5% sucrose for selection with pEU-HSCB vector (Fig. 37.2).
ColiRoller Plating Beads (Novagen, Gibbstown, NJ).

8.1.1. Protocol

1. Remove 10G chemically competent cells from −80 °C freezer and thaw on ice.
2. Aliquot 10 μL of cells into a PCR plate or strip tubes that have been chilled on ice.
3. Add 1.0 μL of the ligation reaction and stir with the pipet tip.
4. Incubate on ice for at least 5 min. Set a thermocycler block to 34 °C.
5. Heat shock at 34 °C for 30 s (per the manufacturer's directions for small reaction volumes).
6. Return the transformation reactions to ice and incubate for 2 min.
7. Remove from ice and add 80 μL of room temp recovery medium.
8. Incubate at 37 °C for 1 h without shaking.
9. While transformations are incubating, label the bottoms of 96 YT plates, containing the appropriate antibiotic, A1–H12 and add 5–10 sterile ColiRoller glass beads to each plate.
10. Apply the entire transformation reaction on each of the correspondingly labeled plates. Shake the plates in a circular motion to spread the

culture. Carefully remove ColiRoller beads by tipping the plate over an appropriate waste container, or, if they will be reused, 100% ethanol.

11. Incubate the plates overnight at 37 °C.

9. Purification of Plasmid DNA

All reagents used for *in vitro* transcription and translation must be RNase-free. Therefore, all glassware used to prepare reagents for cell-free translation reactions must be baked for 3 h at 180 °C to eliminate contaminating RNases. Furthermore, wear gloves, keep one's mouth closed, and avoid sneezing while handling reagents in order to prevent RNase contamination from hands or saliva. All buffers must be sterilized by passage through a 0.2-μm filter, and stored at −20 °C unless otherwise indicated.

Also, unless otherwise stated, 18 MΩ water (Milli-Q water, Millipore, Billerica, MA) is used to prepare all reagents. Diethylpyrocarbonate-treated water should not be used in these protocols as the degradation products of DEPC can inhibit *in vitro* transcription and translation reactions.

9.1. Reagents

The 10× buffer used with proteinase K is 100 m*M* Tris–HCl, pH 8.0, containing 50 m*M* EDTA and 1% (w/v) SDS.

Proteinase K is from Sigma/Aldrich (St. Louis, MO). Prepare a 10× proteinase K solution by addition of 0.5 mg of proteinase K to 1.0 mL of 10× proteinase K buffer. Aliquot the proteinase K solution into 10 μL samples and store them at −80 °C.

9.1.1. Protocol

1. Inoculate 150 mL of 2 × YT medium with a single colony from a pEU-His-FV transformation and grow the culture in a shaking incubator overnight at 37 °C. Recover the cells by centrifugation.
2. Purify the plasmid using a Marligen high-purity plasmid Maxiprep kit (Marligen Biosciences, Ijamsville, MD) and the manufacturer's instructions. Resuspend each separate DNA pellet in 500 μL of Milli-Q water and measure the absorbance at 260 nm to determine the plasmid DNA concentration. A typical yield is 600–900 μg of plasmid DNA.
3. Plasmid DNA prepared by commercially available kits often contains a trace contamination of RNase. This contaminant must be removed for successful transcription and translation. In order to remove trace RNase contamination, treat the purified plasmid with 50 ng/μL of proteinase K for at least 60 min at 37 °C in 1× proteinase K buffer.

4. Extract the remaining protein by adding an equal volume of a 1:1 phenol/chloroform solution to the plasmid preparation, vortex it vigorously, and then centrifuge the preparation at 14,000 rpm (18,000×*g*) and 4 °C in an F2402H rotor and Allegra 21R centrifuge or comparable instrument. Remove the upper aqueous phase to a separate new tube and repeat the extraction, centrifugation of the phenol/chloroform solution, and transfer of the aqueous phase to a new tube.
5. Add 0.1 volume of 3 *M* sodium acetate, pH 5.2, and 2.5 volumes of ethanol to the aqueous phase obtained from step 4 and mix well. Chill each tube at −20 °C for 10 min to facilitate precipitation of the plasmid DNA.
6. Centrifuge each tube at 14,000 rpm (18,000×*g*) and 4 °C in an F2402H rotor and Allegra 21R centrifuge. Wash each pellet with 500 μL of ice-cold 70% ethanol.
7. Centrifuge each tube again for 5 min as above, discard the supernatant, and thoroughly air-dry the pellet, which contains the desired plasmid DNA.
8. Dissolve the plasmid DNA pellets in 400 μL of Milli-Q water and measure the absorbance of each at 260 nm to determine the concentration of each separate plasmid DNA preparation. Adjust the volume of each preparation with Milli-Q water to obtain a DNA concentration for the solution of 1 μg/μL based on the fact that 1 μg/μL of DNA gives an $A_{260} = 20$ (40-fold dilution = 0.5).

10. Preparation of mRNA

In cell-free translation, mRNA is an essential reagent for the protein synthesis reaction. In order to obtain maximal translation efficiency, the mRNA preparations must be added in sufficient quantity to saturate ribosomes present in the translation reaction.

10.1. Reagents

Transcription Buffer plus Mg (5×) is 400 m*M* HEPES-KOH, pH 7.8, containing 100 m*M* magnesium acetate, 10 m*M* spermidine hydrochloride, and 50 m*M* DTT. Store this buffer at −20 °C.

An NTP solution containing 25 m*M* each of ATP, GTP, CTP, and UTP is prepared from 0.2 μm filter-sterilized, 100 m*M* solutions of each NTP prepared in Milli-Q water. NTP solutions are stored at −80 °C.

SP6 RNA polymerase and RNase inhibitor (RNasin) are from Promega.

10.1.1. Protocol

1. Immediately before use, prepare a transcription mixture containing 2× Transcription Buffer plus Mg, 8 m*M* NTPs, 3.2 unit/μL of SP6 RNA polymerase, and 1.6 unit/μL of RNasin.
2. Dispense 2.5 μL of each separate plasmid DNA from Purification of Plasmid DNA, step 8, to a separate well of a PCR plate. To the PCR plate, dispense 2.5 μL of the transcription mixture into each well and mix. This is called the Transcription plate.
3. Tightly cap the wells of the Transcription plate to avoid concentrating the samples by evaporation. Incubate the Transcription plate at 37 °C for 4 h. If the transcription reaction is proceeding correctly, a white precipitate of magnesium pyrophosphate will form and make the transcription solution turbid.
4. Remove the white precipitate from the transcription reaction by centrifugation in the C0650 rotor and Allegra X-22R centrifuge for 5 min at 6230 rpm (4000×*g*) and 26 °C. To avoid coprecipitation of mRNA, the reaction should not be chilled. Transfer the supernatant to a new tube. This clarified solution will be used in the translation reactions as the mRNA solution.

11. Preparation of Liposomes

Unilamelar liposomes are added to the cell-free translation reaction to capture membrane proteins as they are translated. This step provides an alternative to solubilization of membrane proteins by detergents, which often does not allow direct determination of function.

11.1. Materials and reagents

Soybean total extract (20% lecithin) is from Avanti Polar Lipids (Alabaster, AL).

The lipid rehydration buffer is 25 m*M* HEPES, pH 7.5, containing 100 m*M* NaCl.

Track-etch polycarbonate membranes, 0.4 and 0.1 μm, are from Nucleopore (Pleasanton, CA).

A mini-extruder for preparation of unilamelar liposomes is from Avanti Polar Lipids.

11.1.1. Protocol

1. Dissolve 1 g of soybean total extract (20% lecithin) in 3 mL of chloroform.
2. Flush the lipid solution with a stream of N_2 gas in order to remove the bulk of the organic solvent. Dry the remaining lipid further under vacuum for 30 min.

3. Resuspend the dried lipid in 67 mL of lipid rehydration buffer (final concentration of 15 mg/mL). Vortex the lipid solution until homogeneous. Hydration of the lipids can be aided by 3–5 freeze/thaw cycles.
4. Form unilamelar liposomes by extrusion through a mini-extruder. Pass the liposome solution 11 times through a 0.4-μm track-etch polycarbonate membrane and then 11 times through a 0.1-μm membrane.
5. Aliquot the liposomes and flash-freeze them. Liposomes prepared in this way can be stored at -80 °C.

12. Wheat Germ Translation Reaction

Figure 37.3 compares the set up of the bilayer reaction with that of a dialysis reaction. The bilayer reaction separates the extract and mRNA from other reagents by the difference in density of the extract and upper buffer solutions. In this case, diffusion of substrates and products occurs through the entire buffer interface. Because of this simplicity in set up, the bilayer cell-free translation reaction is amenable to automation (Sawasaki *et al.*, 2002a; Tyler *et al.*, 2005; Vinarov *et al.*, 2004, 2006). In our hands, the bilayer reaction has yielded $\sim$ 0.2 mg/mL of various membrane proteins. However, as diffusion dilutes the reaction, the yield is limited.

The dialysis reaction is another method for performing cell-free translation (Fig. 37.3). A 12-kDa cutoff membrane at the bottom of the reactant cup retains the concentrated wheat germ extract, mRNA, and expressed protein. Diffusion replenishes ATP, AAs, cofactors, and other additives required for continued translation and removes inhibitory products. The dialysis method of cell-free translation can yield a 5- to 10-fold higher level of expression than

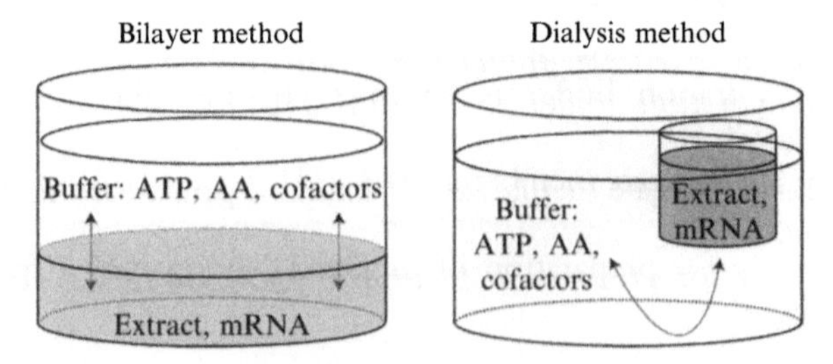

Figure 37.3 Schematic representation of two methods of cell-free translation. In the bilayer method, the extract and mRNA are separated from ATP, amino acids (AA), cofactors, and other buffer additives by the difference in density of the two solutions (also see Fig. 37.1). In the dialysis method, the extract and mRNA are contained within a dialysis membrane. In both cases, diffusion gives transfer of substrates, products, and additives between the extract and buffer.

the bilayer method because the constituents of the extract are not diluted by diffusion, which occurs during the course of the translation in the bilayer reaction. Thus the dialysis method can yield purified membrane proteins in the range of 0.2 mg/mL to greater than 2 mg/mL in the standard reactions (Goren and Fox, 2008). Although not as easily adaptable to robotics, the dialysis reaction can be performed on the benchtop and is scalable from 50 μL to 10 mL or greater with no overall changes in volumetric productivity. In the smaller volumes, this method has utility for simple screening of a few proteins, while in the larger volumes it also has utility for scale-up of the production of proteins whose properties have been investigated at small scale and found to be favorable for more extensive studies. Protocols for cell-free translation in automated batch, bilayer, and dialysis reactions have been published (Endo and Sawasaki, 2006; Sawasaki *et al.*, 2002a; Vinarov *et al.*, 2006). For translation of integral membrane proteins, we have found that translation reactions are productively modified by the addition of unilamellar liposomes (Goren and Fox, 2008; Nozawa *et al.*, 2007).

12.1. Materials and reagents

WEPRO2240 wheat germ extract is from CellFree Sciences, Ltd. (Yokohama, Japan). This preparation has ~ 240 OD_{260}/mL and does not contain AAs. Store the extract at −80 °C, thaw on the bench, and flash-freeze the unused lysate before storing. For addition to translation reactions, the extract is diluted from the concentrated commercial preparation to a final OD_{260} of 60 with 1× Reaction Buffer.

Unlabeled AAs are from Advanced ChemTech (Louisville, KY). A mixture of the 20 unlabeled AAs is prepared in Milli-Q water with each AA present at 2 m*M*. Do not filter these preparations because some AAs will not dissolve in these solutions.

5× Reaction Buffer is prepared from 150 m*M* HEPES–KOH, pH 7.8, and contains 500 m*M* potassium acetate, 12.5 m*M* magnesium acetate, 2 m*M* spermidine hydrochloride, 20 m*M* DTT, 6 m*M* ATP, 1.25 m*M* GTP, 80 m*M* creatine phosphate, and 0.025% (w/v) sodium azide. Store this buffer at −80 °C.

1× Overlay Buffer is prepared by fivefold dilution of 5× Reaction Buffer. The overlay buffer is then amended with AAs solution to give a final concentration of 0.3 m*M* for each AA.

Creatine kinase is from Roche Applied Sciences (Indianapolis, IN). Dissolve in Milli-Q water to make a 50 mg/mL solution and store at −80 °C. Dilute the stock solution to 1 mg/mL prior to use. Avoid multiple cycles of freezing and thawing of the concentrated solution. Discard the diluted solution.

12 kDa molecular weight cutoff (MWCO) dialysis cups are from Cosmo Bio (Tokyo, Japan). Prior to use, check the integrity of the membrane by adding 500 μL of Milli-Q water and monitor for any leakage. If there is no leakage, shake out the water prior to use.

The purified mRNA preparation is from Preparation of mRNA, step 4.

The liposome solution from Preparation of Liposomes, step 6.

24-Deep well pyramid-bottom plate, maximum volume of 10 mL (Artic-white, Bethlehem, PA).

96-Well U-Bottom Plate (Greiner Bio-One, Monroe, NC).

12.1.1. Protocol for a bilayer reaction

1. Prepare a 20 μL translation mixture by mixing 2 μL of a 15 mg/mL liposome solution, 4.25 μL of Milli-Q water, 2.75 μL of 5× Reaction Buffer, 3.75 μL of 2 m*M* unlabeled AAs, 1 μL of 1 mg/mL creatine kinase, and 6.25 μL of WEPRO2240 wheat germ extract. Depending on the number of separate reactions desired, scale these volumes and include ~ 10% extra volume to account for handling losses.
2. Transfer 5 μL of an mRNA preparation to an individual well of a 96-well U-bottom plate.
3. Add 20 μL of translation mixture to the mRNA sample in each well, and mix.
4. Form a bilayer by carefully adding 125 μL of 1× Overlay Buffer. Be careful not to disrupt the layers as that can dilute the extract and reduce protein production.
5. Incubate the reaction for 20 h at 26 °C. Do not disturb the bilayer during this time.
6. Protein translation levels can be determined by denaturing electrophoresis SDS–PAGE with creatine kinase serving as an internal intensity standard.

12.1.2. Protocol for a dialysis reaction

1. Dissolve the purified mRNA pellet in 50 μL of the translation mixture prepared as described in Protocol for Bilayer Reaction, step 1.
2. Place the translation mixture into a 12-kDa MWCO dialysis cup.
3. Prepare the reservoir dialysis buffer by mixing 6.5 mL of Milli-Q water, 2.0 mL of 5× Reaction Buffer, and 1.5 mL of 2 m*M* unlabeled AAs. Sonicate the mixture for 5 min, and then pass the solution through a 0.2-μm filter. Add 2.5 mL of reservoir dialysis buffer to each well of the 24-deep well plate.
4. Suspend the dialysis cup in a reservoir dialysis buffer. Be careful not to trap air bubbles underneath the dialysis cup, which will disrupt replenishment of additives.

5. Cover the 24-deep well plate with Saran Wrap to prevent evaporation of the reservoir dialysis buffer. Incubate the translation reaction at 26 °C for 16 h.
6. Protein levels are determined by denaturing electrophoresis with creatine kinase serving as an internal intensity standard.

13. Purification by Density Gradient Ultracentrifugation

Figure 37.4 shows a schematic representation of the purification of proteoliposomes containing membrane proteins produced by wheat germ cell-free translation. After assembly of the density gradient, the proteoliposomes are separated from the cell-free extract proteins by a centrifugation step. In most cases, the proteoliposomes are sharply concentrated at the interface between the 30% Accudenz solution and the upper buffer.

Figure 37.5 shows a denaturing PAGE analysis of the separation of proteoliposomes obtained by cotranslation of human stearoyl-CoA desaturase (hSCD1) and cytochrome b_5 (cytb5) in the wheat germ extract. hSCD1 accounted for ~ 4% of the total protein present after translation in the presence of liposomes. The density gradient purification yielded 24 μg of hSCD1 from a 50-μL translation reaction with greater than 80% purity. The density gradient purification also gave a 25-fold increase in the specific activity of the enzyme, and provided near complete recovery of the enzyme activity from the extract.

13.1. Materials and reagents

Accudenz is from Accurate Chemical and Scientific (Westbury, NY). Prepare 80% (w/v) and 30% (w/v) solutions in 25 mM HEPES, pH 7.5, containing 100 mM NaCl and 10% (w/v) glycerol. Store these solutions at room temperature.

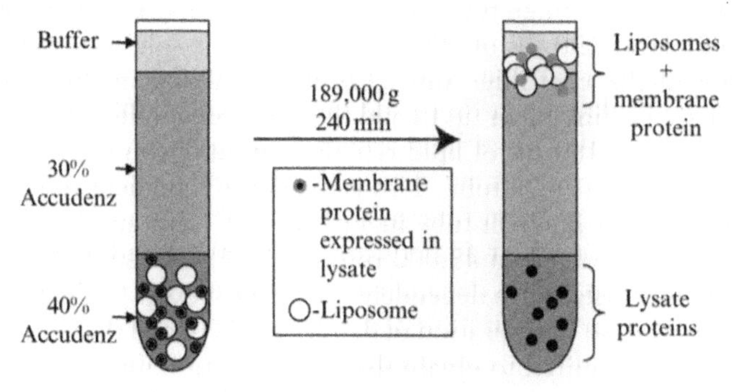

Figure 37.4 Schematic representation of the purification of proteoliposomes containing membrane proteins produced by cell-free translation.

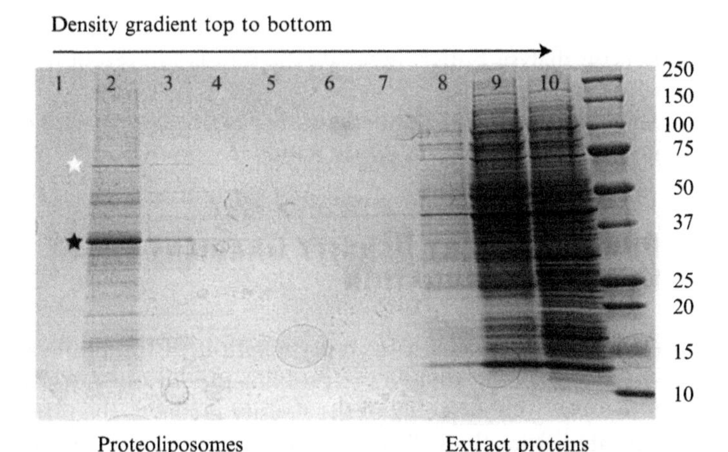

Figure 37.5 Denaturing PAGE analysis of the density gradient obtained after cell-free translation of hSCD1. The order of fractions from the density gradient is indicated. hSCD1 (black star) and a plant Hsp70-like protein (white star) are noted in lane 2, which represents the interface between the upper buffer (lane 1) and the remainder of the 30% Accudenz layer. The majority of wheat germ extract proteins remain in the 40% Accudenz layer. The rightmost lane contains molecular weight markers (kDa, noted alongside image). Adapted from Goren and Fox (2008).

Ultraclear centrifuge tubes (5 mm id × 41 mm h) are from Beckman Coulter (Fullerton, CA).

The lipid rehydration buffer is 25 m*M* HEPES, pH 7.5, containing 100 m*M* NaCl.

13.1.1. Protocol

1. Carefully mix 75 μL of 80% (w/v) Accudenz solution with up to 75 μL of the translation reaction, and place the mixed sample in the bottom of an ultraclear centrifuge tube. This creates the 40% Accudenz layer.
2. Carefully layer 350 μL of 30% (w/v) Accudenz solution on top of the mixture in the centrifuge tube. Minimize mixing of the gradient by using a gel loading pipet tip to add the successive buffer layers.
3. Carefully layer 100 μL of lipid rehydration buffer on top of two other layers in the centrifuge tube. This is the density gradient tube.
4. Spin the density gradient tube in an SW 50.1 rotor and L-60 ultracentrifuge for at least 4 h at 45,000 rpm (189,000×*g*) and 4 °C. The time required for floatation is dependent on the properties of the proteoliposomes created, so optimization of density gradient and the centrifugation time may be required to obtain the best separation for other proteins.
5. Carefully remove 60 μL fractions from the top to the bottom of the gradient in order to fractionate the density gradient. Store the individual

fractions in separate 1.5-mL microfuge tubes. Typically, proteoliposomes will migrate to the interface between the 30% (w/v) Accudenz solution and the liposome rehydration buffer.

6. Use denaturing SDS–PAGE to analyze the individual fractions for protein content.

14. Characterization of Proteoliposomes

Another chapter has elegantly described the process of transfer of solubilized membrane proteins from detergent into liposomes for functional assays (Rigaud and Levy, 2003). In this chapter, the liposome is an integral part of the enzyme production and purification process, so functional assays are immediately feasible without further manipulations of the preparation. The facile recovery of proteoliposomes from the cell-free translation reaction by density gradient centrifugation permits functional assays on highly enriched preparations with a minimal investment of time and effort. This provides some obvious advantages in the discovery and characterization of the membrane proteome (Sawasaki *et al.*, 2002b; Wu *et al.*, 2003), a large and still poorly understood fraction of all proteins.

Figure 37.6 shows results of assays for conversion of ^{14}C-labeled stearoyl-CoA (18:0) to oleoyl-CoA (18:1) obtained from different combinations of proteoliposomes of hSCD1 and cyt5b produced by cotranslation or independently with wheat germ extract (Goren and Fox, 2008). There is no desaturase activity in the cell-free extract only control (*lane* 9). In order to obtain hSCD activity, both cytb5 and Fe^{2+} must be added (*lanes* 1–3 versus *lanes* 4–8). Furthermore, iron and hemin must be optimized to obtain maximal activity (*lanes* 4–8). Furthermore, in this case, cotranslation of hSCD1 and cytb5 versus separate translation and subsequent mixing of the proteoliposomes gave equivalent catalytic activity (*lane* 6 versus *lane* 8).

In many cases, the cell-free extract will not exhibit competing enzymatic reactions that are present in whole cells, microsomes, and other impure natural membrane preparations. This possibility can be experimentally verified by use of the cell-free extract as a control. If verified, the low background reactivity of the cell-free extract can have important implications for discovery of the function of unknown membrane proteins and complexes. Moreover, the absence of competing background reactions may facilitate additional characterizations using site-directed mutagenesis, inhibitors, and other catalytic methods that have been more routinely carried out with purified soluble enzymes.

The availability of functional proteoliposomes also facilitates investigations of the topology of cell-free translated membrane proteins by the combination of proteolytic digestion and mass spectrometry (Speers and

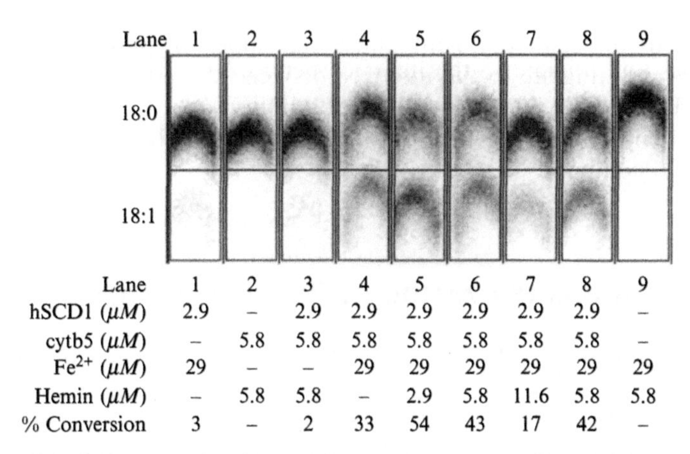

Lane	1	2	3	4	5	6	7	8	9
hSCD1 (μM)	2.9	–	2.9	2.9	2.9	2.9	2.9	2.9	–
cytb5 (μM)	–	5.8	5.8	5.8	5.8	5.8	5.8	5.8	–
Fe^{2+} (μM)	29	–	–	29	29	29	29	29	29
Hemin (μM)	–	5.8	5.8	–	2.9	5.8	11.6	5.8	5.8
% Conversion	3	–	2	33	54	43	17	42	–

Figure 37.6 Catalytic activity observed from various reconstitutions of the complex of human stearoyl-CoA desaturase and human cytochrome b5 produced by cell-free translation and purified by density gradient centrifugation. The soluble domain of human cytochrome b5 reductase was expressed in *E. coli* and added to these assays. The influence of adding Fe^{2+} (required for hSCD1 activity) and hemin (required for cytb5 activity) is also demonstrated. Lane 1, hSCD1 plus iron. Lane 2, cytb5 plus hemin. Lane 3, cotranslation of hSCD1 and cytb5 supplemented with hemin. Lane 4, cotranslation of hSCD1 and cytb5 supplemented with Fe^{2+}. Lane 5, cotranslation of hSCD1 and cytb5 supplemented with Fe^{2+} and 0.5 equiv. of hemin. Lane 6, cotranslation of hSCD1 and cytb5 supplemented with Fe^{2+} and 1 equiv. of hemin. Lane 7, cotranslation of hSCD1 and cytb5 supplemented with Fe^{2+} and 2 equiv. of hemin. Lane 8, mixture of separately translated and purified hSCD1 with separately translated and purified cytb5 supplemented with Fe^{2+} and 1 equiv. of hemin. Lane 5, cotranslation of hSCD1 and cytb5 supplemented with Fe^{2+} and 0.5 equiv. of hemin. Adapted from Goren and Fox (2008).

Wu, 2007; Wu *et al.*, 2003). In some cases, the ability to perform protein synthesis using radiolabeled AAs, unnatural AAs, or to specifically target one or more AAs with diagnostic mass tags can advance these studies.

Further purification likely requires that the proteoliposomes be dissolved with detergents, much as would be required for purification of a membrane protein starting with a microsomal preparation from a living tissue. However, the combination of cell-free translation, liposomes, and density gradient centrifugation allows these additional purification efforts to start at a much higher level of purity. Monitoring the decrease in light scattering from the proteoliposome as detergent is added provides a simple diagnostic for detergent optimization (Seddon *et al.*, 2004; Womack *et al.*, 1983). New approaches using NMR, analytical ultracentrifugation, and gel filtration can provide insight into the ability of a given detergent to yield monodisperse protein-detergent micelles (Maslennikov *et al.*, 2007; Slotboom *et al.*, 2008), which is generally considered to be an indicator of the compatibility of protein and detergent. It is also possible to perform cell-free translation of membrane proteins in the presence of detergents (Klammt *et al.*, 2007;

Nozawa *et al.*, 2007), with the constraints that the detergents used are compatible with the cell-free translation reaction and with solubilization of the nascent membrane protein.

15. Considerations for Scale-Up

With demonstration of suitable behavior using the bilayer or small-volume dialysis reactions, scale-up of protein production is most effectively carried out by use of the dialysis approach. Methods for performing the dialysis reaction in large scale have been published (Madin *et al.*, 2000; Spirin and Swartz, 2008). More recently, a robotic dialysis device has become available to address the need to produce large quantities of proteins using cell-free translation. Wheat germ cell-free translation has been used to solve the NMR structures of numerous soluble proteins (Phillips *et al.*, 2007; Tyler *et al.*, 2005), and more recently an X-ray structure has been solved (Makino *et al.*, 2009). Correspondingly, *E. coli* cell-free translation has been effectively used to produce membrane proteins for refolding and, in the case of EmrE, preparation of a SeMet-labeled sample for structure determination (Klammt *et al.*, 2007; Chen *et al.*, 2007). To prepare sufficient mRNA for a large-scale protein production, carry out a transcription reaction in a 50 mL conical tube with a total volume of 4 mL of 1× Transcription Buffer plus Mg containing 4 m*M* NTPs, 0.05 mg/mL of plasmid DNA, 0.5 unit/μL of SP6 RNA polymerase, and 0.25 unit/μL of RNasin. Incubate the reaction at 37 °C for 3–5 h.

16. Isotopic Labeling for Structural Studies

Because AAs are added to the wheat germ extract as substrates for the protein synthesis, SeMet, ^{2}H-, ^{13}C-, ^{15}N- or other isotopically labeled AAs can be easily substituted. There is no significant metabolism of AAs in wheat germ extract except for alanine transaminase (beta-chloro-L-alanine), aspartate transaminase (aminooxyacetate), and glutamine synthetase (L-methionine sulfoximine), which can be inhibited by the compounds indicated (Endo and Sawasaki, 2006; Morita *et al.*, 2004). Otherwise, the level of isotopic enrichment in translated proteins consistently matches that of the AA precursors used in the translation reactions. ^{15}N and ^{13}C, ^{15}N-labeled AAs are from Cambridge Isotope Laboratories (Andover, MA) or other sources. For NMR studies, the ^{15}N-labeled AAs are prepared at 8 m*M*, and the ^{13}C, ^{15}N-labeled AAs are prepared at 5 m*M*, also in Milli-Q water. These preparations are substituted into the standard protocols to give the same final concentrations as described above. Likewise SeMet (Acros Organics, Morris Plains, NJ) is prepared as a 2-m*M* stock solution and substituted for methionine as described earlier.

17. Conclusions

The process simplifications afforded by the cell-free translation approaches described herein provide significant labor and time advantages, and this conclusion is emphasized when cell-free translation can produce integral membrane proteins and enzymes that cannot be reasonably obtained in a catalytically active state by any other method. By comparison, other methods for the preparation of membrane proteins beginning with expression in living hosts are time-, labor-, and material-intensive. Through application of these methods, uniform samples of membrane proteins can be easily obtained as the starting point for optimization of catalytic assays, purification procedures, antibody production, structure determination, and many other types of studies. This work provides an attractive alternative to the sequence of expressing in a living host, extracting membrane proteins with detergents, and then transferring them back into liposomes or other lipid bilayers for functional studies. It is reasonable to assume that continued application of this approach could open many new avenues to investigations of membrane protein structure and function.

Acknowledgments

This work was supported by NIGMS grant GM50853 to BGF, NIGMS Protein Structure Initiative grant 1U54 GM074901 (J. L. Markley, PI, G. N. Phillips, Jr., and B. G. Fox, Co-Investigators) and NSF East Asia and Pacific Summer Institutes for U.S. Graduate Students to MAG.

References

Abe, M., Hori, H., Nakanishi, T., Arisaka, F., Ogasawara, T., Sawasaki, T., Kitamura, M., and Endo, Y. (2004). Application of cell-free translation systems to studies of cofactor binding proteins. *Nucleic Acids Symp. Ser. (Oxf.)* **48,** 143–144.

Blommel, P. G., and Fox, B. G. (2007). A combined approach to improving large-scale production of tobacco etch virus protease. *Protein Expr. Purif.* **55,** 53–68.

Blommel, P. G., Martin, P. A., Wrobel, R. L., Steffen, E., and Fox, B. G. (2006). High efficiency single step production of expression plasmids from cDNA clones using the Flexi Vector cloning system. *Protein Expr. Purif.* **47,** 562–570.

Boyer, M. E., Stapleton, J. A., Kuchenreuther, J. M., Wang, C. W., and Swartz, J. R. (2008). Cell-free synthesis and maturation of [FeFe] hydrogenases. *Biotechnol. Bioeng.* **99,** 59–67.

CellFree Sciences, Ltd. Yokohama, Japan. http://www.cfsciences.com.

Chen, Y. J., Pornillos, O., Lieu, S., Ma, C., Chen, A. P., and Chang, G. (2007). X-ray structure of EmrE supports dual topology model. *Proc. Natl. Acad. Sci. USA* **104,** 18999–19004.

Drew, D. E., von Heijne, G., Nordlund, P., and de Gier, J. W. (2001). Green fluorescent protein as an indicator to monitor membrane protein overexpression in *Escherichia coli*. *FEBS Lett.* **507,** 220–224.

Endo, Y., and Sawasaki, T. (2006). Cell-free expression systems for eukaryotic protein production. *Curr. Opin. Biotechnol.* **17,** 373–380.

Goerke, A. R., and Swartz, J. R. (2008). Development of cell-free protein synthesis platforms for disulfide bonded proteins. *Biotechnol. Bioeng.* **99,** 351–367.

Goren, M. A., and Fox, B. G. (2008). Wheat germ cell-free translation, purification, and assembly of a functional human stearoyl-CoA desaturase complex. *Protein Expr. Purif.* **62,** 171–178.

Hall, J. F., Ellis, M. J., Kigawa, T., Yabuki, T., Matsuda, T., Seki, E., Hasnain, S. S., and Yokoyama, S. (2005). Towards the high-throughput expression of metalloproteins from the *Mycobacterium tuberculosis* genome. *J. Synchrotron Radiat.* **12,** 4–7.

Kanno, T., Kitano, M., Kato, R., Omori, A., Endo, Y., and Tozawa, Y. (2007). Sequence specificity and efficiency of protein N-terminal methionine elimination in wheat-embryo cell-free system. *Protein Expr. Purif.* **52,** 59–65.

Kiga, D., Sakamoto, K., Kodama, K., Kigawa, T., Matsuda, T., Yabuki, T., Shirouzu, M., Harada, Y., Nakayama, H., Takio, K., Hasegawa, Y., Endo, Y., *et al.* (2002). An engineered *Escherichia coli* tyrosyl-tRNA synthetase for site-specific incorporation of an unnatural amino acid into proteins in eukaryotic translation and its application in a wheat germ cell-free system. *Proc. Natl. Acad. Sci. USA* **99,** 9715–9720.

Kigawa, T., Yabuki, T., Matsuda, N., Matsuda, T., Nakajima, R., Tanaka, A., and Yokoyama, S. (2004). Preparation of *Escherichia coli* cell extract for highly productive cell-free protein expression. *J. Struct. Funct. Genomics* **5,** 63–68.

Klammt, C., Lohr, F., Schafer, B., Haase, W., Dotsch, V., Ruterjans, H., Glaubitz, C., and Bernhard, F. (2004). High level cell-free expression and specific labeling of integral membrane proteins. *Eur. J. Biochem.* **271,** 568–580.

Klammt, C., Schwarz, D., Lohr, F., Schneider, B., Dotsch, V., and Bernhard, F. (2006). Cell-free expression as an emerging technique for the large scale production of integral membrane protein. *FEBS J.* **273,** 4141–4153.

Klammt, C., Schwarz, D., Dotsch, V., and Bernhard, F. (2007). Cell-free production of integral membrane proteins on a preparative scale. *Methods Mol. Biol.* **375,** 57–78.

Madin, K., Sawasaki, T., Ogasawara, T., and Endo, Y. (2000). A highly efficient and robust cell-free protein synthesis system prepared from wheat embryos: Plants apparently contain a suicide system directed at ribosomes. *Proc. Natl. Acad. Sci. USA* **97,** 559–564.

Makino, S.-I., Bingman, C. A., Berge, S., Larkin, A., Fox, B. G., Phillips, G. N. Jr, and Markley, J. L. (2009). *Crystal structure of agmatine iminohydrolase produced by wheat germ cell-free translation* (in preparation).

Maslennikov, I., Kefala, G., Johnson, C., Riek, R., Choe, S., and Kwiatkowski, W. (2007). NMR spectroscopic and analytical ultracentrifuge analysis of membrane protein detergent complexes. *BMC Struct. Biol.* **7,** 74.

Matsuda, T., Koshiba, S., Tochio, N., Seki, E., Iwasaki, N., Yabuki, T., Inoue, M., Yokoyama, S., and Kigawa, T. (2007). Improving cell-free protein synthesis for stable-isotope labeling. *J. Biomol. NMR* **37,** 225–229.

Matsumoto, K., Tomikawa, C., Toyooka, T., Ochi, A., Takano, Y., Takayanagi, N., Abe, M., Endo, Y., and Hori, H. (2008). Production of yeast tRNA (m(7)G46) methyltransferase (Trm8-Trm82 complex) in a wheat germ cell-free translation system. *J. Biotechnol.* **133,** 453–460.

Morita, E. H., Shimizu, M., Ogasawara, T., Endo, Y., Tanaka, R., and Kohno, T. (2004). A novel way of amino acid-specific assignment in ^{1}H–^{15}N HSQC spectra with a wheat germ cell-free protein synthesis system. *J. Biomol. NMR* **30,** 37–45.

Nomura, S. M., Kondoh, S., Asayama, W., Asada, A., Nishikawa, S., and Akiyoshi, K. (2008). Direct preparation of giant proteo-liposomes by *in vitro* membrane protein synthesis. *J. Biotechnol.* **133,** 190–195.

Nosjean, O., and Roux, B. (2003). Anchoring of glycosylphosphatidylinositol-proteins to liposomes. *Methods Enzymol.* **372,** 216–232.

Nozawa, A., Nanamiya, H., Miyata, T., Linka, N., Endo, Y., Weber, A. P., and Tozawa, Y. (2007). A cell-free translation and proteoliposome reconstitution system for functional analysis of plant solute transporters. *Plant Cell Physiol.* **48,** 1815–1820.

Phillips, G. N. Jr., Fox, B. G., Markley, J. L., Volkman, B. F., Bae, E., Bitto, E., Bingman, C. A., Frederick, R. O., McCoy, J. G., Lytle, B. L., Pierce, B. S., Song, J., *et al.* (2007). Structures of proteins of biomedical interest from the Center for Eukaryotic Structural Genomics. *J. Struct. Funct. Genomics* **8,** 73–84.

Rigaud, J. L., and Levy, D. (2003). Reconstitution of membrane proteins into liposomes. *Methods Enzymol.* **372,** 65–86.

Sawasaki, T., Hasegawa, Y., Tsuchimochi, M., Kasahara, Y., and Endo, Y. (2000). Construction of an efficient expression vector for coupled transcription/translation in a wheat germ cell-free system. *Nucleic Acids Symp. Ser.* **44,** 9–10.

Sawasaki, T., Hasegawa, Y., Tsuchimochi, M., Kamura, N., Ogasawara, T., Kuroita, T., and Endo, Y. (2002a). A bilayer cell-free protein synthesis system for high-throughput screening of gene products. *FEBS Lett.* **514,** 102–105.

Sawasaki, T., Ogasawara, T., Morishita, R., and Endo, Y. (2002b). A cell-free protein synthesis system for high-throughput proteomics. *Proc. Natl. Acad. Sci. USA* **99,** 14652–14657.

Schwarz, D., Klammt, C., Koglin, A., Lohr, F., Schneider, B., Dotsch, V., and Bernhard, F. (2007). Preparative scale cell-free expression systems: New tools for the large scale preparation of integral membrane proteins for functional and structural studies. *Methods* **41,** 355–369.

Seddon, A. M., Curnow, P., and Booth, P. J. (2004). Membrane proteins, lipids and detergents: Not just a soap opera. *Biochim. Biophys. Acta* **1666,** 105–117.

Slotboom, D. J., Duurkens, R. H., Olieman, K., and Erkens, G. B. (2008). Static light scattering to characterize membrane proteins in detergent solution. *Methods* **46,** 73–82.

Sobrado, P., Goren, M. A., James, D., Amundson, C. K., and Fox, B. G. (2008). A protein structure initiative approach to expression, purification, and *in situ* delivery of human cytochrome b5 to membrane vesicles. *Protein Expr. Purif.* **58,** 229–241.

Speers, A. E., and Wu, C. C. (2007). Proteomics of integral membrane proteins—Theory and application. *Chem. Rev.* **107,** 3687–3714.

Spirin, A. S., and Swartz, J. R. (eds.) (2008). *In* "Cell-Free Protein Synthesis-Methods and Protocols", WILEY-VCH Verlag Gmbh & Co., Weinheim.

Thao, S., Zhao, Q., Kimball, T., Steffen, E., Blommel, P. G., Riters, M., Newman, C. S., Fox, B. G., and Wrobel, R. L. (2004). Results from high-throughput DNA cloning of *Arabidopsis thaliana* target genes using site-specific recombination. *J. Struct. Funct. Genomics* **5,** 267–276.

Tyler, R. C., Aceti, D. J., Bingman, C. A., Cornilescu, C. C., Fox, B. G., Frederick, R. O., Jeon, W. B., Lee, M. S., Newman, C. S., Peterson, F. C., Phillips, G. N. Jr., Shahan, M. N., *et al.* (2005). Comparison of cell-based and cell-free protocols for producing target proteins from the *Arabidopsis thaliana* genome for structural studies. *Proteins* **59,** 633–643.

Vinarov, D. A., and Markley, J. L. (2005). High-throughput automated platform for nuclear magnetic resonance-based structural proteomics. *Expert Rev. Proteomics* **2,** 49–55.

Vinarov, D. A., Lytle, B. L., Peterson, F. C., Tyler, E. M., Volkman, B. F., and Markley, J. L. (2004). Cell-free protein production and labeling protocol for NMR-based structural proteomics. *Nat. Methods* **1,** 149–153.

Vinarov, D. A., Loushin Newman, C. L., and Markley, J. L. (2006). Wheat germ cell-free platform for eukaryotic protein production. *FEBS J.* **273,** 4160–4169.

Womack, M. D., Kendall, D. A., and Macdonald, R. C. (1983). Detergent effects on enzyme activity and solubilization of lipid bilayer membranes. *Biochim. Biophys. Acta* **733,** 210–215.
Wu, C. C., MacCoss, M. J., Howell, K. E., and Yates, J. R. 3rd. (2003). A method for the comprehensive proteomic analysis of membrane proteins. *Nat. Biotechnol.* **21,** 532–538.
Wuu, J. J., and Swartz, J. R. (2008). High yield cell-free production of integral membrane proteins without refolding or detergents. *Biochim. Biophys. Acta* **1778,** 1237–1250.
Yokoyama, S. (2003). Protein expression systems for structural genomics and proteomics. *Curr. Opin. Chem. Biol.* **7,** 39–43.

SECTION TEN

CHARACTERIZATION OF PURIFIED PROTEINS

CHAPTER THIRTY-EIGHT

Determination of Protein Purity

David G. Rhodes* *and* Thomas M. Laue†

Contents

1. Composition-Based and Activity-Based Analyses 680
   1.1. Problems and pitfalls 681
2. Electrophoretic Methods 681
   2.1. Methods 682
   2.2. Pitfalls and problems 684
3. Chromatographic Methods 685
   3.1. Gel filtration chromatography 685
   3.2. Reversed phase HPLC 686
4. Sedimentation Velocity Methods 687
   4.1. Method 687
5. Mass Spectrometry Methods 687
6. Light Scattering Methods 688
References 689

Abstract

There is no means for directly quantifying the purity of a protein sample. Demonstration of purity of a protein preparation always involves an assessment of the quantity of particular types of impurities, or simply demonstrating their absence. Whether the goal is interpretation of analytical data, demonstration of the quality of a process, or assuring safety of a biopharmaceutical product, determination of sample purity is of central importance. This chapter will focus primarily on methods for detecting protein impurities in protein samples and will describe methods, their capabilities, and their limitations.

To assess the purity of a sample, one must first identify the type of impurity that is to be measured (i.e., nucleic acid, carbohydrate, lipid, unrelated proteins, isozymes, inactive enzyme), and then identify a characteristic property (chemical assay or physical feature) which can distinguish the putative contaminant(s) in the presence of the protein of interest in a specified solution condition. Purity is then the demonstration that the

* Lipophilia Consulting, Storrs, Connecticut, USA
† Department of Biochemistry and Molecular Biology, University of New Hampshire, Durham, New Hampshire, USA

Methods in Enzymology, Volume 463
ISSN 0076-6879, DOI: 10.1016/S0076-6879(09)63038-X

contaminant in question is below a specified level. Note that this specification need not characterize the contaminant. A chromatographic peak may remain an unknown even after a purification procedure has reduced its concentration to levels below detection limits. Clearly, the apparent level of purity will depend on which assay methods are chosen and their sensitivity. Since most isolation procedures are quite good at removing nonprotein contaminants, this chapter will focus primarily on methods for detecting contamination of a protein sample by other proteins. Selective assays for the presence of nucleic acids, lipids, and carbohydrates can be found in other volumes of this series.

Purity of proteins has become a significant issue in regulation of pharmaceutical products as protein-based products become more prevalent. The International Conference on Harmonisation (ICH) quality guidelines for biological products (ICH Q6B, 1998) consider impurities in drug substance (bulk material) and drug product (dosage form or finished product) and recognize the inherent heterogeneity in biological molecules like proteins. The guidelines state: "In addition to evaluating the purity of the drug substance and drug product, which may be composed of the desired product and multiple product-related substances, the manufacturer should also assess impurities which may be present. Impurities may be either process or product-related. They can be of known structure, partially characterized, or unidentified." The guidelines go on to distinguish process-related impurities (derived from the manufacturing process, and including host cell proteins, host cell DNA, cell culture inducers, antibiotics, media components, or the effects of downstream processing) and product-related impurities, including "molecular variants arising during manufacture and/or storage, which do not have properties comparable to those of the desired product with respect to activity, efficacy, and safety." In the guidelines, there is recognition of the analytical capabilities described above. For example, with regard to drug substance, "The absolute purity of biotechnological and biological products is difficult to determine and the results are method-dependent.... Consequently, the purity of the drug substance is usually estimated by a combination of methods. The choice and optimization of analytical procedures should focus on the separation of the desired product from product-related substances and from impurities." The discussion for drug product is similar, but also includes testing for degradation products and product-related impurities due to interaction with excipients.

There are several sensitive methods that may be used to detect the presence of impurities in a sample (Table 38.1). Each method tests for a specific physical property of the molecules. The method of choice depends on the following criteria: (1) the quantity of protein available, (2) the nature of the impurity being evaluated, (3) the accuracy of the estimate needed, (4) the sensitivity of the test needed, and (5) any peculiarities of

Table 38.1 Methods for detecting impurities

Method	Fractionation[a]	Sample recovery[b]	Quantity	Ease[c]	Expense[d]	Property assessed
Electrophoresis						
Denaturing gel	+	±	ng–μg	+	+	Chain length
Native gel	+	+	ng–μg	+	+	Charge/size
Isoelectric focusing	+	+	ng–μg	+	±	Isoelectric pH
Chromatography						
Gel permeation	+	+	μg	+	+	Size, r_g
Ion exchange	+	+	μg	+	+	Net charge
Affinity	+	+	μg	+	±	Specific binding
Reverse phase	+	+	μg	+	±	Polarity
Sedimentation						
Velocity	+	+	μg	+	±	Mass/size
Equilibrium	±	+	μg	±	–	Mass, association
Light scattering						
Static	–	+	μg	+	±	Mass/size
Dynamic	–	+	μg	+	±	Mass/size
Mass spectrometry	+	–	ng–μg	–	–	Mass/charge ratio
Composition	–	–	Variable	+	±	Component content
Activity	–	–	Variable	+	+	Presence of active site

[a] +, Sample fractionated during analysis; separated components may be isolated; ±, incomplete fractionation, no components isolable; –, no sample fractionation.
[b] +, Good recovery of sample following assay; ±, sample may be recovered in favorable cases; –, sample destroyed during analysis.
[c] +, Simple method; ±, somewhat more difficult; –, special equipment or expertise may be required.
[d] +, Equipment and supplies relatively inexpensive; ±, some equipment or supplies may be costly; –, expensive equipment or supplies required.

the protein or its solvent which might interfere with the use of the techniques. It is easiest and most common to carry out one or more fractionation procedures and demonstrate that only one component is detectable. A wide variety of fractionation procedures may be used to detect impurities, with the criterion for purity being the presence of a single detectable component. Orthogonal methods are advisable to guard against situations in which a chosen detection method is incapable of resolving an unknown impurity from the primary analyte. Finally, it is important to handle samples carefully, so that the process of preparing a sample for analysis does not alter the impurity profile. Cleanliness, proper temperature, and containers made of appropriate materials will all contribute to informative analysis.

Many appropriate methods are detailed elsewhere in this volume and will not be reviewed here. Several analytical methods for determining molecular weight or molecular size are outlined elsewhere in this volume, so where appropriate, this discussion will refer to details in that chapter (Rhodes *et al.*, 2009). In this chapter, emphasis will be on the variations of those methods which are better suited to detection of contaminant than to quantitative determination of size, mass, or other molecular parameters. This chapter will also outline criteria for selecting among these methods, some of their limitations and advantages, and some description of assays for purity that do not involve fractionation.

1. Composition-Based and Activity-Based Analyses

Methods that provide molar quantification of amino acids, specific prosthetic groups, or of active sites may be used to assess the purity of a sample. In cases where the specific activity of the pure material is known it may be appropriate to determine the unit activity as a measure of relative purity. These methods are secondary in nature, since reference is always made to the quantity of the analyte that would be expected for a sample that is free of contaminants. Moreover, these methods are appropriate primarily with heterogeneous systems in early phases of purification or molecules like membrane proteins which may require a specific environment to retain function. Typically, two measurements are required: the mass of the protein (or material) in the sample that will be used in the analysis, and quantification of the known activity or of the particular analyte. A calculation is then made of the expected quantity of analyte based on the mass of the material used in the analysis. The purity is then expressed as ratio of the amount measured to the amount expected. Good quantification has been achieved using end-group analysis (Chang, 1983), and quantitative analysis for specific prosthetic groups, and for enzymatic activity (Biggs, 1976). The details for this method are the same as those described in this volume Rhodes *et al.* (2009).

1.1. Problems and pitfalls

Composition-based or activity-based estimates are sensitive only to those impurities being assessed, and typically provide little specific information (e.g., size, charge, etc.) concerning the nature of the contaminants. Activity measurements may provide no information about the contaminant unless the contaminant interferes with the assay or the active protein. Thus, few clues are provided to help the experimenter in removing the contaminant.

2. Electrophoretic Methods

Because the electrophoretic methods provide some of the simplest, least expensive, and often most sensitive approaches to determine the number of protein components in a sample, they are the most commonly used methods. Because of the exceptionally low cost and simplicity, these methods are often the first screen for sample purity, even in early-stage, highly heterogeneous samples. The methods of SDS gel electrophoresis (see Friedman, 2009; Garfin, 2009) and electrophoretic determinations of molecular weight and size are described elsewhere in this volume (Rhodes *et al.*, 2009). Any of these approaches could be used independently to assess the purity of a sample, depending on which characteristic of the protein is to be tested (Table 38.1). If the expected contaminants differ in molecular weight from that of the desired protein, SDS gel electrophoresis would resolve the impurity. Molecules of similar molecular weight but different amino acid composition would generally not be distinguishable using SDS gel electrophoresis, but could have different electrophoretic mobilities in nondenaturing gel electrophoresis. Finally, proteins of almost any molecular weight might be separable using isoelectric focusing techniques. The latter method is detailed elsewhere in this volume (Friedman, 2009), and will not be treated here.

Depending on the type of detection method available, nanogram to microgram quantities of sample are required. Since a contaminant constitutes a fixed weight fraction of a given sample, while the detection of a contaminant depends on the total amount of contaminant, an overloaded gel may provide a better chance of detecting a contaminant. However, an upper limit to the quantity of sample that can be applied to a gel is set by sample solubility and by considerations of resolution (Lunney *et al.*, 1971). The latter limit is usually more restrictive, with band broadening and distortion occurring at higher sample concentrations and with large sample volumes. Bands broadened due to overloading could obscure an impurity with similar electrophoretic mobility. Unfortunately, the most sensitive band detection methods (requiring the smallest amount of sample) do not allow sample recovery. If the protein sample is denatured, or otherwise solubilized under harsh conditions, it is usually difficult to recover functional protein. However, if nondenaturing electrophoresis is used, it is possible to recover the sample from the gel by electrophoretic extraction (Friedman, 2009; Garfin, 2009). If denaturing electrophoresis is performed, it may not be possible to recover the native protein (but see Burgess, 2009). The success or failure of renaturation depends largely on the structure of the protein; a large and or multisubunit protein will be less likely to return to the native conformation than will a small monomeric protein.

2.1. Methods

The details of sample preparation depend on the nature of the electrophoretic method chosen. The reader is referred to the other chapters in this volume for discussion of native gel electrophoresis and isoelectric focusing gels. Most frequently, SDS gels provide the "front-line" method for assessing sample purity. Because purity assessment often is not quantitative, one need not (necessarily) be as concerned about such problems as nonlinearity in mobility at protein size extremes, aberrations in mobility due to protein modifications, or nonuniformities in SDS binding that are discussed in Rhodes *et al.* (2009). On the other hand, one does need to be concerned with consistency and uniformity in sample handling and preparation, and the conditions to which the sample is exposed during fractionation. For example, a homogeneous protein preparation in which the reduction of disulfide is not carried out carefully could appear to be heterogeneous. Likewise, misinterpretation is possible if nonuniformity in a gel cross linking results in mobility differences across the gel, but suitable replicates and controls should minimize such issues.

Because one does not normally know the sizes of all protein components within a sample, it is difficult to predict the gel concentration which will give the optimal fractionation of a set of protein components. Gradient gels, or gels of graded porosity, can cover a very wide range of molecular sizes. Although the conditions of a gradient gel are unlikely to be optimal for a particular fractionation, they will cover the widest possible range of conditions, thus maximizing the probability of identifying a contaminant. Ability to resolve a contaminant will depend largely on the range of gel concentrations chosen for the gradient. The gradient should be designed so that the protein of interest bands at an intermediate gel concentration. The concentration of acrylamide at the top of the gel (low concentration) should be sufficiently low that large molecular weight contaminants may enter the gel matrix. The bottom of the gel (high concentration) should be sufficiently concentrated that small protein is retained in the gel matrix. The lowest concentration commonly available in precast gradient gels is 4% acrylamide and the standard high concentration limit is 20%. If these limits are not adequate, lower concentrations (2%) or higher concentrations (as high as 40%) can be made. Proteins with molecular weights on the order of 1 MDa should be able to penetrate a 2% gel matrix. The high concentration for most gradient gels generally does not exceed 30% acrylamide. Very few polypeptides will penetrate a 30% gel, even after prolonged electrophoresis. The concentration range of a gradient gel determines the resolution of the gel. Thus, a gel with a gradient from 2% to 30% will be able to examine the widest range of molecular weights, but will not provide as much discrimination between two proteins of nearly identical molecular weights as a gel with an 8–16% gradient. Choice of gradient will depend largely on the researcher's knowledge about the system under investigation. If it is known, for example, that no very small or very large protein is present, one may wish to use a narrower range of concentrations in the gradient or a gel of constant concentration.

If a specific gradient is required, the procedure for making gradient gels differs only slightly from that for making conventional polyacrylamide gels or that for making sucrose gradients for sedimentation. The procedure is slightly more difficult than that for conventional polyacrylamide gels only because two gels at different acrylamide concentrations must be made up in parallel. The procedure is more difficult than forming sucrose gradients for sedimentation only because one must work within the time constraints of the acrylamide polymerization. Several commercial gradient-forming devices are available and many variations on the technique may be found by searching the literature.

Electrophoresis procedures for a gradient gel are identical to those used in the analytical methods with conventional gels (Rhodes *et al.*, 2009, and elsewhere in this volume). The staining strategy for analysis of the gel will be determined by the expected contaminant. The potential sensitivity of this approach and the information gained through its use can be enhanced by combining it with other techniques to generate a two-dimensional electrophoretic analysis. While the second dimension would ordinarily be a gradient SDS–polyacrylamide gel electrophoresis, the first dimension could be a nondenaturing gel electrophoresis, an isoelectric focusing gel, or a denaturing gel under different conditions (e.g., without disulfide reduction).

Isoelectric focusing is another technique which can successfully span a wide range of potential contaminants and, since it is orthogonal to electrophoresis, can be combined with these methods. Commercially available precast gels cover the range of p*I* from 3 to 10 in the same gel. The procedures for carrying out isoelectric focusing have been described elsewhere in this volume and will not be repeated here. The wide range of pI can be a considerable advantage, but the sensitivity of the technique to small differences in isoelectric point can be enhanced by covering a narrower range of pI values (e.g., 4–5, 5–8). Aside from the obvious capability of differentiating molecules which differ in pI, this method may be able to resolve contaminants which differ by properties other than isoelectric point.

2.2. Pitfalls and problems

Because gel electrophoresis is such a simple and inexpensive method, it is usually the method of first choice for purity assessment. There are, however, a few potential problems that must be borne in mind when using one of these techniques. For denaturing gels, both false-negative and false-positive artifacts are possible. False negatives can result if a comigrating contaminant is present or if a contaminant is unable to penetrate the gel. For this reason, it is important to stain the entire gel, stacking and running portions, and to examine both for protein-staining material. A band of material left behind in the loading well or at the interface between the stacking gel and the running gel is an indication of either a high-molecular-weight contaminant or a contaminant of limited solubility. Likewise, commonly used protein stains such as Coomassie Brilliant Blue will bind weakly to fibrous proteins or glycoproteins, resulting in an underestimate of these contaminants. False positives may result when covalent modifications are made during preparation of the sample for electrophoresis or when the gel is nonuniform or contains residual oxidant. Similar problems may arise in nondenaturing gels, with the additional problems caused by uncertainties as to the net charge on the protein of interest or the contaminants (see discussion in Rhodes *et al.*, 2009). If the net charge on the contaminants is zero or of opposite sign from that on the protein of interest, the contaminant will not appear in the gel unless it is a center-loaded horizontal gel. It is also helpful to run nondenaturing gels over as wide a range of pH as possible.

Artifacts associated with isoelectric focusing arise mostly from interactions between the proteins and the polyampholyte beads used to create the pH gradient. The results are bands (often smeared) appearing in regions of the gel that may have nothing whatsoever to do with the isoelectric point of the protein of interest. One way to test for this possibility is to isolate protein from one of these bands and to analyze the isolate using the same isoelectric focusing protocol. If the original pattern reflects actual heterogeneity in the sample, the isolate will form only its original spot. However, if the bands are artifacts of the method, the original pattern, including the artifact bands will be regenerated by the isolate.

3. Chromatographic Methods

3.1. Gel filtration chromatography

Gel filtration chromatography is one of the simplest methods for the detection of impurities that differ in size from the molecule of interest. The method is nondestructive and rapid. Samples are diluted as they pass through the column because this is a zonal method, so starting concentrations well above the minimum detectable limit must be used. The exact quantity of material needed will depend on the sensitivity of the assay used to detect for contaminants.

3.1.1. Method

The method outlined in this volume for sample and column preparation for the purposes of protein size determination should be used for assessing impurities (Rhodes *et al.*, 2009). The only difference is that the method of detection must be sensitive to contaminants as well as to the protein of interest. Although it is not necessary to calibrate the column to assess the purity of a sample, it is useful to do so for two reasons. First, by using a calibrated column, the experimenter gains some information concerning the size of the impurity. Second, by using a calibrated column, both the test for impurities and the molecular weight determination can be made simultaneously. To do this, one frequently uses two assays. The first is a nonspecific assay for protein (e.g., absorbance at 280 nm and/or 220 nm) and the second is a specific assay (enzymatic, immunological, mass-spectroscopy, multiangle light scattering, etc.) to confirm the identity of the analyte protein. Impurities are detected as either separate peaks in the chromatogram or as a broadening (e.g., a shoulder) of the elution profile. In principle, the elution profile should be nearly Gaussian (with a very slight skewing of the trailing edge, see Problems and Pitfalls). If the analyte protein peak is strongly skewed, if the peak has a shoulder, or if the specific and nonspecific assay results to not superimpose the sample has impurity present.

3.1.2. Problems and pitfalls

Gel-permeation chromatography is not as sensitive method as gel electrophoresis for the detection of size heterogeneity. Typically, a gel permeation experiment requires larger quantities of material than a gel electrophoresis experiment. Since gel-permeation chromatography usually is conducted under native conditions, it is possible that association behavior will mimic sample heterogeneity. A protein that exists in a variety of stable oligomers (e.g., fatty acid synthase) will appear to be heterogeneous by this method. Likewise, a protein that undergoes rapid, reversible self-association can provide anomolous, concentration-dependent elution profiles. In each case, it is possible to distinguish between heterogeneity and molecular association by further testing. For example, the gel permeation chromatography could be repeated with fractions from different portions of the skewed or broadened peak. Similarly, other methods described in this chapter or elsewhere in this volume could be applied to these fractions. In either case, the experimenter obtains useful information about the molecule of interest.

3.2. Reversed phase HPLC

Another commonly used chromatographic separation method is reversed phase HPLC. In this method, a nonpolar stationary phase such as modified silica is used with an elution gradient of decreasing polarity. Thus, for example, a protein in a buffer could be applied to a C-18 modified silica column equilibrated with aqueous 0.05% trifluoroacetic acid (TFA). The protein will generally bind to the stationary phase by way of interaction between the C-18 and hydrophobic amino acids on the protein. The protein would then be eluted using a gradient from 0.05% aqueous TFA to some level of compatible organic, such as acetonitrile, methanol, or ethanol, typically with the same quantity of TFA. The presence of the organic solvent in the mobile phase decreases the affinity of the protein for the stationary phase and the protein elutes. As with most gradient methods, a shallow gradient will provide optimal resolution, but a fast gradient is a good first step to identify approximate solvent conditions for elution of the primary analyte and to screen for possible impurities with very different affinities for the stationary phase. Detection of the protein is usually based on UV absorbance at 215–220 nm, which will identify peptides as opposed to measurements based on aromatic amino acids at 280 nm. Aside from setup time, a gradient can be completed in well under an hour and most optimized elutions run in 15–20 min.

The simplicity and rapidity of this method, the ready availability of suitable instrumentation, and the flexibility of the method in accommodating proteins with widely differing properties, make this a common method of choice for screening protein samples for purity. With a variety of mobile phases, a sample can be analyzed under a variety of gradients and conditions, if necessary, to confirm that the primary analyte is pure. The method can be combined with other methods such as MS (see below) to confirm the identity of the protein and to obtain information about possible impurities. It should be noted that because of the presence of organic solvent, the proteins that elute cannot be assumed to be native; some degree of denaturation is expected. Denaturation can be minimized by using a stationary phase with C-4 or C-8; because these will have lower affinity for the protein, it will elute at lower concentrations of organic mobile phase. Numerous protocols have been published for specific proteins and manufacturers of HPLC equipment have extensive product-specific information.

4. Sedimentation Velocity Methods

Sedimentation velocity is a simple, rapid, nondestructive technique for evaluating the purity of a protein. The technique is sensitive to the ratio of the molecular weight to the molecular size. An overview of the topic of sedimentation velocity is provided in Rhodes *et al.* (2009). Typically, when using sedimentation velocity as a test of purity, an experimenter is looking for the presence of more than one sedimentable component. The virtue of the technique is that it can be used to test for a wide variety of materials (especially when using one of the refractive optical systems). Its major limitation is that it is not nearly as sensitive to small differences between molecules as the electrophoretic techniques. This is true even when band sedimentation is used.

4.1. Method

Except for band, or zonal, sedimentation, details describing sample preparation, optical systems available, and the interpretation of the experimental results are provided in Rhodes *et al.* (2009) of this volume and reviewed in Cole *et al.* (2008). Consult the manufacturer's instructions for details on how to set up a preparative centrifuge to conduct a rate-zonal sedimentation experiment. For zonal methods in the analytical ultracentrifuge, consult Eason (1984) or Ralston (1993) for details. The analysis of the data is analogous to that above describing gel-permeation chromatography.

Analysis using the differential sedimentation coefficient distribution, $g(s)$, (Cole *et al.*, 2008) is a particularly useful approach to detection of impurity. Because the method is model independent, the presence of multiple components may appear as peaks at different values of s or as broadening of the peak representing the primary analyte. Again, as is the case with gel filtration chromatography, self-associating proteins may exhibit sedimentation behavior with broad peaks in $g(s)$ or multiple peaks, even though they are pure. Analysis using equilibrium ultracentrifugation or another method can help to resolve the problem.

5. Mass Spectrometry Methods

Because mass spectrometry provides a direct measure of the covalent mass distribution in a sample, it is a straightforward and sensitive approach to test for impurities. Not only is the presence of the impurity detected, but the impurity is characterized by its mass and therefore, it is often possible to identify the origin of the impurity. In addition to the direct measurement of the protein molecular mass, it is also possible to determine covalent modification identities and sites using the tandem mass spectrometric (MS/MS or MS^2) methods of Collision Induced Dissociation (CID), Electron Transfer Dissociation (ETD), Electron Capture Dissociation (ECD), and Infrared Multiphoton Dissociation (IRMPD). CID and IRMPD can be used on relatively small proteins (<15 kDa) but ETD and ECD can be used with larger proteins (~ 50 kDa). Beyond that, covalent modifications can be found and site determined by any commonly used proteolytic method such as trypsin or CNBr (Link, 2009). CID can be performed on instruments using a quadrupole collision cell before a Time-Of-Flight mass analyzer (Q-TOF configuration). The ETD, ECD, and IRMPD are usually found on ion-trap analyzers such as linear ion traps, Orbitraps, and Fourier Transform Ion Cyclotron Resonance mass spectrometers.

The MS methods, like the light scattering methods, are often most powerful when used in combination with another method like gel exclusion chromatography. It is possible that an impurity could have the same m/z ratio as the primary analyte, so coupling the MS method to an orthogonal fractionation provides an additional level of confidence in the purity of the sample. If an asymmetric peak is observed in the elution profile from a gel exclusion column, mass detection would provide a means for distinguishing molecules of different sizes from molecules undergoing self-association. Because MS provides a measure of the mass of the protein, as opposed to light scattering measurements, which provide a mass based on a measurement of the radius of gyration, r_g, there is less ambiguity in the result.

6. Light Scattering Methods

As described in Rhodes *et al.* (2009), current instrumentation is capable of static and dynamic light scattering measurements of individual samples or to monitor HPLC or FFF separations. This capability of continuously monitoring size or apparent molecular weight provides enhanced capability to identify the eluting proteins as being the analyte, a form of the analyte such as a multimer, or an impurity. Light scattering methods are simple and nondestructive, and the current instrumentation provides results as a plot of molecular weight as a function of elution time. Thus, MALS (multiple angle light scattering, also known as MALLS, multiple angle laser light scattering) analysis of a peak, which appears to be asymmetric by UV detection may exhibit apparent molecular weight of M in the main portion of the peak and 2M at the leading edge, suggesting the presence of a dimer. It is important to note that the presence of multiples of a molecular weight is not proof of self-association; checking these results with an orthogonal method is advised. Although the results from MALS are not as accurate as those from MS, the equipment is far less expensive and easier to use.

REFERENCES

Biggs, R. (1976). Tests for fibrinolysis, thrombin clotting time test. *In* "Human blood coagulation, hemostasis and thrombosis" (R. Biggs, ed.), 2nd edn. p. 722. Blackwell, Oxford.

Chang, J. Y. (1983). Amino acid analysis in the picomole range by precolumn derivatization and high-performance liquid chromatography. *Methods Enzymol.* **91,** 41–48, this series.

Cole, J. L., Lary, J. W., Moody, T. P., and Laue, T. M. (2008). Analytical ultracentrifugation: Sedimentation velocity and sedimentation equilibrium. *Method Cell Biol.* **84,** 143–179.

Eason, R. (1984). *In* "Centrifugation: A practical approach". (D. Richard, ed.), 2nd edn. p. 251. IRL Press, Washington, DC.

Lunney, J., Chrambach, A., and Rodbard, D. (1971). Pore gradient electrophoresis. *Anal. Biochem.* **40,** 158–173.

Q6B (1998). ICH Q6B Specifications: Test procedures and acceptance criteria for biotechnological/biological products.

Ralston, G. (1993). Introduction to analytical ultracentrifugation Beckman Instruments, Fullerton, CA.

Rhodes, D. G., Bossio, R. E., and Laue, T. M. (2009). Determination of size, molecular weight, and presence of subunits. This volume.

CHAPTER THIRTY-NINE

Determination of Size, Molecular Weight, and Presence of Subunits

David G. Rhodes,* Robert E. Bossio,† *and* Thomas M. Laue‡

Contents

1. Introduction 692
2. Chemical Methods 695
   2.1. Composition 695
   2.2. Colligative properties 697
3. Transport Methods 698
   3.1. Sedimentation equilibrium 698
   3.2. Sedimentation velocity 701
   3.3. Gel-filtration (size-exclusion) chromatography 706
   3.4. Electrophoresis 708
   3.5. Viscosity 711
   3.6. Field-flow fractionation 711
   3.7. Mass spectrometry 712
4. Scattering Methods 716
   4.1. Electron microscopy 718
5. Presence of Subunits 719
References 721

Abstract

The size or apparent molecular weight of a given protein may be the most cited distinguishing characteristic of the molecule. In addition to being the basis of many separation methods, the molecular weight, or simply molecular size, immediately provides the investigator with an idea of the complexity of the molecule, whether it is likely to be difficult to produce in quantity, and whether certain analytical methods are likely to be productive. Knowing whether the polypeptide of interest can self assemble or exists in a heterogeneous complex with other polypeptides may provide valuable information regarding biosynthesis

* Lipophilia Consulting, Storrs, Connecticut, USA
† Department of Chemistry, University of Michigan–Dearborn, Dearborn, Michigan, USA
‡ Department of Biochemistry and Molecular Biology, University of New Hampshire, Durham, New Hampshire, USA

Methods in Enzymology, Volume 463
ISSN 0076-6879, DOI: 10.1016/S0076-6879(09)63039-1

or mechanism. This chapter outlines key methods used to determine the size of proteins, their molecular weight, and whether subunits are present, with a focus on the basis of the determinations, their strengths, and their limitations.

1. Introduction

The size of a protein molecule is the property at the basis of many fractionation methods and is an easy-to-use descriptor for known molecular entities or unknown impurities (e.g., "the 20-kDa product"). Nevertheless, in spite of significant advances in methodology and data-handling techniques, care is required if one requires accurate estimates of molecular size. In this chapter, "size" will refer to the physical dimensions of the protein, as opposed to the molecular weight of the protein, which is related to the mass of the protein. This chapter also considers the asymmetry (or axial ratio) of proteins, as this property normally affects determinations of the apparent size of the molecule.

Many proteins assemble into larger aggregates, with each constituent chain being considered a subunit (Timasheff and Fasman, 1971). The concept of a subunit, however, must be defined in context by the individual investigator, taking into account the system under consideration and the objectives of the work with the system. In this discussion, independent subunits will be defined as those protein moieties which do not have a contiguous polypeptide backbone. Thus, disulfide-linked peptide chains as well as segments associated through noncovalent interactions are considered subunits, even if those subunits may have arisen from proteolytic action on a parent polypeptide.

The methods used to determine the size or molecular weight of a protein can be categorized into three broadly defined areas: *chemical analysis*, such as composition analysis or the effect of the molecule on the properties (e.g., vapor pressure, freezing point, or boiling points) of a solvent; *transport methods*, based on the movement of the molecule in response to some applied force (e.g., electrical, centrifugal, mechanical); and *scattering methods*, which are based upon the interaction of incident radiation (e.g., light, X-rays, neutrons) with the molecule. As these methods are sensitive to different features of the molecule, selection of appropriate methods depends on what one needs to know about the system and with what accuracy. The capabilities, advantages, and limitations of a number of techniques are outlined in Table 39.1. In addition to these criteria, the protein and the method must be compatible with regard to the quantity of protein available, the attainable level of purity (Rhodes and Laue, 2009), solvent requirements, and any peculiarities of the protein that may interfere with a specific technique. The various approaches are useful probes of other molecular

Table 39.1 Comparison of techniques for characterization of proteins[a]

Method	Size	Shape	M_r	Subunits	Primary method	Contaminant sensitivity[b]	Precision	Accuracy	Ease of use	Sample size	References
Chemical											
Composition	×	×	−	×	Y	L, H, N	−	−	−	μg–mg	
Colligative properties											
Vapor pressure	×	×	−	×	Y	L, N	−	−	+	mg	
Osmotic pressure	×	×	+	×	Y	L, N	+	+	+	mg	Kupke (1960)
Transport											
Sedimentation											
Equilibrium	×	×	+	+	Y	L, N	+	+	+	mg	
Velocity	+	+	−	+	Y		+	−	+	mg	
Gel-filtration chromatography	+	−	±	+	N		±	±	+	μg–mg	
Electrophoresis											
SDS denaturing	×	×	+	+	N		±	±	+	μg	
Native	+	−	−	±	N		±	±	+	μg	
Viscosity	+	+	±	−	N	L	±	±	+	mg	Yang (1961)
Mass spectrometry	−	−	+	+	Y		+	+	±		
Field-flow fractionation	+	±	−	−	N		±	±	±	mg	

(continued)

Table 39.1 (*continued*)

Method	Size	Shape	M_r	Subunits	Primary method	Contaminant sensitivity[b]	Precision	Accuracy	Ease of use	Sample size	References
Scattering											
X-ray diffraction											
Crystallography	+	+	+	–	Y		+	+	–	mg	Knox (1972)
Low-angle scattering	+	+	–	–	Y	L, H	–	–	–	mg	Glatter and Kratky (1982)
Rayleigh light scattering	+	+	+	–	Y	H	–	–	–	mg	Timasheff and Townend (1970)
Electron microscopy	+	+	×	–	Y		–	–	±	ng–μg	Topf *et al.* (2008)

[a] ×, not usable; –, not good; ±, acceptable; +, good.
[b] L, sensitive to low M contaminant; H, sensitive to high M contaminant; N, sensitive to nonprotein contaminant.

properties as well, and these will be brought out in discussions of individual techniques. Many techniques are best used in combination with other techniques (e.g., viscosity and sedimentation, composition and SDS gel electrophoresis, HPLC and mass spectrometry). This is not a complete list; many methods not included (such as radiation inactivation and cDNA analysis) may be useful in specific situations.

We have selected for detailed discussion those techniques which are most useful and/or most commonly used. Suggested readings for a more complete treatment of other techniques are listed in Table 39.1.

2. Chemical Methods

2.1. Composition

2.1.1. Overview

Although such methods are used only in unusual circumstances in current practice, measurements that provide molar quantitation of either amino acids and/or specific prosthetic groups may be used to estimate the minimum molecular weight of a protein as $M_{min} = m/n$, where M_{min} is the molecular weight estimate (g/mol) of the protein, m is the mass of the protein used in the analysis, and n is the number of moles of the protein-related component measured in the analysis. Good quantitation has been achieved using amino acids analyses, end-group analysis, and quantitative analysis for specific prosthetic groups (Noble and Bailey, 2009). The quantity of material required for such composition-based molecular weight analyses is most dependent on the sensitivity of the analytical method being used. Since many newer analytical techniques are sensitive in the nanomole to picomole range, only microgram or even smaller quantities of material may be required.

The error in the minimum molecular weight estimate will depend on the errors incurred in both the mass estimate and in the analysis for the particular constituent and, therefore, will depend on the methods chosen for these two measurements. By combining data from analyses for several different constituents, the accuracy can be improved. For this reason, independent estimates based on amino acid analyses, quantitative analysis of the end group, or quantitation of nonprotein cofactors are recommended.

Because composition analysis provides only a minimum molecular weight, it is very useful to combine these data with results from other approaches to obtain the mass of the protein. An accurate composition analysis can be combined with techniques that provide only low-accuracy total molecular weights to yield high-accuracy results. Perhaps, the most often used example of this approach is the combination of amino acid analysis (sequence-based methods as described in Burgess, 2009, are more common now, but see Bartolomeo and Maisano 2006) with SDS gel electrophoresis (Garfin, 2009). In principle, one should be able to estimate *M* by determining the smallest integers that could account for the full amino acid composition. In practice, uncertainty in the concentrations of all 20 individual amino acids makes this approach quite unreliable by itself. However, an approximate molecular weight from SDS gel electrophoresis can guide in the selection of the appropriate absolute concentrations from the relative concentrations, indicating to the investigator whether to "round up" or "round down" individual values.

2.1.2. Method

An estimate of protein mass is made most accurately from a measurement of dry weight when the protein is present in a volatile buffer (e.g., ammonium bicarbonate). Compensation for nonvolatile buffer components can be made from the difference in mass between dried samples of the protein containing solution and the buffer alone, but is difficult to do with great accuracy and should not be attempted with partially volatile buffers. Alternatively, protein concentration measurements are often used in place of dry weight estimates (the mass value used for analysis simply being the product of mass concentration and the volume). Concentration estimates can be made refractometrically, spectrophotometrically, or by chemical assay. The accuracy of such measurement often is limited, even though the precision may be reasonable, because standardization is a major problem (Nozaki, 1986). This is especially true for glycoproteins, lipoproteins, or other proteins with unusual compositions where the method of analysis may be sensitive only to certain constituents of the protein (e.g., peptide bonds). Of the methods listed, refractometry (or differential refractometry) is the most accurate means of estimating concentration. Amino acid analysis also provides a suitable means of quantitation.

2.1.3. Problems and pitfalls

Composition-based minimum molecular weight analyses are sensitive to sample purity. Any contaminant that influences either the mass estimation or the quantitative analysis will affect the accuracy of the determination. Since no fractionation of the starting material is afforded by these analyses, contributions from contamination or heterogeneity will be averaged into the results. Therefore, samples must be purified prior to analysis. The degree of purity required depends on the nature of the contaminant and the sensitivity of the mass estimate or the analytical method to that contaminant. In all of these methods, it must be assumed that there is only one, or an integral number, of moles of the analyzed component per mole of protein. Since no information concerning quaternary structure is available, only a minimum molecular weight is obtained. For example, quantitative analysis of heme iron in either hemoglobin or myoglobin would yield nearly identical apparent molecular weights, despite the fact that the molecular weight of intact hemoglobin is fourfold greater than that of myoglobin.

2.2. Colligative properties

2.2.1. Overview

Dissolution of a solute in a solvent reduces the chemical potential of that solvent and results in a number of observable phenomena known collectively as colligative properties (Tinoco *et al.*, 2002). The relationship at the basis of these phenomena is

$$\mu_a - \mu_a^\circ = RT\ln(\gamma_a X_a), \tag{39.1}$$

where μ_a is the chemical potential of solute "a" (cal/mole), μ_a° is μ_a in the standard state, R is the gas constant (8.3144×10^7 cal/mol ·°K), T is the temperature (K), and γ is the activity coefficient of "a," and X_a is the mole fraction of "a." When X_a is less than 1, μ_a is less than μ_a°. The molecular weight of any solute can, in principle, be determined by the extent to which the solution activity is changed by the presence of a known weight of solute. Thus, the molecular weight of any solute, including proteins, could be determined by measuring freezing point depression, boiling point elevation, osmotic pressure, or vapor pressure. Because changes in freezing temperature or boiling temperature are generally quite small for protein-sized molecules, and occur at extremes of temperature at which the molecules may not be stable, these methods are generally not applicable for study of protein solutions. On the other hand, both vapor pressure and osmotic pressure have been used for measurement of protein molecular weights, but this will not be discussed in detail here.

Vapor pressure, freezing point, and membrane osmometers are all available commercially, but the vapor pressure and freezing point osmometers are most commonly used in clinical and pharmaceutical applications. Vapor pressure and freezing point osmometers reliably measure molecular weights in water to approximately 25,000, whereas membrane osmometers are used for proteins above 20,000. Protein concentrations on the order of 0.1–1 mg/ml and sample volumes of 10–200 μl are required. As with composition methods, accurate determinations of sample mass are necessary for any of these methods. It should also be noted that the molecular weight measured with *any* colligative property is a number average and can be severely affected by contaminating low-molecular-weight species. Additional information can be found in Tinoco *et al.* (2002) and in the manufacturer's literature.

3. Transport Methods

3.1. Sedimentation equilibrium

3.1.1. Overview

Sedimentation equilibrium provides an accurate and powerful method for the determination of the native molecular weight of a protein (Cole *et al.*, 2008). It is a simple, nondestructive, relatively rapid method. All parameters that describe sedimentation equilibrium are either readily measured or easily estimated. It has several unique capabilities, and can provide quantitative estimates of molecular weights, stoichiometries, and association constants for a wide range of chemical systems. For example, by sedimenting in neutrally buoyant (nonsedimenting) detergents, the native molecular weight of detergent-solubilized proteins may be measured (Fish, 1978). Many systems not amenable to analysis by any other technique can be evaluated successfully by sedimentation equilibrium. However, with the advent of powerful sedimentation velocity analysis, because sedimentation equilibrium generally requires sophisticated, expensive equipment, and because other methods are available which provide data that are adequate for many purposes, this procedure is used much less frequently than in the past. Nevertheless, equilibrium sedimentation provides one of the most powerful methods for the quantitative examination of protein–protein associations, and is irreplaceable for the study of protein systems with weak to moderate association constants (Cole *et al.*, 2008).

3.1.2. Method

In an equilibrium sedimentation experiment, the purpose is to produce a measurable protein concentration gradient along the radial axis of the centrifuge cell. The length of time needed to achieve sedimentation equilibrium depends on the length of the solution column, the value of the reduced molecular weight, σ, and the diffusion constant. Because the time to reach equilibrium depends on the square of the column length, many cells and techniques have been developed to examine short columns (0.75 or 3 mm long) (Ralston, 1993). Using these cells, equilibrium can be reached in a matter of minutes to a few hours, depending on the protein. At each rotor speed, the concentration distribution is measured at intervals of 30–60 min and invariance of the distribution over time indicates that equilibrium has been reached. Software (e.g., HeteroAnalysis) helps automate this process (Cole *et al.*, 2008).

The reduced molecular weight, σ, is the measured quantity in a sedimentation equilibrium experiment (Yphantis, 1964). It is defined as:

$$\sigma = \frac{M(1 - \bar{v}\rho)\omega^2}{\mathrm{RT}}, \tag{39.2}$$

where $\bar{v}$ (ml/g) is the partial specific volume of the protein, ρ (g/ml) is the solution density, and ω^2 (s^{-2}) is the square of the angular velocity ($\omega =$ rpm $\cdot\ \pi/30$) of the rotor. (Methods for $\bar{v}$ and ρ are described below, following the description of velocity sedimentation). It can be shown that

$$\frac{\mathrm{d}\ln[c_r]}{\mathrm{d}r^2} = \frac{\sigma}{2}, \tag{39.3}$$

where c_r is the solute concentration at radial position in the rotor, "r." Thus, σ can be determined as the slope of a graph of the natural logarithm of the concentration as a function of $r^2/2$. The slope may be calculated at each of several radial positions (or, correspondingly, at each of several concentrations). Molecular weights determined in this fashion are weight-average values.

Alternatively, σ may be obtained from nonlinear least-squares fitting to equations that describe the concentration distribution (i.e., c as a function of $r^2/2$) that are beyond the scope of this chapter (Johnson *et al.*, 1981). These equations have been derived from thermodynamic first principles for a variety of models that include association and nonideality of the proteins.

A typical equilibrium centrifugation experiment will require approximately 100–200 μl of solution at each of three or four concentrations, which typically vary from 0.1 mg/ml to approximately 3 mg/ml. These solutions must be matched with appropriate buffers of the same composition. The equilibrium centrifugation experiment is normally run at different speeds, from a low speed where the concentration ratio from the top of the cell to the bottom of the cell is about 3–5, to a high speed where the meniscus is depleted of solute.

The concentration distribution at equilibrium can be determined by any number of means, without an analytical centrifuge. If a preparative centrifuge (Attri and Minton, 1986) or an "airfuge" (Bock and Halvorson, 1983) is being used for the measurement, any assay that is proportional to the concentration of protein may be used following fractionation of the content of the cell. Because any method for concentration determination may be used (e.g., enzyme activity), this sort of analysis may be used to determine molecular weights in complex mixtures and concentrated solutions (Howlett *et al.*, 2006). However, it is then necessary to perform several experiments for different lengths of time to ensure that equilibrium has been reached. These techniques suffer a loss of precision due to the collapse of the concentration gradient that occurs as the rotor decelerates and as the cell contents are fractionated. The analytical ultracentrifuge alleviates these problems by permitting the solution contents to be examined optically as the centrifuge is operating and the contents of the cell remain at equilibrium.

Currently, there are three optical systems available for the standard analytical centrifuge, the UV/visible absorbance scanner, the Rayleigh interferometer, and a fluorescence detector. The three systems provide complementary information, with the selectivity and sensitivity of the fluorescence and absorbance optics being useful for some situations, while the precision and accuracy of the Rayleigh system are needed in others. The fluorescence system provides exceptional sensitivity (as low as 100 p*M* fluorescein) but requires a fluorescent molecule or a fluorescently labeled molecule (Kroe and Laue, 2009). The precision of the fluorescence optical system is similar to that of the absorbance optical system, but the accuracy is limited because the quantum yield of the fluorophore is sensitive to the environment of the molecule.

The radius must be calculated with accuracy. The optical systems on the analytical ultracentrifuge provide this information (Ralston, 1993). When using a preparative centrifuge, the radius must be calculated from geometric considerations. The original articles describe how this is done for various rotors and cells (Attri and Minton, 1986); Bock and Halvorson, 1983).

3.1.3. Problems and pitfalls

The principal problem in sedimentation equilibrium analysis is obtaining a sufficient quantity of highly purified protein for analysis. Even for highly purified proteins, it is informative to perform a sedimentation velocity analysis just prior to equilibrium analysis (Cole *et al.*, 2008). A molecular weight may be measured using the Rayleigh interferometer optical system with just 20 μl of solution at a concentration of 1 mg/ml and much lower concentrations are possible for some fluorescent solutes. Although this is no more material than is often used for gel electrophoresis, the presence of minor contaminants can make interpretation of the data more difficult, especially if the contaminant is unknown or unexpected. In cases where a preparative centrifuge is being used and a sensitive assay is available, less material may be required, but lower precision can be expected. It should also be realized that equilibrium sedimentation does not afford the extent of fractionation that is seen with sedimentation velocity, gel filtration, or gel electrophoresis. Even so, the gravitational field does fractionate the solution to some extent, such that very large particles (e.g., dust, flocculant, aggregate) are removed from observation. This means that although proper technique is required, the scrupulous cleanliness of samples necessary for scattering techniques is not needed for analytical centrifugation. Note that buffer components will also be fractionated in the gravitational field. For this reason, buffers that have a $\bar{\nu}$ that is near that of the solvent ($\bar{\nu}\rho = 1$) are preferred. Thus, buffers such as Tris are good choices for sedimentation, while phosphate buffers can pose problems under some circumstances

(Yphantis, 1964). Problems of buffer sedimentation are minimized by extensive dialysis and careful filling of the cell.

Tests are available for detecting impurities in a sample being analyzed by sedimentation equilibrium. One is to generate graphs of the apparent molecular weight (determined as the local slope from the graph of ln*c* vs. $r^2/2$) as a function of concentration. If the $M(c)$ graphs are independent of rotor speed, one can be quite certain that the sample is pure. More sensitive tests for heterogeneity are outlined by Rhodes and Laue (2009).

3.2. Sedimentation velocity

3.2.1. Overview

Sedimentation velocity is a simple, nondestructive technique for the characterization of the hydrodynamic behavior of a protein. It has the advantage over gel chromatography of being a primary technique (not requiring standards) for the determination of hydrodynamic parameters. Because it is based on first principles, sedimentation analysis can be applied to systems that cannot be analyzed by any other means. The principal result from a sedimentation velocity experiment is the sedimentation coefficient (s), which is a measure of the ratio of the buoyant mass of the protein to its frictional coefficient:

$$s = \frac{M(1-\bar{v}\rho)}{N_0 f}, \tag{39.4}$$

Where s is the sedimentation coefficient in Svedberg units ($\times 10^{-13}$ s^{-1}), N_0 is Avogadro's number, and f is the protein frictional coefficient. The diffusion coefficient, D (cm^2/s), defined in Eq. (39.5) also is measurable from boundary spreading in measurements of velocity centrifugation experiments:

$$D = \frac{\mathrm{RT}}{N_0 f}. \tag{39.5}$$

The ratio of s/D [Eq. (39.6)] provides a transport-based measure of the protein molecular weight that is free of any requirements concerning size, shape, or similarity to protein standards:

$$\frac{s}{D} = \frac{M(1-\bar{v}\rho)}{\mathrm{RT}}. \tag{39.6}$$

Alternatively, if an accurate molecular weight is known, measurement of the sedimentation coefficient allows the frictional coefficient, f, to be determined. First, f°, the frictional coefficient expected for an anhydrous sphere of molecular weight and density equal to those of the protein, is calculated by writing the Stokes equation:

$$f^{\circ} = 6\pi\eta r_{\text{sphere}} = \frac{6\pi\eta}{[(M\bar{v}3)/(4\pi N_0)]^{1/3}}, \tag{39.7}$$

where η is the viscosity of the solution and r_{sphere} is the radius of the equivalent sphere. Using Eqs. (39.5)–(39.7), the ratio of f/f° may be calculated and used to estimate the shape of the protein. This ratio, f/f°, is compared to standard and theoretical values to estimate the asymmetry and overall shape of the protein.

With the advent of powerful computer programs to analyze the data, sedimentation velocity has become the more widely used than equilibrium sedimentation (Cole *et al.*, 2008). It is recommended that velocity analysis be performed on any sample prior to sedimentation equilibrium analysis.

3.2.2. Method

Sedimentation velocity experiments are best performed in an analytical ultracentrifuge, although methods of reasonable accuracy have been devised for preparative centrifuges. The advantage of the analytical machine is that its accuracy ($\pm$1–3%) is nearly an order of magnitude better than can be expected from techniques using preparative instruments. Moreover, data are acquired throughout the experiment so that any unusual behavior can be observed and better understood. The advantage of the preparative machines is that sensitive or specific assays, such as ELISA, may be used so that in cases where the purity of the sample is in doubt, a specific assay can identify the component of interest. Detailed methods for using the analytical ultracentrifuge (Ralston, 1993) and for using preparative centrifuges (Freifelder, 1973; Martin and Ames, 1961) are available.

As noted before, the principal measurement in any sedimentation experiment is the concentration as a function of radial position. The sedimentation coefficient may be determined from the slope of a graph of the natural logarithm of the distance that the molecules have sedimented as a function of time (d ln r/dt):

$$s = \frac{(\mathrm{d}\ln r/\mathrm{d}t)}{\omega^2}, \tag{39.8}$$

where r is the distance of the boundary of the molecules from the center of the rotor and t is the time from the start of the experiment, usually taken to be the time at which the rotor reaches two-thirds of the final speed. For experiments in which the sample being examined is a thin zone of molecules, r is taken to be the point of maximum concentration. For broad zone or boundary experiments, r is taken to be the midpoint between zero concentration and the plateau concentration. For proper interpretation and comparison, the sedimentation coefficient must be corrected to standard conditions of water at 20 °C and zero protein concentration ($s^{\circ}_{20,\mathrm{w}}$) using

$$s^{\circ}{}_{20,\mathrm{w}} = s_{\mathrm{observed}}\left(\frac{1-\bar{v}_{20,w}\rho_{20,\mathrm{w}}}{1-\bar{v}_{20,w}\rho_{T,\mathrm{b}}}\right)\left(\frac{\eta_{T,\mathrm{w}}}{\eta_{20,\mathrm{w}}}\right)\left(\frac{\eta_{T,\mathrm{b}}}{\eta_{T,\mathrm{w}}}\right), \tag{39.9}$$

where ν, ρ, and η are the partial specific volume, density, and viscosity, respectively, at the experimental temperature (T) or 20 °C, and in buffer (b) or water (w). Tabulated values for the viscosity of water and measured values for the viscosity of the buffer at T and 20 °C are used to correct for the effect of temperature on the sedimentation coefficient. The density is either measured or calculated from tabulated values.

The diffusion coefficient is determined from the spreading of the boundary as it progresses down the analytical cell. A graph may be made of $[c/(\mathrm{d}c/\mathrm{d}r)_{\max}]^2$ as a function of time, where $(\mathrm{d}c/\mathrm{d}r)_{\max}$ is the concentration gradient at the point of maximum gradient, for most cases the midpoint of the boundary (Ralston, 1993). The slope of this line is used to estimate D, the diffusion coefficient:

$$D = \left(\frac{1}{4\pi}\right)\left(\frac{\mathrm{d}[c/(\mathrm{d}c/\mathrm{d}r)_{\max}]^2}{\mathrm{d}t}\right). \tag{39.10}$$

Although s and D may be determined, as described here, this approach is provided primarily to illustrate the relationships between these parameters, since the advent of powerful analysis programs makes graphing data unnecessary. Instead, sedimentation velocity data are fit directly to solutions of the Lamm equation (Cole *et al.*, 2008). These fitting programs can provide good quantitative analyses of molecular weights and association constants for a variety of models and using a variety of analysis protocols.

The ratio of s/D (Eq. 39.6), or the estimates of M from data analysis programs, are the buoyant, or reduced molecular weight, $M(1-\bar{v}\rho)/\mathrm{RT}$. Determination of M requires other experimental measurements. The solvent density, ρ, is readily measured or calculable from tabulated data. The partial specific volume of the protein, $\bar{v}$, can be measured, but more frequently is calculated from the protein composition using the method of Cohn and Edsall and tabulated values of $\bar{v}$ for the amino acids (McMeekin and Marshall, 1952). Values of $\bar{v}$ for carbohydrates may be found in Gibbons (1972) and Durschlag (1986). Values for other common protein-associated components can be found in Steele *et al.* (1978) or Reynolds and McCaslin (1985). For more accurate work $\bar{v}$ is adjusted for temperature (over the range 4–40 °C) using a coefficient of 4.3×10^{-4} cm^3/g °K, as described by Durschlag (1986). The effects of pH, buffer composition, and preferential hydration on $\bar{v}_{\mathrm{r}}$ are typically neglected, except in cases where a high concentration of denaturant is used. Special provision may be made for estimating the isopotential apparent partial specific volume (Π) of simple proteins in solvents containing 8 *M* urea and 6 *M* guanidinium chloride, using the method given by Prakash and Timasheff (1985). Likewise, Π may be estimated for simple proteins in solvents containing varying amounts of NaCl, Na_2SO_4, $MgSO_4$, glycine, β-alanine, α-alanine, betaine, and $CH_3COO \cdot Na$ using the method of Arakawa and Timasheff (1985). Neither estimate is very accurate, however, if the protein contains significant levels (>10%, w/w) of nonamino acid constituents (e.g., glycoproteins, lipoproteins) or if the axial ratio is greater than 10.

Hydrodynamic interpretation of sedimentation data can be made, once an accurate molecular weight and $\bar{v}$ estimate are available. First, one calculates f° using Eq. (39.7) above. The ratio of the measured frictional coefficient to f° (f/f°) will be a value greater than one. Note that f must be calculated using $s^\circ_{20,\mathrm{w}}$:

$$f = \frac{M(1 - \bar{v}\rho)}{N_0 s^\circ{}_{20,\mathrm{w}}}, \tag{39.11}$$

where the various terms are used as described above, and $s^\circ_{20,\mathrm{w}}$ is determined as described in Eq. (39.9).

There are two reasons that the frictional coefficient ratio, f/f°, will be greater than one. First, proteins have an associated layer of water molecules that move with the protein as it sediments. While these water molecules freely exchange with those in the bulk solvent, the net result of the layer is to increase the effective radius of the protein. As can be seen from Eq. (39.7), any increase in the effective radius will increase f. Second, the frictional coefficient depends on the surface area of the protein accessible to the solvent. For a molecule of given mass and density, a sphere would be the shape that exposes the minimum surface area, so any molecular asymmetry will increase f. The degree of molecular asymmetry is estimated by determining the axial ratio (a/b) of the ellipsoids of revolution, prolate (elongated) or oblate (flattened), that would result in an identical value for f/f°. It should be clear that the asymmetry is being modeled using these ellipsoids of revolution as a mathematical construct, and that the calculated result may not resemble the molecule. However, if one can measure the asymmetry for the isolated subunits of a multimeric protein, as well as for the intact oligomer, it is possible to use changes in a/b to distinguish between possible arrangements of the subunits (Cantor and Schimmel, 1980).

3.2.3. Problems and pitfalls

Sedimentation velocity provides the best and the only primary method for the determination of hydrodynamic parameters available to molecular biologists. Electrophoresis and gel filtration require that molecular weight standards be used, which places restrictions on the data interpretation.

Of all the techniques described in this chapter, the problems and pitfalls of sedimentation analysis are the best documented and most easily overcome.

The most demanding aspect of sedimentation is the availability of sufficient material for analysis. If the standard optical systems on the current (Beckman XL-A) analytical ultracentrifuge are used, 0.5 ml of solution with protein concentrations in the range of 0.1–1 mg/ml is needed to obtain good data.

For proteins that bind significant levels of buffer components (e.g., detergent-solubilized proteins), the interpretation of sedimentation coefficients is made difficult by the fact that the bound components will contribute to the measured s in four terms: M, $\bar{v}$, ρ, and f. Of these terms, M, $\bar{v}$, and f are usually the most affected by bound components, and unlike sedimentation equilibrium, there is no way to "blank out" the contribution of such components. Thus, the measured sedimentation coefficient is for the complex of the protein with the bound component, making it difficult to extract any useful information concerning the protein alone.

3.2.4. Variations

The effect of protein concentration on s is assessed by determining s at each of several protein concentrations. Typically, there is a slight linear decrease in s as the protein concentration is increased.

Another widely used method of estimating molecular weights of proteins is the sucrose gradient sedimentation method introduced by Martin and Ames (1961). In this method, one creates a linear gradient of sucrose in buffer in a swinging bucket centrifuge tube. The sucrose concentration typically ranges from approximately 5% at the top of the tube to 20% at the bottom; the actual range is less important than the linearity and reproducibility of the gradient. The unknown sample is layered onto a gradient and a set of standard proteins of known molecular weight layered onto an equivalent gradient. The centrifugation proceeds for an appropriate interval (typically 12–24 h) and the material on the gradients collected as fractions. The protein concentration in these fractions is determined by spectrophotometric, enzymatic, or other assays.

The basis of the method is the fact that in a linear sucrose gradient, the distance traveled by a molecule should be a linear function of the time of centrifugation at a specified speed. In addition, the distance will depend linearly on s. The ratio of the distance traveled by an unknown protein to that of a standard will be equal to the ratio of their sedimentation coefficients, which will, in turn be approximated by the ratio of the molecular weights to the 2/3 power. This method yields only approximate values of s and M, but is simple, requires no specialized equipment, and can be used to estimate s and M for very small amounts of material if a suitable assay is available.

3.3. Gel-filtration (size-exclusion) chromatography

3.3.1. Overview

Gel-filtration chromatography is one of the most powerful and simplest methods for the estimates of the molecular weight of proteins. Because of the fractionation afforded by the method, and because assays specific for the protein of interest may be used (e.g., enzymatic, immunological), sample purity does not have to be very high. The method is nondestructive, can be fairly rapid, and has moderate accuracy as long as the protein of interest is roughly the same shape as the protein standards used to calibrate the column (Ackers, 1970).

The determination of a molecular weight by gel chromatography relies on the comparison of the elution volume of the unknown with those of several protein standards for which molecular weights are known. The molecular weight of the unknown is estimated from a graph of the logarithm of the molecular weight as a function of elution volume (or K_{av}, as described below) made using the data from the protein standards. (It is worth noting that the actual dependence is on the logarithm of the effective hydrated radius, or "Stokes radius" of the protein, and that the fit of standard proteins to this variable is better.) The elution volume (V_e) for the standards should cover the range from V_0 (the void volume – the volume outside of the stationary phase) to V_i (where V_i is the included volume – the accessible volume within the pores of the stationary phase). Although this discussion refers to elution volumes, most reports of SEC data refer to elution times, which will be proportional to elution volume for a given pumping rate. A column-independent measure of the protein behavior, K_{av} is more useful for comparison of results than simply the elution volume:

$$K_{av} = \frac{V_e - V_0}{V_t - V_0}, \qquad (39.12)$$

where K_{av} is the fraction of the stationary gel volume which is accessible to the protein, and V_t is the total volume of the gel bed. Use of K_{av} is preferred over V_e since, for a given gel type, values of K_{av} will vary only slightly from column to column. The methods described below can be used for both native and denatured proteins and, therefore, provide a means for establishing the presence of subunits. However, gel filtration in denaturing solvents typically requires more material than denaturing gel electrophoresis and is thus not used as often.

3.3.2. Method

The principle of operation and selection of gel media and gel porosity is described in detail elsewhere (Stellwagen, 2009). SEC stationary phase materials are available which will fractionate a very wide range of molecular sizes; these may not provide adequate resolution for specific needs. In some cases, a stationary phase with a narrower range may provide superior resolution, but one must choose a gel in which the protein to be examined is partially included. Normally, choice of the stationary phase is arbitrary, as long as the protein does not bind to the matrix. When there is a choice of bead sizes for a given porosity, the smallest bead size should be used, as this improves the column resolution. Check the manufacturer's recommendations for any limitations on solvents, but in general, nearly any free-flowing aqueous buffer system may be used. It is recommended that buffers of moderate ionic strength be used so that electrostatic interactions between the protein and immobile charges on the matrix are minimized. For best results and highest resolution, a long, narrow column should be used. Preparation of the column and equilibration of the column by washing with buffer should be done according to the manufacturer's specifications. Likewise, flow rates should be chosen in accordance with the manufacturer's specifications. In general, lower flow rates afford better resolution because the solute can fully equilibrate with the gel matrix at all times, but excessive diffusion can limit resolution if a column is run too slowly. All samples should be in the same buffer as that used to equilibrate the column. Sample volumes applied to the column should generally be less than 2% of the column's bed volume. In addition, the flow rate of the column should be kept constant throughout all of the analyses, since flow rate dependence of the elution volume can be expected (Ackers, 1970).

Elution volume (V_e) is the volume eluted from the column, starting once one-half of the sample has penetrated the top of the gel bed and continuing until the maximum (peak) of the protein of interest has eluted. The void volume (V_0) of the column usually can be measured using commercially available, size-graded Blue Dextran ($M = 2{,}000{,}000$), and monitoring the effluent spectrophotometrically at 540 or 280 nm. The included volume can be measured using as a sample some buffer of different pH or conductivity (avoiding extremes that could affect the column) or one that contains a small dye (e.g., bromphenol blue). The included volume is the difference between the elution volume for the small molecule and the void volume measured with a large, excluded molecule. Care should be exercised in the choice of a dye, as many aromatic compounds have an affinity for the stationary phase which results in anomalously high K_{av} and V_i values.

Protein standards should be used that span the full range of sizes that can be analyzed by the stationary phase chosen to assure that the standards bracket the analyte protein. For best results, a minimum of four different standards should be used. Kits containing prestained proteins are commercially available. Any convenient assay for detecting the presence of the protein being analyzed may be used. If, for any reason, the column must be repacked, the calibration must be performed again.

Use of column chromatography to estimate the molecular weight of denatured proteins poses special problems (Fish, 1978). This method assumes that the shape of the unknown protein is identical to that of the standards. This means that the protein must be totally denatured, including reduction of disulfide bonds. Buffers should contain a reducing agent or sulfhydryl groups should be alkylated to prevent reformation of any disulfide bonds. Both the standards and the unknown must be analyzed after the same treatment and the same buffer conditions. Calibration proceeds as described above. This is not a commonly used method.

3.3.3. Problems and pitfalls

The greatest source of error in the use of gel-exclusion chromatography for size determination comes from the requirement that the unknown be similar in shape and density to the protein standards. Since the protein standards used are almost universally compact, globular proteins, this means that fibrous proteins, or protein having fibrous regions, can behave anomalously on gel columns. One indication of such molecular asymmetry is if K_{av} for the unknown increases when analyzed at decreased flow rates, while those for the standards remain unchanged.

Since it is the size of the protein and not the molecular weight that is being assessed by this technique, molecular weight estimates for proteins that are complexed with other molecules (e.g., detergent-solubilized proteins, extensively glycosylated proteins) will be unreliable. Finally, if the protein interacts with the stationary phase through either affinity (protein will appear smaller) or repulsion (protein will appear larger), inaccuracies will result. One can test for column interaction by determining the molecular weight using two chemically different types of stationary phase (e.g., Sepharose and acrylamide) or by changing the buffer conditions. There are extremes of interaction that will be obvious from unacceptable values of K_{av}. If $K_{av} > 1$, then the protein is binding to the column and a different stationary phase should be used. Conversely, if $K_{av} < 0$, the column is "channeling" and must be repoured and recalibrated or replaced.

While the estimation of molecular weights using gel filtration often is inaccurate, the combination of size-exclusion chromatography coupled with light scattering is extremely powerful and widely used (below). In this case, gel filtration is used to fractionate the sample by size (Folta-Stogniew, 2000).

3.4. Electrophoresis

3.4.1. Overview

The most widely used method of evaluating the size of a protein molecule is electrophoresis. The method is simple, inexpensive, rapid, and for a very wide range of proteins, is sufficiently accurate for many needs. For these reasons, it is the method of choice for most protein systems, and almost always included in characterization studies. SDS gel electrophoresis is the most widely used method for determining apparent molecular weights of denatured proteins (Garfin, 2009), but electrophoretic methods for obtaining size, shape, and molecular weight information are not limited to this approach. Despite its popularity, it is not necessary to include SDS in the gel formation; native gel electrophoresis of protein samples may be carried out under almost any buffer condition required. In addition, the sensitivity of current staining procedures allows these approaches to be applied to very small amounts of protein (Garfin, 2009).

3.4.2. Method

The basic procedures are identical to those described earlier (Rhodes and Laue, 2009; Garfin, 2009) for electrophoresis of denatured proteins, except that the buffer composition is determined by the needs of the investigator. There are some restrictions; as with gels under denaturing conditions, the buffer in which the gel is cast cannot contain reducing agents. In addition, because the conditions to which the protein is exposed may affect charge, association, or shape, the composition of the buffer in the gel must be controlled carefully. One must, therefore, be especially cautious about relative proportions of catalyst, so that excess oxidant is not left in the gel. Precast gels are inexpensive and generally have good quality control, but the probability is low that the buffer in the precast gel will be precisely the condition that is needed. It is often easiest to prerun gels to remove any possible by-products of polymerization. Alternatively, if the geometry of the gel apparatus allows the slab to be exposed (e.g., a horizontal slab or a vertical slab in which one of two glass plates can be removed and later replaced), the gel can be dialyzed against the buffer of choice prior to running. Because slabs are generally quite thin, dialysis for several hours is generally sufficient. This dialysis procedure can also be used to introduce reducing reagents postpolymerization, or to use a set of gels cast together (and thus, presumably, uniform in porosity) with a variety of buffer conditions.

The basis of electrophoretic protein size analysis is based on a simple principle: that a charged particle in an electric field is forced through the surrounding medium by a force proportional to the charge on the particle and the strength of the field, and is subject to a frictional force proportional to the velocity, the radius of the particle, and the viscosity of the medium. As with SDS gel electrophoresis, the investigator may control the frictional coefficient by controlling the porosity of the gel matrix. In addition, under nondenaturing conditions, the mobility can be significantly affected by alterations of the intrinsic charge on the protein due to changes of pH at which the electrophoresis is carried out. This distinction is important. In SDS gel electrophoresis, the charge is dominated by the negatively charged

SDS associated with the protein so the sample is applied to the cathode end of the gel and the sample always moves toward the anode. In nondenaturing electrophoresis, the direction of migration will depend on the buffer pH relative to the isoelectric point (pI) of the protein. If the pI of the protein is unknown (and quantities are too limited to allow experimental determination with isoelectric focusing), a horizontal electrophoresis apparatus may be used, and the starting wells placed in the center of the gel. Otherwise, a "best guess" might be made by using the pI of related proteins.

The mobility will also be affected by the chosen buffer condition. Beyond association of the protein or conformational changes associated with changing buffer conditions, the relative mobilities of various proteins at a given pH will usually be fairly constant. The absolute mobility, however, will be strongly affected by the concentrations of counterions.

The analysis of native gel electrophoresis mobility data is analogous to that of SDS gels. The only significant difference is that, in the native system, one cannot assume that the proteins have equal charge density and are all in the shape of long rods. As is necessary in gel-exclusion chromatography, one must compare the mobility of the sample to that of a set of standards (Rodbard and Chrambach, 1971). Ferguson analysis can also be used to identify the feature(s) (size and/or charge) that distinguish two components and to extrapolate to the mobility expected in the absence of sieving effects (Andrews, 1986). The slope of the Ferguson plot (log of the relative mobility, R_f, vs. gel concentration) is proportional to r_s, the Stokes radius. Interpolative estimation of an unknown r_s is more reliable using this relation than using, for example, the slope of the Ferguson plot with molecular weights of the standards. The analysis simply involves running the unknown and several standards in a set (at least five) of gels of different total gel concentrations, and plotting log R_f versus the gel concentration for all standards and for the unknown. The slopes for the standards are then plotted as a function of the (known) r_s, and the r_s of the unknown is derived from this plot and the measured slope. The range of gel concentrations used will depend on the size of the protein under study, but at some upper limit of concentration, the protein will be excluded from the gel. A more closely spaced group of gel concentrations covering a lower range should then be used.

3.4.3. Problems and pitfalls

Although electrophoresis under nondenaturing conditions can provide useful information about the physical characteristics of the protein under a number of different conditions, it is important to be aware that this is a zonal method, and that concentration effects (or dilution effects) can be significant. It is not generally advisable, for example, to use this approach to study association behavior quantitatively.

As with gel-exclusion chromatography, this is not a primary technique; it depends on selection of appropriate standards. These are generally globular, unmodified, soluble proteins, so that highly asymmetric proteins or proteins with unusual nonprotein modifications may yield erroneous results. The Ferguson analysis helps to account for differences in mobility due to charge differences, but exaggerated charge densities may yield anomalous results.

Because the buffer conditions in native gel electrophoresis are selected by the individual investigator, and thus may vary widely, one must be aware of the solution components responsible for carrying current. High ionic strength buffers may result in unacceptably slow protein mobilities, and proteins in low ionic strength buffers may run properly at surprisingly low currents. Because of this variability and because protein integrity is more important, one must be particularly aware of power dissipation and cautious about efficient removal of heat generated in the gel.

3.5. Viscosity

Among the variables upon which the viscosity of a solution depends (T, P, etc.) are the amount and nature of any solute that is present. The response of a particle in a fluid under shear will depend on the frictional coefficient of the particle, which represents the mechanical interaction of the particle with the moving solution (Yang, 1961). The response of the particle will also depend on the mass of the particle, which determines the energy required to attain a given movement (Johnson *et al.*, 1977). Because the intrinsic viscosity of a protein depends very strongly on the asymmetry of the molecule, viscosity measurements are sensitive indicators of protein shape. In addition, it is possible to combine viscosity data with sedimentation data to calculate the molecular weight of a protein. Apparatus for rheological measurements vary widely; their use is generally quite simple, but must be performed under rigorously controlled conditions.

3.6. Field-flow fractionation

Field-flow fractionation (FFF) is a method based on subjecting particles to a force field orthogonal to the direction of laminar flow in a narrow channel (100–500 μm) (Liu *et al.*, 2006). Because the laminar flow in the channel is parabolic, with the fastest flow in the center and slowest at the edge, molecules that are most strongly affected by the force imposed on them will be closer to the edge of the channel and will move more slowly, while molecules that do not move from the center will move faster. The force imposed on the molecules could be centrifugal, electrophoretic, magnetic, or even mechanical flow. The cross-flow approach is the most commonly used method, and is referred to as flow-FFF (fFFF). In this method, a semipermeable membrane allows solvent, but not protein, to flow out the bottom of the channel, thus creating a concentration gradient with higher concentration at the bottom of the channel than at the center or top. Diffusion works against this concentration gradient and allows the more rapidly diffusing small molecules to reach the center of the channel where flow is faster. The larger molecules cannot diffuse as rapidly and move slowly at the edge of the channel.

To its benefit, this method is nondestructive and can be used with any buffer system. By appropriate adjustment of the flow rates, this method is capable of resolving a very wide range of molecular sizes. Current instrumentation utilizes many HPLC components and highly automated systems make the process accessible; a typical experiment runs in approximately 30 min. Nevertheless, the method is technically challenging and will require time and effort for method development. Finally, the method is very good for resolving molecules of different sizes (e.g., monomer, dimer, etc.), but the resulting size distribution is relative and not easily converted to an absolute size. Consequently, FFF often is used as a fractionation method for light scattering to provide absolute molecular weights (Liu *et al.*, 2006).

3.7. Mass spectrometry

Mass spectrometry measures an intrinsic property of the protein, its mass, by measuring the mass-to-charge ratio (m/z) of ions generated from the protein sample. This property is temperature, density, and concentration independent. Advances in electrospray ionization mass spectrometry (ESI-MS) have added significant capabilities to the biochemist's toolbox (Faull *et al.*, 2008). ESI-MS and other MS methods can be combined with liquid chromatographic (LC) methods (i.e., capillary electrophoresis, size-exclusion chromatography, normal and reverse phase LC) to aid in identifying chromatographic peaks and to provide additional information with relatively little sample (1–5 μg). The protein molecular mass obtained by MS analysis can be used not only for protein identification; but also for determining purity (Baldwin *et al.*, 2001) and for analysis of covalent modifications and degradation products, both qualitatively and quantitatively. The ESI time-of-flight (TOF) methods are sufficiently gentle that there is little if any degradation of the protein during analysis. Typical data are obtained as a family of peaks as a function of m/z that represent individual ionization states of the protein. For instance, a 30-kDa protein can acquire upwards of 40 H^+ ions, and is observed as a peak in the mass spectrum at ($30{,}000m/40z = 750\ m/z$) and is labeled as $[M + 40H]^{40+}$. The charge states form a Gaussian-like distribution with many charge states ($[M + 39H]^{39+}$, $[M + 38H]^{38+}$, $[M + 37H]^{37+}$, etc. Deconvolution software (such as that publicly available at http://ncrr.pnl.gov/software/) is used to determine the molecular weight of the parent molecule, and minor peaks reveal the presence of contaminants or degradation products. The protein molecular mass obtained is highly accurate ($\pm$0.01%) and can be used for modified (e.g., glycosylated, alkylated, or PEGylated) proteins. It is important to note, however, that any heterogeneity in the modification, such as variable glycosylation or variable lengths of PEG, will be reflected in the mass spectrum. This is useful in assessing purity (Baldwin *et al.*, 2001) but can complicate the calculation of the molecular mass.

This method has a significant advantage over many other methods in that the sample need not be highly purified. While purity will simplify the analysis, the method is capable of identifying multiple components in a single experiment. On the other hand, if the difficulty in purification arises from affinity of the contaminant for the primary analyte, an analyte-contaminant peak may also be observed, since noncovalent interactions can be preserved in ESI-MS. By diluting the sample, it may be possible to dissociate these noncovalent adducts and the contaminant can be observed separately.

Selection of a suitable buffer requires care, since the buffer must be volatile; traditional buffers like Tris and phosphate are not suitable. Typical buffers include formic acid, trifluoroacetic acid (TFA), acetic acid, ammonium formate, ammonium acetate, ammonium bicarbonate, and heptafluorobutyric acid. Not all of these buffers are suitable for all situations. TFA has been known to suppress ion formation, so only low concentrations (0.1% or less) are acceptable. These are not common buffer systems and changing buffers can affect the protein, so it is important to execute proper controls to assure that results are being interpreted correctly.

Because the analyte protein is delivered to the ESI-MS instrument in a buffer, it is relatively straightforward to combine a chromatographic separation with the mass measurement. The complementarity of a mass measurement and a size measurement that one would obtain with SEC chromatography is particularly valuable in analyzing associating proteins. Although the MS can be used as a primary detector, it is most common to place the MS in combination with a conventional detector such as a UV absorbance system. With this arrangement, an elution profile from the SEC–HPLC can be interpreted by mass-based identification of peaks. For example, a SEC peak skewed to the leading edge may result from an impurity or from association to multimer. An impurity could present as a distinct mass peak, whereas an association might result in peaks with mass values at integral multiples of the mass of the parent peak.

3.7.1. Method

The specific ESI-MS techniques will vary with the individual instrumentation, but some general guidelines are provided here. As mentioned above, samples must be in a volatile buffer if a direct presentation to the mass spectrometer is going to be used. Phosphate- and TRIS-related buffers cause ion suppression and mass shifts due to noncovalent adduct formation between sodium, potassium, counter ions, and the protein. These salts can be dialyzed against a volatile buffer or removed chromatographically. If the sample is analyzed by LC/MS, the small buffer constituents will be fractionated from the larger protein and dialysis is not necessary. In cases where a protein is added to the column in a salt-based buffer, a diverter valve is used to divert the salt peak to waste before the salts arrive at the ESI source. Samples available as a lyophilized powder can be dissolved in a mixture of methanol or acetonitrile and water with formic or acetic acid, usually at a ratio of 1:1 aqueous:organic, with the acid present at 5% or less (typically 0.1%). Higher organic concentrations can be used without difficulty for hydrophobic proteins.

In addition to using the volatile buffers suggested above, it is important to assure that the protein is fully equilibrated to this condition and that any titration of the buffer is with suitable acid or base (ammonia, acetic, formic, or trifluoroacetic). Moreover, for some of these buffers, the concentration limits are relatively low. Flow rates will be determined by the specific instrument limits, but in some cases, a splitter will be required so that only a small portion of the column flow enters the MS. Most LC/MS instruments are capable of operating without a splitter at flow rates as high as 1 ml/min. ESI sources are also available with very small ($\leq$ 25 micrometer inner diameter) capillaries, called nano-ESI. These ESI sources operate at much lower flow rates ($\sim$10–100 nl/min) and are frequently coupled to capillary LC columns for Capillary LC/nano-ESI-MS. Although the sensitivity is improved, these systems can become clogged without scrupulous handling of solvents and samples to avoid dust.

The speed of a protein mass spectral analysis is very rapid when analyzing a single sample (<1 min), and can range up to 1 h for an LC/MS analysis. New ultrahigh-pressure liquid chromatography analysis (UPLC) is showing promise to reduce these analysis times and improve the chromatographic resolution.

Four recently developed ionization modes have allowed for alternate modes of ionization for large proteins. Desorption electrospray ionization (DESI), Direct Analysis in Real Time (DART), Laser Ablation Electrospray Ionization (LAESI) and matrix assisted laser desorption electrospray ionization (MALDESI) have shown the ability to ionize proteins, and may prove useful in cases where ESI and MALDI (matrix-assisted laser desorption ionization) are ineffective (Faull *et al.*, 2008). MALDESI combines the advantages of MALDI, such as the ability to handle high-molecular-weight proteins, with the multiple charging capability of ESI. These ionization methods may be able to overcome some of the limitations encountered in the individual methods.

On the mass spectrometer side, an alternate platform is finding new life in protein analysis, ion mobility spectrometry (IMS). IMS separates ions according to their mobility through an electric field as they collide with a gas (usually He). Protein ions with a large shape/collision cross section travel slower in the electric field than smaller ions with lower collisional cross section. From the cross section information, aggregation information can be determined as well as purity. In some aspects, IMS resembles TOF but reveals information about protein purity through different measures (the gas phase ion shape as opposed to its mass). IMS can also be coupled to mass spectrometers, allowing for interesting multidimensional separations followed mass analysis. While this technique is still in its starting phases, it definitely bears watching as the method matures and finds its place in protein purity analysis.

3.7.2. Problems and pitfalls

The most important factor in interpreting MS data is understanding the origin of the results. Mass shifts of 22, 44, 66, or 88 are due to sodium adduction, as Na^+ ions exchange for H^+ ions, but retain the overall charge state. Similarly, mass shifts of 39 are signals of potassium adduction. Ammonium adduction (mass shift of 17 units) can also be observed. These adducts can be minimized by using ultrapure solvents, but are difficult to completely remove.

Because the method reports a mass-to-charge ratio, in principle, there is no difference between a monomer and a doubly charged dimer. While charge deconvolution algorithms cannot resolve this issue, mass spectrometers other than TOF's can. In instances like these, the ultrahigh mass resolving power of fourier transform ion cyclotron resonance (FTICRMS) or Orbitrap mass spectrometers can resolve these uncertainties by using the isotope peaks of the ions to determine the charge state. These spectrometers unambiguously differentiate singly charged monomers from doubly charged dimers. Dimer formation can also be minimized by diluting the sample to the point where dimers are no longer observed. In cases where tightly bound ligands cannot be separated by dilution, using higher concentrations of organic solvent, acid, or both may dissociate the noncovalent complexes. Acid-labile detergents can also be used to initially dissociate noncovalent complexes, but once the pH is lowered these detergents dissociate into noninterfering volatile monomers.

Similarly, without additional information, a single peak could result from a wide range of charge–mass combinations. Fortunately, for proteins, the multiple charge states that are normally observed can be deconvoluted reliably to yield unambiguous results. The presence of tightly bound ligands can be problematic but can usually be dealt with by changing the pH, increasing the organic content of the solution, or diluting the solution until the complex dissociates.

For some proteins the electrospray process does not work well enough to get usable signal. Very large proteins (>150 kDa) can be difficult to ionize by ESI. ESI is also matrix sensitive; solutions with high salt concentrations can result in reduced signals. In such cases, alternative ionization modes can be used. MALDI is useful in these instances, as this mode can handle very large (>1 MDa) proteins and has a higher tolerance for salts. Chromatographic separation followed by lyophilization of the individual chromatographic peaks allows for LC/MALDI analysis. MALDI also tends to form ions with a single charge state, so that protein ions are largely observed as their $[M + 1H]^+$ peaks. This can simplify spectral interpretation, as dimer peaks will show up at twice the m/z, avoiding the doubly charged dimer issue altogether.

Mass spectrometry is limited due to the fact that proteins are generally not isotopically pure. The presence of ^{13}C, ^{15}N, ^{34}S, and other isotopes occurring in proteins causes broadening of mass spectral peaks in TOF as the mass of the protein increases. This broadening can eventually make m/z envelope of a protein ion so broad that the molecular weight becomes indeterminant. In this instance, peaks due to adduct formation, ligand binding, and di- or multimer formation may become indistinct. Although this issue is generally less significant below about 6–7 MDa, it does provide an upper bound to the method. This can be circumvented in some studies by producing the protein using amino acids or other starting blocks that are isotopically depleted, these have a lower than natural abundance amount of ^{13}C, ^{15}N, ^{34}S or similar isotopes.

It should also be noted that the power of mass spectrometry-based methods comes at a significant cost. As of this writing, TOF instruments range in cost from $250,000 to over $500,000 and FTMS instruments can cost $2 million.

4. Scattering Methods

Light scattering methods are generally used to obtain molecular weights and radii of gyration, but specific scattering methods can provide diffusion coefficients, molecular weight, and thermodynamic parameters (Chu, 1974; Cottingham, 2005; Timasheff and Townend, 1970). Instrumentation for key light scattering methods has been developed which integrates with HPLC and FFF instrumentation and incorporates nearly all of the data analysis functions. This ease of use and accessibility has resulted in an resurgence in the application of these powerful methodologies to protein characterization (Harding *et al.*, 1991).

The basis of light scattering is the interaction of the oscillating electromagnetic field of the incident light with the electrons in the protein to induce oscillating dipoles. These oscillating dipoles radiate light, and the intensity and directionality of the light depends on many factors, including the size and structure of the molecule.

The Zimm relationship forms the basis of what is now referred to as "static" light scattering:

$$\frac{Kc_2}{R(\theta)} = \frac{1}{M_w P(\theta)} + 2A_2 c_2, \tag{39.13}$$

where $R(\theta)$ is the scattering angle- and concentration-dependent Rayleigh ratio, which is proportional to the intensity of light scattered by the solution in excess of that scattered by pure solvent, c_2 is the protein concentration, M_w is the weight-average molecular weight of the protein and any other material in the sample, A_2 is the second virial coefficient, K is a constant $(4\pi^2(dn/dc_2)^2 n_0/N_0\lambda^4)$, dn/dc_2 is the refractive increment of the solute, N_0 is Avogadro's number, n_0 is the refractive index of the solvent, λ is the wavelength of the incident light, and $P(\theta)$ is a parameter related to the angular dependence of the scattered light intensity. $P(\theta)$ can be approximated by:

$$P(\theta) = 1 - \frac{16\pi^2 n_0{}^2}{3\lambda^2}\bar{r}_g^2 \sin^2\frac{\theta}{2} + \ldots, \tag{39.14}$$

where $\bar{r}_g$ is the RMS radius of gyration of the molecule. By plotting $Kc_2/R(\theta)$ as a function of $\sin^2(\theta/2) + Kc_2$ for several values of θ, one can calculate $\bar{r}_g$. This is an approximation obtained at c_2, but is acceptable for many applications and is the data used in many instruments incorporated in HPLC instrumentation. By making measurements at several values of c_2, one can extrapolate to $c_2 = 0$ and to $\theta = 0$, and the intersection of these lines at the ordinate $(Kc_2/R(\theta))$ yields M_w.

The basis for "dynamic" light scattering (also known as photon correlation spectroscopy or quasielastic light scattering) is the time-dependent constructive or destructive interference between scattering centers (proteins) in motion in the solution. The normalized autocorrelation function for the intensity variation with time, $g(\tau)$ can be shown to be

$$g(\tau) = \exp(-\Gamma\tau), \tag{39.15}$$

where Γ is $[(4\pi n_0/\lambda)\sin(\theta/2)]^2 D$. The diffusion coefficient can be related to size and molecular weight by Eqs. (39.5) and (39.7). Data from dynamic light scattering measurements can be presented as distributions in r_g, from which the presence of multiple scattering species can be determined.

With laser illumination and a set of fixed angle detectors, one can simultaneously measure dynamic and static light scattering. From the static light scattering, one can estimate radius of gyration and molecular weight and from the dynamic light scattering, diffusion coefficient, hydrodynamic radius, size distribution, and molecular weight. It is important to be aware of assumptions involved in calculations performed by user-friendly systems; most calculations, for example, assume that all particles are spherical.

As with MS measurements, the power of these light scattering methods is greatly enhanced when the measurements are coupled with a chromatographic separation method. The ability to characterize a protein by orthogonal methods such as SEC or FFF and light scattering provides a ready comparison through which the validity of any assumptions can be tested. The power, reproducibility and ease of use has made SEC–multiple-angle light scattering (SEC-MALS), the principle method for characterizing protein aggregates in the pharmaceutical industry (Wen *et al.*, 1996).

When light scattering methods are not coupled with chromatographic separation, which inherently provides very efficient filtration, it is important to maintain scrupulously conditions to avoid the presence of "dust" particles in the solution. The presence of very small amounts of large particles can easily dominate the scattering signal and compromise the results. On the other hand, this dependence points out the power of this methodology in detecting large assemblies or aggregates of proteins.

4.1. Electron microscopy

Electron microscopy has always been an appealing approach to determining molecular size and shape for large protein molecules or associated complexes of subunits because it is a direct imaging method. Initial work in this area was used to identify the shape of very large complexes like the hemocyanin aggregates (van Bruggen and Wiebenga, 1962), but studies advanced to imaging of crystalline arrays of membrane proteins (Sletyr *et al.*, 1988) and more recently to cryo-EM analysis of protein structures (Topf and Sali, 2005, Topf *et al.*, 2008). This method provides very detailed information about the native size and shape of the molecule and is particularly valuable for analysis of membrane-associated proteins that are incompatible with many other analysis methods. The resolution of electron microscopy is inherently limited, but with cryopreparation and signal averaging from large numbers of images, high-resolution electron density maps can be constructed. These maps can then be combined with structure filling algorithms to obtain detailed structure maps of the protein molecule. The modeling requires knowledge of the protein sequence, so this is not a method by which one would experimentally determine the molecular weight of the protein, but this is an elegant approach to determining the size and shape of the molecule to very good resolution. Although the cryo-EM method is not a common laboratory method for protein size determination, this approach has made great advances in areas of protein science which previously were very difficult to study.

5. Presence of Subunits

To determine whether subunits are present generally involves characterization of one or more properties of the system under conditions which favor association and under conditions which are likely to favor dissociation. Thus, the investigator should have prior knowledge of the conditions under which association and dissociation should be expected, or will need to investigate a variety of conditions which have been shown in other cases to result in dissociation. The investigation to identify these conditions may require significant effort, but if properly planned, will also yield additional information about the nature of the system. Before initiating an extensive search for associating or dissociating conditions, it is worthwhile using extreme conditions as a preliminary evaluation. One normally carries out one fractionation procedure under some (often physiological) buffer condition and a second under strongly denaturing conditions.

One very simple, rapid, and low-cost approach for searching for subunits is to carry out electrophoresis under nondenaturing conditions, using the buffer in which the protein was isolated, followed by a second dimension under dissociating or denaturing conditions. The nature of the denaturing conditions may vary depending on the type of subunit and the association one expects to find. For example, one could run the second electrophoresis at extremes of pH, in the presence of urea or SDS, or in the presence or absence of other buffered components like calcium. Changes in mobility due to changes of buffer condition should be accounted for and understood by analyzing the second dimension. That is, analysis should be based on the apparent size of the molecule(s) under associating and possibly dissociating buffer conditions. If one is investigating the possibility that disulfide-linked subunits are present, SDS gel electrophoresis in the absence of reducing agents may be carried out as the first dimension, and SDS gel electrophoresis in the presence of reducing agents may be carried out in the second dimension. Because in-gel alkylation of disulfides is difficult, it is recommended that mercaptoacetate be included in the second gel running buffer to avoid reoxidation. Differences in the apparent molecular weight or the appearance of multiple components in the second dimension will indicate that disulfide-linked subunits were present. Differences in apparent molecular weight deduced from nondenaturing electrophoresis compared with the apparent molecular weight based on electrophoresis under completely denaturing conditions (SDS–PAGE) could indicate the presence of subunits, but one must consider the possibility that the electrophoretic behavior of the molecule is due to extremes of asymmetry or intrinsic charge.

In the case of subunits which are in equilibrium between associated and dissociated states, the definitive approach to determining the presence of subunits requires experimental determination of the apparent size of the molecule under conditions where the equilibrium is shifted to either associating or dissociating conditions. Because the dissociation implies that the samples being studied will be very dilute, very sensitive methods must be used.

Sedimentation equilibrium is very useful for determining the molecular weight of a native protein, and is therefore useful in determining the stoichiometry of the subunits in the final assembly. This is done by comparison of the native molecular weight with that obtained in separate experiments under denaturing conditions, such as by denaturing gel electrophoresis. If the protein is composed of subunits of a single molecular weight, division of the native molecular weight by the denatured molecular weight will provide the subunit stoichiometry. Likewise, for proteins that contain more than one chain, comparison of the native molecular weight to the sum of the monomer molecular weights often will allow the stoichiometry of the different subunits in the native complex to be determined. In cases where there is a wide discrepancy between the subunit molecular weights or when the native structure contains a large number of monomers, these estimate are imprecise.

While the subject requires too much detail to be presented in full here, it is important to note that equilibrium sedimentation provides one of the most powerful means of determining association constants of mass action-driven macromolecular associations (Johnson *et al.*, 1981). The method of experimentation is essentially as outlined above, except that experiments are done at concentrations that encompass the range where significant mass changes occur due to association. The association constant may be estimated using graphical means, or more accurately using nonlinear least-squares analyses (Johnson *et al.*, 1981).

Velocity centrifugation can also be a simple screen for multiple components in a solution. The integral sedimentation coefficient distribution, $G(s)$, (Demeler *et al.* 1997) is based on an extrapolation of the sedimentation boundaries to eliminate the effect of diffusion during sedimentation. The method is model independent and is a useful diagnostic for heterogeneous distributions in a sample, association behavior, and any nonideality. Similar and related methods have also been developed (Schuck, 2000; Stafford, 1992) from which one may easily compare the apparent sizes of monomers, multimers, or multiple species of dissociated subunits.

Gel-permeation chromatography can also be useful in determining the stoichiometry of the subunits in the final assembly. Again, this is done by comparison of the native molecular weight determined by gel filtration with that obtained under denaturing conditions, such as by SDS gel electrophoresis. One can test for association and, in principle, determine stoichiometry using the ratio of the native molecular weight and the denatured molecular weight.

Gel chromatography is also useful for diagnosing interacting protein systems. If, for example, the protein of interest is a multisubunit molecule undergoing rapidly equilibrating assembly–disassembly, the peak shape is skewed, with the leading edge being hypersharp and the trailing edge being diffuse. Likewise, increase of K_{av} with protein loading concentration usually indicates that such an interaction is occurring, and that more detailed analyses will be required (Ackers, 1970).

REFERENCES

Ackers, G. K. (1970). Analytical gel chromatography of proteins. *Adv. Protein Chem.* **24,** 343–446.

Andrews, A. T. (1986). *Electrophoresis: Theory, Techniques, and Biochemical and Clinical Applications.* Clarendon Press, Oxford.

Arakawa, T., and Timasheff, S. N. (1985). Calculation of the partial specific volume of proteins in concentrated salt and amino acid solutions. *Methods Enzymol.* **117,** 60.

Attri, A. K., and Minton, A. P. (1986). Technique and apparatus for automated fractionation of the contents of small centrifuge tubes: Application to analytical ultracentrifugation. *Anal. Biochem.* **152,** 319–328.

Baldwin, M. A., Medzihradszky, K. F., Lock, C. M., Fisher, B., Settineri, T. A., and Burlingame, A. L. (2001). Matrix-assisted laser desorption/ionization coupled with quadrupole/orthogonal acceleration time-of-flight mass spectrometry for protein discovery, identification, and structural analysis. *Anal. Chem.* **73,** 1707–1720.

Bartolomeo, M., and Maisano, F. (2006). Validation of a reversed-phase HPLC method for quantitative amino acid analysis. *J. Biomol Tech.* **17,** 131–137.

Bock, P., and Halvorson, H. (1983). Molecular weight of human high-molecular-weight kininogen light chain by equilibrium sedimentation in an air-driven ultracentrifuge. *Anal. Biochem.* **137,** 172–179.

Burgess, R. (2009). Use of bioinformatics in planning a protein purification. *Methods Enzymol.* **463,** 21–28.

Cantor, C. R., and Schimmel, P. R. (1980). *In* "Biophysical Chemistry", Part II p. 565. Freeman, San Francisco, CA.

Chu, B. (1974). Laser Light Scattering. Academic Press, New York.

Cole, J. L., Lary, J. W., Moody, T. P., and Laue, T. M. (2008). Analytical ultracentrifugation: Sedimentation velocity and sedimentation equilibrium. *Methods Cell Biol.* **84,** 143–179.

Cottingham, K. (2005). Versatile TOFMS. *Anal. Chem.* **77,** 227–231.

Demeler, B., Saber, H., and Hansen, J. C. (1997). Identification and interpretation of complexity in sedimentation velocity boundaries. *Biophys. J.* **72,** 397–407.

Durschlag, J. (1986). *In* "Thermodynamic Data for Biochemistry and Biotechnology," (H. Hinz, ed.), p. 46. Springer-Verlag, Berlin.

Faull, K. F., Dooley, A. N., Halgand, F., Shoemaker, L. D., Norris, A. J., Ryan, C. M., Laganowsky, A., Johnson, J. V., and Katz, J. E. (2008). An introduction to the basic principles and concepts of mass spectrometry. *Compr. Anal. Chem.* **52,** 1–46.

Fish, W. W. (1978). *In* "Methods in Membrane Biology," (E. E. Korn, ed.), **Vol. 4,** p. 289. Plenum Press, New York.

Folta-Stogniew, E. (2000). Oligomeric states of proteins determined by size-exclusion chromatography coupled with light scattering, absorbance, and refractive index detectors. *Methods Mol. Biol.* **328,** 97–112.

Freifelder, D. (1973). Zonal centrifugation. *Methods Enzymol.* **27,** 140–150.
Garfin, D. (2009). One-dimensional gel electrophoresis. *Methods Enzymol.* **463,** 495–514.
Gibbons, R. A. (1972). *In* "Glycoproteins," Part A, (A. Gottschalk, ed.), p. 31. Elseveier, Amsterdam.
Glatter, O., and Kratky, O. (1982). *Small Angle X-Ray Scattering.* Academic Press, New York.
Harding, S. E., Sattelle, D. B., and Bloomfield, V. A. (1991). Laser light scattering in biochemistry. *Biochem. Soc. Trans.* **19,** 477–516.
Howlett, G. J., Minton, A. P., and Rivas, G. (2006). Analytical ultracentrifugation for the study of protein association and assembly. *Curr. Opin. Chem. Biol.* **10,** 430–436.
Johnson, J. F., Martin, J. R., and Porter, R. S. (1977). *In* "Physical Methods of Chemistry," (A. Weissberger and B. W. Rossiter, eds.), **Vol. 1**, Part VI, p. 63. Wiley, New York.
Johnson, M. L., Correia, J. J., Yphantis, D. A., and Halvorson, H. R. (1981). Analysis of data from the analytical ultracentrifuge by nonlinear least-squares techniques. *Biophys. J.* **36,** 575–588.
Knox, J. R. (1972). Protein molecular weight by x-ray diffraction. *J. Chem. Educ.* **49,** 476–479.
Kroe, R. R., and Laue, T. M. (2009). NUTS and BOLTS: Applications of fluorescence-detected sedimentation. *Anal. Biochem.* **390,** 1–13.
Kupke, D. W. (1960). Osmotic pressure. *Adv. Protein Chem.* **15,** 57–130.
Liu, J., Andya, J. D., and Shire, S. J. (2006). A critical review of analytical ultracentrifugation and field flow fractionation methods for measuring protein aggregation. *AAPS J.* **8,** E580–E589.
Martin, R., and Ames, B. (1961). Method for determining the sedimentation behavior of enzymes: Application to protein mixtures. *J. Biol. Chem.* **236,** 1372–1379.
McMeekin, T. L., and Marshall, K. (1952). Specific volumes of proteins and the relationship to their amino acid contents. *Science* **116,** 142–143.
Noble, J. E., and Bailey, M. J. A. (2009). Quantitation of protein. *Methods Enzymol.* **463,** 73–96.
Nozaki, Y. (1986). Determination of the concentration of protein by dry weight—A comparison with spectrophotometric methods. *Arch. Biochem. Biophys.* **249,** 437–446.
Prakash, V., and Timasheff, S. N. (1985). Calculation of partial specific volumes of proteins in 8 *M* urea solution. *Methods Enzymol.* **117,** 53.
Ralston, G. (1993). *Introduction to Analytical Ultracentrifugation.* Beckman Instruments, Fullerton, CA.
Reynolds, J. A., and McCaslin, D. R. (1985). Determination of protein molecular weight in complexes with detergent without knowledge of binding. *Methods Enzymol.* **117,** 41–53.
Rhodes, D. G., and Laue, T. M. (2009). Determination of protein purity. *Methods Enzymol.* **463,** 691–724.
Rodbard, D., and Chrambach, A. (1971). Estimation of molecular radius, free mobility, and valence using polyacylamide gel electrophoresis. *Anal. Biochem.* **40,** 95–134.
Schuck, P. (2000). Size-distribution analysis of macromolecules by sedimentation velocity ultracentrifugation and Lamm equation modeling. *Biophys. J.* **78,** 1606–1619.
Sletyr, U. B., Messner, P., Pum, D., and Sara, M. (1988). *In* "Crystalline Bacterial Cell Surface Layers." Springer-Verlag, Berlin.
Stafford, W. F. (1992). Boundary analysis in sedimentation transport experiments: A procedure for obtaining sedimentation coefficient distributions using the time derivative of the concentration profile. *Anal. Biochem.* **203,** 295–301.
Steele, J. H. C., Tanford, C., and Reynolds, J. A. (1978). Determination of partial specific volumes for lipid-associated proteins. *Methods Enzymol.* **38,** 11–23.
Stellwagen, E. (2009). Gel filtration. *Methods Enzymol.* **463,** 373–386.

Timasheff, S. N., and Fasman, G. R. (eds.) (1971). *In* "Subunits in Biological Systems" Part A. Dekker, New York.

Timasheff, S. N., and Townend, R. (1970). *In* "Physical Principles and Techniques of Protein Chemistry," (S. J. Leach, ed.), **Part B,** p. 147. Academic Press, New York.

Tinoco, I., Sauer, K., Wang, J., and Puglisi, J. (2002). *Physical Chemistry: Principles and Applications in Biological Sciences.* 4th edn. Prentice-Hall.

Topf, M., and Sali, A. (2005). Combining electron microscopy and comparative protein structure modeling. *Curr. Opin. Struct. Biol.* **15,** 578–585.

Topf, M., Lasker, K., Webb, B., Wolfson, H., Chiu, W., and Sali, A. (2008). Protein structure fitting and refinement guided by cryo-EM density. *Structure* **16,** 295–307.

van Bruggen, E. J. F., and Wiebenga, E. H. (1962). Structure and properties of hemocyanins. I. Electron micrographs of hemocyanin and apohemocyanin from Helix pomatia at different pH values. *J. Mol. Biol.* **4,** 1–7.

Wen, J., Arakawa, T., and Philo, J. S. (1996). Size-exclusion chromatography with on-line light-scattering, absorbance, and refractive index detectors for studying proteins and their interactions. *Anal. Biochem.* **240,** 155–166.

Yang, J. T. (1961). The viscosity of macromolecules in relation to molecular conformation. *Adv. Protein Chem.* **16,** 323–400.

Yphantis, D. A. (1964). Equilibrium Ultracentrifugation of Dilute Solutions. *Biochemistry* **3,** 297–317.

CHAPTER FORTY

Identification and Quantification of Protein Posttranslational Modifications

Adam R. Farley* *and* Andrew J. Link†

Contents

1. Introduction 726
2. Enrichment Techniques for Identifying PTMs 731
   2.1. Phosphorylation 731
   2.2. Glycosylation 737
   2.3. Ubiquitination and sumoylation 738
3. Nitrosative Protein Modifications 740
4. Methylation and Acetylation 741
5. Mass Spectrometry Analysis 744
6. CID versus ECD versus ETD 747
7. Quantifying PTMs 750
8. Future Directions 756
Acknowledgments 758
References 758

Abstract

Posttranslational modifications (PTMs) of proteins perform crucial roles in regulating the biology of the cell. PTMs are enzymatic, covalent chemical modifications of proteins that typically occur after the translation of mRNAs. These modifications are relevant because they can potentially change a protein's physical or chemical properties, activity, localization, or stability. Some PTMs can be added and removed dynamically as a mechanism for reversibly controlling protein function and cell signaling. Extensive investigations have aimed to identify PTMs and characterize their biological functions. This chapter will discuss the existing and emerging techniques in the field of mass spectrometry and proteomics that are available to identify and quantify PTMs. We will

* Department of Biochemistry, Vanderbilt University School of Medicine, Nashville, Tennessee, USA
† Department of Microbiology and Immunology, Vanderbilt University School of Medicine, Nashville, Tennessee, USA

Methods in Enzymology, Volume 463
ISSN 0076-6879, DOI: 10.1016/S0076-6879(09)63040-8

focus on the most frequently studied modifications. In addition, we will include an overview of the available tools and technologies in tandem mass spectrometry instrumentation that affect the ability to identify specific PTMs.

1. Introduction

Protein posttranslational modifications (PTMs) perform essential roles in the biological regulation of a cell. PTMs are enzymatic, covalent chemical modifications of proteins that typically occur after translation from mRNAs. Chemical modifications of proteins are extremely important because they potentially change a protein's physical or chemical properties, conformation, activity, cellular location, or stability. In fact, most proteins are altered by the addition or removal of a chemical moiety on either an amino acid or the protein's N- or C-terminus. Some PTMs can be added and removed dynamically as a mechanism for reversibly controlling protein function. Over 400 specific protein modifications have been identified, and more are likely to be identified (Creasy and Cottrell, 2004). The most commonly identified PTMs to date include phosphorylation, sumoylation, ubiquitination, nitrosylation, methylation, acetylation, sulfation, glycosylation, and acylation (Table 40.1).

Vast amounts of scientific effort have gone toward identifying PTMs and elucidating their biological functions. Perhaps the best studied cases involve eukaryotic histones and the myriad of PTMs that are associated with transcriptional regulation (Berger, 2007; Fuchs *et al.*, 2009; Issad and Kuo, 2008; Olsson *et al.*, 2007; Reid *et al.*, 2009). Histones associate with chromosomal DNA to form nucleosomes that bundle together to form chromatin fibers. The histone code hypothesizes that chromatin–DNA interactions are regulated by combinations of histone PTMs (Jenuwein and Allis, 2001; Strahl and Allis, 2000). The acetylation of lysine was first discovered in histones and is correlated with actively transcribed genes (Roth *et al.*, 2001). Combinations of methylation, acetylation, ADP-ribosylation, ubiquitination, and phosphorylation of histone tails function to regulate specific gene expression programs (Godde and Ura, 2008).

One of the most ubiquitous PTMs, phosphorylation, is the focus of many biochemical investigations. It has been estimated that 30% of the human proteome is phosphorylated (Cohen, 2001; Hubbard and Cohen, 1993). For example, tyrosine kinases and phosphatases transduce signals from ligand-bound receptors on the cell surface to downstream targets in the insulin/IGF-1 signaling pathway (Taniguchi *et al.*, 2006).

Classic approaches to detecting PTMs on proteins involved Edman degradation and thin-layer chromatography (TLC). However, these methods are hampered by the requirement of significant amounts of starting

Table 40.1 Common PTMs encountered in mass spectrometry

PTM	Nominal mass shift (Da)	Stability	Proposed biological function
Phosphorylation			
pSer, pThr	+80	Very labile	Cellular signaling processes, enzyme activity, intermolecular interactions
pTyr	+80	Moderately labile	
Glycosylation			
O-linked	203, >800	Moderately labile	Regulatory elements, O-GlcNAc
N-linked	>800	Moderately labile	Protein secretion, signaling
Proteinaceous			
Ubiquitination	>1000	Stable	Protein degradation signal
Sumoylation	>1000	Stable	Protein stability
Nitrosative			
Nitration, nTyr	+45	Stable	Oxidative damage
Nitrosylation, nSer, nCys	+29	Stable	Cell signaling
Methylation	+14	Stable	Gene expression
Acetylation	+42	Stable	Histone regulation, protein stability
Sulfation, sTyr	+80	Very labile	Intermolecular interactions
Deamidation	+1	Stable	Intermolecular interactions, sample handling artifact
Acylation			
Farnesyl	+204	Stable	Membrane tethering, intermolecular interactions, cell localization signals
Myristoyl	+210	Stable	
Palmitoyl	+238	Moderately labile	
Disulfide bond	−2	Moderately labile	Protein structure and stability
Alkylation, aCys	+57	Stable	Sample handling
Oxidation, oMet	+16	Stable	Sample handling

The associated mass shifts, predicted MS stability, and proposed biological functions are included.

material and an inability to identify rare or substoichiometric PTMs. Because most PTMs result in a concomitant change in the mass of the modified protein, methods that can detect changes in molecular mass, namely mass spectrometry-based proteomics, are now routinely utilized to identify PTMs. Some PTMs, such as phosphorylation or methylation, increase the mass of a protein, while other PTMs, like signal peptide cleavage or disulfide bond formation, decrease the mass. Depending on the mass spectrometry instrument used, the proteomic approaches described in this chapter have the advantage of high sensitivity and can determine molecular masses to an accuracy of less than 1 part per million (<1 ppm) (Makarov *et al.*, 2006; Lu *et al.*, 2008; Scigelova and Makarov, 2006). Instrument manufacturers and academic researchers are constantly improving instrument technologies and developing new methods to identify PTMs (Garcia *et al.*, 2005, 2007). For example, recent advances in the field include such developments as electron transfer dissociation (ETD) for the fragmentation of peptides and dynamic detectors capable of accurately measuring masses of undigested proteins (Coon *et al.*, 2004; Mikesh *et al.*, 2006; Pitteri *et al.*, 2005; Syka *et al.*, 2004).

Enrichment strategies to selectively isolate proteins or peptides with a desired PTM can be coupled with mass spectrometry-based proteomic techniques. These enrichment techniques can alleviate the problems created by rare modifications. Among the variety of affinity enrichment strategies available, there are two major categories. First are approaches that use antibodies to recognize a specific PTM or uniquely modified peptide. For example, antiphosphotyrosine antibodies are used to enrich for peptides with phosphotyrosine residues (Blagoev *et al.*, 2004; Rush *et al.*, 2005; Zhang *et al.*, 2005). Second, there are existing and emerging technologies to enrich for PTMs based on the chemical affinity of a modification for an immobilized resin. Such techniques include immobilized metal affinity chromatography (IMAC) for phosphorylations and lectin chromatography for glycosylations (Ito *et al.*, 2009).

Another widely used proteomic technique useful in the identification of PTMs is two-dimensional gel electrophoresis (2D-GE). PTMs can alter the isoelectric point and electrophoretic mobility of a protein in a 2D-gel experiment. When such a change is detected between different cell types or growth conditions, the protein can be isolated and sequenced to identify its PTMs. For example, the differential phosphorylation of a protein will alter its isoelectric point and may cause charge heterogeneity. The differentially phosphorylated protein may appear as a train of 2D spots with different isoelectric points but with similar molecular weights. This pattern is sometimes referred to as "pearls-on-a-string." Specific protein staining methods for revealing protein PTMs have been devised over the years (Ge *et al.*, 2004; Patton, 2002). These include fluorescent methods for the direct detection of phosphoproteins and glycoproteins in gels (Ge *et al.*, 2004;

Steinberg *et al.*, 2001; Schulenberg *et al.*, 2003a,b, 2004). For these 2D spots, the proteins can be in-gel digested, and the recovered peptides can be analyzed by mass spectrometry to identify, validate, and map the expected PTMs (Hayduk *et al.*, 2004; Steinberg *et al.*, 2003). However, the 2D-GE approaches to identifying PTMs are not trivial since the recovery and analysis of the modified peptides are often problematic.

Multiple tools now exist in proteomics to quantify the absolute or relative abundance of proteins and their specific PTMs (Paoletti and Washburn, 2006). *In vivo* and *in vitro* labeling methods have been developed to use mass spectrometry for quantifying PTMs and precisely measuring their changes during cellular events (Goodlett *et al.*, 2001; Gygi *et al.*, 1999; Oda *et al.*, 1999; Ong *et al.*, 2002; Ross *et al.*, 2004; Zhang *et al.*, 2005). Quantitation can be crucial in determining the biological significance of a given PTM, since simply identifying the presence of a modification may not provide sufficient biological information to model its importance. A study by Matthies Mann and coworkers performed in HeLa cells found that 14% of the identified phosphorylation sites were modulated at least two fold by epidermal growth factor stimulation, demonstating the importance of quantifying PTMs (Olsen *et al.*, 2006).

There are, however, several limitations that must be considered when using mass spectrometry-based proteomics to identify PTMs. Some PTMs, such as phosphorylation of serine, tyrosine and threonine, and either *O*-linked or *N*-linked glycosylation, are labile, and maintaining the modification during sample preparation can be difficult. If ineffective separation techniques are used, unmodified peptides or proteins can mask the modified form during the mass spectrometry analysis. The detection of substoichiometric PTMs is especially difficult when the majority of the protein molecules may be unmodified. For mass spectrometry analysis, there is a multitude of instruments available from different manufacturers that all have benefits as well as disadvantages. Proteins are typically digested into peptides, and the peptides are directly analyzed. Modification of a given peptide can reduce its ionization efficiency, thereby compromising the mass spectrum and preventing sequence determination. Alternatively, there can be ambiguity as to which amino acid is modified, since many peptides contain multiple potential amino acid targets for the PTM. In addition, when using collision induced dissociation (CID) to fragment peptides containing labile PTMs such as phosphorylation and glycosylation, the labile group undergoes a β-elimination reaction to generate a neutral loss fragment. This pathway is more energetically favorable than fragmentation at the amide bonds. The resulting mass spectrum often fails to generate enough sequencing ions to unambiguously identify the peptide or modified amino acid. Finally, the detection of rare PTMs can be challenging, and identification often requires highly sensitive mass spectrometry methods or PTM enrichment strategies.

An important concern for proteomic data generated with any mass spectrometry technique is the volume and complexity of the information obtained. A typical experiment on a heterologous mixture of digested proteins analyzed by liquid chromatography coupled with tandem mass spectrometry can generate tens of thousands of spectra. Within the spectra obtained, there is the possibility of a nearly endless combination of PTMs and peptide adducts responsible for the peaks observed by the mass spectrometer. The large amounts of data necessitate an *in silico* approach involving protein database search algorithms to match theoretical spectra generated from genomic databases to the experimentally observed spectra. Searching the experimental spectra to identify PTMs requires some *a priori* knowledge of the nature of the potential modifications. This arises as a result of a limit to the number of modifications that can be taken into consideration when searching data due to computational resource limitations. For example, when considering only phosphorylation of serine (S), threonine (T), and tyrosine (Y), which adds a nominal mass of 80 Da, the computational resources required are exponentially higher than when searching the same dataset without the mass shift. This is due to the algorithm not only having to consider peptides without any modification but also having to consider all the permutations when the S, T, or Y residues are unmodified or modified in the protein sequences. To address this issue, several groups have been developing software algorithms to identify unanticipated PTMs from mass spectrometry data (Chalkley *et al.*, 2008; Hansen *et al.*, 2001, 2005; Tang *et al.*, 2005). The algorithms calculate mass differences between predicted peptide sequences and experimental MS/MS precursors and localize the mass shift to a sequence position in the peptide (Hansen *et al.*, 2005).

This chapter will highlight the existing and emerging techniques in the field of mass spectrometry and proteomics that are available to identify PTMs. We will focus on some of the most frequently studied modifications, although many of these techniques can be extended well beyond the scope of this chapter. As well as discussing the modifications, we will also include an overview of the available technologies in tandem mass spectrometry instrumentation that affect the ability of the investigator to identify specific PTMs. Previous global analysis studies of PTMs will be considered to emphasize the multitude of difficulties that arise when attempting to discern useful information from the vast array of data generated in these large-scale screens. Example workflows will be presented to assist researchers in developing a method that is most applicable to their current interests and available resources. In addition, we will include an overview of the various tools available that allow further quantification of PTMs. It will become evident that there are a growing number of resources and technologies at the disposal of the investigator, many of which have overlapping applicability to identifying PTMs.

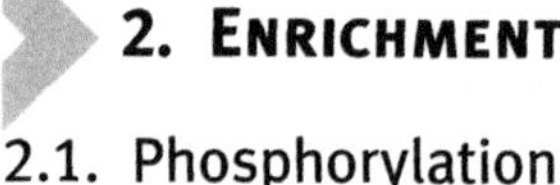

2. Enrichment Techniques for Identifying PTMs

2.1. Phosphorylation

The reversible phosphorylation of serine, threonine, and tyrosine residues is probably the most heavily studied PTM. Protein phosphorylation signaling networks mediate cellular responses to a variety of stressors, growth factors, cytokines, and cellular interactions. Phosphorylation also influences a multitude of cellular processes such as proliferation, apoptosis, migration, transcription, and protein translation (White, 2008). Aberrant regulation of protein kinases and phosphatases has been implicated in cancer, autoimmune diseases, metabolic disorders, and infectious diseases (Blume-Jensen and Hunter, 2001; Gatzka and Walsh, 2007; Sirard *et al.*, 2007; Taniguchi *et al.*, 2006). Single phosphoryation events can have dramatic implications on cellular processes. For example, the dephosphorylation of eIF4E-binding protein is triggered by nutrient deprivation, environmental stress, or infection with some picornaviruses and results in a marked increase in eIF4E-binding activity and an inhibition of translation (Gingras *et al.*, 2001).

The traditional method of identifying phosphorylation sites involves growing cells with ^{32}P-labeled ATP or phosphoric acid. The phosphoproteome incorporates the radioactive phosphate from the growth medium. The radioactive proteins are detected using such techniques as 2D-GE or high-performance liquid chromatography (HPLC). The fractionated proteins are then isolated and hydrolyzed. The resulting phosphopeptides are separated on TLC plates and subsequently sequenced by Edman degradation. However, these techniques are time consuming, require large amounts of relatively pure phosphoproteins, and necessitate the use of significant amounts of radioactive material (McLachlin and Chait, 2001).

Because of these drawbacks, mass spectrometry-based proteomics has quickly emerged as the preferred technique for identifying phosphorylations. A recent proteomic study comparing *Escherichia coli*, *Lactococcus lactis*, and *Bacillus subtilis* found that nearly every enzyme in the glycolytic pathway, which mediates carbon metabolism, were phosphorylated on various serine, threonine, and tyrosine residues (Soufi *et al.*, 2008). Interestingly, they also found that the detected sites of phosphorylation appear far more evolutionarily conserved than the primary sequence (Soufi *et al.*, 2008). A separate study investigating the mouse liver phosphoproteome identified many of the phosphorylation sites previously annotated in the Swiss-Prot mouse database (Pan *et al.*, 2008). However, more than half of the identified sites were novel, suggesting that there are still many more phosphorylation sites yet to be identified in the mouse phosphoproteome (Pan *et al.*, 2008).

Mass spectrometers have the ability to identify specific sites of phosphorylation within complex protein mixtures. This approach does present some problems that are often faced in other proteomic applications, including limited sample amounts, complex samples, and a wide dynamic range of protein concentrations within the sample (White, 2008). To compound the difficulty of these experiments, phosphorylation is often substoichiometric, and therefore the phosphopeptide is at a lower concentration than the other peptides from the same protein. The large population of unmodified peptides suppresses the mass spectrometric response to the phosphopeptides. This phenomenon can lead to a lack of detectable signal for the phosphopeptide of interest (McLachlin and Chait, 2001). Suppression artifacts can be minimized by reducing the fraction of unmodified peptides relative to the PTM variants using any of a variety of enrichment strategies.

One such enrichment strategy applicable to identifying phosphorylations is the use of strong cation exchange (SCX) resin to selectively isolate phosphopeptides. It has been found that, with a pH of less than 2.6, tryptic phosphopeptides are not retained by the SCX particles (Beausoleil *et al.*, 2004). This is likely due to their more anionic nature afforded by the phosphate group. The flow through of the SCX column can be subsequently fractionated with a reverse phase (RP) column and subjected to mass spectrometry analysis (Lim and Kassel, 2006). The separation of the phosphorylated peptides from their unmodified counterparts helps alleviate the problem of suppression during mass spectrometry as discussed above. This enrichment strategy has the advantage of being applicable to both off-line and on-line fractionation approaches. One negative aspect of this approach is that it relies on the inability of the SCX resin to bind the phosphopeptides, as opposed to other strategies that bind the phosphorylated residues and thereby positively enrich for them selectively.

Immobilized Metal Affinity Chromatography (IMAC) is the most widely utilized method for selectively isolating phosphopeptides from complex mixtures of digested proteins. This method typically utilizes a metal chelating agent to bind trivalent metal cations, such as Fe^{3+} or Ga^{3+} (Sykora *et al.*, 2007). The charged resin is subsequently used to bind the phosphorylated peptides. The resin with bound phosphopeptides is washed with a rinse solution, typically an acetic acid solution, to remove unmodified peptides. The phosphopeptides can then be eluted with phosphate buffer or a high pH solution either directly onto a RP column for desalting and subsequent mass spectrometric analysis or off-line for further enrichment or manipulation. The IMAC approach does present some complications, however. If multiply phosphorylated peptides are present in abundance, they can overwhelm the IMAC resin and result in the retention of fewer singly and doubly phosphorylated species. Additionally, other acidic peptides will have an affinity for the IMAC column and will be enriched along with the phosphorylated peptides. This problem is exacerbated by exceedingly complex peptide mixtures such as whole cell extracts. In an attempt to avoid this problem, many investigators methyl esterify the acidic residues and the

C-terminus using a chemical reduction procedure (Ficarro *et al.*, 2002). However, studies have found that the methyl esterification reaction is not quantitative (Cirulli *et al.*, 2008). Thus, the peptide of interest is present as modified, partially modified and unmodified versions. The complexity of the mixture is increased, and the relative amount of a given peptide is actually reduced (Cirulli *et al.*, 2008).

An approach analogous to IMAC is the use of titanium dioxide (TiO_2) as a substitute for the metal chelating resin (Larsen *et al.*, 2005; Pinkse *et al.*, 2004; Thingholm *et al.*, 2006). Just as the phosphate groups of the modified peptides have affinity for the trivalent metal used in IMAC, they also have an affinity for the titanium dioxide molecules under acidic conditions (Larsen *et al.*, 2005; Pinkse *et al.*, 2004; Thingholm *et al.*, 2006). By shifting to a high pH buffer, the phosphopeptides can be eluted from the TiO_2 material. This approach has the advantage of requiring less column preparation time and fewer rinse cycles than the IMAC resin. Studies have shown that, although these two methods use the same principle of affinity, they are complementary in that distinct phosphopeptides are detected by each of the two approaches (Cantin *et al.*, 2007). IMAC is characterized by a higher affinity for multiply phosphorylated peptides while titanium dioxide preferentially binds singly phosphorylated species (Bodenmiller, *et al.*, 2007). This highlights the utility of using both approaches in a single investigation to obtain maximal coverage of the phosphoproteome.

Another method for phosphopeptide enrichment employs chemically introduced affinity tags at the sites of phosphorylation (Goshe *et al.*, 2001; Oda *et al.*, 2001). One such method uses an alkaline environment to remove H_3PO_4 from phosphoserine or phosphothreonine residues in a β-elimination reaction (McLachlin and Chait, 2003; Oda *et al.*, 2001). A dithiol group is then added to the double bond in a Michael-like addition reaction. The thiol is then treated with an alkylating reagent linked to a biotin group. The resulting biotinylated peptides can then be enriched using avidin column chromatography (Fig. 40.1) (McLachlin and Chait, 2003). One difficulty of this approach is the recovery of the tagged species can be problematic due to the high affinity of biotin for avidin. Also, this method enriches for phosphoserine and phosphothreonine, not phosphotyrosine. *O*-linked β-*N*-acetylglucosamine (*O*-GlcNAc) modifications are also succeptible to β-elimination followed by Michael addition and can therefore compete with phosphorylations for enrichment (Wells *et al.*, 2002). Finally, under the alkaline environment of the β-elimination reaction, unwanted side reactions can occur that have unanticipated affects on peptide mass. These side reactions can make it difficult to identify the peptides downstream during mass spectrometry.

During mass spectrometry, specific chemical properties of phosphoproteins must be taken into account. Phosphorylation is a labile addition to S and T amino acids. Traditional fragmentation methods result in the preferential neutral loss of H_3PO_4 or HPO_3 from phosphoserine and phosphothreonine

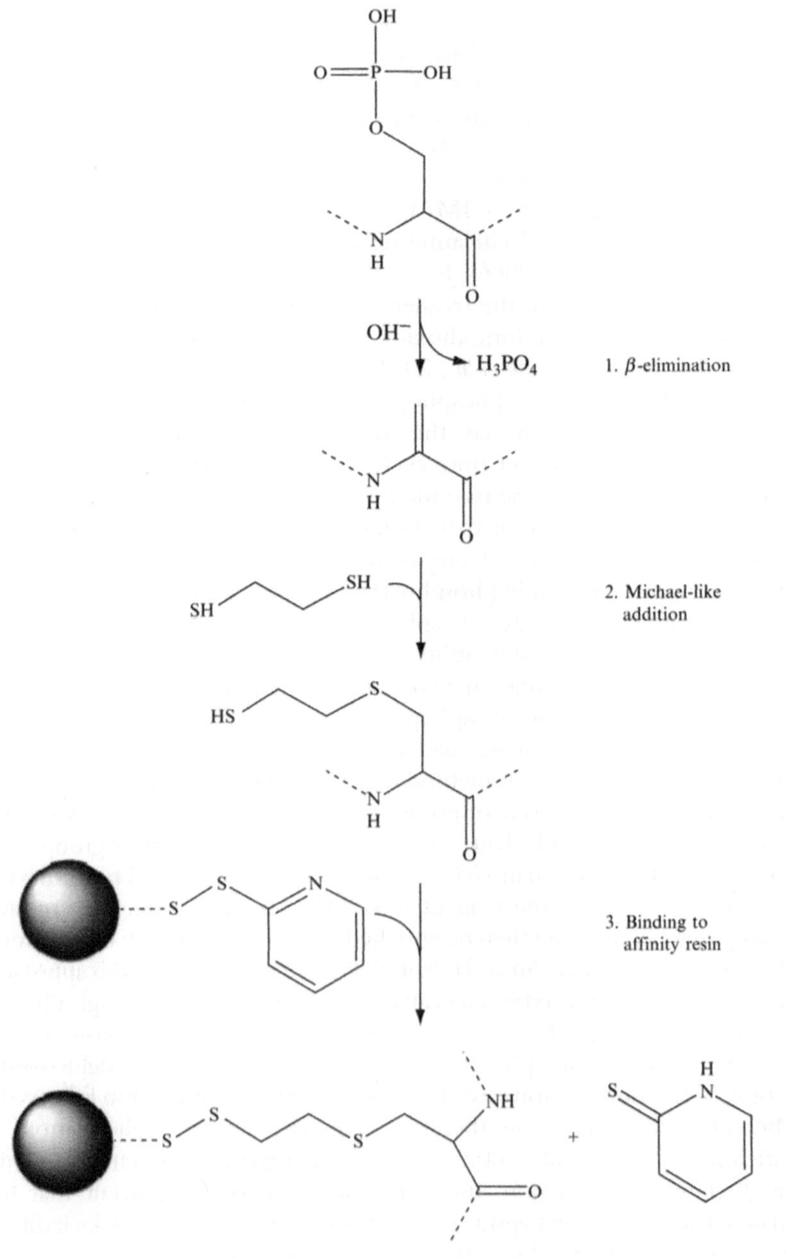

Figure 40.1 The chemical modification and affinity purification of proteins bearing a phosphoserine. Reprinted with permission from McLachlin and Chait (2003).

residues prior to fragmentation along the peptide backbone. The vast majority of the population undergoes β-elimination reaction, leaving only a fraction remaining to fragment along the backbone (Bennett *et al.*, 2002; Lee *et al.*, 2001; Ma *et al.*, 2001). The resulting mass spectrum contains fewer sequencing ions, and these ions are generally lower in intensity. Therefore, there is a low signal-to-noise ratio (Fig. 40.2), which hampers interpreting the peptide sequence. Although capable of losing the phosphate group, phosphotyrosine seldom generates the intense neutral loss ions that are seen for peptides containing phosphoserine or phosphothreonine residues. Some mass spectrometers that use ETD avoid this neutral loss problem and can yield spectra suitable for both sequence analysis and phosphoresidue determination. This alternative fragmentation method will be discussed further when we consider the various instrument types used in the proteomic identification of PTMs.

On the other hand, neutral loss artifacts can be used to the advantage of the investigator. The neutral loss signature is a conspicuous indicator that the peptide contains a site of phosphorylation. The neutral loss peak observed in the mass spectrum is generally of high intensity due to its preferential formation. Therefore, if a loss of 98 m/z (H_3PO_4) or 80 m/z (HPO_3) for a $+1$ charged peptide is observed from the precursor ion, one can infer the presence of a phosphoserine or phosphothreonine reside within the sequence of the peptide. Modern mass spectrometers can monitor for this mass loss. An ion trap mass spectrometer can isolate the neutral loss species and initiate another round of fragmentation (MS^3 scan) (Beausoleil *et al.*, 2004; Gruhler *et al.*, 2005; Olsen and Mann, 2004; Ulintz *et al.*, 2009). This additional analysis step can result in additional sequence coverage to aid in identifying the phosphopeptide and mapping the PTM to a specific amino acid. An alternative approach to an MS3 scan is "Pseudo MS^n" or multistage activation (MSA) for phosphopeptide fragmentation in an ion trap mass spectrometer (Schroeder *et al.*, 2004). The method induces CID of the neutral loss product from the MS/MS scan of the phosphopeptide. The product ions from the neutral loss ions along with the initial MS/MS product ions are trapped together resulting in a composite spectrum. In MSA, the neutral loss product ions are converted into a variety of structurally informative fragment ions, which show improved scores in database search algorithms (Schroeder *et al.*, 2004; Ulintz *et al.*, 2009).

Phosphorylation introduces a negative charge onto the peptide, and this also affects analysis by mass spectrometry. The negative charge can reduce the mass spectrometer response when operated in positive-ion mode (McLachlin and Chait, 2001). This diminished response results in poor quality spectra and makes identifying the phosphopeptide more difficult. Furthermore, this increased negative charge can reduce proteolysis during preparation of protein mixtures prior to mass spectrometry. As a caveat, Hanno Steen and coworkers found no evidence for decreased ionization/

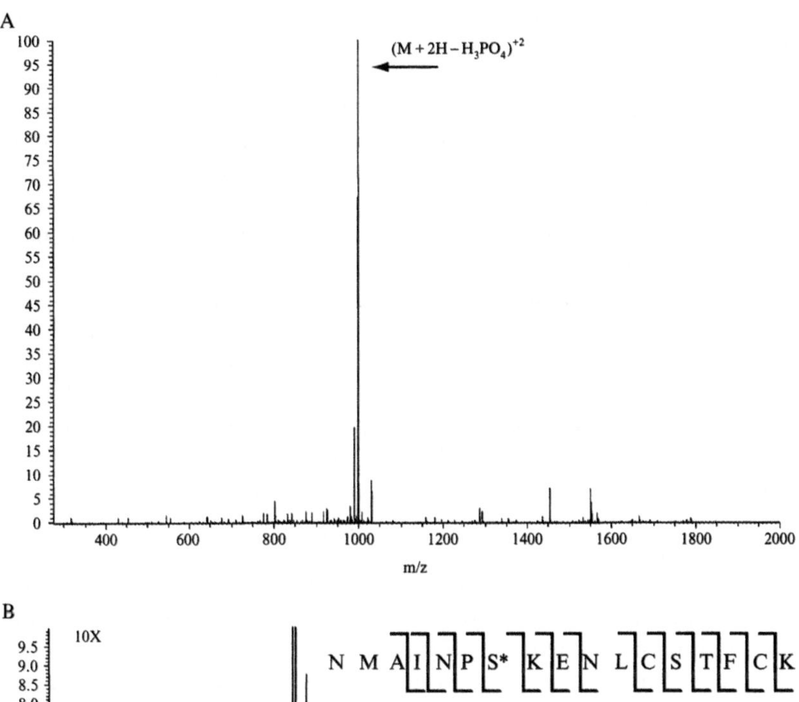

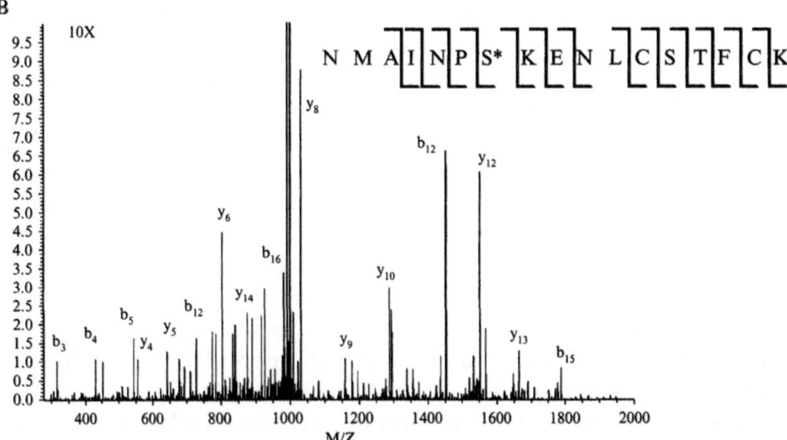

Figure 40.2 Phosphopeptide MS/MS spectrum from a bovine β-casein tryptic digest. Panel A demonstrates the intense neutral loss peak observed at 999.2 m/z from the precursor 1048 at a +2 charge state (−49 Da). This is magnified in Panel B by a factor of 10× to reveal the sequencing ion peaks.

detection efficiencies when selectively investigating phosphopeptides using electrospray ionization (ESI) (Steen *et al.*, 2006).

As an alternative, a precursor ion scanning technique pioneered by Steve Carr and coworkers can be employed to avoid some of the problems generated by performing MS in the positive-ion mode (Carr *et al.*,

1996). Tandem mass spectrometry is performed in negative-ion mode, and phosphorylated residues generate characteristic fragments of −79 Da for PO_3- or −63 Da for PO_2-. When one of these events is detected in the precursor ion scan, MS/MS is preformed on the precursor ion (Zappacosta *et al.*, 2006). This approach can be more sensitive than positive mode mass spectrometry for detecting phosphopeptides (Carr *et al.*, 1996). However, negative-mode spectra are difficult to interpret and not widely studied (Witze *et al.*, 2007). One interesting approach combines both positive-ion mode and negative-ion mode in the same experiment. Precursor ion scanning is performed in the usual negative-ion mode, the instrument's polarity is switched, and MS/MS fragmentation spectra are obtained in the positive-ion mode (Carr *et al.*, 1996). Although identifying sites of phosphorylation is challenging, many proteomic techniques are available that can be customized for characterizing these critically influential PTMs.

2.2. Glycosylation

Glycosylation is a common PTM and has been shown to affect enzyme activity, protein localization, stability, signaling, cell adhesion, and protein interactions (Spiro, 2002). Changes in glycosylation patterns on the surface of cells and on proteins within biological fluids have been discovered in malignant transformation and tumorogenesis (Durand and Seta, 2000). A variety of glycosylations can be present within a proteome. *O*-glycosylated proteins are modified at serine and threonine, while *C*-glycosylation targets tryptophan and *N*-glycosylation targets asparagine. Proteins with *N*-glycosylation share several commonalities such as core structures, prevalence on the extracellular membrane, and enzymes that remove the *N*-linked structures (Medzihradszky, 2005). *N*-glycosylation sites have has an established consensus sequence of N-X-S/T where X is any amino acid except proline (Pless and Lennarz, 1977). Another modification that targets serine and threonine residues is *O*-linked β-*N*-acetylglucosamine (*O*-GlcNAc). Gerald Hart and coworkers have pioneered the study of this dynamic nucleocytoplasmic PTM (Holt and Hart, 1986). This modification targets many of the same proteins selected for phosphorylation. It is postulated that this modification may serve an antagonistic role to protein phosphorylation (Ahmad *et al.*, 2006).

Some of the same problems to identify phosphorylations with mass spectrometry also apply to the study of glycosylations. These modifications tend to be substoichiometric in their abundance relative to unmodified proteins. Glycosylation and modification by *O*-GlcNAc are both labile in the mass spectrometer and often yield poor quality spectra for sequencing. Even though there is a well characterized consensus sequence known for the *N*-glycosylation of proteins, only a subset of the proteins containing this sequence are glycosylated *in vivo*.

In addition to the alternative peptide fragmentation options mentioned for phosphorylation, which also apply to glycosylation, enrichment methods exist for overcoming the issue of low-abundance glycosylations (Mechref *et al.*, 2008). Glycoproteins can be selectively enriched using lectin-affinity chromatography. Lectins are highly specific sugar-binding proteins that can be linked to a solid-phase support. The resin is then exposed to the complex mixture of digested peptides, washed and eluted to isolate an enriched fraction of glycopeptides suitable for proteomic analysis. *O*-GlcNAc can be enriched using β-elimination followed by Michael addition of dithiothreitol or biotin pentylamine in a manner analogous to that previously highlighted for phosphorylations (Wells *et al.*, 2002). Another method is solid-phase extraction of *N*-linked glycopeptides (SPEG) that utilizes hydrazine chemistry to attach the glycopeptides directly to the solid-phase support (Tian *et al.*, 2007). Washes are performed and the glycopeptides are eluted by treatment with peptide-*N*-glycosidase. The lectin-affinity chromatography approach has the advantage of binding a wide array of glycoproteins and can effectively isolate different classes of glycopeptides by eluting with different glycosidases (Yang and Hancock, 2005). There are also specialized columns that contain lectins that uniquely bind specific classes of glycolytic linkages, affording another degree of selectivity in the enrichment (Tian *et al.*, 2007). One caveat of using glycosidases to elute the glycopeptides is that it cleaves the oligosaccharide from the peptide so that any structural information that could be gained by mass spectrometry analysis is lost. However, this does simplify the data analysis since glycosylations can be complex in the nature of their structure and exist in varied polyglycosylated forms.

2.3. Ubiquitination and sumoylation

Ubiquitin is a 76-amino acid protein that is conjugated to an alpha-amino group of a substrate lysine via its C-terminal glycine residue. This process requires the action of E1, E2, and E3 enzymes (Fang and Weissman, 2004). The sequence of ubiquitin contains seven lysines that can also be selected for ubiquitination, leading to the formation of ubiquitin chains (Fang and Weissman, 2004). The internal lysines within ubiquitin that are targeted for further ubiquitination can determine the biological significance of this modification. For example, K48 linked ubiquitination chains signal the attached protein for degradation by the 26S proteosome, and K63 linkages are involved in DNA damage responses, protein trafficking, and signal transduction in NF-κB signaling (Ikeda and Dikic, 2008; Pickart and Fushman, 2004). The turnover of ubiquitinated proteins is very rapid which results in low steady-state levels of this PTM (Peng *et al.*, 2003). To further complicate the prediction of ubiquitination sites, there are little data to support a common sequence motif for this PTM (Sliter *et al.*, 2008).

Another small protein molecule that is similar to ubiquitin is the small ubiquitin-like modifier (SUMO). Although the two proteins are only 18% identical, they show similarity in their 3D structures (Bayer *et al.*, 1998). Sumoylation and ubiquitination share the same E1, E2, and E3 enzymes, which conjugate the SUMO molecule to an internal lysine residue within a conserved motif. Sumoylation occurs preferentially at ψ–K–X–E sequences, where ψ denotes a hydrophobic amino acid (Rodriguez *et al.*, 2001). Sumoylation targets a multitude of proteins that are involved in such diverse cellular processes as transcriptional regulation, nuclear body formation, nuclear pore complexes, and DNA repair (Melchior *et al.*, 2003). Sumoylation can compete with ubiquitination, thereby preventing protein degradation and stabilizing key enzymes needed by the cell. This modification also has the ability to target proteins to specific cellular locations such as the nuclear pore complex (Manza *et al.*, 2004).

Effective enrichment strategies have been developed to isolate proteins modified by ubiquitination and other ubiquitin-like modifications. The most successful of these approaches have used amino-terminal epitope tags incorporated into the genetic sequence of the ubiquitin and ubiquitin-like molecules (Peng *et al.*, 2003; Tagwerker *et al.*, 2006). Polyhistidine tags have been used, as well as tandem-affinity tags composed of both a polyhistidine tag and a FLAG tag (Chang, 2006). The tandem-affinity tag purification method reduces but does not eliminate the rate of false positive discovery. While it is possible to use antibodies generated against the ubiquitin class of molecules, their use in immunoprecipitation reactions to isolate ubiquitinated proteins is still in development. Ubiquitin-binding proteins may also be used to avoid problems associated with using antibodies and affinity tags (Kirkpatrick *et al.*, 2005). Epitope tags are convenient in simple model systems such as *Saccharomyces cerevisiae* because the multiple genes that code for untagged ubiquitin can be deleted so that the only ubiquitin present in the cells contains the epitope tag.

The fact that both ubiquitination and sumoylation are proteinaceous PTMs enables different approaches to identifying the protein targets with mass spectrometry. When trypsin digesting ubiquinated proteins, part of the ubiquitin molecule is cleaved. The remaining tryptic fragment at the site of ubiquination contains a two residue glycine–glycine remnant from the C-terminus of ubiquitin linked to the conjugate lysine through a isopeptide bond (Peng *et al.*, 2003). This peptide causes a nominal mass shift of 114 Da at the lysine residue as well as a missed tryptic cleavage since trypsin cannot recognize the modified lysine. For other ubiquitin-like modifications, there are characteristic mass shifts on lysine associated ubiquitination and ubiquitin-like modifications as shown in Table 40.2 (Denison *et al.*, 2005). All these properties can be taken into consideration when searching spectra for potential PTMs of this type.

Table 40.2 Ubiquitin-like modifications and their associated nominal mass remnants from humans and yeast

Ubiquitin-like modifier	Organism	Nominal mass (Da)	Residual sequence
SUMO-1	Human	2137	(K)-ELGM...QTGG
SUMO-2,3	Human	3552	(R)-FDGQ...QTGG
SUMO	Yeast	485	(R)-EQIGG
URM1	Yeast	2035	(K)-DYILE...LHGG
NEDD8	Yeast	114	(R)-GG
NEDD8	Human	114	(R)-GG
ISG15	Human	114	(R)-GG

One obvious problem with using a mass spectrometry-based proteomics approach to studying this PTM is the redundancy in mass shifts as indicated in Table 40.2. From the spectra of an identified peptide, one may not be able to discern if the modification is ubiquitination or an ubiquitin-like modification since both types can result in the addition of 114 Da to the lysine residue. Sumoylation adds a large mass to the peptide that can affect the ionization of the peptide and reduce the efficiency of ionization. The peptide fragmentation that occurs on the SUMO modification can complicate the resulting mass spectrum to the point that it is uninterpretable. One laboratory took advantage of the latter problem by developing sumoylation pattern recognition software (SUMmOn) to search for fragment ion patterns produced by complex PTMs and to identify modified peptides with their corresponding sites of modification (Pedrioli *et al.*, 2006). The SUMmOn technology allowed them to both identify a user-specified fragment ion series and characterize a modified peptide (Pedrioli *et al.*, 2006).

3. Nitrosative Protein Modifications

Nitration and nitrosylation of tyrosine, tryptophan, methionine, and cysteine side chains comprise the bulk of nitrosative protein PTMs. These additions are mediated by reactive nitrogen species produced during development, oxidative stress, and aging. Increases in reactive nitrogen species result from the reaction of excess or deregulated nitric oxide with reactive oxygen species (Yeo *et al.*, 2008). The reactive nitrogen and oxygen can target DNA, lipids, and proteins (Barnes *et al.*, 2003). Depending on the radical generated, the reactive nitrogen species will preferentially react with tyrosine residues to form either 3-nitrotyrosine or 3-nitrosotyrosine (Beckman *et al.*, 1990). Tyrosine nitrosative modifications are perhaps best characterized by the adducts generated. Increased levels of tyrosine nitration have been implicated in age-related neurodegenerative diseases and can serve as a biomarker for oxidative stress (Yeo *et al.*, 2008). Under normal physiological conditions, the formation of nitrosative tyrosine adducts is prevented by the presence of reducing agents such as glutathione (Chen *et al.*, 2004). Tyrosine nitration shifts the pK_a of the attached hydroxyl group from 10.1 to 7.2 and can have consequences in the structure and function of the modified protein (Sokolovsky *et al.*, 1967).

There are no known selective enrichment strategies for the direct isolation of nitrated peptides or proteins. This is postulated to be a result of the low chemical activity of the nitric oxide groups of this PTM. These fairly inert moieties can be converted to amines under reducing conditions with dithionite to generate amines that are reactive to tagging groups (Yeo *et al.*, 2008). The resulting tagged peptides can then be selectively isolated and analyzed by proteomic methods. The most widely applied approach to identify protein nitrosative modifications is use of a combination of 2D-GE and immunoblot analysis with antibodies specific to a given nitrosative adduct (Zhan and Desiderio, 2004). Proteins are resolved in the 2D-GE experiment, and proteins bearing the modification are visualized using the appropriate antibody. The immunopositive protein spots are then excised from the gel, digested and subjected to mass spectrometry identification (Zhan and Desiderio, 2004). In the case of tyrosine nitrosative modifications, a nominal mass shift of 45 Da for nitration and a shift of 29 Da for nitrosylation is added as a differential modification when searching mass spectrometry data to identify the sites of adduction.

4. Methylation and Acetylation

Lysine and arginine methylation (+14 Da) and lysine acetylation (+42 Da) are two types of PTMs that have come under intense scrutiny, namely for their role in the histone code (Jenuwein and Allis, 2001; Strahl and Allis, 2000). The first instance of finding an acetyl group linked to lysine occurred in a study of histone modifications (Jenuwein and Allis, 2001). Increased levels of histone acetylation are generally associated with local activation of gene transcription (Zhang *et al.*, 2002). Lysine methylation of histone proteins has been shown to localize to the promoters of repressed genes (Berger, 2007). The acetylation events have been found to be reversible and dynamic, while the methylation events are more stable and long-lived (Bernstein and Allis, 2005). Although much scientific effort has gone into studying these PTMs on histones, acetylation and methylation events occur on many proteins

(Glozak *et al.*, 2005; Grewal and Rice, 2004; Sadoul *et al.*, 2008; Yang and Seto, 2008). In eukaryotes, N-terminal acetylation is one of the most common PTMs, occurring on approximately 85% of eukaryotic proteins (Polevoda and Sherman, 2000). The first nonhistone protein found to be regulated by acetylation and deacetylation was the oncogene p53 (Gu and Roeder, 1997).

As with many of the other PTMs, the classical approach to identifying methylation and acetylation was the use of modification-specific antibodies. This technique is limited by cross-reactivity and specificity problems associated with an immunological-based method. In the case of histones, methods exist for their selective isolation from the nuclei of cells (Garcia *et al.*, 2007). A more general approach to identifying sites of acetylation and methylation is the use of an *in vitro* radiolabeling assay. This technique utilizes acetyltransferases or methyltransferases to catalyze the addition of radioactive acetyl or methyl donor group. Thin layer chromatography (TLC) or HPLC separation followed by microsequencing can reveal the radiolabeled amino acids (Sobel *et al.*, 1994). The radiolabeling approach has disadvantages in that it requires significant amounts of pure starting material and utilizes radioactivity to detect modified amino acids.

There are no efficient methods for the selective enrichment of peptides or proteins with methylation or acetylation modifications. Additionally, since acetylation and methylation occur on basic residues, the use of trypsin as a protease can be less effective in cleaving at its usual sites. This can be more dramatic in the case of methylation since the basic residues can be methylated at multiple amine sites. The obvious choice to overcome this problem is selecting an alternative protease. However, the benefit of the missed tryptic cleavages can work to the investigator's advantage, serving as an indicator that the site may be modified. Tandem mass spectral searching algorithms can take into account the average mass shift caused by acetylation and methylation of 42 and 14 Da, respectively, as well as the shift of 28 Da for dimethylation and 42 Da for trimethylation. Some ambiguity can appear when considering trimethylation since the mass shift between it and an acylated peptide is only 0.0363 Da. This mass difference can be resolved with the use of a mass spectrometer with a high enough degree of mass accuracy and resolution (Zhang *et al.*, 2004).

Another diagnostic tool to identify sites of acetylation and methylation is analysis of the immonium ions generated in CID type mass spectrometers. Peptides that harbor an unmodified lysine generally produce a characteristic immonium ion peak at 84 m/z (Trelle and Jensen, 2007). When the lysine is modified, a different immonium is formed at a unique m/z value (see Figure 40.3). Di- and trimethylated forms of the immonium do not form due to constraints of the rearrangement reaction that occurs. However, the trimethylated form produces a unique neutral loss fragment 59 Da less than the precursor due to loss of trimethylamine (Zhang *et al.*, 2004). Figure 40.3 shows the rearrangement reaction that takes place as

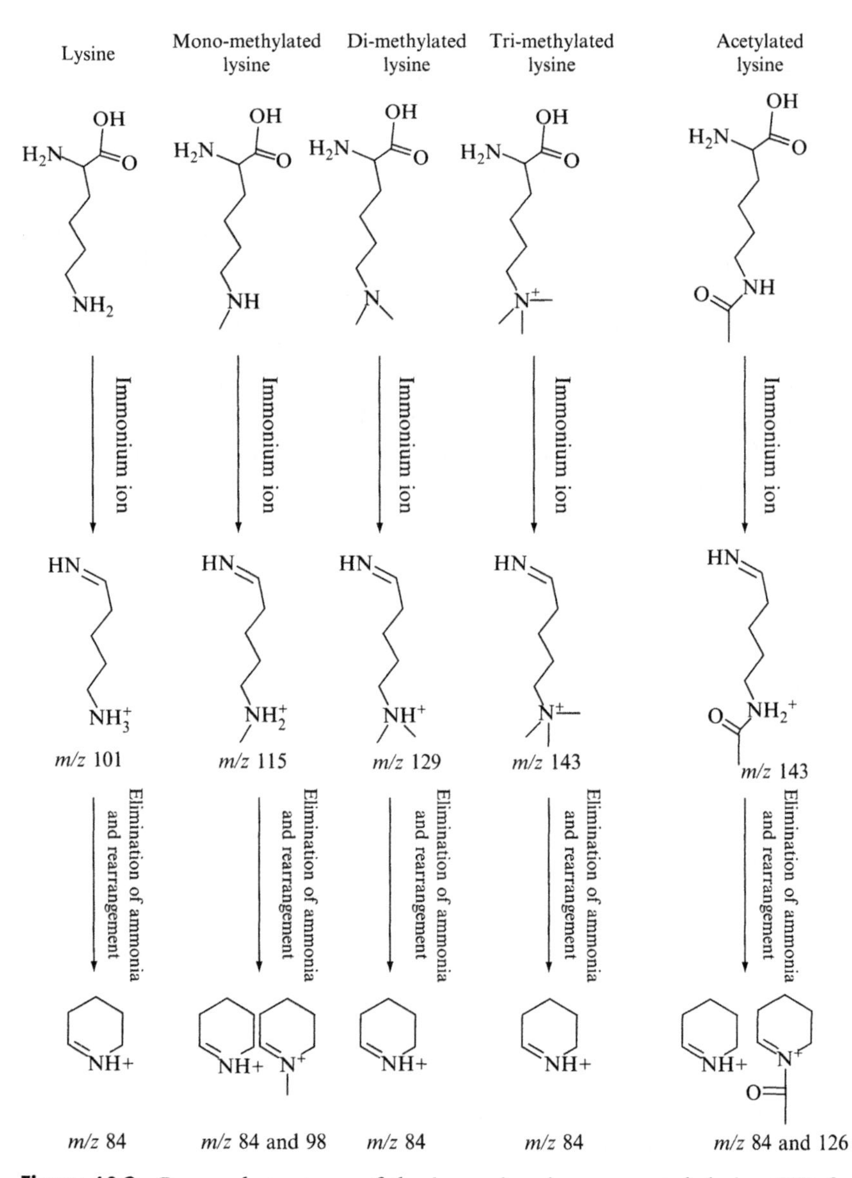

Figure 40.3 Proposed structures of the immonium ions generated during CID for mono-, di-, and trimethylated lysines. Reprinted with permission from Zhang *et al.* (2004).

well as the associated m/z values of the various immonium ions. These unique diagnostic characteristics are only generated in mass spectrometers that are capable of CID.

5. Mass Spectrometry Analysis

A variety of mass spectrometry instrument designs and configurations are available to investigate both the dynamic nature of the proteome and its associated PTMs. For identifying PTMs, two of the most important features of a mass spectrometer are its mass accuracy and resolution. Unlike protein identifications, which are typically based on identifying multiple independent peptides from the same protein, PTMs must be identified based on a single MS/MS spectrum. John Yates and his group have developed approaches using overlapping peptides produced from independent digestions with various proteases to increase the sequence coverage and reduce the ambiguity in mapping modifications (MacCoss *et al.*, 2002; Wu *et al.*, 2003). While the redundant peptide sequences reduce ambiguity, the approach requires sufficient starting material for multiple mass spectrometry analyses.

Since the identification of the PTM and its location on the peptide typically relies on a single MS/MS spectrum, mass spectrometers with high mass accuracy and resolution have a distinct advantage (Clauser *et al.*, 1999; Haas *et al.*, 2006; Mann and Kelleher, 2008). Mass spectrometry instrumentation varies in methods for peptide ionization, peptide fragmentation, ion isolation, and signal detection. Mass spectrometry-based analyses are also often coupled to chromatographic separation techniques to increase the level of sensitivity to detect peptides in a given biological sample and to fractionate complex peptide mixtures. As highlighted previously, many of these chromatographic approaches can enrich for the desired PTM allowing for a more effective method to identify the modifications in a complex protein or peptide mixture. It becomes evident that the combinations of instrument configurations and separation techniques lead to a dizzying array of possibilities for the investigator. In this section, we will attempt to discuss the advantages and possible pitfalls of the mass spectrometers available today.

Perhaps the most widely available and utilized mass spectrometers in proteomics use electrospray ionization (ESI) to introduce peptides into the gaseous phase for downstream analysis by the instrumentation. This class of instruments includes ion traps, triple quadrupoles, time-of-flight (TOF), quadrupole time-of-flight (q-TOF), orbitrap (Makarov, 2000), and Fourier transform ion cyclotron resonance (FT-ICR) designs. Ion trap instruments have the advantage of rapid scan times and high sensitivity. However, they are limited in their mass accuracy (±0.1–1 Da) and resolution. In ion trap mass spectrometers, the mechanism of CID does not allow for trapping of fragment masses below 28% of the precursor mass. As a direct result, the lower 1/3 of the MS/MS data is lost. This is referred to as the "1/3 rule" or "low-mass cut-off" limitation of CID fragmentation. Other mass spectrometers such as q-TOF, triple quadrupole, orbitrap, and Fourier transform instruments retain the low-mass-product ions. The low-mass ions may reveal information about the sequence composition of the peptides that can be very useful for validating PTMs from the database search.

In an attempt to circumvent the limited accuracy and resolution, the linear ion trap can be coupled to an orbitrap analyzer to form the hybrid LTQ-Orbitrap mass spectrometer. The mass accuracy is increased to low ppm (<5 ppm) and the increased resolution resolves the isotopic envelopes of the peptides so the charge state of precursor peptides can be determined. q-TOFs are excellent choices for studying PTMs since they use fragmentation methods that tend to retain labile peptide modifications. This is attributed to their higher energies of fragmentation and the shorter time scale in which the fragmentation occurs. The TOF mass analyzer provides high mass accuracy and resolution to analyze the precursor and fragment ions. For tandem mass spectrometry experiments, a TOF mass analyzer can be separated by a collision cell (TOF–TOF) for routinely generating MS/MS spectra (Medzihradszky *et al.*, 2000; Vestal and Campbell, 2005). FT-ICR instruments have the highest accuracy and resolution of the mass spectrometers available to investigators today. However, their high cost often makes them impractical for many laboratories. In addition, their low scan rate decreases the instrument's sensitivity and necessitates the use of larger amounts of analyte to obtain a measurable signal.

In general, ESI is best suited to identify small peptides that are acidic in nature (Covey *et al.*, 1991). ESI typically generates tryptic peptides bearing a +2 or +3 charge state, as opposed to MALDI techniques that typically produce ions with a +1 charge state. This complexity can make the interpretation of spectra more difficult unless an instrument with high resolution and mass accuracy is used to resolve peptide isotope clusters and determine the precursor or fragment ions, charge states. The routine coupling of these instruments to on-line fractionation using low-flow-rate chromatography allows for a highly sensitive and rapid approach to identifying peptides from a complex protein digest. One significant limitation to these on-line approaches is that spectra can only be obtained for a peptide as it is being eluted from a chromatographic column. That is, once the sample is eluted into the instrumentation, replicate analyses of the peptides are not available, leading to a temporal limit in data acquisition. This is further compounded by peptides that coelute, leading to suppression of the ionization of the relevant peptide. In addition, unmodified peptides and their posttranslationally modified counterparts often have similar elution times, resulting in a diminished signal for the modified peptide.

An alternative to ESI is matrix assisted laser desorption ionization (MALDI). MALDI instruments typically use TOF analyzers but can also be used with other mass analyzers. This ionization technique uses a laser to irradiate a sample consisting of an analyte (peptide) and a UV-absorbing matrix. The matrices are generally aromatic acidic compounds capable of absorbing the energy from the laser. A portion of the energy absorbed by the matrix is transferred to the analyte and both desorb into the gaseous phase where the analyte is ionized and travels downstream in the mass spectrometer for analysis. MALDI can be applied to either peptides or whole proteins. When used with the latter, PTMs on undigested proteins can be identified. This approach does however require some *a priori* knowledge of the modifications present to account for the mass shift in the observed protein from its unmodified counterpart. In addition to being well suited for larger analytes than ESI, MALDI also favors slightly basic peptides (Covey *et al.*, 1991).

The separation of peptides for analysis with MALDI is typically performed off-line of the mass spectrometry, in contrast to the in-line chromatographic separation for ESI. Samples can be fractionated by liquid chromatography and spotted onto MALDI target plates (Lochnit and Geyer, 2004; Pflieger *et al.*, 2008; Zhen *et al.*, 2004). Varying methods exist for spotting the plates, using automated robots specifically designed for the task. The matrix material can be either mixed with the sample and spotted, or spotted directly on the plate before or after applying the sample. The spot size varies according the plate format used and is typically in the microliter to submicroliter range. The off-line nature of this approach removes the temporal restrictions that apply to in-line chromatographic separations. This means that the investigator can return to a sample on a MALDI plate and perform replicate analyses of the same spot until the sample is depleted by the instrument's laser.

The rapid nature of MALDI-TOF instrument and the predominantly +1 charge states for ions makes them a convenient option for quickly measuring the masses of peptides and proteins. A digested protein sample can be spotted on the MALDI plate and a mass spectrum of the peptides generated in a short period time without significant training. The investigator may know from previous data what proteins are present and, therefore, the expected m/z values for the peptides. These known values can be compared with what is experimentally detected, and any differences can be attributed to modified forms of the peptides. This method is simple, fast, and independent of sequencing algorithms. For suspected phosphoproteins, alkaline phosphatase treatment of the sample combined with MALDI-TOF can be used to identify phosphopeptides (Larsen *et al.*, 2001; Zhang *et al.*, 1998). Differences in the peptides maps (80 Da) generated before and after phosphatase treatment can help identify phosphopeptides. Mapping the specific modified amino acid typically requires performing tandem mass spectrometry on the modified peptides using an instrument inherently capable of MS/MS (e.g., ion trap, q-TOF, TOF/TOF).

6. CID versus ECD versus ETD

Mass spectrometry-based proteomic analyses rely on the fragmentation of peptides in the gas phase at low collision energies to generate peaks in the mass spectrum. These resulting peaks are then used to determine the sequence of the peptide, and from this, the associated protein can be inferred. The primary method for peptide fragmentation is collision induced dissociation (CID) (Swaney *et al.*, 2008). LTQ, Orbitrap, q-TOF, MALDI-TOF/TOF, and FT-ICR instruments are all capable of performing CID fragmentation of peptides. CID is best utilized for small, low charged, and unmodified peptide cations (Dongre *et al.*, 1996; Good *et al.*, 2007; Huang *et al.*, 2005). The collision energy transferred to peptides in CID results in the vibrational excitation of the peptides and is distributed among the covalent bonds of the peptide chain. When the internal energy transferred to the peptide in CID exceeds the activation barrier for bond cleavage, the peptide can fragment. These fragments can be detected by the mass spectrometer if they occur within the timescale of the instrument. This fragmentation typically occurs at the amide bonds of the peptide backbone due to the low activation barrier that exists at these locations. This generates the characteristic b- and y-product ions seen in CID spectra (Swaney *et al.*, 2008).

Although CID is effective in generating ion series suitable for peptide sequencing, some caveats exist that diminish its ability to fragment particular peptides. Internal basic residues and proline amino acids can prevent the random protonation of the peptide backbone. These residues direct amide bond dissociation to specific sites and can inhibit the fragmentation of the peptide resulting in an insufficiently diverse set of sequencing ions (Mikesh *et al.*, 2006). The common use of trypsin as a protease alleviates some of these concerns by generating peptides with a basic residue at the C-terminus. Labile PTMs such as phosphorylation and glycosylation also present alternative low-energy fragmentation pathways. In these cases, the aforementioned neutral loss events can occur, generating spectra with insufficient sequencing ion coverage. Some PTMs can also inhibit the random protonation of the peptide backbone and subsequently inhibit fragmentation via CID (Tsaprailis *et al.*, 1999).

An alternative fragmentation method termed electron-capture dissociation (ECD) has the ability to overcome some of the limitations posed by CID. In ECD, protonated peptides are held in the Penning trap of an FT-ICR and exposed to a beam of electrons at thermal or near-thermal energies. The capture of the thermal electron by the protonated peptide results in backbone fragmentation without intramolecular vibrational energy redistribution (Udeshi *et al.*, 2007). One pathway available to account for the fragmentation of the peptide immersed in thermal electrons is depicted in Fig. 40.4. ECD

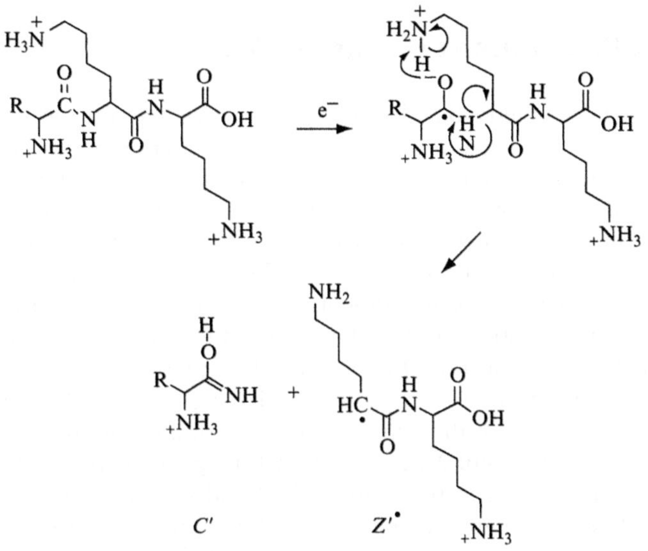

Figure 40.4 Fragmentation scheme for production of ions of type c and $z^{\bullet}$ by reaction of a low-energy electron with a multiply protonated peptide. Reprinted with permission from Udeshi *et al.* (2007).

fragmentation generally produces c- and z-ions series as opposed to the b- and y-ions generated in CID. ECD is a size and sequence independent process and can be used to fragment intact proteins. However, for ECD to be efficient, the sample must be in a dense population of the thermal electrons. This situation is technically difficult to achieve on instruments using electrostatic radiofrequency fields to isolate ions but is easier in FT-ICRs that use static magnetic fields. In addition, ECD spectra suitable for sequencing require the averaging of a large number of scans on a time scale of minutes. This necessitates the introduction of large amounts of analyte into the instrument and precludes the detection of peptides or proteins in a complex sample mixture (Udeshi *et al.*, 2007).

The use of a similar fragmentation method electron transfer disassociation (ETD) overcomes the limitations generated by ECD.

ETD uses ionized radical anions of polyaromatic hydrocarbons to react with multiply protonated peptides in the gas phase. These radical anions are stored in the linear quadrupole ion trap of the mass spectrometer. The radical anions transfer an electron to the ionized peptides and the resulting charge-reduced peptide fragments in a mechanism believed to be analogous to ECD, generating characteristic c- and z-sequencing ions (Syka *et al.*, 2004). The fragmentation patterns generated with ECD and ETD are independent of peptide length, amino acid composition, and the presence of PTMs (Udeshi *et al.*, 2007). This fragmentation process is highly efficient and occurs on a millisecond time scale. Therefore, ETD is sensitive down to femtomole amounts of sample and occurs on a time scale corresponding to current chromatographic separation techniques (Syka *et al.*, 2004). The main benefit of this fragmentation technique is its ability to preserve labile PTMs while still allowing sufficient amide bond cleavage to generate spectra from which sequencing information can be obtained. However, this method does suffer from a decrease in fragment ion production as the precursor ion charge decreases (Swaney *et al.*, 2007).

A recent study has utilized a hybrid approach of ETD and CID (ETcaD) to target the nondissociative electron transfer product ions of ETD with CID. This method resulted in median sequence coverage of 88.9% for ETcaD as opposed to 62.5% for ETD and 77.4% for CID alone (Swaney *et al.*, 2007). Other studies have utilized ETD to identify PTMs, namely phosphorylation. One group used CID based fragmentation to detect over 1000 phosphopeptides but was only able to define 383 sites of phosphorylation, largely due to the neutral loss events on the modified peptides (Udeshi *et al.*, 2007). Using an LTQ modified with ETD, 1252 sites of phosphorylation on 629 proteins of a complex mixture were identified (Udeshi *et al.*, 2007). A separate analysis of the overlap between

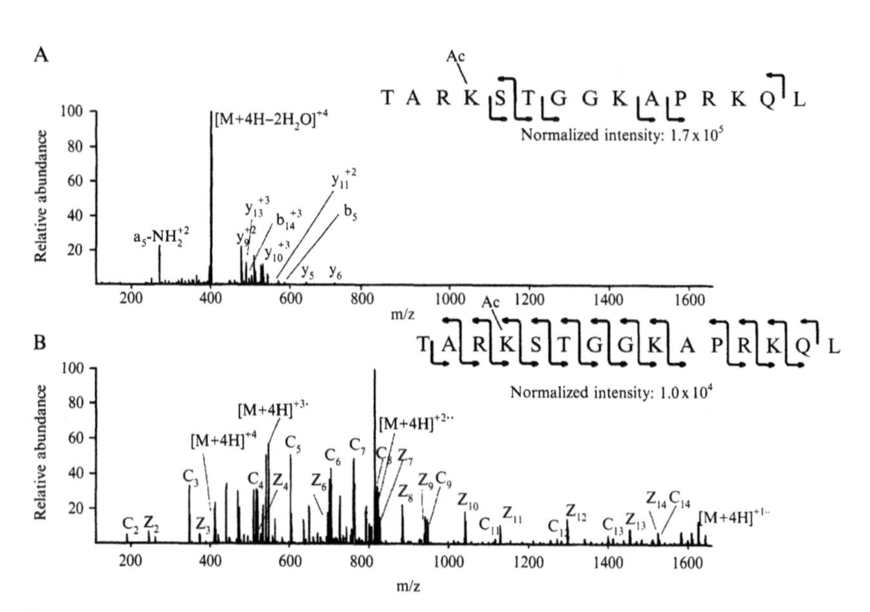

Figure 40.5 A comparison of CID versus ETD spectra for a human histone H3 peptide. Panel A shows the CID spectrum with the corresponding b- and y-ion series. Panel B depicts the ETD spectrum and its corresponding c- and z-ion series. Reprinted with permission from Mikesh *et al.* (2006).

phosphopeptides identified with ETD and CID found that only 17.9% of the sites identified were shared among the two datasets generated (Swaney *et al.*, 2008). This suggests that the best approach to fragmenting and identifying sites of PTMs may be a combination of CID and ETD. Figure 40.5 shows representative spectra generated with both methods of fragmentation. The characteristic b- and y-ions appear in the CID spectrum, while c- and z-type ions appear in ETD.

7. Quantifying PTMs

When investigating the biological significance of PTMs, it is often advantageous to know the relative or absolute abundance of a specific modification or group of PTMs. This allows for the direct comparison of the modification of interest across varied biological samples. An example is comparing the abundance of a PTM in cells or tissues obtained from a normal versus a disease state. Quantifying these changes can lead to insights into the role of PTMs in a myriad of processes from cell growth to diseases and apoptosis. A range of methods are available to quantify PTMs. Traditional methods utilize 2D-GE and differential staining to identify differences in the level of protein expression among samples. However, this approach has a low resolution, and differences in the ability of some proteins to stain can lead to artifacts (Ong and Mann, 2005). The 2D-GE approach is most amenable to abundant protein species (Anderson and Anderson, 1998). More recently, mass spectrometry-based approaches have alleviated some of the problems posed by gel-based approaches. These methods include protein or peptide labeling strategies, tagging approaches and differential proteomic methods using spectral counting (Gygi *et al.*, 1999; Ong *et al.*, 2002; Washburn *et al.*, 2001). While these methods are more widely applied to quantifying protein changes among samples, they can also be used to quantify changes in PTMs from different biological samples.

Quantitative measurements can either be absolute, in terms of concentration or copy number per cell, or relative, in terms of fold change among differentially treated samples. Absolute concentrations are more difficult to obtain, but are more informative and can be used to derive relative measurements. LC-MS/MS experiments allow the plotting of signal intensities as a peptide elutes from the chromatographic column over time. The area under this curve for any given species is directly related to the abundance of the peptide and allows for label-free quantitation. However, the physiochemical properties of peptides such as hydrophobicity, charge, and size vary widely and can lead to differences in the mass spectrometer's detector response. In addition, co-elution of other peptides and variations in elution conditions can be problematic for quantitative investigations with this label-free approach.

Protein abundance estimates using this method can vary by a factor of 3–5-fold from the true abundance (Ong and Mann, 2005).

To circumvent problems associated with ionization efficiencies and MS-detector responses, one can take advantage of approaches based on the stable isotope dilution theory. This theory states that a stable, isotopically labeled peptide is chemically identical to its native counterpart. As a result, isotopically labeled and unlabeled peptides behave identically in chromatography and in MS detection. Since the mass spectrometer can recognize the mass difference between the labeled and unlabeled forms, the relative signal intensities can be determined (Bantscheff *et al.*, 2007). There are four main strategies that exploit the stable isotope dilution theory. First, an isotopically labeled peptide standard corresponding to a PTM of interest can be introduced into the sample (Gerber *et al.*, 2003). Second, cells can be metabolically labeled during growth in media consisting primarily of isotopically pure components (Ong and Mann, 2005). Third, isotopes can be incorporated into the peptide during proteolysis with enzymes such as trypsin (Miyagi and Rao, 2007). Finally, isotopically labeled tags can be chemically linked to specific amino acid side chains (Gygi *et al.*, 1999; Ross *et al.*, 2004). Regardless of which method is chosen, the mass difference between the "heavy" and "light" forms of the peptides should be a minimum of 3–4 Da to avoid isotopic overlapping of the peaks in the mass spectrum. Additionally, deuterated peptides can show different chromatographic elution times in relation to their "light" counterparts (Zhang and Regnier, 2002). This makes using the more expensive ^{13}C- and ^{15}N-reagents necessary.

The most basic approach for using stable isotopes is termed absolute quantitation (AQUA) (Gerber *et al.*, 2003). In this approach, isotopically labeled peptides are generated synthetically and added as internal standards to the peptide mixture. Steve Gygi and his colleagues used this approach for determining the change in the abundance of phosphorylations during the *Xenopus* cell cycle (Stemmann *et al.*, 2001). This method can become cumbersome in that it requires that a labeled synthetic peptide be generated for every PTM to be quantified. Labeled phosphopeptides standards are readily available, but this is not the case for other modifications, limiting broad application of this method. The internal standard is typically added just before or after proteolysis, in which case it is impossible to normalize for variations in sample preparation upstream of the digestion step. This technique requires a prior knowledge of the PTM so the synthetic peptide can be generated. It is also necessary to know the general abundance of the modified peptide so the labeled standard can be added in a similar amount.

To minimize sample preparation errors, cell populations can be grown under different conditions with or without an isotopic label, and the populations pooled prior to sample preparation and analysis. Any errors in sample preparation affect both populations equally. Mathias Mann's laboratory has developed the most popular of these approaches, called stable isotope labeling by amino acids in cell culture (SILAC) (Ong *et al.*, 2002). In the SILAC approach, the culture media contains $^{13}C_6$-arginine and $^{13}C_6$-lysine, which label the sites at which trypsin cleaves. Therefore, excluding the C-terminal fragment, each tryptic product includes one labeled amino acid. The advantage of the SILAC method over total metabolic labeling is that the number of incorporated labels is defined and independent of the amino acid sequence. This simplicity makes data interpretation easier with SILAC than with other methods that label the peptides in more complicated patterns. SILAC limits comparison to a maximum of three culture conditions, which include either unlabeled, $^{13}C_6$-labeled, or $^{13}C_6{}^{15}N_4$-labeled amino acids. Other limitations of SILAC are that it is only useful with certain model organisms and cell lines that can be grown in isotopically defined media and can be prohibitively expensive. However, SILAC's high degree of accuracy makes this method well suited for the study of PTMs (Olsen *et al.*, 2006). A variation of SILAC employs labeling with labeled S-adenosylmethionine for the identification and relative quantification of protein methylation (Ong *et al.*, 2004).

An alternative to *in vivo* metabolic labeling is enzymatic labeling *in vitro*. This labeling can be performed either during protein digestion or after proteolysis with an additional reaction step. Using heavy water ($H_2{}^{18}O$), trypsin or Glu-C will introduce two ^{18}O atoms. The resulting 4 Da shift is sufficient to resolve isotopes and acquire relative quantification data (Miyagi and Rao, 2007). However, some proteases such as endoproteinase Lys-N only incorporate one ^{18}O atom and yield spectra insufficient to resolve isotope peak overlap (Rao *et al.*, 2005). Full enzymatic labeling is rarely achieved and different peptides incorporate the label at different rates thereby complicating data analysis (Ramos-Fernandez *et al.*, 2007). As a result, this method has not found widespread use in quantitative proteomics (Ong *et al.*, 2004).

Efficient chemical tagging approaches can alleviate some of the negative constraints of enzymatic tagging procedures. The first of these was developed by Ruedi Aebersold's laboratory and termed isotope-coded affinity tags (ICAT) (Gygi *et al.*, 1999). The ICAT reagent was initially composed of a thiol reactive group that targets cysteines, a polyether linker region with either zero or eight deuterium atoms, and a biotin group for affinity purification with an avidin column. Since deuterium isotopes can behave differentially in their chromatographic response relative to the light forms, ^{13}C and ^{15}N tagging reagents have been developed (Bottari *et al.*, 2004). The ICAT reaction is performed on a reduced sample to chemically label it with either the heavy or light ICAT reagent. Then, the two samples are mixed and digested with a protease of choice. Labeled peptides are recovered with the affinity tag of the ICAT group and quantified with mass spectrometry. This approach effectively simplifies the peptide mixture by selecting only peptides containing a cysteine. However, since most peptides do not contain cysteine residues, ICAT has not found widespread use for the quantification of PTMs since a majority of PTMs are discarded at the affinity purification step (Ross *et al.*, 2004).

Another set of chemical tagging strategies utilizes the reactivity of the peptide N-terminus and the epsilon-amino group of lysine residues. Isotope-coded protein labeling (ICPL), isotope tags for relative and absolute quantification (iTRAQ), and tandem mass tags (TMT) are all approaches that were developed to exploit the reactivity of amines to specific chemical groups (Ross *et al.*, 2004; Schmidt *et al.*, 2005; Thompson *et al.*, 2003). The most recognized of these approaches, iTRAQ, incorporates isobaric tags that fragment in the tandem mass spectrometer to generate unique reporter ions at *m/z* values of 113–121 (Ong *et al.*, 2004). The peak areas of the low-mass-fragment ions are integrated to determine the quantification values. Mass spectrometers such as quadrupoles and TOF instruments which have the inherent ability to detect low *m/z* fragment ions are typically used for iTRAQ experiments. Commercially available kits allow examination of up to eight states in one single experiment. Since the labeled peptides are isobaric, they should behave similarly in chromatographic separations and reveal quantitative differences in their fragmentation spectra. Using the iTRAQ approach, different samples are isolated under different conditions, trypsin digested, and labeled with the iTRAQ reagents. Samples are then combined and analyzed in a single MS/MS experiment. Figure 40.6 shows an example of tandem mass spectra generated using four iTRAQ reagents in varying ratios (Ross *et al.*, 2004). As with other MS-based approaches, sufficient peptide separation is crucial since coeluting peptides of similar mass can contribute to the observed reporter ion series and interfere with the quantification. Ion trap mass spectrometers are typically not useful for iTRAQ analysis because the reporter ions generated from the fragmentation

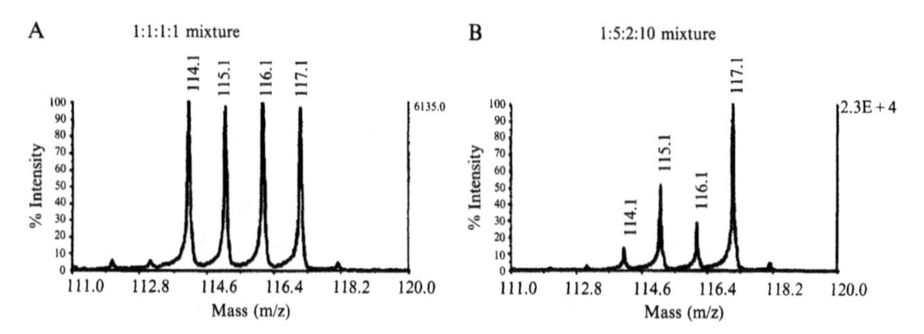

Figure 40.6 iTRAQ reporter ion series for two mixtures of varied ratios of a six protein digest. Reprinted with permission from Ross *et al.* (2004).

of the iTRAQ tags are lost with the lower 1/3 of the MS/MS data. The use of pulsed q-dissociation (PQD) was introduced to alleviate this problem but has found limited use in proteomics (Bantscheff *et al.*, 2008; Cunningham *et al.*, 2006; Griffin *et al.*, 2007).

Using the iTRAQ approach, it is possible to compare the degree of phosphorylation of the proteome under different conditions. Studies performed by Forest White and coworkers used antiphosphotyrosine antibodies followed by IMAC enrichment and iTRAQ labeling to investigate the effect of epidermal growth factor (EGF) stimulation on the phosphorylation state of cells over time (Zhang *et al.*, 2005). This study probed four time points (0, 5, 10, and 30 min) in a single analysis. They were able to quantify relative changes in the phosphorylation of 78 sites on 58 proteins (Zhang *et al.*, 2005). This experimental approach highlights the utility of combining enrichment for PTMs with quantitation to gain additional insight into the biology of cells. It is difficult to determine what fraction of a peptide population is phosphorylated under certain conditions. This is because the unphosphorylated counterpart will behave differently in its MS-detector response. So while it is feasible to compare unmodified forms and other unmodified forms or different phosphorylated forms of the same peptide under different conditions, it is not reliable to compare modified versus unmodified forms based on the ratio of reporter ion peaks generated.

A limited number of additional tagging methods aimed specifically at defined PTMs have been developed. These utilize the selective chemical reactivity of the chemically modified amino acid side chains. For the quantitation of phosphorylations, β-elimination of phosphoric acid, and subsequent Michael addition of ethanedithiol derivatives can be employed to incorporate isotope tags (Tao *et al.*, 2005). Hydrazine chemistry can be applied to glycosylated peptides to replace the carbohydrate group with an isotopically labeled tag (Zhang *et al.*, 2003). These approaches are only effective for a limited number of PTMs. In addition, if the chemical reactions involved are not highly efficient, the resulting peptide mixture can increase in complexity. The increase in complexity can lead to problems in identifying the peptides and their associated PTMs as well as inaccurate quantitative information.

Isotope labeling and tagging strategies can be tedious, cumbersome, and expensive. In addition to the chromatographic peak integration method discussed previously, other label-free quantitation methods exist that can be applied to quantifying PTMs. These methods typically involve some form of spectral counting and normalization for protein length (Washburn *et al.*, 2001). It is debatable whether these methods are truly quantitative, and as a result, these methods are more commonly referred to as differential proteomic techniques. The spectral counting methods are based on the principle that, as the abundance of a protein increases in a given sample, more MS/MS spectra are isolated for peptides derived from that protein. Relative quantitation is inferred by comparing the number of spectra collected for a particular protein between experiments. The utility of this approach is questionable because it does not directly measure any physical property of the peptide and assumes a linear response for every protein (Bantscheff *et al.*, 2007). As previously mentioned, the physiochemical properties of peptides vary widely in a sample digest and their resulting chromatographic and MS-detector behavior can vary greatly. These methods provide a larger dynamic range than labeling and tagging procedures and are most applicable when investigating large or global protein changes among samples. However, the uncertain linear response among peptides and the potentially poor accuracy of spectral counting approaches limits their widespread use (Old *et al.*, 2005).

In an attempt to minimize the variability in label-free quantitation of proteins in a sample, it is possible to compare specific and multiple peptides from the same protein across multiple samples. This has lead to the development of selected reaction monitoring (SRM) and multiple reaction monitoring (MRM) for PTMs (Unwin *et al.*, 2005, 2009; Williamson *et al.*, 2006; Wolf-Yadlin *et al.*, 2007). In an SRM experiment, a unique peptide precursor and its fragment ion m/z value or transition is selectively monitored across multiple experiments, typically using a triple quadrupole mass spectrometer (Lange *et al.*, 2008). For proteins to be quantified with a high degree of statistical confidence, 3–5 peptides and their transitions are typically monitored during an MRM experiment. For these experiments, triple quadrupole mass spectrometers are utilized due to the high degree of selectivity afforded by their first and third quadrupoles to isolate the selected transitions and their high duty cycles. Multiple transitions are monitored over time to produce a set of chromatographic profiles with a retention time and signal intensity for each specific transition. The integrated areas of the transitions are used to quantify the proteins. MRM enables the detection of low abundance proteins in complex mixtures and yields a linear response over a dynamic range up to five orders in magnitude (Lange *et al.*, 2008).

MRM can be used to selectively monitor proteins containing PTMs (Williamson *et al.*, 2006). However, MRM lacks the ability to identify proteins in a mixture, and transitions are typically selected based on previous experimental data or literature searches. Transitions must be optimized for efficient fragmentation in the second quadrupole of the triple quadrupole, and the investigator can be limited by sample availability, although computational tools can compensate for this. In addition, triple quadrupoles use CID to fragment peptides and are subject to neutral loss events for labile PTMs. An extension of MRM includes the use of isotopically labeled synthetic peptides in a manner analogous to AQUA to achieve absolute quantitation of peptides (Wolf-Yadlin *et al.*, 2007). Taken together, this emerging technology in mass spectrometry has the potential to provide quantitative information for many proteins and PTMs across varied biological samples.

8. Future Directions

Traditional proteomic studies employ a bottom-up approach whereby whole proteins are digested with a protease and analyzed using MS/MS strategies to determine the peptide sequences. From these results, the identities of the proteins can be inferred. However, complete coverage of a protein is rarely accomplished. As a result, if peptides harboring the modification are not detected, the PTMs cannot be identified. Top-down investigations aim to determine the entire primary structure of a single protein by performing tandem mass spectrometry on the intact protein to produce fragmentation information. The combination of precursor mass measurement and fragmentation sequencing data allows the identity of the protein to be determined. Top-down methods can be useful in determining specific PTMs and identifying splice variants and degradation products not typically identified in bottom-up approaches (Waanders *et al.*, 2007). Top-down mass spectrometry also can discern gene products with a high degree of sequence identity (Parks *et al.*, 2007; Siuti *et al.*, 2006). Top-down approaches are readily applicable to the identification of PTMs. The initial MS event on the whole protein determines if it has been modified from its native form based upon any mass shift that may be present. The change in the observed mass can be accounted for by only a limited number of modifications. Additional fragmentation data generated by MS/MS can help to further characterize the modification based on the observed mass shift(s) for the peptides.

Top-down proteomics requires the use of instruments with high resolution and high mass accuracy. This has relegated top-down studies to FT-ICR and Orbitrap mass spectrometers with typical resolving powers $> 500{,}000$ and 100,000, respectively (Macek *et al.*, 2006). The high resolving power allows for the accurate determination of charge states, which is essential for measuring the intact protein's mass. The sub- to low-ppm mass accuracy enables the discrimination between PTMs with the same nominal mass, such as acetylation and trimethylation. Recently, studies by Andrea Armirotti and coworkers and Scott McLuckey and his group have demonstrated the ability to perform top-down proteomics on a q-TOF mass spectrometer (Armirotti *et al.*, 2009; Liu *et al.*, 2009). This extending the utility of this approach beyond the generally more expensive FT-ICR and Orbitrap mass spectrometers and opens this methodology to a broader section of investigators. Top-down approaches possess the ability to handle a dynamic range of protein sizes. Parks *et al.* (2007) used such an approach to identify 22 proteins ranging from 14 to 35 kDa in *S. cerevisiae*. One group extended this range to characterize the PTMs of a 115-kDa cardiac myosin-binding protein (Ge *et al.*, 2009).

In addition to being useful for determining the identities of proteins within a biological sample, top-down proteomics is also applicable to studies focusing on quantifying proteins and PTMs. Top-down studies are particularly well suited for quantitating PTMs, since the ionization efficiencies of intact proteins are much less affected by the presence of PTMs as compared to peptides (Ge *et al.*, 2009). ^{15}N labeling strategies were used by Neil Kelleher and coworkers for intact protein quantitation and identification generating 50 abundance ratios (Du *et al.*, 2006). Matthias Mann extended the applicability of his SILAC approach for proteins up to 220 kDa and quantified a 28-kDa signaling protein (Waanders *et al.*, 2007). Mann also determined that the quantitation of a one to one mixture of the 28 kDa species had a standard deviation of 6% and was primarily limited by the signal-to-noise ratio in the protein measurements as compared to peptide measurements using SILAC (Waanders *et al.*, 2007).

Widespread use of top-down proteomics has not yet been achieved. There are technological challenges in measuring and identifying large whole proteins. The implementation of top-down techniques has been slowed by the laborious separations of whole proteins, the requirements for high sensitivity, efficient fragmentation, and the lack of computational tools for data interpretation (Collier *et al.*, 2008; Kelleher *et al.*, 1999; Meng *et al.*, 2002; Siuti and Kelleher, 2007).

The preferred method for fragmenting proteins in bottom-up proteomics utilizes CID of peptides to generate fragmentation information. CID of intact proteins is dependent on protein structure and size (Wu *et al.*, 2004; Zabrouskov *et al.*, 2005). CID produces fragments preferentially at the termini of proteins. One approach to overcome the low internal sequence coverage of CID is to employ a middle-down approach. This technique uses limited proteolysis to generate larger peptides in the 3–20 kDa range. However, the recovery of peptides digested in this manner is unpredictable and as a result has gained little momentum in proteomics (Ge *et al.*, 2009). The use of instruments capable of ECD (FT-ICRs) and, more recently, ETD (Orbitraps), is alleviating some of the problems faced when using CID to fragment whole proteins in top-down proteomics (Bunger *et al.*, 2008; Coon *et al.*, 2005; Ge *et al.*, 2002; McAlister *et al.*, 2007, 2008). As previously mentioned, ECD and ETD are size independent and have the ability to produce fragmentation patterns of whole proteins sufficient to determine protein sequence. For top-down approaches to gain more wide spread use, proteome coverage and instrument sensitivity both need to be improved (Ferguson *et al.*, 2009).

Determining PTMs on proteins can be a daunting task. However, as outlined earlier, there are a variety of tools available that utilize mass spectrometry to assist in identifying and quantifying PTMs. Sample preparation, enrichment strategies, quantitative approaches, instrument configurations, and data analysis tools are all crucial in determining the successfulness of the approach. This chapter has presented many of the techniques and technologies that exist today in the field of proteomics that can be tailored toward the goal of determining PTMs. The chemical modifications of proteins are extremely important because they potentially change many properties of a protein. Therefore, it is important to take advantage of the approaches offered in mass spectrometry to determine sites of PTMs and elucidate their biological function.

ACKNOWLEDGMENTS

We acknowledge Elizabeth M. Link and Kevin Schey for helpful comments and suggestions in the preparation of this manuscript. A. R. F. was supported by NIH training grant T32 CA009385 and GM64779. A. J. L. was supported by NIH grants GM64779 and AR055231.

REFERENCES

Ahmad, I., Hoessli, D. C., Walker-Nasir, E., Choudhary, M. I., Rafik, S. M., and Shakoori, A. R. (2006). *J. Cell. Biochem.* **99,** 706–718.

Anderson, N. L., and Anderson, N. G. (1998). *Electrophoresis* **19,** 1853–1861.

Armirotti, A., Benatti, U., and Damonte, G. (2009). *Rapid Commun. Mass Spectrom.* **23,** 661–666.

Bantscheff, M., Schirle, M., Sweetman, G., Rick, J., and Kuster, B. (2007). *Anal. Bioanal. Chem.* **389,** 1017–1031.

Bantscheff, M., Boesche, M., Eberhard, D., Matthieson, T., Sweetman, G., and Kuster, B. (2008). *Mol. Cell. Proteomics* **7,** 1702–1713.

Barnes, P. J., Shapiro, S. D., and Pauwels, R. A. (2003). *Eur. Respir. J.* **22,** 672–688.

Bayer, P., Arndt, A., Metzger, S., Mahajan, R., Melchior, F., Jaenicke, R., and Becker, J. (1998). *J. Mol. Biol.* **280,** 275–286.

Beausoleil, S. A., Jedrychowski, M., Schwartz, D., Elias, J. E., Villen, J., Li, J., Cohn, M. A., Cantley, L. C., and Gygi, S. P. (2004). *Proc. Natl. Acad. Sci. USA* **101,** 12130–12135.

Beckman, J. S., Beckman, T. W., Chen, J., Marshall, P. A., and Freeman, B. A. (1990). *Proc. Natl. Acad. Sci. USA* **87,** 1620–1624.

Bennett, K. L., Stensballe, A., Podtelejnikov, A. V., Moniatte, M., and Jensen, O. N. (2002). *J. Mass Spectrom.* **37,** 179–190.

Berger, S. L. (2007). *Nature* **447,** 407–412.

Bernstein, E., and Allis, C. D. (2005). *Genes Dev.* **19,** 1635–1655.

Blagoev, B., Ong, S. E., Kratchmarova, I., and Mann, M. (2004). *Nat. Biotechnol.* **22,** 1139–1145.

Blume-Jensen, P., and Hunter, T. (2001). *Nature* **411,** 355–365.

Bodenmiller, B., Mueller, L. N., *et al.* (2007). *Nat. Methods* **4,** 231–237.

Bottari, P., Aebersold, R., *et al.* (2004). *Bioconjug. Chem.* **15,** 380–388.

Bunger, M. K., Cargile, B. J., *et al.* (2008). *Anal. Chem.* **80,** 1459–1467.

Cantin, G. T., Shock, T. R., Park, S. K., Madhani, H. D., and Yates, J. R. 3rd (2007). *Anal. Chem.* **79,** 4666–4673.

Carr, S. A., Huddleston, M. J., and Annan, R. S. (1996). *Anal. Biochem.* **239,** 180–192.

Chalkley, R. J., Baker, P. R., Medzihradszky, K. F., Lynn, A. J., and Burlingame, A. L. (2008). *Mol. Cell. Proteomics* **7,** 2386–2398.

Chang, I. F. (2006). *Proteomics* **6,** 6158–6166.

Chen, Y. R., Chen, C. L., Chen, W., Zweier, J. L., Augusto, O., Radi, R., and Mason, R. P. (2004). *J. Biol. Chem.* **279,** 18054–18062.
Cirulli, C., Chiappetta, G., Marino, G., Mauri, P., and Amoresano, A. (2008). *Anal. Bioanal. Chem.* **392,** 147–159.
Clauser, K. R., Baker, P., and Burlingame, A. L. (1999). *Anal. Chem.* **71,** 2871–2882.
Cohen, P. (2001). *Eur. J. Biochem.* **268,** 5001–5010.
Collier, T. S., Hawkridge, A. M., Georgianna, D. R., Payne, G. A., and Muddiman, D. C. (2008). *Anal. Chem.* **80,** 4994–5001.
Coon, J. J., Syka, J. E. P., Schwartz, J. C., Shabanowitz, J., and Hunt, D. F. (2004). *Int. J. Mass Spectrom.* **236,** 33–42.
Coon, J. J., Ueberheide, B., *et al.* (2005). *Proc. Natl. Acad. Sci. USA* **102,** 9463–9468.
Covey, T. R., Huang, E. C., and Henion, J. D. (1991). *Anal. Chem.* **63,** 1193–1200.
Creasy, D. M., and Cottrell, J. S. (2004). *Proteomics* **4,** 1534–1536.
Cunningham, C. Jr., Glish, G. L., and Burinsky, D. J. (2006). *J. Am. Soc. Mass Spectrom.* **17,** 81–84.
Denison, C., Kirkpatrick, D. S., and Gygi, S. P. (2005). *Curr. Opin. Chem. Biol.* **9,** 69–75.
Dongre, A. R., Jones, J. L., Somogyi, A., and Wysocki, V. H. (1996). *J. Am. Chem. Soc.* **118,** 8365–8374.
Du, Y., Parks, B. A., Sohn, S., Kwast, K. E., and Kelleher, N. L. (2006). *Anal. Chem.* **78,** 686–694.
Durand, G., and Seta, N. (2000). *Clin. Chem.* **46,** 795–805.
Fang, S., and Weissman, A. M. (2004). *Cell. Mol. Life Sci.* **61,** 1546–1561.
Ferguson, J. T., Wenger, C. D., Metcalf, W. W., and Kelleher, N. L. (2009). *J. Am. Soc. Mass Spectrom.* **20,** 1743–1750.
Ficarro, S. B., McCleland, M. L., Stukenberg, P. T., Burke, D. J., Ross, M. M., Shabanowitz, J., Hunt, D. F., and White, F. M. (2002). *Nat. Biotechnol.* **20,** 301–305.
Fuchs, S. M., Laribee, R. N., and Strahl, B. D. (2009). *Biochim. Biophys. Acta* **1789,** 26–36.
Garcia, B. A., Shabanowitz, J., and Hunt, D. F. (2005). *Methods* **35,** 256–264.
Garcia, B. A., Shabanowitz, J., and Hunt, D. F. (2007). *Curr. Opin. Chem. Biol.* **11,** 66–73.
Gatzka, M., and Walsh, C. M. (2007). *Autoimmunity* **40,** 442–452.
Ge, Y., Lawhorn, B. G., *et al.* (2002). *J. Am. Chem. Soc.* **124,** 672–678.
Ge, Y., Rajkumar, L., Guzman, R. C., Nandi, S., Patton, W. F., and Agnew, B. J. (2004). *Proteomics* **4,** 3464–3467.
Ge, Y., Rybakova, I. N., Xu, Q., and Moss, R. L. (2009). *Proc. Natl. Acad. Sci. USA* **106,** 12658–12663.
Gerber, S. A., Rush, J., Stemman, O., Kirschner, M. W., and Gygi, S. P. (2003). *Proc. Natl. Acad. Sci. USA* **100,** 6940–6945.
Gingras, A. C., Raught, B., Gygi, S. P., Niedzwiecka, A., Miron, M., Burley, S. K., Polakiewicz, R. D., Wyslouch-Cieszynska, A., Aebersold, R., and Sonenberg, N. (2001). *Genes Dev.* **15,** 2852–2864.
Glozak, M. A., Sengupta, N., Zhang, X., and Seto, E. (2005). *Gene* **363,** 15–23.
Godde, J. S., and Ura, K. (2008). *J. Biochem.* **143,** 287–293.
Good, D. M., Wirtala, M., McAlister, G. C., and Coon, J. J. (2007). *Mol. Cell. Proteomics* **6,** 1942–1951.
Goodlett, D. R., Keller, A., Watts, J. D., Newitt, R., Yi, E. C., Purvine, S., Eng, J. K., von Haller, P., Aebersold, R., and Kolker, E. (2001). *Rapid Commun. Mass Spectrom.* **15,** 1214–1221.
Goshe, M. B., Conrads, T. P., Panisko, E. A., Angell, N. H., Veenstra, T. D., and Smith, R. D. (2001). *Anal. Chem.* **73,** 2578–2586.
Grewal, S. I., and Rice, J. C. (2004). *Curr. Opin. Cell Biol.* **16,** 230–238.
Griffin, T. J., Xie, H., Bandhakavi, S., Popko, J., Mohan, A., Carlis, J. V., and Higgins, L. (2007). *J. Proteome Res.* **6,** 4200–4209.

Gruhler, A., Olsen, J. V., Mohammed, S., Mortensen, P., Faergeman, N. J., Mann, M., and Jensen, O. N. (2005). *Mol. Cell. Proteomics* **4,** 310–327.

Gu, W., and Roeder, R. G. (1997). *Cell* **90,** 595–606.

Gygi, S. P., Rist, B., Gerber, S. A., Turecek, F., Gelb, M. H., and Aebersold, R. (1999). *Nat. Biotechnol.* **17,** 994–999.

Haas, W., Faherty, B. K., Gerber, S. A., Elias, J. E., Beausoleil, S. A., Bakalarski, C. E., Li, X., Villen, J., and Gygi, S. P. (2006). *Mol. Cell. Proteomics* **5,** 1326–1337.

Hansen, B. T., Jones, J. A., Mason, D. E., and Liebler, D. C. (2001). *Anal. Chem.* **73,** 1676–1683.

Hansen, B. T., Davey, S. W., Ham, A. J., and Liebler, D. C. (2005). *J. Proteome Res.* **4,** 358–368.

Hayduk, E. J., Choe, L. H., and Lee, K. H. (2004). *Electrophoresis* **25,** 2545–2556.

Holt, G. D., and Hart, G. W. (1986). *J. Biol. Chem.* **261,** 8049–8057.

Huang, Y., Triscari, J. M., Tseng, G. C., Pasa-Tolic, L., Lipton, M. S., Smith, R. D., and Wysocki, V. H. (2005). *Anal. Chem.* **77,** 5800–5813.

Hubbard, M. J., and Cohen, P. (1993). *Trends Biochem. Sci.* **18,** 172–177.

Ikeda, F., and Dikic, I. (2008). *EMBO Rep.* **9,** 536–542.

Issad, T., and Kuo, M. (2008). *Trends Endocrinol. Metab.* **19,** 380–389.

Ito, S., Hayama, K., and Hirabayashi, J. (2009). *Methods Mol. Biol.* **534,** 195–203.

Jenuwein, T., and Allis, C. D. (2001). *Science* **293,** 1074–1080.

Kelleher, N. L., Lin, H. Y., Valaskovic, G. A., Aaserud, D. J., Fridriksson, E. K., and McLafferty, F. W. (1999). *J. Am. Chem. Soc.* **121,** 806–812.

Kirkpatrick, D. S., Denison, C., and Gygi, S. P. (2005). *Nat. Cell Biol.* **7,** 750–757.

Lange, V., Picotti, P., Domon, B., and Aebersold, R. (2008). *Mol. Syst. Biol.* **4,** 222.

Larsen, M. R., Sorensen, G. L., Fey, S. J., Larsen, P. M., and Roepstorff, P. (2001). *Proteomics* **1,** 223–238.

Larsen, M. R., Thingholm, T. E., Jensen, O. N., Roepstorff, P., and Jorgensen, T. J. (2005). *Mol. Cell. Proteomics* **4,** 873–886.

Lee, C. H., McComb, M. E., Bromirski, M., Jilkine, A., Ens, W., Standing, K. G., and Perreault, H. (2001). *Rapid Commun. Mass Spectrom.* **15,** 191–202.

Lim, K. B., and Kassel, D. B. (2006). *Anal. Biochem.* **354,** 213–219.

Lochnit, G., and Geyer, R. (2004). *Biomed. Chromatogr.* **18,** 841–848.

Lu, B., Motoyama, A., Ruse, C., Venable, J., and Yates, J. R. 3rd (2008). *Anal. Chem.* **80,** 2018–2025.

Ma, Y., Lu, Y., Zeng, H., Ron, D., Mo, W., and Neubert, T. A. (2001). *Rapid Commun. Mass Spectrom.* **15,** 1693–1700.

MacCoss, M. J., McDonald, W. H., Saraf, A., Sadygov, R., Clark, J. M., Tasto, J. J., Gould, K. L., Wolters, D., Washburn, M., Weiss, A., Clark, J. I., and Yates, J. R. 3rd (2002). *Proc. Natl. Acad. Sci. USA* **99,** 7900–7905.

Macek, B., Waanders, L. F., Olsen, J. V., and Mann, M. (2006). *Mol. Cell. Proteomics* **5,** 949–958.

Makarov, A. (2000). *Anal. Chem.* **72,** 1156–1162.

Makarov, A., Denisov, E., Kholomeev, A., Balschun, W., Lange, O., Strupat, K., and Horning, S. (2006). *Anal. Chem.* **78,** 2113–2120.

Mann, M., and Kelleher, N. L. (2008). *Proc. Natl. Acad. Sci. USA* **105,** 18132–18138.

Manza, L. L., Codreanu, S. G., Stamer, S. L., Smith, D. L., Wells, K. S., Roberts, R. L., and Liebler, D. C. (2004). *Chem. Res. Toxicol.* **17,** 1706–1715.

McAlister, G. C., Berggren, W. T., *et al.* (2008). *J. Proteome. Res.* **7,** 3127–3136.

McAlister, G. C., Phanstiel, D., *et al.* (2007). *Anal. Chem.* **79,** 3525–3534.

McLachlin, D. T., and Chait, B. T. (2001). *Curr. Opin. Chem. Biol.* **5,** 591–602.

McLachlin, D. T., and Chait, B. T. (2003). *Anal. Chem.* **75,** 6826–6836.

Mechref, Y., Madera, M., and Novotny, M. V. (2008). *Methods Mol. Biol.* **424,** 373–396.

Medzihradszky, K. F. (2005). *Methods Enzymol.* **405,** 116–138.
Medzihradszky, K. F., Campbell, J. M., Baldwin, M. A., Falick, A. M., Juhasz, P., Vestal, M. L., and Burlingame, A. L. (2000). *Anal. Chem.* **72,** 552–558.
Melchior, F., Schergaut, M., and Pichler, A. (2003). *Trends Biochem. Sci.* **28,** 612–618.
Meng, F., Cargile, B. J., Patrie, S. M., Johnson, J. R., McLoughlin, S. M., and Kelleher, N. L. (2002). *Anal. Chem.* **74,** 2923–2929.
Mikesh, L. M., Ueberheide, B., Chi, A., Coon, J. J., Syka, J. E., Shabanowitz, J., and Hunt, D. F. (2006). *Biochim. Biophys. Acta* **1764,** 1811–1822.
Miyagi, M., and Rao, K. C. (2007). *Mass Spectrom. Rev.* **26,** 121–136.
Oda, Y., Huang, K., Cross, F. R., Cowburn, D., and Chait, B. T. (1999). *Proc. Natl. Acad. Sci. USA* **96,** 6591–6596.
Oda, Y., Nagasu, T., and Chait, B. T. (2001). *Nat. Biotechnol.* **19,** 379–382.
Old, W. M., Meyer-Arendt, K., Aveline-Wolf, L., Pierce, K. G., Mendoza, A., Sevinsky, J. R., Resing, K. A., and Ahn, N. G. (2005). *Mol. Cell. Proteomics* **4,** 1487–1502.
Olsen, J. V., and Mann, M. (2004). *Proc. Natl. Acad. Sci. USA* **101,** 13417–13422.
Olsen, J. V., Blagoev, B., Gnad, F., Macek, B., Kumar, C., Mortensen, P., and Mann, M. (2006). *Cell* **127,** 635–648.
Olsson, A., Manzl, C., Strasser, A., and Villunger, A. (2007). *Cell Death Differ.* **14,** 1561–1575.
Ong, S. E., and Mann, M. (2005). *Nat. Chem. Biol.* **1,** 252–262.
Ong, S. E., Blagoev, B., Kratchmarova, I., Kristensen, D. B., Steen, H., Pandey, A., and Mann, M. (2002). *Mol. Cell. Proteomics* **1,** 376–386.
Ong, S. E., Mittler, G., and Mann, M. (2004). *Nat. Methods* **1,** 119–126.
Pan, C., Gnad, F., Olsen, J. V., and Mann, M. (2008). *Proteomics* **8,** 4534–4546.
Paoletti, A. C., and Washburn, M. P. (2006). *Biotechnol. Genet. Eng. Rev.* **22,** 1–19.
Parks, B. A., Jiang, L., Thomas, P. M., Wenger, C. D., Roth, M. J., Boyne, M. T. 2nd, Burke, P. V., Kwast, K. E., and Kelleher, N. L. (2007). *Anal. Chem.* **79,** 7984–7991.
Patton, W. F. (2002). *J. Chromatogr. B Analyt. Technol. Biomed. Life Sci.* **771,** 3–31.
Pedrioli, P. G., Raught, B., Zhang, X. D., Rogers, R., Aitchison, J., Matunis, M., and Aebersold, R. (2006). *Nat. Methods* **3,** 533–539.
Peng, J., Schwartz, D., Elias, J. E., Thoreen, C. C., Cheng, D., Marsischky, G., Roelofs, J., Finley, D., and Gygi, S. P. (2003). *Nat. Biotechnol.* **21,** 921–926.
Pflieger, D., Junger, M. A., Muller, M., Rinner, O., Lee, H., Gehrig, P. M., Gstaiger, M., and Aebersold, R. (2008). *Mol. Cell. Proteomics* **7,** 326–346.
Pickart, C. M., and Fushman, D. (2004). *Curr. Opin. Chem. Biol.* **8,** 610–616.
Pinkse, M. W., Uitto, P. M., Hilhorst, M. J., Ooms, B., and Heck, A. J. (2004). *Anal. Chem.* **76,** 3935–3943.
Pitteri, S. J., Chrisman, P. A., Hogan, J. M., and McLuckey, S. A. (2005). *Anal. Chem.* **77,** 1831–1839.
Pless, D. D., and Lennarz, W. J. (1977). *Proc. Natl. Acad. Sci. USA* **74,** 134–138.
Polevoda, B., and Sherman, F. (2000). *J. Biol. Chem.* **275,** 36479–36482.
Ramos-Fernandez, A., Lopez-Ferrer, D., and Vazquez, J. (2007). *Mol. Cell. Proteomics* **6,** 1274–1286.
Rao, K. C., Palamalai, V., Dunlevy, J. R., and Miyagi, M. (2005). *Mol. Cell. Proteomics* **4,** 1550–1557.
Reid, G., Gallais, R., and Metivier, R. (2009). *Int. J. Biochem. Cell Biol.* **41,** 155–163.
Rodriguez, M. S., Dargemont, C., and Hay, R. T. (2001). *J. Biol. Chem.* **276,** 12654–12659.
Ross, P. L., Huang, Y. N., Marchese, J. N., Williamson, B., Parker, K., Hattan, S., Khainovski, N., Pillai, S., Dey, S., Daniels, S., Purkayastha, S., Juhasz, P., *et al.* (2004). *Mol. Cell. Proteomics* **3,** 1154–1169.
Roth, S. Y., Denu, J. M., and Allis, C. D. (2001). *Annu. Rev. Biochem.* **70,** 81–120.
Rush, J., Moritz, A., Lee, K. A., Guo, A., Goss, V. L., Spek, E. J., Zhang, H., Zha, X. M., Polakiewicz, R. D., and Comb, M. J. (2005). *Nat. Biotechnol.* **23,** 94–101.
Sadoul, K., Boyault, C., Pabion, M., and Khochbin, S. (2008). *Biochimie* **90,** 306–312.

Schmidt, A., Kellermann, J., and Lottspeich, F. (2005). *Proteomics* **5,** 4–15.

Schroeder, M. J., Shabanowitz, J., Schwartz, J. C., Hunt, D. F., and Coon, J. J. (2004). *Anal. Chem.* **76,** 3590–3598.

Schulenberg, B., Aggeler, R., Beechem, J. M., Capaldi, R. A., and Patton, W. F. (2003a). *J. Biol. Chem.* **278,** 27251–27255.

Schulenberg, B., Beechem, J. M., and Patton, W. F. (2003b). *J. Chromatogr. B Analyt. Technol. Biomed. Life Sci.* **793,** 127–139.

Schulenberg, B., Goodman, T. N., Aggeler, R., Capaldi, R. A., and Patton, W. F. (2004). *Electrophoresis* **25,** 2526–2532.

Scigelova, M., and Makarov, A. (2006). *Proteomics* **6**(Suppl. 2), 16–21.

Sirard, J. C., Vignal, C., Dessein, R., and Chamaillard, M. (2007). *PLoS Pathog.* **3,** e152.

Siuti, N., and Kelleher, N. L. (2007). *Nat. Methods* **4,** 817–821.

Siuti, N., Roth, M. J., Mizzen, C. A., Kelleher, N. L., and Pesavento, J. J. (2006). *J. Proteome Res.* **5,** 233–239.

Sliter, D. A., Kubota, K., Kirkpatrick, D. S., Alzayady, K. J., Gygi, S. P., and Wojcikiewicz, R. J. (2008). *J. Biol. Chem.* **283,** 35319–35328.

Sobel, R. E., Cook, R. G., and Allis, C. D. (1994). *J. Biol. Chem.* **269,** 18576–18582.

Sokolovsky, M., Riordan, J. F., and Vallee, B. L. (1967). *Biochem. Biophys. Res. Commun.* **27,** 20–25.

Soufi, B., Gnad, F., Jensen, P. R., Petranovic, D., Mann, M., Mijakovic, I., and Macek, B. (2008). *Proteomics* **8,** 3486–3493.

Spiro, R. G. (2002). *Glycobiology* **12,** 43R–56R.

Steen, H., Jebanathirajah, J. A., *et al.* (2006). *Mol. Cell Proteomics* **5,** 172–181.

Steinberg, T. H., Pretty On Top, K., Berggren, K. N., Kemper, C., Jones, L., Diwu, Z., Haugland, R. P., and Patton, W. F. (2001). *Proteomics* **1,** 841–855.

Steinberg, T. H., Agnew, B. J., Gee, K. R., Leung, W. Y., Goodman, T., Schulenberg, B., Hendrickson, J., Beechem, J. M., Haugland, R. P., and Patton, W. F. (2003). *Proteomics* **3,** 1128–1144.

Stemmann, O., Zou, H., Gerber, S. A., Gygi, S. P., and Kirschner, M. W. (2001). *Cell* **107,** 715–726.

Strahl, B. D., and Allis, C. D. (2000). *Nature* **403,** 41–45.

Swaney, D. L., McAlister, G. C., and Coon, J. J. (2008). *Nat. Methods* **5,** 959–964.

Swaney, D. L., McAlister, G. C., Wirtala, M., Schwartz, J. C., Syka, J. E., and Coon, J. J. (2007). *Anal. Chem.* **79,** 477–485.

Syka, J. E., Coon, J. J., Schroeder, M. J., Shabanowitz, J., and Hunt, D. F. (2004). *Proc. Natl. Acad. Sci. USA* **101,** 9528–9533.

Sykora, C., Hoffmann, R., and Hoffmann, P. (2007). *Protein Pept. Lett.* **14,** 489–496.

Tagwerker, C., Flick, K., Cui, M., Guerrero, C., Dou, Y., Auer, B., Baldi, P., Huang, L., and Kaiser, P. (2006). *Mol. Cell. Proteomics* **5,** 737–748.

Tang, W. H., Halpern, B. R., Shilov, I. V., Seymour, S. L., Keating, S. P., Loboda, A., Patel, A. A., Schaeffer, D. A., and Nuwaysir, L. M. (2005). *Anal. Chem.* **77,** 3931–3946.

Taniguchi, C. M., Emanuelli, B., and Kahn, C. R. (2006). *Nat. Rev. Mol. Cell. Biol.* **7,** 85–96.

Tao, W. A., Wollscheid, B., O'Brien, R., Eng, J. K., Li, X. J., Bodenmiller, B., Watts, J. D., Hood, L., and Aebersold, R. (2005). *Nat. Methods* **2,** 591–598.

Thingholm, T. E., Jorgensen, T. J., Jensen, O. N., and Larsen, M. R. (2006). *Nat. Protoc.* **1,** 1929–1935.

Thompson, A., Schafer, J., Kuhn, K., Kienle, S., Schwarz, J., Schmidt, G., Neumann, T., Johnstone, R., Mohammed, A. K., and Hamon, C. (2003). *Anal. Chem.* **75,** 1895–1904.

Tian, Y., Zhou, Y., Elliott, S., Aebersold, R., and Zhang, H. (2007). *Nat. Protoc.* **2,** 334–339.

Trelle, M. B., and Jensen, O. N. (2007). *Expert Rev. Proteomics* **4,** 491–503.

Tsaprailis, G., Nair, H., Somogyi, A., Wysocki, V. H., Zhong, W., Futrell, J. H., Summerfield, S. G., and Gaskell, S. J. (1999). *J. Am. Chem. Soc.* **121,** 5142–5154.
Udeshi, N. D., Shabanowitz, J., Hunt, D. F., and Rose, K. L. (2007). *FEBS J.* **274,** 6269–6276.
Ulintz, P. J., Yocum, A. K., Bodenmiller, B., Aebersold, R., Andrews, P. C., and Nesvizhskii, A. I. (2009). *J. Proteome Res.* **8,** 887–899.
Unwin, R. D., Griffiths, J. R., Leverentz, M. K., Grallert, A., Hagan, I. M., and Whetton, A. D. (2005). *Mol. Cell. Proteomics* **4,** 1134–1144.
Unwin, R. D., Griffiths, J. R., and Whetton, A. D. (2009). *Nat. Protoc.* **4,** 870–877.
Vestal, M. L., and Campbell, J. M. (2005). *Methods Enzymol.* **402,** 79–108.
Waanders, L. F., Hanke, S., and Mann, M. (2007). *J. Am. Soc. Mass Spectrom.* **18,** 2058–2064.
Washburn, M. P., Wolters, D., and Yates, J. R. 3rd. (2001). *Nat. Biotechnol.* **19,** 242–247.
Wells, L., Vosseller, K., *et al.* (2002). *Mol. Cell Proteomics* **1,** 791–804.
White, F. M. (2008). *Curr. Opin. Biotechnol.* **19,** 404–409.
Williamson, B. L., Marchese, J., and Morrice, N. A. (2006). *Mol. Cell Proteomics* **5,** 337–346.
Witze, E. S., Old, W. M., Resing, K. A., and Ahn, N. G. (2007). *Nat. Methods* **4,** 798–806.
Wolf-Yadlin, A., Hautaniemi, S., Lauffenburger, D. A., and White, F. M. (2007). *Proc. Natl. Acad. Sci. USA* **104,** 5860–5865.
Wu, S. L., Jardine, I., *et al.* (2004). *Rapid Commun. Mass Spectrom.* **18,** 2201–2207.
Wu, C. C., MacCoss, M. J., Howell, K. E., and Yates, J. R. 3rd (2003). *Nat. Biotechnol.* **21,** 532–538.
Yang, Z., and Hancock, W. S. (2005). *J. Chromatogr. A* **1070,** 57–64.
Yang, X. J., and Seto, E. (2008). *Mol. Cell.* **31,** 449–461.
Yeo, W. S., Lee, S. J., Lee, J. R., and Kim, K. P. (2008). *BMB Rep.* **41,** 194–203.
Zabrouskov, V., Senko, M. W., *et al.* (2005). *J. Am. Soc Mass Spectrom.* **16,** 2027–2038.
Zappacosta, F., Collingwood, T. S., Huddleston, M. J., and Annan, R. S. (2006). *Mol. Cell. Proteomics* **5,** 2019–2030.
Zhan, X., and Desiderio, D. M. (2004). *Biochem. Biophys. Res. Commun.* **325,** 1180–1186.
Zhang, R., and Regnier, F. E. (2002). *J. Proteome Res.* **1,** 139–147.
Zhang, X., Herring, C. J., Romano, P. R., Szczepanowska, J., Brzeska, H., Hinnebusch, A. G., and Qin, J. (1998). *Anal. Chem.* **70,** 2050–2059.
Zhang, K., Williams, K. E., Huang, L., Yau, P., Siino, J. S., Bradbury, E. M., Jones, P. R., Minch, M. J., and Burlingame, A. L. (2002). *Mol. Cell. Proteomics* **1,** 500–508.
Zhang, H., Li, X. J., Martin, D. B., and Aebersold, R. (2003). *Nat. Biotechnol.* **21,** 660–666.
Zhang, K., Yau, P. M., Chandrasekhar, B., New, R., Kondrat, R., Imai, B. S., and Bradbury, M. E. (2004). *Proteomics* **4,** 1–10.
Zhang, Y., Wolf-Yadlin, A., Ross, P. L., Pappin, D. J., Rush, J., Lauffenburger, D. A., and White, F. M. (2005). *Mol. Cell. Proteomics* **4,** 1240–1250.
Zhen, Y., Xu, N., Richardson, B., Becklin, R., Savage, J. R., Blake, K., and Peltier, J. M. (2004). *J. Am. Soc. Mass Spectrom.* **15,** 803–822.

SECTION ELEVEN

ADDITIONAL TECHNIQUES

CHAPTER FORTY-ONE

Parallel Methods for Expression and Purification

Scott A. Lesley

Contents

1. Introduction 767
2. Strategies Based on End-Use 768
3. Parallel Cloning Strategies for Creating Expression Constructs 771
4. Small-Scale Expression Screening to Identify Suitable Constructs 774
5. Analytical Testing of Proteins for Selection 778
6. Large-Scale Parallel Expression 780
7. Conclusion 783
Acknowledgments 783
References 784

Abstract

Protein properties are highly diverse, making parallel expression and purification a particular challenge. Parallel methods are typically used when a number derivatives of a target protein are desired or when multiple homologs are needed. A typical scenario involves target evaluation, cloning and mutagenesis of the target, expression screening, large-scale expression and purification, and analytical and biophysical testing of the resulting protein. This chapter describes some of the strategies and methods employed for parallel protein expression and purification.

1. Introduction

Parallel protein purification is mainly a challenge of logistics. There are many purification approaches, which can be employed to isolate proteins. Some of these are inherently more amenable to parallel processing than others. (Graslund *et al.*, 2008a; Kim *et al.*, 2004; Lesley *et al.*, 2002b) Most parallel purification strategies employ the use of purification tags to facilitate simple affinity purification with standardization of methods and protein

Genomics Institute of the Novartis Research Foundation, San Diego, California, USA

Methods in Enzymology, Volume 463
ISSN 0076-6879, DOI: 10.1016/S0076-6879(09)63041-X

behavior. Although a single approach using a protease-cleavable his-tag sequence is described in this protocol, other tags and even nontagged approaches can be developed into parallel processes suitable for many applications. Rather than try to be all inclusive, this article attempts to describe a process currently in use in the author's laboratory as a forum to illustrate the key aspects of parallel purification. This process has been applied to thousands of proteins to date and delivers protein suitable for many applications.

2. Strategies Based on End-Use

Successful protein expression and purification typically requires the evaluation of multiple expression constructs. Many iterative truncations, and sometimes orthologous proteins, are required to identify a suitable construct for recombinant expression. Making and characterizing these variants is a significant barrier to success which requires the ability to perform parallel expression screening and protein characterization. The definition of success in this regard depends largely on the end-use application. For use as a simple antigen, expression of sufficient amounts for immunization, possibly even in aggregated form, is sufficient to claim success. More often, demanding applications like biochemical studies or protein structure require not only the production of a purified protein but one which is properly folded, containing the appropriate cofactors or posttranslational modifications and largely homogeneous in nature. Prior to defining any purification strategy or evaluation process, the end-use must be fully understood and protocols designed to achieve the required parameters. One of the most demanding end-use applications is crystallography. Generally speaking, protein which is suitable for crystallization is also sufficient for biochemical, immunological, and other functional studies. Suitable protein should be properly folded (not just soluble), nonaggregated, homogeneous, lacking unstructured regions and expressed at levels sufficient for multiple crystallization trials. In most cases, achieving these criteria is the result of screening multiple derivatives of targets and related targets (Fig. 41.1). This requires parallel processing of purified proteins and the parallel characterization of them.

A bioinformatic evaluation of the protein of interest should be performed before any experimentation. With the vast amount of genome sequence information available, a comparison of proteins with even little to no functional information can provide useful insight used to guide expression design. Software changes so rapidly that there is no point to providing here a detailed method for analysis. Instead, some guidance will be provided as a general approach. The first task, therefore, is to identify

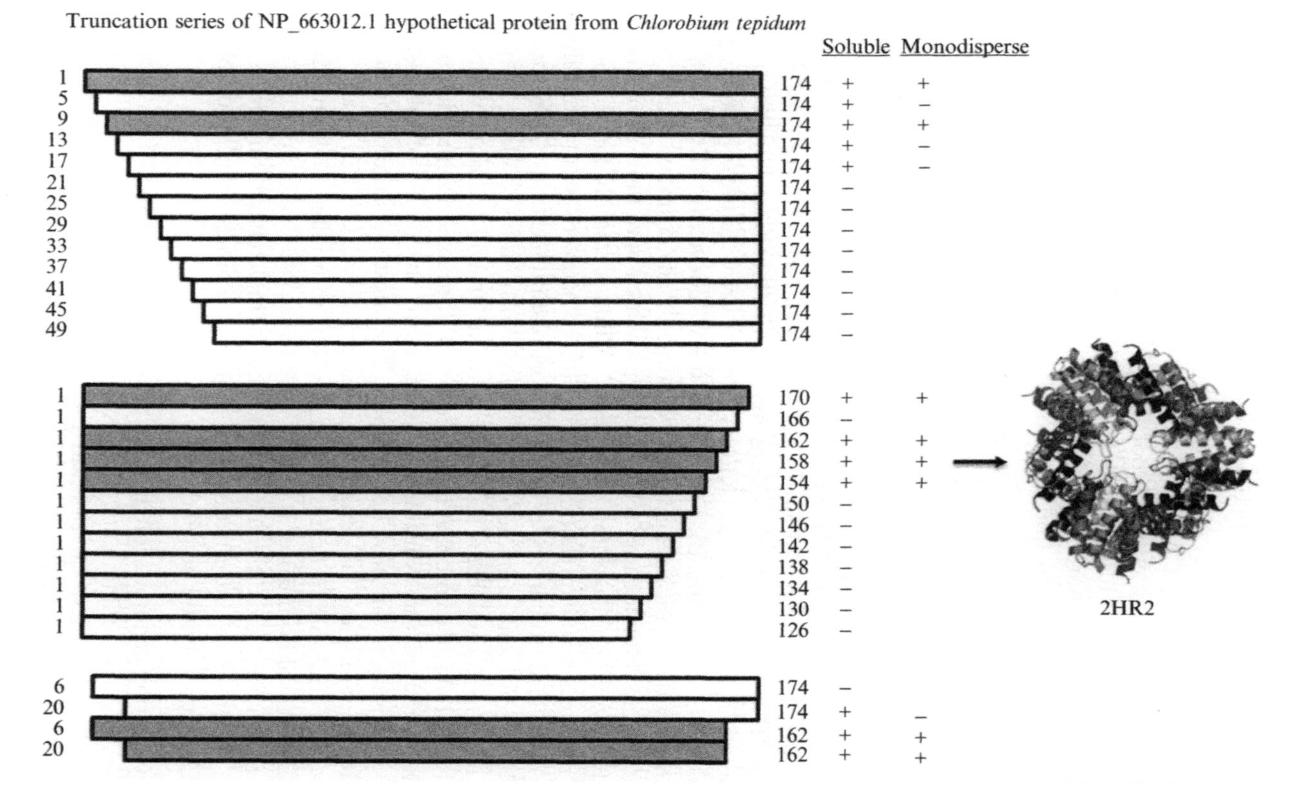

Figure 41.1 Identifying suitable protein expression constructs through parallel protein expression, purification, and characterization. Nested N- and C-terminal truncations were evaluated in parallel for suitability for crystallization trials. Proteins showing both soluble expression with sufficient protein yield and monodisperse behavior by ANSEC are indicated in gray. One such construct was entered into crystallization trials and yielded a high-resolution structure.

related proteins and align them to identify conserved regions. These regions typically constitute the core structural domains which need to be retained within expression constructs and can be used to identify domain boundaries. The basic local alignment search tool (BLAST) (Altschul *et al.*, 1990) is the most common tool for simple retrieval of related sequences. The NCBI (http://blast.ncbi.nlm.nih.gov/Blast.cgi?CMD=Web&PAGE_TYPE=BlastHome) provides many versions of this essential tool along with access to the majority of the genome sequences available to query. A hidden Markov model (HMM) can be developed for a more sophisticated search but a simple PSI-BLAST query is usually sufficient to find the most highly related sequences which are the most useful.

Most proteins have already been aligned and placed into families. Pfam (http://pfam.sanger.ac.uk/) (Finn *et al.*, 2008) is a database of these alignments and provides much useful information on comparative studies of related proteins including domain boundaries and sequence conservation. UniProt (http://www.uniprot.org/) (Consortium, 2008) also provides a useful source of summary data including secondary structure predictions for most proteins. Ligands can also be a useful tool for stabilizing proteins and activity testing. Ligand predictions may be obvious from annotations, but several databases such as KEGG (http://www.genome.jp/kegg/) (Kanehisa *et al.*, 2008) and BioCyc (http://www.biocyc.org) (Karp *et al.*, 2005) can provide additional ligand suggestions. Finally, a simple search of the literature can often identify related proteins which have been expressed previously to give guidance for expression. While seemingly an obvious task, the rush to get an expression construct into whatever system is readily available often leads to cursory or no review of past publications. There are many other useful and easily accessible bioinformatic tools available. The basic advice is to know what is known about your protein and those which are closely related and to use that information for your experimental design.

Having gathered the available information on the target of interest, expression constructs can now be designed. The most successful approach is to evaluate not only the full-length open reading frame (ORF), but also to create and test multiple truncations of this ORF in parallel. Defining the N- and C-terminal boundaries is critical. We have observed many cases of one or a few amino acid differences at the ends completely altering protein behavior. Even with information on sequence conservation, multiple N- and C-termini should be tried.

1. Query existing genome sequences for related proteins using BLAST and perform sequence alignments. Tools such as Pfam can provide a convenient forum for these queries and provides prevetted alignments and domain boundaries. Incorporate information from literature searches.
2. Identify locations of N- and C-terminal sequence conservation as initial truncation boundaries. In some cases, internal domain boundaries may

be preferable. For example, kinase researchers typically study the catalytic domain itself.

3. Starting from these initial boundaries select 5–10 additional N- and C-termini. The final expression construct collection should be a matrix of these termini. We have found that increments of approximately four amino acids provides a useful range of termini without adding excessively to the number of constructs to be screened. A 10×9 matrix (90 constructs) is a convenient number to screen allowing for full-length and empty vector controls.
4. Review the boundary selections with the end-use in mind making sure to retain essential regions.

3. Parallel Cloning Strategies for Creating Expression Constructs

There are many different approaches to cloning and mutagenesis. Individual laboratories have their own preferences based on past experience. Creating large numbers of unique expression clones can be daunting, however. Some cloning approaches are inherently more amenable to parallel processing. A common method utilizes the lambda Int/Xis/IHF recombination at *att* sites to move ORFs into various recipient vectors (Hatley *et al.*, 2000). The commercial embodiment of this protocol, Gateway®, is popular for creating expression vectors with varying tags but typically results in extra amino acids which are encoded by the recombination sites. T-vectors (Harrison *et al.*, 1994), topoisomerase-linked vectors (Shuman, 1994), and cre-lox recombinatorial vectors (Liu *et al.*, 1998) also are amenable to parallel cloning projects, but are not particularly flexible and typically require substantial effort to generate a custom vector. Two methods which have been used in our laboratory to generate thousands of expression clones are ligation-independent-cloning (LIC) (Aslanidis and de Jong, 1990) and polymerase incomplete primer extension (PIPE) (Klock *et al.*, 2008). In both cases these protocols provide a substantial amount of flexibility for cloning ORFs into expression vectors, can be set up to utilize common primers, are rapid and require a minimum of reagent costs. The PIPE method will be described here since it also provides a convenient protocol for creating truncations.

The PIPE method is based on the observation that normal PCR amplifications result in mixtures of products which are not fully extended thereby leaving the 5′ ends of such products variably unpaired. This is similar to LIC which utilizes a proofreading polymerase to remove nucleotides from the 3′-terminus. These unpaired 5′ ends on the PCR products derive from the synthetic amplification primers. Therefore, oligo design defines the

Vector PCR – can be prepared in advance, in bulk

5' annealing site | Negative selection | 3' annealing site

Insert PCR – 5' ends on insert primers include sequences complementary to vector annealing sites

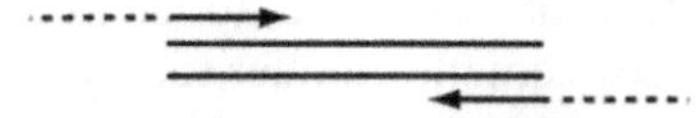

Mix PCR products – unpurified PCR products will anneal directionally across the primer-encoded complementary sequences

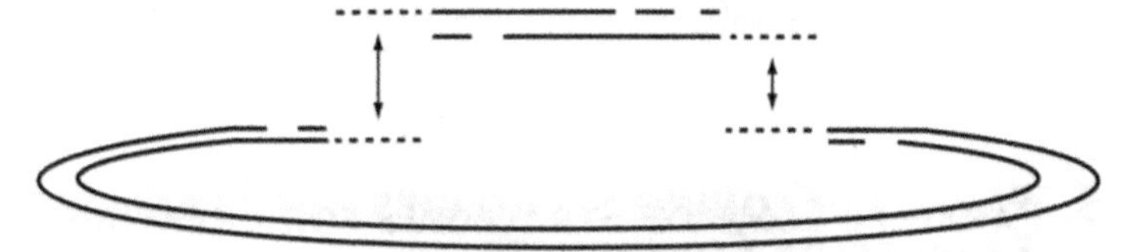

Transform cells – nicks and gaps are repaired *in vivo*
Screen colonies for insert-containing clones by dPCR

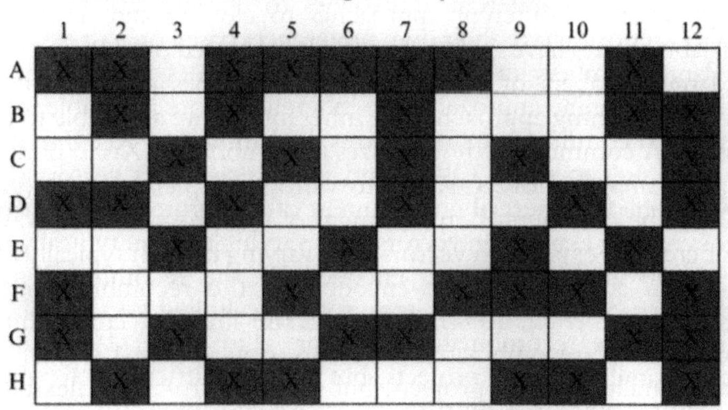

Figure 41.2 *Parallel cloning process.* The PIPE cloning method allows for simple parallel cloning of many expression constructs, including deletions and point mutations for screening. Appropriately designed primers are used for vector amplification and insert amplification. Resulting amplifiers are simply mixed and transformed. Isolated colonies are screened by a dPCR amplification and detection of PCR product by fluorescent dye. Typically, two colonies are screened for each attempted construct which is sufficient to identify insert-containing clones (shaded X) for most targets. Insert-containing clones are then sequenced for confirmation.

resulting 5′-overhangs which can be incorporated into a cloning strategy. By simply preparing PCR products of inserts and vector with complementary 5′ regions of approximately 15 bases, these products will anneal and can be transformed to produce a stable expression plasmid (Fig. 41.2).

1. Design vector and insert primers with appropriate 15 base complementary regions (see Klock *et al.*, 2008) for design criteria. The first 15 bases

on the 5′ ends of the primers are designed to be directionally complementary such that the resultant PCR fragment(s) can anneal as desired and become viable plasmids upon transformation.

2. Set up PCR reactions containing 5 μl of both 10 μM forward and reverse primers, 5 μl of template (20–100 pg per PCR amplification) and 5 μl of 10× Cloned Pfu DNA Polymerase (Stratagene) Reaction Buffer, 2.5 units of PfuTurbo® DNA polymerase, 1 μl of 10 mM dNTPs and water to a volume of 50 μl.
3. Perform PCR by incubating the reaction at 95 °C for 2 min, then 25 cycles of 95 °C for 30 s, 55 °C for 45 s and 68 °C for 14 min, and finally a 4 °C hold.
4. Confirm successful amplifications by gel electrophoresis. PCR products may be faint and smeared, but this is acceptable.
5. Thaw 2 ml aliquot of competent cells on ice for 10–15 min. Chill a 96-well microtiter plate on ice for 10–15 min. Transfer 2 μl of each of the vector and insert PCR reactions into wells of a prechilled microtiter plate and mix by pipetting. Dispense 20 μl of competent cells into each well. Pipette up and down once to ensure DNA has mixed with the cells and incubate for 15 min on ice.
6. Heat shock the cells by floating the microtiter plate in a 42 °C water bath for 45 s. Immediately return the microtiter plate to ice. Dispense 100 μl of LB Broth (no antibiotic) into each well. Recover the transformed cells by incubating at 37 °C while shaking at 250 rpm for 1 h. Dispense 100 or 40 μl of the recovered cells into the respective wells of the selective LB agar trays with glass beads. By hand, gently shake the trays enough to evenly distribute the glass beads across the entire well. Invert the tray to drop the glass beads off of the LB agar and onto the lid and then remove the glass beads. Incubate the inverted trays 12–16 h at 37 °C.

Isolated colonies can be screened by classic procedures such as restriction mapping. Alternatively, large numbers of colonies can be screened by diagnostic PCR (dPCR) or SYBR-PCR.

1. Dispense 200 μl of LB Broth with appropriate antibiotic into the wells of flat-bottom 96-well plate. Using aseptic technique, pick and transfer 1–4 isolated colonies per transformation into unique wells of the microtiter plate. Incubate the plate at 37 °C while shaking at 250 rpm for 3–12 h.
2. Transfer 3 μl samples from each culture into a 96-well PCR plate. Put the remainder of the cultures back into the shaking incubator to continue growth for future glycerol stock archival.
3. Add 47 μl of master mix containing PCR reagents and a gene-specific and a vector-specific primer to each well containing cells in the PCR plate. Primers should be designed such that amplification only occurs when the vector contains the desired insert. Amplify the DNA fragments for 30 cycles using cycling conditions appropriate for the primers.

4. Dispense 50 μl of 10 m*M* EDTA in a flat-bottom 96-well plate. Transfer 5 μl of each PCR product into the wells and dispense 150 μl of SYBR Buffer [40 μl of 10,000× SYBR Green I Dye (Sigma) is diluted in 20 ml of 50 m*M* Tris–HCl (pH 8.0) to make a 20× solution].
5. Using an unamplified sample for a negative control, measure fluorescence of each sample using a microtiter plate fluorescence reader. Excitation: 485 nm. Emission: 525 nm.
6. SYBR results are determined on a relative scale with the positive wells having at least fourfold higher fluorescence than the negatives or the control. dPCR-positive reactions can then be confirmed by gel analysis or direct sequencing procedures.

4. Small-Scale Expression Screening to Identify Suitable Constructs

Small-scale evaluation of clones is an integral part of the clone selection process. Obtaining recombinant protein in good yield, which is in the soluble fraction of a lysate, is often one of the first major hurdles encountered. However, all too often this single criterion dominates the decision process leading to a false sense of success and much more work in the later stages. Screening approaches such as reporter fusions (Lesley *et al.*, 2002a; Waldo *et al.*, 1999) and colony blots (Martinez Molina *et al.*, 2008) provide a means to screen thousands of protein derivatives. Fusion to partners which enhance solubility (Kapust and Waugh, 1999) is another strategy which has been employed to define appropriate construct boundaries for solubility. In many cases, these soluble proteins are aggregates and heterogeneous mixtures of partially folded species (Fig. 41.3A and B). This becomes problematic during purification, often resulting in poor yields and instability, and also presents problems in obtaining high specific activity or crystallizing protein. Partial proteolysis (Fig. 41.3D) can be used to identify stable boundaries. The resulting stable fragments are identified via mass spectrometry and can be used to define subsequent expression constructs. Addition of ligands to the protein can have a stabilizing effect as well (Fig. 41.3C). Ligands can be identified through screening methods such as Thermofluor (Pantoliano *et al.*, 2001) or predicted through annotation and experimentation. To the degree possible, additional protein analytics should be performed at an early stage and the results factored into decisions for selecting expression constructs for larger scale studies. With these caveats in mind, expression screening should be performed at a scale to allow for such tests to be implemented.

Small-scale expression screening is performed using deep-well microtiter plates as a convenient way to grow enough cells to evaluate protein expression. Maximizing cell growth requires optimizing expression conditions. In particular, aeration of cultures can prove limiting. Short-throw plate shakers and gas-permeable plate sealers improve final yields. Care must be taken at each step to minimize volume loss, as even small volume can have a large impact on final yields.

1. Prepare an overnight culture of the desired clones in a 96 deep-well block containing 850 μl of TB containing 2% glycerol and desired antibiotic.
2. In a deep-well block, add 17 μl of overnight culture to 1.25 ml of TB containing 2% glycerol and the desired antibiotic. Incubate 3 h at 37 °C shaking at 900 rpm.

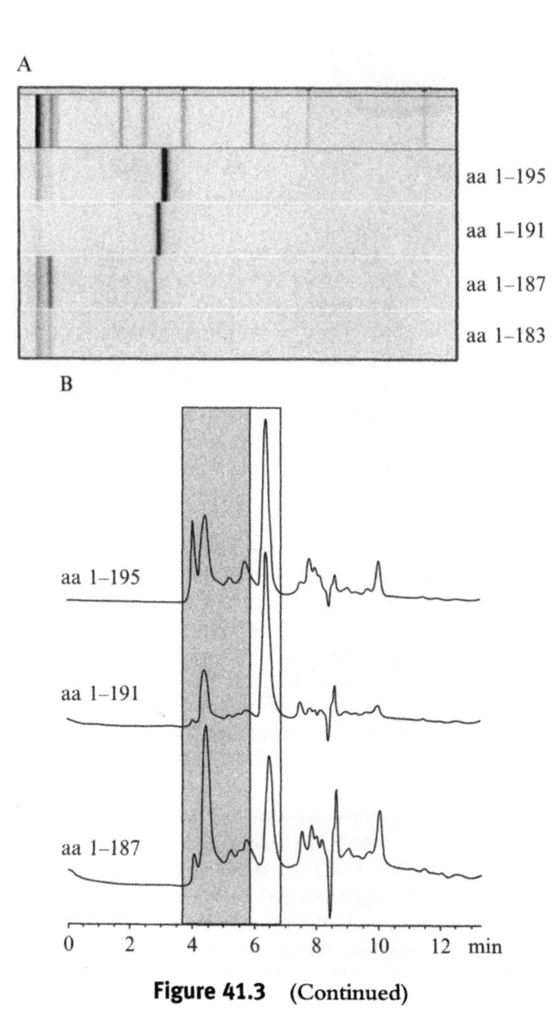

Figure 41.3 (Continued)

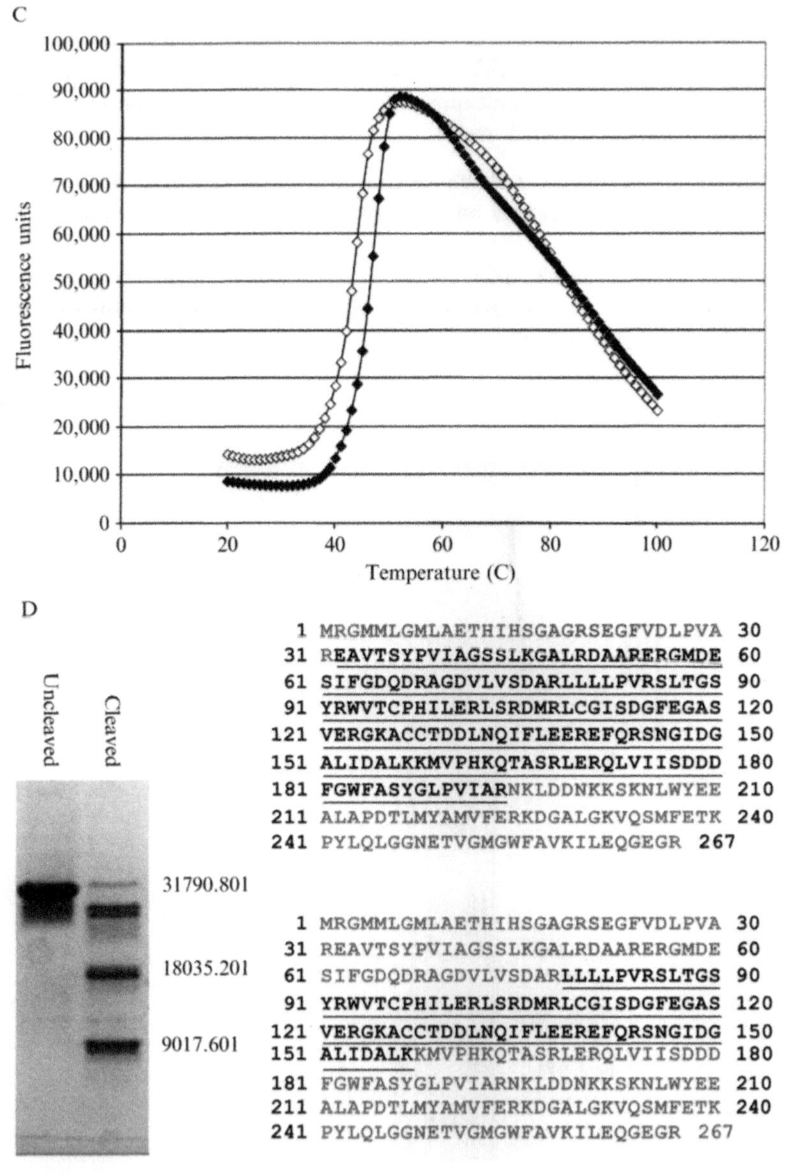

Figure 41.3 Analytical profiling of proteins derived from parallel expression and purification. (A) Capillary electrophoresis results of a truncation series of YP_958225.1. (B) ANSEC profile of a truncation series of YP_958225.1 showing variation in the amount of soluble monomer (white box) and soluble aggregate (gray box) between samples. (C) Thermofluor ligand binding assay showing increased thermostability YP_164977.1 (threonine aldolase) in the presence of pyridoxal phosphate (filled symbols) versus the apo enzyme. (D) Partial trypsin proteolysis of YP_843077.1 shows stable fragments visible by SDS–PAGE (cleaved lane). Mass spectrometry can be used to define stable fragments of the original protein. Two masses (18032.201 and 9017.601 Da) correspond to peptide fragments (underlined) which can be targeted for subsequent subcloning and further expression studies.

3. Add appropriate inducer for protein expression (i.e., 1 m*M* IPTG) and continue incubation at 37 °C with shaking for 5 h. Note the final $OD_{600\,nm}$ of the culture. Another popular method for expression utilized autoinduction media (Studier, 2005) which does not require addition of expression inducer.
4. Pellet cells at 3000 rpm for 15 min. Discard supernatant and freeze cells overnight at – 20 °C. Freezing is important to facilitate cell lysis.
5. Thaw cells by incubating at room temperature briefly then refreeze the pellet for 2 h at – 20 °C. After the second freeze, thaw cells in Extraction Buffer (12 μl [300 kU] Ready Lyse [Epicentre] in 50 ml buffer [50 m*M* Tris pH 7.5, 50 m*M* Sucrose, 1 m*M* EDTA]) using a proportion of 50 μl for every observed $OD_{600\,nm}$ of the original expression culture. Resuspend the pellet well using a multichannel pipette. Incubate 30 min at room temperature.
6. After lysis the viscosity of the solution is often quite high and can prevent efficient transfer of the lysate and protein capture on the resin. To reduce viscosity, a Benzonase is used to degrade nucleic acids which contribute the bulk of the viscosity to the lysate. Prepare Benzonase Solution by diluting 1.5 μl [375 units] Benzonase [Sigma] in 50 ml of buffer [10 m*M* Tris pH 7.5, 50 m*M* NaCl, 1 m*M* EDTA, 10 m*M* $MgCl_2$]. Add Benzonase Solution at 1.25 times the amount of Extraction Buffer used in step 5. Seal the plate and mix by inversion approximately five times or until the viscosity begins to reduce. Incubate an additional 15 min at room temperature. Pellet cells at 4000 rpm for 30 min.
7. Prepare a 96 deep-well plate containing 150 μl of a 50% slurry of affinity resin (Nickel-NTA for his-tag purification) in Equilibration Buffer (50 m*M* HEPES pH 8.0, 50 m*M* NaCl, 10 m*M* imidazole, 1 m*M* TCEP). Add the supernatant from the centrifuged lysate to the resin plate being careful to avoid the residual cell debris pellet. Seal the plate and incubate with mixing for 60 min at 4 °C.
8. Wet a 96 deep-well filter plate with 300 μl of Equilibration Buffer, pulling sufficient vacuum to leave a thin layer of liquid above the frit. Add the bound resin from the previous step and allow the supernatant to flow-through the filter plate. Add 700 μl of Wash Buffer (50 m*M* HEPES pH 8.0, 300 m*M* NaCl, 40 m*M* imidazole, 10% glycerol, 1 m*M* TCEP) to each well and allow to drain. Centrifuge the filter plate at 200*g* for 1 min to remove the remaining Wash Buffer. Place a collection plate beneath the filter plate. Add 150 μl of Elution Buffer (50 m*M* HEPES pH 8.0, 300 m*M* imidazole, 10% glycerol, 1 m*M* TCEP) and incubate for 5 min. Centrifuge the filter plate at 200*g* for 5 min to collect the purified protein. It is important in this step to avoid over-drying the resin which can lead to channeling (cracking) of the resin bed and poor wash and elution of the resin.

5. Analytical Testing of Proteins for Selection

There are many methods for characterizing the quality and homogeneity of a protein prep. Data from these methods are needed to make informed decisions about the suitability of a particular expression construct or purification method. The protein consumption requirements for many of these tests are substantial. However, in some cases miniaturized or abbreviated methods can be used at the level of small-scale screening to provide useful protein characterization. Table 41.1 lists several analytical methods which can be applied to small amounts of protein.

The production of soluble protein is the first step toward successful purification. In many cases alteration of growth conditions such as reduced

Table 41.1 Analytical methods suitable for small-scale protein screens

Method	Protein characteristics	Comments
SDS–PAGE/CE	MW and purity	Capillary electrophoresis methods are rapid, require minimal protein volumes, and provide electronic records for ease of storage and comparison
ANSEC	Protein aggregation, monodispersity, oligomeric state	Trade-offs of resolution versus time and sample
Protease stability	Identification of stable fragments, proper folding	Useful for defining domain boundaries and formulation
Thermal stability	Proper folding, optimizing formulations, ligand binding	Thermofluor™ and Stargazer™ methods utilize small amounts of protein relative to differential scanning calorimetry
Protein concentration	Protein concentration	Various methods with different sensitivities
UV/Vis spectroscopy	Protein concentration, cofactor/ligand predictions	Nanodrop™ instrumentation allows wavelength scans using microliter volumes
Tryptophan fluorescence	Protein folding	Useful for construct comparison

expression temperature or alteration of the expression construct such as fusion to a highly soluble partner like maltose binding protein will greatly improve the soluble yield of the target protein. However, these are often soluble aggregates of the protein which have reduced activity and stability relative to their properly folded counterparts. It is best to identify and avoid these issues as early as possible. Analytical size exclusion chromatography (ANSEC) is a convenient and predictive method which can be employed in a parallel testing mode. By scaling-down column size and increasing flow rates, a screening method with a throughput of 13–18 min per sample can be achieved. By combining this approach with an HPLC autosampler, a full plate of 96 purified proteins produced from the microscale protocol can be screened in a day. A method is presented here for chromatography in a 13.3 min run using an Agilent HP1100 with a refrigerated, 384-well formatted autosampler. A slightly higher resolution method uses an 18 min run, which may also be more gentle on very high molecular weight samples.

1. Equilibrate appropriate size exclusion column (Shodex 8 × 300 mm Protein KW-803 column with 6 × 50 mm Protein KW-G guard column) in 20 m*M* Tris pH 7.5, 200 m*M* NaCl, 0.25 m*M* TCEP, 3 m*M* NaN_3. The flow rate should be 1.0–1.5 ml/min. This will take about 30 min.
2. After the column is appropriately equilibrated and A_{280} baseline has stabilized, 5–15 μl (depending on concentration) injections of sample proteins are performed. Appropriate protein MW standards and buffer blanks should be run prior to the sample sequence and after approximately 50 samples.
3. A comparison of typical results is shown in Fig. 41.3A and B. Although the rapid flow rate and relatively short column do not allow high-resolution size determination, it is quite easy to see the relative proportion of high molecular weight aggregates and the monomeric state of the protein. Generally speaking, proteins which show a uniform and monomeric profile are highly preferable to those containing aggregate, even if the initial purification yield is smaller.

Selecting the correct construct from among many requires understanding the end-use requirements. For example, protein crystallization requires protein to be monodisperse, homogeneous, and lacking unstructured regions. While many constructs might provide sufficient soluble protein to enter into crystal trials, evaluation of SDS–PAGE, ANSEC, ligand binding, and protease stability can effectively narrow the selection to the preferred candidates. The more information that can be gathered at the small-scale screening stage, the better the selection decisions that will be made.

6. Large-Scale Parallel Expression

Large-scale parallel expression and purification of proteins often requires some specialized equipment. Scale should be appropriate for the end-use. One liter bacterial expression is typically enough for most laboratory needs. Simply utilizing multiple 1 l shake flasks is sufficient in most cases, but is difficult to employ when pursuing 10s or 100s of targets. An airlift fermentor (http://www.gnfsystems.com) provides a convenient means to express up to 96 different proteins in parallel at a scale comparable to typical 1 l shake flasks (Fig. 41.4) (Lesley, 2001). Bacterial cultures can be grown to OD_{600} values of 25–30 resulting in a few grams of cell paste for processing.

Lysis and purification of multiple cultures presents the biggest logistical challenge to parallel processing at larger scale. Custom automation can facilitate this effort but much can be accomplished by simplification of the lysis and purification approach. A protocol is given here for purifying his-tagged proteins containing a TEV cleavage site for tag removal. We have found this approach to be reliable and easily scaled to many proteins in parallel, even without the use of automation. Our laboratory processes over 3000 proteins per year by this approach.

1. Cells containing expressed protein directly from fermentation are incubated for 30 min at room temperature with 0.25 mg hen egg white lysozyme per ml of culture, pelleted by centrifugation and the

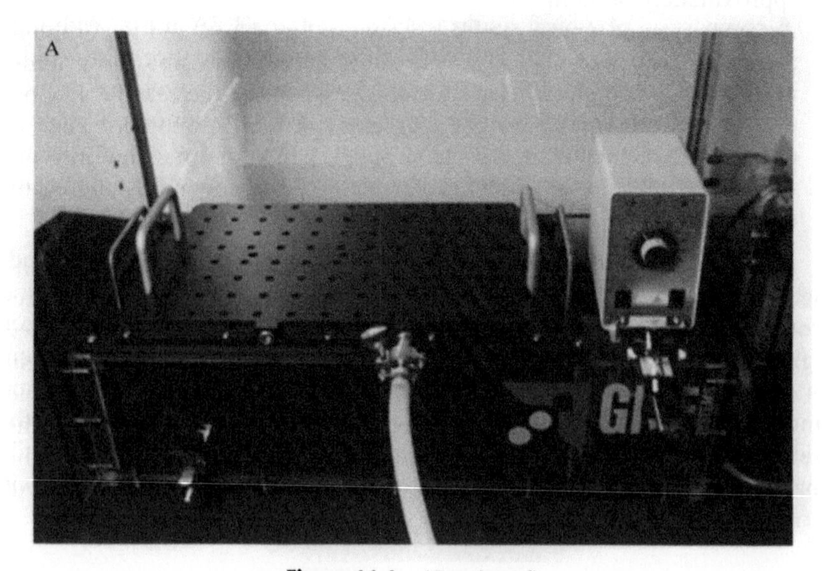

Figure 41.4 (Continued)

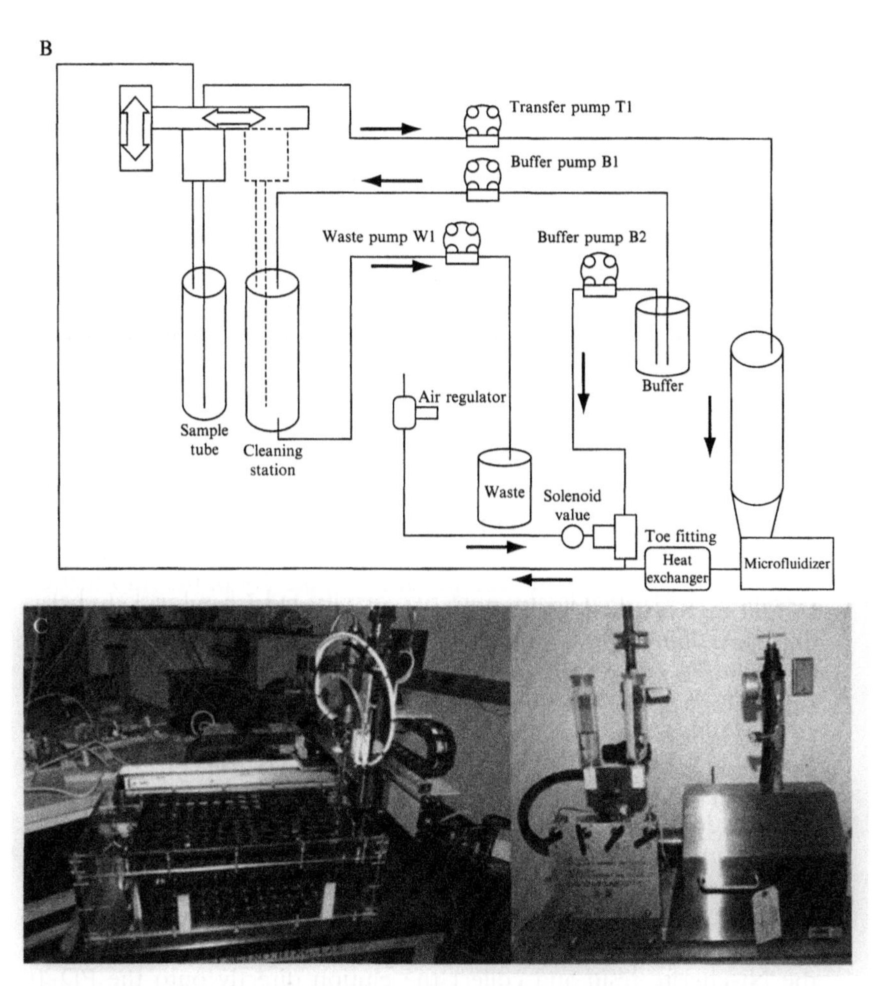

Figure 41.4 Custom instrumentation for parallel expression and purification. (A) GNFermentor (GNF Systems) provides high cell density parallel fermentation of 96 cultures simultaneously. (B) Schematic of an automated cell resuspension and lysis apparatus. (C) Automated suspension and lysis apparatus provides parallel processing of samples expressed in the GNFermentor.

resulting pellets frozen before use. For purification, pellets (~3 g) are thawed in 80 ml of Lysis Buffer (50 m*M* HEPES pH 8.0, 50 m*M* NaCl, 10 m*M* imidazole, 1 m*M* Tris(2-carboxyethyl)phosphine hydrochloride [TCEP]). Each of the cell pellets is lysed as follows.

2. Homogenize the pellet and Lysis Buffer for approximately 80 s at 35,000 rpm (maximum setting) using a homogenizer (Omni International). Each cycle consists of 6 s with the probe near the bottom of the homogenate followed by 3 s with the probe near the surface. These both

resuspend the pellet and effectively lyses the cells. If insufficient lysis is observed, passage through a microfluidizer (Microfluidics) or French press can be used.

3. Centrifuge the lysates for 30 min at 32,000*g*. The clarified lysates are ready for affinity purification.

In parallel purification, conventional purification approaches can be a significant bottleneck. Instrumentation and protocols for multisample processing using HPLC instrumentation has been described (McMullan *et al.*, 2005). For most applications, however, an optimized affinity purification will provide sufficient purity along with the requisite throughput. We have found that a two-step nickel purification combined with a proteolytic removal of the purification tag provides excellent purification and can be performed in parallel for hundreds of samples at a diversity of scale. This approach provides material that has proved suitable for crystallization trials and enzymatic screening studies. Our vector contains a his-tag followed by a TEV protease cleavage site. This protocol can be adapted for other tags and cleavage enzymes.

1. Decant each clarified lysates onto two gravity-fed 1.5-ml nickel-chelating resin columns pre-equilibrated with Lysis Buffer. Collect each flow-through into a waste tray.
2. Add 7.5 ml of Wash Buffer (50 m*M* HEPES pH 8.0, 300 m*M* NaCl, 40 m*M* imidazole, 10% glycerol, 1 m*M* TCEP). Add 0.5 ml of Elution Buffer (20 m*M* HEPES pH 8.0, 300 m*M* imidazole, 10% glycerol, 1 m*M* TCEP). This volume of elution buffer should be no more than one third of the bed resin. The purpose is to push the elution front of the buffer to the bottom of the resin to minimize the elution volume for the next step.
3. Place each column over a PD10 desalting column which has been pre-equilibrated with Digestion Buffer (20 m*M* HEPES pH 8.0, 200 m*M* NaCl, 40 m*M* imidazole, 1 m*M* TCEP). Add 2.5 ml of Elution Buffer to the Ni-chelate resin and collect the elution directly onto the PD-10 column.
4. Elute the protein from the PD-10 column by further addition of 3.5 ml of Digestion Buffer, collecting the elution in a 15 ml disposable tube. Remove a small sample for protein assay and SDS–PAGE analysis.
5. The next step involves removal of the his-tag through proteolytic cleavage. The protease itself is his-tagged. Therefore, subsequent passage of the protein over a nickel resin will remove the protease, as well as uncleaved proteins and those which stick nonspecifically to the nickel resin while the protein of interest passes in the column flow-through. For each 15 mg of protein obtained from the PD-10 elution, add 1 mg of his-tagged TEV protease (Tropea *et al.*, 2009). Bring the total volume of the digestion to 9 ml using Digestion Buffer. Mix by inversion at room temperature for 2 h or overnight at 4 °C.

6. Pour each of the digested protein mixtures onto a gravity-fed 1.5-ml nickel-chelating resin column pre-equilibrated with Digestion Buffer. Collect each flow-through into a 50 ml conical tube. After the digestion mixture has passed through the column, add an additional 1.5 ml of Digestion Buffer to the resin to remove residual cleaved protein and collect into the same tube as above. This 10.5 ml elution volume contains the purified protein with the his-tag sequence removed.

7. Conclusion

Parallel protein expression and purification is necessary to adequately evaluate multiple targets or multiple derivatives of single targets. The methods described here utilize bacterial expression systems. These systems are by far the most convenient and robust platforms for expression, but often suffer problems when applied to mammalian proteins. In many cases, focusing on domains of interest and systematically optimizing expression constructs to identify the proper domain boundaries can result in excellent expression levels of active proteins (Graslund *et al.*, 2008b; Klock *et al.*, 2008). Identifying suitable domain boundaries is one of the best applications of parallel expression and purification. For example, kinase catalytic domains have been defined and expressed in bacteria with great success and have contributed significantly to drug design (Marsden and Knapp, 2008). Despite the difficulties and tedium of mammalian and baculovirus expression systems, these platforms can also be implemented in a parallel processing way (Peng *et al.*, 1993). Platforms such as BacMam (Kost and Condreay, 2002) offer additional flexibility in exploring multiple expression systems. Regardless of the expression platform, the key to successful protein expression is to understand the ultimate requirements of the protein in terms of its physical properties and activities. Too often, simple solubility and total expression levels are used as the sole selection criteria when evaluating expression options. Even very small changes to the domain boundaries of the expression construct can cause drastic changes to protein behavior without visibly changing apparent soluble yield (Klock *et al.*, 2008). Combining parallel expression and purification with parallel analytical techniques and data analysis provides the best opportunity to identify a successful expression construct and purification combination.

Acknowledgments

The protocols described here were developed through the contributions of many people. The author would particularly like to acknowledge Heath Klock, Mark Knuth, Carol Farr, Anna Grzechnik, Julie Feuerhelm, Daniel McMullan, and Dennis Carlton for their contributions to this chapter.

REFERENCES

Altschul, S. F., Gish, W., Miller, W., Myers, E. W., and Lipman, D. J. (1990). Basic local alignment search tool. *J. Mol. Biol.* **215,** 403–410.

Aslanidis, C., and de Jong, P. J. (1990). Ligation-independent cloning of PCR products (LIC-PCR). *Nucleic Acids Res.* **18,** 6069–6074.

Consortium, U. (2008). The universal protein resource (UniProt). *Nucleic Acids Res.* **36,** D190–D195.

Finn, R. D., Tate, J., Mistry, J., Coggill, P. C., Sammut, S. J., Hotz, H. R., Ceric, G., Forslund, K., Eddy, S. R., Sonnhammer, E. L., and Bateman, A. (2008). The Pfam protein families database. *Nucleic Acids Res.* **36,** D281–D288.

Graslund, S., Nordlund, P., Weigelt, J., Hallberg, B. M., Bray, J., Gileadi, O., Knapp, S., Oppermann, U., Arrowsmith, C., Hui, R., Ming, J., dhe-Paganon, S., *et al.* (2008a). Protein production and purification. *Nat. Methods* **5,** 135–146.

Graslund, S., Sagemark, J., Berglund, H., Dahlgren, L. G., Flores, A., Hammarstrom, M., Johansson, I., Kotenyova, T., Nilsson, M., Nordlund, P., and Weigelt, J. (2008b). The use of systematic N- and C-terminal deletions to promote production and structural studies of recombinant proteins. *Protein Expr. Purif.* **58,** 210–221.

Harrison, J., Molloy, P. L., and Clark, S. J. (1994). Direct cloning of polymerase chain reaction products in an XcmI T-vector. *Anal. Biochem.* **216,** 235–236.

Hatley, J. L., Temple, G. F., and Brasch, M. A. (2000). DNA cloning using *in vivo* site-specific recombination. *Genome Res.* **10,** 1788–1795.

Kanehisa, M., Araki, M., Goto, S., Hattori, M., Hirakawa, M., Itoh, M., Katayama, T., Kawashima, S., Okuda, S., Tokimatsu, T., and Yamanishi, Y. (2008). KEGG for linking genomes to life and the environment. *Nucleic Acids Res.* **36,** D480–D484.

Kapust, R. B., and Waugh, D. S. (1999). *Escherichia coli* maltose-binding protein is uncommonly effective at promoting the solubility of polypeptides to which it is fused. *Protein Sci.* **8,** 1668–1674.

Karp, P. D., Ouzounis, C. A., Moore-Kochlacs, C., Goldovsky, L., Kaipa, P., Ahren, D., Tsoka, S., Darzentas, N., Kunin, V., and Lopez-Bigas, N. (2005). Expansion of the BioCyc collection of pathway/genome databases to 160 genomes. *Nucleic Acids Res.* **33,** 6083–6089.

Kim, Y., Dementieva, I., Zhou, M., Wu, R., Lezondra, L., Quartey, P., Joachimiak, G., Korolev, O., Li, H., and Joachimiak, A. (2004). Automation of protein purification for structural genomics. *J. Struct. Funct. Genomics* **5,** 111–118.

Klock, H. E., Koesema, E. J., Knuth, M. W., and Lesley, S. A. (2008). Combining the polymerase incomplete primer extension method for cloning and mutagenesis with microscreening to accelerate structural genomics efforts. *Proteins* **71,** 982–994.

Kost, T. A., and Condreay, J. P. (2002). Recombinant baculoviruses as mammalian cell gene-delivery vectors. *Trends Biotechnol.* **20,** 173–180.

Lesley, S. A. (2001). High-throughput proteomics: Protein expression and purification in the postgenomic world. *Protein Expr. Purif.* **22,** 159–164.

Lesley, S. A., Graziano, J., Cho, C. Y., Knuth, M. W., and Klock, H. E. (2002a). Gene expression response to misfolded protein as a screen for soluble recombinant protein. *Protein Eng.* **15,** 153–160.

Lesley, S. A., Kuhn, P., Godzik, A., Deacon, A. M., Mathews, I., Kreusch, A., Spraggon, G., Klock, H. E., McMullan, D., Shin, T., Vincent, J., Robb, A., *et al.* (2002b). Structural genomics of the *Thermotoga maritima* proteome implemented in a high-throughput structure determination pipeline. *Proc. Natl. Acad. Sci. USA* **99,** 11664–11669.

Liu, Q., Li, M. Z., Leibham, D., Cortez, D., and Elledge, S. J. (1998). The univector plasmid-fusion system, a method for rapid construction of recombinant DNA without restriction enzymes. *Curr. Biol.* **8,** 1300–1309.

Marsden, B. D., and Knapp, S. (2008). Doing more than just the structure-structural genomics in kinase drug discovery. *Curr. Opin. Chem. Biol.* **12,** 40–45.

Martinez Molina, D., Cornvik, T., Eshaghi, S., Haeggstrom, J. Z., Nordlund, P., and Sabet, M. I. (2008). Engineering membrane protein overproduction in *Escherichia coli. Protein Sci.* **17,** 673–680.

McMullan, D., Canaves, J. M., Quijano, K., Abdubek, P., Nigoghossian, E., Haugen, J., Klock, H. E., Vincent, J., Hale, J., Paulsen, J., and Lesley, S. A. (2005). High-throughput protein production for X-ray crystallography and use of size exclusion chromatography to validate or refute computational biological unit predictions. *J. Struct. Funct. Genomics* **6,** 135–141.

Pantoliano, M. W., Petrella, E. C., Kwasnoski, J. D., Lobanove, V. S., Myslik, J., Graf, E., Carver, T., Asel, E., Springer, B. A., Lane, P., and Salemme, F. R. (2001). High-density miniaturized thermal shift assays as a general strategy for drug discovery. *J. Biomol. Screen.* **6,** 429–440.

Peng, S., Sommerfelt, M., Logan, J., Huang, Z., Jilling, T., Kirk, K., Hunter, E., and Sorscher, E. (1993). One-step affinity isolation of recombinant protein using the baculovirus/insect cell expression system. *Protein Expr. Purif.* **4,** 95–100.

Shuman, S. (1994). Novel approach to molecular cloning and polynucleotide synthesis using vaccinia DNA topoisomerase. *J. Biol. Chem.* **269,** 32678–32684.

Studier, F. W. (2005). Protein production by auto-induction in high density shaking cultures. *Protein Expr. Purif.* **41,** 207–234.

Tropea, J. E., Cherry, S., and Waugh, D. S. (2009). Expression and purification of soluble His(6)-tagged TEV protease. *Methods Mol. Biol.* **498,** 297–307.

Waldo, G. S., Standish, B. M., Berendzen, J., and Terwilliger, T. C. (1999). Rapid protein-folding assay using green fluorescent protein. *Nat. Biotechnol.* **17,** 691–695.

[illegible], D. [illegible], and Knapp, S. (2008). [illegible] approaches in [illegible] drug discovery. Curr. Opin. Chem. Biol. 12, [illegible].
Ramirez [illegible], D., [illegible], J., [illegible] (2008). [illegible] Biotechnol. 17, [illegible].
[illegible], H., [illegible], J. M., [illegible], and [illegible], [illegible]. (2011) [illegible] protein [illegible] x-ray crystallography and [illegible] to [illegible] Curr. Opin. [illegible], 13, 43–45.
[illegible], M. W., [illegible], and [illegible], [illegible] (2007). High-throughput [illegible] for drug discovery. [illegible] 6, 1[illegible].
[illegible], S., [illegible], M., [illegible], [illegible], Z., [illegible], and Sanchez, [illegible] (2011). [illegible] expression system. [illegible]
Stone, S. (2000). Novel [illegible] Anal. Chem. 289, [illegible].
[illegible], E. A. (2000). [illegible] Chem. 48, [illegible].
[illegible], and [illegible] (2009). [illegible] 198, [illegible].
[illegible], C., Sanders, S. M., [illegible], and [illegible] (2007). [illegible] [illegible] Curr. [illegible].

CHAPTER FORTY-TWO

Techniques to Isolate O_2-Sensitive Proteins: [4Fe–4S]-FNR as an Example

Aixin Yan* *and* Patricia J. Kiley[†]

Contents

1. Introduction 788
2. Anaerobic Isolation of 4Fe-FNR 790
   - 2.1. Principles and procedures for purifying O_2-labile 4Fe-FNR 790
   - 2.2. Protocols 794
   - 2.3. Isolation of protein for special applications 797
3. Characterization of $[4Fe–4S]^{2+}$ Cluster Containing FNR 799
   - 3.1. Fe and sulfide analysis 800
   - 3.2. ICP-MS analysis of the ^{57}Fe content in the ^{57}Fe labeled 4Fe-FNR 801
   - 3.3. UV–visible spectrophotometric analysis of the $[4Fe–4S]^{2+}$ cluster 802
   - 3.4. Mössbauer spectroscopic analysis 802
   - 3.5. Low-temperature EPR 802
4. Summary 803

References 803

Abstract

Many key enzymes in biological redox reactions require metal centers or cofactors for optimum activity and function. While the metal centers provide unique properties for protein structure and function, some also render protein activity sensitive to environmental O_2 and cause experimental challenges to isolation and biochemical analysis. Iron–sulfur (Fe–S) clusters represent an important class of such metal centers and Fe–S proteins are widely distributed in nature. Here, we utilize FNR, a regulatory Fe–S protein from *Escherichia coli*, as an example to describe the techniques essential to purifying O_2-labile proteins and summarize various approaches for their biochemical analysis. These methods can be readily adapted to purify other O_2-labile proteins and advance our understanding of this interesting class of proteins.

* School of Biological Sciences, The University of Hong Kong, Hong Kong, Hong Kong SAR

† Department of Biomolecular Chemistry, University of Wisconsin, Madison, Wisconsin, USA

Methods in Enzymology, Volume 463

ISSN 0076-6879, DOI: 10.1016/S0076-6879(09)63042-1

1. Introduction

Many key proteins in biology contain O_2-labile metal centers or cofactors. Such proteins are found widely distributed in various physiological processes in nature, such as oxidoreductases in the life-sustaining process of respiration or photosynthesis, nitrogenases for N_2 fixation, dehydratases in metabolic pathways, and DNA helicases in DNA replication and repair, etc. (Beinert *et al.*, 1997; Brzoska *et al.*, 2006; Kiley and Beinert, 2003; Meyer, 2008). While the presence of such metal centers or cofactors provides extraordinary diversity in protein structure and function, it may also render the activity of a protein sensitive to oxidants, such as atmospheric oxygen (O_2) or its reactive oxygen species. These properties present technical challenges in isolating proteins in their active state and in performing subsequent *in vitro* analyses.

Creating and maintaining an anaerobic environment throughout the entire process of protein purification is the key to obtain active O_2-labile proteins. Successful methods for isolating O_2-sensitive proteins in their active state were first reported in the 1960s and employed sealed benchtop glove boxes in which O_2 was removed by evacuating and flushing with an inert gas (Chase and Rabinowitz, 1968; Kajiyama *et al.*, 1969; Vandecasteele and Burris, 1970). This type of device, while significantly advancing the study of O_2-labile metalloproteins, was limited in its ability to maintain an O_2-free environment in procedures requiring sequential transfers of materials into the chamber. These early glove boxes were also limited in the type of equipment that could be enclosed in a chamber, such as those required for centrifugation, chromatography, spectroscopy, etc. As a result, the yield of protein isolated was often low and the purified proteins often exhibited heterogeneity, resulting in some ambiguity in the subsequent biophysical and biochemical characterizations (Gotto and Yoch, 1982; Saari *et al.*, 1984). The invention and construction of the anaerobic laboratory chamber in the late 1970s solved many of these problems by providing greater flexibility in adapting many standard laboratory techniques for anaerobic use (Gunsalus *et al.*, 1980; Poston *et al.*, 1971). Introduction of this device has greatly advanced the isolation and characterization of O_2-labile proteins in the last several decades and has led to an exponential increase in our knowledge of this class of proteins.

In this chapter, we describe the methods used in our laboratory to isolate the O_2-sensitive iron–sulfur (Fe–S) protein, FNR, emphasizing current practices for creating and maintaining an O_2-free environment for the purification and study of O_2-labile proteins. Fe–S proteins are an ancient class of proteins that exist across all biological kingdoms (Beinert *et al.*, 1997; Brzoska *et al.*, 2006; Johnson *et al.*, 2005; Kiley and Beinert, 2003;

Meyer, 2008) and many contain O_2-sensitive Fe–S clusters. Fe–S proteins contain tetrahedral iron usually coordinated with sulfur, provided either by inorganic sulfide or the thiolates of cysteine residues in the polypeptide backbone. The basic types of these metal centers include the mononuclear [Fe(Cys)$_4$] complex and the bi-, tri-, and tetranucleated [2Fe–2S], [3Fe–4S], and [4Fe–4S] clusters (Meyer, 2008), although more complex cluster types can be found in nitrogenases and hydrogenases. While the cluster types are relatively simple in structure, the protein environment surrounding these Fe–S clusters can be remarkably different, resulting in great variation in cluster stability or the redox potential of Fe–S centers (Dey *et al.*, 2007). In addition to their well-known roles in catalyzing electron transfer reactions, the known functions of Fe–S proteins have expanded to also include regulatory factors that mediate transcriptional and translational control of gene expression (Cabiscol *et al.*, 2000; Crack *et al.*, 2008; Kiley and Beinert, 2003). Examples include the well-characterized c-aconitase in eukaryotes, which binds specific RNA sequences (iron regulatory elements) when it is in its apoform (the cytosolic apoform of c-aconitase is also called iron regulatory protein, IRP1) and controls protein translation in response to iron status (Kennedy *et al.*, 1992), and several bacterial transcription factors, such as FNR, IscR, and SoxR. FNR and perhaps IscR appear to exploit the stability of their Fe–S clusters to sense environmental signals and regulate the expression of a large number of genes in *E. coli* (Johnson *et al.*, 2005; Kiley and Beinert, 2003).

In *E. coli*, FNR is a global transcription factor that mediates the transition from an aerobic to anaerobic lifestyle by altering the expression profile of hundreds of genes under anaerobic growth conditions (Constantinidou *et al.*, 2006; Grainger *et al.*, 2007; Kang *et al.*, 2005; Salmon *et al.*, 2003). The ability of FNR to directly and rapidly sense O_2 levels is attributed to its $[4Fe–4S]^{2+}$ cluster that is ligated to FNR through four essential cysteine residues: Cys20, Cys23, Cys29, and Cys120 (Khoroshilova *et al.*, 1995). This form of the protein is referred to as 4Fe-FNR. Under anaerobic conditions, the $[4Fe–4S]^{2+}$ cluster promotes FNR dimerization and subsequent site-specific DNA binding and transcription regulation (Lazazzera *et al.*, 1993, 1996). Upon a switch to aerobic conditions, O_2 rapidly reacts with the $[4Fe–4S]^{2+}$ cluster and converts it to a $[2Fe–2S]^{2+}$ cluster. The resulting protein is referred to as 2Fe-FNR. The 2Fe-FNR is unable to dimerize, and thus is inactive in site-specific DNA binding and transcription regulation (Khoroshilova *et al.*, 1997; Sutton *et al.*, 2004). Since the conversion from a $[4Fe–4S]^{2+}$ cluster to a $[2Fe–2S]^{2+}$ cluster by O_2 is sufficient to inactivate FNR *in vivo* and *in vitro*, isolation of the natively active FNR must be carried out exclusively under anaerobic conditions.

Here, we describe the techniques and protocols necessary to isolate native FNR protein in high yield and cluster occupancy that is suitable for kinetic and various spectroscopic studies. We also summarize various approaches to characterize the purified 4Fe-FNR in the absence of O_2. We expect that these techniques and approaches will be broadly applicable to other O_2-labile proteins and promote further characterization of this class of proteins.

2. Anaerobic Isolation of 4Fe-FNR

2.1. Principles and procedures for purifying O_2-labile 4Fe-FNR

2.1.1. General preparation for achieving and maintaining an oxygen-free environment

Prior to protein purification, common procedures must be adapted to create and maintain an O_2-free environment for all the chemicals, solutions, and apparatus used during protein purification (Sutton and Kiley, 2003). This can be generally achieved by construction of a sparging station and use of an anaerobic glove-box vinyl chamber (Coy Laboratory Products, Model A). A sparging station is usually composed of an argon or nitrogen gas tank, an apparatus to provide copper-scrubbed (Sargent-Welch furnace) gas, and a custom-built manifold (Fig. 42.1). The sparging system is used to remove trace amount of O_2 from chemicals by dispersing inert gas into butyl rubber-sealed glass vials which contain O_2-sensitive solid reagents [such as dithionite (DTH)] or to remove dissolved O_2 from solutions to be used for protein isolation or characterization. For this purpose, the manifold is equipped with multiple lines that are capped by either gas dispersion tubes

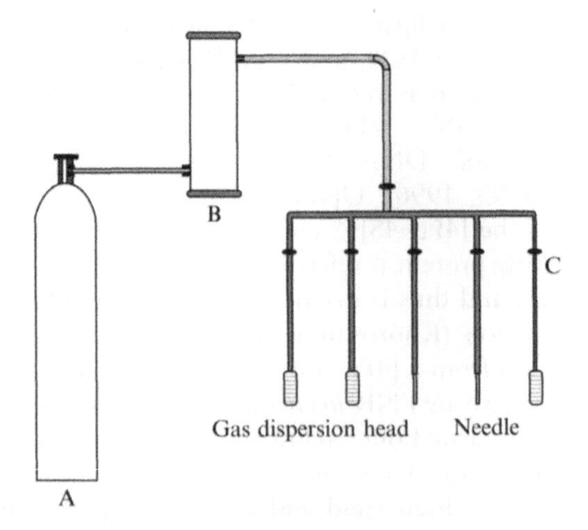

Figure 42.1 Diagram of a sparging station connected to an argon tank. A, argon tank; B, copper-furnace apparatus; C, manifold attached with gas dispersion tubes or needles.

(Upchurch Scientific) or needles (Fig. 42.1). Standard needles are also used to vent gas during the sparging. The glove-box chamber provides sufficient space for storage and operation during the process of protein purification. The vinyl chamber is filled with a gas mixture composed of 80% N_2, 10% CO_2, and 10% H_2 such that any O_2 that enters the chamber will be reduced by H_2 using a palladium catalyst and the generated H_2O will be absorbed by a desiccant. To purify proteins, FPLC or HPLC purification systems are advantageous because the seals are resistant to O_2 diffusion and the newer designs are sufficiently compact to operate within a chamber (assuming minimal heat production). In addition, their largely automated design makes it easy to manipulate within the chamber. However, the design of our older model FPLC requires that most parts including the pump, column, and detection system are placed outside of the chamber and connected to the buffer and fraction collector inside the chamber through PEEK tubing (Upchurch Scientific) (Fig. 42.2), which is impermeable to O_2, so that the entire contents of the system remain anaerobic. The path of solutions through the FPLC is from liquid containers located inside of the chamber, to the pumps and/or column, located outside of the chamber, and then back into the chamber to the fraction collector (Fig. 42.2). Most solutions such as buffers are first sparged with argon gas for a minimum of 20 min before introduction into the anaerobic chamber via the airlock and then equilibrated for at least 12 h in the chamber before use. All glassware and plastic items are equilibrated in the anaerobic chamber for a minimum of 24 h before being used to ensure any O_2 inadvertently introduced into the chamber is reduced.

2.1.2. Expression of FNR and assembly of [4Fe–4S] clusters

E. coli strain PK872 is used to obtain overexpression of 4Fe-FNR. This strain is a derivative of PK22 (BL21(DE3) Δ*fnr*, Δ*crp*) (Lazazzera *et al.*, 1993). It contains plasmid pPK823, carrying the *fnr* coding sequence cloned into

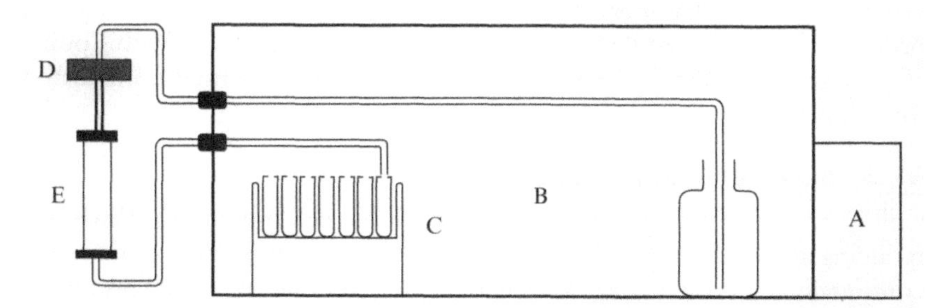

Figure 42.2 Diagram of an anaerobic chamber with an attached FPLC purification system. A, air lock; B, anaerobic chamber; C, fraction collector; D, FPLC pumps and controller (Pharmacia LKB Controller LCC-501 plus); E, column.

the *Nde*I–*Bam*HI sites of pET11a (Novagen) so that *fnr* transcription is under the control of the T7 promoter, and translation is controlled from the gene *10* ribosome-binding site. Cell growth and the synthesis/maturation of 4Fe-FNR follows the procedures described by Sutton and Kiley (2003). Typically, strain PK872 is grown aerobically with shaking at 200 rpm at 37 °C in four 2-l flasks, each of which contains 1 l of M9-minimal medium supplemented with 0.2% glucose, 0.2% casamino acids, 1 m*M* $MgCl_2$, 100 μM $CaCl_2$, 2 μg/ml thiamine, 20 μM ferric ammonium citrate, and 50 μg/ml ampicillin. When the cell culture reaches an OD_{600} of 0.3 (Spectronic 20D+ spectrophotometer), the expression of FNR is induced with 0.4 m*M* isopropyl-β-D-thiogalactoside (IPTG) for 1 h. To promote insertion of the $[4Fe–4S]^{2+}$ cluster by the endogenous Fe–S cluster machinery (Raulfs *et al.*, 2008) into the apo-FNR that largely accumulates during the induction period, these cultures are combined into a 9-l glass carboy and sparged with argon overnight at 4 °C. While we fortuitously discovered that this process is more efficient than expression under steady state anaerobic conditions, our recent discovery that the expression of both Fe–S cluster biogenesis pathways is reduced under anaerobic conditions as compared to aerobic conditions (Mettert *et al.*, 2008) may provide some insight for this observation. The glass carboy for sparging the cultures is fitted with a two-hole stopper containing two pieces of glass tubing: the first piece connects a line from an argon gas tank to a gas dispersion tube, which reaches nearly to the bottom of the carboy, and the second short (3–4 in.) piece permits gas release from the sparging vessel. After the overnight sparging step, cells are quickly poured into a container of suitable size (4-l beaker) to fit within the airlock and transferred immediately into the anaerobic chamber, where the cells are dispensed into 500-ml centrifuge bottles that have been equilibrated in the anaerobic chamber for at least 24 h. The centrifuge bottles are sealed tightly with lids containing O rings (Beckman) before being removed from the anaerobic chamber. Cells are pelleted by centrifugation (Beckman Avanti J-25 centrifuge, JLA10.500 rotor) at 4 °C for 15 min at 8000 rpm. The centrifuge bottles are then returned to the chamber, where the supernatant is poured off and the process is repeated until all of the cells have been collected. At this point, the anoxic cell pellets can either be stored at − 80 °C or used immediately for cell lysis.

2.1.3. Preparation of cell lysate

Cell lysis is achieved by passage of an anaerobic cell suspension containing reducing agents through a French press at 20,000 psi with the collection vial continually gassed with argon, essentially as described by Sutton and Kiley (2003). This step is the most technically challenging for maintaining anoxic conditions. To minimize O_2 exposure to the cells, a chilled French press cell and the bottles containing the anoxic cell pellets are first brought into the anaerobic chamber through two cycles in the airlock. The French press cell is immediately rinsed twice with anaerobic buffer. Typically, the cell pellet is resuspended (to 0.5% of the original volume) in anaerobic buffer A supplemented with dithionite (DTH) and dithiothreitol (DTT) (see Table 42.1) and then pipetted immediately into the rinsed cold French press cell. The French press cell containing the cell suspension is sealed and the sample outlet tubing is fitted with a capped 18-gauge needle before the entire cell is removed from the anaerobic chamber. Once the cell is assembled onto the French press, the needle is inserted into a butyl rubber-stoppered glass vial, previously sparged with argon gas, for collection of the cell lysate. After a single pass through the French press at 20,000 psi, the anaerobic lysate is collected into the sealed glass vial and is immediately brought back into the anaerobic chamber, transferred into ultracentrifuge tubes and sealed with a gas-tight lid before being removed from the chamber. Cell debris is removed by ultracentrifugation (Beckman Optima LE-80K, 70.1 Ti rotor) at 4 °C for 60 min at 45,000 rpm. The samples are then returned to the chamber, where the supernatant is used as the source of 4Fe-FNR for protein isolation.

2.1.4. Protein purification

[4Fe–4S]-FNR is routinely purified using an FPLC (Pharmacia LCC-501 Plus system) equipped with a 5-ml BioRex-70 cation-exchange column (BioRad Laboratories). Buffer solutions are located within the anaerobic chamber and are pumped out to the external FPLC via PEEK tubing through the chamber port. Prior to fractionating the cell lysate, the BioRex-70 column is washed with anaerobic buffers B and A (Table 42.1), containing DTT and DTH for 1 h each and equilibrated in buffer A. The cell lysate containing FNR is first transferred to a superloop and sealed before being brought outside of the anaerobic chamber and connected with the sample loading loop. The 4Fe-FNR is eluted with a gradient from 0.1 to 0.55 *M* KCl at a flow rate of 10 ml/h and is pumped

Table 42.1 Composition of buffers for purifying FNR

Buffer	Composition
A (low salt buffer)	50 m*M* phosphate (pH 6.8), 10% glycerol, 0.1 *M* KCl
B (high salt buffer)	50 m*M* phosphate (pH 6.8), 10% glycerol, 1 *M* KCl
C (no salt buffer)	50 m*M* phosphate (pH 6.8), 10% glycerol
D (elution buffer)	50 m*M* phosphate (pH 6.8), 10% glycerol, 0.4 *M* KCl
E (buffer to dissolve DTH)	0.1 *M* Tris–HCl (pH 7.9)

back into the chamber through PEEK tubing, which is connected to a fraction collector located within the chamber. FNR is eluted at $\sim 0.4\ M$ KCl as measured with a conductivity meter. Because the $[4Fe\text{–}4S]^{2+}$ cluster absorbs visible light maximally at a wavelength of about 420 nm, 4Fe-FNR displays a characteristic green color. The colored fractions are subsequently pooled and concentrated by passing through a gravity-flow column equipped with a 1-ml BioRex-70 cation-exchange resin. The column is stored in anaerobic chamber and is equilibrated with anaerobic buffer A. In addition, since the presence of reducing agents, DTT and DTH, interferes with some subsequent biochemical characterizations of 4Fe-FNR, this step is also used to remove these agents from the protein solution. The 4Fe-FNR obtained from this method typically displays a dark green color and contains $\sim 1.0\ mM$ protein with a cluster occupancy of more than 80%.

2.2. Protocols

A step-by-step protocol for isolating 4Fe-FNR is summarized as follows[1]

Day 1

1. Inoculate 100 ml of Luria Broth containing 1 ml of 20% glucose and 50 μl of 200 mg/ml Ampicillin (Ap) with strain PK872 in the early afternoon and grow overnight at 37 °C.
2. Prepare and autoclave four 2-l flasks containing 1 l of 1× M9 (Miller, 1972) minimal media.

Day 2

1. To each 1-l flask of M9-minimal medium, add:
   - 10 ml 20% glucose
   - 10 ml 20% casamino acids
   - 2 ml vitamin B_1 (thiamine) (2 mg/ml stock)
   - 1 ml $MgSO_4$ (1 M stock)
   - 100 μl $CaCl_2$ (1 M stock)
   - 1 ml ferric ammonium citrate (10 mg/ml stock, filter sterilized)
   - 250 μl Ap (200 mg/ml stock, filter sterilized)
2. For each of the four flasks, inoculate 10 ml of the overnight culture into 1 l of the complete M9 media. Grow at 37 °C with shaking at $\sim$200 rpm to an OD_{600} of 0.3. Induce FNR synthesis by adding 1 ml of 0.4 M IPTG to each flask (final concentration is 0.4 mM) and shaking for 1 additional hour.

[1] This protocol was created and amended by several members in the Kiley group: Lazzaera, B., Khoroshilova, K., Voet, K., Sutton, V., Moore, L., Mettert, E., Yan, A.

3. Pool cells into a 9-l flask and sparge cells with argon gas at 4 °C for 14–16 h.
4. Prepare solutions and glassware that will be used next day and bring them into the anaerobic chamber: (the chamber is normally equipped with standard pipetting devices and tips, a microcentrifuge, a vortex machine and eppendorf tubes.)
   (a) Solutions:
   - *Buffer A*: prepare 600 ml, divide into a 500- and 100-ml screw-capped glass bottles (recipe as shown in Table 42.1)
   - *Buffer B*: prepare 600 ml, divide into a 500- and 100-ml screw-capped glass bottles (recipe as shown in Table 42.1)
   - *Buffer C*: 100 ml in a screw-capped glass bottle (recipe as shown in Table 42.1)
   - *Buffer E*: 100 ml in a screw-capped glass bottle (recipe as shown in Table 42.1)
   - *Water*: 100 ml bottle

Sparge solutions with argon gas for 20 min by adjusting the argon flow to produce steady gas-flow through the custom-built sparging system as described in 2.1 (Fig. 42.1) with the gas dispersion tubes (Upchurch Scientific) placed at the bottom of the buffer solutions.

   (b) Glassware and plastic items to be brought into anaerobic chamber:
   - Two ultracentrifuge tubes and caps
   - Fraction collector tubes
   - Beaker for waste
   - Superloop with piston and cap
   - Glass vials (5–6 small, 2 medium) and butyl rubber caps
   - Glass Pasteur pipettes
   - Four to six centrifuge bottles and caps with O rings
   - 500 ml graduated cylinder
   - 2 and 4-l plastic beakers
   - 25 ml glass pipettes
   - Pasteur pipette bulb
   - Needle and tubing for French Press

Bring these items into anaerobic chamber through air lock, leave the buffers uncapped and equilibrate for 12 h.

Day 3

1. Pour sparged cells from the 9-l flask to a plastic 4-l beaker and immediately cover it with Saran wrap and tape to seal. Bring the beaker into the anaerobic chamber and aliquot 450 ml into each of the 500-ml centrifuge bottles and seal the bottles with the lids.
2. Bring the bottles out of the anaerobic chamber and measure the weight to balance. Unbalanced bottles are brought back to the anaerobic chamber to adjust the weight. Spin at 8000 rpm for 15 min at 4 °C.

3. After centrifugation, bring the bottles back to the chamber and remove the supernatant. Repeat the centrifugation step until all cells are centrifuged. The cell pellet should have an army green color.
4. During the centrifugation steps, weigh 154 mg dithiothreitol (DTT) and 150 mg dithionite (DTH) into individual 1.5 ml eppendorf tubes and seal tightly. Bring them into the chamber together with the bottles after centrifugation.
5. In the chamber, add 1.5 ml buffer E to DTH and 1.0 ml H_2O to DTT. Add 300 μl of DTH (final concentration 1.7 m*M*) and 100 μl of DTT (final concentration 1.0 m*M*) to the 100 ml bottle of buffer A. Resuspend the cells in 20 ml of buffer A containing 1.7 m*M* DTH and 1.0 m*M* DTT. The cell suspension can either be used immediately for cell lysis or transferred into a medium glass vial, sealed with a butyl rubber stopper, frozen on dry ice and then removed from the chamber and stored in a −80 °C freezer until ready for purification.

Day 4

1. To the 500 ml of buffers A and B, add 1.5 ml of DTH and 0.5 ml of DTT (both prepared freshly as described above).
2. Place pump leads into the buffers such that buffer A is connected to pump A and buffer B is connected to pump B.
3. Connect BioRex-70 column to FPLC system, making sure that there are no air bubbles in the column.
4. Run FPLC system program to wash the column with 100% buffer B (contained within the chamber) for 1 h at a flow rate of 0.17 ml/min followed by 100% buffer A for another 1 h.
5. Bring the frozen cells into the chamber and thaw the cell suspension under anaerobic conditions (do not thaw cells outside the chamber as O_2 will be drawn into vial).
6. Bring French press cell into the chamber. Rinse the sample cell twice with anaerobic buffer A containing DTT and DTH. Fill cell with the resuspended cells, seal, and attach sample outlet with the tubing fitted with a capped 18-gauge needle, and remove from chamber along with a butyl rubber-capped glass vial for collection of cell lysate.
7. Assemble cell onto French Press located preferably in a cold room. To exclude O_2 from the lysed extract, insert the needle from the cell into the collection vial and insert two additional needles into the vial, using one to gently flow argon gas and the second to vent the gas. Lyse cells at 20,000 psi at 4 °C.
8. Remove needles and bring the cell lysate into anaerobic chamber and divide the lysate into the two ultracentrifuge tubes. Bring them out of the chamber and weigh the tubes to balance before ultracentrifugation. Unbalanced tubes are brought back into the chamber to adjust the weight.
9. Centrifuge the lysate with 70.1 Ti rotor for 60 min at 45,000 rpm, 4 °C.

10. After centrifuging, bring tubes into the chamber, pour supernatant into superloop and seal. Fill the other end of the superloop with buffer A and seal. Remove the superloop from the chamber and connect superloop to the FPLC system.
11. Run FPLC program to load the sample and initiate the appropriate sequence of buffer solutions to allow FNR to elute from the BioRex-70 column.
12. Program the fraction collector located in the anaerobic chamber to collect the column eluate.

Day 5
Concentration and storage

1. Fractions containing 4Fe-FNR are easily recognized by their green color. Pool these fractions into a medium sized flask and dilute 1:4 with buffer C.
2. Construct a 1-ml BioRex-70 gravity column and wash with 2 column volumes of buffer B followed by 6 column volumes of buffer C.
3. Load protein solution onto column, wash with 2 column volumes of buffer A followed by elution with 0.8 *M* KCl buffer (made from buffers B and C). Collect only colored drops.
4. Aliquot the concentrated 4Fe-FNR into 0.5 ml eppendorf tubes, which have had their caps removed, and place tubes into a small glass vial. The glass vial is topped with butyl rubber stoppers and crimp sealed with aluminum caps (Bellco). The protein samples stored in the glass vial are immediately frozen in a dry ice-ethanol slurry, which is introduced into the anaerobic chamber though the air lock. Samples can be safely stored in a −80 °C freezer for many months using this method. Before thawing, it is critical that the sealed glass vials be placed in dry ice-ethanol slurry during transfer to the anaerobic chamber to prevent O_2 from being drawn into the vial while thawing.

2.3. Isolation of protein for special applications

2.3.1. Removal of trace RNAses from purified 4Fe-FNR for *in vitro* transcription assays

After passage through the two BioRex-70 cation-exchange columns, the purified FNR protein is more than 95% pure as estimated by SDS–PAGE. However, these preparations contain sufficient RNases to degrade RNA products in an *in vitro* transcription assay. RNase activity was subsequently removed by separation via a size exclusion column (HR-12 Superose, GE Healthcare) attached to a Beckman HPLC system (Mettert and Kiley, 2007). The entire HPLC system is housed within an anaerobic chamber and is also fitted with PEEK tubing (Upchurch Scientific). Since FNR isolated this way is to be used in *in vitro* transcription assays, all buffer solutions and fraction collection tubes were treated with diethylpyrocarbonate (DEPC) and/or autoclaved. Purification by size exclusion chromatography also yields ~100% dimeric 4Fe-FNR as judged by sulfide analysis (Moore and Kiley, 2001).

2.3.2. Purification of ^{57}Fe labeled 4Fe-FNR for Mössbauer spectroscopic analysis

Mössbauer spectroscopy is a powerful method that can both qualitatively and quantitatively characterize Fe–S clusters (Beinert *et al.*, 1997). It differentiates Fe–S cluster types based on the hyperfine splitting pattern of Fe nuclear energy levels, as well as energy required for the Fe nuclei to transit from the ground to the excited states in the absorption or emission spectrum of γ-rays. ^{57}Fe is by far the most common element studied with Mössbauer spectroscopy and because of the vital roles of Fe–S containing proteins in biological systems, Mössbauer spectroscopy has been widely used to characterize this class of proteins. In order to prepare ^{57}Fe enriched 4Fe-FNR, we remove the naturally abundant ^{56}Fe from the growth media by passing the 1× M9 media through a Chelex 100 (200–400 mesh, BioRad) column (40 g of resin per 1 l of medium) and then supplement the growth media with ^{57}Fe in the form of ferrous ethylenediammonium sulfate. The protocol for preparing Fe-free growth media and supplementation with ^{57}Fe is as follows:

1. Prepare Chelex 100 column by placing 200 g of Chelex resin and ~200 ml H_2O into a beaker. After incubating at room temperature for 30 min, pour resin into a 250-ml gravity column. Wash the column with 5 column volumes of H_2O before use. To regenerate column, wash with 2 column volumes of 1 N HCl followed by 5 column volumes of H_2O; then wash with 2 column volumes of 1 N NaOH and another 5 column volumes of H_2O till the pH of the column eluate is ~7.0.
2. *Deferrate the growth media*: Pass 5 l of 1× M9 media through the newly prepared or regenerated Chelex 100 column. Collect the flow-through and autoclave. After use, the column is stored in 0.5 *M* ammonium sulfate.
3. Cell growth and isolation of FNR containing ^{57}Fe labeled $[4Fe\text{–}4S]^{2+}$ cluster:
   - Inoculate 100 ml of Luria Broth with strain PK872. Add 1 ml of 20% glucose and 50 μl of 200 mg/ml Ampicillin to the medium and grow overnight at 37 °C.
   - Centrifuge the overnight cell culture to remove the LB media. Add 100 ml of deferrated 1× M9 media to resuspend the cell pellet, centrifuge at 6000 rpm at 4 °C and decant the supernatant.
   - Resuspend the pellet in 100 ml of deferrated 1× M9 media, and inoculate 10 ml of cell suspension to 1 l of deferrated M9-minimal medium containing the same supplements as described earlier, except

that the iron is replaced with 15 μl ^{57}Fe ferrous ethylenediammonium sulfate (1 *M* stock). Cultures are grown as described for the isolation of non-^{57}Fe labeled FNR including sparging cells with argon to allow the assembly of ^{57}Fe labeled [4Fe–4S]$^{2+}$ cluster into FNR.

- Since we have observed that once ^{57}Fe is incorporated into 4Fe-FNR, the exchange between ^{57}Fe in the [4Fe–4S] cluster and ^{56}Fe in solution is unlikely to occur, buffers (A, B, and C) for subsequent protein purification are not deferrated and the purification procedures are the same as those for non-^{57}Fe labeled 4Fe-FNR (section 2.2).

3. Characterization of [4Fe–4S]$^{2+}$ Cluster Containing FNR

Several assays are performed to characterize the purified 4Fe-FNR, including determination of protein concentration, Fe and sulfide content, cluster occupancy and the type, and oxidation state of the cluster. The purity of each protein preparation is estimated by SDS–PAGE and is typically >95%. The protein concentration is measured by the Coomassie Plus protein assay reagent (Pierce) and corrected by dividing by a factor of 1.33, to standardize to the protein concentration determined from amino acid analysis (The Protein Facility Center of the Iowa State University). The Fe and sulfide content of the protein is measured by the methods developed by Beinert and coworkers (Beinert, 1983; Kennedy *et al.*, 1984). The [4Fe–4S]$^{2+}$ cluster content of the purified protein is calculated from the moles of sulfide determined per mole of FNR protein. It can also be determined by the absorbance of [4Fe–4S]$^{2+}$ cluster at 405 nm from the extinction coefficient of 16,125 M^{-1} cm^{-1} (Sutton *et al.*, 2004). The [4Fe–4S]$^{2+}$ cluster occupancy is defined by the percentage of FNR subunits containing a [4Fe–4S]$^{2+}$ cluster. In our measurements of >100 FNR preparations, we have typically observed a slightly higher Fe concentration than the sulfide concentration, which may be due to the presence of a small amount of contaminating iron present in the buffer solutions but not bound in clusters. Therefore, we use the sulfide concentration to determine the concentration of [4Fe–4S]$^{2+}$ cluster in the protein preparation which is unlikely to be introduced artificially.

While the [4Fe–4S]$^{2+}$ cluster exhibits a characteristic visible absorption spectrum (Khoroshilova *et al.*, 1995), the cluster type and the oxidation state of the Fe–S cluster can only be determined by Mössbauer spectroscopy using ^{57}Fe labeled FNR protein (Khoroshilova *et al.*, 1997). Recently, Fourier transform ion cyclotron resonance (FT-ICR) mass spectrometry has been used to identify the apo-, [2Fe–2S]-, and [4Fe–4S]-forms of adenosine 5′-phosphosulfate (APS) reductase isolated from *Mycobacterium tuberculosis* (Carroll *et al.*, 2005), although it is not yet clear how generally applicable this method will be.

3.1. Fe and sulfide analysis[2]

Fe is determined by a rapid procedure described by Kennedy *et al.* (1984) without ashing. The principle of the method is to reduce all the Fe in the protein sample to the Fe^{2+} state and then add the Fe^{2+}-specific chelator ferene (disodium salt of 5,5′-[3-(2-pyridyl)-1,2,4-triazine-5,6-diyl]bis-2-furansulfonic acid, Sigma) to form the Fe^{2+}-ferene complex which has a high extinction coefficient factor at 593 nm. Three solutions, a, b, and c, are prepared prior to the measurement (Table 42.2) (Kennedy *et al.*, 1984). Protein dilution buffer (buffer D in Table 42.1) and semi-micro cuvettes (1-cm path length) are introduced into the anaerobic chamber at least 1 day in advance of the assay. Typically, 5 μl of the protein sample and 95 μl of buffer D are added to the cuvette in the anaerobic chamber; the cuvette is covered with parafilm and then removed from the anaerobic chamber. 100 μl of reagent **a** is then added to the cuvette, mixed by pipetting and then followed by addition of 100 μl of reagent **b**, mixing and incubating the parafilmed cuvette at 30 °C for 15 min. Finally, 5 μl of reagent **c** is added, mixed, and the absorption measured at 593 nm. The molecular absorption of 1 ng atom of Fe is 0.119 with this assay.

Sulfide is determined by the semi-micro analysis method developed by Beinert (1983) using the reagent *N,N*-dimethyl-*p*-phenylenediamine (DMPD) which forms methylene blue with sulfide. In this method, several preventative measures were adopted to avoid the inadvertent loss of S^{2-} in the form of H_2S gas, a side reaction that often occurs in an environment of

Table 42.2 Composition of buffers for Fe analysis

Reagent	Composition
a	1.35 g of sodium dodecyl sulfate dissolved in 30 ml of H_2O and mixed with 0.45 ml of saturated Fe-free sodium acetate
b[a]	270 mg of ascorbic acid and 9 mg of sodium metabisulfate ($Na_2S_2O_5$) dissolved in 5.6 ml of H_2O and mixed with 0.4 ml of saturated acetate
c	18 mg of Ferene in 1 ml of H_2O

[a] Reagent **b** is not stable and has to be remade about every 2 weeks or kept frozen.

[2] The Fe and S analysis of our protein preparations were performed by our long-time collaborator and friend Dr. Helmut Beinert who passed away on December 21, 2007.

strong acid and/or with the presence of an oxidizing agent (such as $FeCl_3$). The preventative measures include: (1) perform the entire procedure in a small glass tube with a 10-mm o.d. and a total volume of 1.7 ml that has a conical bottom and is fitted with a stir bar (10 × 3 × 3 mm) and a glass stopper; (2) use gentle mixing in steps where the solution is acidic or when an oxidizing agent is added. This approach minimizes the agitation of the liquid in the tube and the formation of new air–water interfaces which facilitates the escape of H_2S if it is formed; (3) add a few drops of phenolphthalein in the solution to follow the efficiency of mixing by gentle mixing; (4) strongly acid DMPD solution is added to the zinc hydroxide suspension using a micropipet to underlay this heavy solution under the lighter one. The typical procedure is as follows: prior to the measurement, the glass tubes and H_2O are brought into the anaerobic chamber to equilibrate for 24 h. Then 10 μl of protein sample and 90 μl of H_2O are added to each of the glass tubes, the cap is added and the samples are removed from the anaerobic chamber. With the tubes secured above a stir plate, 300 μl of freshly prepared 1% zinc acetate [$Zn(C_2H_3O_2)_2$] is added and gently stirred. Then 15 μl of 12% NaOH is added immediately and stirred vigorously until the schlieren of NaOH disappears or the phenolphthalein color is homogeneous. After incubation at room temperature for 5 min, 75 μl of 0.1% DMPD solution (in 5 N HCl) is layered under the protein solution. The solution is gently stirred until only the top 2-mm layer of liquid has undissolved zinc hydroxide or pink color. Then 10 μl of 23 mM $FeCl_3$ (in 1.2 N HCl) is added directly to the solution and quickly stirred to achieve a homogeneous (colorless) solution. After incubation for 30 min at 20–25 °C, the tube is centrifuged for 15–20 min until protein is packed. The supernatant is transferred to a cuvette and the absorption values at 670, 710, and 750 nm are measured. The ratio of the absorbance should be approximately A_{670}:A_{750}:A_{710} = 3:2:1. The sulfide content is determined from the extinction coefficient of 34,500 M^{-1} cm^{-1} at 670 nm (Beinert, 1983).

3.2. ICP-MS analysis of the ^{57}Fe content in the ^{57}Fe labeled 4Fe-FNR

Inductively coupled plasma mass spectrometry (ICP-MS) is a type of mass spectrometry that is highly sensitive for analyzing a range of metals and capable of distinguishing isotopic speciation for ions of choice. It is an ideal approach for analyzing the amount of ^{57}Fe enrichment in various forms of natural and laboratory samples. Our protein preparations are currently analyzed with a Thermoscientific ELEMENT2 high resolution ICP-MS housed at the Wisconsin State Laboratory of Hygiene. Typically, 50 μl of a 1-μM protein sample is transferred to a Fe-free tube, sealed and removed from the anaerobic chamber and submitted for analysis. Protein isolation following the protocols mentioned above usually yields 80–90% of cluster occupancy and ~80% of the clusters are incorporated with ^{57}Fe.

3.3. UV–visible spectrophotometric analysis of the $[4Fe–4S]^{2+}$ cluster

UV–visible spectra are routinely obtained employing a Lambda 2 UV–visible spectrophotometer (Perkin-Elmer) (Sutton *et al.*, 2004). Screw-capped quartz cuvettes (1-cm path length) (Starna) are brought into the anaerobic chamber, left uncapped, and equilibrated for 24 h. These cuvettes are manufactured to seal with O-ring adapted screw caps to prevent introduction of O_2 and thus can be used in a standard spectrophotometer. In the anaerobic chamber, a protein sample of the appropriate dilution is placed in a cuvette, sealed with a screw cap and removed from the anaerobic chamber. The sample is scanned from 250 to 700 nm to obtain a full absorption spectrum of the protein and the absorbance at 405 nm is used to calculate the concentration of $[4Fe–4S]^{2+}$ cluster in the protein sample using the extinction coefficient of 16,125 M^{-1} cm^{-1} (Sutton *et al.*, 2004).

3.4. Mössbauer spectroscopic analysis

In order to further identify the type and oxidation state of the Fe–S cluster, Mössbauer spectroscopic analysis is performed at the Department of Chemistry, Carnegie Mellon University (Khoroshilova *et al.*, 1995, 1997; Popescu *et al.*, 1998). The Mössbauer cups, lids, and the tools to tighten the Mössbauer cups are placed into anaerobic chamber 24 h before use. Approximately 400 μl of protein solution is placed into a custom-built Mössbauer cup, the lid is sealed and the entire cup is immediately frozen in liquid nitrogen. Liquid nitrogen was carefully brought into the chamber using the manual cycle on the air lock. After the frozen samples are removed from the chamber, they are shipped for Mössbauer spectroscopic analysis.

3.5. Low-temperature EPR

Electron paramagnetic resonance (EPR) spectroscopy is a technique for studying chemical species that have one or more unpaired electrons, such as organic and inorganic free radicals or inorganic complexes possessing a transition metal ion. It is useful for study of specific oxidations states of Fe–S clusters, which contain unpaired electrons (Cammack and Cooper, 1993). As expected, the $[4Fe–4S]^{2+}$ cluster of FNR is EPR silent, thus EPR is not suitable to characterize the type and oxidation state of the Fe–S cluster in the natively purified 4Fe-FNR protein. However, it is a powerful and widely used tool to study other Fe–S clusters that contain paramagnetic irons such as aconitase (Kennedy *et al.*, 1984). Furthermore, the combination of Mössbauer and EPR allows the detection of Fe–S cluster intermediates that occurs during cluster conversion or destruction of some proteins. EPR samples are prepared similar to those for Mössbauer, except that a special syringe is required to load the EPR tubes because of their long neck. A 0.5-ml glass syringe equipped with a long needle tubing is used to pipette $\sim 200\ \mu l$ of $\sim 100\ \mu M$ protein solution into the tube and then quickly frozen in liquid nitrogen.

4. Summary

In summary, by adapting common procedures to those that can be carried out in the absence of O_2, we routinely isolate 4Fe-FNR from *E. coli* with high purity and cluster occupancy. It is noted that special precautions must be taken during the entire process of the protein purification, including cell growth, protein isolation, and protein storage, in order to achieve a satisfactory yield of this class of O_2-labile proteins. The procedures we routinely use are based upon well-established approaches for eliminating O_2 exposure and maintaining a reduced environment and thus should be applicable for isolation and/or characterization of other O_2-labile proteins. However, modification of this approach may be required in order to optimize expression of a different O_2-labile protein, particularly, if it has a different cofactor or if the cluster cannot be inserted posttranslationally as it is with FNR (Sutton and Kiley, 2003). Other alternatives to studies of O_2-sensitive metal center proteins include purifying the apoform of the protein under aerobic conditions and reconstituting the O_2-labile centers through appropriate *in vitro* systems (Crack *et al.*, 2008) or use of O_2-stable protein variants (Bates *et al.*, 2000; Kiley and Reznikoff, 1991). However, it is an open question whether reconstituted Fe-S protein exhibits all of the same properties of natively isolated protein (Saunders *et al.*, 2008). Nonetheless, there is no doubt that purification and characterization of more O_2-labile proteins and enzymes will help to further broaden our knowledge of this interesting class of proteins.

References

Bates, D. M., Popescu, C. V., Khoroshilova, N., Vogt, K., Beinert, H., Munck, E., and Kiley, P. J. (2000). Substitution of leucine 28 with histidine in the *Escherichia coli* transcription factor FNR results in increased stability of the $[4Fe–4S]^{2+}$ cluster to oxygen. *J. Biol. Chem.* **275,** 6234–6240.

Beinert, H. (1983). Semi-micro methods for analysis of labile sulfide and of labile sulfide plus sulfane sulfur in unusually stable iron–sulfur proteins. *Anal. Biochem.* **131,** 373–378.

Beinert, H., Holm, R. H., and Münck, E. (1997). Iron–sulfur clusters: Nature's modular, multipurpose structures. *Science* **277,** 653–659.

Brzoska, K., Meczynska, S., and Kruszewski, M. (2006). Iron–sulfur cluster proteins: Electron transfer and beyond. *Acta Biochim. Pol.* **53,** 685–691.

Cabiscol, E., Tamarit, J., and Ros, J. (2000). Oxidative stress in bacteria and protein damage by reactive oxygen species. *Int. Microbiol.* **3,** 3–8.

Cammack, R., and Cooper, C. E. (1993). Electron paramagnetic resonance spectroscopy of iron complexes and iron-containing proteins. *Methods Enzymol.* **227,** 353–384.

Carroll, K. S., Gao, H., Chen, H., Leary, J. A., and Bertozzi, C. R. (2005). Investigation of the iron–sulfur cluster in *Mycobacterium tuberculosis* APS reductase: Implications for substrate binding and catalysis. *Biochemistry* **44,** 14647–14657.

Chase, T. Jr., and Rabinowitz, J. C. (1968). Role of pyruvate and S-adenosylmethioine in activating the pyruvate formate-lyase of *Escherichia coli*. *J. Bacteriol.* **96,** 1065–1078.

Constantinidou, C., Hobman, J. L., Griffiths, L., Patel, M. D., Penn, C. W., Cole, J. A., and Overton, T. W. (2006). A reassessment of the FNR regulon and transcriptomic analysis of the effects of nitrate, nitrite, NarXL, and NarQP as *Escherichia coli* K12 adapts from aerobic to anaerobic growth. *J. Biol. Chem.* **281,** 4802–4815.

Crack, J. C., Le Brun, N. E., Thomson, A. J., Green, J., and Jervis, A. J. (2008). Reactions of nitric oxide and oxygen with the regulator of fumarate and nitrate reduction, a global transcriptional regulator, during anaerobic growth of *Escherichia coli*. *Methods Enzymol.* **437,** 191–209.

Dey, A., Jenney, F. E. , Adams, M. W., Babini, E., Takahashi, Y., Fukuyama, K., Hodgson, K. O., Hedman, B., and Solomon, E. I. (2007). Solvent tuning of electrochemical potentials in the active sites of HiPIP versus ferredoxin. *Science* **318,** 1464–1468.

Gotto, J. W., and Yoch, D. C. (1982). Purification and Mn^{2+} activation of *Rhodospirillum rubrum* nitrogenase activating enzyme. *J. Bacteriol.* **152,** 714–721.

Grainger, D. C., Aiba, H., Hurd, D., Browning, D. F., and Busby, S. J. (2007). Transcription factor distribution in *Escherichia coli*: Studies with FNR protein. *Nucleic Acids Res.* **35,** 269–278.

Gunsalus, R. P., Tandon, S. M., and Wolfe, R. S. (1980). A procedure for anaerobic column chromatography employing an anaerobic Freter-type chamber. *Anal. Biochem.* **101,** 327–331.

Johnson, D. C., Dean, D. R., Smith, A. D., and Johnson, M. K. (2005). Structure, function, and formation of biological iron–sulfur clusters. *Annu. Rev. Biochem.* **74,** 247–281.

Kajiyama, S., Matsuki, T., and Nosoh, Y. (1969). Separation of the nitrogenase system of *azotobacter* into three components and purification of one of the components. *Biochem. Biophys. Res. Commun.* **37,** 711–717.

Kang, Y., Weber, K. D., Qiu, Y., Kiley, P. J., and Blattner, F. R. (2005). Genome-wide expression analysis indicates that FNR of *Escherichia coli* K-12 regulates a large number of genes of unknown function. *J. Bacteriol.* **187,** 1135–1160.

Kennedy, M. C., Kent, T. A., Emptage, M., Merkle, H., Beinert, H., and Münck, E. (1984). Evidence for the formation of a linear [3Fe-4S] cluster in partially unfolded aconitase. *J. Biol. Chem.* **259,** 14463–14471.

Kennedy, M. C., Mende-Mueller, L., Blondin, G. A., and Beinert, H. (1992). Purification and characterization of cytosolic aconitase from beef liver and its relationship to the iron-responsive element binding protein. *Proc. Natl. Acad. Sci. USA* **89,** 11730–11734.

Khoroshilova, N., Beinert, H., and Kiley, P. J. (1995). Association of a polynuclear iron–sulfur center with a mutant FNR protein enhances DNA binding. *Proc. Natl. Acad. Sci. USA* **92,** 2499–2503.

Khoroshilova, N., Popescu, C., Münck, E., Beinert, H., and Kiley, P. J. (1997). Iron–sulfur cluster disassembly in the FNR protein of *Escherichia coli* by O_2: [4Fe–4S] to [2Fe–2S] conversion with loss of biological activity. *Proc. Natl. Acad. Sci. USA* **94,** 6087–6092.

Kiley, P. J., and Beinert, H. (2003). The role of Fe–S proteins in sensing and regulation in bacteria. *Curr. Opin. Microbiol.* **6,** 181–185.
Kiley, P. J., and Reznikoff, W. S. (1991). Fnr mutants that activate gene expression in the presence of oxygen. *J. Bacteriol.* **173,** 16–22.
Lazazzera, B. A., Bates, D. M., and Kiley, P. J. (1993). The activity of the *Escherichia coli* transcription factor FNR is regulated by a change in oligomeric state. *Genes Dev.* **7,** 1993–2005.
Lazazzera, B. A., Beinert, H., Khoroshilova, N., Kennedy, M. C., and Kiley, P. J. (1996). DNA binding and dimerization of the Fe–S-containing FNR protein from *Escherichia coli* are regulated by oxygen. *J. Biol. Chem.* **271,** 2762–2768.
Mettert, E. L., and Kiley, P. J. (2007). Contributions of [4Fe–4S]-FNR and integration host factor to FNR transcriptional regulation. *J. Bacteriol.* **189,** 3036–3043.
Mettert, E. L., Outten, F. W., Wanta, B., and Kiley, P. J. (2008). The impact of O_2 on the Fe–S cluster biogenesis requirements of *Escherichia coli* FNR. *J. Mol. Biol.* **384,** 798–811.
Meyer, J. (2008). Iron–sulfur protein folds, iron–sulfur chemistry, and evolution. *J. Biol. Inorg. Chem.* **13,** 157–170.
Miller, J. H. (ed.) (1972). *In* "Experiments in Molecular Genetics", Cold Spring Harbor Laboratory Press, Plainview, NY.
Moore, L. J., and Kiley, P. J. (2001). Characterization of the dimerization domain in the FNR transcription factor. *J. Biol. Chem.* **276,** 45744–45750.
Popescu, C. V., Bates, D. M., Beinert, H., Münck, E., and Kiley, P. J. (1998). Mössbauer spectroscopy as a tool for the study of activation/inactivation of the transcription regulator FNR in whole cells of *Escherichia coli. Proc. Natl. Acad. Sci. USA* **95,** 13431–13435.
Poston, J. M.,, Stadtman, T. C., and Stadtman, E. R. (eds.) (1971). *In* An Anaerobic Laboratory, Academic Press, New York.
Raulfs, E. C., O'Carroll, I. P., Dos Santos, P. C., Unciuleac, M. C., and Dean, D. R. (2008). *In vivo* iron–sulfur cluster formation. *Proc. Natl. Acad. Sci. USA* **105,** 8591–8596.
Saari, L. L., Triplett, E. W., and Ludden, P. W. (1984). Purification and properties of the activating enzyme for iron protein of nitrogenase from the photosynthetic bacterium *Rhodospirillum rubrum. J. Biol. Chem.* **259,** 15502–15508.
Salmon, K., Hung, S. P., Mekjian, K., Baldi, P., Hatfield, G. W., and Gunsalus, R. P. (2003). Global gene expression profiling in *Escherichia coli* K12. The effects of oxygen availability and FNR. *J. Biol. Chem.* **278,** 29837–29855.
Saunders, A. H., Griffiths, A. E., Lee, K. H., Cicchillo, R. M., Tu, L., Stromberg, J. A., Krebs, C., and Booker, S. J. (2008). Characterization of quinolinate synthases from *Escherichia coli, Mycobacterium tuberculosis,* and *Pyrococcus horikoshii* indicates that [4Fe–4S] clusters are common cofactors throughout this class of enzymes. *Biochemistry* **47,** 10999–11012.
Sutton, V. R., and Kiley, P. J. (2003). Techniques for studying the oxygen-sensitive transcription factor FNR from Escherichia coli. *Methods Enzymol.* **370,** 300–312.
Sutton, V. R., Mettert, E. L., Beinert, H., and Kiley, P. J. (2004). Kinetic analysis of the oxidative conversion of the $[4Fe–4S]^{2+}$ cluster of FNR to a $[2Fe–2S]^{2+}$ cluster. *J. Bacteriol.* **186,** 8018–8025.
Vandecasteele, J. P., and Burris, R. H. (1970). Purification and properties of the constituents of the nitrogenase complex from *Clostridium pasteurianum. J. Bacteriol.* **101,** 794–801.

SECTION TWELVE

CONCLUDING REMARKS

CHAPTER FORTY-THREE

Rethinking Your Purification Procedure

Murray P. Deutscher

Contents

1. Introduction 809

1. Introduction

Every protein purification that you undertake should provide you not only with purified material but also with considerable information about the protein. Thus, during the course of purification you most likely will learn about the stability of the protein under a variety of conditions, as well as about its size, charge, and, perhaps, its affinity properties. You will have learned whether the protein can be concentrated, diluted, dialyzed, or exposed to a variety of agents. In addition, you may have prepared an antibody against the protein, subjected it to a limited sequence analysis or mass spec analysis or determined whether it has any covalent modifications. Finally, the sequence information obtained from the purified protein may allow you to identify the gene encoding it, opening up the possibility of overexpression and the generation of mutants for physiological and mechanistic studies.

All of this information can be of great help in deciding whether you have developed an optimal purification scheme. Obviously, in some cases you may not care. However, if this protein is one you plan on studying in some detail, and you can foresee many purifications ahead, a rapid and efficient (meaning high purity and high yield) purification scheme can save you an enormous amount of work in the long run. By taking advantage of what you have learned about the protein, it generally should be possible to streamline and optimize the procedure a great deal. There is a natural tendency, especially after having spent many mouths learning to purify a

Department of Biochemistry and Molecular Biology, University of Miami School of Medicine, Miami, Florida, USA

Methods in Enzymology, Volume 463
ISSN 0076-6879, DOI: 10.1016/S0076-6879(09)63043-3

protein, to go with what you know works. Nevertheless, spending some time rethinking your purification procedure will be a worthwhile exercise.

Clearly, in many situations, overexpressing and tagging the protein of interest will be the simplest route to obtaining additional material. However, in some cases, depending on the organism, regulation of the gene, physiological problems due to overexpression, or negative effects of the tag, this may not be possible or easy to accomplish. While there may be ways to circumvent these particular issues, carefully examining the details of your purification protocol will usually provide benefits.

Some things you might want to consider, with relevant chapters to help you in this process, are as follows: [chapters to be added later, when chapter numbers known]

1. Am I using the best source of material?
   (a) Is the source readily available in large quantities?
   (b) Is the protein associated with a specific subcellular structure that might be a better starting material?
   (c) Would it be possible to develop a better source by cloning the gene for this protein?
   (d) Can the protein be overexpressed from the cloned gene?
2. Are all the steps in the purification scheme necessary and useful?
   (a) Do I extract most of the available activity?
   (b) Can I pool fractions more broadly at a particular step knowing that newly included impurities can be removed at a subsequent fractionation step?
   (c) Do any steps lead to unnecessarily large losses or relatively poor purification, and is there a special reason for including that step?
   (d) If there are large losses at any step, do I know why, and can it be prevented?
   (e) Are all the steps cost effective? Would a cheaper purification medium or procedure suffice?
   (f) Can any time-consuming procedures be eliminated?
3. Are the various purification steps carried out in the most optimal sequence?
   (a) Are procedures best for larger amounts of material in the beginning of the scheme, and those better for smaller amounts later in the procedure?
   (b) Can I avoid a concentration or dialysis step by changing the order of steps? Is it necessary to concentrate a sample prior to a column to which the protein binds?
   (c) Are the solution conditions of the last step in the procedure compatible with storage of the protein, or is a solution change necessary?

4. Should I introduce any new steps into the purification procedure?
   (a) Might a new step that takes advantage of the protein's binding properties (i.e., affinity chromatography) be effective?
   (b) Can the antibody I have prepared be of use for purification?
   (c) Am I taking advantage of all the protein's structural properties in deciding on purification steps?
5. Is the scale of the purification appropriate to the planned uses of the material?
   (a) Can enough material for my needs be obtained in fewer steps by using two-dimensional gel electrophoresis or immunoprecipitation?
6. Am I using the best assay for my protein considering speed, cost, and accuracy? Is a high degree of accuracy necessary during the purification?
7. If I have not already done so, can I learn anything from the literature by examining purification schemes for related proteins?

The answers to these questions will give you a good idea whether modification of your purification procedure or storage conditions might be warranted.

CHAPTER FORTY-FOUR

Important but Little Known (or Forgotten) Artifacts in Protein Biochemistry

Richard R. Burgess

Contents

1. Introduction 814
2. SDS Gel Electrophoresis Sample Preparation 814
   2.1. Proteases at room temperature 815
   2.2. Asp–Pro bond cleavage at 100 °C 815
   2.3. Keratin contamination in sample or SDS sample buffer 816
3. Buffers 816
   3.1. Reducing agent can become an oxidizing agent 816
   3.2. Contaminant in sulfonylethyl buffers 817
4. Chromatography 817
   4.1. EDTA binding to anion-exchange columns 817
   4.2. EDTA in sample can strip nickel from a Ni-chelate affinity column 818
5. Protein Absorption During Filtration 818
6. Chemical Leaching from Plasticware 819
7. Cyanate in Urea 819
References 820

Abstract

Many subtle and sometimes obscure artifacts exist that can have major effects on the outcomes of otherwise carefully performed experiments. This chapter focuses on a few of these artifacts involved in processes such as SDS gel sample preparation, buffers, Ni-chelate affinity and ion-exchange chromatography, urea preparation, use of plasticware, and filtration of protein samples.

McArdle Laboratory for Cancer Research, University of Wisconsin–Madison, Madison, Wisconsin, USA

Methods in Enzymology, Volume 463
ISSN 0076-6879, DOI: 10.1016/S0076-6879(09)63044-5

1. Introduction

Many of us who are experienced in protein biochemistry and protein purification are called on regularly to troubleshoot the problems of our colleagues. I often find that the answer to their problem is a little known bit of knowledge that they have never heard of, but which I have encountered many times. Some of us teach courses in protein purification and add little tidbits of information to our lectures that we feel will help our students avoid mysterious problems with their experiments. Many of us review multiple manuscripts on protein purification and occasionally find that certain artifacts reappear again and again, even when the effect has been seen and reported many years earlier. One of the problems is that while these sorts of artifact are many times reported, they are seldom the subject of the paper, but merely a sentence or paragraph buried deep in the paper or figure legend where they are rarely seen by a wider audience.

Other times, the importance has not been sufficiently emphasized in the original paper, and the novice scientist may have heard of the problem, but has chosen to ignore it or has assumed it really is not a problem most of the time. Finally, young scientists often assume that there is nothing of importance in literature older than a few years old and simply are unaware of issues that were well known 10–20 years ago.

Here, I present a few such potential artifacts and how to avoid them. I hope that some of this obscure and often forgotten information might help readers. These are just a few examples that come to mind, or have made a particularly strong impression on me, but I am sure that there are many, many more that are equally if not more important in various subareas of protein biochemistry.

If you are aware of other such artifacts or have tips, or cautions that you want to share, I would love to hear from you about them (burgess@oncology.wisc.edu).

2. SDS Gel Electrophoresis Sample Preparation

It is quite common to think you have a nearly pure protein and then find multiple bands upon analysis on SDS polyacrylamide gels. In my experience, there are three significant and common ways in which you can actually make your protein look less pure than it really is. While many scientists are quite aware of these problems, it still amazes me how many are not. Some of this material was presented in a short article in the *Novagen Newsletter inNovations* (Grabski and Burgess, 2001).

2.1. Proteases at room temperature

In the early days of SDS gel electrophoresis, it was common to put proteins into a sample buffer consisting of a buffer, glycerol to increase sample density, SDS to denature most proteins, and 2-mercaptoethanol to reduce disulfide bonds. It was soon recognized that many proteases have disulfide bonds and are quite stable and can only be inactivated by heating the sample to better denature and reduce the protease (Pringle, 1975). So, it became standard to heat samples to 95–100 °C for 5 min or more. It was learned a bit more slowly that the heating must be done immediately, since the SDS easily unfolds most proteins, including usually the protein of interest, but not some proteases. As a result, *if a sample is added to sample buffer, but the heating occurs sometime later, then the protease will have time to digest the protein of interest.* To see if this is a problem in your case, a simple experiment is to add a partially purified protein sample to two portions of sample buffer, mix, and immediately heat one. Leave the other at room temperature for 2–4 h and then heat it. Run both samples on an SDS gel and see if there is significant degradation of your protein in the sample not heated immediately. I have heard that as little as 1 pg of protease in a protein sample can cause major degradation of a protein if the sample is added to sample buffer, but not heated immediately.

2.2. Asp–Pro bond cleavage at 100 °C

Another related problem is the inadvertent cleavage of a protein at aspartic acid–proline peptide bonds (DP bonds) due to heating at too high temperature for too long. The DP peptide bond is the most easily cleaved peptide bond by heat or acidic conditions. I work on a protein that, *if heated in sample buffer at 100 °C for 5 min, will undergo about 5% degradation to several smaller bands.* This is particularly evident if the gel is subjected to Western blot analysis, which is easily able to detect a few nanogram of degradation fragments. As a result, I always caution students to do the following experiment. Prepare six identical tubes of your protein in sample buffer. Immediately heat one set of three tubes at 95–100 °C for 5, 20, and 60 min. Heat the other set of three at 75 °C for 5, 20, and 60 min. Put heated samples on ice and run the samples on a SDS gel. Often, but not always, significant cleavage will occur at the higher temperature, especially at the longer time. Of course, one can also analyze the sequence to see if it contains the DP dipeptide. As a result, our lab routinely heats samples for 5 min at 75 °C. We avoid DP peptide bond cleavage and have been able to completely inactivate proteases. The DP bond is known to be particularly labile; 8–20 times more labile that other DX sequences and 100-fold more labile than dipeptide bonds lacking D (Volkin *et al.*, 1995). However, many proteins have been found to be stable in sample buffer for hours at 100 °C (Deutsch, 1976).

2.3. Keratin contamination in sample or SDS sample buffer

Many times someone has shown me a stained SDS gel pattern and lamented that they just cannot seem to remove the impurities from their otherwise pure protein. Often this impurity is a cluster of bands around 55–65 kDa in apparent molecular weight (Ochs, 1983). Keratin is omnipresent in skin, dander, etc. When run on an SDS gel under nonreducing conditions (no 2-mercaptoethanol in the sample buffer), keratin runs near the top of the gel, but under reducing conditions that are most commonly used, *keratin runs as a heterogeneous cluster of bands around 55–65 kDa on the reducing SDS gels*. If you suspect that you have a keratin contamination, you can confirm this suspicion by simply running a nonreducing gel and the characteristic keratin bands should not be seen. The most common problem is that the sample buffer itself becomes contaminated by contact with skin or a flake of dandruff. Usually, this is quite a minor contaminant and can only be seen in silver stained gels. Occasionally, it can be seen on Coomassie Brilliant Blue stained gels. If you suspect this is a problem, simply apply sample buffer alone, with no other added protein, to a gel lane. If you see keratin, it means it is in your sample buffer and the buffer should be remade. We usually prepare batches of 4× SDS sample buffer, aliquot them into 1-ml portions, and store them frozen at −80 °C. Then, we thaw one portion, dilute it appropriately, and use it within a day or two. Keratin is also occasionally seen on Western blots where the antigen used to prepare a polyclonal antibody is contaminated with keratin and reacts with keratin in the proteins transferred from SDS gel to nitrocellulose membranes.

3. Buffers

3.1. Reducing agent can become an oxidizing agent

One very commonly adds a reducing agent to buffers to help prevent air oxidation of enzymes and other proteins. It is typical to add 0.1–1 m*M* 1,4-dithiothreitol (DTT) ($HSCH_2CH_2OHCH_2OHCH_2SH$) (Cleland, 1964). Usually such buffers are stored in the cold and used within a few days, but often buffers are used that are months old. The problem is that slowly over time, *the reducing agent becomes oxidized by oxygen in the air and becomes an oxidizing agent*; effectively, it is now much worse that having no thiol agent in the buffer. How do you know whether a buffer is still good and has effective reducing capability? A simple test is to treat a small portion of the buffer with Ellman's reagent (5, 5′-dithiobis-(2-nitrobenzoic acid) or DTNB) (Ellman, 1959). This chemical reacts with free thiol groups, such as the sulfhydryl of the DTT and releases 2-nitro-5-thiobenzoic acid, which is yellow colored and absorbs light with a molar extinction coefficient at

412 nm of 14,150 M^{-1} cm^{-1} (Riddles *et al.*, 1983). Read the $A_{412\ nm}$ of the solution and a control buffer carefully prepared with fresh DTT. If all the DTT is active then there should be two sulfhydryls per mole of DTT. Buffers should be made fresh if more than 10–20% of the DTT is oxidized. While this effect has been known for decades it is still surprising that this simple test is not more commonly used. The problem is even more pronounced with 2-mercaptoethanol which is more prone than DTT to air oxidation and is usually used at 5–50 times higher concentrations than DTT.

3.2. Contaminant in sulfonylethyl buffers

An unexpected and potentially critical technical observation was made in the Ron Raines lab at the University of Wisconsin–Madison and carefully documented in a paper by Smith *et al.* (2003). In studying ribonuclease A activity in a MES buffer, they noticed that the activity was progressively lower as the salt in the buffer was decreased. What they eventually found was that MES and, in fact, many of the sulfonylethyl group-containing buffers (MES, BES, CHES, and PIPES) contain a contaminant, oligo (vinylsulfonic acid), that is a side product in the chemical synthesis of the chemical buffer. This polyanionic contaminant mimicks RNA and at low salt (0.05 *M* NaCl) binds to and inhibits the enzyme with a K_i of 11 pM. *This has implications to anyone studying nucleic acid-binding proteins.*

4. Chromatography

4.1. EDTA binding to anion-exchange columns

Recently, Robert Chumanov in my lab observed a sharp peak of 215 nm-absorbing material eluting during a salt gradient elution of a MonoQ anion exchange resin (GE Healthcare). It turned out that it was *EDTA, bound quite tightly to MonoQ and eluting at a NaCl concentration of about 250 mM* (R. Chumanov and R. Burgess, unpublished). We routinely monitor column outputs at 215 nm when our protein/peptide lacks absorption at 280 nm or if we are loading low amounts of protein and want more sensitivity (most proteins absorb at least 15 times better at 215 nm than 280 nm). It is not surprising that the di- or trivalent anion EDTA binds to MonoQ column resin. We were surprised that so much bound that it completely depleted EDTA from our low salt equilibration buffer originally containing 0.1 m*M* EDTA. When the salt increased, the EDTA eluted with a peak EDTA concentration of up to 40 m*M*. This could have unexpected effects on many experiments. The fractions where the EDTA elutes will, for example, bind Mg^{2+} needed for many *in vitro* enzyme assays (such as transcription and replication assays) and appear to contain a mysterious enzyme inhibitor. It is also possible that a column nearly saturated with EDTA would have a lower protein-binding capacity, especially for proteins that only bind weakly to the column and elute at NaCl concentrations lower than 250 m*M*. This type of binding had been observed and published over 10 years ago (Sharpe and London, 1997), but we had not seen the paper and had not realized how dramatic this effect could be. Further reflection also suggests that other multivalent anions such as sulfate and phosphate will also bind to and elute from anion exchange resins with potentially artifactual consequences.

4.2. EDTA in sample can strip nickel from a Ni-chelate affinity column

One of the three methods for eluting hexa-His-tagged proteins from Ni^{2+}-chelating columns is by passing EDTA through the column to effectively chelate and strip out the Ni^{2+} and release bound tagged proteins (see Chapter 27). However, many people are used to breaking *Escherichia coli* and other biological materials in lysis buffers containing 0.1–5 m*M* EDTA. In some cases, they centrifuge out insoluble material and load the crude extract directly on an immobilized Ni^{2+} column. The EDTA in the sample, especially at 1 m*M* or higher, could easily decrease or remove completely the chelated Ni^{2+} and thus dramatically decrease the capacity of the column to bind His-tagged protein. When the Ni^{2+} has been stripped from the column, the column will become white instead of the light blue color. Such a column can easily be recharged with Ni^{2+}.

5. Protein Absorption During Filtration

When we refold proteins by dilution out of denaturants, we often get some aggregation that causes light scattering and turbidity of the sample (see Chapter 17). Before we apply the diluted sample onto an ion-exchange column, we filter the sample through a 0.22-μm cellulose acetate filter to remove any particulate material. One time we ran out of cellulose filters and a student used another filter that was in the lab. Unfortunately, it was a cellulose nitrate (or nitrocellulose) filter. *All the soluble protein was lost, due to protein binding to the filter.* This was not too surprising, since proteins generally bind very well to nitrocellulose filters. In fact, this is the principle of protein transfer from SDS gels to nitrocellulose membranes for Western blot analysis. Now we routinely use special PVDF filter units (Stericup-GV 0.22 μm, 500 ml, Millipore #SCGVU05RE) with good recovery of protein. Most filter membranes, no matter how "low protein binding" they are designed to be, will bind μg amounts of protein per cm^2, but some materials, such as cellulose nitrate and polystyrene are able to bind much more material and should be avoided for filtration or storage of protein.

6. Chemical Leaching from Plasticware

It is worth being aware of a recent report (McDonald *et al.*, 2008) that *certain chemicals can leach from common disposable laboratory plasticware into standard aqueous buffers, and in some cases can have strong effects on the results of biological and biochemical experiments.* Some chemicals, like oleamide, are used as lubricating agents in the molding process and others, like certain cationic biocides, are used to help prevent bacterial colonization of the plastic surface. Washing the plasticware in water or better in methanol or DMSO can remove much of this material.

7. Cyanate in Urea

Urea is widely used as a protein-denaturing agent, but it is not widely appreciated that urea solutions contain substantial amounts of ammonium cyanate, which is in chemical equilibrium with urea $(H_2N)_2C{=}O$ in equilibrium with $NH_4^+ + NCO^-$). Isocyanic acid H–N=C=O can react with amino groups on proteins (ε-amino group of lysine and the amino terminus, and to a lesser extent with arginine and cysteine) to form a carbamylated protein (Stark, 1965). This carbamylation can alter charge and in some cases interfere with enzyme function, block certain protease cleavage reactions, and add 43 Da per carbamylation event to the mass as measured by mass spectrometry. How can one diminish or prevent this from happening? One can treat a urea solution with a mixed bed resin such as Bio-Rad AG 501-X8 to remove these contaminant ions. The progress of deionization is easily monitored by measuring conductivity of the solution. Unfortunately, this is a chemical equilibrium and relatively soon (within a few days) the ammonium cyanate builds back up again to levels in the 0.5–3 m*M* range in an 8 *M* urea solution and can reach values of 20 m*M* (Lin *et al.*, 2004). Certain chemical scavengers such as ethylenediamine, glycylglycine, or glycinamide in the 5–25 m*M* range can also reduce cyanate to less than 0.1 m*M* in 8 *M* urea, Tris pH 8 (Lin *et al.*, 2004). Perhaps the best general practice if you want to use urea is to replace some of the NaCl in the buffer with some ammonium salt (such as 25–50 m*M* ammonium chloride) to push the equilibrium back by the common ion effect toward less cyanate. Reaction of isocyanic acid with amino groups in proteins is slowed by lower temperature and by acidic conditions and can be minimized by restricting the exposure of the protein to the urea solution to the shortest time possible.

REFERENCES

Cleland, W. W. (1964). Dithiothreitol, a new protective reagent for SH groups. *Biochemistry* **3,** 480–482.

Deutsch, D. G. (1976). Effect of prolonged 100 °C heat treatment in SDS upon peptide bond cleavage. *Anal. Biochem.* **71,** 300–303.

Ellman, G. L. (1959). Tissue sulfhydryl groups. *Arch. Biochem. Biophys.* **82,** 70–77.

Grabski, A., and Burgess, R. R. (2001). Preparation of protein samples for SDS-polyacrylamide gel electrophoresis: Procedures and tips. *inNovations* **13,** 10–12.

Lin, M.-F., Williams, C., Murray, M. V., Conn, G., and Ropp, P. A. (2004). Ion chromatographic quantification of cyanate in urea solutions: Estimation of the efficiency of cyanate scavengers for use in recombinant protein manufacturing. *J. Chromatogr. B* **803,** 353–362.

McDonald, G., Hudson, A., Dunn, S., You, H., Baker, G., Whittal, R., Martin, J., Jha, A., Edmondson, D., and Holt, A. (2008). Bioactive contaminants leach from disposable laboratory plasticware. *Science* **322,** 917.

Ochs, D. (1983). Protein contaminants of sodium dodecyl sulfate-polyacrylamide gels. *Anal. Biochem.* **135,** 470–474.

Pringle, J. R. (1975). Methods for avoiding proteolytic artefacts in studies of enzymes and other proteins from yeasts. *Methods Cell Biol.* **12,** 149–184.

Riddles, P. W., Blakeley, R. L., and Zerner, B. (1983). Reassessment of Ellman's reagent. *Meth. Enzymol.* **91,** 49–60.

Sharpe, J. C., and London, E. (1997). Inadvertent concentration of EDTA by ion exchange chromatography: Avoiding artifacts that can interfere with protein purification. *Anal. Biochem.* **250,** 124–125.

Smith, B. D., Soellner, M. B., and Raines, R. T. (2003). Potent inhibition of ribonuclease A by oligo(vinylsulfonic acid). *J. Biol. Chem.* **278,** 20934–20938.

Stark, G. R. (1965). Reactions of cyanate with functional groups of proteins: Reaction with amino and carboxyl groups. *Biochemistry* **4,** 1030–1036.

Volkin, D. B., Mach, H., and Middaugh, C. R. (1995). Degradative covalent reactions important to protein stability. (B. A. Shirley, ed.)Protein stability and folding: Theory and practice*Methods Mol. Biol.* **40,** 35–63.

Author Index

A

Abdolzade-Bavil, A., 317
Abe, M., 649
Ackers, G. K., 706–707, 721
Acton, T. B., 456, 460
Afeyan, N. B., 355
Agnew, B., 557, 559
Aguiar, R. W., 142
Ahmad, I., 737
Akermoun, M., 632
Akutsu, H., 164
Al-Ibraheem, J., 293, 298
Al-Rubeai, M., 139
Alegria-Schaffer, A., 573–598
Alexander, C., 424
Ali-Khan, 103
Allen, R. C., 498, 500–501, 506–509, 511–512
Allis, C. D., 726, 741
Alterman, M., 85
Ames, B., 702, 705
Ames, G. F., 162
Anderson, N. G., 517, 750
Anderson, N. L., 517, 750
Anderson, T. R., 247–248
Andersson, C. L., 134
Andersson, D. I., 133
Andersson, L., 444
Andre, N., 632
Andrews, A. T., 498–501, 503, 506–509, 511–512, 710
Andrews, B. A., 286, 290
Anfinsen, C. B., 260
Angal, S., 293
Antal, J., 571
Antharavally, B. S., 586, 614
Anthony, J. R., 478
Anthony, L. C., 269
Antoine, M. D., 582
Anusubel, E. M., 170
Arakawa, T., 273, 704
Armirotti, A., 757
Armstrong, N., 271–272
Arnau, J., 243, 252, 457
Arnold, T., 608, 611–613
Arnold, U., 531
Asenjo, J. A., 286, 290
Asermely, K. E., 89
Ashani, Y., 614
Aslanidis, C. V., 153, 155
Atha, D. H., 118
Attri, A. K., 699–700
Aumiller, J. J., 213
Azzouz, N., 196

B

Bachmair, A., 245
Backliwal, G., 227, 231–232, 234
Bailey, M. J. A., 73–93
Balaji, P. V., 25
Baldi, L., 138, 224–225
Baldwin, M. A., 712–713
Bamber, L., 635
Baneyx, F., 133, 150, 156, 249, 262, 276
Bantan-Polak, T., 89
Bantscheff, M., 751, 753, 755
Barbieri, C. M., 557
Bardwell, J. C., 134
Barnes, P. J., 740
Batard, P., 229
Bates, D. M., 803
Bayer, M. H., 250
Bayer, P., 739
Baynes, B. M., 273
Beausoleil, S. A., 732, 735
Becker, D. M., 176–177, 179
Beckman, T. W., 741
Beidler, D. E., 447
Beinert, H., 788–789, 798–801
Bell, P. J. L., 91, 554
Bellehumeur, T. G., 575
Ben-Naim, A., 407
Bennett, K. L., 735
Benov, L., 293, 298
Bergendahl, V., 478, 481, 490, 575
Berger, I., 196
Berger, S. L., 726, 741
Berggren, K. N., 553
Berkelman, T. R., 555
Berlet, H. H., 86
Bernardi, G., 389
Bernstein, E., 741
Bertschinger, M., 229, 231
Bessette, P. H., 134, 279
Bhalget, M. K., 553
Bhatia, P. K., 139, 144
Bier, M., 498

Biggs, R., 681
Billecke, C., 307, 310
Bjellqvist, B., 515, 519
Blackshear, P. J., 498, 501, 503, 506–508, 511
Blagoev, B., 728
Blakeley, R. L., 83
Blanchard, J. S., 43–56
Blau, H. M., 142
Block, H., 439–468
Blommel, P. G., 160, 651–652, 656
Blumberg, P. M., 428
Blume-Jensen, P., 731
Bock, P., 699–700
Boden, V., 447
Boer, E., 171
Bogyo, M., 543
Bolanos-Garcia, V. M., 246, 449, 455–456
Bonander, N., 633
Booker, C. J., 103
Bornhorst, J. A., 454
Bossio, R. E., 691–721
Bottomley, S. P., 249, 262, 277
Boussif, O., 231
Boyer, H. W., 154
Boyer, M. E., 649
Bradford, M. M., 12, 79, 85, 624
Bradley, M. K., 196, 214
Braun, P., 460
Breier, J., 390
Breitbach, K., 213
Brena, B., 432
Brendel, M. F., 177, 179
Brierley, R., 185
Brinker, N., 439–468
Brisack, K., 387–402
Brondyk, W. H., 131–144
Brooks, C. A., 352
Brown, N., 118
Browne, S. M., 139
Brzoska, K., 788
Buchanan, S. K., 632
Burgess, R. R., 21–27, 29–34, 259–279, 287, 293, 297, 331–342, 475–493, 565–571, 577, 813–819
Burkitt, W. I., 75, 85
Burris, R. H., 788
Bury, A. F., 498, 501
Busso, D., 250, 450
Büssow, K., 460
Butler, M., 139
Butt, T. R., 253
Byrne, B., 463

C

Cabiscol, E., 789
Cabrita, L. D., 244, 249, 262
Calandra, B., 633
Cammack, R., 802
Campbell, D. H., 419
Campbell, J. M., 745
Candiano, G., 547
Cannon, J. G., 586
Cantin, G. T., 733
Cantor, C. R., 704
Carpenter, G., 579
Carr, S. A., 737
Carroll, I. P., 800
Carson, M., 246
Carstens, C. P., 156
Carta, G., 364
Cass, B., 456
Castellanos-Serra, L., 567
Catley, B. J., 289
Catravas, G. N., 614
Cèbe, R., 450
Cecília, A., 428
Celik, E., 136
Chaga, G., 442
Chait, B. T., 731–735
Chalkley, R. J., 730
Chan, J. K., 86
Chance, M. R., 447, 463
Chandran, S. P. 433
Chang, J. Y., 680
Chappel, T., 169–187
Chase, T., Jr., 788
Chasin, L. A., 226
Chatterjee, D. K., 134, 141, 243
Chazenbalk, G. D., 199
Chelikani, P., 635
Chen, C. L., 741
Chen, W., 137, 741
Chen, Y. R., 741
Cheng, Y., 251
Chenuet, S., 229, 232
Cherezov, V., 634
Cheryan, M., 108
Chiari, M., 567
Chirgwin, J. M., 244
Cho, M. S., 225–226
Chong, S., 253
Chopra, I., 162
Chotia, C., 408
Chou, 162
Choudhary, J. S., 244
Choudhary, S., 575
Chow, M. K., 277
Chrambach, A., 498, 500, 506–508, 511–512, 710
Christensen, K., 229
Chu, B., 716
Chu, R., 287, 289
Chumanov, R., 265, 267
Cirulli, C., 733
Clark, P. L., 261–262
Clark, P. M., 559

Clark, W. M., 55
Clauser, K. R., 744
Cleland, J. L., 273
Cleland, W. W., 62, 816
Cline, D. J., 294
Cline, G. B., 311
Clonis, Y. D., 428
Coghlan, D. R., 91, 554
Cohen, P., 726
Cole, J. L., 687, 698, 700, 702–703
Collart, F. R., 149–166
Collier, T. S., 757
Collins-Racie, L. A., 244
Condreay, J. P., 783
Consortium, U., 770
Constantinidou, C., 789
Cook, P. F., 62
Coon, J. J., 728
Cooper, C. E., 802
Cornish-Bowden, A., 62
Costantino, H. R., 114
Cottingham, K., 716
Cottrell, J. S., 726
Covey, T. R., 745–746
Cowieson, N. P., 271–272
Crack, J. C., 789, 803
Cramer, P., 477, 479, 490
Cramer, S. M., 352
Cravatt, B. F., 543
Creasy, D. M., 726
Cregg, J. M., 135, 169–187
Creighton, T., 275
Crowe, J., 455
Cuatrecasas, P., 419
Cull, M., 286
Cummings, L. J., 387–402
Cunningham, F., 244
Curtis, C. A. M., 632, 635
Cussler, E. L., 106
Czerney, P. T., 555

D

Daban, J.-R., 91, 551
Dalbøge, H., 449
Daly, R., 135–136, 141–142
Dang, J. M., 229
Danson, M., 65–66, 68
Das, T. K., 273
Davies, O. R., 449, 455–456
Davis, B. J., 506, 508
Davis, G. D., 250
Davis, T. R., 211
Dawson, R. M. C., 294
De Grip, W. J., 634
de Jong, P. J., 153, 155
De Marco, V., 155
de Marco, A., 249–250
De Mey, M., 293
de Moreno, M. R., 75, 85–86
Dean, D., 224, 226
Deb, J. K. 133
Deber, C. M., 244
D. G.ip, W. J., 613
D. J.ng, J., 455
Delory, G. E., 55
Demeler, B., 720
Dempsey, 44–45, 48, 50
Denison, C., 739
Denisov, G. A., 108
Derewenda, Z. S., 440, 451
Derman, A. I., 154
Desai, S., 584
Desiderio, D. M., 741
Deupi, X., 634
Deutsch, D. G., 815
Deutscher, M. P., 37–42, 121–127, 343–345, 809–811
di Salvo, M. L., 401
Dieckman, L., 150, 152, 156, 286
Dieckman, L. J., 150, 152, 156
Dignam, J. D., 286
diGuan, C., 249
Dikic, I., 738
Dixon, R. A., 162
Döbeli, H., 253
Dong, M. S., 142
Dong, X., 389
Dongre, A. R., 747
Doolittle, R. F., 24
Dorrier, T. E., 83
Douris, V., 194
Drew, D. E., 651
Drews, J., 461
Drott, D., 289
Duellman, S. J., 340, 478, 491–492
Dunbar, B. S., 497–498, 508, 512, 542
Durand, G., 737
Durocher, Y., 138, 227
Durschlag, J., 703
Dyson, M. R., 150, 244, 249–250

E

Eason, R., 687
Edelman, M., 311
Edwards, A. M., 477–478, 490
Einhauer, A., 247
Eisenthal, R., 65–66, 68
Elias, C. B., 211
Eliyahu, H., 229
Ellman, G. L., 816
Endo, Y., 649–650, 663, 669
Englard, S., 332, 335, 341
Engler, C. R., 286
Ernst, W. J., 207

Eschenfeldt, W. H., 153, 156
Eshaghi, S., 301, 462
Esposito, D., 134, 141, 153, 243
Etzel, M., 118
Evans, D. R. H., 97–118

F

Fabis, R., 439–468
Falke, J. J., 454
Fang, S., 738
Farley, A. R., 725–758
Fasman, G. R., 692
Fath-Goodin, A., 210, 213–214
Faull, K. F., 712, 714
Faulmann, E. L., 425
Fausnaugh, J. L., 407
Felsenfeld, G., 324
Ferguson, J. T., 758
Ferguson, W. J., 43, 390
Fernandez, C., 615
Fernandez-Patron, C., 549
Ficarro, S. B., 733
Finn, R. D., 770
Fish, W. W., 698, 708
Fisher, W. D., 311
Flugge, U. I., 117, 531
Foley, K. M., 475–493
Folta-Stogniew, E., 708
Ford, T., 314
Foster, D., 286
Fountoulakis, M., 75, 85
Fox, B. G., 647–670
Franco, A. T., 526
Frank, A. M., 149–166
Franken, K. L. M. C., 453, 460
Freifelder, D., 702
Freitag, R., 390, 457
Frey, P. A., 65
Friedenauer, S., 86
Friedman, D. B., 515–538
Friesen, P. D., 194
Fritze, C. E., 247–248
Fromme, P., 615
Fuchs, S. M., 726
Furth, A. J., 513, 613
Fushman, D., 738
Fussenegger, M., 234
Fux, C., 223–234

G

Gagnon, P., 389, 392, 402
Galbraith, D. J., 234
Gallagher, S. R., 585
García González, L. A., 448
Garcia, B. A., 728, 742
Garfin, D. E., 497–513, 542, 551
Garrett, P. E., 292
Garzia, L., 460
Gasteiger, E., 27
Gatzka, M., 731
Gauchet, C., 433
Ge, Y., 728–729, 757
Gegenheimer, P., 286
Geiser, M., 450
Geisse, S., 223–234
Gellissen, G., 170
Gellman, S. H., 273
Gemmill, T. R. 136
Georgiou, G., 278
Gether, U., 632
Getz, E. B., 294
Ghosh, R., 113
Gibbons, R. A., 703
Giglione, C., 142
Gilbert, H. F., 269
Gill, P., 448
Gill, S. C., 23, 83
Gillies, A. R., 253
Gingras, A. C., 731
Girard, P., 138
Glatter, O., 694
Gleeson, M. A. G., 136, 184
Glimcher, M. J., 392
Glossman, H., 498–499
Glozak, M. A., 741
Godde, J. S., 726
Goerke, A. R., 649
Golks, A., 559
Gomori, G., 52, 54
Good, D. M., 749
Good, N. E., 43, 47
Goodlett, D. R., 729
Gorbunoff, M. J., 388–389–390, 392
Goren, M. A., 647–670
Görg, A., 510, 519
Goshe, M. B., 733
Goto, M., 136
Gotto, J. W., 788
Goustin, A. S., 133
Gräslund, S., 243
Grabski, A. C., 285–300, 814
Grace, T. D. C., 211
Graessmann, A., 226
Graham, F. L., 224
Graham, J., 314
Grainger, D. C., 789
Graslund, S., 261, 456, 767
Grewal, S. I., 741
Grey, M., 177, 179
Gribskov, M., 267
Griffith, D. A., 633
Grisshammer, R., 463, 631–642
Groves, W. E., 82
Gu, T. R., 611
Gu, W., 742

Guarante, L., 176–177, 179
Guarino, L. A., 194–195, 215
Guerini, D., 559
Guerlava, P., 293
Guerrier, L., 402
Guidotti, G., 619–629
Guilbault, G. G., 424
Gunsalus, R. P., 788
Guruprasad, K., 24
Guss, B., 425
Gustafsson, C., 140
Gustavsson, P.-E., 419
Gusman, L. M., 154
Gygi, S. P., 729, 750–752

H

Haas, W., 744
Habig, W. H., 247
Hage, D. S., 419, 422, 424–425, 428
Hager, D. A., 566, 568–570
Hahn, R., 349–370
Haldenwang, W. G., 569
Hale, J. E., 447
Hall, J. F., 649
Halvorson, H., 699–700
Hames, B. D., 498–501, 503, 506–509, 511
Hammer, K. D., 79, 89
Hampsey, M., 477
Han, G. Y., 294
Hancock, 738
Hansen, B. T., 730
Hanson, M. A., 634–636
Harding, I. S., 394
Harding, S. E., 716
Hardwicke, P. I., 211
Hardy, E., 567
Harlow, E., 479
Harrington, M. G., 512, 566, 570
Harris, E. L. V., 293
Harris, T. K., 57–71
Harrison, J. L., 771
Harrison, R. G., 25, 116
Harrison, R. L., 137, 194, 212, 215
Harrison, S. T. L., 286, 290–291
Hart, C., 558
Hart, G. W., 559, 737
Hartinger, J., 516
Hartley, D. L., 265
Hartley, J. L., 155
Hassaine, G., 632
Hatakeyama, H., 526
Hatley, J. L., 771
Haugland, R. P., 552, 556, 558
Haun, R. S., 153, 155
Hawtin, R. E., 208
Hayduk, E. J., 729
Hearn, M. T. W., 135–136, 141–142
Hearon, J., 440
Heerikhuisen, M., 171
Hegeman, A. D., 65
Helenius, A., 608
Hengen, P., 154
Heppel, L. A., 162
Hermanson, G. T., 425, 427, 430, 432
Hernan, R., 248
Higgin, D., 176
Higgin, J. M., 176
Hilbrig, F., 457
Hill, D. R., 210
Hill-Perkins, M. S., 199
Hinkkanen, A., 431
Hinze, W. L., 608
Hirabayashi, J., 419, 427
Hiraide, T., 390
Hirel, P.-H., 449
Hitchcock, A., 262
Hitchman, R. B., 137
Hjelmeland, L. M., 623
Hjertén, S., 409–410
Ho, C. W., 293
Hober, S., 461
Hochuli, E., 440, 442, 449, 451, 458
Hofmeister, F., 117, 407
Hollister, J. R., 213
Holmes, W., 54
Holt, 737
Hopkins, T. R., 286, 290–291, 293
Horvath, C., 407
House, J. E., 58
Hoving, S., 515–538
Howlett, G. J., 699
Hsieh, H.-C., 101
Hsieh, H.-Y., 401
Hu, J., 253
Hu, S. M., 253
Hubbard, M. J., 726
Huber, L. A., 447
Hubert, P., 447
Hulme, E. C., 632, 635, 637
Hulo, N., 245
Hunt, I., 460
Hunter, T., 731

I

Ibel, K., 522
Ichitsuka, T., 388
Idicula-Thomas, S., 25
Ikeda, F., 738
Ikonomou, L., 137
Imagawa, W., 578
Ingham, K. C., 118, 341
Ishihama, Y., 615
Issad, T., 726
Ito, S., 728

J

Jaakola, V. P., 634–635
Jana, S., 133
Janson, J. -C., 354, 406, 411
Jarvis, D. L., 137, 191–218
Jazwinski, S. M., 286
Jendrisak, J. J., 338, 340
Jenkins, N., 137, 139, 142
Jennings, T., 114
Jensen, M. R., 449
Jensen, O. N., 742
Jenuwein, T., 726, 741
Jessani, N., 543
Jesse, J., 179
Jia, W., 101
Jiang, W., 165
Jiang, Y., 477
Jiang, Z., 229
Jin, S., 445
Johnson, J. F., 711
Johnson, M. K., 788–789
Johnson, M. L., 699, 720
Jones, D. D., 408
Jones, L. J., 91
Jordan, M., 229
Jormakka, M., 463
Jovin, T. M., 498, 506
Jungbauer, A., 247, 249, 262, 276, 349–370, 451

K

Kaar, W., 249, 262, 276
Kaba, S. A., 208–209
Kadoya, T., 388
Kadwell, S. H., 211
Kajiyama, S., 788
Kaltenbrunner, O., 365
Kandori, K., 389
Kane, J. F., 140
Kanehisa, M., 770
Kang, C., 555
Kang, Y., 789
Kanno, T., 649–650
Kao, F.-T., 226
Kaplan, W., 247
Kapust, R. B., 155, 249, 252, 774
Karp, N. A., 525
Karp, P. D., 770
Karplus, M., 261
Karuso, P., 91, 554
Kashino, Y., 289
Kaslow, D. C., 458
Kassel, D. B., 732
Kataeva, I., 244, 249–250
Kato, A., 250–251
Kaufmann, S. H., 595
Kawasaki, A., 402
Kawasaki, T., 389
Kelleher, N. L., 744, 757
Kenig, M., 253
Kennedy, M. C. 789, 799–800, 802
Kennedy, S. W., 89
Kersteen, E. F., 269–270
Keshwani, M. M., 57–71
Khoroshilova, N., 789, 799, 802
Kiga, D., 649
Kigawa, T., 649
Kiley, A. Y., 787–803
Kim, C. Y., 246
Kim, D.-M., 457
Kim, J., 446, 457
Kim, M. J., 446, 457
Kim, T.-W., 457
Kim,Y., 767
King, E. J., 55
King, L. A., 199, 215
Kinoshita, E., 557
Kitts, P. A., 200
Kivimäe, S., 226
Klaassen, C. H., 635
Klammt, C., 462, 649, 669
Kleinig, A. R., 291
Klock, H. E., 150, 771–772, 783
Knight, R. D., 463
Knox, J. R., 694
Knuth, M. W., 27, 263, 265–266, 270, 338, 340
Kobilka, B. K., 632, 634–636
Koehn, J., 460
Koike, T., 557
Komoriya, A., 408
Koneracka, M., 423
Kool, M., 216
Kopaciewicz, W., 352
Kornberg, A., 3–6
Korpela, T., 431
Korth, K. L., 142
Kost, T. A., 136, 196, 783
Kovalska, V., 555
Kowal, R., 431
Krane, S. M., 392
Kratky, O., 694
Krause, F., 566, 570
Kresin, L. M., 567
Krigbaum, W. R., 408
Kroe, R. R., 700
Kubicek, J., 439–468
Kuck, U., 171
Kunaparaju, R., 226
Kunitani, M. G., 567
Kunji, E. R., 165
Kuo, M., 726
Kupke, D. W., 83, 693
Kurokawa, M., 313
Kusari, A., 169–187
Kuster, B., 244
Kusumoto, K., 229

Kutter, K., 566
Kuznedelov, K., 490
Kwaks, T. H. J., 139
Kwok, T. T., 579
Kwon, M. S., 137

L

Labahn, J., 439–468
Labrou, N. E., 419, 428
Ladish, R. G., 116
Laemmli, U. K., 498, 501, 506–507, 521–522
Laengle-Rouault, F., 224
Laidler, K. J., 58
Lakowicz, J. R., 66
Lane, D., 479
Langone, J. J., 424
Lanio, T., 460
Largaespada, D. A., 479
Larsen, M. R., 733, 746
Larsson, B., 409
Larsson, P.-O., 419
Lata, S., 445–446
Laue, T. M., 677–689, 691–721
L. V.llie, E. R., 134, 250
Lazazzera, B. A., 789, 791
Le Hir, H., 227
Leach, S. J., 82
Leckband, D. E., 432
Lee, C., 510, 512
Lee, C. D., 246, 251, 253
Lee, C. H., 735
Lee, H., 735
Lee, J., 226
Lee, J. R., 735
Lee, K. A., 735
Lee, S. B., 246, 251, 253
Lee, S. J., 735
Lee, Y. C., 209, 622
Lei, Z., 570
Leibl, H., 402
Leirmo, S., 49
Lennarz, W. J., 737
Leong, K. W., 229
Lerman, L. S., 419
Lesley, S. A., 286, 767–783,
Lettner, H., 366
Levings, C. S., 142
Levinthal, C., 260
Levy, D., 667
Lewinson, O., 462
Laible, P. D., 165
Lichty, J. J., 243
Lieve, L., 289
Lihoradova, O. A., 207
Lilley, K. S., 516, 524–526, 528
Lillie, R. D., 50
Lim, K. B., 732
Lin, M.-F., 819
Lin, S.-H., 619–629
Lin-Cereghino, G. P., 174, 177
Link, A. J., 725
Linke, D., 603–616
Linn, S., 9–19
Listwan, P., 244
Liu, J. W., 244, 251, 711–712
Liu, L., 244, 251
Liu, Q., 771
Lodge, A., 573–598
Londer, Y. Y., 150, 153
London, E., 818
Loo, J. A., 447
Lopez, J. M., 86
Lorenzen, A., 89
Lorow-Murray, D., 179
Los, G. V., 251, 428
Lowe, C. R., 428
Lowry, O. H., 86–87
Lu, A., 194, 207
Lu, Y., 728
Lubs, H. A., 55
Luckow, V. A., 202
Luellau, E., 401
Lungwitz, U., 229
Lunney, J., 681
Lv, G. S., 446
Lynch, N. A., 477
Lynn, D. E., 211

M

Ma, D., 455
Ma, Q. Y., 394
Ma, Y., 735
Mac Coss, M. J., 744
Macek, B., 756–758
MacKenzie, D., 635
Mackintosh, J. A., 554
Madden, E. A., 311, 314
Madden, K., 169–187
Madin, K., 649–651, 669
Madoery, R., 427
Madzak, C., 171
Maertens, B., 439–468
Magnani, F., 634
Majors, B. S., 234
Makarov, A., 728
Makino, S.-I., 647–670
Makrides, S. C., 155, 226, 265
Malakhov, M. P., 253
Maldonado, E., 490
Malhotra, A., 239–254
Mallia, K., 289
Mallik, R., 422, 432
Manavalan, P., 408
Manchenko, G. P., 543, 566

Mann, M., 735, 744, 750–752, 757
Manza, L. L., 739
Marblestone, J. G., 251
Margulies, M. M., 500
Marin, M., 140
Markley, J. L., 649–650
Markwell, J., 75, 93
Marlor, C. W., 344
Marshak, D. R., 289, 293
Marshall, K., 703
Marshall, T., 507
Marston, F. A. O., 265
Martin, R., 702, 705
Martinez Molina, D., 774
Maslennikov, I., 668
Mateo, C., 456
Mathrubutham, M., 578, 590
Matiasson, B., 457
Matsuda, T., 649
Matsumoto, K., 227, 649
Matthey, B., 155
Mattson, D. L., 575
Mc Donald, G., 819
M. C.slin, D. R., 703
M. C.e, J. T., 395–396, 405–413
M. H.nry, C. S., 286
M. I.vaine, T. C., 51
M. L.chlin, D. T., 731–735
M. L.chlin, J. R., 195
M. M.ster, M. C., 67
M. M.ekin, T. L., 703
M. M.llan, D., 782
Mechref, Y., 738
Medzihradszky, K. F., 737, 745
Meiss, G., 290
Meissner, P., 226–227, 229
Melander, W., 407
Melchior, F., 739
Merril, C. R., 508–509, 542, 547
Mészárosová, K., 447
Mettert, E. L., 792, 797
Meyer, J., 788–789
Michael, M. Z., 154
Michaelis, L., 53
Michelsen, U., 305–327
Middelberg, A. P. J., 262, 272, 286, 290–291
Mikesh, L. M., 728, 747, 749
Miles, B. J., 388
Miller, I., 543
Miller, J. V., 326
Miller, L. K., 193–195, 207
Minchiotti, M., 427
Minden, J., 556
Minton, A. P., 699–700
Miroux, B., 154, 164
Mistretta, T. A., 195
Miyagi, M., 751–752
Moffatt, B. A., 154, 156
Mohanty, A. K., 449, 463
Molday, R. S., 635
Monsma, S. A., 137
Mooney, R. A., 612
Moore, L. J., 798
Morita, E. H., 669
Mortensen, K. K., 265
Mueller, E., 354–355
Mueller, U., 244
Mujacic, M., 133, 249, 262, 276
Mukhopadhyay, A., 139, 144
Muller, N., 232
Murby, M., 265
Murhammer, D. W., 215
Muszynska, G., 444

N

Nagel, K., 79, 89
Nagy, P. L., 477
Nakamura, E., 320
Nall, B., 261
Nallamsetty, S., 249, 252
Necina, R., 358
Nelson, P. N., 476
Neophytou, I., 164
Nestle, M., 290
Neu, H. C., 162
Neugebauer, J. M., 289, 605, 608–609
Neuhoff, V., 546
Neville, D. M., Jr., 498–499, 621
Nguyen, H., 287
Nguyen, L., 263, 265
Nguyen, N. Y., 512
Nieba, L., 445, 448
Nielsen, T. B., 499
Nilsson, J., 457
Noble, J. E., 73–93
Nomura, S. M., 652
Norris, J. L., 615
Northcote, D. H., 86
Nosjean, O., 652
Nossal, N. G., 162
Nozaki, Y., 408, 696
Nozawa, A., 647–670
Nuss, J. E., 401

O

O'Farrell, P. H., 515–516, 519
O'Farrell, P. Z., 523
O'Reilly, D. R., 196, 198, 215–216
O'Shannessy, D. J., 432
Oates, M. R., 432
Ochs, D., 510, 816
Oda, Y., 729, 733
Oganesyan, N., 276
Ogawa, T., 390
Ohana, R. F., 428

Ohana, R. F., 251
Ohtaki, T., 636
Old, W. M., 755
Olin, B., 447
Olsen, J. V., 729
Olson, B. J., 75, 93
Olsson, A., 726
Olsson, I., 516, 523–524, 534
Ong, S. E., 729, 750–753
Ooi, B. G., 195
Oprian, D. D., 635
Ornstein, L., 506, 508
Orr, G. A., 328
Osborn, M., 499
Ostrove, S., 419
Otte, A. P., 139
Ozols, J., 75

P

Pabst, T. M., 364–365
Pace, C. N., 23, 83, 266
Padan, E., 445
Påhlman, S., 407
Palumbo, J. L., 156
Pan, C., 731
Panda, A. K., 262, 267, 272, 276–277
Pandolfe, W. D., 291
Pantoliano, M. W., 774
Panvas, T., 251
Parks, B. A., 756–757
Parsons, R. G., 431
Partridge, S. M., 352
Passarelli, A. L., 194
Patonay, G., 582
Patton, W. F., 543, 728
Paulick, M. G., 543
Payne, C. K., 229
Pédelacq, J. D., 243
Pedersen, J., 449–450, 457
Pedrioli, P. G., 740
Peng, J., 739
Peng, S., 783
Pennock, G., 193–194
Perler, F. B., 253
Perrin, D. D., 43–45, 48–50
Pertoft, H., 313
Peter-Katalinic, J., 139
Peterson, G. L., 79, 86–87, 624
Pflieger, D., 746
Pham, P. L., 224–225, 227–,232
Phillips, G. N., Jr., 669
Pickart, C. M., 738
Piehler, J., 445
Pijlman, G. P., 204–205
Pikal, M. J., 114
Pinkse, M. W., 733
Pitteri, S. J., 728
Pless, D. D., 737
Plievaa, F. M., 422
Plumel, M., 52
Podgornik, A., 367
Podgornik, H., 367
Poggeler, S., 171
Poland, J., 547
Polayes, D. A., 457
Polevoda, 742
Ponnuswamy, P. K., 408
Popescu, C. V., 802
Porath, J., 409, 430, 440, 444, 446–447
Posewitz, M. C., 444
Possee, R. D., 199–200, 204, 215
Poston, J. M., 788
Potts, J. T., 248
Prakash, V., 704
Pramauro, E., 608
Presta, L. G., 147
Prickett, K. S., 247
Pringle, J. R., 815
Prinz, B., 456
Prinz, W. A., 156
Prive, G. G., 614
Probasco, M. D., 478, 490
Prokipcak, R. D., 570
Przic, D. S., 114
Puig, O., 244
Puri, N. K., 267

Q

Qasba, P. K., 559
Qing, G., 165
Quronfleh, M. W., 249, 262, 271–272, 276

R

Rabilloud, T., 527–529, 547, 553
Rabinowitz, J. C., 788
Raines, R. T., 248
Ralston, G., 687, 698, 700, 702–703
Ramakrishnan, B., 559
Ramanan, R. N., 291
Ramsby, M. L., 317
Rankl, N. B., 199
Rao, K. C., 751–752
Rao, L., 492
Rapoport, B., 199
Raulfs, E. C., 792
Read, S. M., 86
Reddy, R. C. K., 273
Reeves, P. J., 635
Regnier, F. E., 407, 751
Reichel, A., 446
Reid, G., 726
Renkin, E. M., 420
Reynolds, J. A., 499, 703
Reznikof, W. F., 803
Rhodes, D. G., 677–689, 691–721

Rice, J. C., 741
Richards, F. M., 248
Richardson, C. D., 200, 215
Rickwood, D., 313
Riddles, P. W., 817
Rigaud, J. L., 613, 667
Righetti, P. G., 519
Rinas, U., 262, 267, 269, 272
Roberts, W. K., 290
Robinson, A. S., 402
Rodbard, D., 500, 710
Roeder, R. G., 455, 742
Romanos, M. A., 170
Romero, J. K., 97–118
Rong, W., 110
Roodveldt, C., 244
Rose, G. D., 408
Ross, P. L., 314, 729, 751, 753–754
Rossi, F. M., 142
Rossomando, E. F., 65
Roth, C. B., 634
Roth, M. J., 726
Roulland-Dussoix, D., 154
Roux, B., 652
Rozema, D., 273
Rumbley, J. N., 463
Rush, J., 728
Rydén, L., 354, 406, 411

S

Saari, L. L., 788
Sachdev, D., 244
Sadoul, K., 741
Sadowski, P. G., 315
Sahdev, S., 133, 140, 144, 249
Saiyed, Z., 423
Sakena, S., 113
Saleh, L., 253
Sali, A., 718
Salmon, K., 789
Sambrook, J., 170, 182
Sanchez, J. C., 535
Sapan, C. V., 85
Sarkar, C. A., 634
Sarramegna, V., 632
Sasse, J., 585
Sato, M., 578
Saunders, A. H., 803
Sawasaki, T., 649–650, 652, 662–663, 667, 669
Schäfer, F., 439–468, 458, 460, 466
Schaffner, W., 637, 642
Schagger, H., 516
Scheich, C., 460
Schein, C. H., 272, 278
Scheraga, H. A., 408
Scheraga, H. A., 82
Schimmel, P. R., 704
Schindler, J., 622
Schirch, V., 390, 394
Schlaeger, E.-J., 229
Schley, C., 101
Schmidt, F. R., 164
Schmidt, T. G., 248
Schmitt, J., 454
Schnitzler, G. R., 323
Schroder, E., 394
Schroeder, M. J., 735
Schuck, P., 720
Schulenberg, B., 729
Schumann, C., 401
Schwarz, D., 649
Scigelova, M., 762
Scopes, R. K., 82, 287, 293–294, 332, 341, 394
Scorer, C. A., 186
Scott, M., 137
Seddon, A. M., 668
Seelert, H., 566, 570
Segel, I. H., 62, 65, 68
Sehgal, P. B., 320
Seifter, S., 332, 335, 341
Seo, N. S., 213
Serpa, G., 447
Serrano-Vega, M. J., 634
Seta, N., 737
Seto, E., 741
Shadforth, I., 315
Shah, M. B. 320
Shapiro, A. L., 499
Sharfstein, S. T., 229
Sharpe, J. C., 818
Shaw, A. Z., 164
Shaw, G., 224
Sheen, H., 103
Shen, J. W., 389
Shen, S., 186
Shepard, S. R., 394
Sherman, 742
Sherrier, D. J., 314
Shi, X.-L., 403
Shi, W., 447, 463
Shi, X., 194, 212
Shi, Y., 455
Shiloach, J., 458
Shimamoto, N., 273
Shimeld, S. M., 463
Shinkawa, T., 147
Shirgaonkar, I. Z., 291
Silla, T., 226
Simon, R. H., 324
Simpson, D., 417–435
Sinacola, J. R., 402
Singh, S. M., 262, 267, 272, 276
Sirard, J. C., 731

Sittampalam, G. S., 83
Sivaraman, T., 117
Sjoblom, J., 611
Skartsila, K., 394
Skerra, A., 248
Slack, J. M., 208
Slentz, B. E., 447
Sletyr, U. B., 718
Slotboom, D. J., 668
Smejkal, G. B., 543
Smith, B. D., 817
Smith, G. E., 193–194, 196–197, 211, 214–215
Smith, P. K., 79, 88, 624
Smyth, D. R., 252
Snyder, M. A., 387–402
Sobel, R. E., 742
Sobrado, P., 652
Sokolovsky, M., 741
Song, J., 267
Sorensen, H. P., 265
Sorensen, S. P. L., 50, 53, 55
Soufi, B., 731
Sowell, J., 582
Spanos, N., 394
Speers, A. E., 667
Spencer, D. M., 286
Spencer, J. F. T., 286
Spiess, C., 133
Spinola, S. M., 586
Spirin, A. S., 649, 669
Spiro, R. G., 737
Spriestersbach, A., 439–468
Sridhar, P., 199
Stafford, W. F., 720
Stalder, E. S., 475–493
Stark, G. R., 819
Starkenstein, E., 419
Stasyk, T., 447
Steele, J. H. C., 703
Steen, H., 733, 736
Steen, J., 461
Steinberg, T. H., 541–559, 729, 757
Stellwagen, E., 373–385, 427
Stemmann, O., 751
Stettler, M., 232
Stevens, R. C., 286–287
Stock, A. M., 557
Stoll, S. V., 43–56
Storrie, B., 311, 314
Stoscheck, C. M., 79, 83
Stowers, A. W., 458
Stránská, J., 401
Strausberg, R. L., 135
Strausberg, S. L., 135
Strömberg, P., 461
Strominger, J. L., 428
Studier, F. W., 154, 156, 160, 263, 265, 777
Stuhfelder, C., 401
Suehara, Y., 526
Sugyiama, K., 137
Sulkowski, E., 440
Summers, M. D., 193, 196, 211, 215
Sun, L., 579
Sun, X., 229, 447
Sunga, J., 169–187
Surzycki, S., 292
Sutton, V. R., 789–790, 792, 799, 802–803
Svensson, H., 518
Svensson, J., 450
Swaney, D. L., 747, 749–750
Swart, R., 101
Swartz, J. R., 649, 669
Swiderek, K. M., 289
Swietnicki, W., 262, 276
Syka, J. E., 729, 749
Sykora, C., 732

T

Tait, A. S., 234
Takayama, Y., 164
Tanaka, K., 637
Tanford, C., 407–408
Tang, W. H., 730
Taniguchi, C. M., 726, 731
Tao, W. A., 754
Tate, C. G., 632
Tempst, P., 444
Tennikova, T. B., 355
Terp, K., 150, 156, 243
Thao, S., 652
Thatcher, D. R., 262
Thill, E. R., 185
Thingholm, T. E., 733
Thomas, C. J., 208
Thompson, A. R., 388
Thompson, E. B., 326
Thompson, N., 265, 267
Thompson, N. E., 475–493
Thomsen, D. R., 137
Tian, Y., 738
Tiffany, H. L., 500
Tigges, M., 234
Timasheff, S. N., 390, 692, 694, 704, 716
Timmons, T. M., 498
Tinoco, I., 697
Tipsmark, C. K., 575
Tiselius, A., 388, 399
Tolstorukov, I., 169–187
Tolun, A. A., 249
Tombs, M. P., 13
Tomiya, N., 210
Tonge, R., 556
Topf, M., 694, 718
Towbin, H., 574
Townend, R., 694, 716
Trésaugues, L., 271–272, 274

Trelle, M. B., 742
Trimble, R. B., 136
Tropea, J. E., 252, 782
Tsaprailis, G., 747
Tsien, R. Y., 492
Tsumoto, K., 262
Tucker, J., 463, 634–635, 637–641
Tyler, R. C., 651, 662, 669

U

Udeshi, N. D., 748–750
Uhlén, M., 461
Ulbrich, V., 409
Ulbrich-Hofmann, R., 531
Unger, T., 249
Unlu, M., 516, 524–525
Unwin, R. D., 755
Ura, K., 726
Urh, M., 417–435
Urlaub, G., 226

V

Vaillancourt, P., 249
Vallejo, L. F., 262, 267, 269, 272
van Bruggen, E. J. F., 718
Van Craenenbroeck, K., 226
van Ooyen, A. J., 171
van Veldhoven, P. P., 313
Vancan, S., 447
Vandecasteele, J. P., 788
Varshavsky, A., 24
Vattem, K., 573–598
Vaughn, J. L., 211
Veldkamp, C. T., 262, 276
Ventura, S., 260
Vera, A., 133, 265, 278
Vestal, M. L., 745
Vesterberg, O., 518–519
Vialard, J. E., 198, 200
Vicennati, P., 229
Vijayendran, R. A., 432
Villaverde, A., 260
Vinarov, D. A., 649–651, 662–663
Vincentelli, R., 271–272, 274
Vithayarhil, P. J., 248
Volkin, D. B., 815
von Hagen, J., 305–327
von Hippel, P. H., 23, 83
von Jagow, G., 516

W

Waddell, W. J., 13
Waggoner, A. S., 556
Wagner, S., 633
Waldo, G. S., 244, 774
Walker, G. R., 575
Walker, J. E., 154
Walker, J. M., 543
Walpole, G. S., 51
Walsh, C. M., 731
Walsh, P. S., 444, 448
Wang, K. H., 245
Warne, T., 634–636
Washburn, M. P., 307, 310
Waters, N. J., 107
Waugh, D. S., 134, 155, 243, 249, 774
Waziri, S. M., 107
Weber, G., 307, 310
Weber, K., 499, 566
Weigel, P. H., 144
Weiß, H. M., 632, 634–636
Weissman, A. M., 738
Weissmann, C., 637, 642
Wells, K. S., 733, 738
Wen, J., 718
Wertz, D. H., 408
Wessel, D., 117, 531
Westermeier, R., 515–538, 547
Westoby, M., 97–118
Wetlaufer, D. B., 272
Wheat, T. E., 101
White, F., 731–732, 753
White, J. F., 632, 634–638, 640–641
Wiebenga, E. H., 718
Wiechelman, K. J., 88
Wiener, M. C., 449, 463
Wiggs, J., 569
Wilchek, M., 432
Wilkinson, D. L., 25
Willemsen, O., 310
Williamson, B. L., 755
Willis, M. S., 271–272
Wilson, C. M., 509
Wilson, I. A., 286
Wilt, F. H., 133
Witt, H. T., 615
Woestenenk, E. A., 451
Wolf, B. D., 555
Wolf-Yadlin, A., 755–756
Womack, M. D., 668
Woycechowsky, K. J., 270
Woychik, N. A., 477
Wrobel, R. L., 647–670
Wu, A. M., 87
Wu, C. C., 667–668, 744
Wu, J. C., 87
Wurm, F. M., 138
Wuthrich, K., 615
Wuu, J. J., 649
Wyckoff, M., 501, 506

X

Xia, W., 226–227
Xie, Y., 272
Xing, C., 578
Xu, C.-G., 451

Y

Yamada, S., 557
Yamamoto, S., 352, 354, 361, 365, 412
Yan, A., 787–803
Yang, J. T., 693, 711
Yang, X. J., 738, 741
Yeliseev, A. A., 633, 635
Yeo, W. S., 740–741
Yeung, K. K. C., 103
Yik, J. H. N., 144
Yoch, D. C., 788
Yokoyama, S., 243, 649
You, W. W., 79, 89, 91
Young, J. C., 133
Yphantis, D. A., 698, 701
Yue, S. T., 552

Z

Zachariou, M., 419
Zacharius, R. M., 558
Zagariya, A., 324
Zappacosta, F., 737
Zeheb, R., 328
Zeman, L., 108, 111
Zerbs, S., 149–166
Zerner, B., 83
Zhan, X., 741
Zhang, J., 224–226
Zhang, Y. B., 251, 728–729, 741–743, 746, 751, 754
Zhao, K., 417–435
Zhaohua, H., 445
Zhou, H., 444
Zhu, B., 153
Zhu, H., 286
Zhu, X. D., 388
Zimmerman, J. M., 408
Zydney, A. L., 108, 111

Subject Index

A

Absolute quantitation (AQUA), 751
A431 cells (ATCC), 590
Acetate buffer, 51
Affinity chromatography
  amine reactive linkages, 432
  covalent attachment, 430
  ligand selection
    characterization, 424–425
    covalent affinity chromatography, 428–429
    de novo development, 423
    group specific ligands, 425
    lectins, 426–427
    specificity, 426
  matrix selection
    examples of, 421
    magnetic affinity beads, 422–423
    pore size, 419
    Renkin equation, 420
    specific ligand, 420
    stability, 422
  purification method
    elution, 434–435
    GPCRs, 634–636
    sample preparation, 433
  surface activation, 430
Affinity tags
  basis, 243
  calmodulin-binding peptide (CBP)-tag, 248–249
  epitope, 247–248
  GST, 246–247
  His-tag, 245–246
  S-tag, 248
  STREP-II, 248
Alkaline copper reduction assays. *See* Lowry assay
Amine derivatization assay, 89–91
Amino acid sequence. *See* Bioinformatics, protein purification
2-Amino-2-methyl-1,3-propanediol (ammediol) buffer, 54–55
Amino-terminal epitope tags, 739
Ammonium sulfate (AS) precipitation
  concentration, 335
  principles, 332–334
  problems/solutions, 337
  procedure, 334
  solubility curve, 333
  test, 336
Ampholines®, 519
Analytical size exclusion chromatography (ANSEC), 777
Antibody-affinity chromatography, 627
Arginine, 273
Assay selection quantitation, 75–76
Autoinduction media protocol, 160

B

Bacterial systems, heterologous proteins production
  analysis
    autoinduction method, 159–160
    expression, 159
    solubility, 158–159
  autoinduction media protocol, 160
  cloning
    expression hosts and vectors, 153–155
    ligation independent cloning (LIC) methods, 153
  cytoplasmic expression, 156
  cytoplasmic targets expression and solubility, 157
  Escherichia coli, 150–151
  gram-negative and positive bacterium, 164–165
  osmotic shock protocol, 162–164
  periplasmic targets expression, 160–162
  planning and sequential steps, 151–152
  production scale, 166
  project requirements evaluation, 152
  target analysis, 152–153
  T4 DNA polymerase-treated DNA fragments, 155–156
  vector and induction conditions, 165
Bacteria lysis, 289
BaculoDirect approach, 205–206
Baculovirus–insect cell expression systems
  baculovirus biology and molecular biology, 193–195
  basic protocols
    genomic DNA isolation, 215–216
    insect cell maintenance, 215
    plaque assays, 216–218
  conditions and methods, 211–212
  expression vector, 195–196
  foreign gene deliver and protein encoding, 211

Baculovirus–insect cell expression systems (*cont.*)
genome modifications
accessory genes, 208
BaculoDirect approach, 205–206
β-galactosidase substrate, 207
*flash*BAC approach, 204
linearized/gapped parental viral genome, 201–202
linearized parental viral genome, 200–201
nonessential viral genes, 210
recombinant protein production, 209
transfer plasmid/bacmid, 203
viral chitinase gene, 208–209
insect cell hosts, new generation of
eukaryotic protein process, 212
foreign glycoproteins, 214
transgenic lepidopteran, 213
modification categories, 198
transfer plasmid modifications, 198–199
vector technology
homologous recombination, 197
transfer plasmid, 196–197
Baculovirus/insect cells, recombinant protein, 136–137
Barbital buffer, 53–54
BCA assay. *See* Bicinchoninic acid assay
Bead homogenization, 291
Bead milling, 291
BES. *See* N,N-Bis(2-hydroxyethyl)-2-aminoethanesulfonic acid
Bicinchoninic acid (BCA) assay
procedure and comments, 89
reagents, 88
Bilayer reaction *vs.* dialysis reaction, 662–663
Binding conditions, ion-exchange chromatography
capacity *vs.* concentration, protein, 359
cation exchangers, 356
column volume, 358
isoelectric point, 358, 360
ligand density, 355
prepacked minicolumns, 360
Biogrammatics expression vector map, 174–175
Bioinformatics, protein purification
amino acid sequence
charge *vs.* pH, 22–23
cysteine content, 23–24
E. coli overexpression, solubility, 25
homology and cofactor affinity, 24–25
hydrophobicity and membrane-spanning regions, 24
molar extinction coefficient, 23
polypeptide chain molecular weight, 22
potential modification sites, 25
secondary structure, 24
stability, 24
titration curve, isoelectric point, 22–23
critical problems, 26
denatured state, 27
resources, 27
Biological extracts preparation
cell disruption
buffer composition, 293–294, 296
chemical and enzymes, 287–290
mechanical, 290–292
E. coli cell lysis
gram-scale mechanical disruption, 299–300
microscale protocols, 296–299
Bio-Rad whole gel eluter, 570
N,N-Bis(2-hydroxyethyl)-2-aminoethanesulfonic acid (BES), 390, 401
Blotting and optimization protocols, chemiluminescent substrate
antigen concentration, 596
detection method, 598
enzyme conjugate concentration, 597–598
membrane blocking and washing, 596–597
primary antibody concentration, 597
Western blotting protocol
electrophoretic transfer, 594
primary and secondary antibody, 594
proteins separation, 593
stripping protocol, 595–596
Borax–NaOH buffer, 55
Boric acid–borax buffer, 54
Bradford protein assay. *See* Coomassie blue assay
Broad-range buffers, 50
Buffer capacity (β) *vs.* ΔpH, 45
Buffers, principles and practice
broad-range buffers, 50
preparation, 48–49
selection
pH optimum, 45
pK values, 45–47
temperature-sensitive, 48
stock solutions, recipes
acetate, 51
2-amino-2-methyl-1,3-propanediol (ammediol), 54–55
barbital, 53–54
borax–NaOH, 55
boric acid–borax, 54
cacodylate, 52–53
carbonate–bicarbonate, 55–56
citrate, 50–51
citrate-phosphate, 51–52
glycine–HCl, 50
glycine–NaOH, 55
phosphate, 53
succinate, 52
tris(hydroxymethyl)aminomethane (tris), 54
theory
buffer capacity (β) *vs.* ΔpH, 45
Henderson–Hasselbalch equation, 44
volatile buffers, 49–50
Bulk/batch purifications procedures, 17

C

Cacodylate buffer, 52–53
Calcium-phosphate-mediated transfection, 229
Calmodulin-binding peptide (CBP)-tag, 248–249
Carbonate–bicarbonate buffer, 55–56
Carboxymethyl aspartate (CM-Asp), 442
Catalytic activity, principles
  cell-free translation, 667–668
  chemical kinetics
    basic kinetic mechanism, 58–59
    first-order rate constant, 60–61
    rate law, order, and rate constant, *k*, 59–60
    reactant concentration, 61–62
    reaction velocity, 58, 60
  enzyme kinetics
    initial velocity, 64
    Michaelis constant, 63–64
    steady-state approximation, 63
    turnover number, 62
Cell disruption
  buffer composition
    basic criteria, 294
    protease inhibitors, 295
    volume, 294, 296
  chemical and enzymes, 287–290
    bacteria lysis, 289
    high-throughput (HT) extraction, 287
    lytic enzyme treatments, 290
    reagents, 288
    yeast lysis, 289
  E. coli cell lysis, 296–300
  high-pressure homogenizers mechanism, 290–292
Cell-free translation, integral membrane proteins
  density gradient ultracentrifugation, 665–667
  expression vectors, 651–652
  flexivector and PCR product digestion reaction
    protocol, 657
    reagents, 656
  gene cloning
    materials and reagents, 654
    protocol, 654–655
    sacB-CAT cassette, 652–653
  isotopic labeling, 669
  ligation reaction, 657–658
  liposomes preparation
    materials and reagents, 661
    protocol, 661–662
  mRNA preparation
    protocol, 661
    reagents, 660
  PCR product cleanup
    materials and reagents, 655–656
    protocol, 656
  plasmid DNA purification
    protocol, 659–660
    reagents, 659
  proteoliposomes characterization, 667–668
  scale-up considerations, 669
  steps, 650–651
  transformation reaction
    materials and reagents, 658
    protocol, 658–659
  wheat germ translation reaction
    bilayer reaction protocol, 664
    bilayer reaction *vs.* dialysis reaction, 662–663
    dialysis reaction protocol, 664–665
    materials and reagents, 663–664
Ceramic fluoroapatite (CFT), 391
Ceramic hydroxyapatite type I (CHT I), 391, 397
CFT. *See* Ceramic fluoroapatite
Chaotropic agents, 273
Charged-coupled device (CCD) cameras, 575
Chemical additives, 273
Chemical and enzymatic cell disruption
  bacteria lysis, 289
  high-throughput (HT) extraction, 287
  lytic enzyme treatments, 290
  reagents and enzymes, 288
  yeast lysis, 289
Chemiluminescence signal
  signal capture, 583–584
  signal intensity and duration, 584–585
  Western blot optimizing
    antibodies, 586–587
    blotting membrane, 585
    detection method, 587–588
    signal reduction, chemical treatment, 587
    target protein, 585–586
CHO cell lines. *See* HEK293 cell lines
CHT I. *See* Ceramic hydroxyapatite type I
Citrate-phosphate buffer, 50–52
Cleavage sites, 245
Click chemistry reagents, 433, 559
Cloning, bacterial systems
  expression hosts and vectors, 153–155
  ligation independent cloning (LIC) methods, 153
Cloud point and phase separation, 611–612
CMC. *See* Critical micellization concentration
CNBr. *See* Cyanogen bromide
Collision induced dissociation (CID), 747
Column chromatography, 708
Conacanavalin A (ConA), lectins, 427
Continuous assays, 65–67
Coomassie blue assay
  observation, 86
  procedure, 85–86
  reagents, 85
Coomassie brilliant blue (CBB), 546–547

Copurifying proteins, IMAC
affinity precipitation, 457
bind–wash–elute procedure, 456
chelating amino acids, 455
removal of, 459
size exclusion chromatography (SEC) columns, 455
Core histone purification method, 324–325
Covalent affinity chromatography, 428–429
Critical micellization concentration (CMC), 609–610
Crystallization, 118
Culture media and microbial manipulation techniques, 171–172
Cyanogen bromide (CNBr), 487
Cytoplasmic proteins, expressing method, 140–141
Cytoplasmic targets expression and solubility, 157

D

DDF. *See* Differential detergent fractionation
Denaturing PAGE analysis, cell-free translation, 665–666
Density velocity centrifugation, 311
advantage, 308
mechanical homogenate, 310
procedure, 309–310
proteome complexity reduction
free-flow electrophoresis, 310
multidimensional protein identification technology (MudPIT), 310
Detection methods, immunodetection
assay requirements, 39–40
chemiluminescence detection, 582–583
colorimetric detection, 581
enzyme conjugate, 579
fluorescence detection
single blot and high sensitivity, 581
spectral properties, 582
Detergent-resistant fractionation. *See* Lipid raft enrichment
Detergents, 273
fluorescent detection, 91
GPCRs analysis, 636–637
lipid–receptor, 633
mass spectrometry and nuclear magnetic resonance (NMR) measurements, 615
membrane protein suitable detergent, 613–614
optical spectroscopy, 614
phase diagrams and critical micellization concentration, 609–610
physicochemical parameters, 612–613
properties, 605–609
protein crystallization, 615
reagent lysis, 296–298
removal and exchange, 613
solubility, 611–612
structure, 604–605
temperature effects, 611–612
Dialysis
affinity binding applications and concentration, 107
buffer concentration, 104–105
mechanical design, 106
molecular weight cutoff (MWCO), 104
Difference gel electrophoresis (DIGE), 525–526
Differential detergent fractionation (DDF)
extraction efficacy, 317
procedure, 319–320
reagents and equipment, 318–319
Diffusion coefficient, 703
Diffusion, protein elution
Bio-Rad whole gel eluter, 570
electrophoretic elution, 570–571
gel preparation, 567
location, 567
renaturation, 569
reverse phase HPLC, 570
SDS removal and concentration, 568–569
Discontinuous assays, 67–69
Disposable Wave™ bioreactor, 232
Dithionite (DTH), 790, 793
Dithiothreitol (DTT), 793, 796
Divalent cations, 272–273
DNA preparation, *Pichia pastoris* expression
electroporation procedure, 178–179
protocol, 177–178
solutions prepare, 177
Drip dilution, 267–268
Dye-based protein assay, 75, 80

E

EBNA-1 gene, 224
E. coli cell lysis
gram-scale mechanical disruption
high-pressure homogenization, 300
sonication, 299–300
microscale protocols
detergent-based reagent method, 296–298
freeze-thaw enzymatic method, 298–299
recombinant protein expressing method
disulfide bond formation, 134
fusion partners, 134
inclusion bodies, 133
posttranslational modifications, 134
temperature and molecular chaperones, 133
Electron-capture dissociation (ECD), 747–748
Electron microscopy, 718
Electron transfer dissociation (ETD), 748
Electrophoretic elution, 570–571
Electrophoretic method, 103–104
principle, 709
protein purity determination
contaminants, 681
front-line, 682

gradient gels, 682
isoelectric focusing gel, 683
protein stains, 684
SDS gels, 710
Electroporation procedure and parameters, 178–179
Electrospray ionization (ESI), 744–745
ELISA-elution assay, 480–482
Elution volume (E_v), 707
Endomembrane enrichment
enzyme and substrate markers, 315
material and equipment, 315–316
procedure, 316
workflow, 314
Enrichment techniques
glycosylation
lectin affinity chromatography, 738
O-*β*-*N*-acetylglucosamine (O-GlcNAc), 737
phosphorylation
^{32}P-labeled ATP, 731
mass spectrometry-based proteomics, 731–737
ubiquitination and sumoylation
amino-terminal epitope tags, 739
characteristic mass shifts, 740
Enzyme activity measurement
catalytic activity, principles
chemical kinetics, 58–62
enzyme kinetics, 62–64
continuous assays, 65–67
discontinuous assays, 67–69
reaction assay mixtures formulation, 69–70
Enzyme and substrate markers, 315
Enzyme-catalyzed product, 69
Enzymes purification, 1–5
Epicocconone, 554
Epitope tags, 247–248
Epstein–Barr virus nuclear antigen 1 (EBNA1), 341
Ethanol and acetone precipitation, 341
Expression and purification, parallel methods
analytical testing, proteins
ANSEC, 777
methods, 776
cloning strategies
gel electrophoresis, 771
ligation-independent-cloning (LIC), 769
parallel cloning process, 770
polymerase incomplete primer extension (PIPE), 769
restriction mapping, 771
SYBR results, 772
large-scale parallel expression
airlift fermentor, 778
custom automation, 778–779
digestion buffer, 781
protein expression identification, 767
small-scale expression screening
analytical methods, 776
benzonase, 775
capillary electrophoresis result, 774
Thermofluor ligand, 772, 774
Expression vector technology
homologous recombination, 197
transfer plasmid, 196–197

F

Field-flow fractionation (FFF), 711–712
First-order rate constant, 60–61
FLAG-tag, 247–248
*flash*BAC approach, 204
Flash dilution, 267
Flexivector and PCR product digestion reaction, 656–657
Fluorescent total protein stains
detection sensitivity, 549–550
epicocconone, 554
flamingo and krypton protein gel stain, 555
fluorescein derivatives, 555
LUCY protein gel stains, 555
SYPRO stains
carbazolylvinyl stain, 553
organometallic ruthenium dyes, 553
protein–SDS–dye complex, 552
sulfopropylaminostyryl, 552
[4Fe-4S]-FNR, O_2-sensitive proteins isolation
anaerobic glove-box vinyl chamber, 790
buffer composition, 793
cell lysate preparation, 792–793
dithionite (DTH), 790, 793
dithiothreitol (DTT), 793, 796
57Fe labeled 4Fe-FNR purification, 798–799
Fourier transform ion cyclotron resonance (FT-ICR), 799
FPLC purification system, 791
ICP-MS analysis, 801–802
low-temperature EPR, 802–803
Mössbauer spectroscopic analysis, 802
PK872, *E. coli* strain, 791
protein purification, 793–794
protocols, 793–797
sparging station, 790
trace RNAses removal, 797–798
UV–visible spectrophotometric analysis, 802
Foreign gene deliver and protein encoding, 211
Forgotten artifacts, protein biochemistry
buffers
reducing agent *vs.* oxidizing agent, 816–817
sulfonylethyl buffers, 817
chemical leaching, plasticware, 819
chromatography, EDTA
anion-exchange columns, 817–818
Ni-chelate affinity column, 818
cyanate, urea, 819

Forgotten artifacts, protein biochemistry (*cont.*)
  protein absorption, filtration, 818–819
  SDS gel electrophoresis
    Asp–Pro bond cleavage, 815
    keratin contamination, 816
    proteases, 815
Fourier transform ion cyclotron resonance (FT-ICR), 799
Fractionation requirements, laboratory setting, 40–42
Free-flow electrophoresis, 310
Freeze-thaw plus enzymatic lysis, 298–299
Frictional coefficient ratio (f/f^0), 703–704
Fusion protein. *See* Protein expression tagging

G

Gel filtration
  conventional matrix
    absorbance, 383
    column, 382
  desired protein disappearance, 385
  effluent fractions, 381
  hydrodynamic diameter, 374
  low flow rate, 384
  matrices
    chromatograhic solvents, 380
    conventional, 376
    high-performance, 376
    packed column, 379
    parameters, 377
    powdered and suspended parameters, 378–379
    sample preparation, 380
  resolution, 384
  scaling upward, 383
  semilogarithmic plot, 375
  size-exclusion matrix, 375, 381
  skewed peaks, 384–385
  zero elution volume, 374
Gel-filtration chromatography, 99
  method
    column chromatography, 708
    elution and void volume, 707
  problems and pitfalls, 708
  stationary gel volume fraction, 706
Gene cloning, cell-free translation, 654–655
Genetic strain construct, *Pichia pastoris*, 172–174
Genome modifications, baculovirus
  accessory genes, 208
  BaculoDirect approach, 205–206
  *flash*BAC approach, 204
  β-galactosidase substrate, 207
  linearized/gapped parental viral genome, 201–202
  linearized parental viral genome, 200–201
  nonessential viral genes, 210
  recombinant protein production, 209
  transfer plasmid/bacmid, 203
  viral chitinase gene, 208–209
Genomic DNA isolation, baculovirus, 215–216
Glutathione-S-transferase (GST) tag, 246–247
Glycerol, 273
Glycine
  HCl buffer, 50
  NaOH buffer, 55
Glycoprotein and phosphoprotein detection
  acid fuchsin dye, 558
  glycosylation state, 558–559
  O-GlcNAc detection, 559
  Phos-tag phosphoprotein stains, 557
  Pro-Q diamond phosphoprotein gel stain, 557
Glycosylation
  lectin affinity chromatography, 738
  O-β-*N*-acetylglucosamine (O-GlcNAc), 737
GPCRs stability, detergent solution, 634
G-protein-coupled receptors (GPCRs). *See* Recombinant G-protein-coupled receptors
Gram-scale cell disruption. *See* Mechanical cell disruption
Green fluorescent protein (GFP), 492
GuHCl, 266

H

Halotag technology, 251, 428–429
HEK293 cell lines
  cultivation, 228
  EBNA-1 gene, 224
  expression vectors, 226–227
  HKB-11 cell line, 226
  overview, 225–226
  SV40 enhancer, 224, 226
  transfection methods, 228–234
Heterologous protein production, *E. coli*, 150–151
HIC. *See* Hydrophobic interaction chromatography
High-performance liquid chromatography (HPLC), 67
High-pressure homogenization lysis method, 300
High-pressure homogenizers, 291–292
High-resolution ion-exchange chromatography, 268–269
High-throughput (HT) cell extraction, 287
His tagged protein purification, IMAC, 245–246
  applications
    detection and immobilization, 445–446
    protein fractions purification, 446–448
  carboxymethyl aspartate (CM-Asp), 442
  chemical compatibility, 452–453
  copurifying proteins
    affinity precipitation, 457
    bind–wash–elute procedure, 456
    chelating amino acids, 455
    removal of, 459

size exclusion chromatography (SEC) columns, 455
detergent screening, 462
iminodiacetic acid (IDA), 440, 442
industrial-scale protein production, 458–460
interaction model, 441
membrane protein purification, 461–463
metal-ion stripping, 467
nitrilotriacetic acid (NTA), 440–441
protein expression, 448–449
protein purification protocols
denaturing conditions, 466
native conditions, 465
reducing conditions, 454
sanitization, 466–467
sequences, 449
n-terminal tags, 449
zinc-finger proteins purification, 463–464
HKB-11 cell line, 226
Homologous recombination vector technology, 197
Homology and cofactor affinity, 24–25
Hydrophobic and affinity chromatography, 100–102
Hydrophobic interaction chromatography (HIC)
adsorbents
choice of, 409
feed/load preparation, 410
gradient elutions, 412
isocratic elution, 412–413
$\pi-\pi$ interactions, 409
preparation, 410
product elutions, 410–412
properties of, 408
regeneration and sanitization, 413
chaotropes, 407
Hofmeister series, 407
lyotropes, 407
polarity, 407
proteins *vs.* hydrophobic ligand, 406
Hydroxyapatite (HA) columns, protein chromatography
BES, 390, 401
calcium ions, 396
chemical characteristics
protein separation, 392
pyrophosphate (PPi), 392
effluent pH, 395
hydroxyapatite sources, 399
mechanism
CHT I and CFT, elution time, 391
desorption, 390
P-and C-sites, 389
zero net charge, 389
MES, 394
metal adsorption, 393–394
phosphate gradient, 401
process-scale columns, 399–401
rEPO, 401
solubility constant, 394
voids, HA solubility, 394

I

Iminodiacetic acid (IDA), 440, 442
Immobilized metal affinity chromatography (IMAC), 732–733
applications
detection and immobilization, 445–446
protein fractions purification, 446–448
carboxymethyl aspartate (CM-Asp), 442
chemical compatibility, 452–453
copurifying proteins
affinity precipitation, 457
bind–wash–elute procedure, 456
chelating amino acids, 455
removal of, 459
size exclusion chromatography (SEC) columns, 455
detergent screening, 462
His tagged protein purification
protein expression, 448–449
sequences, 449
N-termianl tags, 449
iminodiacetic acid (IDA), 440, 442
industrial-scale protein production, 458–460
interaction model, 441
media
His-tag, 245–246
tag use, 243
membrane protein purification, 461–463
metal-ion stripping, 467
nitrilotriacetic acid (NTA), 440–441
protein purification protocols
denaturing conditions, 466
native conditions, 465
reducing conditions, 454
sanitization, 466–467
zinc-finger proteins purification, 463–464
Immobilized pH gradient (IPG), 519–520
Immunoaffinity reagents enrichment
antibodies, 326
material and equipment, 326
procedure, 326–327
Immunodetection
blotting and optimization protocols, chemiluminescent substrate
antigen concentration, 596
detection method, 598
enzyme conjugate concentration, 597–598
membrane blocking and washing, 596–597
primary antibody concentration, 597
Western blotting protocol, 593–596
chemiluminescence signal
signal capture, 583–584
signal intensity and duration, 584–585

Immunodetection (*cont.*)
Western blot optimizing, 585–588
detection methods
chemiluminescence detection, 582–583
colorimetric detection, 581
enzyme conjugate, 579
fluorescence detection, 581–582
problems and explanations
bands/entire blot glowing, darkroom, 590
brown/yellow bands, 589
ghost/hollow bands, 590–593
high background, 589
no signal, 588
signal fades quickly, 588–589
substrate different, 589
Western blotting
advances, 575
charged-coupled device (CCD) cameras, 575
direct and indirect, 575–577
far-Western, 577
semiquantitative, 577–580
Inclusion body (IB) proteins, 241
characterization, 270
disulfide bonds
formation methods, 269–270
reoxidation, 269
folding process, 261
high-resolution ion-exchange chromatography, 268–269
native state, 260
overexpression, 264–265
procedure steps, 262
protocol steps, 263–264
refolding
alkaline pH shift, 276–277
catalysts, 276
characterization, 270
database, 277
dilute ways, 267–268
high-pressure, 276
on-column, 275–276
problem, 262
rational approach, 274–275
systematic screens, 271
variables, 271–274
solubilizing, 265–267
washing, 265
Inductively coupled plasma mass spectrometry (ICP-MS), 801–802
Insect cell hosts, baculovirus
eukaryotic protein processing, 212
foreign glycoproteins, 214
transgenic lepidopteran, 213
Insect cell maintenance, 215
$\pi-\pi$ Interactions, 409
Ion-exchange chromatography, 99
binding conditions
capacity *vs.* concentration, protein, 359
cation exchangers, 356
column volume, 358
isoelectric point, 358, 360
ligand density, 355
prepacked minicolumns, 360
column operation conditions
equilibration, 363
gradient volume, 364
oxidizing agents, 366
pH gradient elution, 365
salt loading, 363
complex protein mixture separation
cation exchange resins, 366
milk, pretreatment, 366
elution conditions
diplacement chromatography, 361
α-lactoglobulin isoform separation, 362
resolution, 362
monolithic column, high-resolution separation
bovine whey proteins purification, 368
MP production, 367
pH *vs.* concentration gradient, 370
Superdex 200, 369
principle
isocratic elution, 352
pH gradients, 352
protein retention *vs.* salt concentration, 353
salting-in effect, 350
stationary phases
binding site *vs.* molecular mass, 354
monoliths, 355
pore size, 354
Ion trap mass spectrometers, 744
Isoelectric focusing (IEF)
high resolution 2D gel, 517
pH gradients types
carrier ampholytes, 518–519
different configurations, 521
IPG strip technology, 519–520
principle, 517–518
Isoelectric precipitation, 341
Isotope-coded affinity tags (ICAT), 752–753
Isotope tags for relative and absolute quantification (iTRAQ), 753–754
Isotopic labeling, cell-free translation, 669

K

Krafft point, 611–612

L

Laboratory setting
chemicals, 38
detection and assay requirements, 39–40
disposables, 38
equipment and apparatus, 39
fractionation requirements, 40–42

glassware and plasticware, 38
small equipment and accessories, 39
β-Lactamase expression, 181
Lectin affinity chromatography, 626–627, 738
Lectins, ligand selection, 426–427
Ligand-affinity chromatograph, 627
Ligation independent cloning (LIC) methods, 153
Ligation reaction, cell-free translation, 657–658
Linear map of pEU-HSCB, 652–653
Lipid raft enrichment
material and equipment, 320
mild detergents, 321
procedure, 322
Lipofection. *See* Small-scale transfection methods
Liposomes preparation, 661–662
LOPIT. *See* Endomembrane enrichment
Loss of activity, protein stability, 125–127
Lowry assay
comments, 88
procedure, 87
reagents, 87
Lyophilization, 114–115

M

mAb 8RB13, 490
Maltose-binding protein (MBP) tag, 249–250
Mammalian expression methods, 138–139
Manganese peroxidase (MP) production, 367
Marker proteins, 510–511
Mass spectrometry
analysis
electrospray ionization (ESI), 744–745
features, 744
matrix assisted laser desorption ionization (MALDI), 746
detergents, 615
methods, 687–688, 713–715
Matrices, gel filtration
chromatograhic solvents, 380
conventional, 376
high-performance, 376
packed column, 379
parameters, 377
powdered and suspended parameters, 378–379
sample preparation, 380
Matrix assisted laser desorption ionization (MALDI), 746
Mechanical cell disruption
instrumentation
bead milling, 291
microfluidics microfluidizer, 292
mechanisms, 290–291
Medium-scale transfection methods, 231–232
Membrane and secreted proteins, expressing method, 141
Membrane fractionation methods, 621
Membrane proteins purification
antibody-affinity chromatography, 627
detergent removal and exchange, 627–628
lectin-affinity chromatography, 626–627
ligand-affinity chromatograph, 627
precautions, 625–626
preparation
advantage, 622
membrane fractionation methods, 621
plasma membranes isolation, 620
recombinant integral membrane proteins, expression and purification, 628–629
solubilization, native membrane proteins
detergents choice, 623
preparing membrane fractions, 624–625
screening, 624
Membranes preparation
advantage, 622
membrane fractionation methods, 621
plasma membranes isolation, 620
2-Mercaptoethanol/dithiothreitol, 499
MES. *See* 4-Morpholinepropanesulfonic acid
Metalloproteome, IMAC, 447
Methylation and acetylation
antibodies, 742
histone, 741
immonium ions, 742–743
Michaelis constant, 63–64
Microfluidics microfluidizer, 292
Micro-scale cell disruption. *See* Chemical and enzymatic cell disruption
Mitochondria enrichment
density gradient formulations, 313
drawback, 311
material and equipment, 312
procedure, 312–313
Molar absorbance extinction coefficients (ε), 65
Molar extinction coefficient, 23
4-Morpholinepropanesulfonic acid (MES), cobuffer, 394–396
Mössbauer spectroscopic analysis, 798–799, 802
mRNA preparation, cell-free translation, 660–661
MudPIT. *See* Multidimensional protein identification technology
Multidimensional protein identification technology (MudPIT), 310
Multiple angle light scattering (MALS), 688–689
Multiple reaction monitoring (MRM), 755
Multiple tags. *See* Tandem tags
Multisubunit features, critical problems, 26
M11V3 cultivation medium
large-scale transfection, 232–233
medium scale transfection, 231–232

N

Native membrane proteins solubilization, 623
Nitrilotriacetic acid (NTA), 440–441
Nitrosative protein modifications, 740–741

Nonionic polymer precipitation. *See* Polyethylene glycol precipitation
Nonultracentrifugation-based organelle enrichment technique. *See* Differential detergent fractionation (DDF)
NTS1 fusion protein
analysis, 641–642
purification
immobilized metal affinity chromatography, 640
neurotensin column, 640–641
solubilization, 639–640
Nuclear extract enrichment, 325
Nuclear magnetic resonance (NMR) measurements, 615
Nuclei and histone enrichment
core histone purification, 324–325
washed nuclear pellet
materials and equipment, 323
mechanism, 322
procedure, 323–324
Nucleic acid removal, affi-gel blue
Cibacron Blue F3GA dye, 344
column buffer, 345
NusA tag, 250

O

O-β-*N*-acetylglucosamine (O-GlcNAc), 737
One-dimensional gel electrophoresis
marker proteins, 510–511
2-mercaptoethanol/dithiothreitol, 499
molecular weight determination, 511
polyacrylamide gels, 500
preparative electrophoresis, 511–513
principle, 501
protein detection, gel
Coomassie Brilliant Blue R-248, 508–509
copper staining, 510
silver staining, 509–510
SDS–PAGE procedure
casting gels, 503–505
catalyst, 503
electrode buffer, 503
electrophoresis, 505–506
2-mercaptoethanol, 507
Ornstein–Davis system, 506
sample preparation, 505
stock solutions, 502
Optical spectroscopy, detergents, 614
Ornstein–Davis system, 506
O_2-sensitive proteins isolation
dithionite (DTH), 790, 793
dithiothreitol (DTT), 793, 796
4Fe-FNR, anaerobic isolation
anaerobic glove-box vinyl chamber, 790
buffer composition, 793
cell lysate preparation, 792–793
57Fe labeled 4Fe-FNR purification, 798–799
FPLC purification system, 791
PK872, *E. coli* strain, 791
protein purification, 793–794
protocols, 793–797
sparging station, 790
trace RNAses removal, 797–798
$[4Fe–4S]^{2+}$ cluster characterization
Fourier transform ion cyclotron resonance (FT-ICR), 799
ICP-MS analysis, 801–802
low-temperature EPR, 802–803
Mössbauer spectroscopic analysis, 802
UV–visible spectrophotometric analysis, 802
Osmotic shock protocol, 162–164

P

Parallel methods, expression and purification
analytical testing, proteins
ANSEC, 777
methods, 776
cloning strategies
gel electrophoresis, 771
ligation-independent-cloning (LIC), 769
parallel cloning process, 770
polymerase incomplete primer extension (PIPE), 769
restriction mapping, 771
SYBR results, 772
large-scale parallel expression
airlift fermentor, 778
custom automation, 778–779
digestion buffer, 781
protein expression identification, 767
small-scale expression screening
analytical methods, 776
benzonase, 775
capillary electrophoresis result, 774
Thermofluor ligand, 772, 774
PCR product cleanup, 655–656
Periplasmic targets expression, 161–162
Phase diagrams, 609–610
Phosphate buffer, 53
Phosphorylation
mass spectrometry-based proteomics
chemical affinity purification, 733–734
immobilized metal affinity chromatography (IMAC), 732–733
neutral loss artifacts, 735
precursor ion scanning technique, 736–737
strong cation exchange (SCX) resin, 732
^{32}P-labeled ATP, 731
Physicochemical parameters, detergents, 612–613
Pichia pastoris expression
Arxula adeninivorans, 171

culture media and microbial manipulation techniques, 171–172
DNA preparation
electroporation procedure, 178–179
protocol, 177–178
solutions preparation, 177
expression cassette, multiple copies
general approaches, 185–186
selection procedure, drawback, 186–187
gene preparation and vector selection
biogrammatics expression vector map, 174–175
nucleotide sequences, 176
genetic strain construct
mating, creating diploids, 172–173
random spore analysis, 173–174
Hansenula polymorpha, 170–171
Kluyveromyces lactis, 171
Pichia methanolica, 170–171
posttranslational modification, 184–185
strains, recombinant protein production
cell extracts preparation, 182
β-lactamase expression, 181
plate activity assay, 180
transformation, electroporation, 176
Yarrowia lipolytica, 171
Yeastern blot assay development, 182–184
Pichia pastoris, recombinant protein expressing method, 135–136
Plaque assays, 216–218
Plasma membranes isolation, 620
Plasmid/bacmid transposition, 203
Plasmid DNA purification, 659–660
Plate activity assay, 180
Polyethylene glycol precipitation, 341
Polyethyleneimine (PEI) precipitation, 337
vs. AS precipitation, 339–340
DNA binding capacity, 340
procedure, 338–339
test, 340
usage strategies, 338
PEI-mediated transfection, 230. *See also* Large-scale transfection methods; Medium-scale transfection methods
Polymin P. *See* Polyethyleneimine (PEI) precipitation
Polyol-responsive monoclonal antibodies (PR-mAbs)
antibody purification, 485–487
continuous culture, mAbs productions
cell compartment harvest, 484
CELLline flask 350 (CL 350), 482–483
Dulbecco's modified eagle medium (DMEM), 482
total viable cell count, 485
ELISA-elution assay
characterization, 481
false-positive, 482
o-phenylenediamine (OPD), 480
specific antibody production, 480–481
immobilization, 487–488
mAbs source, 479
properties of, 478–479
protein purification
epitopes, 491–492
RNA polymerase, 488
SDS–PAGE, 489
using cross-reacting PR-mAbs, 490–491
Posttranslational modifications (PTMs)
collision induced dissociation (CID), 747
electron-capture dissociation (ECD), 747–748
electron transfer dissociation (ETD), 748
enrichment techniques
glycosylation, 737–738
phosphorylation, 731–737
ubiquitination and sumoylation, 738–740
identification, 184–185
mass spectrometry analysis
electrospray ionization (ESI), 744–745
features, 744
matrix assisted laser desorption ionization (MALDI), 746
methylation and acetylation
antibodies, 742
histone, 741
immonium ions, 742–743
nitrosative, 740–741
quantification
meaurement forms, 750
stable isotope dilution theory approach, 751–756
Precipitation
critical problems, 26
proteins concentration, 116–118
Precursor ion scanning technique, 736–737
Primary and secondary drying, lyophilization, 114
PR-mAbs. *See* Polyol-responsive monoclonal antibodies
PR-mAbs, immunoaffinity chromatography
antibody purification, 485–487
continuous culture, mAbs productions
cell compartment harvest, 484
CELLline flask 350 (CL 350), 482–483
Dulbecco's modified eagle medium (DMEM), 482
total viable cell count, 485
ELISA-elution assay
characterization, 481
false-positive, 482
o-phenylenediamine (OPD), 480
specific antibody production, 480–481
immobilization, 487–488
mAbs source, 479
properties of, 478–479
protein purification
cross-reacting PR-mAbs, 490–491

PR-mAbs, immunoaffinity chromatography (*cont.*)
 epitopes, 491–492
 RNA polymerase, 488
 SDS–PAGE, 489
Protease inhibitors, 125, 295
Proteinases and glycosylation, 184–185
Protein chromatography, HA columns
 BES, 390, 401
 calcium ions, 396
 chemical characteristics
  protein separation, 392
  pyrophosphate (PPi), 392
 effluent pH, 395
 hydroxyapatite sources, 399
 mechanism
  CHT I and CFT, elution time, 391
  desorption, 390
  P-and C-sites, 389
  zero net charge, 389
 MES, 394
 metal adsorption, 393–394
 phosphate gradient, 401
 process-scale columns, 399–401
 rEPO, 401
 solubility constant, 394
 voids, HA solubility, 394
Protein colligative phenomena, 697
Protein composition
 analysis, 695
 method, 696
 problems and pitfalls, 696–697
Protein crystallization, 615
Protein elution, gels
 electrophoretic elution, diffusion, 570–571
 gel preparation, 567
 location, 567
 renaturation, 569
 reverse phase HPLC, 570
 SDS removal and concentration, 568–569
Protein expression tagging
 affinity tags
  basis, 243
  calmodulin-binding peptide (CBP), 248–249
  common tags, 241
  epitope, 247–248
  glutathione-S-transferase (GST), 246–247
  His-tag, 245–246
  S-tag, 248
  STREP-II, 248
 designing
  affinity and solubility, 241–243
  choice, 243–244
  N-/C-terminal, 244–245
  tags removal, 245
  tandem tags, 244
 endoproteases, 242
 schematics, 242
 solubility tags
  basis, 242–243
  common tags, 241
  maltose-binding protein (MBP), 249–250
  NusA, 250
  smaller solubility enhancement tags (SETs), 250–251
  solubility enhancement peptide (SEP) tags, 250–251
  thioredoxin (Trx), 250
 tags removal, 251–253
Protein gel staining methods
 colorimetric total protein stains
  Coomassie brilliant blue (CBB), 546–547
  silver staining, 547–548
  Zn^{2+}reverse staining, 548–549
 fluorescent total protein stains
  detection sensitivity, 549–550
  epicocconone, 554
  flamingo and krypton protein gel stain, 555
  fluorescein derivatives, 555
  LUCY protein gel stains, 555
  SYPRO stains, 552–554
 glycoprotein detection
  acid fuchsin dye, 558
  glycosylation state, 558–559
  O-GlcNAc detection, 559
 instrumentation, 543
 phosphoprotein detection
  Phos-tag phosphoprotein stains, 557
  Pro-Q Diamond phosphoprotein gel stain, 557
 preelectrophoresis sample labeling, 556
 zymography, 543
Protein inactivation, causes, 121–122
Protein posttranslational modifications (PTMs)
 CID *vs.* ETD, 749–750
 collision induced dissociation (CID), 747
 ECD *vs.* ETD, 749
 electron-capture dissociation (ECD), 747–748
 electron transfer dissociation (ETD), 748
 identification
  enrichment techniques, 731–740
  mass spectrometry analysis, 744–746
  methylation and acetylation, 741–743
  nitrosative, 740–741
 proteomic techniques
  bottom-up, 757–758
  mass spectrometry-based, 727–729
  top-down, 756–757
 quantification
  meaurement forms, 750
  stable isotope dilution theory approach, 751–756
Protein precipitation techniques
 ammonium sulfate
  concentration, 335
  principles, 332–334

problems/solutions, 337
procedure, 334
solubility curve, 333
test, 336
ethanol and acetone, 341
isoelectric and thermal, 341
polyethylene glycol, 341
polyethyleneimine (PEI)
vs. AS precipitation, 339–340
DNA binding capacity, 340
procedure, 338–339
test, 340
usage strategies, 338
procedures, 341–342
Protein purifications, strategies and considerations
assays, 11–13
bulk/batch procedures, 17
contaminating activities, 15
extracts preparation, 16–17
properties and sensitivities, 10
refined procedures
high-capacity steps, 18–19
intermediate-capacity steps, 19
low-capacity steps, 19
source
overexpressed protein, 16
sequence information, 15–16
storage, protein solutions, 14–15
suspension, storage, and assay buffers, 13–14
use, 10–11
Protein purity determination
chromatographic methods
gel filtration method, 685–686
reversed phase HPLC, 686
composition and activity based analyses, 680–681
electrophoretic method
contaminants, 681
front-line, 682
gradient gels, 682
isoelectric focusing gel, 683
protein stains, 684
light scattering methods, 688–689
mass spectrometry methods, 687–688
methods for, 679
sedimentation velocity methods, 687
Protein quantitation
amine derivatization, 89–91
assay selection, 75–76
bicinchoninic acid (BCA)
procedure and comments, 89
reagents, 88
Coomassie blue
observation, 86
procedure, 85–86
reagents, 85
detergent-based fluorescent detection, 91
dye-based protein assay, 75, 80
flow chart, purification procedures, 75–76
instructions
cuvettes, 91–92
interfering substrates, 92–93
microwell plates, 92–93
Lowry assay
comments, 88
procedure, 87
reagents, 87
substance compatibility, 75, 77–79
ultraviolet absorbance, 205 nm
concentration standards, 83–58
dye-based protein assays, 83
extinction coefficient (ε) calculation, 82–83
ultraviolet absorption spectroscopy, 280 nm, 80–81
Proteins and solutes removal, concentration
chromatography
applications, manufacturing-scale, 102–103
gel filtration, 99
hydrophobic and affinity, 100–102
ion-exchange, 99
reversed phase, 100
crystallization, 118
dialysis
affinity binding applications and concentration, 107
buffer concentration, 104–105
mechanical design, 106
molecular weight cutoff (MWCO), 104
electrophoresis, 103–104
lyophilization
process and key system, 114–115
steps, 114
precipitation, 116–118
ultrafiltration
convective and diffusive solute transport, 108–109
devices, 110–112
membranes, 109–110
modeling approach, 108
purification applications, 112–113
Protein size determination
categorization, 692
chemical methods
colligative properties, 697
composition, 695–697
scattering methods
dynamic, 717
electron microscopy, 718
static, 716–717
subunits
associated and dissociated states, 719
stoichiometry, 720–721
techniques comparison, 693–694
transport methods
electrophoresis, 708–711
field-flow fractionation (FFF), 711–712

Protein size determination (*cont.*)
gel-filtration chromatography, 706–708
mass spectrometry, 712–716
sedimentation equilibrium, 698–701
sedimentation velocity, 701–705
viscosity, 711
Protein stability
concentration and solvent conditions, 122–123
inactivation causes, 121–122
loss of activity, 125–127
procedures, 122
proteolysis and protease inhibitors, 124–125
trials and storage conditions, 123–124
Proteoliposomes
characterization, 667–668
purification, 664–665
Proteolysis and protease inhibitors, 124–125
Proteomic techniques
bottom-up, 757–758
mass spectrometry-based, 727–729
top-down, 756–757
Purification procedures
affinity chromatography
amine reactive linkages, 432
characterization, 424–425
covalent affinity chromatography, 428–429
covalent attachment, 430
de novo development, 423
elution, 434–435
examples of, 421
group specific ligands, 425
lectins, 426–427
ligands and specificity, 426
magnetic affinity beads, 422–423
pore size, 419
Renkin equation, 420
sample preparation, 433
specific ligand, 420
stability, 422
surface activation, 430
amino acid sequence, bioinformatics
charge *vs.* pH, 22–23
cysteine content, 23–24
E. coli overexpression, solubility, 25
homology and cofactor affinity, 24–25
hydrophobicity and membrane-spanning regions, 24
molar extinction coefficient, 23
polypeptide chain molecular weight, 22
potential modification sites, 25
secondary structure, 24
stability, 24
titration curve, isoelectric point, 22–23
critical problems, 26
denatured state, 27
protein purifications
assays, 11–13
bulk/batch procedures, 17
contaminating activities, 15
extracts preparation, 16–17
properties and sensitivities, 10
refined procedures, 18–19
source, 15–16
storage, protein solutions, 14–15
suspension, storage, and assay buffers, 13–14
use, 10–11
resources, 27
summary table
footnotes, 32
mistakes and problems, 32–34
SDS–polyacrylamide gel analysis, 32
steps, 30–31

R

Reactant concentration, 61–62
Reaction assay mixtures formulation, 69–70
Receptor-specific ligand affinity chromatography, GPCRs, 636
Recombinant erythropoietin (rEPO), 401–402
Recombinant G-protein-coupled receptors
maltose-binding protein (MBP) fusion approach, 637
purification
affinity purification, 634–636
detergent-solubilized GPCRs analysis, 636–637
GPCRs stability, detergent solution, 634
receptor-specific ligand affinity chromatography, 636
purified NTS1 analysis, 641–642
solubilization, 633–634
Recombinant integral membrane proteins, expression and purification, 628–629
Recombinant protein expressing method
advantages and disadvantages, 143–144
applications, 142, 144
baculovirus/insect cells, 136–137
Escherichia coli
disulfide bond formation, 134
fusion partners, 134
inclusion bodies, 133
posttranslational modifications, 134
temperature and molecular chaperones, 133
mammalian cells, 138–139
Pichia pastoris, 135–136
protein characteristics
cytoplasmic proteins, 140–141
E. coli and codon usage, 140
membrane and secreted proteins, 141
toxic proteins, 141–142
Recombinant protein production
criteria, 224
transient gene expression (TGE)
HEK293 and CHO cell lines, 224–228
transfection methods, 228–234

Redox agents, 272
Refolding inclusion body (IBs)
alkaline pH shift, 276–277
catalysts, 276
characterization, 270
database, 277
dilute ways, 267–268
high-pressure, 276
on-column, 275–276
problem, 262
rational approach, 274–275
systematic screens, 271
variables
arginine, glycerol and sugars, 273
chemical additives, 273
detergents and chaotropic agents, 273
divalent cations, 272–273
pH and temperature, 272
salt concentration and redox agents, 272
target protein-specific additives, 274
Relative/fold purification, 31
Renkin equation, 420
Reverse dilution, 267
Reversed phase chromatography, 100

S

Salt concentration, 272
Scattering methods
dynamic, 717
electron microscopy, 718
static, 716–717
SDS-PAGE. *See* Sodium dodecyl sulfate-polyacrylamide gel electrophoresis
Sedimentation coefficient(s), 701–703
Sedimentation equilibrium
capabilities, 698
method
equilibrium centrifugation experiment, 699–700
reduced molecular weight (σ), 698–699
problems and pitfalls, 700–701
Sedimentation velocity
diffusion coefficient, 703
frictional coefficient ratio (f/f^0), 703–704
*vs.*gel chromatography, 701
methods, 687, 702–704
s/D ratio, 703–704
variations, 705
Selected reaction monitoring (SRM), 755
Size exclusion chromatography (SEC), 457. *See also* Gel–filtration chromatography
Smaller solubility enhancement tags (SETs), 250–251
Small-scale expression screening
analytical methods, 776
benzonase, 775
capillary electrophoresis result, 774
Thermofluor ligand, 772, 774
transfection methods, 229–230
Small ubiquitin-like modifier (SUMO) protein, 251
Sodium dodecyl sulfate-polyacrylamide gel electrophoresis (SDS-PAGE)
casting gels, 503–505
catalyst, 503
difference gel electrophoresis (DIGE)
quantitative analytical algorithms, 525–526
software analytical tools, 524–525
staining and detection, 524–525
electrode buffer, 503
electrophoresis, 505–506, 511–513
gel types, 522–523
marker proteins, 510–511
2-mercaptoethanol/dithiothreitol, 499, 507
molecular weight determination, 511
Ornstein–Davis system, 506
polyacrylamide gels, 500
principle, 501
protein detection, gel
Coomassie Brilliant Blue R-248, 508–509
copper staining, 510
silver staining, 509–510
resolution enhancement, 523–524
cathodic drift, 523
hydroxyethyl disulfide (HED), 524
IPG strip, 524
sample preparation, 505
separation principle, 522
stock solutions, 502
vertical electrophoresis systems, 523
Solid-phase peptide synthesis, 641
Solubility enhancement peptide (SEP) tags, 250–251
Solubility tags
basis, 242–243
common tags, 241
maltose-binding protein (MBP), 249–250
NusA, 250
smaller solubility enhancement tags (SETs), 250–251
solubility enhancement peptide (SEP) tags, 250–251
thioredoxin (Trx), 250
Soluble multimer, 262
Sonication, 299–300
Stable isotope labeling by amino acids in cell culture (SILAC), 752
S-tag, 248
Stationary gel volume fraction, 706
Strains, recombinant protein production
cell extracts preparation, 182
β-lactamase expression, 181
plate activity assay, 180
STREP-II-tag, 248
Strong cation exchange (SCX) resin, 732

Subcellular organelles
  complexity, 307
  density and dimensions, 309
  enzyme and substrate markers, 315
  extraction and prefractionation
    centrifugation, 308–311
    core histone purification, 324–325
    differential detergent fractionation (DDF), 316–320
    enrichment, 311–327
  isolation
    complexity, 307
    core histone purification, 307
    density gradient, centrifugation, 311
    density velocity, 308–310
    differential detergent fractionation (DDF), 307
    endomembrane, enrichment, 314–316
    immunoaffinity reagents, 326–327
    lipid raft, 320–322
    mitochondria, 311–313
    nuclear extract, 325
    nuclei and histone, 322–324
  schematic workflow, 308
Subproteomes. *See* Subcellular organelles
Subunits
  associated and dissociated states, 719
  stoichiometry, 720–721
Succinate buffer, 52
Sucrose gradient sedimentation method, 705
Sucrose gradient separation method. *See* Mitochondria enrichment
Sumoylation. *See* Ubiquitination and sumoylation
SV40 enhancer, 224, 226

T

Tags removal
  chemical cleavage, 253
  cleavage sites, 252
  endo and exo proteases, 252
Tandem affinity purification (TAP), 244
Tandem tags, 244
Target protein-specific additives, 274
T4 DNA polymerase-treated DNA fragments, 155–156
Tetramethylethylenediamine (TEMED), 500
Thermal precipitation, 341
Thioredoxin (Trx) tag, 250
Three-dimension structure, critical problems, 26
Time-of-flight (TOF) mass analyzer, 745
Toxic proteins, expressing method, 141–142
Transfection methods, large-scale
  modifications, 234
  process parameters, 232
  protocol, 232–234
Transfer plasmid modifications, 198–199
Transformation, electroporation, 176
Transformation reaction, 658–659
Transgenic lepidopteran, 213
Transient gene expression (TGE)
  HEK293 and CHO cell lines
    cultivation, 228
    EBNA-1 gene, 224
    expression vectors, 226–227
    HKB-11 cell line, 226
    overview, 225–225
    SV40 enhancer, 224, 226
  transfection methods
    large-scale approach, 232–234
    medium scale approach, 231–232
    small-scale approach, 229–230
    reagents, 229
Transport methods
  electrophoresis, 708–711
  field-flow fractionation (FFF), 711–712
  gel-filtration chromatography, 706–708
  mass spectrometry, 712–716
  sedimentation equilibrium, 698–701
  sedimentation velocity, 701–705
  viscosity, 711
Tris(hydroxymethyl)aminomethane (tris) buffer, 54
Two-dimensional gel electrophoresis, isoelectric focusing
  acidic range IPG gels, 531–534
  alkaline range IPG gels, 534–535
  difference gel electrophoresis (DIGE)
    quantitative analytical algorithms, 525–526
    software analytical tools, 524–525
    staining and detection, 524–525
  equilibration, IPG gels, 535–536
  gel types, 522–523
  high resolution 2D gel, 517
  materials
    equipment, 527
    solutions and reagents, 527–528
  pH gradients types
    carrier ampholytes, 518–519
    different configurations, 521
    IPG strip technology, 519–520
  principle, 517–518
  protein sample preparation, 528–530
  resolution enhancement, 523–524
    cathodic drift, 523
    hydroxyethyl disulfide (HED), 524
    IPG strip, 524
  sample cleanup and precipitation, 530–531
  separation principle, 522
  vertical electrophoresis systems, 523

U

Ubiquitination and sumoylation
  amino-terminal epitope tags, 739
  characteristic mass shifts, 740

Ultrafiltration
convective and diffusive solute transport, 108–109
devices, 110–112
membranes, 109–110
modeling approach, 108
purification applications, 112–113
Ultraviolet absorbance, 205 nm
concentration standards, 83–58
dye-based protein assays, 83
extinction coefficient (ε) calculation, 82–83
Ultraviolet absorption spectroscopy, 280 nm, 80–81
Ultraviolet–visible absorption spectroscopy, 66
Urea, 266

V

Viral chitinase gene, 208–209
Viscosity, 711
Void volume (E_0), 707

W

Washed nuclear pellet preparation, 322–324
Western blotting
advances, 575
charged-coupled device (CCD) cameras, 575
direct and indirect, 575–577
far-Western, 577
semiquantitative
advantageous, 577
assay methods, 578
densitometric analysis, results, 579
Wheat germ translation reaction
bilayer reaction protocol, 664
bilayer reaction *vs.* dialysis reaction, 662–663
dialysis reaction protocol, 664–665
materials and reagents, 663–664

X

6xHis-tag. *See* His-tag

Y

Yeastern blot assay procedure, 182–184
Yeast expression. *See Pichia pastoris* expression
Yeast lysis, 289

Z

Zymography, 543

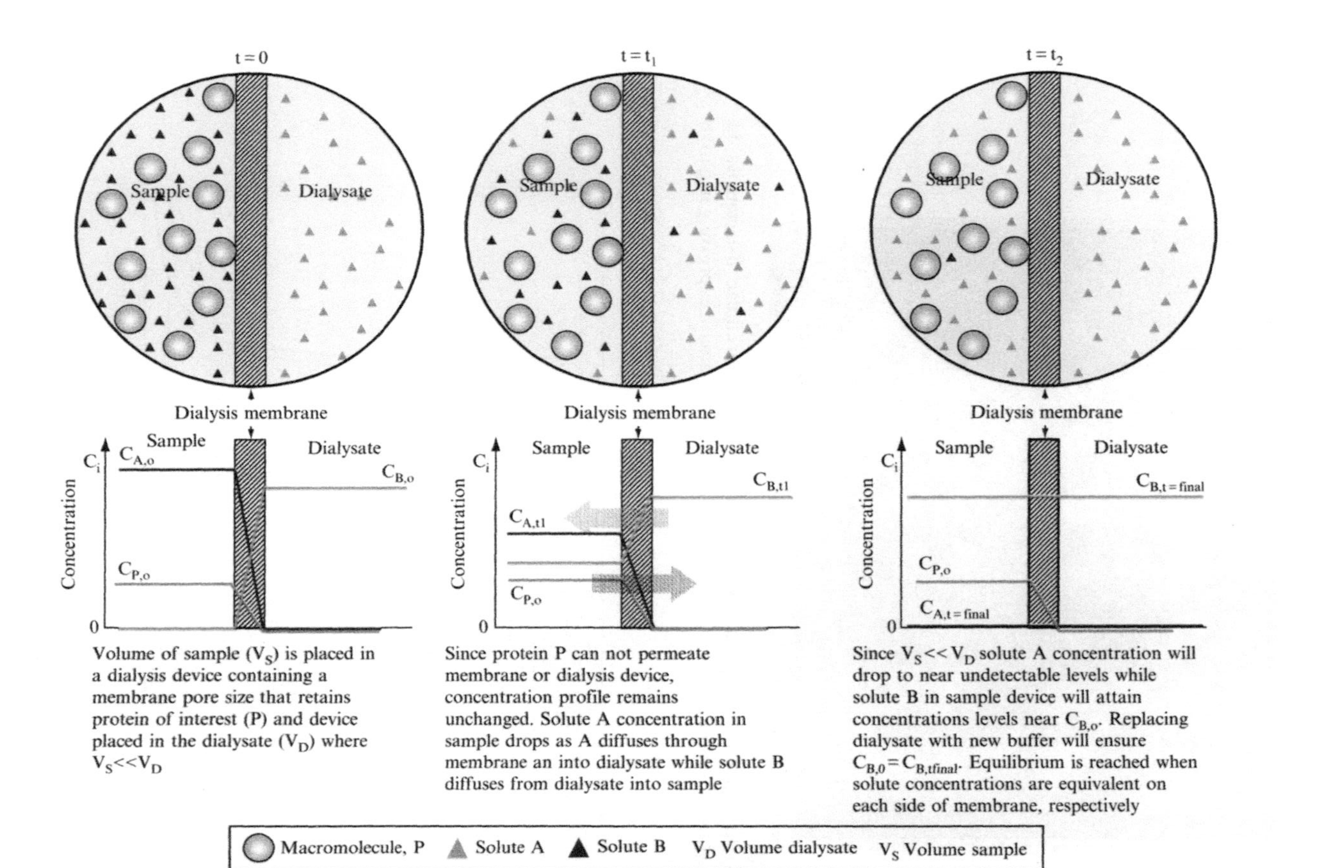

David R. H. Evans *et al.*, Figure 9.1 Schematic of the dialysis process. Top circular diagrams provide graphic representation of solute distributions within "sample" and "dialysate" pools throughout dialysis process. The bottom graphs show the respective concentration profiles from each pool.

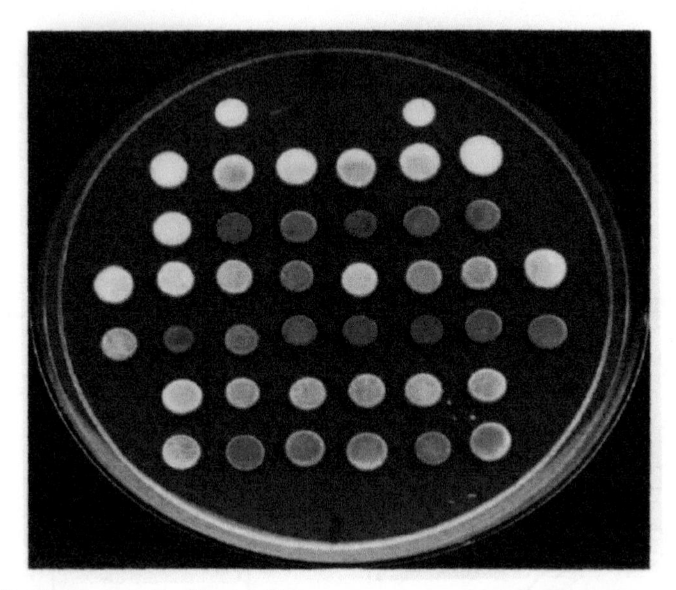

James M. Cregg *et al.*, Figure 13.3 Expression of intracellular β-lactamase in *P. pastoris*. The two spots at the top of the plate are non-β-lactamase expressing negative control strains.

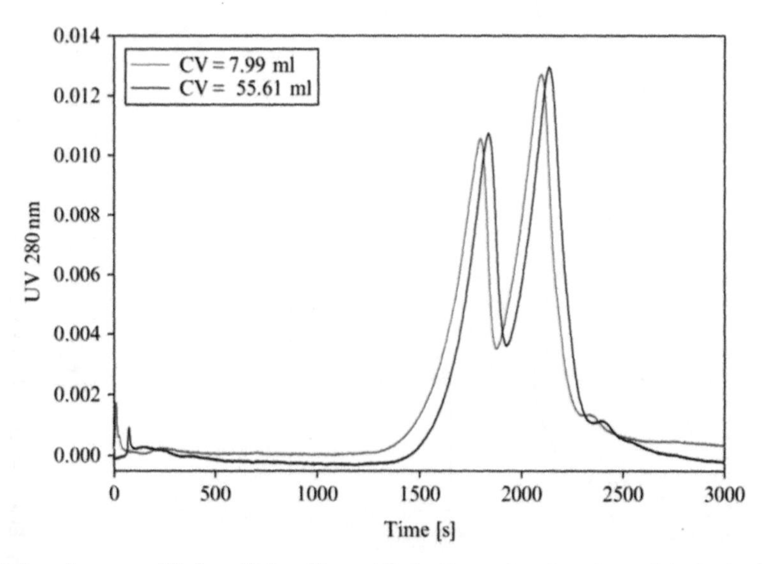

Alois Jungbauer and Rainer Hahn, Figure 22.6 Example of a α-lactoglobulin isoform separation on Source30Q (GE Healthcare, Uppsala, Sweden). The flow rate, gradient length, and load are scaled to the column volume thus consequently γ was kept constant. The flow rate is 30 CV/h. Column dimensions are: 1×10 and 2×17.7 cm. Reproduced by kind permission of Jorgen Mollerup.

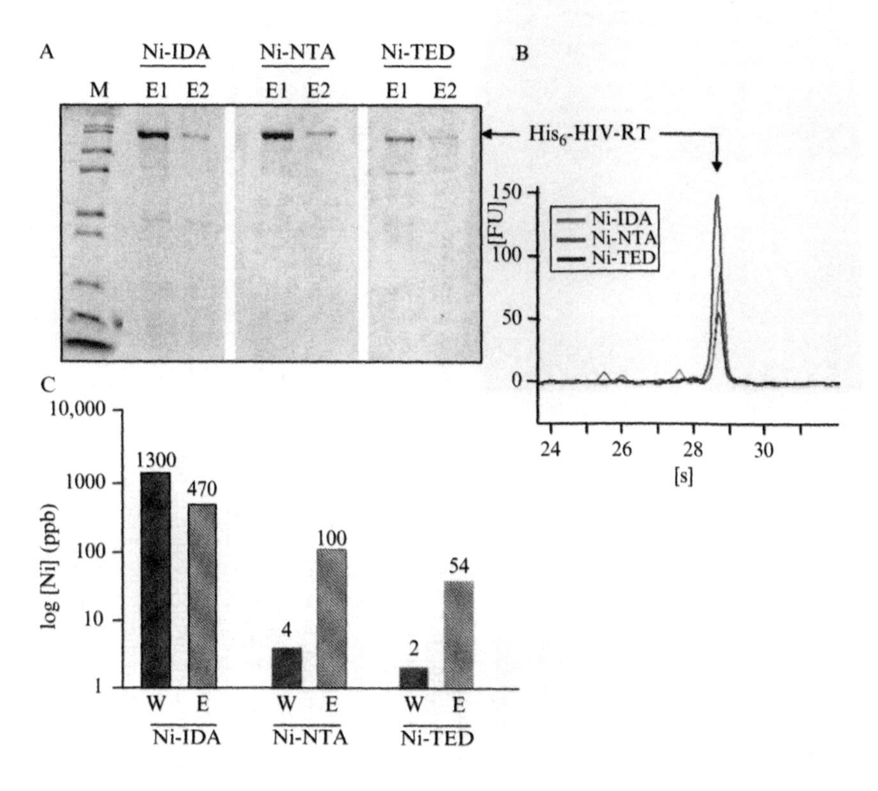

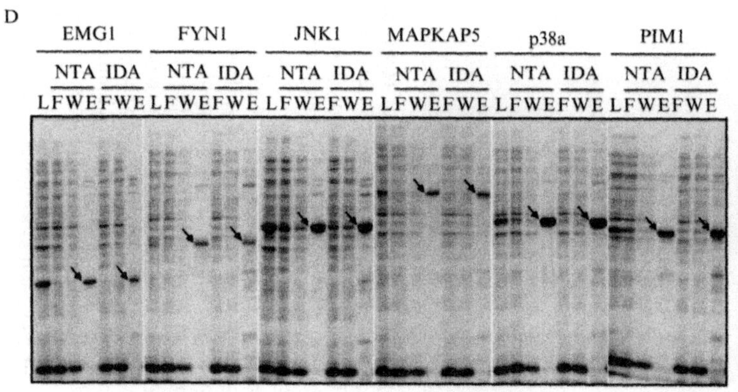

Helena Block *et al.*, Figure 27.2 (Continued)

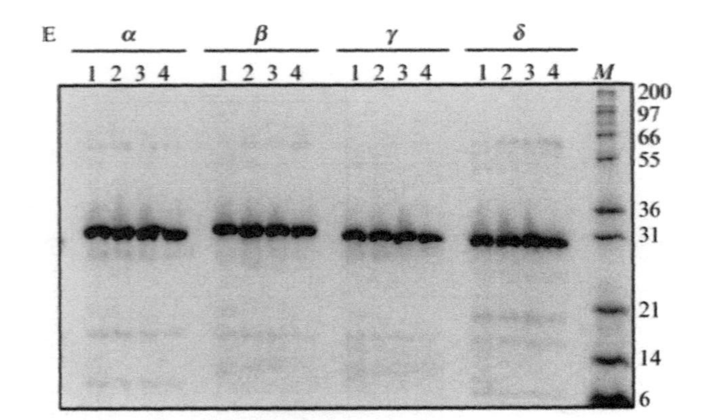

Helena Block *et al.*, Figure 27.2 Purification of His-tagged proteins with NTA, IDA, and TED. (A) H_6-HIV-RT was expressed in *E. coli* BL21(DE3) and purified via Ni-IMAC in the presence of 1 m*M* DTT under standard conditions (IDA, NTA; see Section 2.10 for standard conditions) or according to the manufacturer's recommendations (TED). Corresponding aliquots of the IMAC elution fractions (E1, E2) were analyzed by SDS–PAGE and Coomassie staining. (B) Bioanalyzer 2100 lab-on-a-chip analysis of pooled elution fractions. The peaks from the electropherograms corresponding to H_6-HIV-RT were overlayed. Peak areas directly correlate to the protein amount in the respective pool fraction. (C) Determination of the nickel content in wash (W) and pooled elution fractions (E) of the H_6-HIV-RT purifications described in (A) and (B). Nickel was measured by ICP-MS (intercoupled plasma mass spectrometry) at Wessling Laboratories, Bochum, Germany, and values are provided in μg/l (parts per billion, ppb). (D) QIAgene constructs carrying optimized human genes were used for expression of the indicated proteins in *E. coli* BL21(DE3) LB cultures. Cleared lysates were divided for purification of His-tagged proteins via Ni-NTA (NTA) and Ni-IDA (IDA), respectively. Fractions were analyzed by SDS–PAGE and Coomassie staining as follows: L, cleared lysate; F, IMAC flow-through fraction; W, wash fraction; E, peak elution fraction. (E) Fragments of H_6-tagged proteins named α, β, γ, and δ expressed in *E. coli* were purified under standard conditions using NTA and Cm-Asp tetradentate ligands loaded with Ni^{2+} or Co^{2+} as follows: 1, Ni-NTA; 2, Co-NTA; 3, Co-CmAsp; 4, Ni-CmAsp. Aliquots from the peak elution fractions (2 μl each) were analyzed by SDS–PAGE and Coomassie staining.

David B. Friedman *et al.*, Figure 30.5 Fluorescence image of a multiplexed DIGE gel, 24 cm pH 4–7 (12% second dimension) in false-color representation. Samples had been labeled prior to separation with Cy2 (blue; labeling 100 μg human AGS cells), Cy3 (green; labeling 100 μg human AGS cells infected with *H. pylori* strain 26695), and Cy5 (red; labeling 100 μg human AGS cells infected with *H. pylori* strain 26695 deleted for the cag pathogenicity island). Individual Cydye signals were acquired using mutually exclusive excitation and emission spectra at 100 μm resolution, and recorded at 16-bit depth using a gray-scale intensity from 1 to 100,000. Using Cy2 to label a mixed-sample internal standard enables a statistically powered differential expression experiment by registering independent-replicate samples (separately labeled with Cy3 and Cy5) from multiple experimental conditions across several matched DIGE gels.

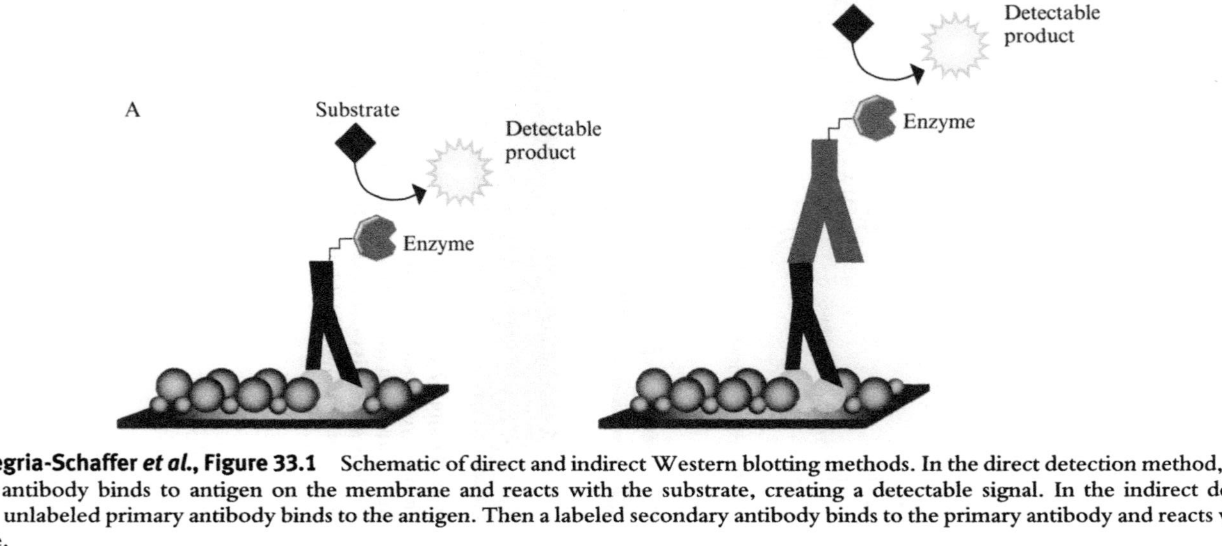

Alice Alegria-Schaffer *et al.*, Figure 33.1 Schematic of direct and indirect Western blotting methods. In the direct detection method, labeled primary antibody binds to antigen on the membrane and reacts with the substrate, creating a detectable signal. In the indirect detection method, unlabeled primary antibody binds to the antigen. Then a labeled secondary antibody binds to the primary antibody and reacts with the substrate.

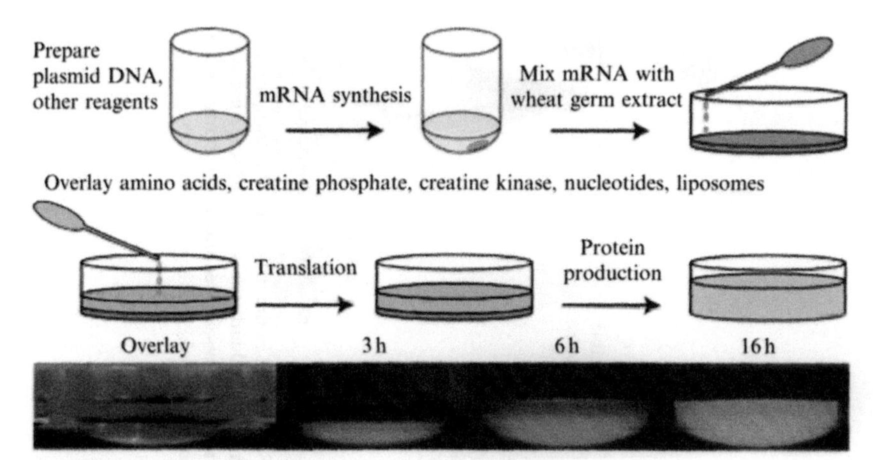

Michael A. Goren *et al.*, Figure 37.1 A schematic representation of the steps used for cell-free translation in a bilayer mode reaction. A customized plasmid is used to prepare reagent mRNA, which is added to the translation mixture, overlaid with amino acids, other substrates, and desired additives, and then the reaction can begin. The photos show an experimental demonstration of GFP translation over a 16 h period.

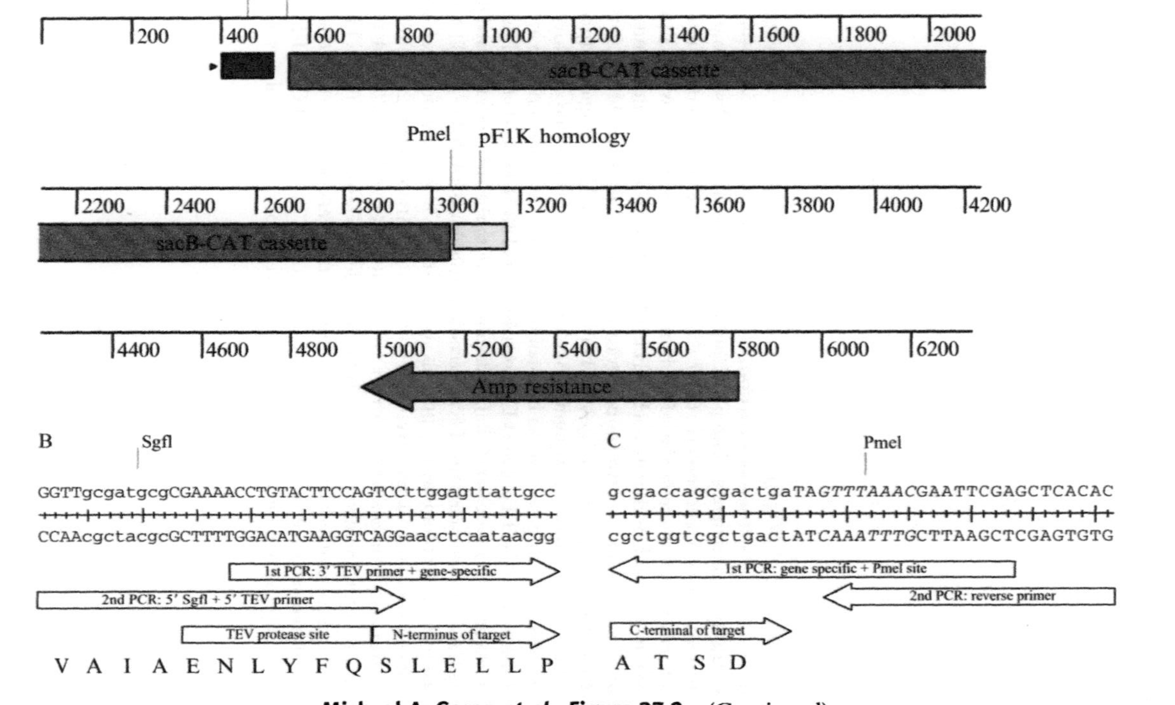

Michael A. Goren *et al.*, Figure 37.2 (Continued)

Michael A. Goren *et al.*, Figure 37.2 (A) Linear map of pEU-HSCB showing important features of the vector construction. The TMV omega sequence (blue box) enhances translation efficiency. The pF1K homology sequence (yellow box) prevents ligation of two plasmid backbones. (B) Example of the 5′ primer designed for the cloning of *Halobacterium salinarium* bacteriorhodopsin (GenBank M11720.1). The first-step PCR forward primer is 5′-ACCTGTACTTCCAGTCCttggagttattgcc, with the upper case nucleotides corresponding to the 3′ TEV primer and the lower case nucleotides corresponding to the gene-specific sequence. The second-step universal forward primer is 5′-GGTTgcgatcgcCGAAAACCTGTACTTCCAGTCC, with the SgfI site in lower case. There is an 18 bp overlap between the first-step forward primer and universal forward primer. The TEV protease recognition sequence is ENLYFQ/S, with proteolysis between Q and S. (C) Example of the 3′ primer designed for the cloning of *Halobacterium salinarium* bacteriorhodopsin (GenBank M11720.1). The first-step PCR reverse complement primer is 5′-GCTCGAATTC*GTTTAAAC*TA-tcagtcgctggtcgc, with the upper case, italic nucleotides corresponding to the PmeI site and the lower case nucleotides corresponding to the gene-specific sequence. The second-step universal reverse primer is 5′-GTGTGAGCTCGAATTCGTTTAAAC. There is an 18 bp overlap between the first-step reverse primer and universal reverse primer.

Printed and bound by CPI Group (UK) Ltd, Croydon, CR0 4YY

07/10/2024

01041951-0002